Springer Geology

The book series Springer Geology comprises a broad portfolio of scientific books, aiming at researchers, students, and everyone interested in geology. The series includes peer-reviewed monographs, edited volumes, textbooks, and conference proceedings. It covers the entire research area of geology including, but not limited to, economic geology, mineral resources, historical geology, quantitative geology, structural geology, geomorphology, paleontology, and sedimentology.

More information about this series at http://www.springer.com/series/10172

Abderrahmane Bendaoud
Zakaria Hamimi · Mohamed Hamoudi
Safouane Djemai · Basem Zoheir
Editors

The Geology of the Arab World—An Overview

Springer

Editors
Abderrahmane Bendaoud
University of Science and Technology
Houari Boumediene
Algiers, Algeria

Zakaria Hamimi
Department of Geology
Benha University
Benha, Egypt

Mohamed Hamoudi
Faculty of Earth Sciences
University of Science and Technology
Houari Boumediene
Algiers, Algeria

Safouane Djemai
Faculty of Earth Sciences
University of Science and Technology
Houari Boumediene
Algiers, Algeria

Basem Zoheir
Department of Geology
Benha University
Benha, Egypt

ISSN 2197-9545 ISSN 2197-9553 (electronic)
Springer Geology
ISBN 978-3-030-07253-7 ISBN 978-3-319-96794-3 (eBook)
https://doi.org/10.1007/978-3-319-96794-3

Softcover re-print of the Hardcover 1st edition 2019

This Springer imprint is published by the registered company Springer Nature Switzerland AG
The registered company address is: Gewerbestrasse 11, 6330 Cham, Switzerland

To My late Father, Prof. Mohamed Bendaoud, who gave me the curiosity to try to understand the universe.

To my daughter, Maya Ines and my son Rostom with whom I share this passion.

Abderrahmane Bendaoud

To my family: my wife, my children and my grandchildren.

Mohamed Hamoudi

To my wife, my three children and my parents.

Safouane Djemai

Preface

April 2013, the Arabian Geosciences Union (ArabGU) has been established, as a nonprofit international association, similar in scope to the European Geosciences Union (EGU), the American Geophysical Union (AGU), the Japanese Geosciences Union (JpGU), and the Asia Oceania Geosciences Society (AOGS). ArabGU aims to promote, disseminate, and contribute to geosciences as a whole, with emphasis on the Arab World, through organizing an annual international meeting bringing distinguished scientists and professionals together to discuss topics of broad interest. February 17–22, 2016, the first ArabGU International Congress (AIC-1) was co-organized by ArabGU, the University of Science and Technology Houari Boumediene (USTHB), the Center for Research in Astronomy, Astrophysics and Geophysics (CRAAG) and the Algerian Society of Geophysics (SAG). More than 150 participants from 15 Arab and Western countries (Morocco, Algeria, Tunisia, Egypt, Sudan, Yemen, the Emirates, Libya, Canada, Spain, France, Italy, Austria, Japan) have taken a part in this international event. The topics covered include structural geology, tectonics, remote sensing, GIS, stratigraphy, seismology, metallogeny, Precambrian geology, Phanerozoic geology, geophysics and modeling, planetology and geoheritage, along with many other multidisciplinary items. This book is the result of the work of this first international congress of the ArabGU.

From the geological point of view, Arab region (13,132,327 km^2) is dominated by a wide variety of lithologic units ranging from Archean to Quaternary. Since the amalgamation of Gondwana, this region formed a continuous ensemble consisting of Archean and/or Palaeoproterozoic cratonic terranes and Neoproterozoic Pan-African belts surrounded by Paleozoic basins. It was affected to a greater or lesser extent by various Phanerozoic Caledonian, Hercynian, and Alpine orogeneses. The countries of the Arab world also have the particularity to have their economic growth mainly driven by exports to the world market of hydrocarbons (oil and natural gas) and mineral resources or by tourism thanks in particular to their geoheritage.

This book, comprising 14 chapters, is aimed at readers who are already familiar with the Geology of the Arab Region. Some chapters are research articles and others are reviews that embrace a wide range of Earth sciences. The volume as a whole represents a valuable basic reference for junior/senior geoscientists who are keen interest with the aspects to be addressed.

In Chapter "Tectonics of the Eastern Desert of Egypt: Key to Understanding the Neoproterozoic Evolution of the Arabian–Nubian Shield (East African Orogen)", Hamimi et al. highlighted the tectonic setting of the Egyptian Nubian Shield in an attempt to decipher the Neoproterozoic tectonic evolution of the Arab-Nubian Shield and the entire East African Orogen.

In the next contribution, Liégeois presents and discusses his geological map of the Tuareg Shield map that served as the basis for this region in the new 1/10,000,000 scale geological map of Africa (BRGM, 2016). It synthesizes all the knowledge acquired to date on this Shield, and to which the author has greatly contributed through his work over the past 30 years. It differentiates between terranes whose ages may be Archean, Paleoproterozoic or Neoproterozoic and the orogenic events that affected them during the Paleoproterozoic or the Neoproterozoic. This chapter also proposes a global evolution model for the amalgamation of these terranes during the Pan-African.

Chapter "The 600 Ma-Old Pan-African Magmatism in the In Ouzzal Terrane (Tuareg Shield, Algeria): Witness of the Metacratonisation of a Rigid Block", is a discussion given by Fezaa et al. on the geochronological, geochemical and isotopic characteristics of Pan-African granitic plutons intruding the Archean-Paleoproterozoic terrane of In Ouzzal in the Hoggar Shield (Algerian part of the Tuareg Shield). The authors indicated that this pluton was formed by a process of metacratonization of an old lithosphere at a late stage of the Pan-African Orogeny.

Chapter "An Overview of the Plutons Magnetic Fabric Studies in the Hoggar Shield: Evolution of the Major Shear Zones During the Pan-African", by Henry et al., is essentially a synthesis of their work on magnetic fabrics obtained on different granitic plutons of the Hoggar Shield. The authors highlighted different stages of the Pan-African Orogeny, responsible for deformation along megashears traversing the shield. Changes in the direction of continental convergence during the Pan-African Orogeny are also dealt with in this chapter.

Chapter "Electrical Conductivity Constraints on the Geometry of the Western LATEA Boundary from a Magnetotelluric Data Acquired Near Tahalgha Volcanic District (Hoggar, Southern Algeria)", by Bouzid et al., is a magnetotelluric study that attempts to constrain the geometry of the lithosphere in a key region of the Hoggar Shield. The region encompasses a shear zone separating the archean-paleoproterozoic terranes of the central Hoggar from the Neoproterozoic juvenile terranes of the western Hoggar.

Chapter "Regional Geology and Petroleum Systems of the Main Reservoirs and Source Rocks of North Africa and the Middle East", by Lučić and Bosworth, presents various paleomagnetic studies carried out on the western Saharan basins allowing the determination of the position of several paleomagnetic poles mainly between the Bashkirian and the Autunian and between the Middle Triassic and the Lias. Results indicate a post-Lias regional tectonics affecting the Paleozoic coverage of these basins.

In the next chapter, Lučić and Bosworth provide a remarkable and very edifying review of the major regional geological events that contributed to the genesis of large hydrocarbon fields in North Africa and Arabia. This is achieved in particular by high-quality illustrations that synthesize a large amount of information. In particular, they highlight the similarities that existed over geological time in this vast region as it evolved as contiguous parts of Gondwana.

In Chapter "Archaeoseismology in Algeria: Observed Damages Related to Probable Past Earthquakes on Archaeological Remains on Roman Sites (Tel Atlas of Algeria)", Deroin outlines of the main applications of satellite sensors in geological mapping, tectonics and structural geology, hydrogeology, mining geology, geoarchaeology, oil and gas exploration, earthquake and seismicity, landslides, and coastal erosion. He attested that remote sensing techniques are particularly suitable for geological studies in arid regions, such as the Arab world. Three examples of very different applications all located in Arab countries are given in this chapter.

Chapter "A Glimpse at the History of Seismology in Algeria", by Harbi et al., represents the first attempts to discuss the archaeoseismology in Algeria at the international scale. The authors demonstrate that the seismically active Tellian Atlas of Algeria preserves numerous archaeological evidence of ancient earthquakes ranging from the Roman period (BC 146–429) to the Vandal and Byzantine period (AD 429–533). They interpret damage to some monuments as being caused by strong earthquakes or landslides.

Chapter "Active Tectonics and Seismic Hazard in the Tell Atlas (Northern Algeria): A Review", by Maouche et al., is a contribution to the knowledge of the history of seismology in Algeria. The authors, after having inventoried what we know about the great seismic events in Algeria during the Middle Ages until the French colonization, which are very little documented, show the different stages of the evolution of seismology during the colonial period in Algeria, with notably the creation of the Algiers observatory and the first studies and measurements of seismic activity. Then, the postindependence period is detailed with the creation of the CNAAC (currently CRAAG). The authors have taken this opportunity to highlight the mastery and great scientific qualities of Algerian actors, during this period, in the field of seismology.

In Chapter "Seismicity of the Algerian Tell Atlas and the Impacts of Major Earthquakes", Ousadou and Bezzeghoud review the active tectonics and seismic hazard in the Tellian Atlas of Algeria, which has experienced several destructive earthquakes in the past. The authors note that seismicity is not randomly distributed but directly related to active geological structures, which correspond mainly to faulted folds. They also show the fundamental role of studies on the El Asnam and Zemmouri faults at the source of two of the most destructive earthquakes known in Algeria, in understanding seismogenic active tectonics in the region.

In the next Chapter "An Overview on 40 Years of Remote Sensing Geology Based on Arab Examples", Deroin, appraise the seismicity in northern Algeria and carry out a detailed analysis of the main earthquakes that have occurred in the Tell Atlas since 1980. Throughout this revision, the authors characterize the impacts of several major earthquakes that occurred between 1364 and 2015 in terms of seismic energy.

Chapter "Meteorite Impact Structures in the Arab World: An Overview", by Chabou, records the confirmed, proposed, or refuted impact craters in the Arab countries. The author did not just present a bibliographical work but he also combines it with the examination of satellite images and a critical synthesis of available geological, petrographic, and geochemical data. He proposes certain structures that could be affiliated to impact craters. Moreover, he came to the conclusion that the assessment of impact structures in the Arab region is still incomplete, given the large size of the territory and the low concentration of confirmed structures compared to other better studied regions.

In the last Chapter "Holocene Climate Development of North Africa and the Arabian Peninsula", Lüning and Vahrenholt examine the hydroclimatic and temperature changes in the Arab region over the past 15,000 years by correlating and integrating all available case studies. The authors describe the passage of a wet green Sahara, between 15,000 and 9000 years BP, under conditions currently 6500–3500 years BP, depending on the regions. They suggest that the Holocene climate history of North Africa and Arabia is closely linked to global development and that significant temperature changes have also occurred in subtropical climate belts.

In conclusion, we would like to express our thanks to the contributors who provided manuscripts to this textbook. We are also indebted to the following list of reviewers (arranged in alphabetical order) for their valuable comments and constructive criticism.

Abdelhakim Ayadi, Algeria
Ahmed N. El Barkooky, Egypt
Abderrahmane Bendaoud, Algeria
Mourad Bezzeghoud, Portugal
Roger Bilham, USA
Jean Boissonnas, France
Jean Chorowicz, France
Michel Cottin, France
Tari Gabor, Hungary
Mohamed Hamoudi, Algeria
Peter Johnson, USA
Assia Harbi, Algeria
Jean-Paul Liegeois, Belgium
Djamel Machane, Algeria
Merzouk Ouyed, Algeria
Neil Roberts, USA
Wolf Uwe Reimold, Germany

Hakim Saibi, UEA
Nicola Scafetta, Italy
Martin Schmieder, Germany
Stathis Stiros, Greece
Abdelkrim Yelles-Chaouche, Algeria

The Editors

Algiers, Algeria	Abderrahmane Bendaoud
Benha, Egypt	Zakaria Hamimi
Algiers, Algeria	Mohamed Hamoudi
Algiers, Algeria	Safouane Djemai
Benha, Egypt	Basem Zoheir

Contents

Tectonics of the Eastern Desert of Egypt: Key to Understanding the Neoproterozoic Evolution of the Arabian–Nubian Shield (East African Orogen)

Z. Hamimi, M. A. Abd El-Wahed, H. A. Gahlan and S. Z. Kamh

Abstract The tectonic evolution of the Arabian–Nubian Shield (ANS), the northern continuation of the East African Orogen (EAO), is enigmatic and a matter of controversy. The EAO is observed as a N–S trending major suture zone separating East and West Gondwanaland. It documents a prolonged tectonic history bracketed by the fragmentation of Rodinia Supercontinent and the amalgamation of Gondwana. The ANS is dominated by Neoproterozoic juvenile continental crust (i.e., crust formed directly from the mantle), formed by magmatic arc accretion and subsequent post-tectonic magmatism, and includes a mosaic of tectonic terranes juxtaposed along ophiolite-decorated megashears (suture zones). Among them is the Eastern Desert terrane (namely, Aswan or Gerf terrane in some literatures) which is regarded as the western extension of Midyan terrane in Western Arabian and shows most of the polydeformed history of the ANS. This chapter is devoted to discuss the Neoproterozoic crustal evolution of the Pan-African belt of the Eastern Desert terrane in an attempt to understand the tectonic setting of the ANS. Main points to be discussed in this chapter are: (1) infracrustal–supracrustal rocks, (2) thrusting, shearing, and folding relations; (3) gneiss domes versus metamorphic core complexes; (4) the conjugate pairs of Najd-related shears; (5) role of Najd Fault System in tectonic evolution of gneiss domes; (6) rates and transport directions of metaultramafic nappes; (7) the voluminous intrusives in northern Eastern

Z. Hamimi (✉)
Department of Geology, Faulty of Science, Benha University, Benha 13518, Egypt
e-mail: yahiahamimi@gmail.com

M. A. Abd El-Wahed · S. Z. Kamh
Geology Department, Faculty of Science, Tanta University, Tanta 31527, Egypt

H. A. Gahlan
Geology and Geophysics Department, College of Sciences, King Saud University, Riyadh 11451, Saudi Arabia

H. A. Gahlan
Department of Geology, Faculty of Science, Assiut University, Assiut 71516, Egypt

A. Bendaoud et al. (eds.), *The Geology of the Arab World—An Overview*,
Springer Geology, https://doi.org/10.1007/978-3-319-96794-3_1

Desert; (8) the post-amalgamation Hammamat sediments and their relation to Dokhan Volcanics; and (9) the northward decrease in intensity of deformation in the entire Eastern Desert.

1 Introduction

A general agreement is established that rifting of the Red Sea since the Oligocene and younger times led to uplifting and exhumation of the Neoproterozoic Arabian–Nubian Shield (ANS) (Hamimi et al. 2015b). The Egyptian Eastern Desert occupies the northwestern part of the ANS and is dominated by crystalline basement complex considered by many workers as the mirror image of the Neoproterozoic belt exposed at Western Arabia (Arabian Shield).

This chapter addresses the tectonic setting of the Egyptian Eastern Desert as a key to deciphering the Neoproterozoic tectonic evolution of the ANS and consequently the entire East African Orogen (EAO). Depending upon outstanding differences in exposed lithologies and remarkable contrast in physiographic features, Abdel Khalek (1979), Stern and Hedge (1985) and El-Gaby et al. (1988) subdivided the Egyptian Eastern Desert into three main provinces; Northern Eastern Desert (NED), Central Eastern Desert (CED) and Southern Eastern Desert (SED). Such threefold division has been one of the few broadly accepted concepts in the tectonic analysis of the Egyptian Neoproterozoic basement (Fowler and Osman 2009). The provinces juxtapose along two major structural elements; Qena-Safaga Shear Zone separates NED from CED, and Idfu-Mersa Alam Shear Zone splits CED from SED (Fig. 3). Fowler and Osman (2009) considered the northerly dipping Sha'it–Nugrus shear zone as the boundary separating the CED from the SED. The NED is dominated by voluminous granitoids, together with slightly deformed–unmetamorphosed Dokhan Volcanics and post-amalgamation Hammamat volcano-sedimentary sequence. The CED encompasses gneisses–migmatites-sheared granitoids and remobilized equivalents outcropping as elliptical domal-like structures, in addition to volcano-sedimentary succession (mainly volcanogenic metagraywackes and metamudstones) and ophiolitic metaultramafics. Lithologic units encountered in the SED resemble those of the CED with the exception of high percentage of gneisses and migmatitic gneisses, as well as the ophilitic metaultramafics that form conspicuous tectonically transported nappes. Gneissic and migmatitic rocks of the SED occupy much larger, more complexly shaped areas associated with batholiths of foliated granodiorite (Fowler and Osman 2009). Moreover, the typical greenschist facies of volcanogenic metagraywackes and metamudstones of the CED are also comparable to the high-grade schistose-metasediments of the SED (El-Gaby et al. 1988; Hermina et al. 1989; Hassan and Hashad 1990). Besides the lithologic differences, the previously mentioned provinces of the Egyptian Eastern Desert display remarkable differences in structural architecture (Hamimi et al. 2015b). The NED is dominated by fault/joint systems and marked by younger granitoid intrusions. Fold-related faults are dominant in the CED and are commonly associated with pull-apart basins linked to Najd Fault System. In the SED, fold-thrust belts prevail and thrusts are first-order kinematics that are later overprinted by map-scale transpression.

2 Proposed Models for the Arabian–Nubian Shield, East African Orogen

According to the concept of supercontinental cycle, continental blocks and cratons converge into single supercontinent, split and disperse into numerous continents, then amalgamate once again. This scenario, which is estimated to be 300–500 Ma long, occurred more than once throughout earth's history giving Pangaea and preceding supercontinents such as Rodinia and Colombia (Hoffman 1999). Several lines of evidence indicated that rifting and fragmentation of Rodinian Supercontinents lasted over 300 Ma and reassembly of the old continental blocks resulted in the formation of Gondwana supercontinent. Gondwana comprises two continental blocks juxtapose along a 3000 km stretched EAO (Stern 1994). The EAO extends in a N- to NNW-direction from Sinai Peninsula to Mozambique. The East African Orogen has long been considered the best exposed bowels of former mountain building that there are results of continent–continent collision and the bulldozing together of many oceanic arcs and remnants of oceanic lithosphere that once separated the cratons (Drury 2013). It consists of several Neoproterozoic juvenile island arc terranes accreted on the westward Saharan Metacraton (Fig. 1), during a long orogenic cycle started with the breakup of Rodinia (Li et al. 2004) and continued to the final amalgamation of Gondwana (Pisarevsky et al. 2003; Collins and Pisarevsky 2005).

Several models have been proposed to discuss the tectonic evolution of the ANS which represents the northern extension of the EAO and upper crustal equivalent of the Mozambique belt (Hamimi et al. 2013a, b).

(1) Infracrustal orogenic model, whereby ophiolites and island arc volcanics and volcaniclastics were thrusted over an old craton consisting of high-grade gneisses, migmatites, and remobilized equivalents during Neoproterozoic time (e.g., Akaad and Noweir 1980; El-Gaby et al. 1988; Abdel Khalek et al. 1992; Khudeir and Asran 1992).
(2) Turkic-type orogenic model, whereby much of the ANS formed in broad fore-arc complexes (Şengor and Natal'in 1996).
(3) Hot-spot model, whereby much of the ANS is due to accretion of oceanic plateau formed by upwelling mantle plumes (Stein and Goldstein 1996).
(4) Arc assembly (arc accretion) model, whereby EAO juvenile crust was generated around and within a Pacific-sized ocean (Mozambique Ocean). This model was proposed first by Vail (1985) and Stoeser and Camp (1985), and modified by Stern (1994).

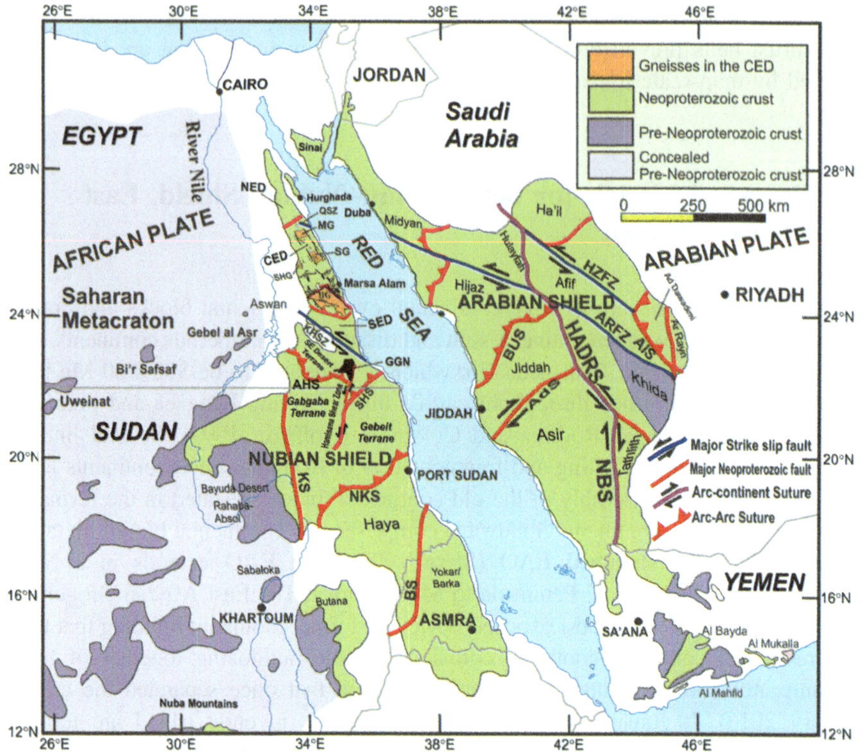

Fig. 1 Relationship between the Arabian–Nubian Shield (ANS) to the adjacent older continental crust. The border with the Saharan Metacraton, on the west, is the effective contact between ANS and West Gondwana: a border, on the east, with putative East Gondwanan crust, is not certain. NED = North Eastern Desert; CED = Central Eastern Desert; SED = South Eastern Desert; QSZ = Queih Shear Zone; MG = Meatiq Gneisses; SG = Sibai Gneisses; MG = Hafafit Gneisses; SHG = Shalul Gneisses; KHSZ = Kharit-Hodein shear zone: SHS = Sol Hamed Suture; AHS = Allaqi–Heiani Suture; GGN = Gabal Gerf Nappe; NKS = Nakasib Suture; KS = Keraf Suture; BS = Baraka Suture; AIS = Al Amar Suture; ADS = Ad Damm Suture; BUS = Bir Umq Suture; YS = Yanbu Suture; HADRS = Hulayfah-Ad Dafinah-Ruwah; NBS = Nabatih Suture; HZFZ = Halaban-Zarghat Fault Zone; ARFZ = Al Rika Fault Zone (compiled from Johnson et al. 2011; Abdeen and Abdelghaffar 2011; Fritz et al. 2013; Abd El-Wahed et al. 2016)

3 Rock Succession in the Egyptian Nubian Shield

The Precambrian igneous–metamorphic complex (basement) of Egypt, covers $\sim 100{,}000\ km^2$, crops out along the Red Sea hills in the Eastern (Arabian) Desert and southern Sinai Peninsula as well as limited areas in the south Western (Libyan) Desert (Oweinat area) (Fig. 2) (El- Gaby et al. 1990; Hassan and Hashad 1990). The basement complex of Egypt is part of the Arabian–Nubian Shield (ANS). The ANS is the largest tract of juvenile continental crust of Neoproterozoic age on Earth (Patchett and Chase 2002). It forms the suture between East and West

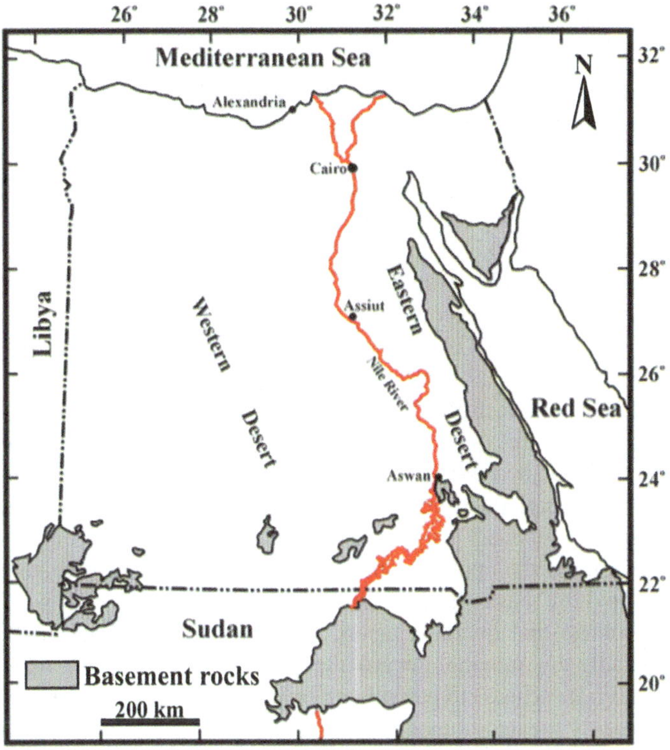

Fig. 2 Distribution of the Precambrian basement rocks in Egypt (El-Gaby et al. 1990). Rocks of Archean age are confirmed in the Western/Libyan Desert only

Gondwana at the northern end of the East African Orogen. The ANS has a complex history including a record of the breakup of Rodinia at circa 900–800 Ma, and the evolution of numerous arc systems, oceanic plateaux, oceanic crust, and sedimentary basins (Stern 1994; Stein and Goldstein 1996; Johnson 1998; Kusky et al. 2003; Kusky 2004). The ANS can be represented by a complex amalgam of arc, ophiolite and micro-continental terranes that had been resulted from the Neoproterozoic closure of the Mozambique Ocean (e.g., Kusky et al. 2003). The shield was subsequently over buried by the Phanerozoic sediments. In the Oligocene and younger times, the shield had been exposed by uplift and erosion on the Red Sea flanks (Stern et al. 2004). Accordingly, the basement complex is well exposed along the Eastern Desert of Egypt and western Arabia.

Generally, the ANS/basement of Egypt comprises three main tectonostratigraphic units (e.g., Ali et al. 2012) from bottom upward as follows: (1) high-grade gneisses and migmatites, (2) arc-type volcanic/volcano-sedimentary units, along with dismembered ophiolites, and (3) the Ediacaran Hammamat and Dokhan supracrustal sequences. And granitoids intrude all the three units.

3.1 Gneisses and Migmatites

Gneisses and migmatites constitute the infrastructural rocks of the Egyptian basement complex, as well as the ANS. They comprise 7% of the surface outcrops of the Egyptian basement. Pohl (1979) inferred that gneiss domes represent the northern continuation of the Mozambique belt. They show tectonic windows or continental basement being exposed in the uplifted regions of the Pan-African belt (e.g., El-Gaby et al. 1984). Among the most famous gneiss domes of the Egyptian Eastern Desert: Meatiq, Um Had, Sibai, El Shalul, Hafafite, Migif, Beitan, and Fiqo from the North- to South Eastern Desert (Fig. 3) (e.g., El-Gaby et al. 1984; Hamimi et al. 1994; Hamimi 1996; Gahlan 2006; Fowler et al. 2007; Abd El-Wahed 2007, 2008; Andersen et al. 2009, 2010; Ali et al. 2010, 2012; Lundmark et al. 2012, among others). They are structurally below arc and ophiolite rocks.

Briefly, the infrastructural rocks comprise schists, gneisses, and migmatites constituting the so-called *Tier-I* (e.g., Bennet and Mosely 1987; Greiling et al. 1994). They were derived from an ancient sedimentary succession accumulated along a passive continental margin, being affected by low-*P* and high-*T* regional metamorphism up to the upper amphibolite facies. The sedimentary sequence was also intercalated by basic igneous sills (ortho-amphibolites) as well as marls (para-amphibolites) and quartzite layers. The metamorphic grade and grain size change gradually by moving toward the core of the antiform, from common metamorphites (i.e., schists) to partially migmatized metatexites before reaching the high-grade weakly foliated homogenized mesocratic to leucocratic diatexites which occupy the core of the antiform. The homogenized diatexites comprise tonalite to granodiorite gneisses. The gneiss domes are commonly surrounded folded thrust belt of low-grade supracrustal rock assemblage, namely, island arc metavolcanic and metavolcano-sedimentary sequences, and ophiolites.

A variety of ages have been proposed for the Egyptian Eastern Desert gneisses; pre-Neoproterozoic/pre-Pan-African (e.g., El-Ramly and Akaad 1960; El-Ramly 1972; El-Gaby et al. 1984, 1990; Hamimi et al. 1994; Khudeir et al. 2008) or juvenile/Pan-African (El-Ramly et al. 1984; Greiling et al. 1984, 1988; Kröner et al. 1994; Andresen et al. 2009, 2010; Ali et al. 2010, 2012; Augland et al. 2012; Lundmark et al. 2012). The isotopic dating of gneiss domes at Meatiq (631 ± 2 Ma, Andersen et al. 2009), El Shalul (631 ± 6 Ma, Ali et al. 2012), Hafafit (659 ± 5, Lundmark et al. 2012) and Beitan ($\sim 725 \pm 9$ Ma, Ali et al. 2015) has denied the pre-Neoproterozoic age for those gneisses. Although the pre-Neoproterozoic zircons are recognized in the juvenile Eastern Desert basement rocks (Ali et al. 2015, and references therein), which indicate a contribution from an older crustal component to the Eastern Desert juvenile rocks. Isotopic data (U–Pb zircon and Sm–Nd whole rock) further indicate that most of the Eastern Desert curst was derived from a depleted mantle source (e.g., Moghazi et al. 1998; Moussa et al. 2008; Andresen et al. 2009; Ali et al. 2009, 2010, 2012; Liégeois and Stern 2010; Augland et al. 2012; Lundmark et al. 2012).

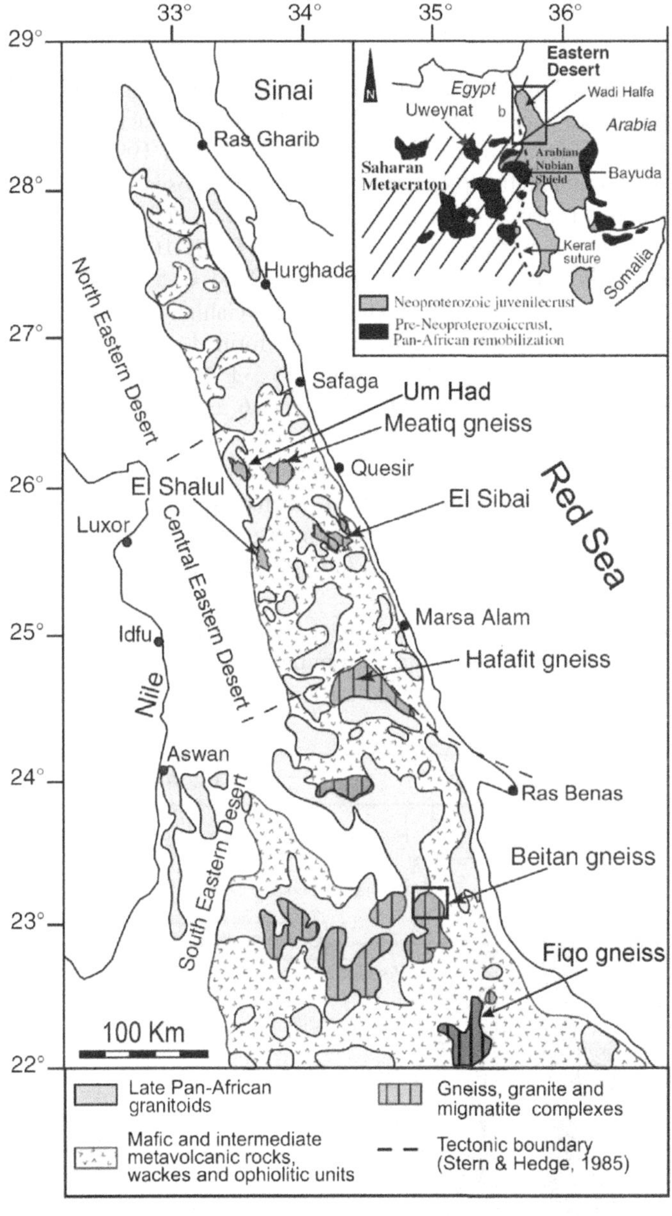

Fig. 3 Distribution of the most famous gneiss domes in the Central and South Eastern Desert of Egypt, modified after Ali et al. (2015) and locality of the Fiqo gneisses is after Gahaln (2006). The inset is a general geological map of the Arabian–Nubian Shield, Saharan Metacraton, and Archean and Paleoproterozoic crust that were remobilized during the Neoproterozoic era

3.2 Relict Oceanic Lithosphere

It is widely accepted that the ANS ophiolites represent remnants of Neoproterozoic oceanic lithosphere (~870–690 Ma) obducted along destructive plate boundaries during the Pan-African orogeny (~750–650 Ma) (Stern 1994; Ali et al. 2010). A variety of tectonomagmatic scenarios have been proposed for the Egyptian ophiolites, including: (i) back-arc basins (e.g., Ahmed et al. 2001; Farahat et al. 2004b; Abd El-Rahman et al. 2009a); (ii) mid-ocean ridges (MOR) (e.g., Zimmer et al. 1995; Khalil 2007); and/or (iii) forearcs (e.g., Stern et al. 2004; Gahlan 2006; Azer and Stern 2007; Abd El-Rahman et al. 2009b; Ahmed 2013; Gahlan et al. 2015).

The Egyptian ophiolites seem to be formed at ~730–750 Ma (e.g., Allaqi ~730 Ma, Ali et al. 2010; Fawakhir 736.5 ± 1.5 Ma, Andersen et al. 2009; Gerf ~750 Ma, Zimmer et al. 1995; Ghadir 746 ± 19 Ma, Kröner et al. 1992; Gerf 741 ± 21 Ma; Kröner et al. 1992) (Fig. 4).

According to the Penrose Conference ophiolite model (Anonymous 1972), the Egyptian ophiolites are variably dismembered due to tectonic disruption. A pseudo-lithostratigraphic column for the Egyptian ophiolites can display a sequence from mantle section, upward through mafic crust to the overlying mafic volcanic rocks (Fig. 5) (Gahlan 2006).

The mantle section is represented by sheet-like bodies of serpentinized ultramafics dominated by harzburgite and dunite. Chromitite pods are commonly small-scale and observed in the shallowest parts of the mantle section. The uppermost part of mantle

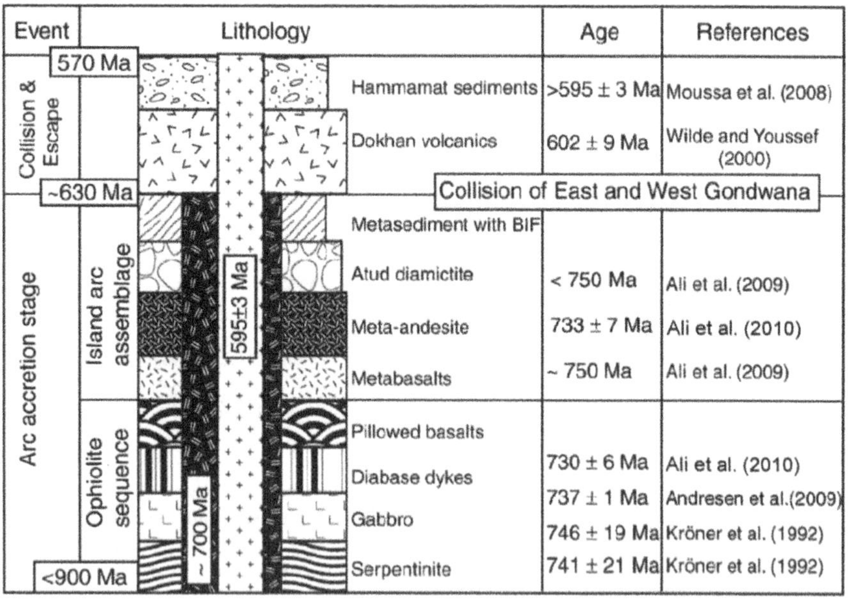

Fig. 4 An evolutionary diagram shows major tectonic events of the Neoproterozoic basement complex of Egypt (Ali et al. 2010, and references therein)

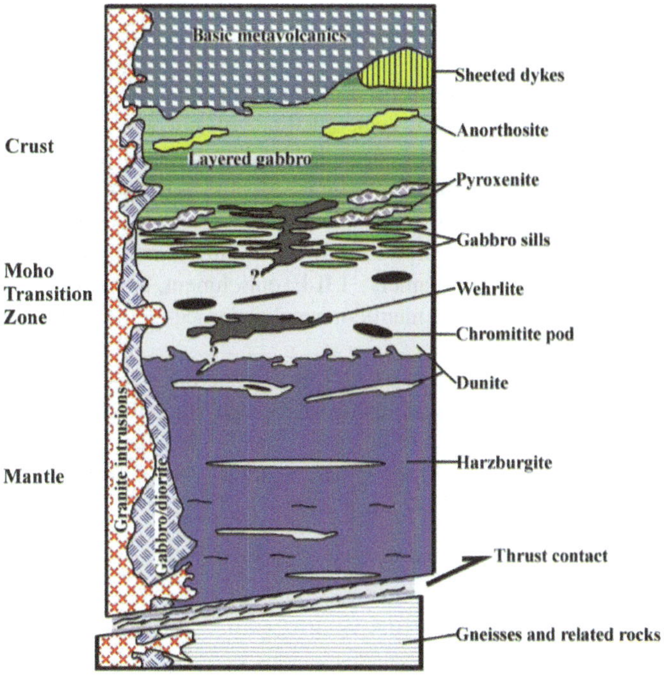

Fig. 5 A general pseudo-lithostratigraphic column of the Egyptian ophiolites, an example from the Abu Dahr ophiolite (Gahlan et al. 2015)

sometimes contains large masses of serpentinized dunite containing concordant gabbro sills or layers of serpentinized ultramafic cumulates, representing the so-called Moho Transition Zone (MTZ) (Fig. 5). The ultramafic cumulates are represented by wehrlites and pyroxenites, being restricted to the Moho transition zone and the lower crustal sequence (e.g., Ras Salatit ophiolite, Gahlan et al. 2012) (Fig. 5). Toward the sole thrust and along shear zones, the serpentinized ultramafics are transformed into schistose serpentinites and talc-carbonates.

The crustal section is represented by metagabbro and metavolcanic rocks. The metagabbros range from medium-grained to appinitic pyroxene metagabbro, Hb-metagabbro, and meta-anorthosite. Isomodal-type igneous layering and foliation are observed. The sheeted dykes were locally observed in the best cases as massive 100% dykes with chilled margins discriminate the contacts between them (e.g., Gerf ophiolite, Gahlan 2006; Sol Hamed ophiolite, Gahlan et al. submitted). The metavolcanics are represented by basalts and basaltic andesites. They are locally pillowed and associated with ophiolitic metasedimentary rocks. They stratigraphically and structurally overlie the metagabbros. Locally, near and along thrust contacts, the metavolcanic rocks are transformed into chlorite and actinolite schists.

3.3 Island Arc Assemblage

The island arc rock assemblage of the Egyptian basement includes (1) an intrusive gabbro–diorite–tonalite suit, and (2) calc-alkaline volcanic and volcano-sedimentary rocks, with rare sedimentary iron formation and carbonates (e.g., El-Gaby et al. 1984; Mohamed and Hassanen 1996; Abu El-Ela 1997; Kharbish 2010; Azer et al. 2016, and references therein). The island arc rocks are commonly regionally deformed, sheared, and metamorphosed up to the lower amphibolite facies. They show a calc-alkaline nature; namely, LILE enrichment, negative Nb and P_2O_5 anomalies, and jagged primitive mantle-normalized trace-element pattern compared to the smooth MORB pattern.

According to Be'eri-Shlevin et al. (2009b, 2011), the calc-alkaline magmatism in the ANS can be divided into two stages; the first stage (625–650 Ma) includes deformed syn- to late-orogenic island arc intrusive and the associated extrusive rocks, and the second stage (590–625 Ma) includes less deformed post-collision calc-alkaline magmatic rocks.

3.4 Dokhan Volcanics

The Dokhan volcanics are K-rich calc-alkaline volcanic rocks that characterize the stage after cratonization of the shield in the Neoproterozoic ($\sim$590–610 Ma) (Fig. 4). They record the second major volcanic episode in the Neoproterozoic crust of the ANS (e.g., Abdel Wahed et al. 2012). The Dokhan volcanics are largely restricted to southern Sinai, North, and Central Eastern Desert of Egypt, in decreasing order of abundance (Fig. 5). Additionally, Gahlan (2003) recorded the Dokhan volcanics in the southernmost Eastern Desert, along the Egyptian–Sudanese boarder, at Wadi Soaorib.

They form a thick sequence of variegated stratified andesitic to rhyolitic lava flows and their pyroclastics as well as ignimbritic rhyolites of Ediacaran age (615 $\pm$ 4 Ma, Breitkreuz et al. 2010; Basta et al. 1980). The Dokhan volcanics are characterized by K-rich nature, great abundance of acidic volcanics, common presence of ignimbrites and welded tuffs, LILE enrichment relative to HFSE, low-pressure fractionation, low Nb and Ta, depletion of Sr and Ti, and high total REEs with LREE enriched over HREE (e.g., El-Gaby et al. 1991; Abdel Wahed et al. 2012). Moreover, these volcanics provide evidences of fractional crystallization and crustal contamination magma processes.

A variety of origins have been proposed for the Dokhan volcanic: (1) compressional tectonomagmatic regime/active continental margin (e.g., Basta et al. 1980; El-Gaby et al. 1989; Abdel Rahman 1996); (2) extensional tectonomagmatic regime/rift system (e.g., Stern et al. 1984); (3) and/or transitional tectonomagmatic regime between extensional and compressional (e.g., Mohamed et al. 2000; Eliwa et al. 2006).

3.5 Hammamat Molasse Sediments

The Hammamat sediments are late-orogenic molasse-type fluvial sediments deposited in foreland, strike-slip, intermontane basins (e.g., El-Gaby et al. 1984; Fritz and Messner 1999; Abd El-Wahed 2010). They commonly crop out in N–S extent along the North- and Central Eastern Desert of Egypt (e.g., Hassan and Hashad 1990). They mark a significant change in the Pan-African tectonics, the end of compressional and the onset of extensional regime (Stern and Hedge 1985).

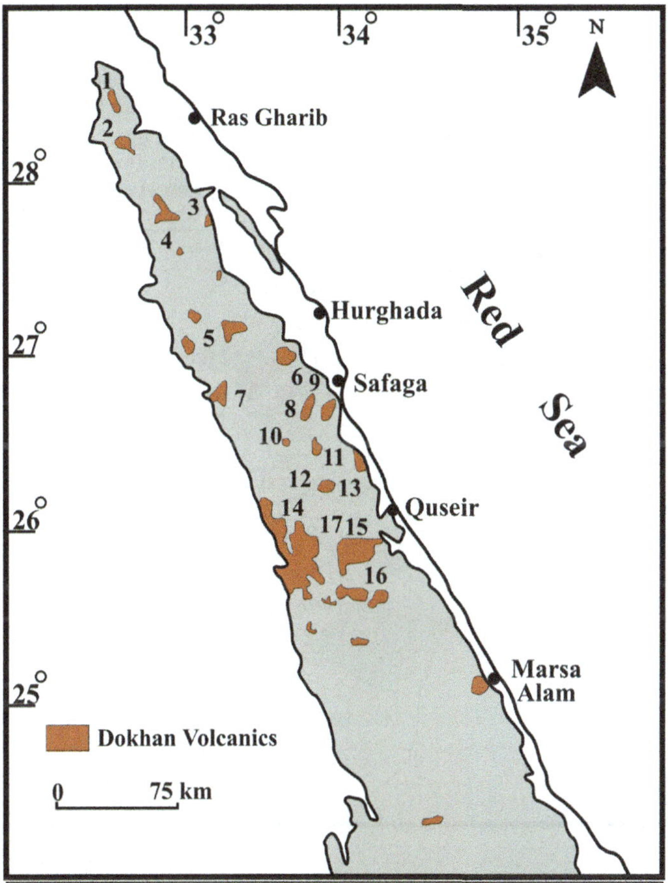

1. Um Arta	7. W. Fatira	13. W. Hamrawein
2. G. Gharib-G. Sager El-Abd	8. W. Um Taghir	14. W. Hammamat
3. Esh El Malaha	9. G. Nugara	15. W. kareim
4. G. Abu Had	10. W. Wassif	16. W. Arak
5. G. Dokhan (Type locality)	11. W. Queih	17. W. Ambagy
6. G. El-Shayib	12. W. Sodmein	

Fig. 6 Distribution of the most famous Dokhan volcanic outcrops in the Eastern Desert of Egypt (El Ramly 1972; Hashad 1980)

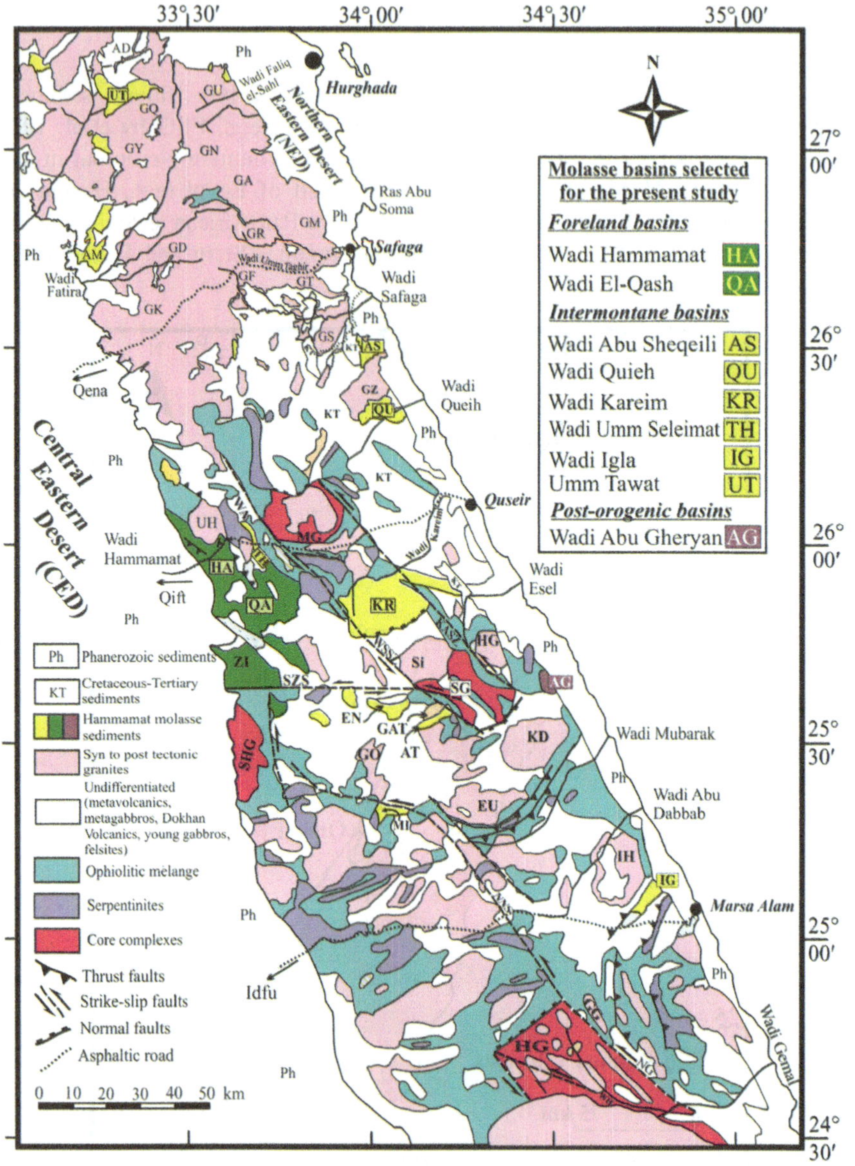

The lower part of the Hammamat sequence comprises red-colored clastics (Igla formation), which passes upward into greenish gray subgraywackes, volcanic arenites, and conglomerate banks representing fanglomerates (El-Gaby et al. 1984). According to Willis et al. (1988), the lower part of the Hammamat sequence comprises polymictic breccia and the upper part comprises sandstone and shale.

◂**Fig. 7** Simplified geological map of the Central and part of the Northern Eastern Desert of Egypt showing distribution of Hammamat molasse basins (after Abd El-Wahed 2010; Compiled from the Geological Map of Egypt (El-Ramly 1972) and the geological map of Quseir (Klitzch et al. 1987). Major structures are after Fritz et al. (1996), Bregar et al. (2002), Shalaby et al. (2005) and Abd El-Wahed 2008). HG, Hafafit Gneiss; WH, Wadi Hafafit; NG, Wadi Nugrus; GG, Gebel (G.) Nugrus; IG, Wadi Igla molasse basin; IH, Igl Al-Ahmar monzogranite, GX, G. Umm Nagat; NNS, Um Nar-Nugrus shear zone; EU, Gebel El-Umra older granite; MI, El-Miyah molasse basin; KD, Kadabora monzogranite pluton; SHG, El Shalul gneisses; AT, Atawi molasse basin; GAT; Gebel Atawi Alkali feldspar granite; EN, Andiya molasse basin; AG, Abu Gheryan molasse basin; Si, Gebel Sibai alkali feldspar granite; SG, Sibai gneisses; WSSZ, Wadi Sitra shear zone; KASZ, Kab Ahmed Shear zone, HG; Homrat Ghaunam alkali feldspar granite; ZI, Wadi Zeidun molasse basin; SZS, Wadi Zeidoun-Wadi El-Shush strike-slip fault; QA, Wadi El-Qash molasse basin; KR, Kareim molasse basin; TH, Um Esh-Um Seleimat molasse basin; HA, Hammamat molasse basin; UH, Um Had granite pluton; WA, Wadi Atalla; MG, Meatiq gneisses; QU, Wadi Quieh molasse basin; AS, Abu Sheqeili molasse basin; GZ, G. Umm Zarabit; GK, G. Kafari; GS, G. Gasus; GD, G. el-Dob; GF, G. Abu Furad; GT, G. Umm Taghir; GR, G. Ras Barud; GM, G. el-Magal; GA, G. Umm Anab; GY, G. Samyuk; GN, G. Shayib el Banat; GQ, G. Qattar; GU, G. Umm Araka; UT, Umm Tawat molasse basin and AD, Gebel Dokhan

They are typically well bedded and commonly overlie the Dokhan volcanics. The Hammamat sequence is largely derived from the erosion of the Dokhan volcanics (El-Gaby et al. 1984). And the sequence in turn is intruded by the syn- to late-orogenic granitoids (younger granites). Alteration is manifested by hematitization, epidotization, and kaolinitization.

Most of the Eastern Desert molasse basins were evolved between 650 and 580 Ma (Fig. 6) in individual basins with different individual tectonic settings (e.g., Shalaby et al. 2006; Abd El-Wahed 2010). Geochemistry revealed that the Hammamat molasse sedimentary rocks were originated from felsic to intermediate igneous sources formed in a continental arc setting.

Abd El-Wahed (2010) concluded that the deformational history of the molasse basins includes two main thrusting events overprinted by an event of sinistral shearing along the Najd Fault System (NFS). The NNW-directed thrusts and the NE-, ENE-, and WSW-trending folds are the structures related to NNW–SSE shortening (650–640 Ma). The SW- and NE-directed thrusts are due to ENE–ESE constriction during oblique convergence and arc accretion around 640–620 Ma. Fritz and Messner (1999) and Abd El-Wahed (2010) classified the Hammamat molasse basins in the Central and part of the Northern Eastern Desert (Fig. 7) into: (i) The foreland basins (e.g., Wadi Hammamat basin, Wadi El-Qash basin); (ii) Intermontane basins (e.g., Wadi Queih basin, Wadi Abu Sheqeili basin, Wadi Kareim basin, Wadi Igla basin, Wadi Um Seleimat basin, Gebel Umm Tawat basin); and (iii) Post-orogenic basins (e.g., Wadi Abu Gheryan basin). The basins which are characterized by prevalence of thrust-related structures include foreland basins (Wadi Hammamat, Wadi El-Qash) and two of the intermontane basins (Wadi Umm Tawat, Wadi Abu Sheqeili basins) (Abd El-Wahed 2010). The other intermontane basins are markedly pull-apart basins characterized by structures related to thrusting overprinted by strike-slip faulting (e.g., Wadi Queih, Wadi Igla, Wadi Um Seleimat). Some basins exhibit structures developed by the sinistral shearing only (e.g., Wadi Kareim, Wadi Atawi, and Wadi El-Miyah basins) (Abd El-Wahed 2010).

3.6 Granites

The Egyptian granites can exclusively be divided into three major groups: (1) older gray granites and (2) younger pink granites, which include (3) A-type (alkaline/peralkaline) granites; all possibly constitute one granite series from less to more evolved (e.g., El-Ramly and Akaad 1960; El-Gaby 1975; Hussein et al. 1982; Farahat et al. 2004a, 2007; Ali et al. 2013).

(1) The older gray granites (615–820 Ma, Stern and Hedge 1985) comprise ~27% of the cropping out Egyptian basement (Stern 1979). They show a composition ranges from qz-diorite to granodiorite. They are syn- to late-orogenic, subduction-related, and calc-alkalic I-type granitoids (e.g., Rise et al. 1983; Hassan and Hashad 1990; Kroner et al. 1994; Moghazi 2002; El Mahallawi and Ahmed 2012). Hussein et al. (1982) identified these granites as G_1 granites. Most of the older gray granites are pre-collision of arc assemblages (650–580 Ma, Stern 1994) and variably deformed.
(2) The younger pink granites (590–610 Ma, Moussa et al. 2008; Ali et al 2012, and reference therein) comprise ~16% of the cropping out Egyptian basement (Stern 1979). They show a composition ranges from alkali feldspar granite to normal granite (e.g., Ali et al. 2012; El-Bialy and Omar 2015, and references therein). They are late-orogenic, suture related, and highly fractionated calc-alkalic I-type granitoids (e.g., El-Bialy and Omar 2015, and references therein). Hussein et al. (1982) identified these granites as G_2 granites.
(3) The A-type granites (~590–610 Ma, Be'eri-Shlevin et al. 2009a) are recorded in Sinai, North Eastern Desert, and South Eastern Desert, in decreasing order of abundance (e.g., Stern 1979). They show a composition ranges from alkali feldspar granite to syenite (e.g., Gahlan et al. 2016, and references therein). They are alkaline/peralkaline, anorogenic, and within-plate granitoids (e.g., Abdel Rahman and El-Kibbi 2001). Hussein et al. (1982) discriminated these granites as G_3 granites.

A variety of origins have been proposed for the aforementioned granite groups; (a) partial melting of the crust (e.g., Moghazi et al. 2001; Farahat et al. 2011; El Mahallawi and Ahmed 2012) or (b) fractionation of a mantle derived magma (e.g., Finger et al. 2008; Moussa et al. 2008; Ali et al. 2012).

4 Landsat- and ASTER-Based Mapping of the Egyptian Nubian Shield

The Egyptian Arabian–Nubian Shield consists of extensive outcrops of metamorphosed gneissic domes, ophiolitic-related assemblages, island arc metavolcanics, and their volcaniclastic associations together with clastic Molasse-type sediments, which they intruded by suites of mafic, syn-late to post-tectonic volcanics and granitoids. Remotely sensed processed data including Enhanced Landsat Thematic Mapper (ETM+) and Advanced Spaceborne Thermal Emission Reflection Radiometer (ASTER) have been used by many workers in lithological discrimination and mineral exploration in the Egyptian Nubian Shield (e.g., Kusky and Ramadan 2002; Sadek 2004, 2005; Gad and Kusky 2006; Amer et al. 2010; Sadek and Hassan 2012; Gabr et al. 2015; Abou El-Magd et al. 2013; Asran et al. 2013; Hassan and Ramadan 2014; Hassan et al. 2015; Sadek et al. 2015), where these data were utilized in the geological mapping at various scales and have shown a great success (Qiu et al. 2006; Gaber et al. 2010; Zoheir and Emam 2014 and Kumar et al. 2015). From the lunch of spaceborne multispectral sensors, particularly Landsat (TM and ETM+) in 1982 and 1999 with five and eight spectral channels, respectively, they have been well employed for geological applications (Gad and Kusky 2006). Moreover, the launch of ASTER in December 1999 with three spectral bands in the Visible/near Infrared (VNIR) region, six spectral bands in the Shortwave Infrared (SWIR) region and five spectral bands in the Thermal Infrared (TIR) region with 15, 30 and 90 m spatial resolution has provided a new prospective of investigating the geological materials (Rowan and Mars 2003; Gad and Kusky 2007; Gaber et al. 2010).

Khan et al. (2007) pointed out that the remote sensing data has been used increasingly during the last two decades for mapping, structural analysis, and mineral exploration. There is a wide acceptance that these data conveys useful information for mapping different rock types and their alteration products (Torres-Vera and Prol-Ledesma 2003; Ramadan and Kontny 2004). The recent developments in sensor technology have enabled remote sensing to become an increasingly important tool for mapping lithologies, structures, and ore deposits, particularly for remote areas with little or no access, or areas that lack detailed topographic or geologic base maps (Qaoud 2014). Numerous and success trials for mapping different areas of the Egyptian Nubian shield using Landsat (ETM+) and ASTER were performed by many scientists, especially after the free availability of the Landsat images through the Global Land Cover Facility (GLCF) (www.landcover.org) which authorized by the National Aeronautics and Space Administration (NASA). These studies produced new geologic maps as well as improved pervious geologic maps for these areas, where the resulted maps were verified by the fieldwork, geochemical and pertographical studies.

4.1 Remote Sensing Techniques and Lithological Discrimination

To carry out lithologic mapping and to identify mineral deposits in arid and semi-arid environments, remote sensing data have been analyzed using several digital image processing techniques such as image enhancement, fusion, band rationing, and principal component transformation (e.g., Sultan et al. 1986; Kusky and Ramadan 2002; Gad and Kusky 2007; Youssef et al. 2009; Madani and Emam 2011; Ali-Bik et al. 2012 and Zoheir and Emam 2012). Thus, the development of the specific remote sensing methods and data is considered as a critical part of geological mapping over the last years. Many scientists have statistically selected

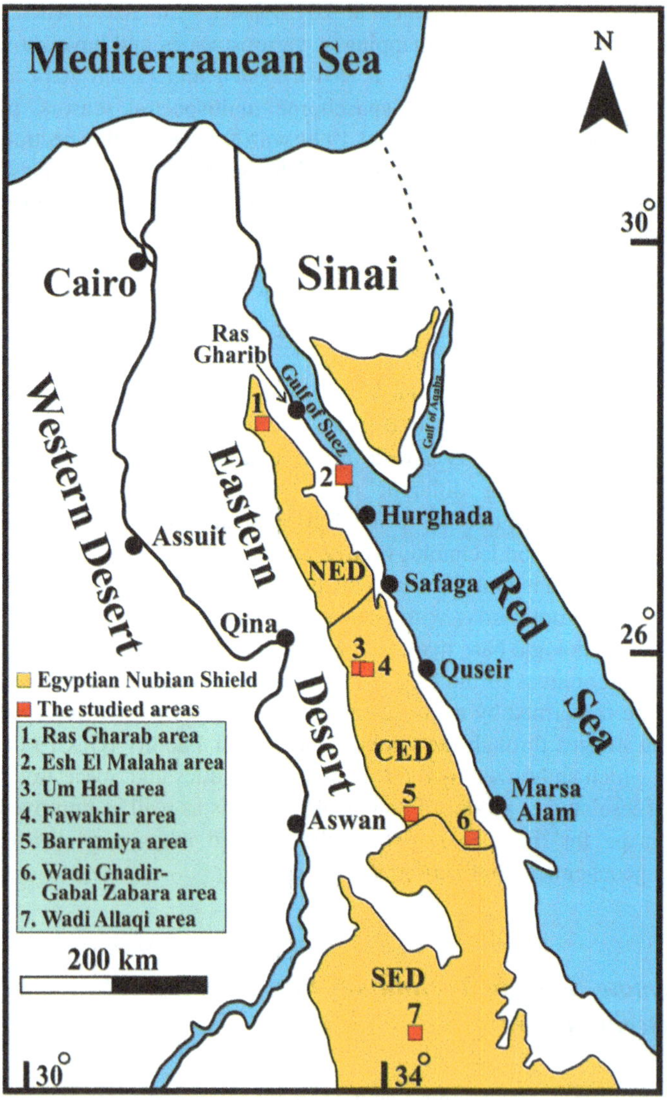

Fig. 8 Location map for seven areas in the Egyptian Nubian shield had been subjected to Landsat and ASTER geological mapping. NED = Northern Eastern Desert, CED = Central Eastern Desert, and SED = Southern Eastern Desert

Landsat TM enhanced false-color composite to map geological terranes with high degree of confidence (Souza Filho and Drury 1997).

Remote sensing techniques usually offer a medium for a successful discrimination and mapping of exposed rocks and associated weathering products (Yenne et al. 2015), providing information that is relevant to maps of bedrock geology (Rowan et al. 2005 and David and Wooil 2012). The reflectance characteristics of individual lithological classes (i.e., rock or soil types) are mainly used as a function of the presence of minerals to be utilized (Yenne et al. 2015). It is essential to note that Landsat bands 4, 5, and 7 are most successfully used to discriminate between major rock types (Rothery 1987) which achieved through several processes.

The enhanced resolutions, high Signal-to-Noise Ratio (SNR), the effective spectral coverage and global data availability makes ASTER more suitable particularly for operational geological applications (Kumar et al. 2015). The three subsystems of ASTER sensor, i.e., VNIR, SWIR, and TIR, has different roles to play in spectroscopy for geological applications such as the VNIR region provides spectral features of transition metals such as iron, SWIR region is very effective for analyzing spectral characteristics of carbonate, hydrate, and hydroxide minerals, and TIR region is effective for characterization of silicates (Clark 1999; Gad and Kusky 2007). The ASTER sensor acquires earth's surface imagery in the VNIR, SWIR, and TIR wavelength regions and has offered a great opportunity of using these data sets for mapping of various lithological units (Aboelkheir et al. 2010; Tangestani et al. 2011), minerals (Salem et al. 2014; Salem and Soliman 2015).

Various image enhancement techniques such as Principle Component Analysis (PCA), Minimum Noise Fraction (MNF), Band Ratios (BRs), Band Combinations (BCs) and Spectral Indices (SIs) as well as image classifications can be used (Gad and Kusky 2006, 2007). The present study offers some of these remote sensing techniques to discriminate the various lithological units in selected areas in the Egyptian Nubian Shield (Fig. 8). Actually, aforementioned image processing techniques will be applied after the needed preprocessing techniques (e.g., geometric, atmospheric, and radiometric corrections) performed to the Landsat and ASTER images.

4.2 False-Color Composites (FCC)

Color composite is an image produced by displaying multiple spectral bands as colors different from the spectral range in which they were taken. This method is commonly used for displaying multiband (multichannel) imagery (Kalelioğlu et al. 2009). This is usually achieved by assigning three of the image bands to the fundamental colors red (R), green (G) and blue (B), the combination of which results in a RGB (false) color composite image. The selected band combination should be the most informative one and has the highest sum of standard deviation and lowest correlation among band pairs. Seleim and Hammed (2016) selected the False-Color Composite (FCC) of RGB 742 of Landsat 8 as the best one to

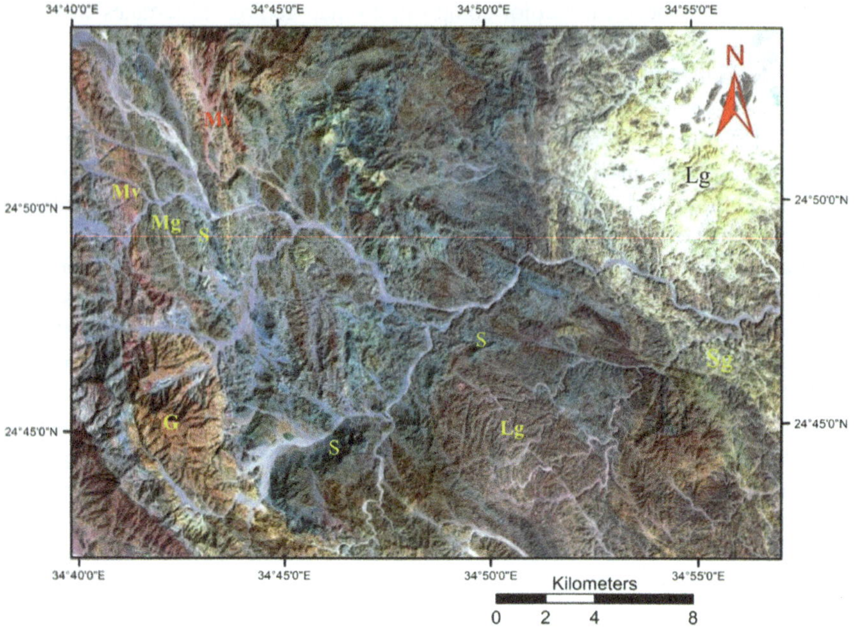

Fig. 9 False-color composite of Landsat 7 (ETM+) image RGB 742 of Wadi Ghadir-Gabal Zabara area at the central Eastern Desert of Egypt (after Kamel et al. 2016). Gneiss (Gn), serpentinites (Sp), metagabbroic (Mgb), metavolcanics (Mv), syn-tectonic granite (Sgr), late-tectonic gabbro (Lgb), and late-tectonic granite (Lgr)

discriminate the rock units in Esh El Malaha area at the North Eastern Desert of Egypt. Where the FCC of RGB 742 of Landsat 7 success to separate the rock types in Wadi Ghadir-Gabal Zabara area and Wadi Allaqi area at the central and southern Eastern Desert of Egypt, respectively (Salem and Soliman 2015; Kamel et al. 2016). Kamel et al. (op. cit) pointed out that the RGB 742 band combination success to discriminate the serpentinites, metavolcanics, metagabbros, and syn- and late-tectonic granites rocks in Wadi Ghadir-Gabal Zabara area obviously (Fig. 9). The FCC RGB 731 of ASTER has the ability to differentiate between granitic rocks and serpentinites at Fawakhir area (Abou El-Maged et al. 2015; Fig. 10) and to discriminate the different rock types in the Barramiya district (Salem et al. 2014) at the central Eastern Desert of Egypt.

4.3 *Principal Component Analysis (PCA)*

The principal component analysis (PCA) is the image processing technique commonly used for analysis of correlated multidimensional data. It is multivariate statistical method used to compress multispectral data sets by calculating a new

Fig. 10 False-color composite of ASTER image RGB 731 of Fawakhir area at the central Eastern Desert of Egypt (after Abou El-Maged et al. 2015)

coordinate system and redistribute data onto a new set of axes in multidimensional space (Drury 1987; Gupta 1991). This technique has several advantages, where most of the variance in a multispectral data set is compressed into one or two PC images. Moreover, noise may be relegated to the less correlated PC images and the spectral differences between materials may be more apparent in PC images than in individual bands (Sabins 1997). Seleim and Hammed (2016) adopted three PCA RGB combinations of PC2, PC1, PC3; PC5, PC3, PC2 and PC4, PC5, PC2 of Landsat 8 to discriminate the lithologic units in Esh El Malaha area at the North Eastern Desert of Egypt. They success to separate the metavolcanics, Dokhan volcanics, and granitoid rocks as well as sedimentary cover using the fore mentioned PCs images. Where the PCA RGB combination of PC2, PC1, PC4 of Landsat 7 is the good one to differentiate between the different rocks assemblage of Wadi Ghadir-Gabal Zabara area at the central Eastern Desert of Egypt (Kamel et al. 2016). Moreover, Qaoud (2014) recognized the gneisses, serpentinites, Hammamat sediments, and granitoids in Um Had area at the central Eastern Desert of Egypt using PCA RGB combination of PC4, PC2, PC3 of Landsat 7 (Fig. 11). In addition, different PCs combinations of ASTER images played a role to differentiate between

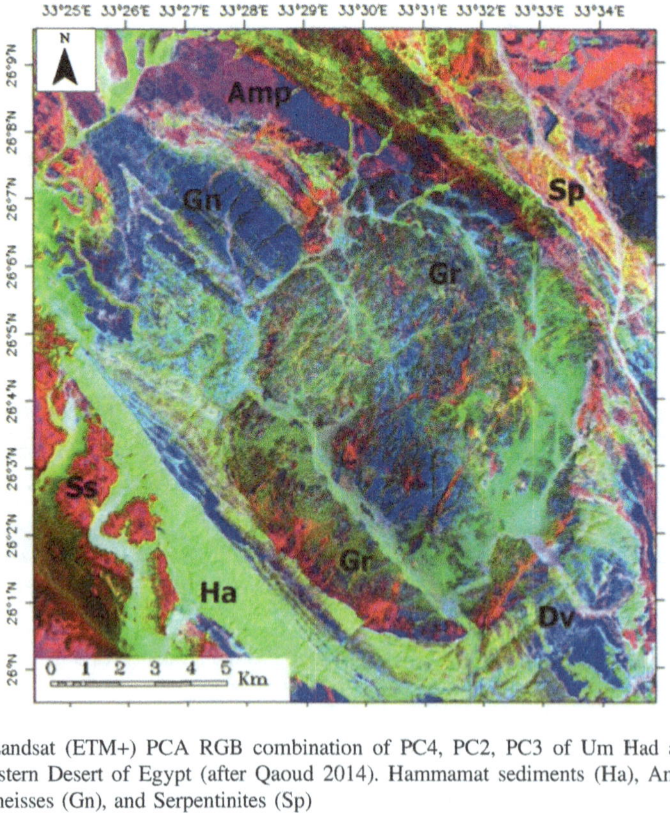

Fig. 11 Landsat (ETM+) PCA RGB combination of PC4, PC2, PC3 of Um Had area at the central Eastern Desert of Egypt (after Qaoud 2014). Hammamat sediments (Ha), Amphibolites (Amp), Gneisses (Gn), and Serpentinites (Sp)

Fig. 12 ASTER PCA RGB combination of PC5, PC7, PC6 of Ras Gharib area at the North Eastern Desert of Egypt (after Jakob et al. 2015)

rock types in different areas of the Eastern Desert of Egypt. Jakob et al. (2015) used PCA RGB combination of PC5, PC7, PC6 to recognize the rock types in Ras Gharib area (Fig. 12) but Salem et al. (2014) adopted PC3, PC4, PC2 PCA RGB combination to differentiate between rock units in the Barramiya district.

4.4 Band Ratioing

The ratio images were prepared simply by dividing the Digital Number (DN) values of each pixel in one band by the DN values of another band (Drury 1993). Band ratio technique enhances the objects based on the differences in reflectivity between the numerator and denominator spectral bands. There are some effective factors controlling the lithological mapping using remote sensing techniques including the increased concentration of minerals relative to the background in the locality and the mineral assemblage characteristics (Frei and Jutz 1990). The main advantage of band ratio images is that they used to reduce the variable effects of illumination condition, thus suppressing of the expression of topography (Crane 1971). Band ratios have been used successfully in lithological mappings for the Arabian–Nubian shield and for other areas worldwide (Gad and Kusky 2006), where the band selection for the different ratio images used is based on the spectral signature of these rocks.

The band ratio transformation of Landsat ETM+ and ASTER data is useful for qualitative detection of hydrothermal alteration minerals (Di Tommaso and Rubinstein 2007), and also has wide acceptance in geological mapping in the Eastern Desert of Egypt (e.g., Qiu et al. 2006; Amer et al. 2010; Aboelkhair et al. 2010; Madani and Emam 2011). For example, Landsat ETM+ band ratios (5/7, 5/1, 5/4 * 3/4) in RGB coloring mode have been used for mapping serpentinites in the Egyptian Nubian Shield (Sultan et al. 1986). They concluded that these band ratios can be used to distinguish serpentinites from the surrounding mafic rocks with high amounts of magnetite and hydroxyl-bearing minerals in arid regions. Gad and Kusky (2006) used the false-color combinations of ETM+ band ratios (5/3, 5/1, 7/5) and (7/5, 5/4, 3/1) for mapping serpentinites in Barramiya area in the central Eastern Desert of Egypt (Fig. 13). They suggested that these band ratios can be used as well as the Sultan et al. (1986) band ratios for mapping serpentinites in the Eastern Desert of Egypt. Kamel et al. (2016) adopted Landsat ETM+ band ratios

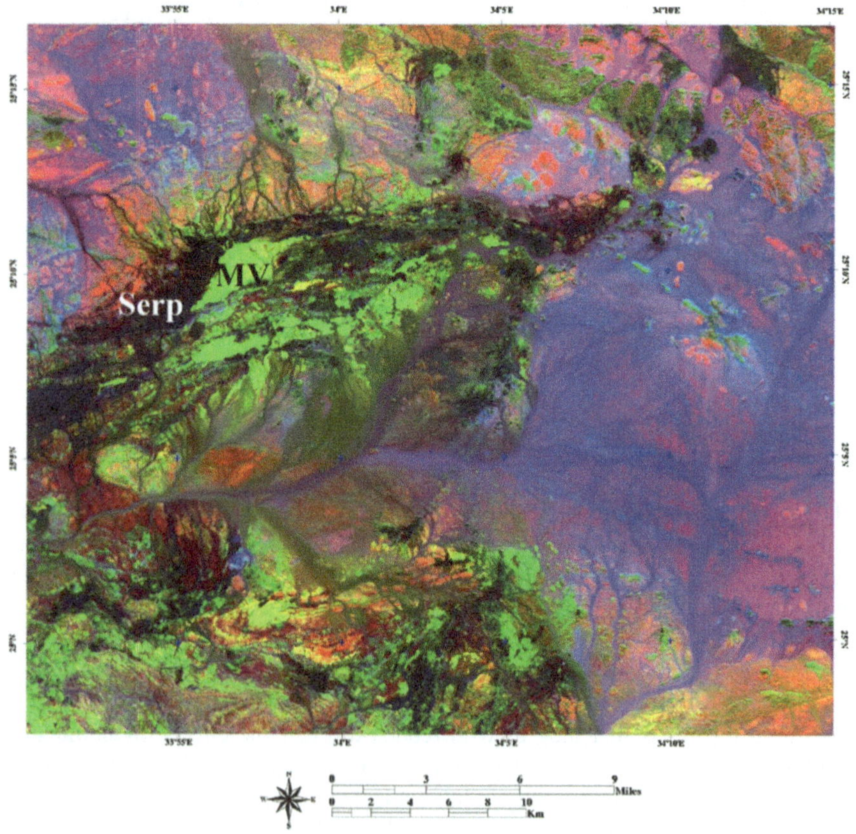

Fig. 13 Landsat TM RGB band ratio image (7/5, 5/4, 3/1) for the Barramiya area at the central Eastern Desert of Egypt (after Gad and Kusky 2006). Serpentinites (Serp) and Metavolcanics (Mv)

(5/1, 3/2, 7/2) and (3/1, 4/2, 5/7) to differentiate serpentinites and granites from their surrounding rocks, respectively. Moreover, Youssef et al. (2009) used the false-color combination of Landsat ETM+ band ratios (5/3, 5/1, 7/5) to discriminate between different granitic phases of the Kadabora granitic intrusion in the central Eastern Desert of Egypt.

Abdeen et al. (2001) used band ratios (4/7, 3/4, 2/1) of ASTER image, which are equivalent to (5/7, 4/5, 3/1) Landsat TM image, also named as Abram's combination, for mapping units of the Neoproterozoic Allaqi Suture of southeast Egypt, mainly serpentinites, marble, and granite. Amer et al. (2010) used ASTER band ratios ((2 + 4)/3, (5 + 7)/6, (7 + 9)/8) to distinguish between ophiolites and granites and for general lithological mapping of arid areas. Zoheir and Emam (2012) found that the band ratio (4/7, 2/4, 6/8) of ASTER images, the most efficient in differentiating all lithologic units and hydrothermally altered zones around Gabal Egat within the Wadi Allaqi region at the southern Eastern Desert (Fig. 14). Moreover, Jakob et al. (2015) considered the band ratios of (7/6, 2/1, 4/6) and (8/6, 2/1, 4/8) of ASTER images produced images of high contrast and good discrimination between different lithological units of Ras Gharib area in the north of the Eastern Desert of Egypt.

4.5 Image Classification Techniques

Classification is the process by which pixels having similar spectral characteristics are consequently assumed to belong to the same class that can be identified and assigned a unique color. The base of image classification is in comparing it to predefined class, which requires definition of the classes and methods for comparison. Definition of the predefined classes is an interactive process and is carried out during the training process or collecting the spectral signature. After the training sample sets have been defined, classification of the image can be carried out by applying a classification algorithm. The choice of a particular algorithm depends on the purpose of the classification, the characteristics of the image and training data. There are two types of image classification can be identified unsupervised and supervised. Where the first one needs a prior knowledge about the study area, but the second one needs to build training areas from the fieldwork or old geologic maps of the study area to use them as a base to classification.

Several workers were produced geological maps for a selected areas at the Egyptian Nubian Shield based on the supervised classification of Landsat (ETM+) and ASTER images with satisfied accuracies. Jakob et al. (2015) differentiated between the rock units in Ras Gharib area using the Support Vector Machine (SVM) algorithm to Landsat and ASTER images and produced a geologic map with overall accuracy 99.81% (Figs. 12 and 15). They prepared geological map shows clear improvements and variations to the earlier version of Breitkreuz et al. (2010) by set up new borders between lithologic units and adding three new rock types to the pervious map. Kamel et al. (2016) performed the maximum likelihood

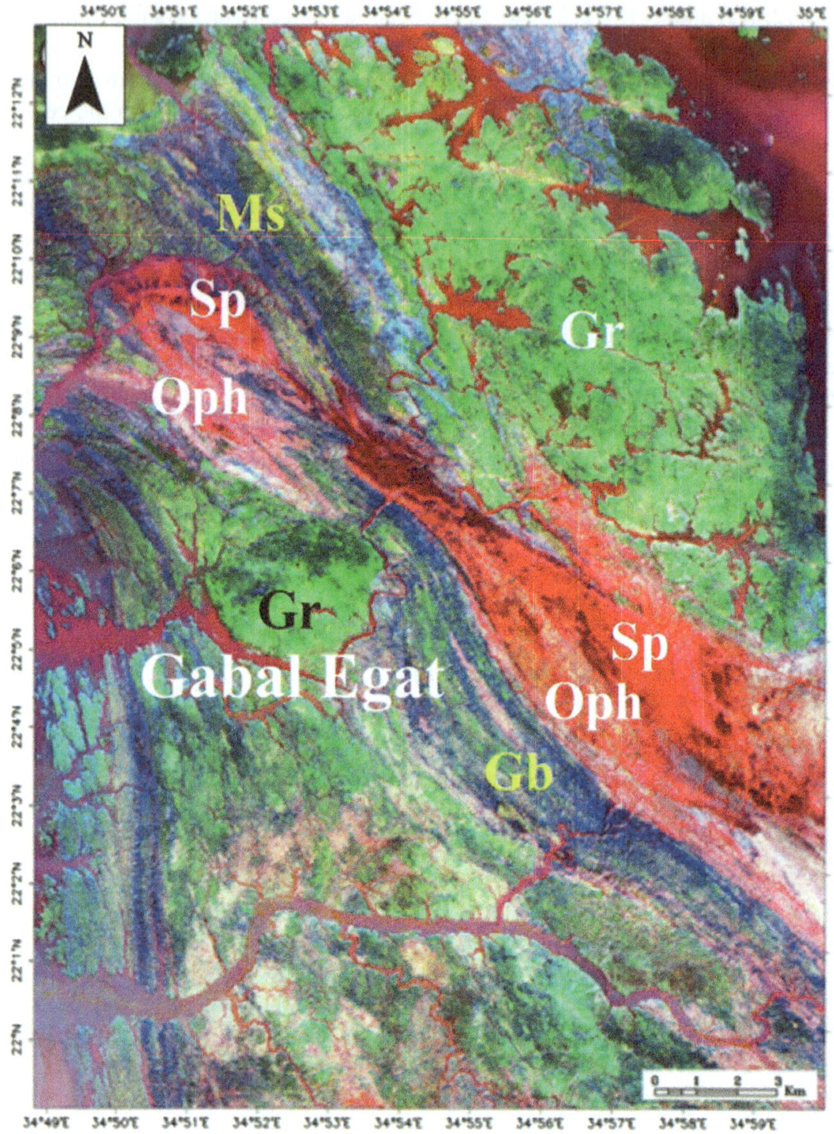

Fig. 14 ASTER RGB band ratio image (4/7, 2/4, 6/8) for Wadi Allaqi region at the South Eastern Desert of Egypt (after Zoheir and Emam 2012). Serpentinites (Sp), Ophiolites (Oph), MetaGabggros (Gb), Metasediments (Ms), and Granites (Gr)

algorithm to Landsat 7 image to discriminate between the different rocks in Wadi Ghadir-Gabal Zabara area at the central Eastern Desert of Egypt. They displayed the defined rock units of granite gneiss, serpentinites, metagabbros, metavolcanics, Melange, and late-tectonic gabbro syn- and late-tectonic granites in a thematic map

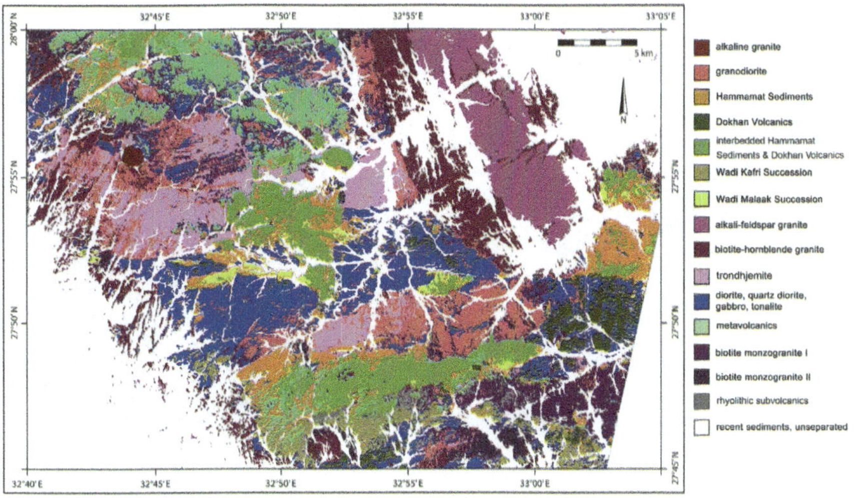

Fig. 15 The classified geologic map of Ras Gharib area, North Eastern Desert, Egypt (after Jakob et al. 2015)

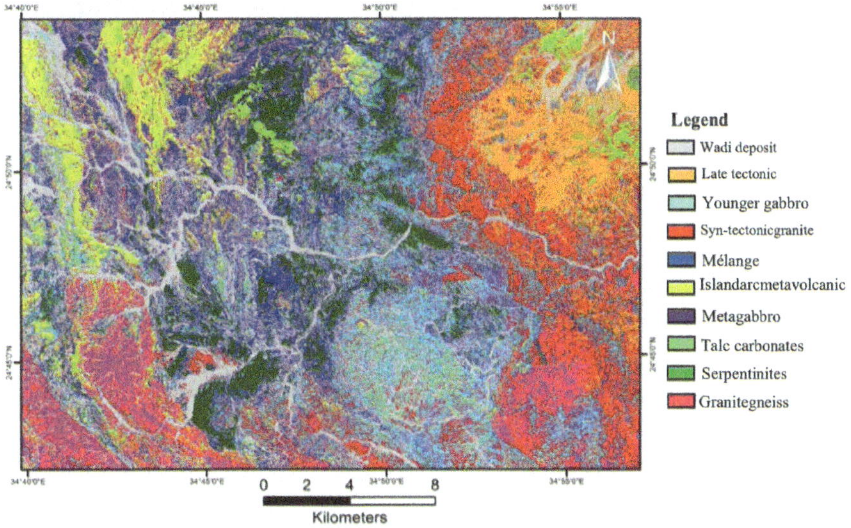

Fig. 16 The classified geologic map of Wadi Ghadir-Gabal Zabara area at the central Eastern Desert of Egypt (after Kamel et al. 2016)

(Fig. 16). Moreover, Seleim and Hammed (2016) produced a new geologic map of Esh El Malaha area at the north of the Eastern Desert of Egypt, with 13 classes using Mahalanobis distance algorithm to Landsat 8 image (Fig. 17). They added that the new lithologic map produced from the remote sensing data shows more

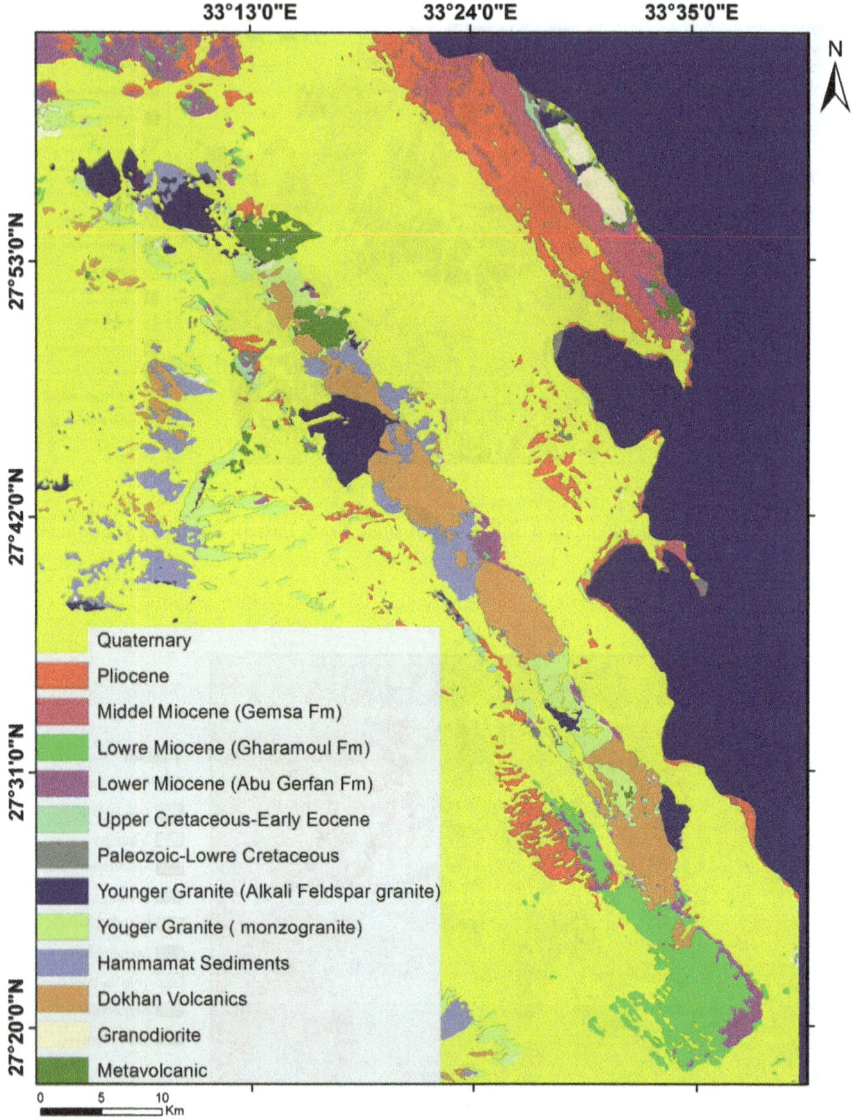

Fig. 17 The classified geologic map of Esh El Malaha area at the North Eastern Desert of Egypt (after Seleim and Hammed 2016)

clear improvement and variations than the previously published maps and can discriminate easily between the granitic rock phases at the north of Gabal El Zeit.

Finally, Landsat and ASTER multispectral remote sensing imagery provides a potentially useful data source that can be used for discriminate between the geological units in the Egyptian Nubian Shield. The processing techniques for these data enable the rapid and inexpensive mapping of surface geological and mineralogical features, especially in arid regions. Several enhancements of Landsat and ASTER images such as band ratios, false-color composites, and principal component analysis as well as image classification are used and evaluated to obtain new geologic maps or improve the previous geologic ones. Where many studies stated that the maps derived from remote sensing data agree well with the field derived maps and show details unobtainable from conventional ground-based mapping.

5 Structural Succession and Sequence of Tectonic Events

Detailed field investigations of structural fabrics encountered in the Neoproterozoic crystalline basement outcropping in the Egyptian Eastern Desert, along with their overprinting relations, allow us to establish a plausible structural succession comprising three successive deformations ($D_1 \rightarrow D_3$). The D_1 deformation, termed here as Syn-accretion phase, resulted in the formation of syn-accretionary structures and embraces two progressive stages; D_1a and D_1b. Both stages took place at greenschist facies condition as indicated by minerals forming slip lineations and slickenlines. The D_1a was a shortening stage, concurrent with the E–W assembly of Gondwanalands, led to the formation of W- and WSW-propagated thrusts such as those encountered in Wadi Beitan, Allaqi–Heiani Belt (Abdeen and Abdelghaffar 2011), and Wadi Ghadir in the SED (Greiling et al. 1994). W- and WSW-propagation of D_1a-thrusting was responsible for the formation of obvious thrust-related folds, as in Wadi Beitan and Um Shilman areas. The D_1b was a progressive stage, concomitant with the N-directed escape of the entire ANS, and resulted in the formation of N- (to NNW-) propagated thrusts and thrust-propagation folds which are best typified in Wadi El Mayet, Wadi Mubarak, and Wadi El-Umra in the CED (Shalaby et al. 2005; Abd El-Wahed and Kamh 2010). The D_2 deformation, termed here as post-accretion phase, was also a shortening phase, played a noteworthy role in the structural shaping of the basement complex of the Egyptian Eastern Desert. A wide variety of post-accretionary transpressive structures affiliated to this phase, including the sinistral transcurrent shearing along the NNW-directed Najd Shear Corridor (e.g., Nugrus and Atalla Shear Zones), the dextral transcurrent shearing along the NE-directed megashears (e.g., Idfu-Mersa Alam and Qena-Safaga Shear Zones), and the post-accretionary shear zone-related gneiss domes (e.g., Meatiq, Sibai, Shalul, and Hafafit gneiss domes) (Fritz et al. 1996, 2002, 2013; Loizenbauer et al. 2001; Abd El-Wahed 2008, 2014; Abdeen et al. 2014; Abd El-Wahed et al. 2016; Stern 2017). Deposition of the Hammamat Volcano-sedimentary Sequence in fault-controlled down sags and pull aparts occurred during also during the D_2 deformation. The D_3 deformation, was an extensional long-lasting phase, associated the crustal relaxation followed the Gondwana assembly and the subsequent Red Sea rifting since the Oligocene and younger times. Rejuvenation of the preexisting Neoproterozoic fractures and shear zones, as well as rift-related structures are related to this deformation phase. Also, the ring complexes of the Eastern Desert of Egypt are connected to this phase. These complexes represent the northward continuation of the East African chain of ring complexes, range in age from Cambrian (554 Ma) to Late Cretaceous (89 Ma), and include a wide variety of rock types, ranging from basic to acidic and from undersaturated to quartz-bearing (El-Ramly and Hussein 1983). However, the previously mentioned structures will be dealt with in a way or another in the following sessions.

6 Thrusting and Thrust Duplex System

The Eastern Desert of Egypt, the northwestern part of the Arabian–Nubian Shield, has experienced a polyphase deformation history involving many successive events. The oldest deformation event (D_1) in the Eastern Desert is recorded in its southern part, where fold-thrust belts prevail and thrusts are the first-order kinematic later overprinted by map-scale transpression. The Southern Eastern Desert in general seems to generally represent a deeper level of exposure than the Central Eastern Desert and is less affected by Najd shearing. The major structures in the south are manifested by accretion-induced sutures developed at ca. 750–720 Ma. $D_{1a,\ b}$ is represented by NNW or SSW direction of thrust transport in the Allaqi–Heiani shear zone (Abdeen and Abdelghaffar 2011). The Allaqi–Heiani Shear Zone (Figs. 18 and 19) involves an early N–S shortening followed by a late E–W shortening. The N–S shortening is associated with accretion and obduction of the South Eastern Desert island arc rocks over the Gabgaba island arc rocks to its south (Abdeen and Abdelghaffar 2011). The late E–W shortening is associated with a post-accretion phase probably related to the latest stage of the Pan-African orogen or to the Najd orogen (Abdeen and Abdelghaffar 2011).

The tectonic evolution of the Allaqi–Heiani shear zone is characterized by a polyphase deformation history involving, at least, two phases of deformation (Abdeen and Abdelghaffar 2011). D_1 phase involves N–S to NNE–SSW shortening due to collision between the South Eastern Desert (Aswan) terrane and the Gabgaba terrane (Figs. 18 and 19) between ca. 800 and 700 Ma (Kröner et al. 1992). D_1 structures include WNW–ESE trending imbricate thrust sheets dipping gently toward the NNE indicating a nappe transport toward SW, megascopic, and mesoscopic folds verging to SSW, shear foliations along thrust planes, and SSW plunging stretching lineation. The ENE–WSW collision led to the formation of D_2 structures including NNW–SSE-oriented folds and reactivation of older thrust faults as D_2 transpressional shear zones.

Fold-related faults are dominant in the Central Eastern Desert and are commonly associated with pull-apart basins linked to Najd System (Hamimi et al. 2015b). The most recent subdivision of deformation events in the Central Eastern Desert (Makroum 2001; Fritz et al. 1996, 2002; Loizenbauer et al. 2001; Shalaby et al. 2005; Abd El-Wahed 2008, 2014; Abd El-Wahed and Kamh 2010; Abdeen et al. 2014; Hamimi et al. 2015b; Abd El-Wahed et al. 2016) arranged as follows: (1) D_1

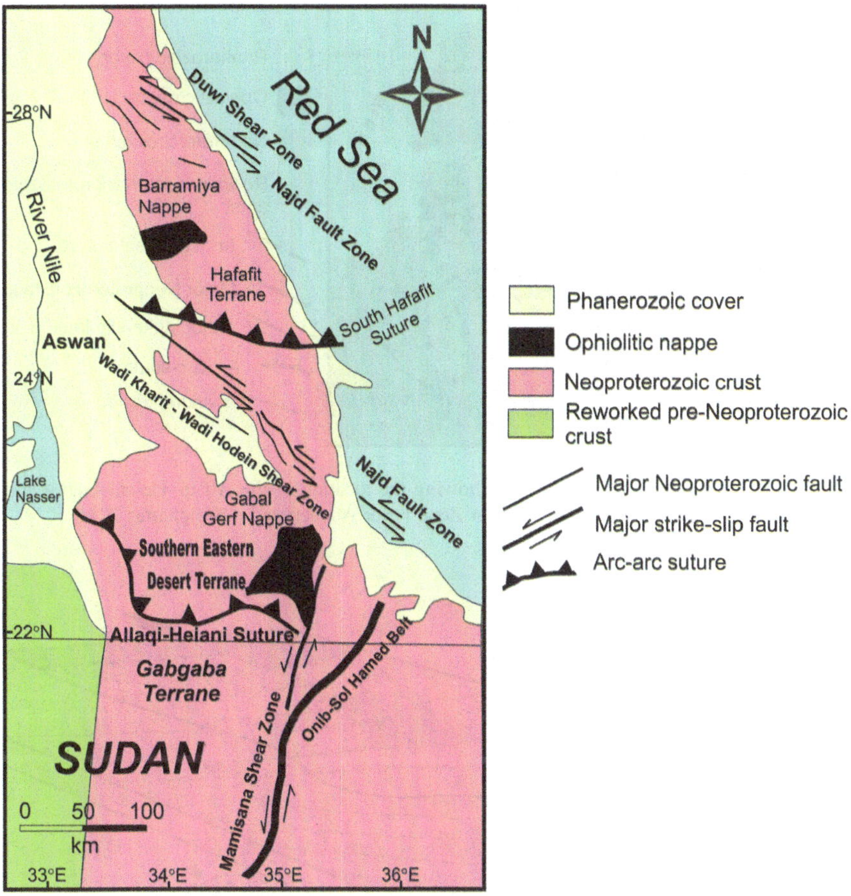

Fig. 18 Simplified tectonic sketch map showing location of the Allaqi–Heiani and the South Hafafit sutures (after Greiling et al. 1994; Abdeen and Abdelghaffar 2011)

linked to NNW- and WSW-propagated thrusts. D_1 deformation took place at greenschist facies conditions (Neumayr et al. 1998) and related to the Pan-African deformation associated with oblique convergence of the arc and back-arc assemblage onto the Nile craton around 620–640 Ma (Fritz et al. 1996) (3) D_{2a} attributed to sinistral movement along the NW-trending shear zones of the Najd Fault System (Fig. 19). (4) D_{2b} associated with dextral movement along NE-trending shear zones. D_2 is linked to the formation of crustal-scale northwest–southeast sinistral shear zones of the Najd Fault System (Stern 1985) followed by exhumation of core complexes within orogen-parallel extension around 620–580 Ma. D_3 deformation occurred much later and may be related to brittle deformation associated with the Red Sea rifting.

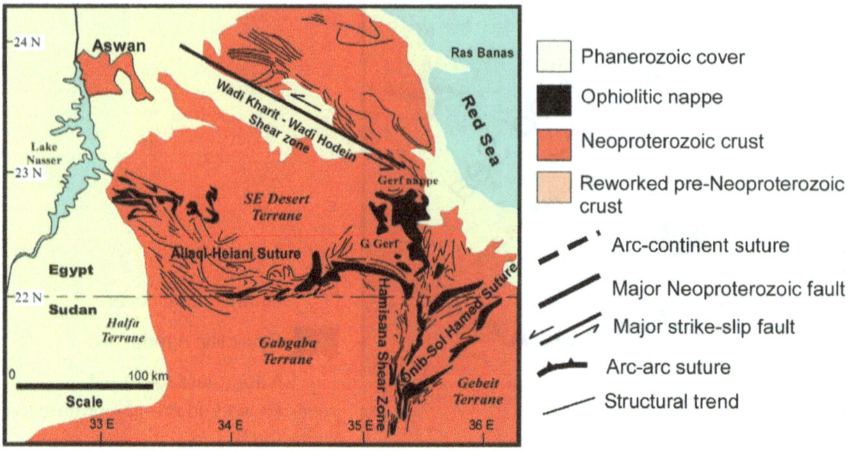

Fig. 19 Simplified tectonic sketch map showing the location of the Allaqi–Heiani–Gerf–Onib–Sol Hamed suture and the Hamisana Shear Zone (after Abdeen and Abdelghaffar 2011)

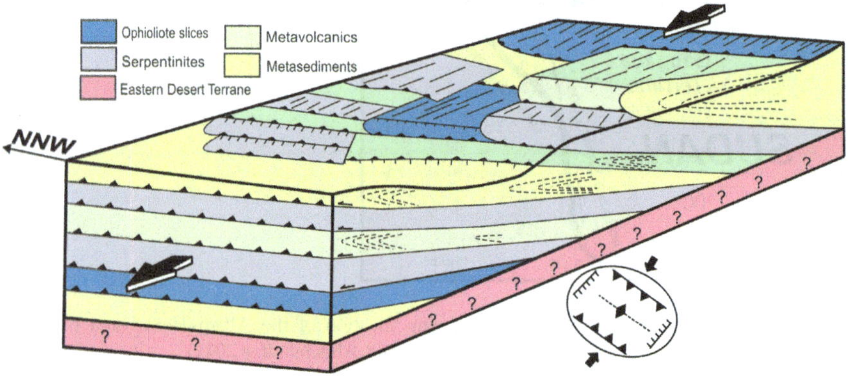

Fig. 20 Early NNW–SSE shortening associated with accretion of relict island arcs and thrusting of ophiolites over old continent (680–660 Ma)

The early NNW–SSE shortening event (D_1) is related to oblique island arc accretion (740–680 Ma) and characterized by stacking and imbrications of Pan-African nappe from SSW to NNW (Fig. 20). NNW–SSE shortening is documented by top-to-NNW large-scale thrusting and folding, distributed over the Eastern Desert of Egypt (Greiling et al. 1994; Abd El-Wahed 2014; Abdeen et al. 2014). A regional nappe transport toward the NW is documented in the Central Eastern Desert from Meatiq, Sibai, Hafafit core complexes (Greiling 1997; Fritz et al. 1996; Loizenbauer et al. 2001; Abd El-Wahed 2008) and from Wadi Mubarak belt (Shalaby et al. 2005; Abd El-Wahed and Kamh 2010). The structure associated with these events includes imbricated shear zones (Fig. 21) forming thrust duplex

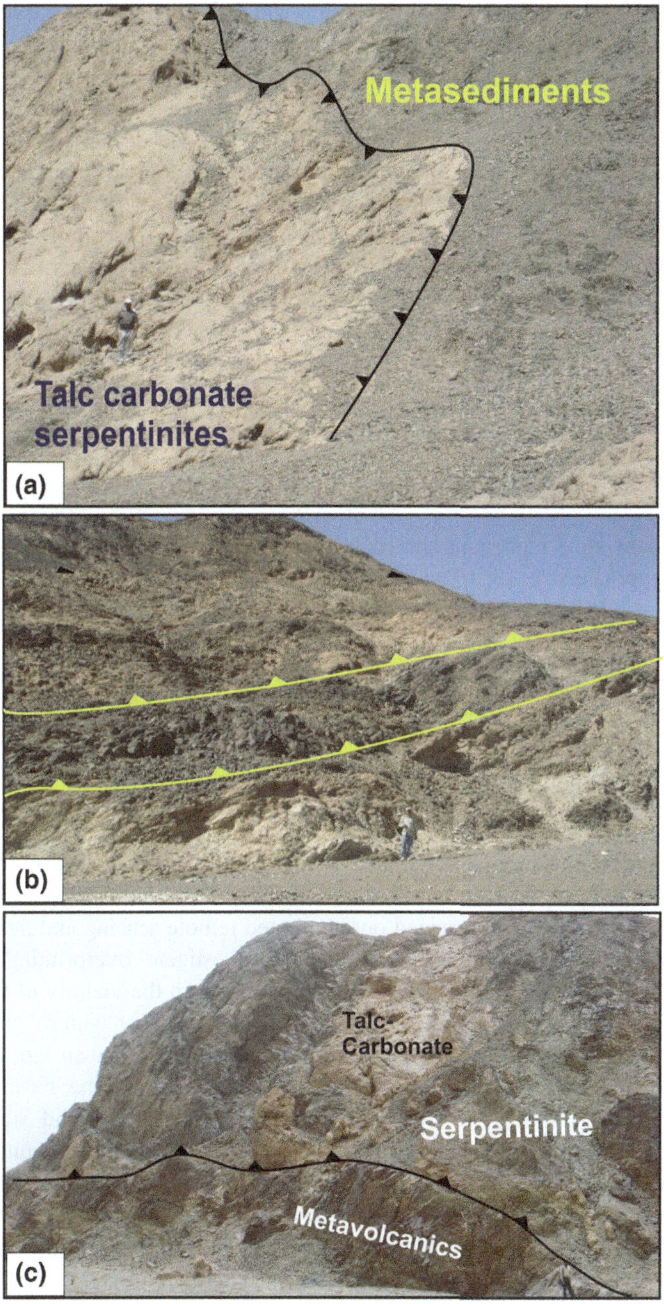

Fig. 21 Example of imbricate thrusts from the Central Eastern Desert of Egypt **a** imbricate thrusts between talc-carbonate serpentinites and volcaniclastic metasediments, looking WSW (After Abd El-Wahed and Kamh 2010); **b** imbricate thrusts between serpentinites and talc–carbonate pods, Wadi Barramiya, looking NW; and **c** EW-oriented imbricate thrusts along Wadi Dungash

structures (e.g., thrust duplexes and the ramp which formed the summit of Gabel El Mayet, Shalaby et al. 2005), E- and NE-trending bedding and fracture cleavage, NE- and ENE-trending map-scale, and mesoscopic folds.

The Northern Eastern Desert is very different than either the Central the Southern Eastern Deserts. Gneisses, ophiolites, and Najd deformation are absent. Faults and joint systems are dominant, and thrusts where observed are attenuated as most of the sector is masked by younger granitoid intrusions. Brittle deformation in the northern sector is evidently late. The nature and overprinting relationships between the pervasive structures at all scales in the Northern Eastern Desert point toward structural younging toward the north (Stern 2017).

7 Suture Zones

7.1 Allaqi–Heiani Suture

Allaqi–Heiani zone represents the most obvious structural feature in the extreme SED of Egypt (Figs. 1, 18, 19 and 22). This zone was described for the first time by Taylor et al. (1993) and has been mentioned in numerous studies later on (e.g., Greiling et al. 1994; EGSMA 1996; El-Kazzaz and Hamimi 1999; Hamimi and El-Kazzaz 2000). It extends over 200 km (average width $\approx$ 3 km) from Gabal Um Shilman and probably to Nasser Lake in the west to the NNE-trending Hamisana Shear Zone in the east. It could be traced easily on the satellite imagery and aerial photomosaics (scale 1:50.000) covering Gabgaba-Elba Topographic Sheets (scale 1:250.000). The strike of this zone is remarkably variable from E–W, NW–SE and N–S. Such strike variation makes it to be perpendicular to the main Wadi Allaqi to the west and align the southern flank of the same wadi to the east, where it is apparently cut by the NE-oriented Hamisana Shear Zone (Greiling et al. 1994). Kusky and Ramadan (2002) carried out integrated remote sensing and field work to distinguish exposed lithologic units, and to investigate overprinting relations between geologic structures along Allaqi–Heiani zone in the vicinity of Gabal Um Shilman. They considered the zone as an arc/arc collision suture zone (750–720 Ma) formed when the Gerf terrane (Eastern Desert or Aswan terrane) in the north overrode the Gabgaba terrane in the south, prior to the closure of the Mozambique Ocean (830–720 Ma). In this context, Abdelsalam and Stern (1996) proposed four Neoproterozoic deformations ($D_1 \rightarrow D_4$) for the development of the Allaqi–Heiani suture zone. D_1 and D_2 are associated with early collisional stages between the Gerf terrane in the north and Haya and Gabgaba terranes in the south, whereas D_3 and D_4 are associated with later stages of collision. Abdelsalam et al. (2003) believed that the Neoproterozic Allaqi–Heiani suture is the western extension of the Allaqi–Heiani–Onib–Sol Hamed–Yanbu suture (Fig. 22) that represents one of arc–arc sutures in the Arabian–Nubian Shield. From the authors opinion, this suture is worthy to understand Neoproterozoic evolution of the Arabian–Nubian Shield because: (1) It is the northernmost linear ophiolitic belt that defines an

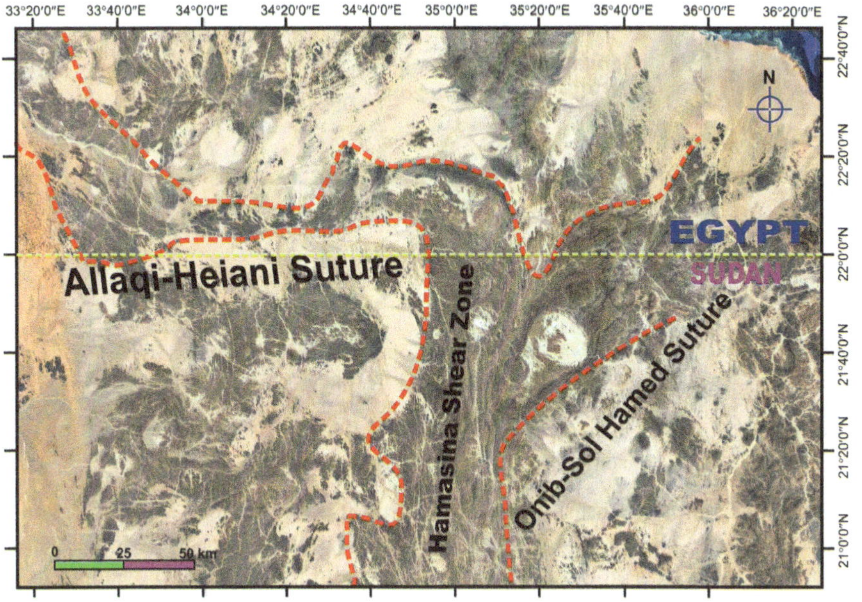

Fig. 22 ETM+ false-color image combination of bands 7, 4, 2 in (RGB) showing the Allaqi–Heiani and Onib–Sol Hamed suture zones

arc–arc suture in the Arabian–Nubian Shield (Kröner et al. 1987a; Stern 1994; Abdelsalam and Stern 1996); (2) It is the only suture in the ANS where a complete ophiolite is preserved at Gabal Gerf (Zimmer et al. 1995); (3) The suture extends in a general east-west direction and its western end is at a high angle to the proposed N-trending, western margin of the ANS; and (4) Recent tectonic models have resulted in conflicting views about the continuity of the Allaqi–Heiani suture, its structural style, and the overall tectonic transport direction involved. The authors concluded that the Allaqi–Heiani Suture is an E–W to NW–SE trending fold/thrust belt forming three allochthons and one autochthonous block. Such conclusion is inconsistent with that given by El-Kazzaz and Taylor (2000) who used facing direction and folded thrust patterns to demonstrate north-verging and top-to-north transport direction. The Allaqi–Heiani suture zone shows sinistral sense of shear indicated by shear band mylonitic foliation, mineral and mica fish, and S-C fabrics. Progressive shearing produced a complex history of folding with development of planar and nonplanar refolded sheath folds.

Abdeen and Abdelghaffar (2011) subdivided the Allaqi–Heiani belt into three structural domains. The western domain (I) is characterized by NNE dipping thrusts and SSW-vergent folds. The central domain (II) includes upright tight to isoclinal NNW–SSE-oriented folds and transpressional faults. The eastern domain (III) shows NNW–SSE-oriented open folds. Structural analysis indicates that the area has a polyphase deformation history involving at least two events. Event D_1 was an N–S to NNE–SSW regional shortening generating the SSW-verging folds

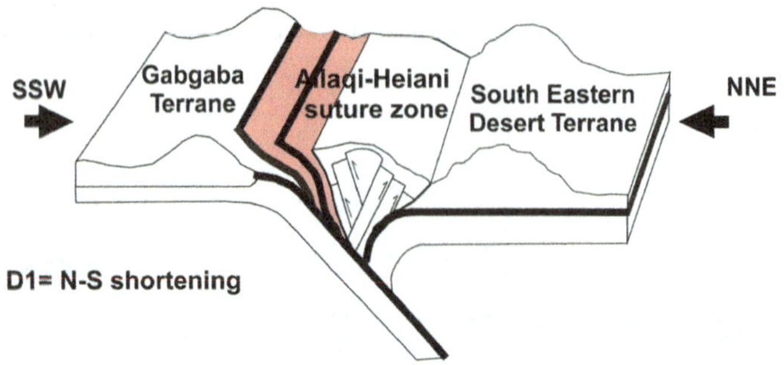

Fig. 23 Schematic sketch showing the tectonic evolution of the central Allaqi–Heiani suture zone during the Neoproterozoic East African Orogen. D1 phase showing the results of the N–S compression between the South Eastern Desert terrane and the Gabgaba terrane (after Abdeen and Abdelghaffar 2011)

and the NNE dipping thrusts (Fig. 23). Event D_2 was an ENE–WSW shortening producing NNW–SSE-oriented folds in the central and eastern parts of the study area and reactivating older thrusts with oblique-slip reverse fault movement.

7.2 *South Hafafit Suture (?)*

Suture is simply an ophiolite-decorated shear zone defining closure of ocean or back-arc basin. In orogenic belts, indications for suturing are multiple but the most significant at all is the distribution of ophiolitic serpentinites in a linear pattern. In the Eastern Desert of Egypt, the complete ophiolite sequence was described for the first time by El-Sharkawi and El-Bayoumi (1979) in Wadi Ghadir area. Since that time, the ophiolitic affinity of the Egyptian serpentinites has been the subject matter of detailed discussion. Some workers (e.g., Akaad 1997; Akaad and Abu El Ela 2002) raised doubt about the ophiolitic nature and tend to consider these suites as intruded magams. However, a number of criteria demonstrate the tectonic obduction of the Egyptian serpentinites in convergent zone. Enigmatic point, from our opinions, do all serpentinites suites could be considered as suture zone? Greiling et al. (1994) introduced the term South Hafafit Suture to the south of the Wadi Hafafit island arc terrane (Fig. 18), which juxtaposes Hafafit terrane to the north and Aswan terrane to the south. They believed that such suture restores for a lateral transport distance of at least 300 km to the SE to form the eastward continuation of the previously mentioned Allaqi–Heiani suture zone. The Aswan terrane is considered to be the equivalent and westward continuation of the Hafafit terrane, and the terrane to the south of the South Hafafit suture, is regarded to extend as far south as the Onib–Sol Hamed Belt, is an equivalent of the Gabgaba terrane. Nevertheless, such restoration needs to be confirmed elsewhere in the CED and SED of Egypt.

8 Major Shear Zones in the Egyptian Eastern Desert

Shear zones (high-strain zones) are known to form important mechanical weaknesses that affect the rheology of the continental lithosphere and its kinematic response of deformation (Bulter et al. 1995). They are the deep counterpart or extension of faults, where both are strain localization structures, both involve displacement parallel to the walls, and both tend to grow in width and length during displacement accumulation (Fossen 2010). These zones play the noteworthy roles as deformation-buffer zones in the upper and lower crust of continents and island arcs, accommodating the crustal thinning (extension) and thickening (shortening) (Coward 1990). The accommodation patterns of crustal thinning and thickening depend on the preexisting anisotropy (geometry of preexisting fracture zones), spatial variation in temperature and spatial distribution of the various rock materials with various mechanical properties, and stress state and its variation in time and space (Sakakibara 1995). Based on ductility, shear zones are subdivided into brittle, brittle–ductile (semi-brittle), ductile–brittle and ductile shear zones, and based on deformation mechanism they are discriminated into brittle/frictional or plastic shear zones. In brittle shear zones, the deformation is localized in a narrow fracture surface separating the wall rocks, whereas in ductile shear zones the deformation is spread out through a wider zone, the deformation state varying continuously from wall to wall (Ramsay and Huber 1987). In the last decades, a large number of studies have been focused on kinematic analysis of shear zones, such as the sense of shear (e.g., Hanmer and Passchier 1991; Hippertt 1993; Greiling et al. 1994; Doblas et al. 1997; Greiling 1997; Hamimi 1999; de Wall et al. 2001; Abdelsalam et al. 2003; Hamimi et al. 2012a, b), non-coaxiality or kinematic vorticity number (e.g., Ratchbacher et al. 1991; Ishii 1992; Hamimi et al. 2015a), finite strain value and its direction (e.g., Hudleston 1983; Treagus 1983), three dimensional shape and finite strain (*k*-value) (e.g., Gapais et al. 1987) and its strain path (e.g., Lacassin and Van Den Driesshce 1983; El-Kazzaz and Taylor 2000).

In the Eastern Desert of Egypt, the shear zones are the most prominent tectonic features that playing a major role in the structural shaping of the Neoproterozoic Pan-African belt. They have given much more attention since the work of El-Gaby (1983) who considered Qena-Safaga Zone as a conspicuous right-lateral shear zone juxtaposes remobilized older continental crust and infolded, locally metamorphosed, Dokhan Volcanics and molasses Hammamat Sediments to the north, and ophiolites, island arc metavolcanics, and metavolcaniclastics to the south. Further studies investigated and referred to megashear zones in Eastern Desert include: Dixon et al. (1987), El-Gaby et al. (1988), Sultan et al. (1988), Greiling et al. (1994), Abdelsalam and Stern (1996), El-Kalioubi and Osman (1996), Fritz et al. (1996), Abdel Khalek et al. (1999), Hamimi and El-Kazzaz (2000), Wallbrecher et al. (1999), El-Kazzaz and Taylor (2000), Fowler and El-Kalioubi (2002), Shalaby et al. (2005), Fowler and Osman (2009), Abd El-Wahed (2008, 2014) Abd El-Wahed and Kamh (2010), Abdeen et al. (2014), Hamimi et al. (2015b), Abd El-Wahed et al. (2016) and Stern (2017). However, the eye-catching shear trend in the Eastern Desert is the NW- (to NNW-) trend. This trend is typified by Wadi Kharit–Wadi Hodein, Nugrus, and Atallah Shear Zones. Other obvious shear zones are NE-, ENE-, N- to NNE-oriented, such as Qena-Safaga, Idfu-Mersa Alam, and Hamisana Shear Zones. The N- to NNE-oriented shear trend is less pronounced compared to the other trends.

8.1 Wadi Kharit–Wadi Hodein Shear Zone

Wadi Kharit–Wadi Hodein Shear Zone (≈186 km long) is a distinctive high angle NW-oriented transcurrent shear zone (Figs. 1 and 18) in the SED of Egypt exhibiting sinistral sense of shear confirmed by various kinematic indicators such as veins, deflected markers, S-C structures, microscale foliations, porphyroclasts, mica fish and mineral fish. Opinions differ about the tectonic affinity of this shear zone, where some workers (e.g., El-Gaby et al. 1988; Stern et al. 1990) proved that it is a Najd-related shear system (analog of the 655–540 Ma, Najd Fault System in the Egyptian Eastern Desert), others (e.g., Fritz et al. 1996; Fowler and El Kalioubi 2004) considered it as a youngest major structural element in the Egyptian Eastern Desert, or even a transpressional corridor (Greiling et al. 1994; Nano et al. 2002). Ramadan and Kontny (2004) reported gold mineralization in listvenite-type wall-rock alteration at Gabal El-Anbat in the vicinity of this shear zone. Greiling et al. (1994) supposed that the Wadi Kharit–Wadi Hodein Shear Zone has connected the —perhaps once continuous—previously mentioned Allaqi and South Hafafit sutures (before being overprinted by the Hamisana Shear). Hamimi et al. (2016) reported a dextral sense of shear along the main Wadi Kharit overprinting the main sinistral shearing, which may demonstrate switching in tectonic regime from sinistral to dextral along the Najd Shear Corridor in the Egyptian Eastern Desert.

8.2 Nugrus Shear Zone

Nugrus Shear Zone (Fig. 24) is a NW-oriented high-strain zone reaching up to 750 m in maximum width, and constitutes the boundary between Wadi Ghadir area (East) and Hafafit Core Complex (West). It is steeply dipping toward the NE direction and separates hanging wall low-grade ophiolitic metaultramafic nappes and volcanogenic metasediments from Migif-Hafafit high-grade gneisses in its footwall.

Geochronologically, the data suggests that the emplacement of low-grade nappes above gneisses occurred at around 680 Ma. The timing of the Nugrus Shear Zone shearing is a distinctly younger shearing event at around 600 Ma (Fowler and Osman 2009). This corroborates the conclusion by (Greiling et al. 1994) that extensional collapse in the region began at about 600 Ma, and accelerated during the period 595–575 Ma. Greiling (1985) stated that Nugrus Shear Zone age is

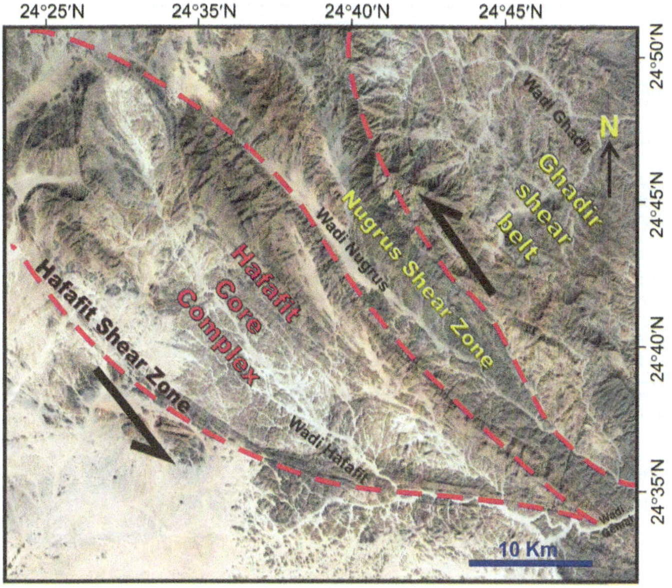

Fig. 24 ETM+ false-color image showing Hafafit Core Complex and Nugrus shear zone

bracketed between 680 Ma (the age of sheared trondhjemite) to 595 Ma (age of post-tectonic granite). Relatively, Mohamed (1993) mentioned the activity time on the Nugrus shear zone was bracketed between the intrusion of the older granitoids and the younger granitoids. Rb/Sr whole rock ages of 610 ± 20 Ma and 594 ± 12 Ma for leucogranites intruded into the schists bordering the Sha'it-Nugrus Shear Zone was given by (Moghazi et al. 2004).

A wide variety of structures are recorded along strike of the shear zone including mylonitic foliation, shear bands, S-C foliations and deformed objects with monoclinic symmetry (Fig. 24). These structures reflect the sinistral sense of shearing which is also confirmed under the microscope by sigmoidal structure and mineral fish. These structures overprint arc collision-related nappe structures (≈680 Ma) and are therefore post-arc collision (Fowler and Osman 2009). The structural architecture of Wadi Nugrus was and still is a matter of debate. It was interpreted in terms of (a) thrust duplexes (Greiling et al. 1988; El-Ramly et al. 1993), (b) Najd-related Shear Zone with sinistral sense of shear (Fritz et al. 1996; Makroum 2003; Abd El-Wahed et al. 2016), and (c) as part of a northerly dipping Sha'it–Nugrus shear zone which is a post-arc collision low-angle normal ductile shear zone separating CED from SED (Fowler and Osman 2009). The steep NE dip of shear-related schistosity and low-pitching slip lineations along Wadi Nugrus are due to NW–SE folding of the Sha'it–Nugrus shear zone, and do not indicate a sinistral strike-slip shear zone (Fowler and Osman 2009).

The footwall of NSZ is Migif-Hafafit gneisses deemed to be of high-temperature–low-pressure amphibolite facies (El-Ramly et al. 1984, 1993).

Fig. 25 σ-type serpentinite porphyroclasts indicating sinistral sense of shear, northern part of Nugrus shear zone

Numerically, the metamorphic conditions were estimated by (Asran and Kabesh 2003) as 720–740 °C for Migif-Hafafit amphibolites and 800–820 °C for associated migmatites, both under pressures of 6–7 kbars. The pressure conditions were confirmed as 6–8 kbars by (Abd El-Naby and Frisch 2006) (Fig. 25).

8.3 Atalla Shear Zone

Atalla Shear Zone is a NW-oriented steeply dipping zone showing sinistral transcurrent regime (Figs. 26 and 27). It is marked by the Atalla felsite mass (28.4 km long by 7.2 maximum width) (Akaad and Noweir 1977; Akaad 1996). The Atalla felsite was considered by many workers as pertaining to the "post-Hammamat felsites" (e.g., Noweir 1968). Conversely, Essawy and Abu Zeid (1972) considered the Atalla felsite and the associated siliceous metatuffs and acidic flows as rocks belonging to ummetamorphosed old volcanics of the Dokhan type. Akaad and

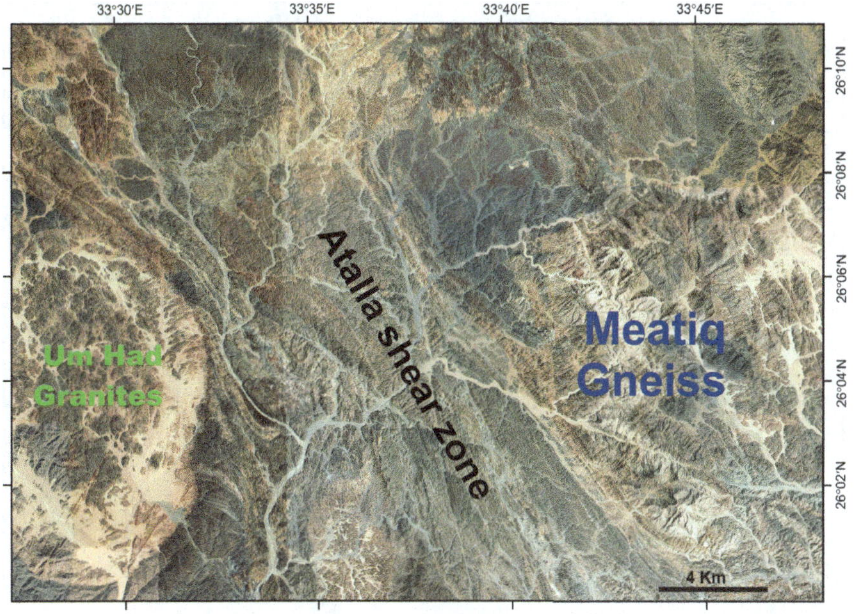

Fig. 26 ETM+ false-color image showing Atalla shear zone

Noweir (1977) and Akaad et al. (1979) considered the Atalla felsite as older than Um Had granite pluton, and this is proved in the field where the felsite extruded the mélange rocks (serpentinites, metasediments metavolcanics, and acidic tuffs) and both of them are intruded by Um Had granite. Because of the effect of Atalla Shear Zone, the huge felsite mass and the mélange rocks are strongly sheared and cataclased, exhibiting stretching lineations and slickenlines particularly at their margins. Kinematic indicators with monoclinic symmetry that encountered at outcrop scale reflect sinistral sense of shearing.

Akawy (2003, 2007) concluded that the structural pattern of the Atalla Shear Zone consists of thrust faults dipping toward the NE with a transport direction toward the SW. Folds developed upon the thrusting represent the second generation of structures. The fold axes trend in a NW–SE direction. The folds occur on centimeter-to-kilometer scales. The thrust faults and folds were cut by four sets of faults, which trend NW–SE, E–W, N–S and NE–SW. The oldest faults are mainly compressive (strike-slip faults) which were superseded by extensional (normal faults).

Fig. 27 Landsat image showing kinematic indicator with monoclinic symmetry reflecting sinistral shearing along Atalla Shear Zone (after Sultan et al. 1988)

8.4 Qena-Safaga Shear Zone

Qena-Safaga Shear Zone (Fig. 28) is a semi-ductile–semi-brittle subvertical shear zone (5 km average width), running along Qena-Safaga line in a NE–SW direction, at nearly right angle to the Red Sea. It passes through two lithologically different domains; Pan-African basement complex in the east and Phanerozoic sediments in the west. In the basement domain, the shear zone juxtaposes Dokhan Volcanics and post-amalgamation Hammamat volcano-sedimentary sequence to the north from ophiolites, arc metavolcanics, and metavolcaniclastics to the south. Qena-Safaga Shear Zone has a long-lasting tectonic history since the Precambrian time and shows dextral sense of movement proved by the remarkable displacement of the basement rocks, particularly near 22 km from Safaga Coastal City. Dextral shear

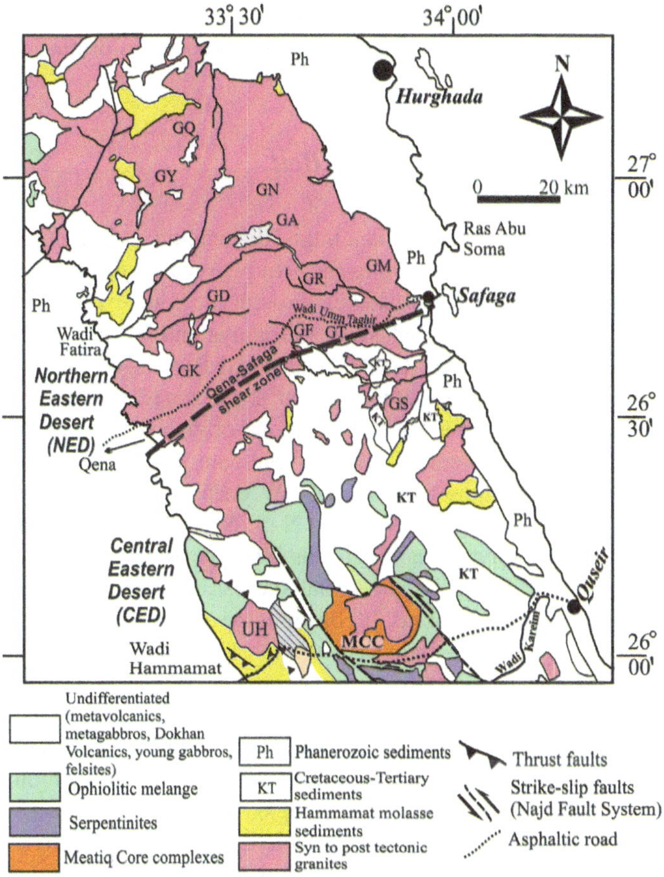

Fig. 28 Simplified geological map of northern part of the Central Eastern Desert and the southern part of the Northern Eastern Desert of Egypt (After Abd El-Wahed and Abu Anbar 2009). Compiled from the Geological Map of Egypt (El-Ramly 1972) and the geological map of Quseir (Klitzch et al. 1987). Major structures are after Fritz et al. (1996). UH, Um Had granite pluton; MCC, Meatiq core complex; GK, G. Kafari; GS, G. Gasus; GD, G. El-Dob; GF, G. Abu Furad; GT, G. Umm Taghir; GR, G. Ras Barud; GM, G. El-Magal; GA, G. Umm Inab; GY, G. Samyuk; GN, G. Shayib El Banat; GQ, G. Qattar. The thick gray line is the Qena-Safaga line of El-Gaby et al. (1994) and represents the approximate boundary between CND and NED

criteria are also recorded some 2 km to the south of the Qena-Safaga asphaltic road. In places, sinistral sense of movement is detected, reflecting reactivation along the shear zone. El-Gaby et al. (1988) suggested that the pronounced reactivation which was during the Tertiary had been responsible for the creation of Qena Bend in the course of the River Nile. In this context, some authors (e.g., El Kazzaz 1999) believed that Qena-Safaga Shear Zone is an active tectonic zone till the present time. The presence of tensile and en echelon opening fractures within the recent sediments, especially between 20 and 30 km to the east of Qena City, together with high magnitude seismicity (up to 3.9) are rather evidence conforming this opinion.

8.5 Mubarak–Barramiya Shear Belt

The Mubarak–Barramiya shear belt is a huge NE-trending shear belt (about 90 km in length and 40 km in width) occupies the area between Wadi Mubarak and Wadi Ghadir on the Red Sea coast and extends through the whole width of the Eastern Desert to include the area between Wadi Barramiya and Wadi Sha'it to the west (Fig. 29). Hassaan and El-Sawy (2009) considered Barramiya, Atud–Sukari; Um Khariga-Wadi Abu Dabbab as sutures and/or shear zones. Greiling et al. (1994) considered E–W trending Idfu-Mersa shear zone originated as extension collapse during a post-collision event. However, Salloum et al. (1989) considered Idfu-Mersa Alam Shear Zone as a major E–W deep seated fault, overprinted by several thrusts and strike-slip faults having N–S, NE–SW and NW–SE directions

The Mubarak–Barramiya Shear Belt runs NE–SW to ENE–WSW in the CED (Figs. 29 and 30) and deforms supracrustal successions and structures associated with the NW-trending shear fabric. It constitutes well-defined ophiolite-decorated linear belt where serpentinites represent the most characteristic lithological unit. The geology of The Mubarak–Barramiya shear belt is commonly described in terms of three major lithotectonic units, namely, (i) ophiolite slices and ophiolitic mélange, (ii) island arc metavolcanic and metasedimentary successions, and (iii) syn- to post-orogenic gabbroic to granitic intrusions. The ophiolites display imbricate thrust sheets and slices of dismembered ophiolite suites distributed along several localities within the Mubarak–Barramiya shear belt (Shalaby et al. 2005; Abd El-Wahed and Kamh 2010; Abd El-Wahed 2014; Abd El-Wahed et al. 2016).

The Mubarak–Barramiya shear belt is post-accretionary deformational belts in the Arabian–Nubian Shield and characterize by the following evolutionary sequence (Abd El-Wahed 2014): (1) Early NW–SE shortening (D_1) associated with accretion of island arcs and obduction of ophiolites over old continent. D_1 produced NNW-directed thrusts and ENE–WSW-oriented folds in the CED. (2) an E–W-directed shortening deformation was superimposed due to oblique collision between the Arabian–Nubian Shield and the Nile Craton (D_2) this produced NW-trending upright folds, NE-dipping and SW-dipping thrusts, and discreet NW–SE tending shear zones in the CED. NNW-directed thrusts belonging to D_1 were folded around NNW–SSE trending fold axes. Continuing E–W shortening rotated the folded thrust to steeply dipping orientations and initiation of major NW-trending sinistral shear zones and culminated in the initiation of major dextral strike–slip shear zones (D_3) as conjugate sets with the NW-trending sinistral shear zones at c. 640–540 Ma ago. The structures associated with the NW-sinistral shear zones are strongly superimposed by the NE-trending transpressional deformation of the Mubarak–Barramiya Shear Belt.

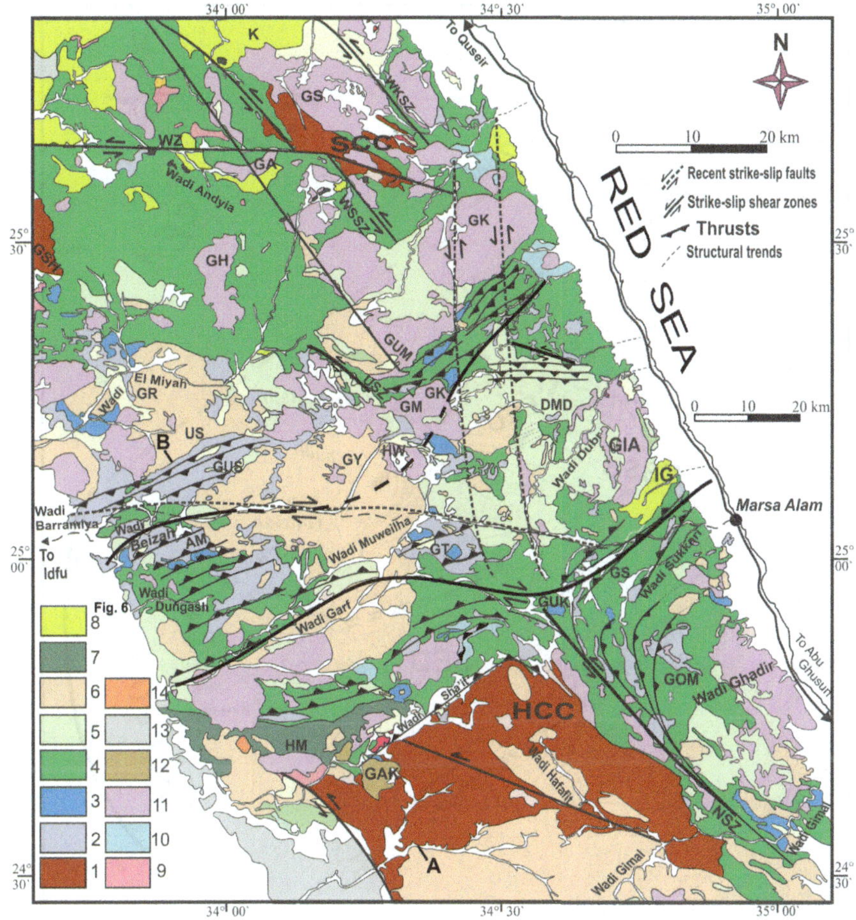

Fig. 29 Geological map of the southern part of the Central Eastern Desert of Egypt (modified after Klitzsch et al. 1987). 1; core complexes 2; serpentinites, 3; ophiolitic metagabbros, 4; metavolcanics and metasediments, 5; syn-tetonic intrusive metagabbros, 6; syn-tectonic granite, 7; Dokhan volcanics, 8; molasse sediments, 9; felsites, 10; gabbros, 11; post- to late-tectonic granites; 12; ring complex, 13; Natash volcanics, and 14; trachyte plugs. GAK; Gebel Abu Khruq, HG; Hafafit gniess, GOM; Wadi Ghadir ophiolitic mèlange, HM; Hamash gold mine, NSZ; Wadi Nugrus shear zone, GS; Gebel Sukkari and Sukkari gold mine, GUK; Gebel Um Khariga, IG; Igla molasse basin, DMD; Dubr metagabbro–diorite complex, GIA; Gebel Igl Al-Ahmar, HW, Gebel Homrat Waggad, GY, Gebel El-Yatima, GUS; Gebel Umm Salim, US, Gebel Umm Saltit, GK; Gebel Abu Karanish, GM, Gebel Al Miyyat, USZ; Um Nar shear zone, GUM; Gebel El-Umra, GK; Gebel Kadabora, GA; GH; Gebel El-Hidilawi, GU; Gebel Umm Atawi, SHG; El Shalul gneiss, GR; Gebel El Rukham; SG; Sibai gneiss, GS; Gebel Sibai, WZ; Wadi Zeidon, WSSZ; Wadi Sitra shear zone, WKSZ; Wadi Kab Ahmed shear zone, and K; Kareim molasse basin. The major structures are after Akaad et al. (1993), Fritz et al. (1996), Helmy et al. (2004), Shalaby et al. (2005), Abd El-Wahed (2008) and Abd El-Wahed and Kamh (2010)

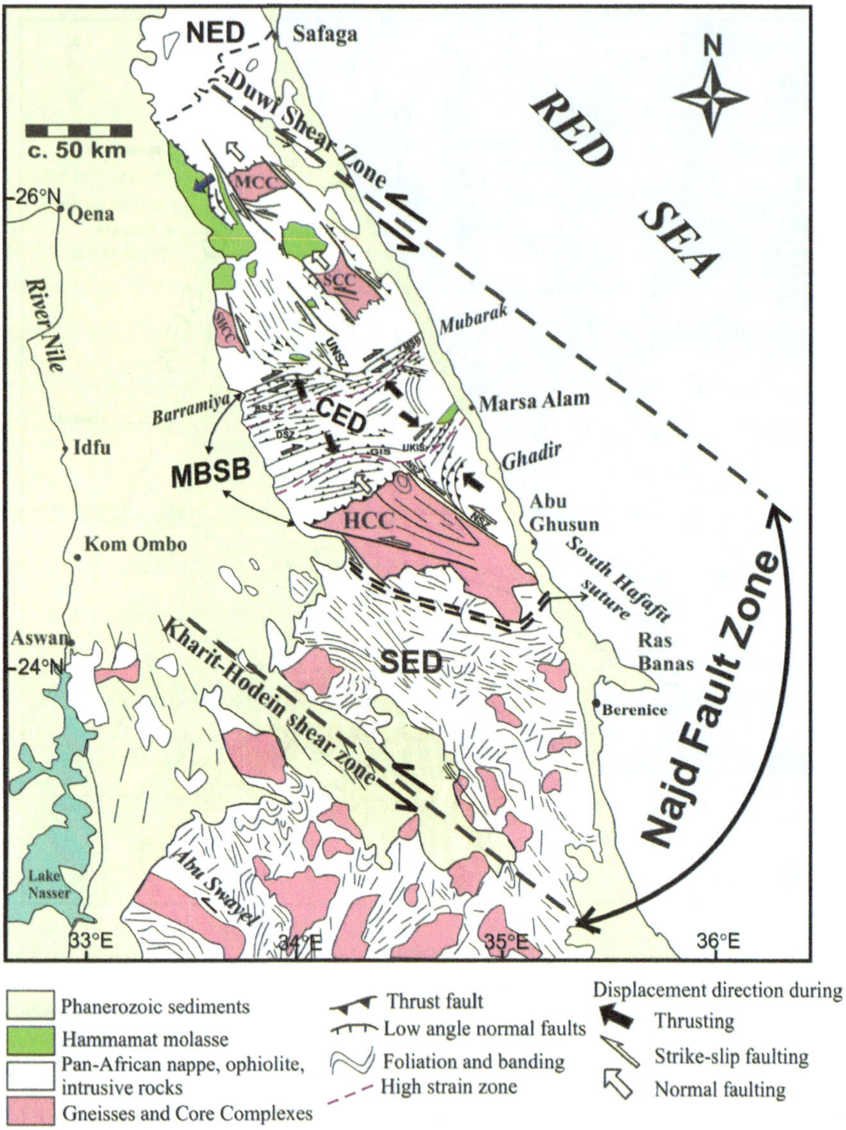

Fig. 30 Major structures in the Central Eastern Desert (CED). The Najd Fault Zone in the Eastern Desert is enclosed between Kharit-Hodein shear zone in the south and Duwi Shear Zone to the north. SED; South Eastern Desert, CED; Central Eastern Desert, NED; Northern Eastern Desert, NSZ, Nugrus shear zone, UNSZ, Um Nar shear zone, HCC; Hafafit Core Complex, SCC; Sibai Core Complex; MCC; Meatiq Core Complex; and MBSB, Mubarak–Baramiya shear belt. This map is compiled from Greiling et al. (1994), Fritz et al. (1996), de Wall et al. (2001) and (Abd El-Wahed 2014)

8.6 Hamisana Shear Zone

Hamisana Shear Zone (660–610 Ma) belongs to the post-accretionary structures (Fig. 22) that formed in the ANS post-dating arc–arc collision and distinct from the assembly between the ANS and Gondwana fragments (Abdelsalam and Stern 1996). It extends in N- to NNE–direction from northern Sudan and because of its sinistral sense of shear causes remarkable dragging of the Allaqi–Heiani Suture Zone and goes further north bounding the Gerf metaultramafic nappe from the east, then continues to meet the Red Sea Coast. Abdelsalam and Stern (1996) proposed dextral sense of shear along the Hamisana Shear Zone based on the deteral off-setting of the Yoshgah Suture (Stern et al. 1989, 1990). The geometric aspect and kinematic history of the Hamisana Shear Zone have been the subject matter of controversy where it is interpreted in terms of a major high-strain zone, a suture zone, and a large-scale transpressional transcurrent zone. Miller and Dixon (1992) argued that transpression in itself is polyphase and this is consistent with the geometric relationship between the eastern extension of the Allaqi Suture Zone and the Hamisana Shear Zone. de Wall et al. (2001) carried out integrated field and AMS studies, and demonstrated that deformation in the Hamisana Shear Zone is dominated by pure shear under upper greenschist/amphibolite grade metamorphic conditions, producing E–W shortening, but with a strong N–S-extensional component. The authors demonstrated that deformation was responsible for folding of regional-scale thrusts (including the base of Gerf and Onib ophiolitic nappes) and indicate that high-strain deformation is younger than ophiolite emplacement and suturing of arc–arc terranes. The obtained data led these authors to conclude that the Hamisana Shear Zone is dominated by late-orogenic compressional deformation and cannot be related to either large-scale transpressional orogeny or major escape tectonics.

Hamisana Shear Zone (Fig. 22) is cut by NW-striking faults and shear zones related to the Najd Fault System, although the amount of displacement and extent of deformation associated with the Najd system is controversial (e.g., Sultan et al. 1988; Smith et al. 1999; Kusky and Ramadan 2002). Four deformational phases have been identified (Stern et al. 1990). The oldest (D_1) records emplacement of the ophiolitic rocks. This produced a complex imbrication of ophiolitic and metavolcanic sequences. D_2 folding around north-trending axes produced a regional cleavage (S_2), subhorizontal intersection lineation (L_2), and tight, upright to inclined fold (F_2). D_3 is coaxial with D_2 and refolds S_2, locally producing pencil structures and crenulations in the western Hamisana. The resultant pervasive northsouth fabric is truncated by narrow, north-northeast-trending D_3 dextral shear zones. These become more dominant in the extreme south as the Hamisana shear zone turns southwest. Later kinks (D_4) and brittle faults have variable movement sense and account for limited regional strain. Thus, the principal ductile deformation in the Hamisana is characterized by nearly coaxial folding about a northsouth axis, indicating shortening normal to the zone.

9 Najd Shear Corridor

The first requirement for comprehensive understanding of the Eastern Desert regional tectonic setting is the entire fathoming of the Najd Fault System (NFS) and its complex arrays of secondary structures. The NFS has a great importance due to its great size, role in the exhumation of metamorphic core complexes and prominence in Gondwana cratonization. Moore (1979) described this system as a major transcurrent (strike-slip) fault system, of Proterozoic age in the Arabian Shield (Fig. 1). He suggested a similarity of NFS to many of the world's major transcurrent fault systems, including the San Andreas (USA) and Alpine (New Zealand) faults in terms of its length (possible length of more than 2000 km), and added the system is a braided complex of parallel and curved en echelon faults. For the NFS and especially close to the terminations of some major faults a complex association of secondary structures including strike-slip, oblique-slip, thrusts, and normal faults, in addition to folds and dike swarms are usually present forming an intricate array. Therefore, we should keep in our mind that the importance and complexity of NFS is augmented by this array of secondary structures that give an allusion to synchronous compressional and extensional conditions in various parts of the fault zones.

The NFS was identified originally as a huge late Proterozoic and early Phanerozoic NW-trending brittle–ductile shear zone with 300 km width and length over 1100 km extending across the northern part of the ANS (Brown and Jackson 1960; Stern 1985; Johnson et al. 2011). The displacement along the strike of the NFS was reported by (Brown 1972) as 240 km cumulative displacement but field displacements can be demonstrated as only tens of kilometers for faults (Johnson et al. 2011). The NFS and other NW-trending strike-slip faults in the ANS are regarded also as post-accretionary structures and were interpreted to be the result of the squeezing of the Arabian–Nubian Shield between East and West Gondwana (Berhe 1990; Stern 1994; Abdelsalam 1994; Abdelsalam and Stern 1996; Abdelsalam et al. 2003). The formation of NFS is a result of simple shear that allowed the Nubian and southern Arabian shield to move several hundred kilometers sinistrally with respect to northern Arabia (Moore 1979).

From the Arabian Shield, the northwestern extensions are probably in the Eastern Desert of Egypt and to the southeast, the line of faulting coincides with structures in the south Yemen coast and in the bed of the Arabian Sea (Brown 1972). In southern Jordan rocks, the shear zone is inferred to be present and disrupted by much younger Cenozoic slip on the Dead Sea Transform (El-Rabaa et al. 2001). In the Mozambique Belt in Kenya and Madagascar, Similar NW-trending shear zones were identified (Raharimahefa and Kusky 2010). The influence of the NFS also continued to southeast into parts of India and the Lut block of Iran as was reported by (Al-Husseini 2000) from Magnetic and gravity data. Thus, the NFS attains more than 2000 km (Moore 1979) total possible length making this system one of the greatest shear systems known on Earth.

Geochronologically, the absolute radiometric ages obtained from small intrusions denoted that the faults were active from late Proterozoic into early Phanerozoic times, 580–530 Ma ago (Fleck et al. 1979). The late stages (630–535 Ma) of the Pan-African event witnessed the NFS development in the form of huge shear zone system striking NW–SE (Stern 1985; Johnson et al. 2011).

Geophysically, subsurface aeromagnetic maps interpretation denoted a continuity of the NFS beneath the surface faults arrays concluding that this system is broader at depth than the outcropping fault complex (Moore 1979). At depth and under amphibolite facies prevailing conditions, an early shear ductile activity of the NFS is prevailing which is turned into brittle shearing at the shallower levels (Johnson et al. 2011; Fritz et al. 2013). Hydrothermal activity, in turn was pervasive indicating abnormally high heat transfer in the time of faulting. The hydrothermal alteration is probably also a reflection of the mechanical importance of fluid pressure in the mechanism of faulting at this structural level (Phillips 1972).

Deformation was followed by brittle failure during the main faulting episodes. The currently exposed structural level in the Southern Najd was that of ductile to semi-brittle deformation (Moore 1979). In addition to strike-slip movement, minor thrusting, oblique-slip faulting, and normal faulting have occurred in some areas. Thrust faults and folds adjacent to terrane containing normal faults can be seen. These unusual assemblages of minor structures define local areas of anomalous compression or dilation within the fault belt. The compressional regimes, in which thrusting and folding accompany second-order wrench faulting, occur on the northeast side of the northwest terminations and on the southwest side of southeast terminations and in the "overlap" between en echelon major faults. The compressional regimes are complemented by "dilation" on the opposite side of the master fault, marked by normal faulting and extension fissure formation. The dilational areas offer the most mechanically favorable loci for dike intrusion during the faulting events (syn-tectonically) and, subsequently, for hydrothermal vein emplacement.

Strictly speaking, the NFS is dominated by NW-trending shears, and best typified in the Egyptian Eastern Desert by the previously mentioned Kharit-Hodein, Nugrus, and Atalla Shear Zones (Fig. 30). On the other hand, the theoretical complements (major NE-trending dextral faults) to the main system are rare (Moore 1979). The NE- (to ENE-) trending dextral shears are regarded as the conjugate pairs of the NW trend and are represented in the Eastern Desert by Qena-Safaga and Mubarak–Baramiya shear belt (Fig. 30). It is worth mentioning to denote that the NFS is a coherent structure that can be explained by a single regional event (Moore 1979). The sinistral strike–slip shearing along the NW-oriented Najd-related Shear Zones was accompanied by transpressional and transtensional tectonic regimes (Fritz et al. 1996; Abd El-Wahed 2014; Abd El-Wahed et al. 2016; Stern 2017).

10 Shear Zone-Related Gneiss Domes

As mentioned before, the Eastern Desert of Egypt is characterized mainly by the prevalence of a NW-trending tectonic fabric marking the NW–SE sinistral shear zone (Fig. 30) of the NFS (Abd El-Wahed and Kamh 2010). Another important structural feature in the Eastern Desert is presence of a series of gneiss domes

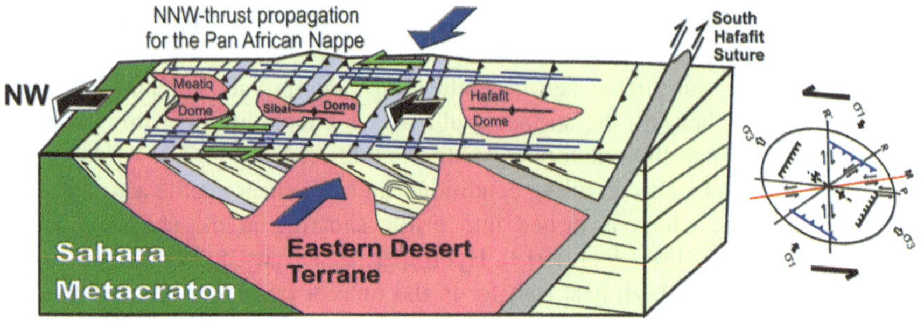

Fig. 31 E–W shortening associated with a post-accretion phase produced sinistral shearing along NW–SE shear zones (660–560 Ma)

(Figs. 30 and 31) or core complexes (e.g., Meatiq, Sibai, El- Shalul, and Hafafit) (Hamimi et al. 1994; Hamimi 1996, Fritz et al. 1996, 2002, 2013; Loizenbauer et al. 2001; Abd El-Wahed 2008; Abdeen et al. 2014). These domes are bounded in the northeast and southwest by NW-trending sinistral strike–slip shear zones related to the NFS such as Nugrus Shear Zone to the east of Hafafit Culmination (Fig. 24). Gneiss domes have either been interpreted as (1) antiformal stacks formed during thrusting (e.g., Greiling et al. 1994), (2) core complexes formed during orogen-parallel crustal extension (e.g., Fritz et al. 1996; Bregar et al. 2002; Abd El-Wahed 2008), and (3) interference patterns of sheath folds (Fowler and El Kalioubi 2002). Geochronology suggests that extension and exhumation of gneiss domes commenced around 620–606 Ma (Fritz et al. 2002; Andresen et al. 2009). The gneiss domes in the Eastern Desert (e.g., Meatiq, Sibai, Shalul, and Hafafit) are distributed along a NW-trending major zone that extends for more than 400 km between Wadi Hodien in the south to Wadi Quieh in the north (Figs. 30 and 32).

Fig. 32 Field photographs showing the southern part of El Shalul Dome, looking E (after Hamimi et al. 1994)

11 Transpressional Regime in the Egyptian Eastern Desert

Transpressional deformation has played an important role in the late Neoproterozoic evolution of the Eastern Desert of Egypt and the ANS as a whole. Transpression is strike-slip deformation that deviate from simple shear because of a component of, respectively, shortening or extension orthogonal to the deformation zone (Harland 1971; Tikoff and Greene 1997; Dewey et al. 1998). It is a combination of a strike-slip component and a shortening component orthogonal to the deformation zone and includes a deformation developed between two undeformed blocks resulting from both simple and pure shear (Sanderson and Marchini 1984; Fossen et al. 1994; Fossen and Tikoff 1998; Jones et al. 2004).

Transpressional shear zones are characterized by an association of structures that suggest zone-normal shortening and zone-parallel shearing. In wrench-dominated transpression, stretching lineations are either horizontal (low strain) or vertical (high strain), whereas they are always vertical in a pure shear dominated transpression. Vertical stretching lineations within a vertically oriented shear zone, perpendicular to the simple shear component of deformation and the direction of tectonic movement, were first interpreted to be the result of transpressional deformation by Hudleston et al. (1988). Theoretical models of heterogeneous transpression (Jiang and Williams 1998) interpret these lineations as can range from horizontal to vertical continuously, depending upon the value of finite strain.

Transpression occurs in a wide variety of tectonic settings and scales during deformation of the earth's lithosphere such as the Archean North Caribou greenstone belt (Gagnon et al. 2016), the Pan-African Kaoko belt in Namibia (Goscombe and Gray 2008; Knopásek et al. 2005), Southern Uplands of SE Scotland (Tavarnelli et al. 2004), Kushtagi schist belt, India (Matin 2006), Sanandaj–Sirjan metamorphic belt, Zagros mountains, Iran (Sarkarinejad and Azizi 2008; Shafiei Bafti and Mohajjel 2015), Al Jabal Al Akhdar, Libya (Abd El-Wahed and Kamh 2013), Central Asian Orogenic Belt (Li et al. 2016), Cauvery shear zone, southern Granulite Terrain, India (Chetty and Bhaskar Rao 2006), Dom Feliciano Belt, Uruguay (Oriolo et al. 2016) Salem–Attur shear zone, south India (Kumar and Prasannakumar 2009), Egyptian Eastern Desert (Fritz et al. 1996, 2002, 2013; Bregar et al. 2002; Shalaby et al. 2005; Abd El-Wahed 2008, 2010, 2014; Abd El-Wahed and Kamh 2010; Zoheir 2008, 2011; Zoheir and Weihed 2013; Abd El-Wahed et al. 2016), and Arabian Shield (Hamimi et al. 2013a, b, 2014).

In the Eastern Desert of Egypt, transpressional deformation is recognized as post-collisional structures (Greiling et al. 1994; Fritz et al. 1996; Abd El-Wahed and Kamh 2010; Abdeen and Abdelghaffar 2011) identified by localized transpressional strike-slip shear zones (Greiling et al. 1994; Fritz et al. 1996, 2013; Abd El-Wahed and Kamh 2010) deforming the early compressional structures. The best examples for transpressional strike-slip shear zones in the Eastern Desert of Egypt are the Allaqi–Heiani shear zone (Figs. 18, 19 and 22), the NW-oriented Wadi Kharit–Wadi Hodein (Figs. 18, 19 and 30), and the N-oriented Hamisana in southern Eastern Desert (Greiling et al. 1994), and the Najd Fault Zones in the CED. The sinistral strike–slip shearing along the NFS was accompanied by transpressional and transtensional tectonic regimes (Fritz et al. 1996). Transpression resulted in formation of ESE- to ENE-shortening that produced NNW- to N-trending folds throughout the CED. These are superimposed on early NNW-directed thrusts and related structures (Makroum 2001; Shalaby et al. 2005; Abd El-Wahed and Kamh 2010; Abd El-Wahed 2014; Abdeen et al. 2014; Abd El-Wahed et al. 2016).

Early transpression in the Eastern Desert of Egypt produced the Allaqi Allaqi–Heiani shear belt and final transpression is documented in the Wadi Kharit–Wadi Hodein Zones (Greiling et al. 1994). Abdeen and Abdelghaffar (2011) described two major deformation events from Allaqi–Heiani Shear Belt; (i) D_1 was an N–S to NNE–SSW regional shortening generating the SSW-verging folds and the NNE dipping thrusts, and (ii) D_2 was an ENE–WSW shortening producing NNW–SSE-oriented folds and reactivating older thrusts with oblique-slip reverse fault movement. D_1 is related to the terrane accretion during the early Pan-African orogen and associated with collision between the Eastern Desert terrane and the Gabgaba terrane along the Allaqi–Heiani Shear Belt, whereas D_2 is related to Najd Orogen due to collision between East- and West-Gondwanalands at ca. 750–650 Ma associating the closure of the Mozambique Ocean. Collision between East- and West- Gondwanalands deformed the Allaqi–Heiani Shear Belt along N–S trending shortening zones and produced NW–SE and NE–SW-oriented sinistral and dextral transpressional faults, respectively (Abdeen and Abdelghaffar 2011).

Allaqi–Heiani shear belt is apparently linked by the high angle sinistral strike-slip Wadi Kharit–Wadi Hodein Shear Zone with a tectonic transport of about 300 km toward the W/NW. The Wadi Kharit–Wadi Hodein Shear Zone is characterized by compressional/transpressional deformation and interpreted as an equivalent of the NFS (Greiling et al. 1994; Zoheir 2011). The Hamisana Shear Zone is one of the largest transpressional shear zone in NE Africa, covers an area of about 15,000 km^2. It has been interpreted as a Precambrian suture, as a zone of strike-slip displacement, or as a zone of crustal shortening (e.g., Stern et al. 1989; Miller and Dixon 1992; de Wall et al. 2001; Sakran et al. 2001; Takla et al. 2002; Ali-Bik et al. 2014). The Hamisana Shear Zone represents a transpressional shearing event post-dating the terrane accretion (Smith et al. 1999). On the contrary, a post-accretion origin is postulated for the Hamisana Shear Zone by many workers (e.g., Stern et al. 1989, 1990; Miller and Dixon 1992; de Wall et al. 2001). Stern et al. (1989) bracketed the shearing event of the Hamisana Shear Zone between 660 and 510 Ma post-dating the terrane accretion. They related the Hamisana Shear Zone to the Najd Tectonic System.

The most important transpressional shear zones in the CED of Egypt are Nugrus Shear Zone, Mubarak–Baramiya shear belt and the major shear zones bounding the gneiss domes. Nugrus Shear Zone interpreted as thrust accommodating westward or SW-ward ophiolitic material transport over the continental margin (El-Gaby et al. 1990; El-Bayoumi and Greiling 1984; El-Ramly et al. 1984) or as roof thrust (Greiling et al. 1994 and Greiling 1997). Many workers considered Nugrus Shear zone as a Najd-related transpressional strike-slip shear zone (e.g., Fritz et al. 1996; Shalaby et al. 2006; Abd El-Wahed et al. 2016).

Nugrus Shear Zone and Mubarak–Baramiya Shear Belt are dominated NW-sinistral and NE-dextral shear zone characterized by wrench-dominated transpressional structures such as oppositely dipping thrusts, subhorizontal lineation (Fig. 33a), pop-up structures, and flower-like cleavage (Figs. 33b and 34). The Mubarak–Baramiya Shear Belt is composed of a network of anastomosed shear

Fig. 33 **a** Subhorizontal slickensides overprinted by steeply dipping striae in sheared metavolcanics along Wadi Al Alam, Central Eastern Desert, looking NW; **b** flower-like cleavage in serpentinites, Wadi Barramiya, Central Eastern Desert, looking SW

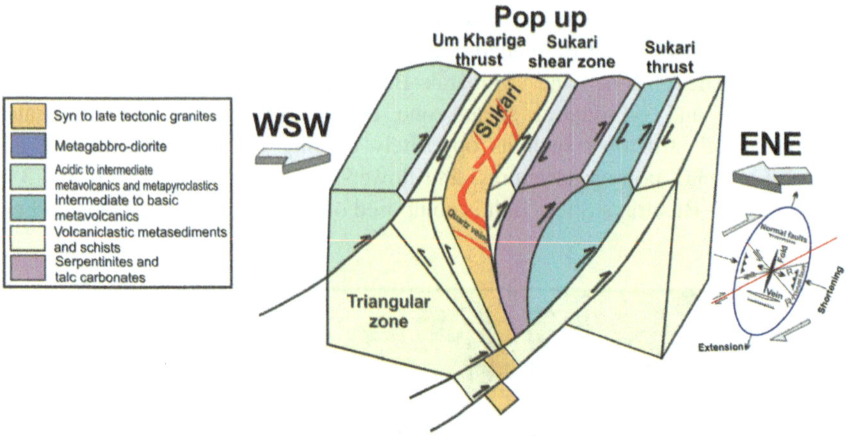

Fig. 34 A schematic block diagram showing the geometry and kinematics of pop-up structure along Sukari shear zone, northern Nugrus shear zone (after Abd El-Wahed et al. 2016)

zones arising from conjugate shears with a prevalent dextral sense (Abd El-Wahed and Kamh 2010). The transpression-related sinistral shear regime is superimposed by the dominant dextral transpression along NE–SW trending Wadi El-Umra Shear Zone. This dextral shearing is characterized by development of NNE- to NE-trending cleavage, strike-slip duplex, NNE- and NE-trending folds, and NNW-directed thrusts. These two events represent a single progressive phase associated with sinistral transpressional deformation, which is related to a younger E–W shortening event (Abd El-Wahed 2014). E–W-directed shortening is due to oblique convergence between East and West Gondwana along the Mozambique belt. Transpressional structures in the NE–SW trending Mubarak–Barramiya Shear Belt (Figs. 29 and 30) indicate highly oblique convergence leading to wrench-dominated dextral transpression and development of a major flower structure between Wadi Mubarak and Hafafit dome occupying the whole width of the CED (Abd El-Wahed 2014).

Transpression has also been used to explain the geometry and kinematics of gold-bearing quartz veins in the Eastern Desert of Egypt (Hassaan et al. 2009; Zoheir 2008, 2011; Zoheir and Lehmann 2011; Abd El-Wahed 2014). The syn-orogenic gold mineralization in the Eastern Desert relates to transpressional NW-sinistral (Fig. 34) and NE-dextral strike-slip shear zones related to the NFS and hosted by volcaniclastic metasediments and altered ophiolitic serpentinites (Abd El-Wahed et al. 2016). Structural analysis of the shear fabrics in Sukari gold mine area indicates that the geometry of the mineralized quartz veins and alteration patterns are controlled by the regional NNW- and NE-trending conjugate zones of transpression (Abd El-Wahed et al. 2016). Gold-bearing quartz veins are located within steeply dipping NW- and SE-dipping thrusts and NE- and NS-oriented dextral and sinistral shear zones around Sukari mine area, and along E-dipping back thrusts and NW–SE and N-S fractures in Sukari granite (Abd El-Wahed et al. 2016).

The high grade of gold mineralization in Sukari is mainly controlled by SE-dipping back thrusts branched from the major NW-dipping Sukari Thrust. The gold mineralization in Sukari gold mine and neighboring areas in the Eastern Desert of Egypt is mainly controlled by the conjugate shear zones of the NFS. In the SED, many of the mineralized quartz veins and alteration patterns are controlled by the regional, NNW-trending zone of transpression, such as the Wadi Kharit–Wadi Hodein Shear Zone, which is related to the 655–540 Ma, Najd Shear Corridor (Zoheir 2011).

12 The Active Seismotectonic Zone in the Egyptian Eastern Desert

The abovementioned fractures and shear zones traversing the Eastern Desert of Egypt show no indications for reactivation at present time. The only known active seismotectonic zone in the entire Pan-African belt of the Eastern Desert is Abu Dabbab zone which represents one of five seismotectonic zones in Egypt (Hamimi and Hagag 2017). Abu Dabbab zone is an ENE-oriented narrow zone, lying some 30 km north of Marsa Alam City along the Red Sea Coast. This zone could easily be affiliated to the ENE Najd-related Shear Zones, such as Idfu-Mersa Alam Shear Zone, in terms of direction and sense of shear, but it differs where it shows daily recorded microearthquakes with local magnitudes ($ML < 2.0$). Besides, November 12, 1955 and July 2, 1984, two giant earthquakes were recorded with magnitudes 5.6 and 5.2, respectively (Fairhead and Girdler 1970; Badawy et al. 2008). The recorded seismic activity from Abu Dabbab region by the Egyptian National Seismic Network (ENSN) ranges from 10 to 15 events/day to more than 60 events/day, and sometimes attained 100 events/day during swarms (Badawy et al. 2008; Mohamed et al. 2013). Such enigmatic seismic record has attracted the attention of many workers (e.g., Fairhead and Girdler 1970; Daggett et al. 1986; Hassoup 1987; Kebeasy 1990; El-Hady 1993; Ibrahim and Yokoyama 1998; Badawy et al. 2008; Hosny et al. 2009, 2012; Azza et al. 2012; Mohamed et al. 2013) to decipher origin of the earthquakes. The magmatic origin of the seismicity and associated shallow and deep earthquakes is promoting most of the publications dealt with the tectonic setting and seismic activity of Abu Dabbab area. Sabet et al. (1976) suggested that the tectonic evolution of the area was associated with volcanic activity, whereas Daggett et al. (1986) attributed Abu Dabbab seismicity to the subsurface volcanic environment of a cooling pluton. Meanwhile, Hassoup (1987) interpreted this seismicity in the light of the subsurface structural heterogeneity. Hosny et al. (2009) proposed a structural model for the area based on seismic velocity tomography, and related the P and S-wave velocity anomaly to magmatic intrusion. Recently,

El Khrepy et al. (2015) reveal strong arguments for the tectonic origin of the seismicity of Abu Dabbab area in despite of the prevalent magmatic origin.

Hamimi and Hagag (2016) proposed a new tectonic model for Abu Dabbab seismogenic zone based on integrated field-structural investigations, and EMR/ seismic data. The obtained results led the authors to indicate a present-day faulting activity in the area and to determine the depth of the brittle–ductile transition zone underlies the Abu Dabbab area. The transition zone is estimated to be existed at a relatively shallow depth (10–12 km) depending upon the following main criteria: (1) the absence of a large seismic main shock, (2) the periodically recorded swarm's hypocenters of focal depths not deeper than 16 km, (3) the high *Vp*/*Vs* ratio (from seismic tomography) until 12 km depth, (4) the occurring of tensile earthquakes of high compensated linear vector dipole (CLVD) ratios, and (5) the high heat flow rates (about 92 $mW/m^2 \pm 10$, which is more than twice the average value of Egyptian Eastern Desert; 47 mW/m^2). The authors came to the conclusion that there is a mechanical decoupling between the shallow and deep crustal levels of Abu Dabbab Neoproterozoic basement succession, where the maximum principal stress axis ($\sigma 1$) rotates from a subhorizontal position at the uppermost crustal levels practicing transpressional deformation to a near vertical attitude in the deeper levels, where the transtensional deformation predominated.

13 Effect of Gravitational Tectonics

The structural architecture of the crystalline basement complex of the Eastern Desert is attributed to the Pan-African Orogeny that led to the formation of a wide variety of tectonic structures, such as thrusting and thrust-related structures, and shearing and shear-related structures. Nevertheless, effect of gravitational collapse in fracturing and deforming some domains is worth mentioning. Best example of this effect could be detected in Abu Fas area in Wadi Allaqi District in the extreme SED. Lithologic units outcropping in this area comprise metavolcanics, metagabbros, metaultramafics, and metasediments, intruded by tonalite, layered gabbro and muscovite-biotite granite. Striking features in Abu Fas gabbroic mass are margin-parallel arcuate faulting and inward symmetric dipping of layering and the remarkable decrease in the amount of dip from very steep layers in the outermost part of the intrusion to mildly dipping layers toward the center (Fig. 35). El-Kazzaz and Hamimi (2000) attributed the arcuate faulting to the gravitational forces, and believed that the inward dipping of foliation in the enveloping rocks close to the layered gabbro is related to a sagging of the preexisting structure post-dating the gabbro emplacement. Such observations may give a clue to the role of gravitational tectonics and crustal relaxation in the structural history of the Egyptian Eastern Desert.

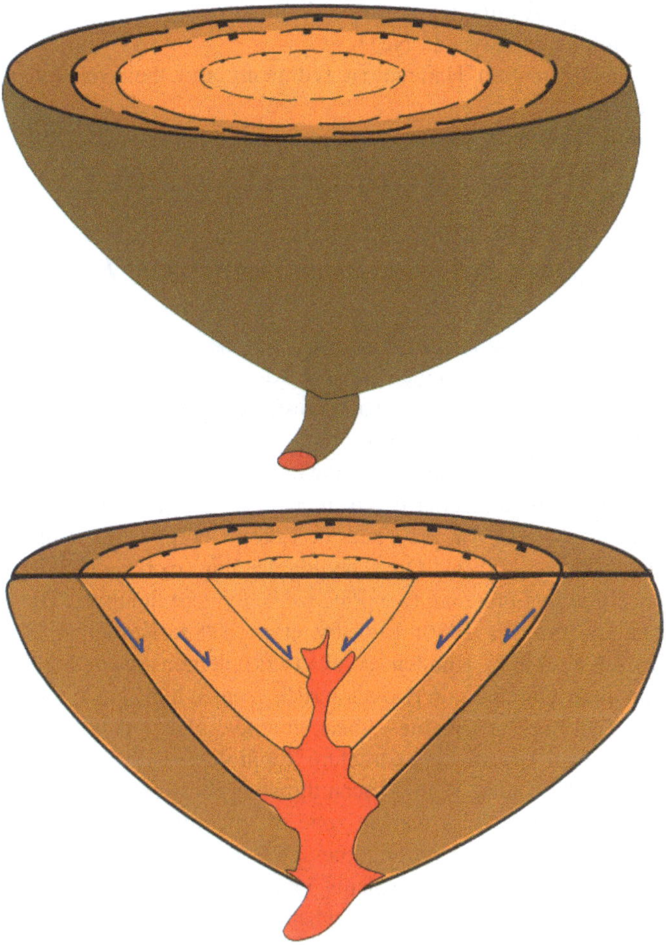

Fig. 35 Block diagrams showing arcuate faulting in Abu Fas layered gabbro, Gabal Muqsim area (after Hamimi et al. 2012c)

14 Discussion

In discussing the Neoproterozoic crustal evolution of the Pan-African belt and the tectonics of the Eastern Desert of Egypt the following enigmatic points are worthy to be addressed.

- Do we have pre-Pan-African infracrustal rocks?
- Thrusting, shearing, and folding relations.
- Gneiss domes vs. metamorphic core complexes.
- The conjugate pairs of Najd-related shears.
- Role of Najd Fault System in tectonic evolution of Gneiss domes.

- Rates and transport directions of metaultramafic nappes.
- The voluminous intrusives in NED.
- The post-amalgamation Hammamat sediments and their relation to Dokhan Volcanics.
- Northward decrease in intensity of deformation.

14.1 Do We Have Pre-pan-African Infracrustal Rocks?

The existence of the pre-Pan-African rocks in the Egyptian Eastern Desert is controversy and a matter of much debate. Also, the nature of the Pan-African orogeny and characteristics of the early crustal complex, are more controversial (Dixon 1981). Evaluating the extent of pre-ANS crustal input into the shield's core region carries important implications for the ongoing debate regarding its origin and evolution, because this affects critically models of crustal accretion rates and the interpretation of the Nd, Sr, Pb, and O isotope compositions of ANS rocks (Be'eri-Shlevin et al. 2009b). On scanning leading works done by influential scholars, once find that there are two principal schools of thoughts. The first school (e.g., Akaad and Noweir 1980; El-Gaby et al. 1988; Kroner et al. 1987a, b; Abdelkhalek et al. 1992; Khudeir et al. 1995) demonstrates that the exposed lithologic units are affiliated into two main rock groups; high-grade pre-Pan-African infrastructure (gneisses, migmatites, shear granites, and remobilized equivalents) and low-grade Pan-African suprastructure (ophiolites island arc volcanics and volcaniclastics). The contact (Pan-African thrusting) between the infrastructure and suprastructure is marked by a narrow zone of mylonite and intensive degree of shearing and cataclases, along with tightness of foliation. Affiliation of the gneisses, migmatites, and remobilized equivalents into the pre-Pan-African age is based mainly on the deformation and shearing, where these rocks are highly deformed and sheared compared to the ophiolites, island arc volcanics, and volcaniclastics. The Pan-African dating obtained for some gneisses and sheared granites elsewhere in the Eastern Desert are regarded to represent the age of metamorphism and not the protolith age. In this context, El-Gaby (1983), El-Gaby et al. (1988) and Hassan and Hashad (1990) believed that the mid-Proterozoic continental crust extends into the Eastern Desert and crops out in gneiss domes underneath overthrusted Pan-African island arc volcanics and volcaniclastics and associated ophiolites. Khudeir et al. (1995) contributed to the debate on the existence of pre-Pan-African continental crust in the Eastern Desert by presenting new data on the geology, petrography, and geochemistry of deformed granites exposed within the Sibai swell. The obtained results showed that the granite masses within the Sibai swell are intensely deformed and are older than the thrusting event which emplaces the overlying ophiolites and island arc volcanics and volcaniclastics of Pan-African age. The second school considered that (1) the rocks of the Eastern Desert have oceanic affinity characterized by mafic to intermediate volcanics and thick sequences of immature sediments and volcaniclastic material (e.g., Engel et al. 1980), and (2) the gneisses are regarded to represent the highly deformed and metamorphosed equivalent of metavolcanics and metasedimentary rocks. Such opinion is in complete harmony with the arc assembly (arc accretion) model that shows, as mentioned before, that the EAO juvenile crust was generated around and within a Pacific-sized ocean (Mozambique Ocean) (Vail 1985; Stoeser and Camp 1985; Stern 1994).

14.2 Thrusting, Shearing, and Folding Relations

Structural styles and overprinting relations in the Egyptian Eastern Desert Shield rocks offer a good opportunity to decipher the relationship between folding, on the one hand, and thrusting and shearing on the other hand. Our investigations of key areas in the Eastern Desert indicate that these structures are geometrically and kinematically related. In many areas, a complete transition from slightly deformed to highly deformed rocks are potential in deducting such geometric and kinematic relations. Propagation of thrusting often took place according to "footwall-nucleating–footwall vergent rule". Accordingly, earlier hanging walls are carried forward in a piggyback manner, and newly formed thrusts grow in the footwalls of the older thrusts. Out-of-sequence thrusts are infrequently observed. The propagation of thrusting led to the formation of thrust-related folds which are best typified by Beitan major structure (Abdelkhalek et al. 1992; Abdeen et al. 2008). Other types of folds are shear zone-related. Overprinting relations between thrust-related folds and shear zone-related folds demonstrate that the former are in most case the older as documented in many deformed belts in the Eastern Desert, such as Mubarak–Baramiya belt, Um Nar-Gabal Elhadid Belt, and Wadi Khuda Belt.

14.3 Gneiss Domes Versus Metamorphic Core Complexes

The main structural feature of the Eastern Desert terrane comprises structural basement (lower and higher grade gneissic domes) overthrusted by structural cover nappes (lower grade volcaniclastic metasediments) (Sabet 1961; Akaad and El-Ramly 1960; El-Ramly 1972; El-Gaby 1983; El-Gaby et al. 1984, 1994; Kamal El Din et al. 1992; Khudeir et al. 1992, 1995; Kamal El Din 1993; Greiling et al. 1994; Akaad et al. 1996; El-Sayed et al. 1999, 2002; Ibrahim and Cosgrove 2001; Fritz et al. 1996, 2002, 2013; Fowler and Osman 2001; Bregar et al. 2002; Abdeen 2003; Abdeen and Greiling 2005; Fowler et al. 2007; Abd El-Wahed 2008; Youssef et al. 2009; Amer et al. 2010; Johnson et al. 2011; Abdeen et al. 2014; Abu Enen et al. 2016). The directions of nappe transport reported from the CED vary from top to the NE (e.g., El-Bayoumi and Greiling 1984), top to the NW (e.g., Ries et al. 1983; Greiling 1987), top to the SE (e.g., Kamal El Din et al. 1992), and top to the

SW (e.g., Abdeen et al. 2002; Abdelsalam et al. 2003). One remarkable feature of the Eastern Desert is the presence of a series of gneiss domes (e.g., Meatiq, Sibai, El-Shalul, Hafafit) surrounded by low-grade nappes (Fig. 34). Also, there are several gneissic windows in Sinai (e.g., Feiran–Solaf metamorphic complex: Abu Alam and Stuwe 2008; Kid area: Abu El-Enen et al. 2003; Blasband et al. 1997, 2000; Brooijmans et al. 2003; Eliwa et al. 2008; Taba area: Abu El-Enen et al. 2004; Eliwa et al. 2008). Abu Alam and Stuwe (2009) proposed that the fold-and-thrust deformation and exhumation of the Feiran–Solaf complex was a result of transpressional deformation and the NW-striking Najd Fault System exhuming the complex in an oblique transpressive regime.

The gneiss domes in the Eastern Desert terrane have been interpreted as metamorphic core complexes exhumed in extensional settings. The origin and mode of deformation and exhumation of these gneissic domes and their relation to the Najd Fault System are the subjects of many publications (e.g., Fritz et al. 1996, 2002, 2013; Loizenbauer et al. 2001; Bregar et al. 2002; Shalaby et al. 2005; Fowler et al. 2007; Abd El-Wahed 2008, 2014; Andresen et al. 2010; Fowler and Osman 2009; Abu Alam and Stüwe 2009; Abd El-Wahed and Kamh 2010; Shalaby 2010; Johnson et al. 2011; Abu Alam et al. 2014; Abd El-Wahed et al. 2016; Makroum 2017; Stern 2017; Hassan et al. 2016). Gneiss domes in Egypt are mostly bordered by NW-striking sinistral shear zones and low-angle normal faults (Fritz et al. 1996). Geochronology suggests that extension and exhumation of gneiss domes commenced around 620–606 Ma (Fritz et al. 2002; Andresen et al. 2009)

Within the Eastern Desert Terrane of Egypt, there is a general agreement on (i) significant NW–SE extension and crustal-scale thinning, (ii) NNW-thrust propagation and juxtaposition of low-grade volcano-sedimentary sequences against high-grade gneisses along extensional shears, and (iii) intense magmatic activity and emplacement of extension-related granitoids (Blasband et al. 2000; Fowler and El Kalioubi 2004; Fowler and Osman 2009; Andresen et al. 2010). But, the presence or absence of core complexes (e.g., Meatiq, Sibai, El-Shalul, and Hafafit) and the surrounding sinistral shear zones in the Eastern Desert is a matter of debate. Based on structural arguments and the succession of magmatic events, the gneiss domes were interpreted as a remobilized early Proterozoic older continental crust (Sturchio et al. 1983; El-Gaby et al. 1990, 1994; Hassan and Hashad 1990). Kinematically, four tectonic models were proposed to decipher origin, evolution, and exhumation of the gneiss-cored domes in the Eastern Desert: (1) development of fault-bend fold "antiformal stacks" (e.g., Hafafit domal structure; Greiling et al. 1988), (2) orogen-parallel crustal extension as a consequence of sinistral shearing along the NW-trending shear zones of the Najd Fault System (e.g., Hafafit, Sibai, El-Shalul and Meatiq domal structures; Wallbrecher et al. 1993; Fritz et al. 1996, 2002, 2013; Bregar et al. 2002; Loizenbauer et al. 2001; Abd El-Wahed 2008; Khudeir et al. 2008), (3) emplacement within regional domal structures (Ibrahim and Cosgrove 2001) followed by extension parallel to their fold axes (e.g., Sibai dome, Fowler et al. 2007), and (4) interference patterns of sheath folds (e.g., Hafafit domal structure, Fowler and El Kalioubi 2002). Shalaby (2010) assigned the development of the northern dome of Wadi Hafafit culmination to oblique convergence in a scissor-like wrench corridor model not to regional-scale lithospheric extension associated with formation of core complexes.

14.4 The Conjugate Pairs of Najd-Related Shears

The structures in the Najd Fault zone were developed in response to a sinistral transpressional and transtensional tectonic regimes, with the axis of maximum compressional stress oriented at oblique angles to the NW-trending orogenic front (Fritz et al. 1996). Transpression resulted in formation of ESE- to ENE-shortening that produced NNW- to NW-trending folds throughout the Central Eastern Desert (Abd El-Wahed 2014). These are superimposed on early NNW-directed thrusts and related structures (Abdeen et al. 2014). The structures associated with the NW-sinistral shear zones are strongly superimposed by the NE-trending structures of the Mubarak–Barramiya shear belt that extends from Wadi Mubarak in the East to Wadi Barramiya in the west across the whole width of the CED forming a remarkable structural feature in the eastern desert of Egypt. This belt constitutes well-defined ophiolite-decorated linear belt where serpentinites represent the most characteristic lithological unit. It is oriented orthogonal to the major NW-trending fabric characterizing the CED. In the D_{2b}, D_{2b} is the NE-trending dextral transpression along Wadi Mubarak overprinted on D1 and D2a structures. D_{2b} structures include the NNE- to NE-trending WUSZ, NNE- and NE-trending (NNW-verging) folds, NNW-directed thrusts (Shalaby et al. 2005; Abd El-Wahed and Kamh 2010; Abd El-Wahed 2014). D_{2a} and D_{2b} represent a non-coaxial progressive event formed in a dextral NE- over NW-sinistral shear zone during a partitioned transpression in response to E–W-directed compression during oblique convergence between East and West Gondwana developed due to closure of the Mozambique Ocean and amalgamation of the Arabian–Nubian Shield in Cryogenian-early Ediacaran time (Abd El-Wahed and Kamh 2010; Abd El-Wahed 2014). Superimposition of the NW Najd-related shear zones by the NE-trending transpressional deformation (Fig. 30) demonstrates that both trends are not conjugate. In this framework, the entire Pan-African Belt of the Eastern Desert represents a megashear, called in the present study as the Eastern Desert Megashear (Fig. 36). This megashear exhibits sinistral sense of movement and the NE-trending shears were formed by the same way as the domino structure.

14.5 Role of Najd Shear System in Tectonic Evolution of Gneiss Domes

Another controversy in the Eastern Desert is about the role of sinistral shearing and transpression related to the Najd Fault System in the exhumation of these gneiss

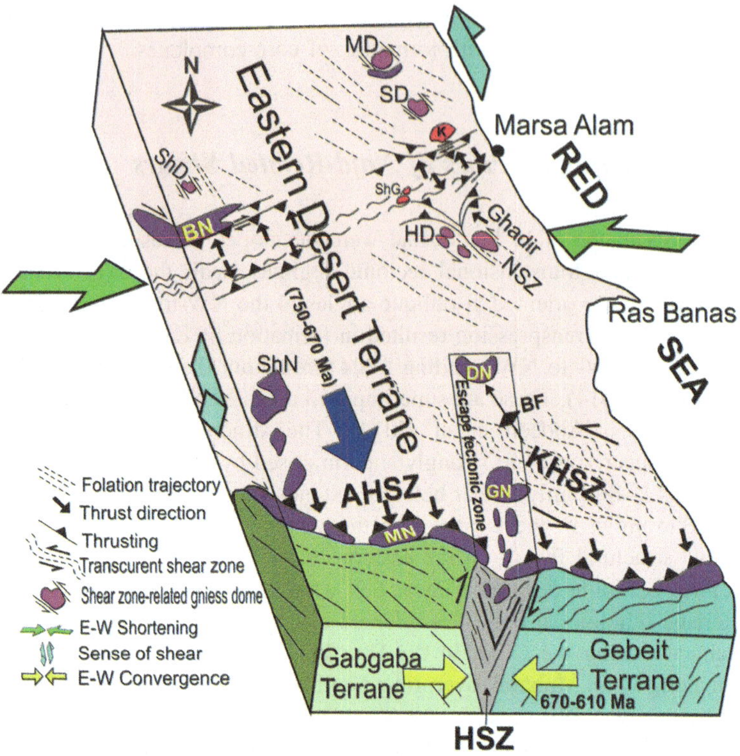

Fig. 36 Tectonic style sketch of the Southern and Central Eastern Desert. MD, Meatiq dome; SD, Sibai dome; HD, Hafafit dome, ShD, Shalul dome; K, Kadabbora granite; ShG, Sheikh Salem granite; BN, Barramiya nappe; NSZ, Nugrus shear zone; DN, Abu Dahr nappe; BF, Beitan fold; KHSZ, Kharit-Hodein shear zone; Gn, Gerf nappe; Mn, Moqsem nappe; ShN, Shilman nappe; AHSZ, Allaqi–Heiani shear zone; and HSZ, Hamisana shear zone

domes and in the crustal structure of the Central Eastern Desert. Northwest-striking shear zones and faults of the Najd Fault System are the dominant structural elements within the Afif–Hijaz and Midyan terranes of northwestern Saudi Arabia and the Eastern Desert terrane of Egypt. The Najd Fault System consists of brittle–ductile shears in a zone as much as 300 km wide and more than 1100 km long, extending across the northern part of the Arabian Shield and developed during the interval 540–620 Ma (Stern 1985). The Najd Fault System in Saudi Arabia extends northwest in the Central Eastern Desert of Egypt affecting the area between Duwi shear zone to the north and Kharit-Hodein shear zone to the south (Sultan et al. 1993; Abd El-Wahed 2014). The major shear zones bounding the gneiss domes have been interpreted as sinistral strike-slip shear zones combined with extensional shears that formed during exhumation of domes (e.g., Fritz et al. 1996, 2002) or as remnants of NW-directed thrusts (Andresen et al. 2010). Nowadays, sinistral shearing along the NW-trending shear zones of the Najd Fault System is genetically linked with deposition of sediments, exhumation of gneiss domains, and emplacement of syn-tectonic granitoids (e.g., Fritz et al. 1996, 2002, 2013; Bregar et al. 2002; Shalaby et al. 2005; Abd El-Wahed 2008, 2010; Abd El-Wahed and Kamh 2010; Abdeen et al. 2014; Abd El-Wahed et al. 2016; Makroum 2017).

There are some points support the role of the Najd shear zones in the evolution and exhumation of core complexes in the Eastern Desert: (i) The large-scale, oblique transpressive shear zones of the Najd Fault System in the Eastern Desert and Sinai was developed during the second tectonometamorphic event (D2) occurred between 680 and 640 Ma (Johnson et al. 2011). During D_2 the Arabian–Nubian Shield collided with the Sahara Metacraton and moved toward the Paleo-Tethys ocean (Stern 1994), (ii) the gneiss domes is bounded by transtensional marginal shears linked by low-angle normal faults (Fritz et al. 1996, 2002, 2013), (iii) The oblique setting of the gneiss domes to the main NW-trending shear zones (Abu Alam et al. 2014), (iv) Exhumation history of the core complexes is accompanied by crustal thickening, development of molasse sedimentary basins (e.g., Kareim Basin to the north of Sibai core complex; Abd El-Wahed 2010), and (v) There is relationship between change of the Najd shear kinematics from transpressive to transtension through time and emplacement of transpressive- and transtension-related granitoids (655 and 645 Ma, respectively, Bregar et al. 2002; Abu Enen et al. 2016).

14.6 Rates and Transport Directions of Metaultramafic Nappes

Several metaultramafic nappes have been the subject matter of detailed field-structural studies in the Egyptian Eastern Desert. The largest nappe, not only in the Eastern Desert but also in the ANS as a whole, is the Gerf nappe which exists in a remarkable escape tectonic zone (Fig. 36) includes the southern tip of Hamisana Shear Zone and extends further north to Abu Dahr nappe that ranks the second after the Gerf nappe by volume. Other eye-catching nappes area is observed in Moqsem, Shilman, and Barramiya areas. Both Moqsem and Shiman nappes form part of an ophiolite-decorated belt defining the Allaqi–Heiani Suture which is itself regarded to represent the western extension of the greater Allaqi–Heiani–Onib–Sol Hamed–Yanbu (Abdelsalam and Stern 1996; Abdelsalam et al. 2003). However, the rates and directions of transport of these metaultramafic nappes are debatable, although there is a general agreement that these nappes follow two main directions in their transportation; W- (to WSW) and N- (to NNW-) direction. The first transport direction is mostly a consequence of the final assembly and accretion of the entire ANS to the Saharan Metacraton (SM) (Abdelsalam et al. 2011; Liégeois et al. 2013) which was concurrent with the assembly of eastern and western Gondwana during the late Cryogenian–Ediacaran (650–542 Ma). The second transport direction is linked to the N-directed tectonic escape of the ANS, and proved by N- (to NNW-) trending stretched lineations. The rates of transportation of the metaultramafic nappes vary considerably from deformed belt to another in the Eastern Desert depending upon several factors, such as nappe size, shearing rate, and enveloping lithologies.

14.7 The Voluminous Intrusives in NED

Dissimilar to the Arabian Shield, the Nubian Shield had been invaded by anorogenic/within-plate magmatism (500–30 Ma) (El-Ramly and Hussein 1983; Abdel-Rahman and Martin 1987). The Pan-African basement of the NED of Egypt differs from the southern continuation by (1) voluminous calc-alkaline batholithic magmatism (635–590 Ma) and alkaline A-type magmatism ($\sim$600 Ma), (2) scarcity of older metasedimentary, metavolcanic, and ophiolitic rocks, (3) the absence of igneous activity younger than 475 Ma, and (4) the absence of megashear zone like the Najd-related shears (e.g., El-Ramly 1972; Engel et al. 1980; Ries et al. 1983; Bentor 1985; Stern and Hedge 1985; Abdel-Rahman and Martin 1987; Beyth et al. 1994; Cosca et al. 1999; Garfunkel 1999).

The voluminous magmatism of the NED includes: (1) batholithic post-collisional high-K calc-alkaline plutons ($\sim$635–590 Ma) of granodiorite and monzogranite, as well as some minor gabbros and quartz-diorites; and (2) within-plate alkaline to peralkaline high-level plutons and the associated bimodal (mafic–felsic) volcanic rocks and dyke swarms of similar age ($\sim$608–580 Ma) and geochemical affinity (e.g., Garfunkel 1999; Jarrar et al. 2003, 2004; Katzir et al. 2006, 2007a, b; Be'eri-Shlevin et al. 2009b). These magmas intruded the older deformed island arc complexes of c. 820–740 Ma (e.g., Be'eri-Shlevin 2009, and references therein). In the post-collisional calc-alkaline suite, the voluminous pulse of granodiorite to granite intrusions (610–600 Ma) characterized the change from mafic to felsic magmatism in most of the northeastern African region. Contemporaneously, the alkaline magmatism commenced at the peak of the latter calc-alkaline activity at $\sim$608 Ma and had continued to $\sim$580 Ma (Be'eri-Shlevin 2009).

The calc-alkaline magmatic suite of the NED is generally characterized by low initial $^{87}Sr/^{86}Sr$ ratios, low-Nb contents, and progression from intermediate to felsic calc-alkaline magmatism. Accordingly, the NED crustal segment can be compared to the modern continental arc magmatic suits, such as the Cordilleran magmatic suite (e.g., Abdel-Rahman and Martin 1987). The cease of subduction is characterized by alkaline/within-plate magmatism that emplaced into the orogenic calc-alkaline suite. Accordingly, a continental margin tectonomagmatic model (e.g., western North America) can be adopted for the NED crustal evolution.

The voluminous post-collisional magmatism in the NED/northernmost ANS is possibly the result of cratonization of the juvenile crust. According to Jackson (1986), the post-collisional magmatic activity in the ANS has generated large volumes of melt from crust and mantle. Černý (1991) proposed that peraluminous

A-type granitoids were generated by partial melting of undepleted (possible the ANS oceanic crust), previously unmelted, crustal sources. The first volumetrically important melting event of the ANS crust was most likely during the post-collisional magmatic activity (610–530 Ma) (e.g., Kuster 2009). The voluminous felsic magmatism took place during a period of ~40–45 Ma and was interpreted to represent the migration of melting zone from mantle to lower crustal levels (Be'eri-Shlevin et al. 2009). The latter felsic magmatism could lead to the internal differentiation of the ANS juvenile crust and cratonization of the nascent shield (e.g., Kuster 2009). The post-collisional setting of the NED voluminous magmatism can be supported by (1) the absence of arc formation in the northernmost ANS at ca 620–600 Ma, and (2) the absence of subduction-related features (e.g., accretionary prisms, paired metamorphic belts and linear alignment of calc-alkaline intrusions and volcanic rocks) (e.g., Be'eri-Shlevin et al. 2009).

14.8 The Post-amalgamation Hammamat Sediments and Their Relation to Dokhan Volcanics

Post-amalgamation depositional basins (each of which covers a surface area from 200 to 72,000 km^2) occur in some forty areas in the ANS (e.g., Abdeen et al. 1992; Johnson 2003; Matsah and Kusky 1999, 2001; Willis et al. 1988; Johnson and Woldehaimanot 2003; Abdeen and Greiling 2005; Eliwa et al. 2006, 2010; Hamimi et al. 2012b, 2013a, b, 2014; Hamimi and Kattu 2014). These basins encompass slightly to moderately metamorphosed, and at the same time variably deformed, volcano-sedimentary successions that were deposited after 650 Ma over newly amalgamated arc terranes (Johnson et al. 2011). They are commonly structurally controlled (fault-controlled down sags, pull aparts, rifts, half-grabens, thrusting, normal faulting, magmatic doming, etc.) (e.g., Abdeen and Greiling 2005; Shalaby et al. 2006; Hamimi et al. 2014), and autochthonous as attested by their unconformable basal contacts vs. older basement rocks. Depending on their carbonate succession, relative abundances of gray–green and red–purple rocks and other sedimentary structures, Johnson and Woldehaimanot (2003) subdivided these basins into marine, terrestrial and mixed terrestrial–marine basins. Marine post-amalgamation basins are prominent in the eastern part of the ANS, such as in Murdama, Bani Ghayy, Fatima, and Ablah areas.

In NED and CED, the post-amalgamation Hammamat basins (590–585 Ma depositional age, Rice et al. 1993) are epicontinental and showing mostly NW–SE and N–S directions (Rice et al. 1993; Abdeen and Greiling 2005). The sedimentary section of the Hammamat sediments in the type locality reaches up to 4000 m. It comprises polymictic conglomerates, gritstone, sandstone, siltstone, claystone, and rare limestone intercalated with volcaniclastic layers (Akaad and Now-eir 1969, 1980). The Hammamat sediments have been deposited in three main types of basins (Abd El-Wahed 2010) including foreland, intermontane (El-Gaby et al. 1990), and strike-slip basins (Fritz and Messner 1999). They were deposited by alluvial fan and braided stream systems in intermontane (Grothaus et al. 1979) and foreland basins (El-Gaby et al. 1988) formed during the late stage of the Pan-African orogeny. The Hammamat sediments were deposited in late Precambrian after the eruption of subduction-related Dokhan Volcanics of andesitic to rhyolitic composition (Eliwa et al. 2006) and prior to the emplacement of the post-orogenic granitoids (Akaad and Noweir 1980). Hamimi et al. (2014) believed that the Hammamat basins are fault-bounded basins affected by a NW–SE- to NNW–SSE-oriented shortening phase just after the deposition of the molasse sediments, proved by NW- to NNW-verging folds and SE- to SSE-dipping thrusts that were refolded and thrusted in the same direction. The shortening phase in the Hammamat was followed by a transpressional wrenching phase related to the Najd Shear System, which resulted in the formation of NW–SE sinistral-slip faults associated with positive flower structures that comprise NE-verging folds and SW-dipping thrusts.

The Hammamat sediments show nearly the same deformation patterns and underwent the same deformation history of the post-amalgamation marine sediments in Western Arabia (Hamimi et al. 2014). The only exception is the absence of the earlier-formed structures (D_1) that predating the conspicuous transpressional phase (D_2) and formed due to the effect of an early E–W (to ENE–WSW) shortening phase accompanied with the convergence between East and West Gondwana. Such a conclusion is attested by MRL and CESS quantitative strain calculations. A general agreement is established that the deposition of the Hammamat was after the eruption of subduction-related Dokhan Volcanics (El-Gaby et al. 1988) and prior to the emplacement of the post-orogenic granitoids (Akaad and Noweir 1980).

14.9 Northward Decrease in Intensity of Deformation

One of the most amazing issues in dealing with the tectonics of the Pan-African Belt of the Egyptian Eastern Desert is the northward decrease in intensity of deformation. Although this point will be the subject matter of detailed investigation and will be addressed in the near future, the wealth of collected data from key areas in CED and SED attribute such variation in the intensity of deformation to several factors such as (1) presence of the Allaqi–Heiani suture zone at the southern end of the SED, (2) the NNE-SSW convergence of the Eastern Desert terrane (750–670 Ma) with the Gabgaba and Gebeit terranes, and (3) the effect of the N- to NNE-oriented Hamisana Shear Zone which in fact concomitant with the N-directed escape of the entire ANS.

References

Abd El-Naby H, Frisch W (2006) Geochemical constraints from the Hafafit Metamorphic Complex (HMC): evidence of Neoproterozoic back-arc basin development in the central Eastern Desert of Egypt. J Afr Earth Sci 45:173–186

Abd El-Rahman Y, Polat A, Dilek Y, Fryer BJ, El-Sharkawy M, Sakran S (2009a) Geochemistry and tectonic evolution of the Neoproterozoic Wadi Ghadir ophiolite, Eastern Desert, Egypt. Lithos 113:158–178

Abd El-Rahman Y, Polat A, Dilek Y, Fryer BJ, El-Sharkawy M, Sakran S (2009b) Geochemistry and tectonic evolution of the Neoproterozoic incipient arc–forearc crust in the Fawakhir area, Central Eastern Desert of Egypt. Precambr Res 175:116–134

Abd El-Wahed MA (2007) Pan-African strike-slip tectonics of Wadi El-Dabbah area, north Sibai core complex, Central Eastern Desert, Egypt. Ann Egypt Geol Surv 29:1–36

Abd El-Wahed MA (2008) Thrusting and transpressional shearing in the Pan-African nappe southwest El-Sibai core complex, Central Eastern Desert, Egypt. J Afr Earth Sci 50:16–36

Abd El-Wahed MA (2010) The role of the Najd fault system in the tectonic evolution of the Hammamat molasse sediments, Eastern Desert, Egypt. Arab J Geosci 3:1–26

Abd El-Wahed MA (2014) Oppositely dipping thrusts and transpressional imbricate zone in the Central Eastern Desert of Egypt. J Afr Earth Sci 100:42–59

Abd El-Wahed MA, Abu Anbar MM (2009) Syn-oblique convergent and extensional deformation and metamorphism in the Neoproterozoic rocks along Wadi Fatira shear zone, Northern Eastern Desert, Egypt. Arab J Geosci 2:29–52

Abd El-Wahed MA, Kamh SZ (2010) Pan-African dextral transpressive duplex and flower structure in the Central Eastern Desert of Egypt. Gondwana Res 18:315–336

Abd El-Wahed MA, Kamh SZ (2013) Evolution of conjugate strike-slip duplexes and wrench-related folding in the Central part of Al Jabal Al Ahkdar, NE Libya. J Geol 12 (2):173–195

Abd El-Wahed MA, Harraz HZ, El-Behairy MH (2016) Transpressional imbricate thrust zones controlling gold mineralization in the Central Eastern Desert of Egypt. Ore Geol Rev 78:424–446

Abdeen MM (2003) Tectonic history of the Pan-African orogeny in Um Gheig area, central Eastern Desert, Egypt. Egypt J Geol 47(1):239–254

Abdeen MM, Abdelghaffar AA (2011) Syn- and post-accretionary structures in the Neoproterozoic Central Allaqi-Heiani suture zone, Southeastern Egypt. Precambr Res 185:95–108

Abdeen MM, Greiling RO (2005) A quantitative structural study of late Pan-African compressional deformation in the Central Eastern Desert (Egypt) during Gondwana assembly. Gondwana Res 8(4):457–471

Abdeen MM, Dardir AA, Greiling RO (1992) Geological and structural evolution of the Wadi Queih Area (N Quseir), Pan-African basement of the Eastern Desert of Egypt. Zbl Geol Palfiontol Teil I:2653–2659

Abdeen MM, Allison TK, Abdelsalam MG, Stern RJ (2001) Application of ASTER band-ratio images for geological mapping in arid regions; the Neoproterozoic Allaqi Suture, Egypt. Abstr. Program Geol. Soc. Am. 3:289

Abdeen MM, Abdelsalam MG, Dowidar HM, Abdelghaffar AA (2002) Varying structural style along the Neo-Proterozoic Allaqi-Heiani suture, Southern Egypt. In: 19th Colloquium of African Geology. El Jadida, Morocco. (Abstract)

Abdeen MM, Sadek MF, Greiling RO (2008) Thrusting and multiple folding in the neoproterozoic pan-African basement of Wadi Hodein area, south Eastern Desert, Egypt. J Afr Earth Sci 52:21–29

Abdeen MM, Greiling RO, Sadek MF, Hamad SS (2014) Magnetic fabrics and Pan-African structural evolution in the Najd Fault corridor in the Eastern Desert of Egypt. J Afr Earth Sci 99:93–108

Abdel Khalek ML (1979) Tectonic evolution of the basement rocks in the Southern and Central Eastern Desert of Egypt. Bull Inst Appl Geol 3(1):53–62. (King Abdel Aziz University, Jeddah)

Abdel Khalek ML, Takla MA, Sehim A, Hamimi Z, El Manawi AW (1992) Geology and tectonic evolution of Wadi Beitan area, south Eastern Desert, Egypt. In: International Conference Geology. The Arab World, Cairo University, Egypt (GAW1), pp 369–393

Abdel Khalek ML, Takla MA, Sehim A, Abdel Wahed M, Hamimi Z, Sakran Sh (1999) Geology and tectonic evolution of the shield rocks, East Wadi Beitan area, south Eastern Desert, Egypt. Egypt J Geol 43(1):1–15

Abdel Rahman AM (1996) Pan-African volcanism: petrology and geochemistry of the Dokhan volcanic suite in the northern Nubian shield. Geol Mag 133:17–31

Abdel Rahman AM, El-Kibbi MM (2001) Anorogenic magmatism: chemical evolution of the Mount El-Sibai A-type complex (Egypt), and implications for the origin of within-plate felsic magmas. Geol Mag 138:67–85

Abdel Rahman AM, Martin RF (1987) Late Pan-African magmatism and crustal development in northeastern Egypt. Geol J 22:281–301

Abdel Wahed AA, Ali KG, Khalil MMA, Abdel Gawad A (2012) Dokhan volcanics of Gabal Monqul area, North Eastern Desert, Egypt: geochemistry and petrogenesis. Arab J Geosci 5:29–44

Abdelsalam MG (1994) The Oko shear zone: post-accretionary deformation in the Arabian-Nubian shield. J Geol Soc, London 151:767–776

Abdelsalam MG, Stern RJ (1996) Sutures and shear zones in the Arabian-nubian shield. J Afr Earth Sci 23:289–310

Abdelsalam MG, Abdeen MM, Dowaidar HM, Stern RJ, Abdelghaffar AA (2003) Structural evolution of the Neoproterozoic Western Allaqi-Heiani suture, southeastern Egypt. Precambr Res 124:87–104

Abdelsalam MG, Gao SS, Liégeois J-P (2011) Upper mantle structure of the Saharan metacraton. J Afr Earth Sci 60:328–336

Aboelkhair H, Yoshiki N, Yasushi W, Isao S (2010) Processing and interpretation of ASTER TIR data for mapping of rare-metal-enriched albite granitoids in the Central Eastern Desert of Egypt. J Afr Earth Sci 58(1):141–618

Abou El-Magd K, Emam A, Ali-Bik MW (2013) Chemostratigraphy, petrography and remote sensing characterization of the Middle Miocene-Holocene sediments of Ras Banas peninsula, Red Sea Coast, Egypt. Carpathian J Earth Environ Sci 8(3):27–42

Abou El-Magd I, Mohy H, Basta F (2015) Application of remote sensing for gold exploration in the Fawakhir area, Central Eastern Desert of Egypt. Arab J Geosci 8(6):3523–3536

Abu El-Ela FF (1997) Geochemistry of an island-arc volcanic suit: Wadi Dabr intrusive complex, Eastern Desert, Egypt. J Afr Earth Sci 24:473–496

Abu El-Enen MM, Okrusch M, Will TM (2003) Metapelitic assemblages in the Zariq schists, central western Kid Belt, Sinai Peninsula, Egypt. Neues Jahrbuch Mineralogie Abhandlungen 178:277–306

Abu El-Enen MM, Okrusch M, Will TM (2004) P-T evolution of the Taba metamorphic belt, Egypt: constraints from the metapelite assemblages. J Afr Earth Sci 38:59–78

Abu El-Enen MM, Abu-Alam TS, Whitehouse MJ, Ali K, Okrusch M (2016) P-T path and timing of crustal thickening during amalgamation of East and West Gondwana: a case study from the Hafafit Metamorphic Complex, Eastern Desert of Egypt. Lithos 263:213–238

Abu-Alam TS, Stüwe K (2008) Metamorphic evolution of the Wadi Feiran-Solaf core complex, Sinai, Egypt. J Alpine Geol 49:2. (PANGEO-Austria 2008)

Abu-Alam TS, Stüwe K (2009) Exhumation during oblique transpression: the Feiran-Solaf region, Egypt. J Metamorph Geol 27:439–459

Abu-Alam TS, Hassan M, Stüwe K, Meyer S, Passchier C (2014) Multistage tectonism during Gondwana collision: Baladiyah Complex, Saudi Arabia. J Petrol 55:1941–1964

Ahmed AH (2013) Highly depleted harzburgite–dunite–chromitite complexes from the Neoproterozoic ophiolite, south Eastern Desert, Egypt: a possible recycled upper mantle lithosphere. Precambr Res 233:173–192
Ahmed AH, Arai S, Attia AK (2001) Petrological characteristics of podiform chromitites and associated peridotites of the Pan African ophiolite complexes of Egypt. Miner Deposita 36:72–84
Akaad MK (1996) Rock succession of the basement: an autobiography and assessment. In: The Egyptian geological survey, centennial conference, paper no 71, p 87
Akaad MK (1997) On the behavior of serpentinites and its implications. Geol Surv Egypt, Paper, p 74
Akaad MK, Abu El Ela AM (2002) Geology of the basement rocks in the eastern half of the belt between latitudes 25 30′ and 26 30′ N Central Eastern Desert. Egypt, Geol Surv Egypt, p 78
Akaad MK, El Ramly MF (1960) Geological history and classification of the basement rocks of the central Eastern Desert of Egypt. Geol Surv Egypt 9:24
Akaad MK, Noweir AM (1969) Lithostratigraphy of the Hammamat-Umm Seleimat district, Eastern Desert of Egypt. Nature 223:284–285
Akaad MK, Noweir AM (1977) The post Hammamat felsites and their age relation with the younger granites of Egypt. Proc Acad Sci 30:163–168
Akaad MK, Noweir AM (1980) Geology and lithostratigraphy of the Arabian Desert orogenic belt of Egypt between latitudes 25 35′ S and 26°30′ N. Inst Appl Geol Jeddah Bull 3:127–135
Akaad MK, Noweir AM, Kotb H (1979) Geology and petrochemsitry of the granite association of the Arabian Desert Orogenic Belt of Egypt between 25 35′ and 26°30′. Delta J Sci 3:107–151
Akaad MK, Abu El Ela AM, El Kamshoshy HI (1993) Geology of the region West of Mersa Alam, Eastern Desert, Egypt. Ann Egypt Geol Surv 19:1–18
Akaad MK, Nowei AM, Abu El-Ela AM (1996) Geology of the Pan-African basement rocks of the Jabal al Hadid–Wadi Mubarak district, Eastern Desert, Egypt. Geol Surv Egypt 73:78
Akawy A (2003) Fault characterization and paleostress analysis in the Precambrian rocks west of the Meatiq Dome, Central Eastern Desert, Egypt. Neues Jahrbuch Mineralogie Abhandlungen 228:135
Akawy A (2007) Geometry and texture of quartz veins in Wadi Atalla area, Central Eastern Desert. J Afr Earth Sci 47:73–87
Al-Husseini M (2000) Origin of the Arabian plate structures: Amar collision and Najd rift. Geo Arabia 5:527–542
Ali KA, Stern RJ, Manton WI, Kimura J-I Khamees HA (2009) Geochemistry, Nd isotopes and U–Pb SHRIMP dating of Neoproterozoic volcanic rocks from the Central Eastern Desert of Egypt: new insights into the 750 Ma crust-forming event. Precambr Res 171:1–22
Ali KA, Azer MK, Gahlan HA, Wilde SA, Samuel MD, Stern RJ (2010) Age constraints on the formation and emplacement of Neoproterozoic ophiolites along the Allaqi-Heiani Suture, South Eastern Desert of Egypt. Gondwana Res 18:583–595
Ali KA, Andresen A, Stern RJ, Manton WI, Omar SA, Maurice AE (2012) U-Pb zircon and Sr–Nd–Hf isotopic evidence for a juvenile origin of the c 634 Ma El-Shalul Granite, Central Eastern Desert, Egypt. Geol Mag 149:783–797
Ali K, Wilde S, Stern RJ, Moghazi A, Mahbubul Ameen S (2013) Hf isotopic composition of single zircons from Neoproterozoic arc volcanics and post-collision granites, Eastern Desert of Egypt: implications for crustal growth and recycling in the Arabian-Nubian Shield. Precambr Res 239:42–55
Ali KA, Kröner A, Hegner E, Wong J, Li S, Gahlan HA, Abu El Ela F (2015) U-Pb zircon geochronology and Hf–Nd isotopic systematics of Wadi Beitan granitoid gneisses, South Eastern Desert, Egypt. Gondwana Res 27:811–824
Ali-Bik MW, Taman Z, El Kalioubi B, Abdel Wahab W (2012) Serpentinite-hosted talc–magnesite deposits of Wadi Barramiya area, Eastern Desert, Egypt: Characteristics, petrogenesis and evolution. J Afr Earth Sci 64:77–89

Ali-Bik MW, Sadek MF, Ghabrial DS (2014) Late Neoproterozoic metamorphic assemblages along the Pan-African Hamisana Shear Zone, southeastern Egypt: Metamorphism, geochemistry and petrogenesis. J Afr Earth Sci 99:24–38

Amer RM, Kusky TM, Ghulam A (2010) New methods of processing ASTER data for lithological mapping. Examples from Fawakhir, Central Eastern Desert of Egypt. J Afr Earth Sci 56:75–82

Andresen A, Abu El-Rus MA, Myhre PI, Boghdady GY, Corfu F (2009) U-Pb TIMS age constraints on the evolution of the Neoproterozoic Meatiq Gneiss Dome, Eastern Desert, Egypt. Int J Earth Sci (Geol Rundsch) 98:481–497

Andresen A, Augland LE, Boghdady GY, Elnady OM, Lundmark AM, Hassan MA, Abu El-Rus MA (2010) Structural constraints on the evolution of the Meatiq Gneiss Dome (Egypt), East African Orogen. J Afr Earth Sci 57:413–422

Anonymous (1972) Penrose field conference on ophiolites. Geotimes 17:24–25

Asran AM, Kabesh M (2003) Evolution and geochemical studies on a stromatic migmatite-amphibolite association in Hafafit area CED Egypt. Egypt J Geol 47:1–24

Asran AM, Rkh Bekir, Abdel Rahman E, Abdel El-Rashed A (2013) Petrology, geochemistry and remote sensing data of island arc assemblage along Wadi Abu Marawat, Central Eastern Desert, Egypt. Arab J Geosci 6:2285–2298

Augland LE, Andresen A, Boghdady GY (2012) U-Pb ID-TIMS dating of igneous and metaigneous rocks from the El-Sibai area: time constraints on the tectonic evolution of the Central Eastern Desert, Egypt. Int J Earth Sci 101:25–37

Azer MK, Stern RJ (2007) Neoproterozoic (835–720 Ma) serpentinites in the Eastern Desert, Egypt: fragments of fore-arc mantle. J Geol 115:457–472

Azer MK, Obeid M, Gahlan HA (2016) Late Neoproterozoic layered mafic intrusion of arc-affinity in the Arabian-Nubian Shield: a case study from the Shahira layered mafic intrusion, southern Sinai, Egypt. Geol Acta 14:237–259

Azza M, Hosny A, Gharib A (2012) Rupture process of Shallow Earthquakes of Abou-Dabbab Area, South East of Egypt. Egypt J Appl Geophys 11(1):85–100

Badawy A, El-Hady Sh, Abdel-Fattah AK (2008) Microearthquakes and neoteconics of Abu-Dabbab, Eastern Desert of Egypt. Seismol Res Lett 79(1):55–67

Basta EZ, Kotb H, Awadallah MF (1980) Petrochemical and geochemical characteristics of the Dokhan Formation at the type locality, Gabal Dokhan, Eastern Desert, Egypt. Bull Inst Appl Geol 4(3):121–140 King Abdel Aziz University, Jeddah

Be'eri-Shlevin Y, Katzir Y, Whitehouse MJ, Kleinhanns IC (2009a) Contribution of pre Pan-African crust to formation of the Arabian Nubian Shield: new secondary ionization mass spectrometry U-Pb and O studies of zircon. Geol Soc Am 37(10):899–902

Be'eri-Shlevin Y, Katzir Y, Whitehouse M (2009b) Postcollisional tectono-magmatic evolution in the northern Arabian-Nubian Shield (ANS): time constraints from ion probe U-Pb dating of zircon. J Geol Soc London 166(1):71–8

Be'eri-Shlevin Y, Samuel MD, Azer MK, Rämö OT, Whitehouse MJ, Moussa HE (2011) The Ediacaran Ferani and Rutig volcano-sedimentary successions of the northernmost Arabian–Nubian Shield (ANS): new insights from zircon U-Pb geochronology, geochemistry and O-Nd isotope ratios. Precambr Res 188:21–44

Bennett JD, Mosley PN (1987) Tiered-tectonics and evolution, Eastern Desert and Sinai. In: Matheis G, Schandelmeier H (eds) Current research in African earth sciences. Balkema, Rotterdam, pp 79–82

Bentor YK (1985) The crustal evolution of the Arabian-Nubian Massif with special reference to the Sinai Peninsula. Precambr Res 28:1–74

Berhe SM (1990) Ophiolites in northeast and east Africa: implications for Proterozoic crustal growth. J Geol Soc 147:41–57

Beyth M, Stern RJ, Altherr R, Kroner A (1994) The late Precambrian Timna igneous complex, Southern Israel: evidence for comagmatic-type sanukitoid monzodiorite and alkali granite magma. Lithos 31:103–124

Blasband B, Brooijmans P, Dirks P, Visser W, White S (1997) A pan-African core complex in the Sinai, Egypt. Geologie en Mijnbouw 76:247–266

Blasband B, White S, Brooijmans P, De Brooder H, Visser W (2000) Late Proterozoic extensional collapse in Arabian-Nubian Shield. J Geol Soc 157:615–628
Bregar M, Bauernhofer A, Pelz K, Kloetzli U, Fritz H, Neumayr P (2002) A late Neoproterozoic magmatic core complex in the Eastern Desert of Egypt; emplacement of granitoids in a wrench-tectonic setting. Precambr Res 118:59–82
Breitkreuz C, Eliwa H, Khalaf I, El Gameel K, Bühler B, Sergeev S, Larionov A (2010) Neoproterozoic SHRIMP U-Pb zircon ages of silica-rich Dokhan Volcanics in the northeastern Desert, Egypt. Precambr Res 182:163–174
Brooijmans P, Blasband B, White SH, Visser WJ, Dirks P (2003) Geothermobarometric evidence for a metamorphic core complex in Sinai, Egypt. Precambr Res 123:249–268
Brown GB (1972) Tectonic map of the Arabian Peninsula. Saudi Arabian Directorate General of Mineral Resources Map AP-2, Scale 1:4,000,000
Brown GF, Jackson RO (1960) The Arabian shield. In: XIX international geological congress, Pt. 9, pp 69–77
Bulter CA, Holdsworth RE, Strachan RA (1995) Evidence for Caledonian sinistral strike-slip motion and associated fault zone weakening, Outer Hebrides Fault Zone, NW Scotland. J Geol Soc 152:743–746
Černý P (1991) Fertile granites of Precambrian rare-element pegmatite fields: is geochemistry controlled by tectonic setting or source lithologies? Precambr Res 51:429–468
Chetty TRK, Bhaskar Rao YJ (2006) The Cauvery Shear Zone, Southern Granulite Terrain, India: a crustal-scale flower structure. Gondwana Res 10(1–2):77–85
Clark RN (1999) Spectroscopy of rocks, and minerals and principles of spectroscopy. In: Rencz AN, Ryerson RA (eds) Remote sensing for the earth sciences, manual of remote sensing, 3rd edn. Wiley, New York, NY, pp 3–58
Collins AS, Pisarevsky SA (2005) Amalgamating eastern Gondwana: the evolution of the Circum-Indian Orogens. Earth Sci Rev 71(3–4):229–270
Cosca MA, Shimron A, Caby R (1999) Late Precambrian metamorphism and cooling in the Arabian-Nubian Shield: petrology and 40Ar/39Ar geochronology of metamorphic rocks of the Elat area (Southern Israel). Precambr Res 98:107–127
Coward MP (1990) Shear Zones at the Laxford Front, NW Scotland and their significance in the interpretation of lower crust structure. J Geol Soc 147:279–286
Crane RB (1971) Preprocessing techniques to reduce atmospheric and sensor variability in multispectral scanner data. In: Proceedings of 7th International Symposium Remote Sensing Environment Ann Arbor Michigan 2, pp 1345–1355
Daggett PH, Morgan P, Boulos FK, Hennin SF, El-Sherif AA, El Sayed AA, Basta NZ, Melek YS (1986) Seismicity and active tectonics of the Egyptian Red Sea margin and the northern Red Sea. Tectonophysics 125:313–324
David WL, Wooil MM (2012) Landsat-TM-Based discrimination of lithological units associated with the Purtuniq Ophiolite, Quebec, Canada. Remote Sens 4:1208–1231
de Wall H, Greiling RO, Sadek MF (2001) Post-collisional shortening in the late Pan-African Hamisana high strain zone, SE Egypt: field and magnetic fabric evidence. Precambr Res 107:179–194
Dewey JF, Holdsworth RE, Strachan RA (1998) Transpression and transtension zones. In: Holdsworth RE, Strachan RA, Dewey JE (eds) Continental transpressional and transtensional tectonics. J Geol Soc 135:1–14
Di Tommaso I, Rubinstein N (2007) Hydrothermal alteration mapping using ASTER data in the Infiernillo porphyry deposit, Argentina. Ore Geol Rev 32:275–290
Dixon TH (1981) Age and chemical characteristics of some pre-Pan-African rocks in the Egyptian shield. Precambr Res 14:119–133
Dixon TH, Stern RJ, Hussein IM (1987) Control of red sea rift geometry by Precambrian Structures. Tectonics 6:551–571
Doblas M, Faulkner D, Mahecha V, Aparicio A, Lopez-Ruiz J, Hoyos M (1997) Morphologically ductile criteria for the sense of movement on slickensides from an extensional detachment fault in Southern Spain. J Struct Geol 19:1045–1054

Drury S (1987) Image interpretation in geology. Allen & Unwin, London, p 243
Drury S (1993) Image interpretation in geology, 2nd edn. Chapman and Hall, London
Drury S (2013) The East African orogen: neoproterozoic tectonics on display. Research news from the Earth Sciences, Earth-pages. https://earth-pages.co.uk/2013/09/19/the-east-african-orogen-neoproterozoic-tectonics-on-display/
Egyptian Geological Survey and Mining Authority (EGSMA) (1996) Geological map of Wadi Jabjahah Quadrangle, Egypt, Scale 1:250.000
El Khrepy S, Koulakov I, Al-Arifi N (2015) Crustal Structure in the area of the cannon earthquakes of Abu Dabbab (Northern Red Sea, Egypt), from seismic tomography inversion. Bull Seismo Soc Am 105(4): 13. http://dx.doi.org/10.1785/ 0120140333
El Mahallawi MM, Ahmed AF (2012) Late Proterozoic older granitoids from the North Eastern desert of Egypt: petrogenesis and geodynamic implications. Arab J Geosci 5:15–27
El-Bayoumi RM, Greiling R (1984) Tectonic evolution of a Pan-African plate margin in southeastern Egypt—a suture zone overprinted by low angle thrusting? In: Klerkx J, Michot J (eds), African geology. Tervuren, pp 47–56
El-Bialy MZ, Omar MM (2015) Spatial association of Neoproterozoic continental arc I-type and post-collision a-type granitoids in the Arabian-Nubian Shield: the Wadi Al-Baroud older and younger granites, North Eastern Desert, Egypt. J Afr Earth Sci 103:1–29
El-Gaby S (1975) Petrochemistry and geochemistry of some granites from Egypt. Neues Jahrbuch für Mineralogie Abhandlungen 124:147–189
El-Gaby S (1983) Architecture of the Egyptian basement complex. In: Riad lS, Baars DL (eds) Proceedings of 5th international conference on basement tectonics, Cairo University, Egypt, pp 1–8
El-Gaby S, El Nady O, Khudeir A (1984) Tectonic evolution of the basement complex in the central Eastern Desert of Egypt. Geol Rundsch 73:1019–1036
El-Gaby S, List FK, Tehrani R (1988) Geology, evolution and metallogenesis of the pan-African Belt in Egypt. In: El-Gaby S, Greiling RO (eds) The pan-African belt of Northeast Africa and adjacent areas. Vieweg & Sohn, Weisbaden, pp 17–68
El-Gaby S, Khudier AA, El Taky M (1989) The Dokhan volcanics of wadi Queh area, Central Eastern Desert, Egypt. In: 1st conference on geochemistry, Alexandria, Egypt, pp 42–62
El-Gaby S, List FK, Tehrani R (1990) The basement complex of the Eastern Desert and Sinai. In: Said R (ed) The geology of Egypt. Balkema, Rotterdam, pp 175–184
El-Gaby S, Khudeir AA, Abdel Tawab M, Atalla RF (1991) The metamorphosed volcanosedimentary succession of Wadi Kid, Southeastern Sinai, Egypt. Ann Geol Surv Egypt 17:19–35
El-Gaby S, El Aref M, Khudeir A, El Habbak G (1994) Geology and genesis of banded iron formation at Wadi Kareim, Eastern Desert, Egypt. 7th Ann Meeting, Min Soc, Egypt. Abstract
El-Hady ShM (1993) Geothermal Evolution of the red sea margin and its relation to Earthquake Activity. M.Sc. thesis, Cairo University, Cairo, Egypt
Eliwa HA, Kimura JI, Itaya T (2006) Late Neoproterozoic Dokhan Volcanics, N Eastern Desert, Egypt: geochemistry and petrogenesis. Precambr Res 151:31–52
Eliwa HA, Abu El-Enen MM, Khalaf IM, Itaya T, Murata M (2008) Metamorphic evolution of Neoproterozoic metapelites and gneisses in Sinai, Egypt: insights from petrology, mineral chemistry and K-Ar age dating. J Afr Earth Sci 51:107–122
Eliwa H, Breitkreuz C, Khalaf I, El Gameel K (2010) Depositional styles of early Ediacaran terrestrial volcanosedimentary succession in Gebel El Urf area, N Eastern Desert, Egypt. J Afr Earth Sci 57:328–344
El-Kalioubi BA, Osman AF (1996) Petrological and structural aspects of the mylonitized felsite at Wadi Atalla, Central Eastern Desert, Egypt. Proc Geol Surv Egypt 159–184
El-Kazzaz YA (1999) Active faulting along along Qena-Safaga road. In: The first international conference on the geology of Africa, p 2
El-Kazzaz YA, Hamimi Z (1999) Polyphase deformation of Wadi Murra, south Eastern Desert, Egypt. Int Conf Geol Afr 1(1):353–368 Assiut University

El-Kazzaz YA, Hamimi Z (2000) Temporal and spatial relationships of Abu Fas geologic structures, south Eastern Desert, Egypt. In: Fifth international conference of geology of Arab world, Cairo University, Egypt, (GAW5), pp 619–634

El-Kazzaz YA, Taylor WE (2000) Tectonic evolution of the Allaqi Shear Zone and implications for the pan-African terrain amalgamation in the South Eastern Desert of Egypt. J Afr Earth Sci 33(2):177–197

El-Rabaa SM, Al-Shumaimri MS, Al-Mishwat AT (2001) Fate of the Najd fault system in northwestern Saudi Arabia and southwestern Jordan. Gondwana Res 4:164–165

El-Ramly MF (1972) A new geological map for the basement rocks in the Eastern and Southwestern Deserts of Egypt. Ann Geol Surv Egypt 2:1–18

El-Ramly MF, Akaad MK (1960) The basement complex in the CED of Egypt between lat. 24°30′ and 25°40′. Geol Surv Egypt 8:33

El-Ramly MF, Hussein AAA (1983) The ring complexes of the Eastern Desert of Egypt. J Afr Earth Sci 3(1/2):77–82

El-Ramly MF, Greiling R, Kroner A, Rashwan AA (1984) On the tectonic evolution of the Wadi Hafafit area and environs, Eastern Desert of Egypt. Bull Fac Earth Sci 6:114–126 King Abdulaziz University, Jiddah

El-Ramly MF, Greiling RO, Rashwan AA, Ramsy AH (1993) Explanatory note to accompany the geological and structural maps of Wadi Hafafit area, Eastern Desert of Egypt. Ann Geol Surv Egypt 9:1–53

EL-Sayed MM, Furnes H, Mohamed FH (1999) Geochemical constraints on the tectonomagmatic evolution of the late Precambrian Fawakhir ophiolite, Central Eastern Desert, Egypt. J Afri Earth Sci 29:515–533

EL-Sayed MM, Mohamed FH, Furnes HS, Kanisawa S (2002) Geochemistry and petrogenesis of the Neoproterozoic granitoids in the Central Eastern Desert, Egypt. Chem Erde 62:317–346

El-Sharkawy MA, El-Bayoumi RM (1979) The ophiolites of Wadi Ghadir area Eastern Desert, Egypt. Ann Geol Surv Egypt 9:125–135

Engel AEJ, Dixon TH, Stern RJ (1980) Late Precambrian evolution of Afro-Arabian crust from ocean to craton. Geol Soc Am Bull 91:699–706

Essawy MA, Abu Zeid KM (1972) Atalla felsite intrusion and its neighbouring rhylitic flows and tuffs, Eastern Desert, Egypt. Ann Geol Surv Egypt 2:271–280

Fairhead JD, Girdler RW (1970) The seismicity of the Red Sea, Gulf of Aden and Afar triangle. Phil Trans R Soc A 267:49–74

Farahat ES, Abdel Ghani MS, Ahmed AF (2004a) Mineral chemistry as a guide to magmatic evolution of I- and A-type granitoids, Eastern Desert, Egypt. In: Proceedings of the 6th international conference on geochemistry, Alexandria University, Egypt, pp 1–23

Farahat ES, El Mahalawi MM, Hoinkes G, Abdel Aal AY (2004b) Continental back-arc basin origin of some ophiolites from the Eastern Desert of Egypt. Miner Petrol 82:81–104

Farahat ES, Mohamed HA, Ahmed AF, El Mahallawi MM (2007) Origin of I- and A-type granitoids from the Eastern Desert of Egypt: implications for crustal growth in the northern Arabian-Nubian shield. J Afr Earth Sci 49:43–58

Farahat ES, Zaki R, Hauzenberger C, Sami M (2011) Neoproterozoic calc-alkaline peraluminous granitoids of the Deleihimmi pluton, Central Eastern Desert, Egypt: implications for transition from late- to post-collisional tectonomagmatic evolution in the northern Arabian-Nubian Shield. Geol J 46:544–560

Finger F, Dorr W, Gerdes A, Gharib M, Dawoud M (2008) U-Pb zircon ages and geochemical data for the monumental granite and other granitoid rocks from Aswan, Egypt. Implications for geological evolution of the western margin of the Arabian-Nubian Shield. Miner Petrol 93:153–183

Fleck RJ, Greenwood WR, Hadley DG, Anderson RE, Schmidt DL (1979) Rubidium-Strontium geochronology and plate-tectonic evolution of the southern part of the Arabian Shield. US Geological Survey Saudi Arabian Mission Report 245, p 105

Fossen H (2010) Structural geology. Cambridge University Press, London, p 457

Fossen H, Tikoff B (1998) Extended models of transpression/transtension and application to tectonic settings. In: Holdsworth RE, Strachan RA, Dewey JF (eds) Continental transpressional and transtensional tectonics. J Geol Soc London, Special Pub 135, pp 15–33

Fossen H, Tikoff B, Teyssier C (1994) Strain modeling of transpressional and transtensional deformation. Norsk Geol Tidsskr 74:134–145

Fowler A, El Kalioubi B (2002) The Migif-Hafafit gneissic complex of the Egyptian Eastern Desert: fold interference patterns involving multiply deformed sheath folds. Tectonophysics 346:247–275

Fowler AR, El-Kalioubi B (2004) Gravitational collapse origin of shear zones, foliations and linear structures in the Neoproterozoic cover nappes, Eastern Desert, Egypt. J Afr Earth Sci 38:23–40

Fowler TJ, Osman AF (2001) Gneiss-cored interference dome associated with two phases of late Pan-African thrusting in the Central Eastern Desert, Egypt. Precambr Res 108:17–43

Fowler AR, Osman AF (2009) The Sha'it-Nugrus shear zone separating Central and South Eastern Deserts, Egypt: a post-arc collision low-angle normal ductile shear zone. J Afr Earth Sci 53:16–32

Fowler A, Khamees H, Dowidar H (2007) El Sibai gneissic complex, Central Eastern Desert, Egypt: folded nappes and syn-kinematic gneissic granitoid sheets-not a core complex. J Afr Earth Sci 49(4):119–135

Frei M, Jutz S (1990) Use of thematic mapper data for the detection of gold bearing formations in the Eastern Desert of Egypt. In: 7th thematic conference on remote sensing for exploration geology—methods, integration, solutions, Calgary, Canada, pp 1157–1172

Fritz H, Messner M (1999) Intramontane basin formation during oblique convergence in the Eastern Desert of Egypt: magmatically versus tectonically induced subsidence. Tectonophysics 315:145–162

Fritz H, Wallbrecher E, Khudier AA, Abu El Ela F, Dallmeyer RD (1996) Formation of Neoproterozoic metamorphic core complexes during oblique convergence, Eastern Desert, Egypt. J Afr Earth Sci 23:311–329

Fritz H, Dalmeyer DR, Wallbrecher E, Loizenbauer J, Hoinkes G, Neumayr P, Khudeir AA (2002) Neoproterozoic tectonothermal evolution of the Central Eastern Desert, Egypt: a slow velocity tectonic process of core complex exhumation. J Afr Earth Sci 34:543–576

Fritz H, Abdelsalam M, Ali KA, Bingen B, Collins AS, Fowler AR, Ghebreab W, Hauzenberger CA, Johnson PR, Kusky TM, Mace P, Muhongo S, Stern RG, Viola G (2013) Orogen styles in the East African Orogen: a review of the Neoproterozoic to Cambrian tectonic evolution. J Afr Earth Sci 86:65–106

Gabr S, Ghulam A, Kusky T (2010) Detecting areas of high-potential gold mineralization using ASTER data. Ore Geol Rev 38(1):59–69

Gabr SS, Hassan SM, Sadek MF (2015) Prospecting for new gold-bearing alteration zones at El-Hoteib area, South Eastern Desert, Egypt, using remote sensing data analysis. Ore Geol Rev 71:1–13

Gad S, Kusky TM (2006) Lithological mapping in the Eastern Desert of Egypt, the Barramiya area using Landsat thematic mapper (TM). J Afr Earth Sci 44:196–202

Gad S, Kusky TM (2007) ASTER spectral ratioing for lithological mapping in the Arabian-Nubian shield, the Neoproterozoic Wadi Kid area, Sinai, Egypt. Gondwana Res 11(3):326–335

Gagnon É, Schneider D, Kalbfleisch T, Habler G, Biczok J (2016) Characterization of transpressive deformation in shear zones of the Archean North Caribou greenstone belt (NW Superior Province) and the relationship with regional metamorphism. Tectonophysics 693(B):261–276

Gahlan HA (2003) Geology of the area of Wadi Fiqo, South Eastern Desert, Egypt. M.Sc. thesis, Assiut University, Assiut, Egypt, p 141

Gahlan HA (2006) Petrological characteristics of the mantle section in the Proterozoic ophiolites from the Pan-African Belt. Ph.D. thesis, Kanazawa University, Kanazawa, Japan, p 227

Gahlan H, Arai S, Abu El-Ela F, Tamura A (2012) Origin of wehrlite cumulates in the Moho Transition zone of the Neoproterozoic Ras Salatit Ophiolite, Central Eastern Desert, Egypt. Contrib Mineral Petrol 163:225–241

Gahlan HA, Azer MK, Khalil AES (2015) The Neoproterozoic Abu Dahr ophiolite, South Eastern Desert, Egypt: petrological characteristics and tectonomagmatic evolution. Mineral Petrol 109:611–630

Gahlan HA, Azer MK, Asimow P, Al-Kahtany K (2016) Late Ediacaran post-collisional a-type syenites with shoshonitic affinities, northern Arabian-Nubian Shield: a possible mantle-derived a-type magma. Arab J Geosci 9:603. https://doi.org/10.1007/s12517-016-2629-x

Gapais D, Bale P, Choukroune P, Cobbold PR, Mahjoub Y, Marquer D (1987) Bulk kinematics from shear zone patterns: some field examples. J Struct Geol 9:635–646

Garfunkel Z (1999) History and paleogeography during the Pan-African orogen to stable platform transition: reappraisal of the evidence from Elat area and the northern Arabian-Nubian Shield. Israel J Earth Sci 48:135–157

Goscombe BD, Gray R (2008) Structure and strain variation at mid-crustal levels in a transpressional orogen: a review of Kaoko Belt structure and the character of West Gondwana amalgamation and dispersal. Gondwana Res 13(1):45–85

Greiling RO (1985) Thrust tectonics in Pan-African rocks of SE Egypt. Terra Cognita 5

Greiling RO (1997) Thrust tectonics in crystalline domains: the origin of a gneiss dome. Proc Ind Acad Sci (Earth Planet Sci) 106:209–220

Greiling RO, Kröner A, El-Ramley MF (1984) Structural interference patterns and their origin in the Pan-African basement of the Southeastern Desert of Egypt. In: Kröner A, Greiling RO (eds) Precambrian tectonics illustrated. Schweitzerbart'sche Verlagsbuchhandlung, Stuttgart, Germany, pp 401–412

Greiling RO, Kröner A, El-Ramly MF, Rashwan AA (1988) Structural relationship between the southern and central parts of the Eastern desert of Egypt: details of a fold and thrust belt. In: El-Gaby S, Greiling RO (eds) The Pan-African belt of Northeast Africa and adjacent areas. Vieweg & Sohn, Weisbaden, Germany, pp 121–146

Greiling RO, Abdeen MM, Dardir AA, El Akhal H, El Ramly MF, Kamal El Din GM, Osman AF, Rashwan AA, Rice AH, Sadek MF (1994) A structural synthesis of the Proteorozoic Arabian-Nubian Shield in Egypt. Geol Rundsch 83:484–501

Grothaus B, Eppler D, Ehrlish R (1979) Depositional environments and structural implications of the hammamat formation, Egypt. Ann Geol Surv Egypt 9:564–590

Gupta RP (1991) Remote sensing geology. Springer, Heidelberg, p 356

Hamimi Z (1996) Tectonic evolution of the shield rocks of Gabal El Sibai area, Central Eastern Desert, Egypt. Egypt J Geol 40:423–453

Hamimi Z (1999) Kinematics of some major shear zones in the Eastern Desert of Egypt. Egypt J Geol 43(10):53–71

Hamimi Z, El-Kazzaz YA (2000) Structure and metamorphism of Rawd El-Bil area, south Eastern Desert, Egypt. Egypt J Geol 43(1):297–310

Hamimi Z, Hagag W (2016) A new tectonic model for Abu-Dabbab seismogenic zone (Eastern Desert, Egypt): evidence from field-structural, EMR and seismic data. Arab J Geosci (2017) 10:11. https://doi.org/10.1007/s12517-016-2786-y

Hamimi Z, Hagag W (2017) A new tectonic model for Abu-Dabbab seismogenic zone (Eastern Desert, Egypt): evidence from field-structural, EMR and seismic data. Arab J Geosci 10:11. https://doi.org/10.1007/s12517-016-2786-y

Hamimi Z, Kattu G (2014) Post-Accretionary structures in the Ediacaran Ablah Group volcanosedimentary sequence, Asir Terrane, Saudi Arabia. Gondwana 15, Madrid, Spain, 14–18 July 2014

Hamimi Z, El Amawy MA, Wetait M (1994) Geology and structural evolution of El Shalul Dome and environs, central Eastern Desert, Egypt. Egypt J Geol 38(2):575–595

Hamimi Z, El-Shafei M, Kattu G, Matsah M (2012a) Transpressional regime in southern Arabian Shield: insights from Wadi Yiba Area, Saudi Arabia. Mineral Petrol 107(5):849–860. http://dx.doi.org/10.1007/s00710-012-0198-6

Hamimi Z, Matsah MI, El-Shafei M, El-Fakharani A, Shujoon A, Al-Gabali M (2012b) Wadi Fatima thinskinned foreland FAT belt: a post amalgamation marine basin in the Arabian Shield. Open J Geol 2:271–293. http://dx.doi.org/10.4236/ojg.2012.24027

Hamimi Z, El-Kazzaz Y, El-Fakharani A (2012c) Deformation history of Jabal Muqsim Ophiolitic Nappe and Environs, Allaqi District: implications for the tectonic evolution of the South Eastern Desert Tectonic Terrane In NE Nubian Shield. American Geophysical Union, Fall Meeting 2012, abstract id. V33B-2863

Hamimi Z, Matsah M, Shujoon A, Al-Jabali M (2013b) The NE-Oriented Wadi Fatima Fault Zone, Near Jeddah, Saudi Arabia: a possible arc-arc suture in western Arabian shield. The abstract submitted to the CAG24, Addis Ababa, January 8–14

Hamimi Z, El-Fakharani A, Abdeen M (2014) Polyphase deformation history and strain analysis of the post-amalgamation depositional basins in the Arabian-Nubian Shield: Evidence from Fatima, Ablah and Hammamat basins. In: Fowler AR, Greiling RO and Abdeen MM (eds) Special issue: Arabian-Nubian Precambrian basement geology—progress and developments, J Afr Earth Sci 99(1):64–92

Hamimi Z, Kassem OMK, El-Sabrouty MN (2015a) Application of kinematic vorticity techniques for mylonitized rocks in Al Amar Suture, Eastern Arabian Shield, Saudi Arabia. Geotectonics 49(5):439–450

Hamimi Z, Zoheir BA, Younis MH (2015b) Polyphase deformation history of the Eastern Desert tectonic terrane in northeastern Africa. In: XII international conference "new ideas in earth sciences", Muscow, April 2015 (Abstract)

Hamimi Z, Ghazaly MK, Mohamed EA, Younis AM (2016) Structural history of Wadi Kharit transpressional Shear Zone, Eastern Desert tectonic terrane, Northern Nubian Shield. Ann Meet Egypt Geol Soc (Abstract)

Hanmer S, Passchier C (1991) Shear sense indicators: a review. Geology Survey, Canada, paper no 90, p 71

Harland WB (1971) Tectonic transpression in Caledonian Spitsbergen. Geol Mag 108:27–42

Hashad AH (1980) Present status of geochronological data on the Egyptian basement complex. In: Al-Shanti AMS (eds) Evolution and mineralization of the Nubian-Arabian Shield vol 3, pp 31–46. King Abdulaziz University, Jeddah, Saudi Arabia

Hassaan MM, El-Sawy E (2009) Tectonic environments and distribution of gold deposits in the Pan African Nubian Shield, Egypt. Aust J Basic Appl Sci 3(2):797–809

Hassaan MM, Ramadan TM, Abu El Leil I, Sakr SM (2009) Lithochemical surveys for ore metals in arid region, Central Eastern Desert, Egypt: using Landsat ETM + imagery. Aust J Basic Appl Sci 3:512–528

Hassan MA, Hashad AH (1990) Precambrian of Egypt. In: Said R (ed) The geology of Egypt. Balkema, Rotterdam, pp 201–245

Hassan SM, Ramadan TM (2014) Mapping of the late Neoproterozoic Basement rocks and detection of the gold-bearing alteration zones at Abu Marawat-Semna area, Eastern Desert, Egypt using remote sensing data. Arab J Geosci 8(7):4641–4656

Hassan SM, Sadek MF, Greiling RO (2015) Spectral analyses of basement rocks in El-Sibai-Umm Shaddad area, Central Eastern Desert, Egypt using ASTER thermal infrared data. Arab J Geosci 8(9):6853–6865

Hassan M, Stüwe K, Abu-Alam TS, Klötzli U, Tiepolo M (2016) Time constraints on deformation of the Ajjaj branch of one of the largest Proterozoic shear zones on Earth: the Najd Fault System. Gondwana Res 34:346–362

Hassoup A (1987) Microearthquakes and magnitude studies on Earthquake activity at Abu Dabbab region, Eastern Desert Egypt. M.Sc. thesis. Faculty of science, Cairo University, pp 1–169

Helmy H, Kaindl R, Fritz H, Loizenbauer J (2004) The Sukari gold mine, Eastern Desert, Egypt: structural setting, mineralogy and fluid inclusion study. Miner Deposita 39:495–511

Hermina M, Klitzsch E, List FK (1989) Stratigraphic lexicon and explanatory notes to the geological map of Egypt 1:500,000. Conoco Inc, Cairo, Egypt

Hippertt J (1993) "V"-pull-apart microstructures: a new shear sense indicator. J Struct Geol 15 (12):1393–1403

Hoffman PF (1999) The break-up of Rodinia, birth of Gondwana, true polar wander and the snowball earth. J Afr Earth Sc 17:17–33

Hosny A, El Hady SM, Mohamed AA, Panza GF, Tealeb A, El Rahman MA (2009) Magma intrusion in the upper crust of Abu Dabbab area, south east of Egypt, from Vp and Vp/Vs tomography. Rend Fis Acc Lincei 20:1–19. http://dx.doi.org/; https://doi.org/10.1007/s12210-009-0001-8

Hosny A, El-Hady SM, Guidarelli M, Panza GF (2012) Source moment tensors of the earthquake Swarm in Abu-Dabbab area, South-East Egypt. Rend Fis Acc Lincei 23:149–163

Hudleston PJ (1983) Strain patterns in an ice cap and implications for strain variations in shear zones. J Struct Geol 5:455–463

Hudleston PJ, Schultz-Ela DD, Southwick DL (1988) Transpression in an Archean greenstone belt, northern Minnesota. Can J Earth Sci 25:1060–1068

Hussein AA, Aly MM, El Ramly MF (1982) A proposed new classification of the granites of Egypt. J Volcanol Geoth Res 14:187–198

Ibrahim S, Cosgrove J (2001) Structural and tectonic evolution of the Umm Gheig/El-Shush region, central Eastern Desert of Egypt. J Afr Earth Sci 33:199–209

Ibrahim ME, Yokoyama I (1998) Probable origin of the Abu Dabbab earthquakes swarms in the Eastern Desert of Egypt. In: Bull IISEE 32

Ishii K (1992) Partitioning of non-coaxiality in deforming layered rock masses. Tectonophysics 210:33–43

Jackson NJ (1986) Petrogenesis and evolution of Arabian felsic plutonic rocks. J Afr Earth Sci 4:47–59

Jakob S, Bühler B, Gloaguen R, Breitkreuz C, Eliwa HA, El Gameel K (2015) Remote sensing based improvement of the geological map of the Neoproterozoic Ras Gharib segment in the Eastern Desert (NE-Egypt) using texture features. J Afr Earth Sci 111:138–147

Jarrar G, Stern RJ, Saffarini G, Al-Zubi H (2003) Late- and postorogenic Neoproterozoic intrusions of Jordan: implications for crustal growth in the northernmost segment of the East African Orogen. Precambr Res 123:295–319

Jarrar G, Saffarini G, Wachendorf H, Baumann A (2004) Origin, age and petrogenesis of neoproterozoic composite dykes from the Arabian-Nubian Shield, SW Jordan. Geol J 39:157–178

Jiang D, Williams PF (1998) High-strain zones: a unified model. J Str Geology 20:1105–1120

Johnson PR (1998) Tectonic map of Saudi Arabia and adjacents areas (scale 1–4,000,000). Technical report USGSOF-98-3.Saudi Arabian Deputy Ministry for Mineral Resources, Saudi Arabia, Jeddah, p 2

Johnson PR (2003) Post-amalgamation basins of the NE Arabian shield and implications for Neoproterozoic tectonism in the northern East African Orogen. Precambr Res 123:321–337

Johnson PR, Woldehaimanot B (2003) Development of the Arabian–Nubian shield: perspectives on accretion and deformation in the northern East African Orogen and the assembly of Gondwana. In: Yoshida M, Windley BF, Dasgupta S (eds), Proterozoic East Gondwana: supercontinent assembly and breakup. Geol Soc London, Special Publ 206, pp 290–325

Johnson PR, Andresen A, Collins AS, Fowler AR, Fritz H, Ghebreab W, Kusky T, Stern RJ (2011) Late Cryogenian-Ediacaran history of the Arabian-Nubian Shield: a review of depositional, plutonic, structural, and tectonic events in the closing stages of the northern East African Orogen. J Afr Earth Sci 61:167–232

Jones RR, Holdsworth RE, Clegg P, McCaffrey K, Travarnelli E (2004) Inclined transpression. J Struct Geol 30:1531–1548

Kalelioglu O, Zorlu K, Kurt MA, Gul M, Guler C (2009) Delineating compositionally different dykes in the Ulukıs¸la basin (Central Anatolia, Turkey) using computer-enhanced multi-spectral remote sensing data. Int J Remote Sens 30(11):2997–3011

Kamal El, Din GM (1993) Geochemistry and tectonic significance of the pan-African El Sibai Window, Central Eastern Desert, Egypt. Unpubl. Ph.D. thesis. Forschungzentrum Julich GmbH, 19, p 154

Kamal El Din GM, Khudeir AA, Greiling RO (1992) Tectonic evolution of a pan-African gneiss culmination, Gabal El Sibai area, Central Eastern Desert, Egypt. Zbl Geol Palaont Teil I 11:2637–2640

Kamel M, Youssef M, Hassan M, Bagash F (2016) Utilization of ETM + Landsat data in geologic mapping of wadi Ghadir-Gabal Zabara area, Central Eastern Desert, Egypt. J Remote Sens Space Sci 19:343–360

Katzir Y, Litvinovsky BA, Jahn BM, Eyal M, Zanvilevich AN, Vapnik Y (2006) Four successive episodes of Late Pan-African dykes in the central Elat area, southern Israel. Israel J Earth Sci 55:69–93

Katzir Y, Eyal M, Litvinovsky BA, Jahn BM, Zanvilevich AN, Valley JW, Beeri Y, Pelly I, Shimshilashvili E (2007a) Petrogenesis of A-type granites and origin of vertical zoning in the Katharina pluton, area of Gebel Mussa (Mt. Moses), Sinai, Egypt. Lithos 95(3–4):208–228

Katzir Y, Litvinovsky BA, Jahn BM, Eyal M, Zanvilevich AN, Valley JW, Ye Vapnik, Beeri Y, Spicuzza MJ (2007b) Interrelations between coeval mafic and A-type silicic magmas from composite dykes in a bimodal suite of southern Israel, northernmost Arabian-Nubian Shield: geochemical and isotope constraints. Lithos 97(3–4):336–364

Kebeasy RM (1990) Seismicity. In: Said R (ed) Geology of Egypt. Balkema, Rotterdam, pp 51–59

Khalil KI (2007) Chromite mineralization in ultramafic rocks of the Wadi Ghadir area, Eastern Desert, Egypt: mineralogical, microchemical and genetic study. Neues Jahrbuch für Mineralogie Abhandlungen 183:283–296

Khan T, Murata M, Karim T, Zafar M, Ozawa H, Rehman H (2007) Cretaceous dike swarm provides evidence of a spreading axis in the back-arc basin of the Kohistan paleo-island arc, northwestern Himalaya, Pakistan. J Asian Earth Sci 29:350–360

Kharbish S (2010) Geochemistry and magmatic setting of Wadi El-Markh island-arc gabbro–diorite suite, central Eastern Desert, Egypt. Chemie der Erde 70:257–266

Khudeir AA, Asran AH (1992) Back-arc Wizr Ophiolites at Wadi Um Gheig district, Eastern Desert, Egypt. Bull Fac Sci 21:1–22 Assiut University

Khudeir AA, El-Gaby S, Greiling RO, Kamal El Din GW (1992) Geochemistry and tectonic significance of polymetamorphosed amphibolites in the Gebel Sibai window, Central Eastern Desert, Egypt. Geology of the Arab world conference. Cairo University, Egypt, pp 461–476

Khudeir AA, Kamal El-Gaby S, El Din GM, Asran AMH, Greiling RO (1995) The pre-Pan-African deformed granite cycle of the Gabal El-Sibai swell, Eastern Desert, Egypt. J Afr Earth Sci 21:395–406

Khudeir AA, Abu El-Rus MA, El-Gaby S, El-Nady O, Bishara WW (2008) Sr–Nd isotopes and geochemistry of the infrastructural rocks in the Meatiq and Hafafit core complexes, Eastern Desert, Egypt: evidence for involvement of Pre-Neoproterozoic crust in the growth of Arabian-Nubian shield. Island Arc 17:90–108

Klitzch C, List F, Pöhlmann A (1987) Geological Map of Egypt (CONOCO), NG 36NE. Qusier, scale 1:500000

Knopásek J, Kröner S, Kitt S, Passchier CW, Kröner A (2005) Oblique collision and evolution of large-scale transcurrent shear zones in the kaoko belt, NW Namibia. Precambr Res 136:139–157

Kröner A, Greiling RO, Reischmann T, Hussein IM, Stern RJ, Durr St, Kruger J, Zimmer M (1987a) Pan-African crustal evolution in northeast Africa. In: Kröner A (ed) Proterozoic lithospheric evolution. American geophysical union, geodynamic series 17, pp 235–257

Kröner A, Stern RJ, Dawoud AS, Compston W, Reischmann T (1987b) The Pan-African continental margin in northeast Africa: evidence from a geochronological study of granulites at Sabaloka, Sudan. Earth Planet Sci Lett 85:91–104

Kröner A, Todt W, Hussein IM, Mansour M, Rashwan AA (1992) Dating of late Proterozoic ophiolites in Egypt and the Sudan using the single grain zircon evaporation technique. Precambr Res 59:15–32

Kröner A, Krüger J, Rashwan AA (1994) Age and tectonic setting of granitoid gneisses in the Eastern Desert of Egypt and south-west Sinai. Geol Rundsch 83:502–513

Kumar RS, Prasannakumar V (2009) Fabric evolution in Salem-Attur Shear Zone, South India, and its implications on the kinematics. Gondwana Res 16:37–44

Kumar C, Shetty A, Raval S, Sharma R, Ray C (2015) Lithological discrimination and mapping using ASTER SWIR data in the Udaipur area of Rajasthan, India. Proc Earth Planet Sci 11:180–188

Kusky TM (2004) Precambrian ophiolites and related rocks. Elsevier, Amsterdam, p 748

Kusky TM, Ramadan TM (2002) Structural controls on Neoproterozoic mineralization in the South Eastern Desert, Egypt: an integrated field, Landsat TM, and SIR C/X SAR approach. J Afr Earth Sci 35(1):107–121

Kusky TM, Abdelsalam M, Tucker R, Stern RJ (2003) Evolution of the East African and related Orogens, and the assembly of Gondwana. Spec Iss Precambr Res 123:81–344

Kuster D (2009) Granitoid-hosted Ta mineralization in the Arabian-Nubian Shield: ore deposit types, tectono-metallogenetic setting and petrogenetic framework. Ore Geol Rev 35:68–86

Lacassin R, Van Den Driessche J (1983) Finite strain determination of gneiss: application of Fry's to porphyroid in Southern Massif Central (France). J Struct Geol 5:245–253

Li ZX, Evans DAD, Zhang S (2004) A 90o Spin on Rodinia: possible causal links between the Neoproterozoic supercontinent, superplume, true polar wander and low-latitude glaciation. Earth Planet Sci Lett 220:409–421

Li S, Wilde SA, Wang T, Xiao WJ, Guo QQ (2016) Latest Early Permian granitic magmatism in southern inner Mongolia, China: implications for the tectonic evolution of the southeastern Central Asian orogenic belt. Gondwana Res 29:168–180

Liégeois JP, Stern RJ (2010) Sr–Nd isotopes and geochemistry of granite–gneiss complexes from the Meatiq and Hafafit domes, Eastern Desert, Egypt: no evidence for pre-Neoproterozoic crust. Int J Earth Sci 57:31–40

Liégeois J-P, Abdelsalam MG, Ennih N, Ouabadi A (2013) Metacraton: nature, genesis and behavior. Gondwana Res 23:220–237

Loizenbauer J, Wallbrecher E, Fntz H, Neumayr P, Khudeir AA, Kloetzil U (2001) Structural geology, single zircon ages and fluid inclusion studies of the Meatiq metamorphic core complex. Implications for Neoproterozoic tectonics in the Eastern Desert of Egypt. Precambr Res 110:357–383

Lundmark AM, Andresen A, Hassan MA, Augland LE, Abu El-Ru MA, Boghdady GY (2012) Repeated magmatic pulses in the East African Orogen of Central Eastern Desert, Egypt: an old idea supported by new evidence. Gondwana Res 22:227–237

Madani AA, Emam AA (2011) SWIR ASTER band ratios for lithological mapping and mineral exploration: a case study from El Hudi area, southeastern desert, Egypt. Arab J Geosci 4(1–2):45–52

Makroum FM (2001) Pan-African tectonic evolution of the Wadi El Mayit Area and its environs, Central Eastern Desert, Egypt. Second Int Conf Geol Afr 2:219–233 Assiut University, Egypt

Makroum FM (2003) Lattice preferred orientation (LPO) study of the orogen-parallel Wadi Nugrus and Wadi Um Nar shears, Eastern Desert-Egypt, using EBSD-technique. Third Int Conf Geol Afr 1:213–232

Makroum F (2017) Structural Interpretation of the Wadi Hafafit Culmination: a pan African gneissic dome in the central Eastern Desert. Lithosphere (in press), Egypt

Matin A (2006) Structural anatomy of the Kushtagi schist belt, Dharwar craton, south India-an example of archaean transpression. Precambr Res 147:28–40

Matsah MI, Kusky TM (1999) Sedimentary facies of the Neoproterozoic Al-Jifn basin, NE Arabian shield: relationships to the Halaban-Zarghat (Najd) faults system and the closure of the Mozambique Ocean. In: Greiling R (ed) Pan-African of northern Africa–Arabia. Forschungszentrum Jülich. Geologisch Palaeontoloigsches Institut, Ruprecht-Karls-Universitaet, Heidelberg, proceedings of a workshop, Oct 22–23, 1998

Matsah MI, Kusky TM (2001) Analysis of Landsat ratio imagery of the Halaban Zarghat fault and related Jifn basin, NE Arabian shield: implications for the kinematic history of the Najd fault. Gondwana Res 4(2):182

Miller MM, Dixon TH (1992) Late Proterozoic evolution of the north part of the Hamisana zone, northeast Sudan: constraints on Pan-African accretionary tectonics. J Geol Soc London 149:743–750

Moghazi AM (2002) Petrology and geochemistry of Pan-African granitoids, Kab Amiri area, Egypt-implications for tectonomagmatic stages of the Nubian Shield evolution. Miner Petrol 75:41–67

Moghazi AM, Andersen T, Oweiss GA, El Bouseily AM (1998) Geochemical and Sr–Nd–Pb isotopic data bearing on the origin of Pan-African granitoids in the Kid area, southeast Sinai, Egypt. J Geol Soc London 155:697–710

Moghazi AM, Hassanen MA, Hashad MH, Mohamed FH (2001) Garnet-bearing leucogranite in the El-Hudi area, southern Egypt: evidence of crustal anatexis during Pan-African low pressure regional metamorphism. J Afr Earth Sci 33:245–259

Moghazi AM, Hassanen MA, Mohammed FH, Ali S (2004) Late Neoproterozoic strongly peraluminous leucogranites, South Eastern Desert, Egypt—petrogenesis and geodynamic significance. Miner Petrol 81:19–41

Mohamed FH (1993) Rare metal-bearing and barren granites, Eastern Desert of Egypt: geochemical characterization and metallogenetic aspects. J Afr Earth Sci 17:525–539

Mohamed FH, Hassanen MA (1996) Geochemical evolution of arc-related mafic plutonism district, Eastern Desert of Egypt. J Afr Earth Sc 22:269–283

Mohamed FH, Moghazi AM, Hassanen MA (2000) Geochemistry, petrogenesis and tectonic setting of late Neoproterozoic Dokhan type volcanic rocks in the Fatira area, eastern Egypt. Int J Earth Sci 88:764–777

Mohamed AS, Hosny A, Abou-Aly N, Saleh M, Rayan A (2013) Preliminary crustal deformation model deduced from GPS and earthquakes' data at Abu-Dabbab area, Eastern Desert, Egypt. NRIAG J Astron Geophys 2(1):67–76

Moore JM (1979) Tectonics of the Najd transcurrent fault system, Saudi Arabia. J Geol Soc London 136:441–454

Moussa EMM, Stern RJ, Manton WI, Ali KA (2008) SHRIMP zircon dating and Sm/Nd isotopic investigations of Neoproterozoic granitoids, Eastern Desert, Egypt. Precambr Res 160:341–356

Nano L, Kontny A, Sadek MF, Greiling RO (2002) Structural evolution of metavolcanics in the surrounding of the gold mineralization at El Beida, South Eastern Desert, Egypt. Ann Geol Surv Egypt 25:11–22

Neumayr P, Hoinkes G, Puhl J, Mogessie A, Khudeir AA (1998) The Meatiq Dome, (Eastern Desert, Egypt) a precambrian metamorphic core complex: petrological and geological evidences. J Metamorph Geol 16:259–279

Noweir AM (1968) Geology of the Hammamat-Um Seleimat District, Eastern Desert, Egypt. Ph. D. thesis, Assiut University, Egypt

Oriolo S, Oyhantçabal P, Wemmer K, Basei MAS, Benowitz J, Pfänder J, Hannich F, Siegesmund S (2016) Timing of deformation in the Sarandí del Yí Shear Zone, Uruguay: implications for the amalgamation of the Western Gondwana during the Neoproterozoic Brasiliano–Pan-African Orogeny. Tectonics 35(3):754–771

Patchett PJ, Chase CG (2002) Role of transform continental margins in major crustal growth episodes. Geology 30:39–42

Phillips WJ (1972) Hydraulic fracturing and mineralization: geological society of London 128:337–361

Pisarevsky SA, Wingate MTD, Powell CM, Johnson S, Evans DAD (2003) Models of Rodinia assembly and fragmentation. In: Yoshida M, Windley Dasgupta BFS (eds) Proterozoic East Gondwana: supercontinent assembly and breakup. J Geol Soc London Special Publ 206:35–55

Pohl W (1979) Metallogenic/minerogenic analysis—contribution to the differentiation between Mozambiquian basement and Pan-African superstructure in the Red Sea region. Ann Geol Surv Egypt 9:32–44

Qaoud N (2014) Utilization of spaceborne imagery for lithologic mapping: a case study from Um Had area, Central Eastern Desert, Egypt. J Geogra Geol 6(2):113–123

Qiu F, Abdelsalam M, Thakkar P (2006) Spectral analysis of ASTER data covering part of the Neoproterozoic Allaqi-Heiani suture, southern Egypt. J Afr Earth Sci 44:169–180

Raharimahefa T, Kusky TM (2010) Structural and remote sending analysis of the Betsimisaraka Suture in northeastern Madagascar. Gondwana Res 15:14–27

Ramadan TM, Kontny A (2004) Mineralogical and structural characterization of alteration zones detected by orbital remote sensing at Shalatin District area, SE Desert, Egypt. J Afr Earth Sci 40:89–99
Ramsay JG, Huber MI (1987) The techniques of modern structural geology. Academic Press, New York, I & II, p 700
Ratschbacher L, Wenk HR, Sintubin M (1991) Calcite textures: an example from nappes with strain-path partitioning. J Struct Geol 13:369–384
Rice AHN, Greiling RO, Dardir AA, Rashwan AA, Sadek MF (1993) Pan-African extensional structures in the area S of the Hafafit Antiform, Eastern Desert of Egypt. Zentralblatt fur Geologie und Palaeontolgie Teil I 11:2641–2651
Ries AC, Shackelton RM, Graham RH, Fitches WR (1983) Pan-African structures, ophiolites and melange in the Eastern Desert of Egypt: a traverse at 26° N. J Geol Soc London 140:75–95
Rothery DA (1987) Improved discrimination of rock units using landsat thematic mapper imagery of the Oman ophiolite. J Geol Soc London 144:587–597
Rowan LC, Mars JC (2003) Lithologic mapping in the mountain pass, California, area using advanced spaceborne emission and reflection radiometer (ASTER) data. Remote Sens Environ 82:350–366
Rowan LC, Mars JC, Simpson CJ (2005) Lithologic mapping of the Mordor, NT, Australia ultramafic complex by using the advanced spaceborne thermal emission and reflection radiometer (ASTER). Remote Sens Environ 99:105–126
Sabet AH (1961) Geology and mineral deposits of Gebel El Sibai area, Red Sea hills, Egypt. Ph.D. thesis, Leiden State University, Netherlands, p 188
Sabet AH, Tsogoev V, Sarin LP, Azazi SA, Bedewi MA, Ghobrial GA (1976) Tin tantalum deposits of Abu Dabbab. Ann Geol Surv Egypt 6:93–117
Sabins F (1997) Remote Sensing: principles and interpretation, 3rd edn. Wiley Inc, New York, p 494
Sadek MF (2004) Discrimination of basement rocks and alteration zones in Shalatein area, Southeastern Egypt using Landsat Imagery data, Egypt. J Remote Sens Space Sci 7:89–98
Sadek MF (2005) Geology and spectral characterization of the basement rocks at Gabal Gerf area Southeastern Egypt. J Remote Sens Space Sci 8:109–128
Sadek MF, Hassan SM (2012) Application of remote sensing in lithological discrimination and geological mapping of Precambrian basement rocks in the Eastern Desert of Egypt. In: Proceedings of 33rd Asian conference remote sensing, Pattaya, Thailand, 26–30 Nov 2012
Sadek MF, Ali-Bik MW, Hassan SM (2015) Late Neoproterozoic basement rocks of Kadabora-Suwayqat area, Central Eastern Desert, Egypt: geochemical and remote sensing characterization. Arab J Geosci 8(12):10459–10479
Sakakibara N (1995) Structural evolution of multiple ductile shear zone system in the Ryoke Belt, Kinki Province. J Sci Hiroshima Univ Ser C 10(2):267–332
Sakran ShM, Hussein AA, Takla MA, Makhlouf AA (2001) Structural evolution of Um Radam area, south Eastern desert, Egypt. Ann Geol Surv Egypt XXII I(2):1–14
Salem SM, Soliman NM (2015) Exploration of gold at the east end of Wadi Allaqi, South Eastern Desert, Egypt, using remote sensing techniques. Arab J Geosci 8(11):9271–9282
Salem SM, Soliman NM, Ramadan TM, Greilling RO (2014) Exploration of new gold occurrences in the alteration zones at the Barramiya District, central Eastern Desert of Egypt using ASTER data and geological studies. Arab J Geosci 7:1717–1731
Salloum GM, Yehia MA, Tolba ME (1989) Tectonics of the Idfu-Marsa Alam area, central Eastern Desert, Egypt. In: The first conference on geochemistry, Alexandria University, pp 225–237
Sanderson DJ, Marchini WRD (1984) Transpression. J Struct Geol 6:449–458
Sarkarinejad K, Azizi A (2008) Slip partitioning and inclined dextral transpression along the Zagros thrust system, Iran. J Struct Geol 30:116–136
Seleim AM, Hammed MS (2016) Applications of remote sensing in lithological mapping of east Esh El Malaha area, southwest Gulf of Suez, Egypt. Int J Sci Eng Res 7(12):691–701
Sengör AMC, Natal'in BA (1996) Turkic-type orogeny and its role in the making of the continental crust. Ann Rev Earth Planet Sci 24:263–337

Shafiei Bafti S, Mohajjel M (2015) Structural evidence for slip partitioning and inclined dextral transpression along the SE San-andaj–Sirjan zone, Iran. Int J Earth Sci 104:587–601

Shalaby A (2010) The northern dome of Wadi Hafafit culmination, Eastern Desert, Egypt: structural setting in tectonic framework of a scissor-like wrench corridor. J Afr Earth Sci 57:227–241

Shalaby A, Stüwe K, Makroum F, Fritz H, Kebede T, Klotzli U (2005) The Wadi Mubarak belt, Eastern Desert of Egypt: a neoproterozoic conjugate shear system in the Arabian-Nubian Shield. Precambr Res 136:27–50

Shalaby A, Stüwe K, Fritz H, Makroum F (2006) The El Mayah molasse basin in the Eastern Desert of Egypt. J Afr Earth Sci 45:1–15

Shebl A (2018) Contribution of remote sensing and GIS in deciphering the structural and geomorphological attributes of Wadi Ghadir environs, Eastern Desert, Egypt. Unpublished M. Sc thesis, Geology Department, Faculty of Science, Tanta University, Egypt, 402 pp

Smith M, O'Connor E, Nasr BB (1999) Transpressional flower structures and escape tectonics: a new look at the pan-African collision in the Eastern desert, Egypt. In: de Wall H, Greiling RO (eds) Aspects of pan-African tectonics. International Cooperation, Bilateral Seminars Int Bureau, Forschungszentrum Jülich, Germany, pp 81–82

Souza Filho CR, Drury SA (1997) Remote sensing strategies for lithological mapping of Pan African assemblages in arid environments—a case study in Eritrea, NE Africa. Bol./G-USP. Ser Gent 28:1–22

Stein M, Goldstein SL (1996) From plume head to continental lithosphere in the Arabian Nubian shield. Nature 382:773–778

Stern RJ (1979) Late precambrian ensimatic volcanism in the Central Eastern Desert of Egypt. Ph. D. thesis, University of California, USA

Stern RJ (1985) The Najd fault system, Saudi Arabia and Egypt: a late precambrian rift-related transform system. Tectonics 4:497–511

Stern RJ (1994) Arc assembly and continental collision in the neoproterozoic East African orogen: implications for the consolidation of Gondwanaland. Ann Rev Earth Planet Sci 22:319–351

Stern RJ (2017) Neoproterozoic formation and evolution of Eastern Desert continental crust—the importance of the infrastructure-superstructure transition. J Afr Earth Sci (in press). http://dx.doi.org/10.1016/j.jafrearsci.2017.01.001

Stern RJ, Hedge CE (1985) Geochronologic and isotopic constraints on late precambrian crustal evolution in the Eastern Desert of Egypt. Am J Sci 258:97–127

Stern RJ, Gottfried D, Hedge CE (1984) Late Precambrian rifting and crustal evolution in the North Eastern Desert of Egypt. Geology 12:168–172

Stern RJ, Kroner A, Manton WI, Reischmann T, Mansour M, Hussein IM (1989) Geochronology of the late Precambrian Hamisana shear zone, Red Sea Hills, Sudan and Egypt. J Geol Soc London 146:1017–1030

Stern RJ, Nielsen KC, Best E, Sultan M, Arvidson RE, Kroner A (1990) Orientation of late precambrian sutures in the Arabian-Nubian shield. Geology 18:1103–1106

Stern RJ, Johanson PR, Kroner A, Yibas B (2004) Neoproterozoic ophiolites of the Arabian-Nubian shield. In: Kusky TM (ed) Precambrian Ophiolites and Related Rocks. Elsevier, Amsterdam, pp 95–128

Stoeser DB, Camp VE (1985) Pan-African microplate accretion of the Arabian shield. Geol Soc Am Bull 96:817–826

Sturchio NC, Sultan M, Batiza R (1983) Geology and origin of Meatiq Dome, Egypt: a precambrian metamorphic core complex? Geology 11:72–76

Sultan M, Arvidson RE, Sturchio NC (1986) Mapping of serpentinites in the Eastern Desert of Egypt by using Landsat thematic mapper data. Geology 14(12):995–999

Sultan M, Arvidson RE, Duncan IJ, Stern RJ, El Kalioubi B (1988) Extension of the Najd shear system from Saudi Arabia to the central Eastern Desert of Egypt based on integrated field and Landsat observations. Tectonics 7:1291–1306

Sultan M, Becker R, Arvidson RE, Shore P, Stern RJ, El Alfy Z, Attia RI (1993) New constraints on red-sea rifting from correlations of Arabian and Nubian neoproteozoic Outcrops. Tectonics 12:1303–1319

Takla MA, Hussein AA, Sakran ShM, Makhlouf AA (2002) Tectonometamorphic history of Um Radam area, south Eastern Desert, Egypt. Egypt. Ann Geol Surv Egypt XXIV:41–61

Tangestani MH, Jaffari L, Vincent RK, Sridhar BM (2011) Spectral characterization and ASTER-based lithological mapping of an ophiolite complex: a case study from Neyriz ophiolite. SW Iran Remote Sens Environ 115(9):2243–2254

Tavarnelli E, Holdsworth RE, Clegg P, Jones RR, McCaffrey KJW (2004) The anatomy and evolution of a transpressional imbricate zone, Southern Uplands, Scotland. J Struct Geol 26:1341–1360

Taylor WEG, El Hamad YA, El-Kazzaz YA, Rashwan AA (1993) An outline of the tectonic framework for the Pan-African orogeny in the vicinity of Wadi Um Relan, SE Desert, Egypt. In: Thorweihe U, Shandelmeier H (eds) Geoscientific research in NE Africa. Balkema, Rotterdam, pp 31–34

Tikoff B, Greene D (1997) Stretching lineations in transpressional shear zones: an example from the Sierra Nevada Batholith, California. J Struct Geol 19:29–39

Torres-Vera MA, Prol-Ledesma RM (2003) Spectral enhancement of selected pixels in thematic mapper images of the Guanajuato district (Mexico) to identify hydrothermally altered rocks. Int J Remote Sens 24:4357–4373

Treagus S (1983) A theory of finite strain variation through contrasting layers and its bearing on cleavage refraction. J Struct Geol 5:351–368

Vail JR (1985) Pan-African (late Precambrian) tectonic terrains and the reconstruction of the Arabian-Nubian shield. Geology 13:839–842

Wallbrecher E, Fritz H, Khudeir AA, Farahat F (1993) Kinematics of pan-African thrusting and extension in Egypt. Geosci Res Northeast Afr 27–30

Wallbrecher E, Loizenbauer J, Fritz H, Bauernhofer A, Hauzenherger C (1999) Convergence of East and West Gondwana-constraints from Egypt and East Africa. In: The first international conference on the geology of Africa (Abstract)

Willis KM, Stern RJ, Clauer N (1988) Age and geochemistry of Late Precambrian sediments of the Hammamat series from the Northeastern Desert of Egypt. Precambr Res 42:173–187

Yenne EY, Bala DA, Abalaka IE, Ozoji TM, Nimze LW (2015) Utilization of Landsat and field data in geological mapping of Vom-Kuru Area, North-central Nigeria. J Environ Earth Sci 5 (16):95–114

Youssef AM, Hassan AM, Mohamed MA (2009) Integration of remote sensing data with the field and laboratory investigation for lithological mapping of granitic phases: Kadabora pluton, Eastern Desert, Egypt. Arab J Geosci 2(1):69–82

Zimmer M, Kröner A, Jochum KP, Reischmann T, Todt W (1995) The Gabal Gerf complex: a Precambrian N-MORB ophiolite in the Nubian Shield, NE Africa. Chem Geol 123:29–51

Zoheir BA (2008) Characteristics and genesis of shear zone-related gold mineralization in Egypt: a case study from the Um El Tuyor mine, south Eastern Desert. Ore Geol Rev 34:445–470

Zoheir BA (2011) Transpressional zones in ophiolitic mélange terranes: potential exploration targets for gold in the SouthEastern Desert, Egypt. J Geochem Explor 111:23–38

Zoheir BA, Emam A (2012) Integrating geologic and satellite imagery data for high-resolution mapping and gold exploration targets in the South Eastern Desert, Egypt. J Afr Earth Sci 66:22–34

Zoheir BA, Emam A (2014) Field and ASTER imagery data for the setting of gold mineralization in Western Allaqi-Heiani belt, Egypt: a case study from the Haimur deposit. J Afr Earth Sci 99 (1):150–164

Zoheir BA, Lehmann B (2011) Listvenite-lode association at the Barramiya gold mine, Eastern Desert, Egypt. Ore Geol Rev 39:101–115

A New Synthetic Geological Map of the Tuareg Shield: An Overview of Its Global Structure and Geological Evolution

J.-P. Liégeois

Abstract Here is presented a geological map of the Tuareg Shield (Algeria, Mali, Niger), based on the various available geological maps, which are often old and uneasily available, and on the various papers published during the last decades that give mapping or geochronological information. This work has been initiated by the new geological map of Africa at the scale of 1/10,000,000 for which a largely simplified version has been used. The present map focuses on the age of the large magmatic units and of their basement, including the age of the major reactivations that may have affected them. The publication here of this georeferenced map at an appropriate resolution will be useful for future studies at both local and regional scales but also allows to synthetize and discuss the different concepts that have been applied to the Tuareg Shield: the terrane structure, for which some modifications are proposed, the localization of the different Paleoproterozoic orogenies, of the Neoproterozoic juvenile terranes and of the Pan-African reactivation of the old blocks that generated metacratonic terranes or associations of terranes. This map allows also to visualize the eastern margin of the West African craton and the western margin of the Saharan metacraton and the influence of the Murzukian orogeny. Finally, different provinces, separated by oceans prior to the Pan-African orogeny are proposed as well as a global model for their amalgamation.

Keywords Trans-Saharan belt · Tuareg terrane structure · Geodynamical model Metacratons

J.-P. Liégeois (✉)
Geodynamics and Mineral Resources Unit, Royal Museum for Central Africa, B-3080 Tervuren, Belgium
e-mail: jean-paul.liegeois@africamuseum.be

A. Bendaoud et al. (eds.), *The Geology of the Arab World—An Overview*, Springer Geology, https://doi.org/10.1007/978-3-319-96794-3_2

1 Introduction

A comprehensive map of the Hoggar has been published by Bertrand and Caby (1977). Since that time numerous geological and geochronological data have been acquired and several geodynamical models have been proposed. A major advance was the elaboration of a model with a structure in 23 terranes for the entire Tuareg Shield (Black et al. 1994). The 1977 map was limited to the Algerian territory and did not include the Malian and Nigerian parts of the Tuareg Shield (Iforas and Aïr regions) that must be considered for a comprehensive understanding of this key Saharan Precambrian territory. Later, various sketch maps of the terrane structure with more or less geological information have been built on the basis of the Fig. 3 of Black et al. (1994) but their aims were not to provide a detailed map.

The aim of this work is to provide a synthetic geological map of the Tuareg Shield with as much detail as possible, taking into account the current knowledge and the scale of this map. This map has been built following the method used for the recent geological map of Africa (Thiéblemont et al. 2016), by focusing on the ages of the lithologies and considering, if needed, the existence of major reactivation events. This synthetic geological map differentiates the lithological ages at the system rank (IUGS international chronostratigraphic chart; Cohen et al. 2013, updated in 2017) of the Archean/Paleoproterozoic metamorphic basement for which geochronological data are still sparse, with additional information about lithologies for the Neoproterozoic, through the use of additional colors.

This representation allows to view the general structure of the Tuareg Shield and to understand the different geodynamical models proposed for the Pan-African evolution of this part of the Trans-Saharan Belt. Let us note that the Tuareg Shield offers the entire width of an orogen (1400 km) between two major cratons that acted as rigid boundaries after oceans closed.

2 Regional Geological Setting

The Tuareg Shield is a Cenozoic swell with a surface of c. 550,000 km^2 made of Precambrian rocks surrounded by Paleozoic sediments deposited after the end of the Pan-African orogeny. This present swell results from a widespread Eocene exhumation of the shield (Rougier et al. 2013; English et al. 2017) and was followed by intraplate volcanic activity (from 35 Ma to recent times; Liégeois et al. 2005 and references therein). The Tuareg Shield belongs to the Trans-Saharan Belt that runs from the Atlantic Ocean (Nigerian Shield) in the south towards the Alpine Atlas Belt to the north (Fig. 1). It is partly covered by large and often thick Phanerozoic sedimentary basins, with the Iullemmeden basin between the Nigerian and Tuareg Shield and the oil-rich north Saharan sedimentary basins between the latter and the Atlas Belt (Fig. 1). The Trans-Saharan Belt was one of the first loci where plate tectonics was demonstrated to have been at work in Precambrian times

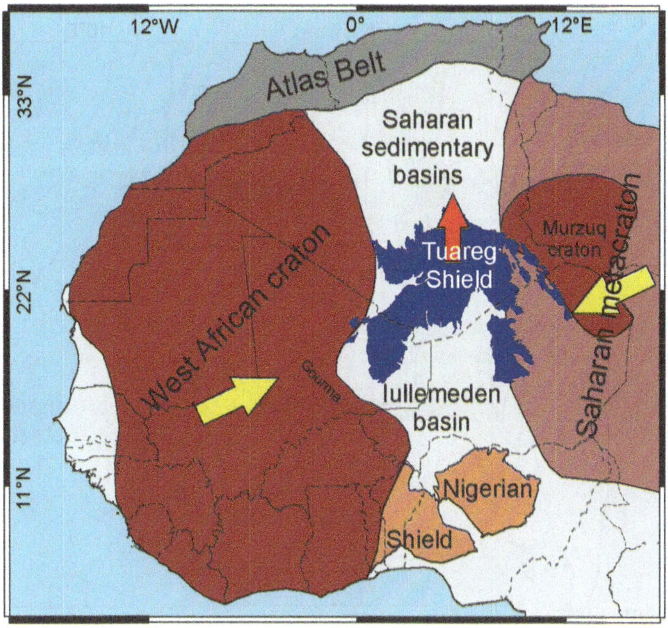

Fig. 1 Map of west Africa with the major geodynamical entities. The Trans-Saharan Belt corresponds to the band in the middle, with the Nigerian Shield and the Tuareg Shield and the basement below the Iullemmeden and Saharan sedimentary basins

(Black et al. 1979). Indeed, it has been squeezed between the West African craton to the west and the Saharan metacraton to the east during the Pan-African orogeny due to the closure of a large ocean (Caby et al. 1981). This cratonic convergence resulted in a general northward tectonic escape of the Tuareg shield terranes (Black et al. 1994). This intracontinental escape can be bracketed between 630 and 580 Ma (Liégeois et al. 2003; Abdallah et al. 2007).

Twenty-three terranes have been identified in the Tuareg Shield, separated either by subvertical mega-shear zones or by thrust fronts (Fig. 2); these terranes differ by lithologic, metamorphic, magmatic or tectonic characteristics (Black et al. 1994). Some of these terranes are juvenile Neoproterozoic terranes, mostly oceanic island arcs bearing sometimes ophiolites, with building ages between 870 and 635 Ma (Caby et al. 1982, 1989; Béchiri-Benmerzoug et al. 2011) while others are Archean to Paleoproterozoic terranes variably reactivated during the Pan-African orogeny. The latter can bear large well-preserved tracts of the old basement lithologies (Bendaoud et al. 2008). These old variably reactivated regions have been described as metacratons (Liégeois et al. 2003, 2013a).

In the center of the shield, four metacratonic terranes share enough common characteristics to be considered as displaced terranes belonging to the same pre-Neoproterozoic passive margin; they have been grouped under the term "LATEA metacraton", LATEA being the acronym of their names (Laouni–Azrou

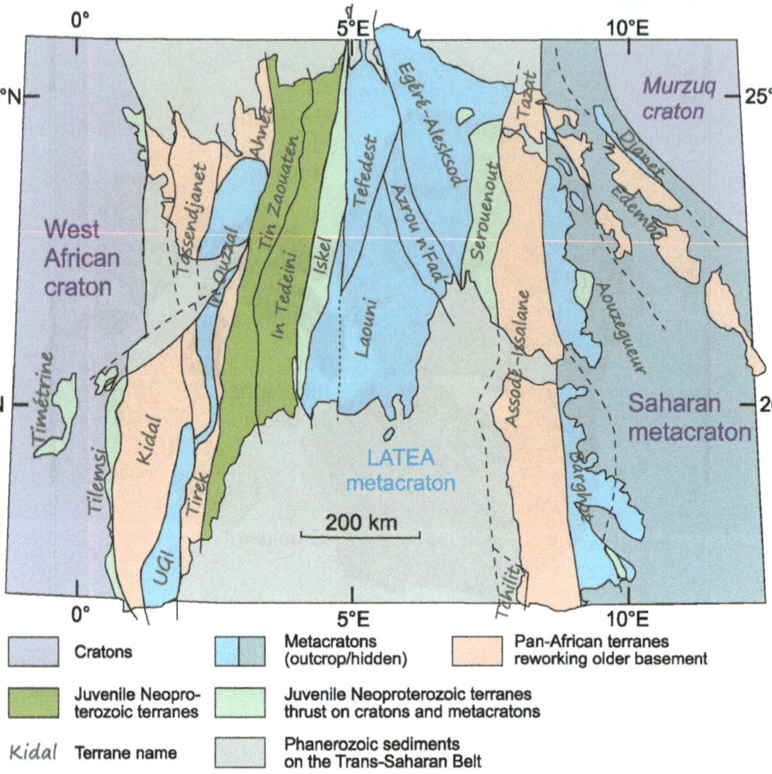

Fig. 2 The Tuareg Shield with the 23 terranes of Black et al. (1994) with a geodynamical attribution, modified from Liégeois et al. (2003)

n'Fad–Tefedest–Egéré-Aleksod; Liégeois et al. 2003; Fig. 2). LATEA corresponds to the western part of the former "Polycyclic Central Hoggar" (Bertrand 1974; Bertrand and Caby 1978; Vitel 1979; Caby et al. 1981). LATEA is composed of a Paleoproterozoic granulite–amphibolite facies basement mostly 2.1 and 1.9 Ga in age with some local Archean relics (Peucat et al. 2003; Bendaoud et al. 2008) variably reactivated during the Pan-African orogeny. The latter orogeny generated large movements along mega-shear zones accompanied by high-temperature greenschist to amphibolite metamorphisms (Bendaoud et al. 2004, 2008), locally up to eclogite-facies (Sautter 1986; Liégeois et al. 2003; Doukkari et al. 2014, 2015), and intrusion of large granitoid batholiths during the 630–580 Ma period (Acef et al. 2003; Liégeois et al. 2003; Abdallah et al. 2007; Bendaoud et al. 2008).

To the west of LATEA, there are three mostly juvenile Neoproterozoic terranes, Iskel, In Tedeini (referred incorrectly as In Teidini by Black et al. 1994) and Tin Zaouatene (Fig. 2), the former of which is considered to lie on LATEA (Azzouni-Sekkal et al. 2003; Liégeois et al. 2003). They correspond to the central Pharusian Belt (Gravelle 1969; Bertrand 1974; Bertrand and Caby 1978; Caby et al. 1981).

More to the west, there is exists a series of terranes comprising Archean and Paleoproterozoic lithologies variably affected by the Pan-African orogeny. The In Ouzzal in Algeria and UGI in Mali ("Unité granulitique des Iforas") are mostly made of Archean rocks affected by a very high-temperature metamorphism that occurred at c. 2 Ga (Peucat et al. 1996; Ouzegane et al. 2003; Bosch et al. 2016). The Pan-African orogeny ended with brittle faults that can have induced local retrogression under greenschist facies conditions, controlled the intrusion of high-level subcircular alkali-calcic to alkaline plutons in the 600–580 Ma age period (Hadj-Kaddour et al. 1998; Fezaa et al. 2018) and the external shape of these terranes. In Ouzzal and UGI are considered as Pan-African metacratons (Aït-Djafer et al. 2003; Adjerid et al. 2008; Fezaa et al. 2018). Around the In Ouzzal and UGI terranes, four terranes (Ahnet, Tirek, Tassendjanet, Kidal) have a Paleoproterozoic basement largely affected by the Pan-African orogeny (Caby 2003; Bosch et al. 2016). A Pan-African active margin developed in the Tassendjanet and Kidal terranes with the emplacement of continental arc andesites (Chikhaoui et al. 1978), of volcano-sedimentary sequences (Fabre et al. 1988), of eclogites (Berger et al. 2014) and of a huge post-collisional composite batholith displaying a high-K-calc-alkaline to alkaline transition (Kidal terrane; Liégeois and Black 1987; Liégeois et al. 1987, 1998). The Kidal and Tassendjanet terranes are not considered as metacratonic because if they could have become cratonic during the Mesoproterozoic, as no orogenic events occurred during nearly 1 Ga, their behavior during the Pan-African orogeny does not reveal cratonic characteristics. For instance, the Kidal assemblage, which is representative of the basement of the Kidal terrane, is a tectonic mixing of an Eburnean granulitic basement, its metamorphic Neoproterozoic sedimentary cover (with some volcanic rocks) intruded by various Pan-African magmatic rocks (Champenois et al. 1987). This can be attributed to the active continental margin environment, able to entirely decratonize small cratonic areas (Liégeois et al. 2013a). So if these terranes were cratonic prior to the Pan-African orogeny, they evolved beyond the metacratonic state, becoming decratonized areas. The contact with the West African craton is made through the juvenile Pan-African Tilemsi island arc terrane (Caby et al. 1989) and the Timétrine terrane that belongs to the WAC passive margin (Caby 2014). Equivalents of these terranes are known to the south in the Gourma area with the presence of an island arc root (Berger et al. 2008, 2011) and of active and passive margin nappes including high-pressure rocks (Caby 1994; Jahn et al. 2001; Caby et al. 2008).

LATEA is bordered to the east by the mostly juvenile Neoproterozoic Serouenout terrane (Fig. 2) that comprises white schist and eclogitic assemblage lithologies reminiscent of an ophiolitic mélange, metamorphosed and strongly deformed under eclogite-facies conditions (Bitam-Derridj et al. 2010; Adjerid et al. 2012, 2015). Further to the east is the Assodé-Issalane terrane, marked by a high-temperature amphibolite facies and the intrusion of a regional Pan-African anatectic leucogranite (Renatt granite; Liégeois et al. 1994), unknown elsewhere in the Tuareg Shield (Liégeois et al. 1998), contemporaneous with classical high-K calc-alkaline batholiths (Dabaga-West) at c. 600 Ma (Liégeois et al. in prep.) and not at c. 660 Ma (Liégeois et al. 1994). Having the same age, the Renatt and

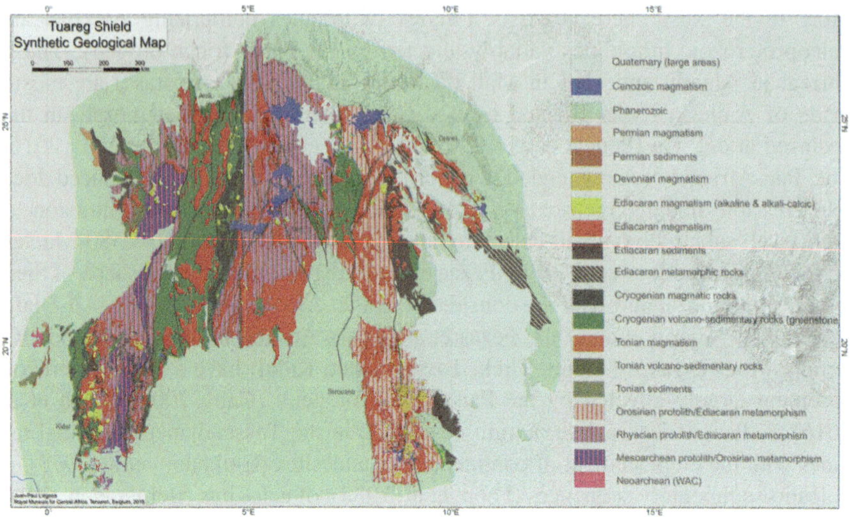

Fig. 3 Georeferenced synthetic geological map of the Tuareg Shield. See text for the geological map and the method that have been used for its elaboration. A high-resolution map (PDF, 465 Mb) is given in annex. The correspondence between the geological map and the DEM (digital elevation model) is good but not perfect everywhere. This is due to the map referential from the fifties and the map projection that should be corrected in details. This is out of the scope of this article and the differences are not a problem for its purpose and considering the scale at which the map is interpreted. Obtaining a very detailed geological map well superposed on a DEM is clearly a main objective for the near future

Dabaga-West granitoids are not differentiated in Figs. 3 and 4 but are distinctly mapped in Liégeois et al. (1994). The emplacement of the Renatt granite and of the generation of the accompanying high-temperature amphibolite facies metamorphism has been ascribed to lithospheric delamination of the terrane, bringing the asthenosphere close to the lower crust, which induced a regional partial melting of the medium and lower crusts (Liégeois et al. 1994, 1998). Its eastern boundary is the Raghane mega-shear zone, also the western boundary of the Saharan metacraton (SmC; Liégeois et al. 1994; Abdelsalam et al. 2002, formerly the East Saharan ghost craton, Black and Liégeois 1993). To the north, the small Tazat terrane is marked by distinctive metasedimentary sequences (Blaise 1967; Bertrand et al. 1968) lying on a poorly known granite–gneiss basement but similar to that of the Assodé-Issalane terrane (BRMA 1961) of which it probably constitutes the northern prolongation.

Just to the southwest of Assodé-Issalane terrane, in Aïr, there is the Tchilit exotic terrane, made of a Paleoproterozoic amphibolite facies bimodal volcanic assemblage (Navez et al. 1999) whose origin is unknown due to a very limited outcropping (Fig. 2).

To the east of the Raghane shear zone, the eastern part of the Hoggar was first considered to have been stabilized early at 730 Ma (Caby and Andreopoulos-Renaud 1987). However, more comprehensive studies showed that Eastern Hoggar

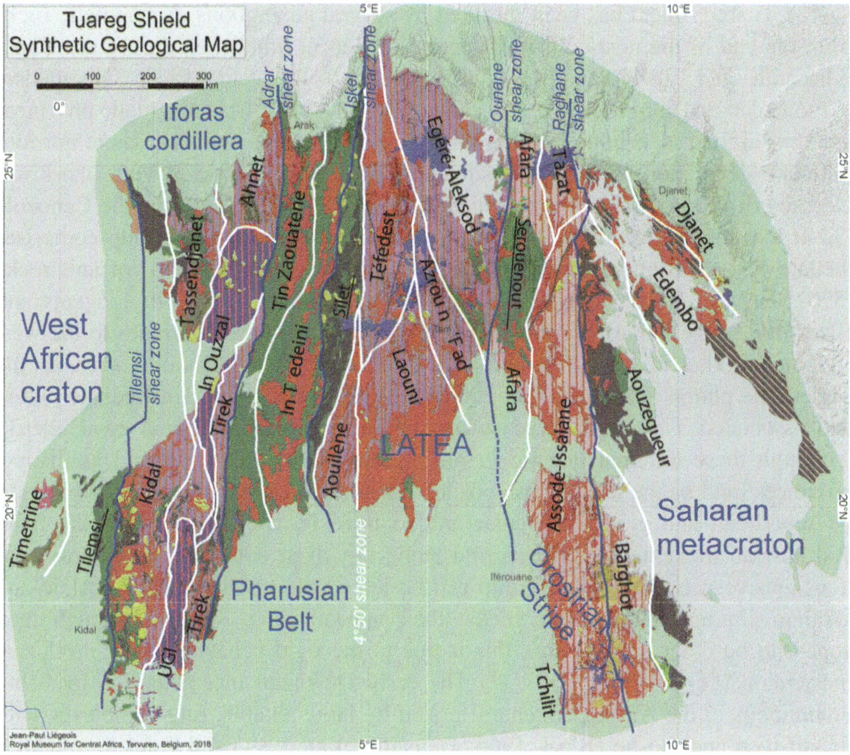

Fig. 4 Synthetic geological map of the Tuareg Shield with the delimitation of the 25 terranes (separated by white or blue lines). 2 terranes have been added (Aouilène and Afara terranes) to those of Black et al. (1994). The name of the major shear zones and the delineation of the major entities (separated by blue lines) have also been added. See text for more explanation (Sect. 4). For the legend of colors and symbols, see Fig. 3

(Aouzegueur, Edembo, and Djanet terranes) was a stable area belonging to the SmC, overthrust by ophiolitic complexes (Boullier et al. 1991) and covered by the Pan-African molasse until 580 Ma. It was lately metacratonized (intracontinental orogeny) during the 575–545 Ma period (Murzukian orogeny; Fezaa et al. 2010). The western boundary of the SmC, the Raghane mega-shear zone, is marked by the intrusion of several generations of granitoids from 790 to 550 Ma (Henry et al. 2009). The presence of old basement has been demonstrated in one point (1.9 Ga; Arokam area, Aouzegueur terrane; Nouar et al. 2011). Pan-African granitoids in the Edembo terrane display inherited zircons at 1.9, 2.1 and 2.7 Ga (Fezaa et al. 2010). To the south, the Barghot terrane was thought to have been formed at c. 700 Ma (Liégeois et al. 1994) but new LA-ICP-MS zircon ages indicate that its evolution is also at c. 600 Ma (Liégeois et al. in prep.).

The Tuareg Shield has been intruded by several generations of magmatic rocks from the end of the post-collisional period to Recent times, i.e., during the Upper Ediacaran and the Phanerozoic. Compared to the Pan-African magmatism (Cryogenian and above all, late Ediacaran, 630–580 Ma), these magmatic provinces are very limited in volume but are critical as geodynamic markers. There are four different generations: the composite late Ediacaran (570–520 Ma), the lower Devonian (420–400 Ma), the Permian–Jurassic (270–130 Ma) and the Cenozoic (35–0 Ma) magmatic groups: (1) The Late Ediacaran composite group comprises the late post-collisional alkali-calcic and alkaline plutons, dykes and volcanic rocks (570–520 Ma) of the Adrar des Iforas (linked to a slab break-off; Liégeois and Black 1987), the Taourirt province in LATEA and in adjacent terranes to the west (Azzouni-Sekkal et al. 2003) and additional more isolated bodies such as the Tisselliline pluton (Liégeois et al. 2003) in NE LATEA or the Tin Bedjane pluton and associated Tin Amali dyke swarm in Eastern Hoggar (Fezaa et al. 2010). Although these different late Ediacaran plutons could be considered as distinct provinces (and so the use of "composite" for this group), they have in common to be superficial and globally circular and to have been emplaced after the main Pan-African movements that occurred along the shear zones during extensive or trans-tensive periods. To that group can be joined the In Ouzzal alkali-calcic and alkaline plutons and volcanic rocks: their appearance is similar although they appear to be older (c. 600 Ma), this is due to the rigid behavior of the in Ouzzal metacraton (Fezaa et al. 2018); (2) The early Devonian magmatism is local but voluminous in the Aïr region, characterized by large alkaline ring-complexes with massif-type anorthosite (Black 1965; Demaiffe et al. 1991) and volcanic products (Black et al. 1967) emplaced under a transtensional tectonic regime (Moreau et al. 1994). The other known manifestations are the large Arrikine sill having an enriched mantle source (Derder et al. 2016) and magmatically generated (phreatic) sand injections (Moreau et al. 2012), all being located in the Edembo and Djanet terranes at the periphery of the Murzuq craton (Derder et al. 2016). This early Devonian group could be related to the Caledonian event that prevented the deposition of lower Devonian sediments in the Murzuq area (Ghienne et al. 2013); (3) The Permian to Jurassic magmatic event is limited to the Tadhak area (Liégeois et al. 1983, 1991), to the west of the Adrar Iforas in Mali (Timetrine terrane; Fig. 2). It is a carbonatite–nepheline syenite association whose source was a Dupal-type asthenospheric mantle (Weis et al. 1987) pointing, as for the Arrikine sill, to a sublithospheric source. This implies the need for the reactivation of pre-existing lithospheric structures linked to the end of Variscan stress and to the initiation of the Atlantic Ocean (Liégeois et al. 1991); (4) The Cenozoic magmatism is volumetrically limited but geographically widespread in NW Africa and as a whole related to the Africa–Europe collision (Liégeois et al. 2005). It followed some Pan-African mega-shear zones, allowing the magneto-telluric method to highlight the >100 km depth of their source (Bouzid et al. 2015).

3 Sources and Design of the Synthetic Tuareg Shield Geological Map

The first regional map of the Hoggar was proposed by Lelubre (1952) who established, together with Kilian (1947), the general Precambrian framework of the Hoggar geology. On this basis, the first comprehensive geological maps were realized by the BRMA (1961) at 1/500,000 scale. These maps constitute the foundation used for Hoggar map in the present work. Available more recent but more focused maps were also used: Tazat area (1/50,000 map; Blaise 1967), Northwest Ahaggar (1/500,000; Caby 1970), which actually replaced the BRMA Ouallen-Bidon V map, Egéré area (1/200,000 scale; Duplan 1974), Aleksod-Tazoulet area (1/100,000 map; Bertrand 1974), Gour Oumelalen area (1/68,000 map; Latouche 1978), Assekrem area (1/200,000 scale; Vitel and Girod 1983), Tamanrasset area (1/200,000 scale; Zeghouane 2002). In Mali, the Iforas geological map (1/500,000 scale; Fabre et al. 1982) and in Niger, the Aïr geological map (1/500,000 scale; Black et al. 1967) were used as basis for the map here presented. In addition, the references cited in the above geological setting were used especially for the age of the lithologies, unknown when most of the above maps were established. For a general overview of the available geochronological data, see Béchiri-Benmerzoug et al. (2017). The resulting high-resolution georeferenced map was superposed on the regional DEM (Digital Elevation Model) and is shown in Fig. 3.

A fundamental feature of the Tuareg Shield is the coexistence of juvenile Neoproterozoic terranes and Archean and Paleoproterozoic terranes more or less reactivated during the Pan-African orogeny (metacratons). Information on the pre-Pan-African evolution of these terranes is still sparse and variable depending on the area but has been rapidly growing these last years. Especially, it appeared that different Eburnian orogenic periods occurred in the Tuareg Shield, affecting different regions of the shield. The evolution of some areas comprises three periods of activity, i.e. during the Archean, Paleoproterozoic and Pan-African but where this is the case, one of these activities is of minor importance and has not been represented on the map. For instance, there is sparse evidence for Archean events in the NE of LATEA or for the Pan-African brittle reactivation of the In Ouzzal terrane not highlighted in the map. Based on the current knowledge, no Mesoproterozoic rocks are known in the Tuareg Shield, as globally for West Africa (Ennih and Liégeois 2008). Recent studies indicate the existence of an early sedimentation in the Taoudeni basin in the late Mesoproterozoic (c. 1.1 Ga; Rooney et al. 2010) and of Statherian (late Paleoproterozoic) magmatic events (dykes and sills at c. 1.7 Ga; Youbi et al. 2013; Ikenne et al. 2017) but these are local and volumetrically minor anorogenic events, even if they have an important geodynamic meaning. In the Tuareg Shield, a "Kibaran" event (actually a Stenian event at 1.1–1.0 Ga) was postulated early (Bertrand and Lasserre 1976) on the basis of Rb–Sr data but this has not been confirmed until now.

The rationale of this map is to focus on geological mapping but also to highlight the geodynamical contrasts existing between terranes by using colors adapted to this purpose. This means that the metacratonic terranes have as base color that of the age of their oldest phase (Mesoarchean, Rhyacian, Orosirian) with a hatching showing the color of the main reactivation (Orosirian or Pan-African). As the Pan-African reactivation is largely associated with the intrusion of large batholiths, the Pan-African hatching has the same color as these batholiths (red). For a better contrast, the Neoproterozoic proto-Pan-African Tonian and Cryogenian sediments and volcano-sedimentary series are represented in green colors while the Ediacaran syn-Pan-African sediments (molasse) are represented in brown. A special hatching has been used for the amphibolite facies rocks of the Edembo terrane, as their protolith is the Ediacaran molasse (Fezaa et al. 2010), which is a unique case in the Tuareg Shield.

Based on the current available data, all the non-metamorphic magmatic rocks are Neoproterozoic or Phanerozoic in age and are represented by plain colors. They are orange for Tonian, dark green for Cryogenian, red for Ediacaran (high-K) calc-alkaline, yellow for late Ediacaran alkali-calcic and alkalic, dark yellow for Devonian, brown for Permian and blue for Cenozoic magmatism. Phanerozoic sediments around the shield (and locally in the Shield, small Cretaceous basins) are in light green, except large Quaternary areas that are in light gray due to the information they convey.

4 Global Structure of the Tuareg Shield

The terrane structure is represented on the new synthetic geological map in Fig. 4. It is basically that of Black et al. (1994) but some modifications have been made to include several major entities defined and are added, together with others here proposed.

In the center of the shield is located the LATEA metacraton (Liégeois et al. 2003). It is an old gneissic basement (Arechchoum, Aleksod series) with a complex structure (e.g., Bertrand 1974), comprising an important major metasedimentary sequence (Egéré series; Duplan 1974). Geochronological data are still sparse but the major event is at 2.1 Ga with a regional granulite- to amphibolite facies metamorphism (Bendaoud et al. 2008 and references therein), i.e., a Rhyacian orogenic period.

LATEA has been defined as being composed of four terranes, Laouni, Azrou n'Fad, Tefedest and Egéré-Aleksod. The LATEA portion west of the 4°50′ shear zone was shown to be similar to that to the east (Bertrand et al. 1986; Lapique et al. 1986) and so included in the Laouni terrane by Black et al. (1994). However, its Pan-African structure, similarly to the Tefedest terrane, is much more intense (Lapique et al. 1986), does not bear eclogites, is limited to the east by the major 4° 50′ shear zone, which suggests a likely major movement relatively to the western part of the Laouni terrane. This justifies considering this portion of LATEA as a distinct terrane, which we propose to name the Aouilène terrane, echoing the Aouilène domain of Lapique et al. (1986). This means that LATEA is now the acronym of the five constituting terranes, Laouni, Aouilène, Tefedest, Egéré-Aleksod, Azrou n'Fad. It must be noted that the relative displacement of the LATEA terranes estimated considering a unique eclogite formation (min. 500 km relative movement; Liégeois et al. 2003) is valid only for the Laouni, Egéré-Aleksod and Azrou n'Fad terranes. The more deformed character of the Tefedest and Aouilène terranes suggests that a larger movement could be envisaged, including relative movements between the Tefedest and Aouilène terranes themselves. The eastern boundary of the Tefedest terrane is not unequivocally defined, mostly because this terrane is poorly studied. To the south, the Tamanrasset shear zone has been chosen because of its major reactivation during recent times (Liégeois et al. 2005) but the large shear zone located 25 km to the west (Fig. 3) is also a good candidate for separating the Tefedest and Laouni terranes. To the north, a large Quaternary deposit blurs the boundary between the Tefedest and the Egéré-Aleksod terranes. Due to a large shear zone present to the south, the boundary has been located along a large batholith (Fig. 3) but lithological arguments within the Eburnian basement suggests that the boundary could be 10 km more to the east (Black et al. 1994). This boundary needs more work to be definitely determined.

During the Pan-African orogeny, LATEA was a continental passive margin that was first subducted towards the west and then dissected along major shear zones during the general tectonic escape towards the north of the whole Tuareg Shield. This resulted in the obduction of a Neoproterozoic terrane on its western margin, initially named Iskel (Black et al. 1994) and now named after the main oasis of the area, Silet (Béchiri et al. 2016), because the Iskel pluton lies outside the so-called Iskel terrane. By contrast, Iskel is a suitable name for the shear zone bounding the terrane to the west (Fig. 4). The presence of the old LATEA basement at depth is attested by the isotopic signature of the cross-cutting Taourirt plutons (Azzouni-Sekkal et al. 2003) and by the excellent preservation of the Tonian and Cryogenian plutons of this terrane (Caby et al. 1982; Béchiri-Benmerzoug et al. 2011), almost unaffected by the main Pan-African phase except in the main shear zones. This means that the Silet terrane is a crustal terrane, having lost its initial lithospheric mantle during its obduction (Liégeois et al. 2003). At the surface, the western boundary of LATEA is the western boundary of the Tefedest and Aouilène terranes but at depth, in the lithosphere, it is the Iskel shear zone that corresponds to the western boundary of LATEA (Fig. 4). The preservation of the Silet juvenile terrane is probably due to limited uplift of this terrane during the recent swell formation.

To the northeast, the eastern boundary of LATEA has been modified. It follows now the Ounane shear zone (Fig. 4). This is justified by a younger metamorphic basement to the east of this shear zone, dated at 1.9 Ga (Peucat et al. 2003), than to the west (c. 2.1 Ga, see above). In addition, structurally, the shape of the Egéré-Aleksod terrane of Black et al. (1994; Fig. 2) was problematic. To the south, LATEA was bordered by the Serouenout terrane, considered to be juvenile (Fig. 2). The geological map presented here shows that to the south, an old basement is also present in which preliminary zircon ages give also Orosirian ages (Baziz et al. 2005; Baziz et al. in prep.). This means that the juvenile Serouenout terrane lies actually upon this basement and, as Silet, is a crustal terrane. We propose here to name the underlying lithospheric terrane the "Afara terrane" (Fig. 4), from the name of a major oued present in the northern part of the terrane. The allochthonous position of the Serouenout terrane agrees with the presence of high-pressure rocks (eclogites, white schists) in this terrane in the Ti-N-Eggoleh area (Bitam-Derridj et al. 2010; Adjerid et al. 2012, 2015), interpreted as a high-pressure meta-ophiolitic series consisting mainly of oceanic metasediments formerly subducted (Adjerid et al. 2015). As for the Silet terrane, the preservation of the Serouenout crustal terrane is probably due to a less important recent uplift of the Afara terrane, which is indeed typified by a thicker lower Paleozoic sedimentary sequence (Beuf et al. 1971) than in adjoining terranes, indicating that the Afara terrane was already a depression at that time. It is also on the Arafa terrane, just to the north of the Serouenout terrane, that Cretaceous sediments have been preserved (Fig. 3).

The Afara terrane is grouped with the Assodé-Issalane terrane, whose basement is also c. 1.9 Ga (Liégeois et al. in prep.) and with its northern prolongation, the poorly known Tazat terrane. Together they constitute what is called the Orosirian stripe (Fig. 4), that may have participated in the formation of the Columbia supercontinent (maximum packing at 1.8 Ga; Meert 2012) but this has to be investigated. This Orosirian Stripe is bounded to the east by the Raghane shear zone and to the west by the Ounane shear zone (Fig. 4) and corresponds to the eastern part of the former Central Polycyclic Hoggar. A northward movement of c. 1000 km of the Assodé-Issalane terrane (and so of the Orosirian Stripe) along the Raghane shear zone has been proposed (Liégeois et al. 1994). This was facilitated by a lithospheric delamination that softened the Assodé-Issalane terrane (Liégeois et al. 1994); such a delamination did not occur in the Afara terrane.

To the east of the Orosirian stripe, on the other side of the Raghane shear zone, is the Saharan metacraton that behaved as a rigid and low block during the Pan-African orogeny and was the receptacle of the Hoggar molasse (Fezaa et al. 2010). At that period, this area was still a craton, having not been metacratonized yet, meaning that, strictly speaking, before 580 Ma, we should use the term "Saharan craton" and "Saharan metacraton" only after that date. Indeed, this area was metacratonized during the Murzukian intracontinental orogenic phase, which generated the three terranes of Djanet (greenschist facies molasse), Edembo (amphibolite facies molasse) and Aouzegueur (mostly unmetamorphosed molasse). The margins of these terranes are reactivated Paleoproterozoic or Archean boundaries (Fezaa et al. 2010). Detrital zircons with mostly 1.9 Ga zircons as Paleoproterozoic inheritage (Fezaa et al. 2010) suggest that the source of the molasses is mainly the Orosirian stripe. In the northern part of the Aouzegueur terrane, the molasse was deformed along the Raghane shear zone (Bertrand et al. 1978) during the Murzukian event at c. 550 Ma (Henry et al. 2009). Its western boundary, corresponding also to the western boundary of the Saharan metacraton, was repeatedly intruded by plutonic rocks during the Neoproterozoic (Henry et al. 2009; Nouar et al. 2011). The only large part of the Saharan craton that was

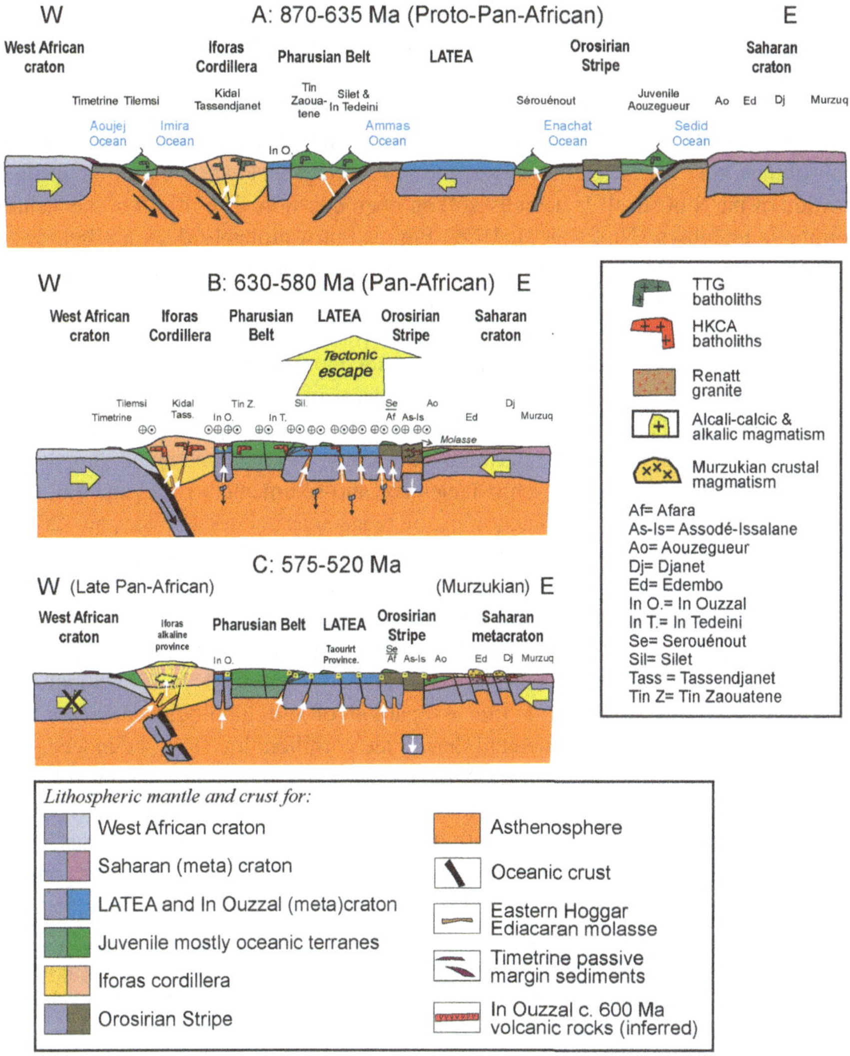

Fig. 5 Sketched E–W cross-sections through the Tuareg Shield at a latitude of 19°N from 1°W to 2°E (Adrar des Iforas) and of 23°N from 2°E to 14°E. This means that the Ahnet, Tchilit and Barghot terranes are not present in the figure. These cross-sections combined snapshots of different times included in the time range. **a** Situation at 870–635 Ma, i.e. in Proto-Pan-African times, a period dominated by subduction of several oceanic domains. The Imira Ocean should have been represented much wider. **b** Situation at c. 630–580 Ma during the general intracontinental northward tectonic escape of the Tuareg terranes due to the continuing convergence of the West African craton and of the Saharan craton. **c** Situation at 575–520 Ma when the West African craton convergence stopped but when the metacratonization of the Saharan craton also occurred. All events represented here are not strictly contemporaneous but all belong to that period. See text for explanations (Sect. 5)

metacratonized during the main Pan-African phase (630–580 Ma) is the Barghot terrane in Niger intruded by c. 600 Ma high-K calc-alkaline batholiths (Fig. 4; Liégeois et al. 1994; Liégeois et al. in prep.). There, metacratonization preserved the earlier ophiolitic complex and the contemporaneous molasse in the Aouzegueur terrane just to the east (Liégeois et al. in prep.). Whether the Barghot terrane is a part of the Orosirian Stripe thrust over the SmC (Liégeois et al. 1994) or an uplifted portion of the SmC itself is actually still an open question. The origin of the Tchilit terrane is unknown (Navez et al. 1999; Fig. 5) but it probably does not belong to the Orosirian Stripe, the Afara terrane closes most probably more to the north (Fig. 4) and the Ounane shear zone that constitutes most likely the eastern boundary of the Laouni terrane to the south (Fig. 4) may be correlated with the eastern Tchilit shear zone but could also passes more to the west.

To the west of lithospheric LATEA, i.e., west of the Iskel shear zone, are the poorly known In Tedeini and Tin Zaouatene terranes, corresponding to the former "Central Pharusian Branch" (Bertrand and Caby 1978), constituted largely by Neoproterozoic juvenile greenschist facies volcano-sedimentary rocks and granitoid plutons associated with the functioning of mega-shear zones (Boissonnas 2008). We propose here to restrict the notion of "Pharusian belt" to these two terranes, being bracketed between the Iskel and Adrar shear zones (Fig. 4).

To the west of the Adrar fault and east of the Tilemsi shear zone, there is a complex association of terranes, which is named here the "Iforas Cordillera" in the sense of an active continental margin eventually involved in collisional and post-collisional processes, with the association of old lithologies more or less reactivated and lithologies generated during the cordillera life (Fig. 4). Indeed, this domain is located close to the West African craton and was largely modeled by the oceanic subduction towards the east that led to the collision between the WAC and the Tuareg Shield (Black et al. 1979). The Kidal and Tassendjanet terranes show typically the association of a strongly reactivated Paleoproterozoic basement and of a major input from the subducting slab in pre-, syn- and post-collisional settings (Chikhaoui et al. 1978; Liégeois et al. 1987; Bosch et al. 2016). To the east, the In Ouzzal terrane displays very well preserved c. 2 Ga high-temperature granulitic parageneses (Peucat et al. 1996; Ouzegane et al. 2003) but it has been shaped during the Pan-African orogeny, becoming very narrow to the south where it is relayed by the UGI. The Ahnet and Tirek terranes are also old terranes reactivated during the Pan-African orogeny. The exact relative behavior of these two terranes during the Pan-African orogeny must be yet established.

To the west of the Tilemsi shear zone is the Tilemsi terrane, which is a late Tonian–early Cryogenian island arc assemblage (Caby et al. 1989) thrust over the WAC. To this assemblage can be associated the ultramafic rocks marking the gravimetric suture between the WAC and the Tuareg Shield (Bayer and Lesquer 1978).

Finally, at the extreme West, the West African craton basement is outcropping in a small area in the Timetrine terrane with remnants of its passive margin (Caby 2014). Later this area has been affected by Permian–Jurassic carbonatites, nepheline syenite ring-complexes and associated graben (Tadhak province; Liégeois et al. 1991).

Late Ediacaran alkali-calcic and alkalic magmatism is well developed in the Iforas cordillera (Liégeois and Black 1987; Fezaa et al. 2018), to the west of the LATEA metacraton (Silet terrane) and locally elsewhere in LATEA (Azzouni-Sekkal et al. 2003) as well as in the Djanet terrane (Fezaa et al. 2010). Devonian magmatism is very well developed in the Aïr area (Moreau et al. 1994) along the western boundary of the Saharan metacraton and locally in the Djanet terrane, along the western boundary of the Murzuq craton (Figs. 1, 2; Derder et al. 2016). The Permian–Jurassic magmatism is restricted to the Tilemsi terrane, close to the eastern boundary of the WAC (Liégeois et al. 1991). Finally, the Cenozoic volcanism (Liégeois et al. 2005) is present in the LATEA metacraton, especially along the eastern boundary of the Tefedest terrane, at the boundary between the SmC and the Orosirian Stripe, both to the south (Aïr) and to the north (Tazat area) and along the western boundary of the Murzuq craton, in the Djanet terrane (Fig. 4).

5 Geological Evolution of the Tuareg Shield

The Tuareg Shield resulted from the interplay between the subduction of several oceanic domains that generated island arcs and continental cordilleras and the differential behaviors of rheologically contrasted juvenile, cratonic or metacratonic terranes during a general northward tectonic escape between the closing large and rigid blocks that are the WAC and the SmC. A major ocean with an eastern subducting oceanic plate was present between the WAC and the Tuareg shield (Black et al. 1979; Liégeois et al. 1987) with the presence of the Tilemsi oceanic arc in between (Caby et al. 1989). The life of this arc spanned from 750 to 700 Ma ending with more felsic products (Caby et al. 1989) suggesting that it was thrust over the WAC at c. 700 Ma, closing the Aoujej Ocean (Aoujej = far in Tuareg language) and forming the Tilemsi crustal terrane (Fig. 5). To the south in the Gourma area, the preserved Amalaoulaou island arc root (Berger et al. 2008, 2011) belongs probably to the same arc system. The subduction period lasted later under the Iforas cordillera producing TTG and volcano-sedimentary series (Fabre 1982; Liégeois et al. 1987; Bosch et al. 2016) with the closure of the Imira Ocean (Imira = large) at c. 625 Ma, the first post-collisional granitoids of the Iforas batholith (Liégeois et al. 1987) and, to the south in the Gourma area, the thrusting of eclogitic thrust nappes (Jahn et al. 2001; Caby et al. 2008) with additional eclogitic remnants in the Iforas cordillera itself (Berger et al. 2014). The WAC lithosphere began to block the subduction and provoked the beginning of the general northward tectonic escape of the Tuareg terranes (Fig. 5b).

To the east other oceanic subductions existed but were probably smaller than the Imira Ocean considering the less important remnants and the less important post-collisional magmatism, even if this has still to be more deeply investigated. In the middle of the orogen, the Ammas Ocean (Ammas = inside) westerly subduction generated the Silet, In Tedeini and Tin Zaouatene island arcs (Fig. 5a). Its closure caused the obduction of the Silet crustal terrane over LATEA (Béchiri-Benmerzoug et al. 2011) whose western margin was subducted (Liégeois et al. 2003), while the In Tedeini and Tin Zaouatene terranes were preserved as lithospheric terranes. It is considered here that the Silet terrane is the "thrust on LATEA" equivalent of the In Tedeini terrane. Being not thrust on the rigid LATEA, the Tin Zaouatene and In Tedeini terranes were squeezed between LATEA and the rear of the Iforas cordillera comprising the rigid In Ouzzal metacraton (Fig. 5b) and invaded by large post-collisional granitoid batholiths (Gravelle 1969; Boissonnas 2008). It must be noted that the Tin Zaouatene terrane comprises lithologies reminiscent of continental areas (felsic gneiss, migmatites associated to crustal granites, …) in addition to the greenschist facies volcano-sedimentary series (Black et al. 1994) suggesting a more complex evolution than that of the In Tedeini terrane.

The correlation between the Kandi shear zone marking the eastern boundary of the WAC and the 4°50′ shear zone is often proposed because they are roughly in continuation (Ganade de Araujo et al. 2014 and references therein), which may suggest that this was the Ammas Ocean the largest ocean. However, this Kandi—4° 50′ alignment, only suggested through geometric considerations, is a post-collisional feature linked to the northern tectonic escape due to the WAC embayment existing in front of the Gourma region (Fig. 1). Remnants of the main Imira Ocean are marked by heavy gravimetric anomalies following the eastern boundary of the WAC (Bayer and Lesquer 1978). So we consider that the Imira Ocean was the major ocean that separated the Trans-Saharan Belt from the WAC in Rodinia reconstructions during Tonian and early Cryogenian times (Li et al. 2008; Merdith et al. 2017). Nevertheless, rocks with subduction signature in the range 870–635 Ma for the Ammas Ocean (Caby et al. 1982; Béchiri-Benmerzoug et al. 2011) indicate it was also a significant ocean.

More to the east, the westerly subducting Enachat Ocean (Enachat = active, efficient) generated the Serouenout island arc (Fig. 5a) whose closure led to the Serouenout crustal terrane on the Orosirian stripe (Arafa terrane; Fig. 5b) with spectacular high-pressure lithologies (Adjerid et al. 2008, 2012, 2015). Finally the Sedid Ocean (Sedid = meagre) westerly subduction produced the Aouzegueur island arc (Fig. 5a) whose closure generated allochthonous nappes or slivers made of juvenile oceanic material onto the SmC in Aïr (ophiolitic association; Boullier et al. 1991) and in Hoggar (Caby and Andreopoulos-Renaud 1987) (Fig. 5b). A crustal juvenile terrane has not been distinguished here because these remnants appear to be dispersed and of limited extension. Similarly, the presence of juvenile or potentially juvenile thrust material (including eclogites; Liégeois et al. 2003; Doukkari et al. 2014, 2015) on LATEA has not been considered as a distinct terrane for the same reason, but that could be discussed and revised.

Even if some diachronisms are likely, the closure of the Imira, Ammas, Enachat and Sedid oceans occurred roughly at the same time, i.e., just before c. 630 Ma, the age of the obduction of high-pressure rocks (collision climax; Liégeois 1998) and of the first post-collisional granitoids in the shield. Probably, the closure of the large Imira Ocean and the beginning of the great WAC push contributed to the closure of the narrower oceans more to the east. This generated the general post-collisional northward tectonic escape that complicated the image by segmenting the more rigid blocks in metacratonic terranes by stretching the more ductile lithospheric terranes (Pharusian belt and Assodé-Issalane terrane from the Orosirian Stripe) and by generating huge shear zones accompanied by high-temperature metamorphism and abundant granitoid batholiths and plutons (Fig. 5b). This is especially the period of the genesis of the LATEA metacraton with the individualization of its five terranes (Laouni, Aouilène, Tefedest, Egéré-Aleksod, Azrou n'Fad) and the lithospheric delamination of the Assodé-Issalane terrane generating its spectacular high-temperature metamorphism and regional leucogranites. The latter effect probably facilitated the global tectonic escape of the Tuareg Shield between the WAC and the SmC, the Assodé-Issalane terrane acting as a lubricant.

During that period, to the east of the Raghane shear zone, all the region was stable and flat and received the molasse (Tiririne, Proche-Ténéré, Djanet groups; Fezaa et al. 2010), resulting from the erosion of the Tuareg orogen, mainly from the Orosirian Stripe (Fig. 5b). This region corresponds to the western side of the Saharan craton, not yet metacratonic at the time, at least in its western part.

The end of the WAC push is attributed to slab break-off below the Iforas Cordillera (Fig. 5c), with the consequent uprise of the asthenosphere whose decompression melting generated the huge alkaline province of the Adrar des Iforas, the volcanic products of which probably covered all the area (60,000 km^2; Liégeois and Black 1987) similarly as the Ouarzazate volcanic Supergroup in the Anti-Atlas (Morocco; Thomas et al. 2004; Ennih and Liégeois 2008).

At the extreme east by contrast, major intracontinental metacratonization occurred, reactivating the Paleoproterozoic shear zones, allowing the emplacement of largely crustal granitoids, under greenschist facies in Djanet terrane and under amphibolite facies in Edembo terrane, called the Murzukian intracontinental episode (575–545 Ma; Fezaa et al. 2010; Bendaoud et al. 2016). The stress was transmitted by the Murzuq cratonic nucleus (Fig. 1) originating from a plate boundary event to the NE. This event has still to be identified but could be related to an impingement of the Siberian craton (Liégeois et al. 2013b) but this hypothesis needs to be validated. During that period and until c. 520 Ma, in the Orosirian Stripe and specifically in the LATEA metacraton, a series of shallow and subcircular plutons were emplaced. These are the Taourirt province (Azzouni-Sekkal et al. 2003), the Tisselliline pluton (Liégeois et al. 2003), and other isolated late plutons. They can have emplaced due to the vanishing stress generated by the end of the convergence with the WAC or to a far-distant Murzukian effect for the younger phases, when the stress coming from the west was negligible (Fig. 5c), even if younger than 540 Ma. Indeed, building a more accurate model would require more age data on this composite magmatic group on the one hand and on the Murzukian episode for knowing its actual end, on the other hand.

During the Phanerozoic, several periods of magmatic activity, Devonian, Permian, Jurassic and Cenozoic correspond to the reactivation of faults having the right orientation due to the stress applied at plate margins respectively due to the Caledonian event (Derder et al. 2016), the Variscan event (Liégeois et al. 1991; Derder et al. 2016), the Atlantic Ocean opening (Liégeois et al. 1991) and the Africa–Europe convergence (Liégeois et al. 2005).

6 Conclusions

The Tuareg Shield presents the outstanding particularity to offer the entire width of an orogen from a craton to another, including the margins of the latter, i.e., a current 1400 km width. The new synthetic geological map of the Tuareg Shield presented here is based on mostly old geological maps especially those of BRMA (1961) but enhanced by some more recent ones and uses the literature of the last decades furnishing geological and/or geochronological constraints.

This map conforms to the available geological constraints but has been conceived to highlight the geodynamical structure of the shield and particularly the coexistence of juvenile Neoproterozoic terranes and metacratons, with an Archean and/or Paleoproterozoic basement more or less reactivated during the Pan-African orogeny, taking advantage of the absence of Mesoproterozoic events.

The metacratonic terranes have a base color corresponding to the age of their formation (Mesoarchean, Rhyacian or Orosirian) with a hatching related to the main reactivation age (Orosirian or Pan-African). Neoproterozoic terranes are distinguished as are the various Neoproterozoic and Phanerozoic magmatic events.

The original terrane structure (Black et al. 1994) is partly revised by adding two terranes, Aouilène, and Afara, at the expense of the Laouni and Egéré-Aleksod terranes, respectively. LATEA is now the acronym standing for Laouni, Aouilène, Téfedest, Egéré-Aleksod, Azrou n'Fad. These 25 Tuareg terranes are grouped in several larger entities that are from west to east: the West African craton, the Iforas cordillera, the Pharusian belt, the LATEA metacraton, the Orosirian Stripe and the Saharan metacraton.

A comprehensive evolution of the Tuareg Shield is proposed on the basis of the new map consisting of (1) the closure of five oceans (from west to east, Aoujej, Imira, Ammas, Enachat, and Sedid oceans; Imira Ocean being the largest) (900–630 Ma); (2) the general northward tectonic escape of the Tuareg terranes squeezed between the West African craton and the Saharan craton (630–580 Ma). This main Pan-African phase did not affect the Saharan craton east of the Raghane shear, which was the receptacle of an important molasse resulting from the erosion mostly of the Orosirian Stripe; this phase stopped with a slab break-off to the west, marked by the huge Iforas alkaline province blocking the West African craton ramming; (3) by contrast, the intracontinental metacratonization that lately occurred (575–545 Ma), may be due to the impingement of the Siberian craton on the Saharan craton, eventually becoming the Saharan metacraton.

The contrasted rheology of the different terranes and entities is a key parameter for understanding the Tuareg Shield Neoproterozoic evolution, including the magmatism. The late Ediacaran and Phanerozoic magmatism is the result of the interplay between the Pan-African terrane structure and the orientation of the stress applied at plate boundaries.

The aim of the map presented here is to illustrate this complicated but exciting story. *Electronic files of the different maps are available from the author.*

Acknowledgements Mohamed Laghmouch (RMCA) is warmly thanked for his skilled help in the GIS processing. Reviewers Jean Boissonnas and Abderrahmane Bendaoud are thanked for their careful and expert suggestions that nicely enhanced the streamline and the clarity of the paper. I would like to warmly thank all the Algerian, French and other colleagues, most often friends, with whom I worked, and discussed so passionately and so efficiently, especially with the USTHB team of Algiers with which I had an extremely tight and friendly collaboration. I cannot cite all of them here, they are too numerous but a series of them are present in the references, even if all are not cited. Obviously, this paper is based on many papers made by generations of geologists since the pioneering work of C. Kilian and M. Lelubre until the work of the young enthusiastic generation. However, I would like to cite here colleagues and close friends who did outstanding work in the Tuareg Shield and sadly have already left us, Russell Black, Jimmy Bertrand and Louis Latouche. They belong, of course, to the team that developed the terrane concept in the Tuareg Shield a quarter of a century ago. Finally, I am eager to see the developments that the current skillful young Algerian geologists will accomplish concerning the amazing Tuareg Shield geology the following years, hoping that this contribution will be a useful tingle.

References

Abdallah N, Liégeois JP, De Waele B, Fezaa N, Ouabadi A (2007) The Temaguessine Fe-cordierite orbicular granite (Central Hoggar, Algeria): U–Pb SHRIMP age, petrology, origin and geo-dynamical consequences for the late Pan-African magmatism of the Tuareg shield. J Afr Earth Sci 49:153–178

Abdelsalam M, Liégeois JP, Stern RJ (2002) The Saharan meta-craton. J Afr Earth Sci 34:119–136

Acef K, Liégeois JP, Ouabadi A, Latouche L (2003) The Anfeg post-collisional Pan-African high-K calc-alkaline batholith (Central Hoggar, Algeria), result of the Latea microcontinent metacratonisation. J Afr Earth Sci 37:295–311

Adjerid Z, Ouzegane K, Godard G, Kienast JR (2008) First report of ultrahigh-temperature sapphirine + spinel + quartz and orthopyroxene + spinel + quartz parageneses discovered in Al–Mg granulites from the Khanfous area (In Ouzzal metacraton, Hoggar, Algeria). In: Ennih N, Liégeois J-P (eds) The boundaries of the west African craton, vol 297. Geological Society, London, Special Publications, pp 147–167

Adjerid Z, Ouzegane K, Godard G, Bitam-Derridj A, Kienast JR (2012) Le Sérouènout: un fragment de lithosphère océanique subducté à haute pression, exhumé puis granulitisé à haute température. Bull du Serv Géol Natl [algérien] 23:199–217

Adjerid Z, Godard G, Ouzegane K (2015) High-pressure whiteschists from the Ti-N-Eggoleh area (Central Hoggar, Algeria): a record of Pan-African oceanic subduction. Lithos 226:201–216

Aït-Djafer S, Ouzegane K, Liégeois JP, Kienast JR (2003) An example of post-collisional mafic magmatism: the gabbro-anorthosite complex from the Tin Zebane area (western Hoggar, Algeria). J Afr Earth Sci 37:313–330

Azzouni-Sekkal A, Liégeois JP, Béchiri-Benmerzoug F, Belaidi-Zinet S, Bonin B (2003) The "Taourirt" magmatic province, a marker of the very end of the Pan-African orogeny in the Tuareg Shield: review of the available data and Sr–Nd isotope evidence. J Afr Earth Sci 37:331–350

Bayer R, Lesquer A (1978) Les anomalies gravimétriques de la bordure orientale du craton ouest africain: géométrie d'une suture pan-africaine. Bull Soc Géol Fr 6:863–876

Baziz K, Liégeois JP, Azzouni-Sekkal A (2005) La carte géologique Asséo au 1/200 000: structure, Pétrologie et Géochronologie de deux "Terranes" Issalane et Sérouènout dans le S. E. Hoggar (Algérie). Séminaire de géologie et de métallogenie des massifs du Hoggar et des Eglab, Tamanrasset, Algérie

Béchiri H, Azzouni-Sekkal A, Béchiri-Benmerzoug F, Bonin B, Liégeois JP, Kheloui R, Matukov DI, Sergeev SA (2016) La faille de Tin-Dahar, un élément clé dans la structuration du bloc de Silet (ex Iskel) Hoggar occidental, Algérie. In: 1st Arab geoscience union international conference (AIC-1), Algiers, Algeria, p 4

Béchiri-Benmerzoug F, Liégeois JP, Bonin B, Azzouni-Sekkal A, Bechiri H, Kheloui R, Matukov DI, Sergeev SA (2011) The plutons from the cryogenian Iskel composite oceanic island arc (Hoggar, Tuareg Shield, Algeria): U–Pb on zircon SHRIMP geochronology, geochemistry and geodynamical setting. In: Seventh Hutton symposium on granites and related rocks, Avila, Spain, 4–9 July 2011, p 17

Béchiri-Benmerzoug F, Bonin B, Bechiri H, Khéloui R, Talmat-Bouzeguela S, Bouzid K (2017) Hoggar geochronology: a historical review of published isotopic data. Arab J Geosci 10:351–383

Bendaoud A, Derridj A, Ouzegane K, Kienast JR (2004) Granulites of the Laouni terrane, (Tamanrasset, Tidjenouine, Tin Begane). J Afr Earth Sci 39:187–192

Bendaoud A, Ouzegane K, Godard G, Liégeois JP, Kienast JR, Bruguier O, Drareni A (2008) Geochronology and metamorphic P–T–X evolution of the Eburnean granulite-facies metapelites of Tidjenouine (Central Hoggar, Algeria): witness of the LATEA metacratonic evolution. In: Ennih N, Liégeois J-P (eds) The boundaries of the west African craton, vol 297. Geological Society of London, Special Publication, pp 111–146

Bendaoud A, Liégeois JP, Sergeev S, Kienast JR, Drareni A, Ouzegane K (2016) The Edembo Murzukian (575–545 Ma) monophase high-temperature amphibolite metamorphism (Eastern Hoggar, Algeria): a key for understanding the development of the Saharan metacraton during the Gondwana supercontinent amalgamation. In: 1st Arab geoscience union international conference (AIC-1), Algiers, Algeria, p 103

Berger J, Caby R, Liégeois JP, Mercier JC, Demaiffe D (2008) Dehydration, melting and related garnet growth in the deep root of the Amalaoulaou Neoproterozoic magmatic arc (Gourma, NE Mali). Geol Mag 146:1–14

Berger J, Caby R, Liégeois JP, Mercier JCC, Demaiffe D (2011) Deep inside a Neoproterozoic intra-oceanic arc; growths, differentiation and exhumation of the Amalaoulaou complex (Gourma, Mali). Contrib Miner Petrol 162:773–796

Berger J, Ouzegane K, Bendaoud A, Liégeois JP, Kiénast JR, Bruguier O, Caby R (2014) Continental subduction recorded by Neoproterozoic eclogite and garnet amphibolites from western Hoggar (Tassendjanet terrane, Tuareg Shield, Algeria). Precambr Res 247:139–158

Bertrand JML (1974) Evolution polycyclique des gneiss du Précambrien de l'Aleksod (Hoggar central, Sahara algérien). Aspects structuraux, pétrologique, géochimiques et géochronologiques. Centre National de la Recherche Scientifique Paris, France, Série Géologique 19, p 350

Bertrand JML, Caby R (1977) Carte géologique du Hoggar, Algérie à l'échelle 1 million (2 feuilles). Direction des Mines et de la Géologie, SONAREM, Algeria

Bertrand JLM, Caby R (1978) Geodynamic evolution of the Pan-African orogenic belt: a new interpretation of the Hoggar shield (Algerian Sahara). Geol Rundsch 67:357–388

Bertrand JM, Lasserre M (1976) Pan-African and pre-Pan-African history of the Hoggar (Algerian Sahara) in the light of new geochronological data from the Aleksod area. Precambr Res 3:343–362

Bertrand JM, Caby R, Fabriès J, Vitel G (1968) Sur la structure et l'évolution orogénique du Tazat (Ahaggar oriental). C R Somm Soc Géol Fr 8:257–258

Bertrand JML, Caby R, Ducrot J, Lancelot JR, Moussine-Pouchkine A, Saadallah A (1978) The late Pan-African intracontinental linear fold belt of the eastern Hoggar (Central Sahara, Algeria): geology, structural development, U/Pb geochronology, tectonic implications for the Hoggar shield. Precambr Res 7:349–376

Bertrand JM, Meriem D, Lapique F, Michard A, Dautel D, Gravelle M (1986) Nouvelles données radiométriques sur l'âge de la tectonique pan-africaine dans le rameau oriental de la chaîne pharusienne (région de Timgaouine, Hoggar, Algérie). C R Acad Sci Paris 302:437–440

Beuf S, Biju-Duval B, De Charpal O, Rognon P, Gariel O, Bennacef A (1971) Les grès du Paléozoïque inférieur au Sahara. Publication IFP, collection "Science et techniques du pétrole", Paris, p 464

Bitam-Derridj A, Ouzegane K, Adjerid Z, Godard G, Kiénast JR (2010) Les éclogites granulitisées de Ti-N-Eggoleh (terrane du Sérouènout, Hoggar central): étude métamorphique et conséquence géodynamique. Bull Serv Géol Natl [Algér] 21:117–136

Black R (1965) Sur la signification pétrogénétique de la découverte d'anorthosites associées aux complexes annulaires subvolcaniques du Niger. C R Acad Sci Paris 260:5829–5832

Black R, Liégeois JP (1993) Cratons, mobile belts, alkaline rocks and continental lithospheric mantle: the Pan-African testimony. J Geol Soc London 150:89–98

Black R, Jaujou M, Pellaton C (1967) Carte géologique de l'Aïr et sa notice explicative à l'échelle 1/500,000, direction nationale des mines et de la Géologie, Niger. Direction Nationale des Mines et de la Géologie, Niger

Black R, Caby R, Moussine-Pouchkine A, Bayer R, Bertrand JML, Boullier AM, Fabre J, Lesquer A (1979) Evidence for late Precambrian plate tectonics in west Africa. Nature 278:223–227

Black R, Latouche L, Liégeois JP, Caby R, Bertrand JM (1994) Pan-African displaced terranes in the Tuareg shield (central Sahara). Geology 22:641–644

Blaise J (1967) Le Précambrien du Tazat. Publications Centre de Recherche Zones Arides, série géologiques, n 7, Paris CNRS

Boissonnas J (2008) Cisaillements ductiles et mise en place de plutons granitiques dans le nord de la chaîne panafricaine du Sahara central: le secteur de Tinnirt (Mouydir, Hoggar du NW, Algérie). Bull Serv Géol Natl (Algér) 19:101–113

Bosch D, Bruguier O, Caby R, Buscail F, Hammor D (2016) Orogenic development of Adrar des Iforas (Touareg Shield, NE Mali): new geochemical and geochronological data and geodynamic implications. J Geodyn 96:104–130

Boullier AM, Rocci G, Cosson Y (1991) La chaîne pan-africaine d'Aouzegueur en Aïr (Niger): un trait majeur du bouclier touareg. C R Acad Sci Paris 313:63–68

Bouzid A, Bayou B, Liégeois J-P, Bourouis S, Bougchiche SS, Bendekken A, Abtout A, Boukhlouf W, Ouabadi A (2015) Lithospheric structure of the Atakor metacratonic volcanic swell (Hoggar, Tuareg Shield, southern Algeria): electrical constraints from magnetotelluric data. In: Foulger GR, Lustrino M, King SD (eds) The interdisciplinary earth: a volume in honor of Don L. Anderson, vol 514. Geological Society of America special paper, pp 239–255

BRMA (Bureau de Recherches Minières en Algérie) (1961) Cartes géologiques du Sahara, 9 sheets of Hoggar: Amguid-Arak, Fort de Polignac, Ouallen-Bidon V, Fort Laperrine, Tazrouk, Fort Charlet-In Ezzane, In Ouzzal, Ti-M-Misaou, In Azaoua. Bureau de Recherches Géologiques et Minières (BRGM), Rambault et Guiot imprimerie, Paris

Caby R (1970) La chaîne pharusienne dans le NW de l'Ahaggar (Sahara central, Algérie): sa place dans l'orogenèse du Précambrien supérieur en Afrique. Thesis, University of Montpellier, France and 1983 Publication de la Direction des Mines et de la Géologie, Algiers, Algeria, vol 47, p 289

Caby R (1994) First record of Precambrian coesite from northern Mali: implications for late Proterozoic plate tectonics around the west African craton. Eur J Mineral 6:235–244

Caby R (2003) Terrane assembly and geodynamic evolution of central-western Hoggar: a synthesis. J Afr Earth Sci 37:133–159

Caby (2014) Nature and evolution of Neoproterozoic ocean-continent transition: evidence from the passive margin of the west African craton in NE Mali. J Afr Earth Sci 91:1–11

Caby R, Andreopoulos-Renaud U (1987) Le Hoggar oriental, bloc cratonisé à 730 Ma dans la chaîne pan-Africaine du Nord du continent africain. Precambr Res 36:335–344

Caby R, Bertrand JM, Black R (1981) Oceanic closure and continental collision in the Hoggar-Iforas pan-African segment. In: Kröner A (ed) Precambrian plate tectonics. Elsevier, pp 407–434

Caby R, Andreopoulos-Renaud U, Gravelle M (1982) Cadre géologique et géochronologie U/Pb sur zircon des batholites précoces dans le segment pan-Africain du Hoggar central (Algérie). Bull Soc Géol Fr 24:677–684
Caby R, Andreopoulos-Renaud U, Pin C (1989) Late Proterozoic arc-continent and continent-continent collision in the pan-African trans-Saharan belt of Mali. Can J Earth Sci 26:1136–1146
Caby R, Buscail F, Dembélé D, Diakité, Sacko S, Bal M (2008) Neoproterozoic garnet-glaucophanites and eclogites: new insights for subduction metamorphism of the Gourma fold- and thrust belt (eastern Mali). In: Ennih N, Liégeois J-P (eds) The boundaries of the west African craton, vol 297. Geological Society, London, Special Publication, pp 204–216
Champenois M, Boullier AM, Sautter V, Wright LI, Barbey P (1987) Tectonometamorphic evolution of the gneissic Kidal assemblage related to the Pan-African thrust tectonics (Adrar des Iforas, Mali). J Afr Earth Sci 6:19–27
Chikhaoui M, Dupuy C, Dostal J (1978) Geochemistry of late Proterozoic volcanic rocks from Tassendjanet area (N.W. Hoggar, Algeria). Contrib Miner Petrol 66:157–164
Cohen KM, Finney SC, Gibbard PL, Fan JX (2013) (updated 2017) The ICS international chronostréatigraphic chart. Episodes 36, pp 199–204
Demaiffe D, Moreau C, Brown WL, Weis D (1991) Geochemical and isotopic (Sr, Nd and Pb) evidence on the origin of the anorthosite-bearing anorogenic complexes of the Aïr Province, Niger. Earth Planet Sci Lett 105:28–46
Derder MEM, Maouche S, Liégeois JP, Henry B, Amenna M, Ouabadi A, Bellon H, Bruguier O, Bayou B, Bestandji R, Nouar O, Bouabdallah H, Ayache M, Beddiaf M (2016) Discovery of a Devonian mafic magmatism on the western border of the Murzuq basin (Saharan metacraton): Paleomagnetic dating and geodynamical implications. J Afr Earth Sci 115:159–176
Doukkari SA, Ouzegane K, Arab A, Kienast JR, Godard G, Drareni A, Zetoutou S, Liégeois JP (2014) Phase relationships and P-T path in NCFMASHTO system of the eclogite from the Tighsi area (Egere terrane, Central Hoggar, Algeria). J Afr Earth Sci 99:276–286
Doukkari SA, Ouzegane K, Godard G, Diener JFA, Kienast JR, Liégeois JP, Arab A, Drareni A (2015) Prograde and retrograde evolution of eclogite from Adrar Izzilatène (Egéré-Aleksod terrane, Hoggar, Algeria) determined from chemical zoning and pseudosections, with geodynamic implications. Lithos 226:217–232
Duplan L (1974) Carte géologique de la chaîne de l'Eg'éré, échelle 1/200,000ème. Publication du Service de la carte géologique de l'Algérie, Société nouvelle de Cartographie, Paris
English KL, Redfern J, Bertotti G, English JM, Cherif RY (2017) Intraplate uplift: new constraints on the Hoggar dome from the Illizi basin (Algeria). Basin Res 29:377–393
Ennih N, Liégeois JP (2008) The boundaries of the west African craton, with a special reference to the basement of the Moroccan metacratonic Anti-Atlas belt. In: Ennih N, Liégeois J-P (eds) The boundaries of the west African craton, vol 297. Geological Society, London, Special Publications, pp 1–17
Fabre J (1982) Pan-African volcano–sedimentary formations in the Adrar des Iforas (Mali). Precambr Res 19:201–214
Fabre J, Ba H, Black R, Caby R, Leblanc M, Lesquer A (1982) La chaîne Pan-Africaine, son avant-pays et la zone de suture au Mali. Carte géologique et gravimétrique de l'Adrar des Iforas au 1/500 000, notice and map. Direction nationale de la Géologie et des Mines, Bamako, p 85
Fabre J, Ait Kaci A, Bouima T, Moussine-Pouchkine A (1988) Le cycle molassique dans le Rameau trans-saharien de la chaine panafricaine. J Afr Earth Sci 7:41–55
Fezaa N, Liégeois JP, Abdallah N, Cherfouh EH, De Waele B, Bruguier O, Ouabadi A (2010) Late Ediacaran geological evolution (575–555 Ma) of the Djanet Terrane, Eastern Hoggar, Algeria, evidence for a Murzukian intracontinental episode. Precambr Res 180:299–327
Fezaa N, Liégeois JP, Abdallah N, Bruguier O, De Waele B, Ouabadi A (2018) The 600 Ma-old Pan-African magmatism in the In Ouzzal terrane (Tuareg Shield, Algeria): witness of the metacratonisation of a rigid block. In: Bendaoud A, Hamimi Z, Hamoudi M, Djemai S, Zoheir B (eds) The geology of the Arab world—an overview. Springer, this volume

Ganade de Araujo CE, Rubatto D, Hermann J, Cordani UG, Caby R, Basei MAS (2014) Ediacaran 2,500-km-long synchronous deep continental subduction in the west Gondwana Orogen. Nat Commun 5(5198):1–8

Ghienne JF, Moreau J, Degermann L, Rubino JL (2013) Lower Palaeozoic unconformities in an intracratonic platform setting: glacial erosion versus tectonics in the eastern Murzuq Basin (southern Libya). Int J Earth Sci 102:455–482

Gravelle M (1969) Recherches sur la géologie du socle Précambrien de l'Ahaggar centro-occidental dans la région de Silet Tibehaouine. PhD thesis, Paris, p 781 (unpublished)

Hadj-Kaddour ZH, Liégeois JP, Demaiffe D, Caby R (1998) The alkaline–peralkaline granitic post-collisional Tin Zebane dyke swarm (Pan-African Tuareg shield, Algeria): prevalent mantle signature and late agpaitic differentiation. Lithos 45:223–243

Henry B, Liégeois JP, Nouar O, Derder MEM, Bayou B, Bruguier O, Ouabadi A, Belhai D, Amenna M, Hemmi A, Ayache M (2009) Repeated granitoid intrusions during the Neoproterozoic along the western boundary of the Saharan metacraton, Eastern Hoggar, Tuareg shield, Algeria: an AMS and U-Pb zircon age study. Tectonophysics 474:417–434

Ikenne M, Söderlund U, Ernst RE, Pin C, Youbi N, El Aouli EH, Hafid A (2017) A c. 1710 Ma mafic sill emplaced into a quartzite and calcareous series from Ighrem, Anti-Atlas – Morocco: evidence that the Taghdout passive margin sedimentary group is nearly 1 Ga older than previously thought. J Afr Earth Sci 127:62–76

Jahn BM, Caby R, Monié P (2001) The oldest UHP eclogites of the World: age of UHP metamorphism, nature of protoliths and tectonic implications. Chem Geol 178:143–158

Kilian C (1947) Du Précambrien d'Afrique. C R Somm Soc Géol Fr 224:350–352

Lapique F, Bertrand JM, Meriem D (1986) A major Pan-African crustal decoupling zone in the Timgaouine area (western Hoggar, Algeria). J Afr Earth Sci 5:617–625

Latouche L (1978) Etude pétrographique et structurale du Précambrien de la région des Gour Oumelalen (NE Hoggar, Algérie). Unpublished thesis, Paris, France, p 225

Lelubre M (1952) Recherche sur la géologie de l'Ahaggar central et occidental (Sahara central). Thèse d'Etat, Paris, France tome 1, p 354, tome 2, p 385

Li ZX, Bogdanova SV, Collins AS, Davidson A, De Waele B, Ernst RE, Fitzsimons ICW, Fuck RA, Gladkochub DP, Jacobs J, Karlstrom KE, Lul S, Natapovm LM, Pease V, Pisarevsky SA, Thrane K, Vernikovsky V (2008) Assembly, configuration, and break-up history of Rodinia: a synthesis. Precambr Res 160:179–210

Liégeois JP (1998) Some words on the post-collisional magmatism. Lithos 45:xv–xviii

Liégeois JP, Black R (1987) Alkaline magmatism subsequent to collision in the Pan-African belt of the Adrar des Iforas (Mali). In: Fitton JG, Upton BJG (eds) Alkaline igneous rocks, vol 30. Geological Society Special Publication, London, pp 381–401

Liégeois JP, Bertrand H, Black R, Caby R, Fabre J (1983) Permian alkaline undersaturated and carbonatite province and rifting along the west African craton. Nature 305:42–43

Liégeois JP, Bertrand JM, Black R (1987) The subduction- and collision-related Pan-African composite batholith of the Adrar des Iforas (Mali): a review. In: Kinnaird J, Bowden P (eds) African geology review. Wiley, London, pp 185–211 and Geol J 22(S2):185–211

Liégeois JP, Sauvage JF, Black R (1991) The Permo-Jurassic alkaline province of Tadhak, Mali: geology, geochronology and tectonic significance. Lithos 27:95–105

Liégeois JP, Black R, Navez J, Latouche L (1994) Early and late Pan-African orogenies in the Aïr assembly of terranes (Tuareg Shield, Niger). Precambr Res 67:59–88

Liégeois JP, Navez J, Hertogen J, Black R (1998) Contrasting origin of post-collisional high-K calc-alkaline and shoshonitic versus alkaline and peralkaline granitoids. Lithos 45:1–28

Liégeois JP, Latouche L, Boughrara M, Navez J, Guiraud M (2003) The LATEA metacraton (Central Hoggar, Tuareg shield, Algeria): behaviour of an old passive margin during the Pan-African orogeny. J Afr Earth Sci 37:161–190

Liégeois JP, Benhallou A, Azzouni-Sekkal A, Yahiaoui R, Bonin B (2005) The Hoggar swell and volcanism: reactivation of the Precambrian Tuareg shield during Alpine convergence and west African Cenozoic volcanism. In: Foulger GR, Natland JH, Presnall DC, Anderson DL

(eds) Plates, plumes and paradigms, vol 388. Geological Society of America special paper, pp 379–400
Liégeois JP, Abdelsalam MG, Ennih N, Ouabadi A (2013a) Metacraton: nature, genesis and behavior. Gondwana Res 23:220–237
Liégeois JP, Bendaoud A, Fezaa N, Abdallah N, Bonin B, Bruguier O, Passchier C, Sergeev S, Ouzegane K, Ouabadi A (2013b) The Djanet and Edembo terranes (Saharan metacraton, Eastern Hoggar, Algeria): a tale of late Pan-African intracontinental metacratonization due to impingement of the Siberian craton? In: 24th colloquium of African geology, Addis-Abeba, Ethiopia, p 46
Meert JG (2012) What's in a name? The Columbia (Paleopangaea/Nuna) supercontinent. Gondwana Res 21:987–993
Merdith AS, Collins AS, Williams SE, Pisarevsky S, Foden JD, Archibald DB, Blades ML, Alessio BL, Armistead S, Plavsa D, Clark C, Müller RD (2017) A full-plate global reconstruction of the Neoproterozoic. Gondwana Res 50:84–134
Moreau C, Demaiffe D, Bellion Y, Boullier AM (1994) A tectonic model for the location of Paleozoic ring-complexes in Aïr (Niger, West Africa). Tectonophysics 234:129–146
Moreau J, Ghienne JF, Hurst A (2012) Kilometre-scale sand injectites in the intracratonic Murzuq Basin (South-west Libya): an igneous trigger? Sedimentology 59:1321–1344
Navez J, Liégeois JP, Latouche L, Black R (1999) The Palaeoproterozoic Tchilit exotic terrane (Aïr, Niger) within the Pan-African collage of the Tuareg shield. J Geol Soc London 156:247–259
Nouar O, Henry B, Liegeois JP, Derder MEM, Bayou B, Bruguier O, Ouabadi A, Amenna M, Hemmi A, Ayache M (2011) Eburnean and Pan-African granitoids and the Raghane mega-shear zone evolution: Image analysis, U–Pb zircon age and AMS study in the Arokam Ténéré (Tuareg shield, Algeria). J Afr Earth Sci 60:133–152
Ouzegane K, Kienast JR, Bendaoud A, Drareni A (2003) A review of Archaean and Paleoproterozoic evolution of the In Ouzzal granulitic terrane (western Hoggar, Algeria). J Afr Earth Sci 37:207–227
Peucat JJ, Capdevila R, Drareni A, Choukroune P, Fanning CM, Bernard-Griffiths J, Fourcade S (1996) Major and trace element geochemistry and isotope (Sr, Nd, Pb, O) systematic of an Archaean basement involved in 2.0 Ga very high-temperature (1000 °C) metamorphic event. In Ouzzal Massif, Hoggar, Algeria. J Metamorph Geol 14:667–692
Peucat JJ, Drareni A, Latouche L, Deloule E, Vidal P (2003) U–Pb zircon (TIMS and SIMS) and Sm–Nd whole rock geochronology of the Gour Oumalelen granulitic basement, Hoggar massif, Tuareg Shield, Algeria. J Afr Earth Sci 37:229–239
Rooney AD, Selby D, Houzay JP, Renne PR (2010) Re–Os geochronology of a Mesoproterozoic sedimentary succession, Taoudeni basin, Mauritania: implications for basin-wide correlations and Re–Os organic-rich sediments systematics. Earth Planet Sci Lett 289:486–496
Rougier S, Missenard Y, Gautheron C, Barbarand J, Zeyen H, Pinna R, Liégeois JP, Bonin B, Ouabadi A, Derder MEM, Frizon de Lamotte D (2013) Eocene exhumation of the Tuareg Shield (Sahara, Africa). Geology 41:615–618
Sautter V (1986) Les eclogites de l'Aleksod (sud algérien): des témoins in situ d'une collision intracontinentale. J Afr Earth Sci 5:345–357
Thiéblemont D, Liégeois JP, Fernandez-Alonso M, Ouabadi A, Le Gall B, Maury R, Jalludin M, Vidal M, Ouattara Gbélé C, Tchaméni R, Michard A, Nehlig P, Rossi P, Chêne F (2016) Geological map of Africa at 1:10 M scale, CGMW-BRGM
Thomas RJ, Fekkak A, Ennih N, Errami E, Loughlin SC, Gresse PG, Chevallier LP, Liégeois JP (2004) A new lithostratigraphic framework for the Anti-Atlas Orogen, Morocco. J Afr Earth Sci 39:217–226
Vitel G (1979) La région Tefedest-Atakor du Hoggar central (Sahara). Evolution d'un complexe granulitique précambrien. Unpublished thesis, University of Paris, France, p 324
Vitel G, Girod M (1983) Carte géologique de l'Algérie, feuille Assekrem au 1/200,000. Direction de la Géologie et des Mines, Ministère de l'Industrie Lourde, Alger, Algérie

Weis D, Liégeois JP, Black R (1987) Tadhak alkaline ring-complex (Mali): existence of U–Pb isochrons and "Dupal" signature 270 Ma ago. Earth Planet Sci Lett 82:316–322

Youbi N, Kouyaté D, Söderlund U, Ernst RE, Soulaimani A, Hafid A, Ikenne M, El Bahat A, Bertrand H, Rkha Chaham K, Ben Abbou M, Mortaji A, El Ghorfi M, Zouhair M, El Janati M (2013) The 1750 Ma magmatic event of the west African craton (Anti-Atlas, Morocco). Precambr Res 236:106–123

Zeghouane H (2002) Carte géologique de l'Algérie, feuille Tamanrasset au 1/200,000. ORGM, Office National de Recherche Géologiques et Minières, Ministère de l'Energie et des Mines, Boumerdès, Algérie

The 600 Ma-Old Pan-African Magmatism in the In Ouzzal Terrane (Tuareg Shield, Algeria): Witness of the Metacratonisation of a Rigid Block

N. Fezaa, J. P. Liégeois, N. Abdallah, O. Bruguier, B. De Waele and A. Ouabadi

Abstract The high-level sub-circular North Tihimatine granitic pluton, intrusive in the In Ouzzal terrane, has been dated at 600 ± 5 Ma (LA-ICP-MS U–Pb zircon) and at 602 ± 4 Ma (SHRIMP U–Pb zircon). At this time, while Tihimatine intruded a brittle In Ouzzal without major metamorphism, large high-K calc-alkaline granitoid batholiths emplaced in the adjacent terranes under ductile conditions and regional amphibolite facies metamorphism. Outside In Ouzzal, high-level plutons emplaced under brittle conditions are known only at c. 580 Ma. The In Ouzzal terrane (500 km × 80 to 5 km), made of c. 2 Ga very high-temperature granulitic lithologies with Archean protoliths, is the sole terrane within the Tuareg Shield to have been largely unaffected by the Pan-African orogeny. The field, petrographic, geochemical and isotopic characteristics of the In Ouzzal granitic plutons studied herein, give keys for the understanding of the atypical behavior of the In Ouzzal terrane. The In Ouzzal Pan-African granitoids present chemical compositions varying from medium-K to high-K calc-alkaline to alkaline compositions. This is recorded by the Sr and Nd radiogenic isotopes

N. Fezaa (✉) · N. Abdallah · A. Ouabadi
Laboratoire de Géodynamique, Géologie de l'Ingénieur et Planétologie, USTHB/FSTGAT, USTHB, Algiers, Algeria
e-mail: fezaanassima@yahoo.fr

J. P. Liégeois
Geodynamics and Mineral Resources, Royal Museum for Central Africa, 3080 Tervuren, Belgium

O. Bruguier
Géosciences Montpellier, Université de Montpellier, CNRS-France, Montpellier, France

B. De Waele
SRK Consulting, Level 1, 10 Richardson Street, West Perth, WA 6005, Australia

B. De Waele
Curtin University, Bentley, WA, Australia

A. Bendaoud et al. (eds.), *The Geology of the Arab World—An Overview*, Springer Geology, https://doi.org/10.1007/978-3-319-96794-3_3

($-4 < \varepsilon_{Nd} < -30$; $0.704 < I_{Sr} < 0.713$), pointing to a mixing between a heterogeneous and old Rb-depleted source, the Eburnean granulitic In Ouzzal crust, and a Pan-African mantle. The latter is represented by the nearby bimodal Tin Zebane dyke swarm ($\varepsilon_{Nd} = +6.2$, $I_{Sr} = 0.7028$; Hadj Kaddour et al. in Lithos 45:223–243, 1998), emplaced along the mega-shear zone bounding the In Ouzzal terrane to the west. Trace element composition and Sr–Nd isotope modeling indicate that 20–40% of different crustal lithologies outcropping in the In Ouzzal terrane mixed with mantle melts. At least two, most probably three, Eburnean granulitic reservoirs with Archean protoliths are needed to explain the chemical variability of the In Ouzzal plutons. The Pan-African post-collisional period is related to a northward tectonic escape of the Tuareg terranes, including the rigid In Ouzzal terrane, bounded by major shear zones. Blocking of the movement of the In Ouzzal terrane, which occurred 20 Ma earlier (at 600 Ma) on the western side than on the eastern side, induced its fracturing along oblique faults inside the terrane. This process allowed asthenosphere to rise and to locally melt the In Ouzzal crust, giving rise to the studied plutons. This corresponds to a metacratonization process. The In Ouzzal terrane demonstrates that a relatively small rigid block can survive within a major orogen affected by a post-collisional tectonic escape at the cost of a metacratonization, particularly at depth along faults.

1 Introduction

The Tuareg Shield (Central Sahara) is composed of old Archean/Paleoproterozoic terranes variably reactivated during the Pan-African orogeny (Black et al. 1994; Peucat et al. 1996, 2003; Ouzegane et al. 2003; Bendaoud et al. 2008) and of juvenile Neoproterozoic oceanic and continental terranes, mostly formed during the 870–630 Ma period (Caby et al. 1982, 1989; Liégeois et al. 1987, 1994, 2003; Bechiri-Benmerzoug et al. 2011). These two contrasting types of terranes are variably crosscut by magmatic intrusions emplaced along numerous shear zones, which affected the Tuareg shield during the 630–580 Ma period. This period corresponds to the Pan-African phase s.s. and is related to the collision between the Tuareg Shield and the West African craton (Liégeois et al. 1987, 2003; Jahn et al. 2001; Caby 2003). To the West of the shield, the numerous Pan-African intrusions can be related to a cordilleran environment (Liégeois et al. 1987; Bosch et al. 2016), while in the center of the shield, they correspond to the metacratonisation of the LATEA microcontinent (Liégeois et al. 2003, 2013). More to the east, a later intracontinental phase (Murzukian, 575–545 Ma; Fezaa et al. 2010) generated high-K calc-alkaline granitoids along the Murzuk craton.

To the west of the Tuareg shield, the In Ouzzal terrane is strongly N–S elongated (500 km × 30 km in the north, 5 km in the south), extending to the south in Mali with the Iforas granulitic unit (UGI) terrane (Fig. 1). The In Ouzzal terrane is characterized by the excellent preservation of Archean rocks affected by an Eburnean (c. 2 Ga; Peucat et al. 1996) high-temperature granulitic metamorphism

(>1000 °C and 10–11 kbar; Ouzegane et al. 2003). In the In Ouzzal terrane, except along its eastern and western boundaries, the Pan-African orogeny is only responsible for brittle tectonics, localized greenschist facies metamorphism and the intrusion of some high-level sub-circular plutons, subject of this paper. The UGI terrane was affected by a more widespread Pan-African greenschist facies metamorphism, but the much older granulitic paragenesis are well preserved (Boullier and Barbey 1988; Black et al. 1994), and dated in this paper at c. 2 Ga (Bosch et al. 2016).

The In Ouzzal terrane is bound by two N–S mega-shear zones, the East Ouzzalian shear zone, separating it from the Tin Zaouatene and Tirek terranes to the east, and the West-Ouzzalian shear zone, separating it from the Tassendjanet terrane to the west. To the north, it is partly thrust over the Ahnet terrane (Fig. 1). The ductile deformation related to the East Ouzzalian shear zone has been constrained to the 627–575 Ma age range by $^{40}Ar/^{39}Ar$ laser-probe dating of ferrotschermakite, biotite and hydrothermal muscovite (Ferkous and Monie 2002). Movements along the East Ouzzalian shear zone and associated fluid circulation were responsible for gold concentration (Marignac et al. 1996; Ferkous and Monie 2002), mined in the Tirek and Amesmessa areas (Mobbs 2004 and references therein).

In contrast to the In Ouzzal terrane, the adjacent terranes have been generated or largely affected by the Pan-African orogeny, with regional amphibolite facies, pre-collisional cordilleran volcano-sedimentary rocks and plutonism (Caby et al.

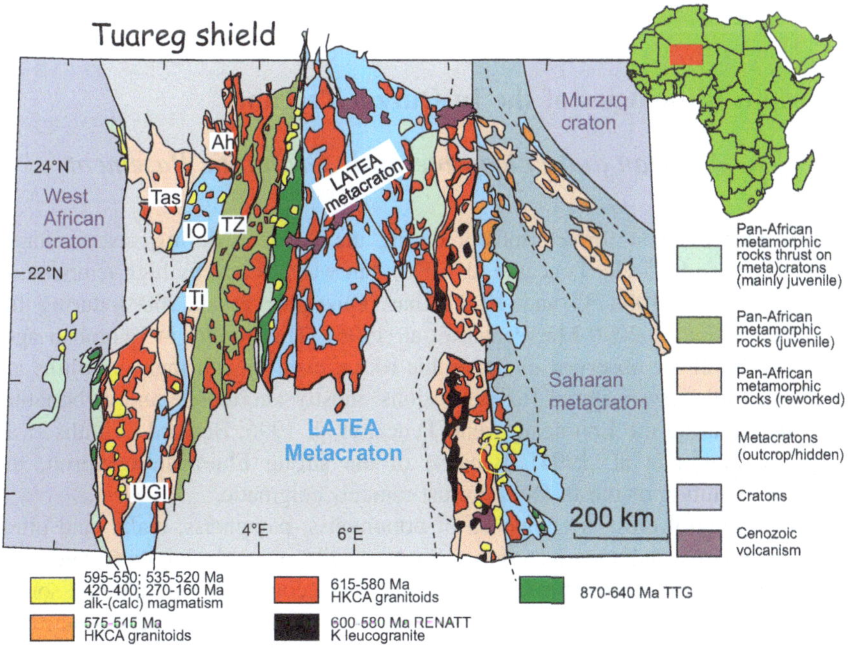

Fig. 1 Simplified geological map of the Tuareg shield

1977; Chikhaoui et al. 1978), large post-collisional high-K calc-alkaline batholiths (Liégeois and Black 1987; Liégeois et al. 1987) and major transcurrent tectonics (Boullier et al. 1986) during the 630–580 Ma period (Caby et al. 1985; Liégeois et al. 1987; Caby and Andreopoulos-Renaud 1985, 1989; Abdallah et al. 2007; Berger et al. 2014; Bosch et al. 2016).

To the south, in Mali, the UGI has been interpreted either as a lithospheric-scale block (Bertrand and Caby 1978) or as a rather superficial basement nappe (Boullier et al. 1978). In the case of the In Ouzzal terrane, magnetotelluric data show distinctly that the west- and east Ouzzalian shear zones are sub-vertical and reach at least the base of the crust (Bouzid et al. 2008), thus suggesting that the In Ouzzal terrane is a distinct lithospheric block. Magnetotelluric data also show the existence of a deeply rooted major fault in the middle of the terrane (Bouzid et al. 2008). This is in agreement with the major late NE–SW shear zones depicted by satellite radar mapping (Deroin et al. 2014). The age of these structures are unknown and could be late Eburnean or Pan-African, or both.

The aim of this paper is to study eight granitic plutons intrusive in the In Ouzzal terrane through U–Pb geochronology on zircon, geochemistry and Sr–Nd isotope geochemistry. These eight plutons are North Tihimatine, South Tihimatine, In Hihaou, Ihaouhouene, Oued Ihaouhouene, In Eher and Tin Chik-Chik outcropping in the north of the terrane, and Lozher, which outcrops more to the south (Fig. 2). Characterizing their emplacement time and origin allow to propose that the process at work corresponds to the metacratonization (Abdelsalam et al. 2002; Liégeois et al. 2013) of an old Archean microcraton.

2 Main Lithologies of the In Ouzzal Terrane

2.1 *The Archean and Paleoproterozoic Granulitic Basement*

The In Ouzzal crust formed mostly during the Archean during several stages between 3.3 and 2.5 Ga (Peucat et al. 1996), but was affected by high-temperature metamorphism (>1000 °C and 10–11 kbar; Ouzegane et al. 2003) during the Eburnean event (c. 2000 Ma; Peucat et al. 1996). This results in Eburnean ages (c. 2 Ga) in all the observed metamorphic rocks, even though their protoliths are Archean. Little new material (rare intrusions, mostly anorthosite and carbonatite) was added during the Eburnean event (Peucat et al. 1996; Bernard Griffiths et al. 1996; Haddoum et al. 1994). Because of the strong Eburnean overprint, the Archean evolution of the In Ouzzal crust remains enigmatic.

The In Ouzzal crust is composed of orthogneiss, paragneiss, mafic, and ultramafic granulites and rare intrusions. Some details on their lithology and geochemistry are summarized here, as these lithologies constitute an important source for the Pan-African granitoids. For a model of their genesis, the reader is referred to Ouzegane et al. (2003).

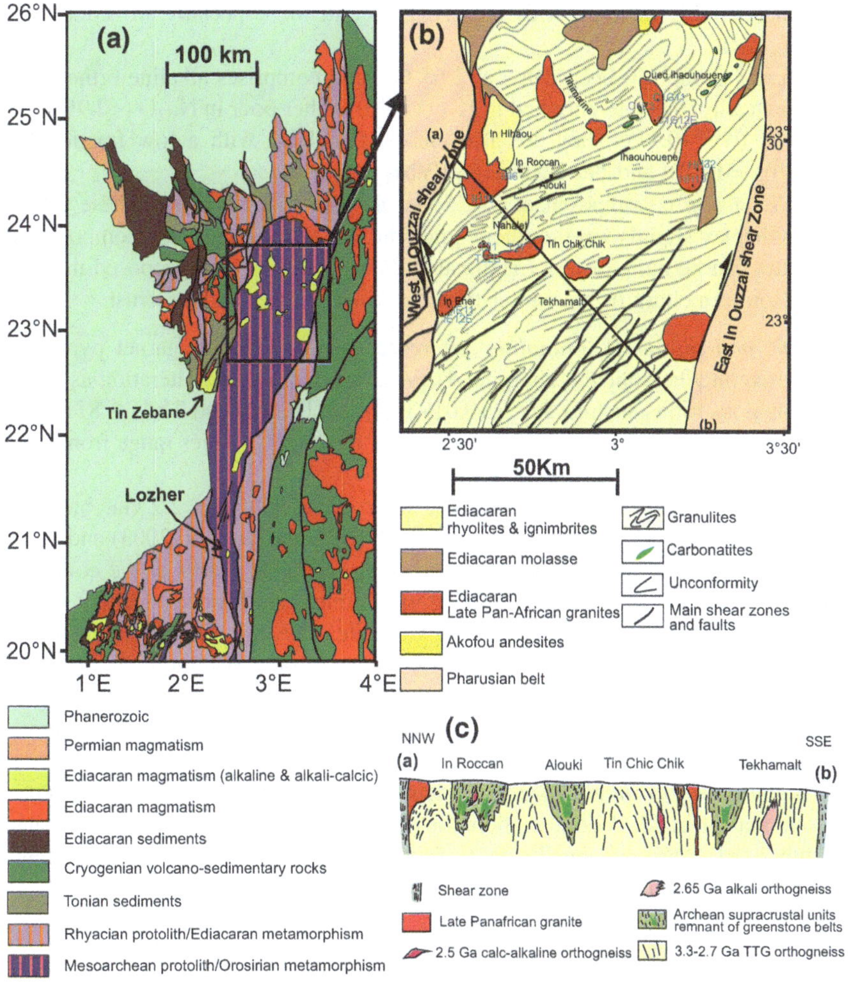

Fig. 2 **a** Geological map of the In Ouzzal terrane and surrounding region. **b** The northern In Ouzzal terranes with the studied plutons. Foliations from Caby (2003), main internal faults from Deroin et al. (2014). **c** Cross-section **a–b** represented in **b** following Ouzegane et al. (2003)

Most of the In Ouzzal granulites are comprised of *orthogneiss*. Despite a rather homogeneous aspect in the field, they display a wide range of mineralogical and geochemical compositions. Based on their ages and geochemistry, three groups can be distinguished (Peucat et al. 1996).

The first group comprise the oldest rocks (3.2–2.7 Ga) that constitute a TTG series (tonalite–trondjhemite–granodiorite) with $K_2O/Na_2O < 0.5$ and typical geochemical patterns: negative Ta, Nb, P and Ti anomalies, a positive Pb anomaly and highly fractionated REE patterns with low HREE contents, especially in Yb as well as in Y.

The second group has been dated at 2.65 Ga, and comprises alkaline orthogneiss. Geochemically, they are enriched in K_2O (>6%), rather poor in Na_2O (<2.9%) with low X_{Mg} (0.07–0.11). They are also enriched in REE with a low fractionation ($(La/Yb)_n = 6–10$) and flat HREE ($(Gd/Yb)_n < 1.7$).

The third group is composed of c. 2.5 Ga old calc-alkaline rocks. These orthogneisses display slightly peraluminous granodioritic to monzogranitic compositions. The latter can be more or less fractionated. Peucat et al. (1996) proposed that they result from remelting of the earlier tonalitic lower to intermediate crust.

Mafic and ultramafic granulites are respectively metanorite/garnet pyroxenite and pyroxenite/lherzolite/harzburgite. They suffered an intense alteration as shown by anomalous ratios such as Rb/Sr (0.01–3.44), La/Th (<6) and Th/U (<8) (Peucat et al. 1996). Their precise age is not known but their NdT_{DM} ages range from 3.1 to 4.2 Ga.

The protoliths of the *metasediments* were deposited after 2.7 Ga, based on detrital zircon ages (Bernard Griffiths et al. 1996; Ouzegane et al. 2003) and before 2.65 Ga, the age of the alkaline intrusions that generated skarns at the contact of marbles (Ouzegane et al. 2003). They are light quartzite, banded magnetite quartzite, marble, Al–Fe granulite and Al–Mg granulite. They are often interlayered with the mafic and ultramafic granulites (Ouzegane 1987; Haddoum et al. 1994; Ouzegane et al. 2003). The Al–Mg granulites are enriched in Cr, Ni, and Co, suggesting that their protoliths comprised mafic and ultramafic rocks (Bernard Griffiths et al. 1996).

Paleoproterozoic magmatism is represented by anorthosite (c. 2.0 Ga, U–Pb on zircon; Peucat et al. 1996), carbonatite (c. 1.99 Ga; U–Pb on zircon; Bernard Griffiths et al. 1996) and small stocks of cordierite-bearing monzogranitic granulite (c. 1.98 Ga, U–Pb on zircon; Peucat et al. 1996).

2.2 The Pan-African Plutons

The studied plutons, always display sharp contacts with their country-rocks, and typically form inselbergs, especially in the north of the terrane, over the Eburnean granulites, which develop a flat surface (Fig. 3a). They can also be smaller intrusions, linearly associated, and are often crosscut by granitic or granodioritic dykes.

They are mostly granites, often with megacrysts (Fig. 3b) and with frequent microgranular enclaves (Fig. 3c, d), attesting to mixing/mingling magma processes. Two main facies are observed: an amphibole–biotite granite (Fig. 3b) and a biotite granite. A spotted biotite–muscovite granite (Fig. 3d) has been locally observed in the South Tihimatine pluton.

The two main facies have been studied in detail in the North Tihimatine pluton. The two facies are fine to medium-grained and often porphyritic (K-feldspar

Fig. 3 Field pictures: **a** Tihimatine inselberg within the flat granulitic In Ouzzal basement; **b** Typical porphyritic granite (Lozher pluton); **c** Mafic microgranular enclaves (North Tihimatine pluton); **d** Spotted biotite–muscovite granite (South Tihimatine pluton)

megacryst). (1) The amphibole–biotite facies is composed of quartz, K-feldspar, less abundant oligoclase (An_{10-25}), biotite and amphibole (magnésio-hornblende). Accessory minerals are apatite, zircon, titanite, allanite, and oxides, mostly magnetite. Chemically, the biotite belongs to the alkali-calcic and alkaline series (Nachit et al. 1985). (2) The biotite facies is composed of quartz, K-feldspar, much less abundant oligoclase-albite (An_{0-20}) and biotite. Accessory minerals are apatite, zircon, allanite and oxides, mostly magnetite. Chemically, the biotite belongs also to the alkali-calcic and alkaline series (Nachit et al. 1985).

Both types of granites contain microgranular enclaves with mafic to felsic compositions.

The felsic enclaves have a mineralogical composition very close to the host granite and probably represent a disrupted chilled margin dropped into the magma. Enclaves of intermediate composition are enriched in biotite and especially in amphibole. They often contain quartz ocelli representing xenocrysts from the magma, around which amphibole from the mafic magma crystallized preferentially. Such ocelli attest to mixing/mingling processes between a mafic and a granitic magma.

These two facies are also found in the In Hihaou, Ihaouhouene, In Eher and Tin Chik-Chik plutons. In the Oued Ihaouhouene (West In Ouzzal) and Lozher (South in Ouzzal), only the biotite–amphibole facies has been found. In the Lozher pluton, a peculiar cumulative gabbroic enclave, composed exclusively of amphibole and plagioclase has been found.

The third facies, a biotite–muscovite granite, intrudes the amphibole-biotite granite in the South Tihimatine pluton, close to its southern boundary. This is a metric-scale outcrop (Fig. 3d) and no contact can be seen. This granite is more aluminous, which is shown by the presence of the muscovite and by a biotite belonging to the Al–K calc-alkaline series (Nachit et al. 1985). This could be correlated with the presence of cordierite-bearing enclaves, which could correspond to country-rock blocks collapsed in the magma and partially melted (Abdallah et al. 2007).

3 Geochronology

3.1 U–Pb Zircon Ages

The North Tihimatine pluton, outcropping in the center of the northern part of the In Ouzzal terrane, has been dated on zircon by two different methods (LA-ICP-MS and SHRIMP; Tables 1 and 2) whose analytical techniques can be found in Appendix.

LA-ICP-MS: 24 zircon from sample TH10 have been dated. 20 zircon are sub-concordant and provide a mean $^{206}Pb/^{238}U$ age of 600 ± 5 Ma (MSWD = 0.87) (Fig. 4a). The six most concordant zircon (MSWD of concordance = 0.16) give a similar Concordia age within error limits of 595 ± 9 Ma but with a larger error. Four inherited zircon are present. Only one is concordant giving a Concordia age of 651 ± 14 Ma. The three others are too discordant to give reliable age information but suggest the presence of older material in this granite linked to the end of the subduction period that characterized the Western Hoggar and Adrar des Iforas at this time (Bosch et al. 2016), maybe through subduction of a sedimentary basin towards In Ouzzal followed by an assimilation process. More data are needed for assessing such a scenario.

SHRIMP: 13 zircon from the same sample (TH10) have been dated. 10 sub-concordant zircon define a mean $^{206}Pb/^{238}U$ age of 602 ± 4 Ma (MSWD = 0.37) (Fig. 4b). The seven most concordant zircon (MSWD of concordance = 0.96) give a similar Concordia age within error limits of 600 ± 8 Ma but with a larger error. The three more discordant spots cannot deliver reliable age information.

Table 1 U–Pb zircon LA-ICP-MS results for sample TH10 (North Tihimatine)

Spot name	Pb* (ppm)	U (ppm)	Th (ppm)	$^{232}Th/^{238}U$	$^{207}Pbr/^{206}Pbr$	1σ err	$^{207}Pbr/^{235}U$	1σ err	$^{206}Pbr/^{238}U$	1σ err	Err corr	$^{206}Pb/^{238}U$ age	1σ err
odZr37	29.3	331	246	0.74	0.0651	0.0020	0.8474	0.0360	0.0944	0.0027	0.68	582	16
odZr38	10.5	118	108	0.91	0.0624	0.0016	0.8337	0.0435	0.0969	0.0044	0.87	596	26
odZr39	12.8	138	104	0.75	0.0668	0.0007	0.9168	0.0224	0.0996	0.0022	0.91	612	13
odZr40	47.1	557	615	1.10	0.0617	0.0010	0.8084	0.0246	0.0951	0.0024	0.85	585	14
odZr41	17.9	201	151	0.75	0.0628	0.0003	0.8544	0.0253	0.0987	0.0029	0.99	607	17
odZr4	13.4	150	127	0.84	0.0596	0.0019	0.8021	0.0360	0.0975	0.0031	0.70	600	18
odZr44″	30.5	338	386	1.14	0.0611	0.0008	0.8331	0.0315	0.0989	0.0035	0.94	608	21
odZr45	33.8	392	277	0.71	0.0627	0.0009	0.8241	0.0193	0.0953	0.0018	0.81	587	11
odZr46	33.0	321	184	0.57	0.0905	0.0043	1.3207	0.0770	0.1059	0.0035	0.57	649	20
odZr47	25.8	277	166	0.60	0.0612	0.0016	0.8394	0.0299	0.0995	0.0024	0.68	611	14
odZr48	53.3	603	417	0.69	0.0621	0.0013	0.8293	0.0347	0.0968	0.0035	0.87	596	21
odZr49	56.9	678	496	0.73	0.0638	0.0011	0.8361	0.0351	0.0951	0.0037	0.92	586	22
odZr50	70.5	854	1157	1.36	0.0553	0.0023	0.7292	0.0322	0.0957	0.0011	0.27	589	7
odZr51	41.3	424	405	0.95	0.0577	0.0026	0.7602	0.0467	0.0956	0.0040	0.68	588	23
odZr53	26.8	268	259	0.96	0.0613	0.0011	0.8972	0.0270	0.1062	0.0026	0.81	651	15
odZr54	42.0	486	343	0.71	0.0628	0.0008	0.8466	0.0374	0.0977	0.0042	0.96	601	24
odZr56	17.4	160	172	1.07	0.0674	0.0028	1.0017	0.0426	0.1078	0.0009	0.20	660	5
odZr57	41.5	250	346	1.38	0.0626	0.0006	0.8647	0.0127	0.1001	0.0012	0.79	615	7
odZr58	33.7	385	356	0.93	0.0637	0.0015	0.8697	0.0338	0.0990	0.0031	0.80	609	18
odZr60	32.7	239	310	1.30	0.0709	0.0010	1.1300	0.0311	0.1156	0.0027	0.86	705	16
odZr61	52.3	468	561	1.20	0.0645	0.0015	0.8522	0.0236	0.0959	0.0015	0.57	590	9
odZr62	31.8	358	350	0.98	0.0614	0.0001	0.8242	0.0152	0.0974	0.0018	1.00	599	10
odZr63	22.3	271	235	0.87	0.0597	0.0012	0.8007	0.0383	0.0972	0.0043	0.92	598	25
odZr64	23.0	418	243	0.58	0.0595	0.0015	0.8113	0.0241	0.0989	0.0014	0.49	608	8

*Pb** Total radiogenic Pb; *Pbr* Radiogenic Pb of the considered isotope

Table 2 U–Pb zircon SHRIMP results for sample TH10 (north Tihimatine)

Spot name	^{206}Pb (% comm)	U (ppm)	Th (ppm)	$^{232}Th/^{238}U$	$^{207}Pbr/^{206}Pbr$	1σ % err	$^{207}Pbr/^{235}U$	1σ% err	$^{206}Pbr/^{238}U$	1σ % err	Err corr	$^{206}Pb/^{238}U$ age	1σ err
th10-1	0.15	282	225	0.82	0.0695	2.36	0.9073	2.72	0.0947	1.36	0.50	583	8
th10-2	0.67	235	177	0.78	0.0576	2.74	0.7713	3.03	0.0971	1.29	0.43	597	7
th10-3	0.31	520	726	1.44	0.0612	1.31	0.8279	1.74	0.0980	1.14	0.65	603	7
th10-4	0.02	566	391	0.71	0.0638	1.22	0.8693	1.66	0.0988	1.13	0.68	607	7
th10-5	0.20	492	752	1.58	0.0589	1.35	0.7974	1.77	0.0982	1.15	0.65	604	7
th10-6	0.09	939	2154	2.37	0.0640	0.83	0.8604	1.36	0.0975	1.08	0.79	600	6
th10-7	0.20	520	637	1.26	0.0608	1.96	0.8198	2.27	0.0978	1.15	0.51	602	7
th10-8	0.25	370	236	0.66	0.0580	1.21	0.7756	1.82	0.0969	1.35	0.74	596	8
th10-9	0.40	331	237	0.74	0.0800	2.13	1.1116	2.45	0.1007	1.21	0.49	619	7
th10-10	1.63	363	351	1.00	0.0722	3.13	0.9516	3.36	0.0956	1.22	0.36	589	7
th10-11	0.04	498	713	1.48	0.0620	1.01	0.8340	1.53	0.0976	1.15	0.75	600	7
th10-12	−0.15	237	249	1.09	0.0614	1.66	0.8206	2.11	0.0970	1.30	0.62	597	7
th10-13	0.02	460	412	0.93	0.0636	0.98	0.8684	1.51	0.0990	1.16	0.76	609	7

Pbr Radiogenic Pb of the considered isotope

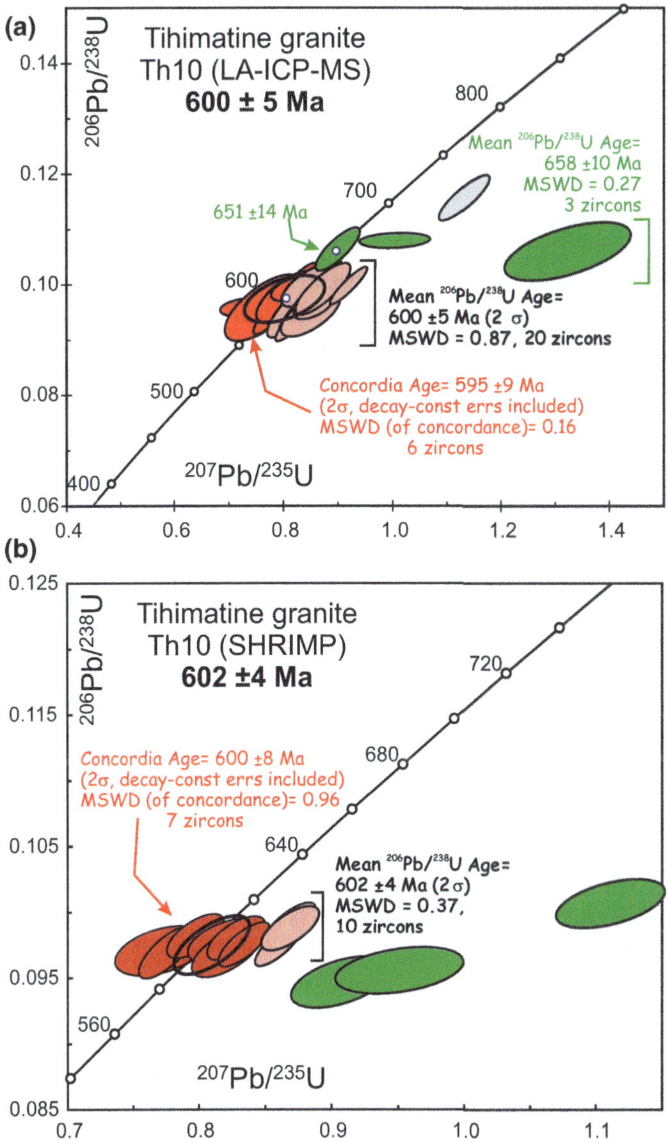

Fig. 4 Concordia U–Pb diagrams for Tihimatine zircons: **a** LA-ICP-MS method; **b** SHRIMP method

Combining the LA-ICP-MS and SHRIMP analyses results in an age of 601 ± 4 Ma, which is confidently taken as the age of emplacement of the North Tihimatine pluton.

3.2 Rb–Sr Isochron Ages

The Rb–Sr whole-rock dating method is less robust than the U–Pb zircon dating method. However, coupled with the latter, it can provide additional information as it concerns the whole-rock system and not only one mineral (Liégeois and Stern 2010; Stern et al. 2010).

Thirty-four whole rocks, from the eight studied plutons have been analyzed for the Sr isotopes (Table 3). Three sub-parallel isochrons, statistically distinct and so with different initial ratios (I_{Sr}), have been obtained (Fig. 5). These isochrons are based on distinct plutons and, include nearly all analyzed samples. Only five samples out of 34 were not included in the calculation.

The first isochron is based on the North Tihimatine, South Tihimatine, Oued Ihaouhouene, and In Eher (sample IE11) plutons, and gives an age of 579 ± 11 Ma (I_{Sr} = 0.70799 ± 0.00018, MSWD = 2.3, 15 samples). Most samples of this isochron belong to the neighboring South and North Tihimatine plutons. Restraining the isochron calculation to these two plutons gives rise to a similar age (576 ± 9 Ma), slightly more precise, with a more accurately defined initial ratio and with a lower MSWD (0.708165 ± 0.000083, MSWD = 0.62, 11 samples). Two samples from South Tihimatine (THZ1 and THZE) plot slightly below this isochron.

The second isochron is based only on the southern In Ouzzal Lozher pluton, which yields a younger age (510 ± 11 Ma) and a lower I_{Sr} (I_{Sr} = 0.706677 ± 0.000028, MSWD = 1.8, five samples). Two samples are statistically below and above this isochron (AT111 & AT106) and one sample (sample AT110), actually a cross-cutting dyke of fine-grained granite, with a very low Sr content (inducing a high Rb/Sr ratio) plots largely above all isochrons.

The third isochron is based on the In Hihaou, Ihaouhouene, In Eher (sample IE12) and Tin Chik-Chik plutons and on one sample from North Tihimatine (TH30). This isochron yields an age of 581 ± 28 Ma (I_{Sr} = 0.7130 ± 0.0024, MSWD = 2.1, nine samples).

These isochrons are defined by rocks from different plutons and give younger ages than the emplacement of the North Tihimatine pluton (601 ± 4 Ma), although the third one (581 ± 28 Ma) is compatible within errors. This indicates that these isochrons either (1) represent the age of the closure of the Rb–Sr isotopic system. This implies that these plutons remained blanketed for about 20 Myrs by a cover that could be the volcanic sequences now found in In Hihaou. This suggests that this In Ouzzal magmatism lasted at least 20 Myrs or (2) result from late fluid movements, most probably along faults that enable homogenization of the Sr initial ratios of plutons rapidly emplaced at c. 601 Ma. The second isochron is based only on the Lozher pluton that is located much further to the south, in a place where the In Ouzzal terrane is narrower and where large fluid movements are documented along faults (Marignac et al. 1996). In this case, the younger apparent age could be related to effect from large fluid movements. Ar–Ar ages in the range 627–575 Ma for ductile deformation along shear zone (Ferkous and Monie 2002) can be

Table 3 Sr and Nd isotopic compositions for the studied In Ouzzal Pan-African granitoids

In Ouzzal Massifs	Type	Sample	Rb (ppm)	Sr (ppm)	$^{87}Rb/^{86}Sr$	$^{87}Sr/^{86}Sr$	2σ	Sr_i (600 Ma)	Sm (ppm)	Nd (ppm)	$^{147}Sm/^{144}Nd$	$^{143}Nd/^{144}Nd$	2σ	ε_{Nd} (600 Ma)	T_{CHUR} (Ma)	T_{DM} (Ma)
Tihimatine North	Bi-Amph syenogranite	TH30	130	90	4.197	0.748776	0.000009	0.712866	8.15	39.11	0.1261	0.511450	0.000008	−17.8	2551	2798
Tihimatine North	Bi-Amph monzogranite	TH31	109	116	2.725	0.730357	0.000012	0.707039	3.88	26.44	0.0887	0.511683	0.000009	−10.4	1346	1656
Tihimatine North	Bi-Amph syenogranite	TH10	106	110	2.795	0.730870	0.000010	0.706955	3.60	26.63	0.0816	0.511678	0.000009	−9.9	1271	1573
Tihimatine North	Bi syenogranite	TH43	108	96	3.264	0.735582	0.000015	0.707650	4.16	28.91	0.0869	0.511667	0.000011	−10.5	1346	1652
Tihimatine North	Bi syenogranite	TH82	110	65	4.917	0.749521	0.000018	0.707446	4.75	31.47	0.0913	0.511665	0.000016	−10.9	1406	1714
Tihimatine North	Bi syenogranite	TH81	105	99	3.077	0.733507	0.000012	0.707179	4.10	28.08	0.0882	0.511661	0.000007	−10.8	1371	1676
Tihimatine North	Dyke granodiorite	THR	99	306	0.937	0.715672	0.000009	0.707655	3.21	24.32	0.0797	0.511614	0.000008	−11.0	1333	1624
Tihimatine North	Enclave granodiorite	TH34E	84	116	2.099	0.725526	0.000009	0.707564	4.69	28.33	0.1001	0.511688	0.000011	−11.1	1496	1816
Tihimatine South	Bi-Amph monzogranite	TH92	100	123	2.357	0.727337	0.000010	0.707168	3.75	29.87	0.0758	0.511667	0.000010	−9.7	1223	1519
Tihimatine South	Bi-Amph alkaline granite	TH90	112	176	1.844	0.723165	0.000012	0.707384	3.54	24.09	0.0889	0.511689	0.000008	−10.3	1340	1651
Tihimatine South	Bi-Amph monzogranite	THZ1	95	247	1.114	0.716008	0.000010	0.706477	2.76	21.48	0.0777	0.511683	0.000009	−9.5	1222	1522
Tihimatine South	Bi-Mus granodiorite	STH1	61	457	0.386	0.711326	0.000007	0.708020	3.68	32.89	0.0676	0.511343	0.000014	−15.4	1526	1772
Tihimatine South	Dyke granodiorite	THF9	44	517	0.246	0.710207	0.000010	0.708099	5.01	34.73	0.0872	0.511390	0.000007	−16.0	1733	1992
Tihimatine South	Enclave qz-monzodiorite	THZE	87	429	0.587	0.711404	0.000011	0.706381	3.56	22.65	0.0949	0.511672	0.000009	−11.1	1444	1757
In Eher	Bi-Amph alkaline granite	IE12E	130	56	6.757	0.766637	0.000009	0.708824	8.46	49.83	0.1026	0.511458	0.000013	−15.8	1906	2177
In Eher	Bi syenogranite	IE11	109	105	3.011	0.732688	0.000010	0.706921	3.84	27.77	0.0837	0.511651	0.000011	−10.6	1329	1629
Ihaouhouene	Bi-Amph monzogranite	HH32	75	212	1.025	0.721323	0.000009	0.712551	4.55	30.84	0.0892	0.511413	0.000008	−15.7	1732	1995

(continued)

Table 3 (continued)

In Ouzzal Massifs	Type	Sample	Rb (ppm)	Sr (ppm)	$^{87}Rb/^{86}Sr$	$^{87}Sr/^{86}Sr$	2σ	Sr_i (600 Ma)	Sm (ppm)	Nd (ppm)	$^{147}Sm/^{144}Nd$	$^{143}Nd/^{144}Nd$	2σ	ε_{Nd} (600 Ma)	T_{CHUR} (Ma)	T_{DM} (Ma)
Ihaouhouene	Bi alkaline granite	HH10	109	46	6.898	0.767892	0.000011	0.708872	9.43	45.94	0.1241	0.511473	0.000009	−17.2	2434	2691
In Hihaou	Bi-Amph syenogranite	IH 10	168	115	4.245	0.749319	0.000009	0.712999	6.69	41.29	0.0980	0.511362	0.000009	−17.3	1964	2218
In Hihaou	Bi syenogranite	IH 6	172	65	7.709	0.776398	0.000009	0.710434	5.08	27.05	0.1135	0.511513	0.000010	−15.6	2054	2334
Tin Chik - Chik	Bi-Amph alkaline granite	T21	95	32	8.657	0.785697	0.000011	0.711624	15.67	89.02	0.1064	0.511336	0.000007	−18.5	2190	2435
Tin Chik-Chik	Bi syenogranite	T27	104	25	12.167	0.815868	0.000008	0.711765	15.40	72.75	0.1280	0.511378	0.000011	−19.3	2779	3000
Tin Chik-Chik	Enclave alkaline granite	T22E	114	55	6.031	0.763159	0.000009	0.711556	11.30	56.98	0.1200	0.511299	0.000009	−20.3	2645	2864
Oued Ihaouhouene	Bi-Amph granodiorite	C1G11	76	331	0.665	0.713123	0.000009	0.707435	4.85	28.97	0.1011	0.511092	0.000012	−22.9	2454	2658
Oued Ihaouhouene	Enclave tonalite	C1E3	83	257	0.935	0.715682	0.000009	0.707679	7.17	42.96	0.1009	0.511415	0.000006	−16.5	1939	2202
Oued Ihaouhouene	Enclave qz-monzodiorite	C1G12E	69	303	0.659	0.713278	0.000010	0.707636	10.69	45.79	0.1412	0.511562	0.000009	−16.8	2937	3172
Lozher	Bi-Amph monzogranite	AT108	115	284	1.172	0.714965	0.000009	0.704938	4.78	29.33	0.0986	0.511709	0.000010	−10.6	1441	1764
Lozher	Bi-Amph monzogranite	AT107	114	271	1.218	0.715415	0.000010	0.704994	4.37	24.49	0.1079	0.511781	0.000007	−9.9	1469	1815
Lozher	Bi-Amph monzogranite	AT116	143	270	1.527	0.718169	0.000010	0.705100	4.35	28.30	0.0928	0.511580	0.000008	−12.7	1550	1843
Lozher	Bi-Amph monzogranite	AT111	122	224	1.581	0.720165	0.000016	0.706633	4.59	29.51	0.0939	0.511282	0.000011	−18.6	2005	2245
Lozher	Dyke syenogranite	AT110	185	37	14.700	0.852022	0.000014	0.726240	2.60	7.33	0.2145	0.512176	0.000011	−10.4	−4017	−6796
Lozher	Enclave tonalite	AT105	107	232	1.338	0.716410	0.000011	0.704962	4.44	23.09	0.1163	0.511767	0.000008	−10.8	1648	1996
Lozher	Enclave qz-monzodiorite	AT106	125	348	1.042	0.712151	0.000007	0.703238	6.42	33.61	0.1155	0.512114	0.000010	−4.0	984	1439
Lozher	Enclave gabbro	AT112	23	864	0.078	0.707241	0.000010	0.706577	7.87	45.11	0.1055	0.510744	0.000006	−30.0	3144	3287

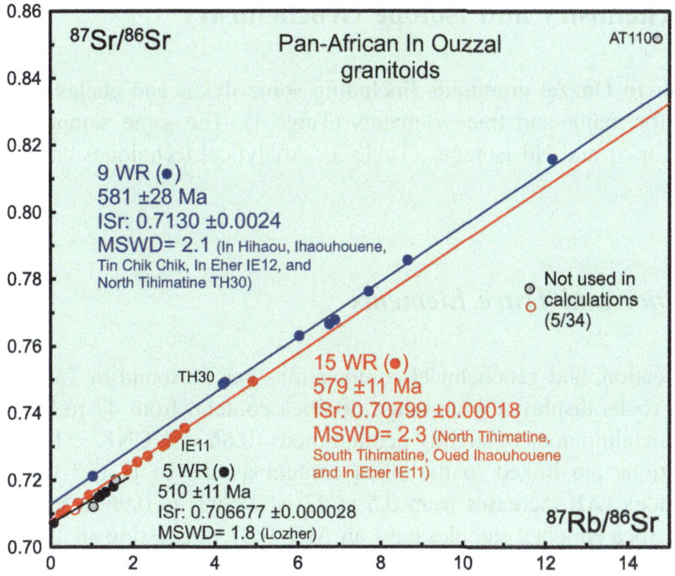

Fig. 5 Rb–Sr isochron diagram for the studied In Ouzzal Pan-African granitoids

reconciled with both hypotheses. Only additional U–Pb zircon ages could resolve the alternatives. But, whatever the right interpretation, it has far-reaching consequences demonstrating that the Rb–Sr isotopic system, if based on enough samples and on a reasonable interpretation, can provide complementary major information that are not documented by more robust geochronometers.

Another important consequence of these data is that the Cambrian age attributed to the In Ouzzal rhyolites et ignimbrites and associated sediments must be re-evaluated. Indeed, this assertion is based on an old Rb–Sr age of 520 ± 10 Ma on a In Hihaou rhyolite interstratified within the Série Pourprée sediments (Allègre and Caby 1968, giving few analytical information). This Rb–Sr age cannot be taken with confidence as the emplacement age of this volcanic layer. Late plutons in Hoggar, generally located on brittle faults, have often generated Rb–Sr isochron ages younger than their emplacement ages dated by the U–Pb zircon method (e.g., Azzouni-Sekkal et al. 2003; Fezaa et al. 2010). As a consequence, until proven otherwise, we consider that the Pan-African magmatism in the In Ouzzal terrane is Ediacaran in age, probably older than 580 Ma, the age of the end of the movements along the Ouzzalian shear zones (Ferkous and Monie 2002).

4 Geochemistry and Isotope Geochemistry

Thirty-four In Ouzzal granitoids (including some dykes and enclaves) have been analyzed for major and trace elements (Table 4). The same samples have been analyzed for Sr and Nd isotopes (Table 3). Analytical techniques can be found in Appendix.

4.1 Major and Trace Elements

Nature, location, and geochemical compositions can be found in Table 4. Studied magmatic rocks display a wide range of silica content, from 47 to 76 wt% SiO_2. They are metaluminous to slightly peraluminous (0.66 < A/CNK < 1.05) (Fig. 6a). The variations are linked to the silica content as shown in Fig. 6b, where the agpaitic index (AI) increases from 0.5 at 47 wt% SiO_2 to 0.96 at >70 wt% SiO_2. The most silica enriched samples have an AI > 0.87, suggesting an alkaline affinity (Liégeois and Black 1987). One Tin Chik-Chik intermediate rock (66 wt% SiO_2) has a particularly high AI, close to 1 (0.96; Fig. 6b), in agreement with that suggestion. The globally high AI is mainly due to K. Most of the studied samples have the composition of high-K calc-alkaline granitoids (Fig. 6c). The most K-enriched rocks are actually alkali-calcic (Fig. 6d). The high AI intermediate rock even falls in the shoshonitic field (Fig. 6c) but not in the alkaline field (Fig. 6d). Most of the enclaves are less enriched in K and are in the medium-K calc-alkaline field (Fig. 6c) or in the calc-alkalic trend (Fig. 6d). The two-mica granite (STH1) has also a lower K_2O content (Fig. 6c). The variations observed in these diagrams indicate that the studied samples cannot be linked simply by a unique magmatic differentiation.

The sliding normalization was proposed for distinguishing potassic series from true alkaline series (Liégeois et al. 1998). This technique normalizes studied samples to an interpolated rock from a reference series having the same silica content to limit the differentiation bias. Normalizing to the alkali-calcic Yenchichi-Telabit series (NYTS) from the Iforas region (Liégeois et al. 1998) is particularly discriminating. Both K- and ALK series can show similar enrichments in one given incompatible element but never in all considered elements. In K-series, enrichment in LILE (large ion lithophile elements) is much more important than in HFSE (high field strength elements), the inverse being observed in ALK series. In this diagram (Fig. 7), the In Ouzzal granites determine two trends. The first one displays a large variation along the *Y* axis. This group comprises the samples defining isochrons 1 and 2, i.e., those with the lower I_{Sr} (red and black dots in Fig. 5). They are grouped hereafter in the IOPAG (In Ouzzal Pan-African granitoid) group I. The second trend displays a large variation along the *X* axis (Fig. 7) and corresponds to the samples forming the Rb–Sr isochron with the highest I_{Sr} (blue dots in Fig. 5). This group will be hereafter called the IOPAG group II. The enrichment in REE, Zr and Y (HFSE) cannot be

Table 4 Major and trace element concentrations for the studied In Ouzzal Pan-African granitoids

Type	Tihimatine North								Tihimatine South					
	Syenogr.	Monzgr.	Syenogr.	Syenogr.	Syenogr.	Syenogr.	Dyke	Enclave	Monzgr.	Alkali. gr.	Monzgr.	Granodi.	Dyke	Enclave
Sample	TH30	TH31	TH10	TH43	TH82	TH81	THR	TH34E	TH92	TH90	THZ1	STH1	THF9	THZE
SiO_2	72.33	72.53	73.64	73.48	73.72	74.03	71.11	69.17	73.25	73.50	69.59	71.46	65.53	64.53
TiO_2	0.29	0.14	0.13	0.12	0.10	0.11	0.26	0.25	0.13	0.14	0.22	0.15	0.36	0.46
Al_2O_3	13.24	13.86	13.00	14.03	13.42	13.81	15.14	15.17	13.84	12.78	14.25	15.23	17.03	16.18
Fe_2O_{3t}	2.45	1.89	1.85	1.59	1.40	1.65	2.63	2.84	1.67	1.99	2.23	2.01	4.11	4.28
MnO	0.05	0.05	0.05	0.05	0.06	0.05	0.08	0.14	0.06	0.06	0.05	0.05	0.10	0.15
MgO	0.18	0.25	0.08	0.20	0.23	0.21	0.52	0.97	0.30	0.28	1.93	0.51	0.78	1.93
CaO	1.35	1.25	1.15	1.03	0.85	0.89	2.20	1.89	1.09	1.49	1.93	2.23	3.97	3.49
Na_2O	4.22	4.56	4.13	4.62	4.64	4.62	4.24	5.80	4.56	4.38	4.73	4.65	4.56	4.82
K_2O	4.46	3.99	4.17	3.94	4.08	4.05	3.13	2.19	3.91	4.59	3.75	2.76	2.54	2.59
P_2O_5	0.04	0.02	0.01	0.01	0.01	0.01	0.07	0.05	0.01	0.04	0.05	0.05	0.20	0.12
P.F.	0.66	0.78	1.00	0.98	0.74	0.93	1.03	1.42	0.71	0.80	0.90	1.12	1.23	1.41
Total	99.27	99.32	99.20	100.05	99.25	100.37	100.40	99.87	99.52	100.05	99.64	100.22	100.39	99.95
V	13.09	9.10	7.12	5.11	2.40	5.90	14.25	27.59	6.56	12.67	17.70	4.21	1.74	49.08
Rb	118.8	96.4	91.9	93.8	95.3	92.4	84.5	69.1	85.7	97.7	78.6	46.7	40.1	73.8
Sr	90	116	110	96	65	99	306	116	123	176	247	457	517	429
Y	63.7	34.0	29.6	32.1	32.2	30.2	23.9	32.6	28.1	32.7	23.1	19.2	24.5	23.6
Zr	284	125	123	129	141	158	157	161	127	135	119	146	240	103
Nb	15.24	9.15	9.04	11.09	12.91	9.79	7.98	8.09	9.72	11.27	7.81	9.19	7.40	10.52
Ba	543	1260	1185	1259	1177	1246	1583	588	1349	1302	1449	1928	1604	1112

(continued)

Table 4 (continued)

Type	Tihimatine North								Tihimatine South					
	Syenogr.	Monzgr.	Syenogr.	Syenogr.	Syenogr.	Syenogr.	Dyke	Enclave	Monzgr.	Alkali. gr.	Monzgr.	Granodi.	Dyke	Enclave
Sample	TH30	TH31	TH10	TH43	TH82	TH81	THR	TH34E	TH92	TH90	THZ1	STH1	THF9	THZE
La	52.0	41.3	44.5	46.2	52.9	44.3	39.6	38.7	53.4	43.9	37.8	50.5	45.0	27.6
Ce	96.4	69.2	77.2	81.3	86.2	78.3	67.5	70.9	89.5	63.9	63.2	88.0	81.5	51.3
Pr	11.00	7.81	8.18	8.75	9.27	8.40	7.18	8.11	9.32	7.10	6.54	9.67	9.47	6.11
Nd	39.1	26.4	26.6	28.9	31.5	28.1	24.3	28.3	29.9	24.1	21.5	32.9	34.7	22.7
Sm	8.15	3.88	3.60	4.16	4.75	4.10	3.21	4.69	3.75	3.54	2.76	3.68	5.01	3.56
Eu	0.57	0.21	0.20	0.18	0.18	0.17	0.48	0.33	0.21	0.28	0.33	0.86	1.53	0.63
Gd	8.40	3.70	3.13	3.91	4.12	3.69	3.01	3.88	3.29	3.39	2.58	2.63	4.24	3.01
Dy	8.72	3.42	2.83	3.56	3.57	3.42	2.39	3.41	2.90	3.26	2.18	1.22	2.77	2.29
Ho	1.94	0.74	0.61	0.77	0.74	0.73	0.47	0.73	0.60	0.70	0.44	0.16	0.47	0.43
Er	5.35	2.11	1.67	2.21	2.15	2.09	1.34	2.03	1.74	2.13	1.27	0.38	1.03	1.23
Yb	5.31	2.26	1.89	2.39	2.32	2.41	1.19	2.20	1.89	2.31	1.21	0.29	0.71	1.09
Lu	0.74	0.32	0.26	0.35	0.34	0.34	0.19	0.30	0.26	0.32	0.19	0.04	0.08	0.16
Hf	8.98	4.05	3.76	4.36	4.69	4.95	4.36	4.54	4.01	4.53	3.68	4.03	5.74	3.18
Ta	1.21	0.74	0.71	0.94	0.99	0.85	0.77	0.65	0.77	1.02	0.74	0.57	0.50	0.74
W	0.76	0.50	0.34	0.75	1.13	0.87	0.96	1.51	0.48	1.00	0.69	0.76	0.61	2.16
Pb	28.4	14.1	15.9	17.7	17.7	19.6	11.2	14.0	15.1	18.0	15.1	22.6	15.0	7.8
Th	11.78	11.43	10.29	13.63	12.98	12.98	11.30	10.40	12.01	12.23	12.28	7.53	4.84	10.52
U	1.46	2.06	2.02	2.06	2.23	2.59	2.12	1.97	2.12	2.46	2.48	1.03	0.75	2.73

(continued)

Table 4 (continued)

In Eher		Ihaouhouene		In Hihaou		Tin Chik-Chik		
Alkali. gr.	Syenogr.	Monzgr.	Alkali. gr.	Syenogr.	Alkali. gr.	Alkali. gr.	Alkali. gr.	Enclave
IE12E	IE11	HH32	HH10	IH10	IH6	T21	T27	T22E
71.17	74.34	72.10	75.57	72.70	74.71	74.38	75.71	66.28
0.38	0.12	0.37	0.11	0.24	0.17	0.24	0.15	0.75
13.98	13.55	13.92	12.66	13.16	13.07	12.23	12.25	12.82
2.92	1.55	2.97	1.61	2.10	1.65	2.69	1.98	5.74
0.05	0.05	0.06	0.03	0.05	0.04	0.05	0.04	0.12
0.24	0.20	0.24	0.31	0.19	0.06	0.01	0.06	0.84
1.21	0.86	2.25	0.79	1.26	0.83	0.76	0.58	2.70
4.80	4.51	3.61	3.87	3.68	3.65	3.61	3.38	3.83
4.63	4.26	4.12	4.85	4.72	5.27	5.06	4.78	5.55
0.05	0.01	0.08	0.01	0.02	0.02	0.01	0.01	0.14
1.06	0.91	0.97	0.65	1.02	0.71	1.21	1.24	1.37
100.48	100.35	100.68	100.46	99.13	100.18	100.24	100.18	100.13
11.10	5.79	18.58	3.24	11.57	6.20	2.01	−0.30	42.38
102.6	105.1	75.5	100.0	164.5	168.8	80.2	90.1	113.2
56	105	212	46	115	65	32	25	55
52.7	35.4	34.0	76.0	54.7	52.6	74.7	95.4	70.9
271	136	195	203	233	172	413	309	346
10.79	10.46	8.49	16.01	13.86	11.35	18.04	20.22	17.48
437	1242	1481	493	971	599	291	159	529
76.5	45.5	45.0	62.7	67.2	39.2	112.3	86.5	66.3
136.9	79.2	79.1	103.8	116.8	71.9	210.0	163.6	127.3
14.82	8.58	8.76	12.52	12.50	8.01	24.40	19.62	15.43
49.8	27.8	30.8	45.9	41.3	27.1	89.0	72.8	57.0
8.46	3.84	4.55	9.43	6.69	5.08	15.67	15.40	11.30
0.48	0.19	1.06	0.58	0.73	0.50	1.02	0.45	1.61
7.55	3.44	4.36	9.59	6.47	5.10	14.13	15.81	10.87

(continued)

Table 4 (continued)

In Eher		Ihaouhouene		In Hihaou		Tin Chik-Chik		
Alkali. gr.	Syenogr.	Monzgr.	Alkali. gr.	Syenogr.	Alkali. gr.	Alkali. gr.	Alkali. gr.	Enclave
IE12E	IE11	HH32	HH10	IH10	IH6	T21	T27	T22E
6.77	3.19	3.83	10.00	6.29	5.68	11.67	14.77	10.12
1.49	0.67	0.81	2.25	1.39	1.31	2.56	3.26	2.29
3.96	1.94	2.26	6.27	3.85	3.93	6.60	8.56	5.96
3.93	2.25	2.26	6.30	4.14	4.27	6.34	7.67	5.77
0.57	0.33	0.32	0.88	0.60	0.61	0.96	1.14	0.91
7.09	3.98	5.06	7.21	6.04	5.28	11.23	9.81	9.09
0.79	0.85	0.75	1.16	1.18	1.22	1.10	1.33	1.26
0.63	0.73	0.50	0.75	0.44	1.09	1.25	0.93	2.35
27.5	24.7	20.2	25.6	25.9	25.4	19.3	40.2	30.7
9.91	11.64	5.16	10.60	12.87	18.52	12.09	13.00	9.71
1.24	2.25	0.60	1.37	2.08	2.91	1.24	1.49	1.49

Oued Ihaouhouene			Lozher							
Granodi.	Enclave		Monzgr.	Monzgr.	Monzgr.	Monzgr.	Dyke	Enclave		
C1G11	C1E3	C1G12E	AT108	AT107	AT116	AT111	AT110	AT105	AT106	AT112
63.69	61.46	55.54	68.22	69.94	70.80	71.44	77.70	66.91	57.64	47.79
0.65	1.34	0.91	0.38	0.36	0.30	0.24	0.04	0.61	0.77	1.48
15.60	14.80	16.64	15.62	15.40	15.10	14.08	12.81	16.30	17.32	15.90
5.58	8.73	9.17	4.06	3.03	2.43	1.95	0.55	4.51	7.54	10.48
0.10	0.13	0.21	0.07	0.07	0.06	0.04	0.02	0.09	0.19	0.21
1.85	1.89	3.31	0.97	0.95	0.93	0.50	0.06	1.48	2.78	6.41
4.42	4.33	6.43	2.57	2.59	2.36	1.69	0.82	3.59	5.41	8.73
3.63	3.48	4.19	3.78	3.92	3.71	3.67	3.83	4.75	4.33	4.27
3.21	2.10	2.06	3.88	3.72	4.28	4.47	4.30	1.70	2.85	0.91
0.26	0.70	0.34	0.17	0.17	0.14	0.06	0.01	0.21	0.43	0.50
1.60	1.34	1.48	0.58	0.59	0.78	0.68	0.21	0.67	0.83	1.82
100.57	100.30	100.27	100.29	100.73	100.89	98.81	100.35	100.83	100.10	98.50

(continued)

Table 4 (continued)

In Eher			Ihaouhouene		In Hihaou		Tin Chik-Chik			
Alkali. gr.		Syenogr.	Monzgr.	Alkali. gr.	Syenogr.	Alkali. gr.	Alkali. gr.		Alkali. gr.	Enclave
IE12E		IE11	HH32	HH10	IH10	IH6	T21		T27	T22E
56.52	85.19	101.11								
63.9	70.4	55.3	115.1	114.1	142.6	122.4	185.5	107.2	125.1	23.1
331	257	303	284	271	270	224	37	232	348	864
35.7	41.0	65.4	18.0	21.1	13.4	15.6	25.2	20.2	24.3	23.4
202	286	219	162	154	152	107	42	140	130	151
10.15	16.76	15.84	47.18	17.64	10.14	10.08	20.82	19.13	18.98	14.39
1405	853	851	1047	925	924	1136	36	383	580	3410
40.1	61.1	26.3	44.6	34.3	50.0	48.7	3.6	35.0	38.8	46.6
70.8	108.1	69.4	80.4	64.1	85.7	85.7	11.3	62.1	75.6	96.9
7.86	11.63	10.17	8.53	6.91	8.66	8.79	1.58	6.88	9.21	12.05
29.0	43.0	45.8	29.3	24.5	28.3	29.5	7.3	23.1	33.6	45.1
4.85	7.17	10.69	4.78	4.37	4.35	4.59	2.60	4.44	6.42	7.87
1.19	1.11	1.71	1.02	0.89	0.98	0.98	0.38	0.88	1.72	2.53
4.72	6.48	10.39	4.15	4.09	3.63	3.92	3.04	4.18	5.64	6.87
4.20	5.48	9.94	3.17	3.45	2.64	2.94	4.17	3.52	4.43	4.82
0.93	1.16	2.25	0.68	0.79	0.56	0.63	1.06	0.69	0.87	0.87
2.58	3.01	6.20	1.91	2.27	1.54	1.81	3.43	2.10	2.56	2.32
2.61	2.97	6.08	1.89	2.85	1.66	1.81	4.70	2.07	2.56	1.97
0.40	0.45	0.93	0.31	0.48	0.27	0.28	0.79	0.32	0.40	0.29
6.09	8.41	7.01	4.51	5.06	4.80	3.59	4.08	3.84	3.84	3.70
0.66	0.93	1.07	1.03	3.97	0.83	0.90	5.25	1.42	1.42	0.77
0.33	0.54	0.52	0.37	0.63	0.21	0.21	0.48	0.48	0.18	0.32
17.0	11.9	12.6	17.0	16.1	20.4	26.7	29.6	14.7	17.5	15.1
6.38	7.99	5.47	7.25	6.87	9.75	6.43	18.18	4.29	5.16	2.61
1.45	1.29	1.44	1.03	3.87	2.02	1.23	17.81	0.67	1.55	0.94

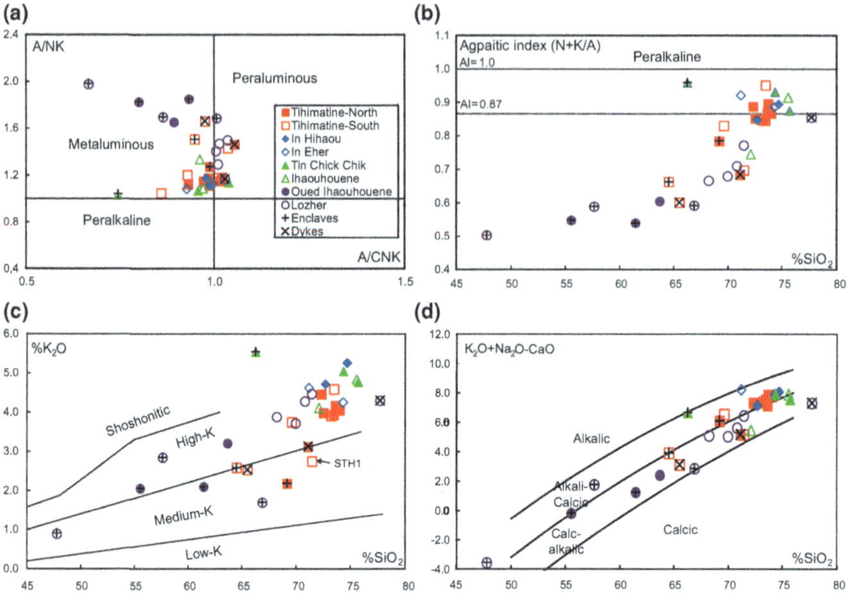

Fig. 6 Major element diagrams for the in Ouzzal Pan-African granitoids: **a** Inverse agpaitic index A/NK (molar Al/(Na + K)) versus peraluminous index A/CNK (molar Al/(Na + K+2Ca)). **b** Agpaitic index (molar (Na + K)/Al) versus wt% SiO_2. The line at 0.87 separates the alkaline granitoids (above) from the sub-alkaline granitoids (Liégeois and Black 1987). **c** wt% K_2O versus wt% SiO_2. **d** MALI (Modified Alkali Lime Index) diagram (% Na_2O + K_2O–CaO vs. % SiO_2) from Frost et al. 2001). In these diagrams, each pluton is represented by a given symbol. When it is an enclave or a dyke, a cross is superimposed

linked directly to the Rb–Sr isotopic systematics and so points to a difference in the source rather than to a magmatic process. Most of the IOPAG group I samples display values lower than the alkali-calcic Yenchichi-Telabit series, the reference (coordinates 1, 1) indicating that these granitoids are not so enriched. In this group, there are several really enriched enclaves, the fine-grained granitic dyke AT110 and the porphyritic granite AT107. Representative gneisses from the In Ouzzal terrane are reported in Fig. 7, with four groups of orthogneisses: TTG-type, granodioritic CA (calc-alkaline)-type, monzogranitic CA-type and ALK (alkaline)-type (Peucat et al. 1996). They are all relatively depleted in LILE, which can be attributed to the granulitic metamorphism (Peucat et al. 1996). The ALK gneisses are in the same range as the IOPAG group II.

Two REE main patterns, with two outlier groups can be defined. Group 1 corresponds to most of the IOPAG group I (17 samples on 25, the eight remaining forming the two outlier groups) and share similar REE patterns but with variable Eu negative anomaly (Fig. 8a). It can be divided in a 1a and 1b group. Group 1a includes the South and North Tihimatine plutons and the In Eher IE11 sample and is characterized by deep Eu anomalies. Group 1b includes the Lozher pluton and

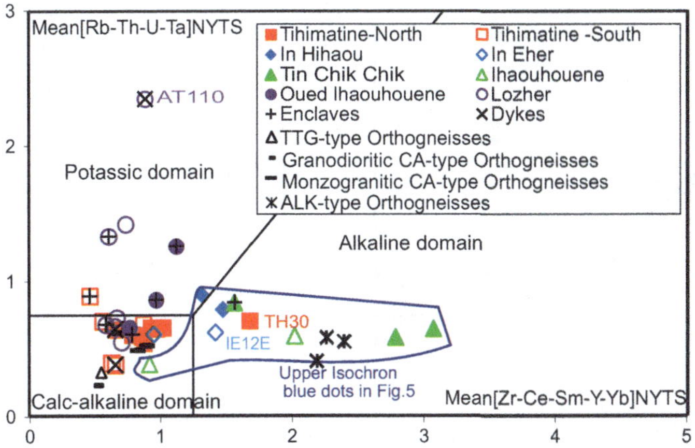

Fig. 7 Sliding mean [Zr–Ce–Sm–Y–Yb] versus mean [Rb–Th–U–Ta] normalized values to the Yenchichi-Telabit reference series (Liégeois et al. 1998) for the In Ouzzal Pan-African granitoids

Oued Ihaouhouene sample C1E3. Group 2 corresponds to all IOPAG group 2 (In Hihaou, Ihaouhouene, Tin Chik-Chik, + North Tihimatine TH30 and In Eher IE12E samples; Fig. 8b). Outliers 1 group comprises samples from South Tihimatine, Lozher and Oued Ihaouhouene plutons (Fig. 8c) and outliers 2 group comprises samples from Lozher and Oued Ihaouhouene plutons (Fig. 8d).

Group 1a is constituted by 12 samples, mostly granitic (SiO_2 from 69 to 74 wt% and one sample at 64.5 wt%). These samples have ΣREE (Fig. 9a) in the range 120–200 ppm, moderately fractionated REE patterns (La_N/Lu_N: 13–21), with most of the fractionation occurring in the LREE (La_N/Sm_N: 4.9–9), the HREE being rather flat (Gd_N/Lu_N: 1.3–2.4). Eu anomalies are variable, but always highly negative (Eu/Eu*: 0.12–0.58). *Group 1b* is made of five samples of intermediate to acid composition (61–71 wt% SiO_2). Samples from this group resemble those from Group 1a but with less pronounced negative Eu anomalies. Sample C1E3 is more enriched (Fig. 8a). Overall, these samples are characterized by ΣREE in the range 145–250 ppm, La_N/Lu_N: 11–19, La_N/Sm_N: 4.9–7.2, Gd_N/Lu_N: 1.6–1.8, and Eu/Eu*: 0.49–0.74. The richest Group 1 sample (C1E3) is close to the Iforas HKCA mean pattern, both in the shape and abundance. The Iforas ALK pattern is strongly different (Fig. 8a)

Compared to Group 1, *Group 2* (nine samples; Fig. 8c) is granitic (72–76 wt% SiO_2 and one sample at 66 wt%). It is, as a mean, richer in REE (ΣREE in the range 170–500 ppm), moderately but less fractionated (La_N/Lu_N: 6.7–14.5), most of the fractionation occurring also in the LREE but with lower values (La_N/Sm_N: 3.5–6.3), the HREE being even flatter (Gd_N/Lu_N: 1.05–1.83). Eu anomalies are highly variable, moderately to highly negative in these granitoids (Eu/Eu*: 0.09–0.72), in a similar way as in group 1 (Fig. 8a). The Group 2 REE patterns are similar to the ALK Iforas pattern, especially for HREE in the richest samples. The Iforas LREE are less fractionated. As a whole, the Iforas pattern appears to be more affected by

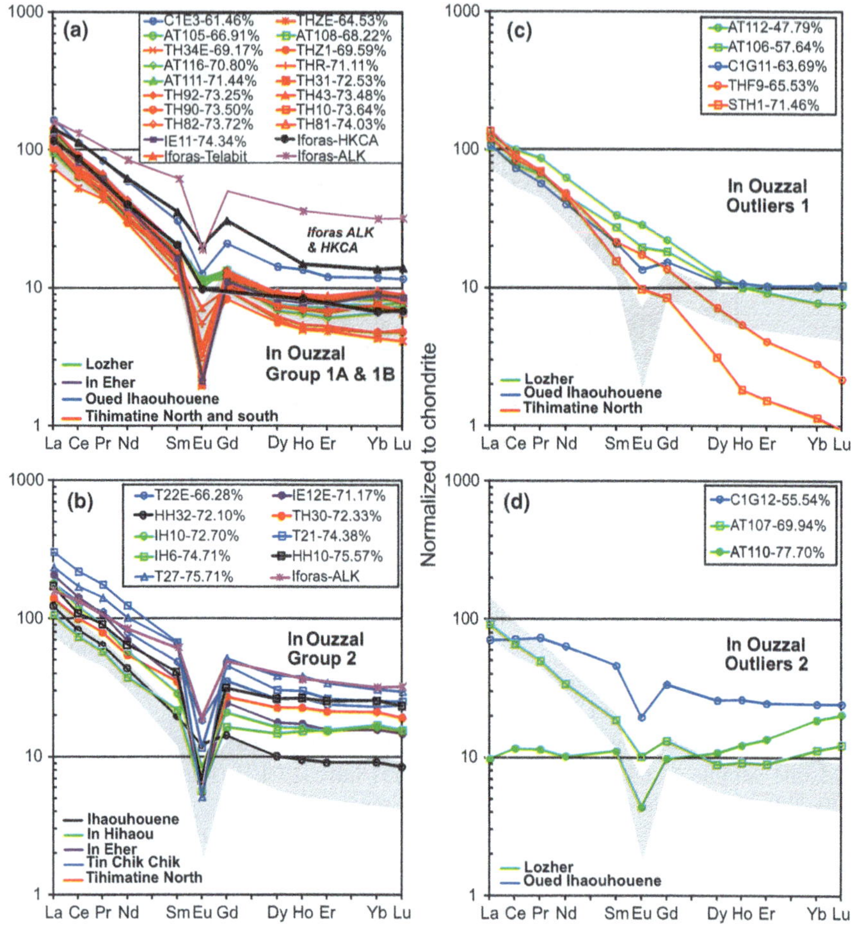

Fig. 8 Chondrite-normalized REE (rare earth elements) diagram for the In Ouzzal Pan-African granitoids. Chondrite values from Sun and McDonough (1989)

late fluids rich in F–Cl (seagull pattern) than those from In Ouzzal Group 2 but they can be considered as fairly similar.

The outlier 1 group (five samples, Fig. 8c) is highly variable in silica (48–71 wt%). It is characterized by weak Eu anomalies (Eu/Eu*: 0.75–1.03), similar LREE (Ce: 74–101 ppm) but different HREE (Yb: 9–31 ppm). This leads to close ΣREE (169–228 ppm), large REE fractionation (La_N/Lu_N: 10–145), with larger HREE fractionation (Gd_N/Lu_N: 1.46–9.08) than LREE fractionation (La_N/Sm_N: 3.8–8.6). Samples STH1 (two-mica granite) and THF9 (South Tihimatine, 71 and 66 wt% SiO_2) display TTG-like spectra (high La_N/Lu_N = 145 and 56 that can be attributed to low HREE, Lu_N = 0.9 and 2.2) and slight Eu negative anomalies (Eu/Eu* = 0.81–0.99).

The outlier 2 group (three samples; Fig. 8d) is also highly variable in silica (56, 70, and 78 wt%). These samples are characterized by lower REE fractionation (La_N/Lu_N: 0.5–7.5), especially in HREE (Gd_N/Lu_N: 0.5–1.4) and moderate negative Eu anomalies (Eu/Eu*: 0.42–0.64). In this group, the granitic dyke AT110 (78 wt% SiO_2) is particularly low in REE (ΣREE = 44 ppm), and displays an apparent flat pattern, actually enriched in HREE (La_N/Lu_N = 0.48) with a marked negative Eu anomaly (Eu/Eu* = 0.42).

This means that, globally, REE do not show a positive correlation with SiO_2 (Fig. 9a) as it should be if considering the usual incompatible behavior of REE. The REE content is in the range 150–200 ppm from 47 to 75 wt% SiO_2, except in Group 2, where REE are higher but still without correlation with silica (Fig. 9a). This suggests different sources and mixing rather than magmatic differentiation. The enrichment in REE linked to the alkaline character of Group 2 is shown by the agpaitic index (Fig. 9b), as well as by the NYTS parameter (Fig. 9c).

In the MORB-normalized diagrams (Fig. 10), elements are ordered with the LILE to the left and the HFSE to the right, with increasing incompatibility towards the center of the diagram, for both element families. Group 1a (South and North Tihimatine; Fig. 10a) displays rather homogeneous patterns with greater variations in P and Ti due to differentiation of apatite and titanite. When compared to Iforas high-K calc-alkaline batholiths and alkali-calcic Telabit dyke swarm (Liégeois et al. 1998; to the SW of In Ouzzal, Fig. 1), Group 1a displays a similar pattern, except for P and Ti, but is less enriched except in Sr, K, Ba and Y–Yb. Group 1b (mostly Lozher; Fig. 10a) is similar to Group 1a but poorer in Ba, Th, Y, Yb, and less-depleted in P and Ti. The Iforas alkali group (Liégeois et al. 1998) is strongly enriched in HFSE.

In contrast, IOPAG group 2 displays a pattern similar to the Iforas alkali group (Fig. 10b), confirming its distinct, more alkaline character.

In the IOPAG outliers 1 group (Fig. 10c), the TTG-like samples (THF9 and STH1) have a pattern quite like Group 1a, except for Sr and Ba, higher, and Yb, much lower. The other members of this group display generally higher element concentrations, although in a variable and different way than the alkali Group 2 (Fig. 10c).

Two samples of IOPAG outliers 2 group (C1G12 and AT107) display similar patterns as Group 1a, but with higher values, especially C1G12 enclave (Fig. 10d). The dyke AT110 represents a highly differentiated granitic liquid. It yields high K, Rb, Th, pointing to poor K-feldspar effect, and low Sr, Ba reflecting plagioclase fractionation. Marked troughs in P, Ti, and Zr are related to apatite, titanite, and zircon fractionation respectively. Moreover, interaction with late fluids is reflected by low Ta and Ce, high Yb, and seagull-like REE patterns (Figs. 8d and 10d). The two other samples, especially AT107 share most of these characteristics, indicating that interaction with fluids is a main trait of this group.

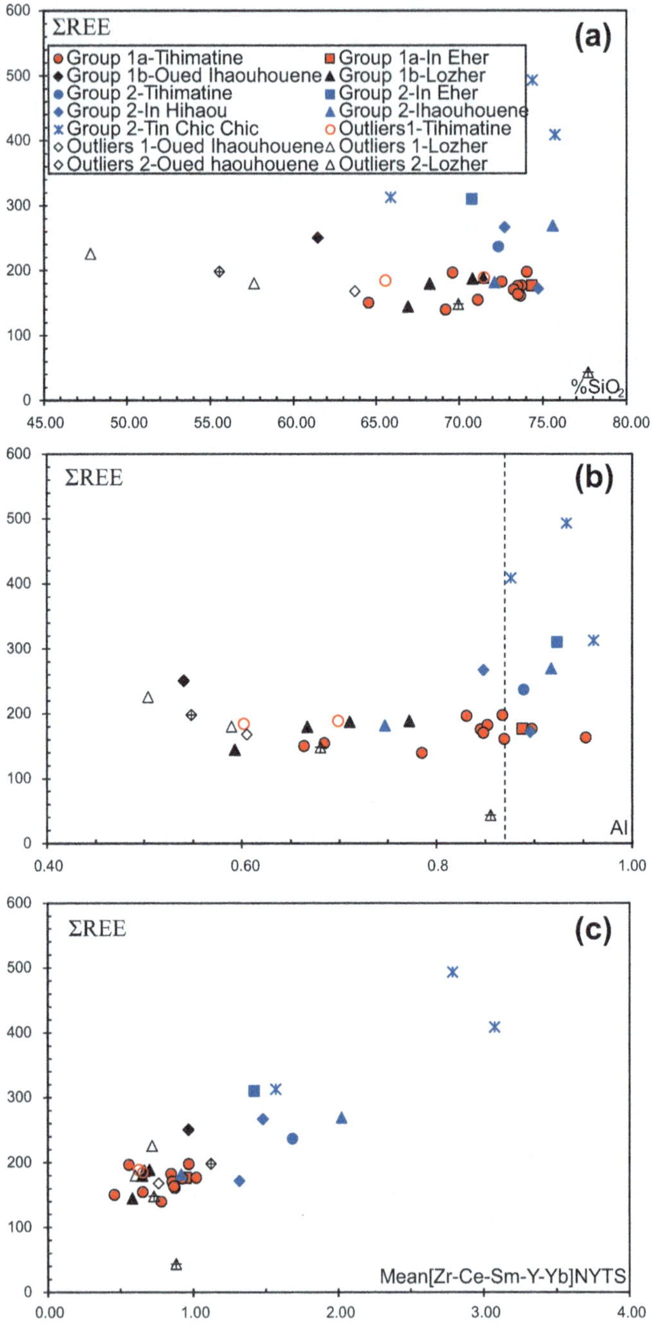

Fig. 9 Diagrams ΣREE versus **a** wt% SiO_2, **b** AI, agpaitic index (molar (Na + K)/Al), **c** mean [Zr–Ce–Sm–Y–Yb] normalized values to the Yenchichi-Telabit reference series (NYTS; Liégeois et al. 1998)

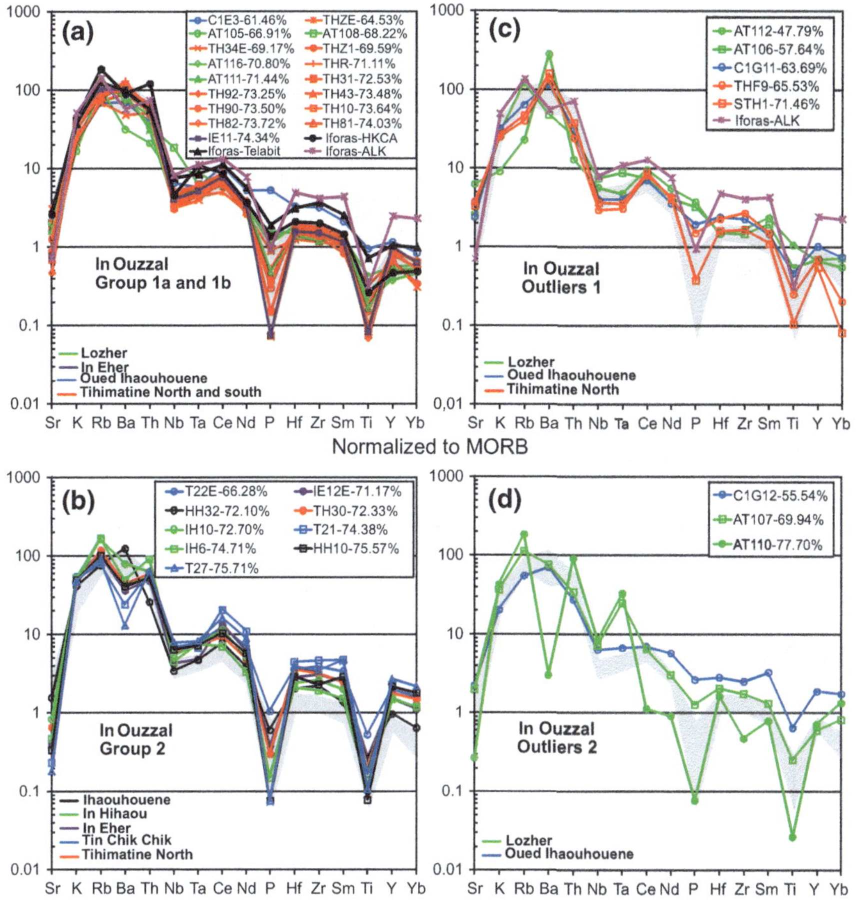

Fig. 10 MORB-normalized incompatible element spidergrams for the In Ouzzal Pan-African granitoids. MORB values from Sun and McDonough (1989). The Iforas HKCA (high-K calc-alkaline) and Iforas ALK (alkaline–peralkaline) are mean values of Adma and Aoukenek plutons and of N–S dykes and Kidal ring complex, respectively (Liégeois et al. 1998)

4.2 Sr and Nd Isotopes

The geochemical variability of the In Ouzzal Pan-African granites is mirrored by the Sr and Nd isotopic ratios. Recalculated at 601 Ma, ε_{Nd} range from −30 to −4 and I_{Sr} from 0.704 to 0.713. In consequence, NdT_{DM} model ages vary from 1440 to 3290 Ma (Fig. 11a, b). However, all these values are intermediate between the juvenile mantle source represented by the nearby and sub-contemporaneous (592 Ma) Tin Zebane dyke swarm (ε_{Nd} = + 6.2, I_{Sr} = 0.7028; Hadj Kaddour et al. 1998; Fig. 2a) and the various In Ouzzal Archean lithologies, especially those depleted in Rb due to the c. 2 Ga granulitic metamorphism (Peucat et al. 1996;

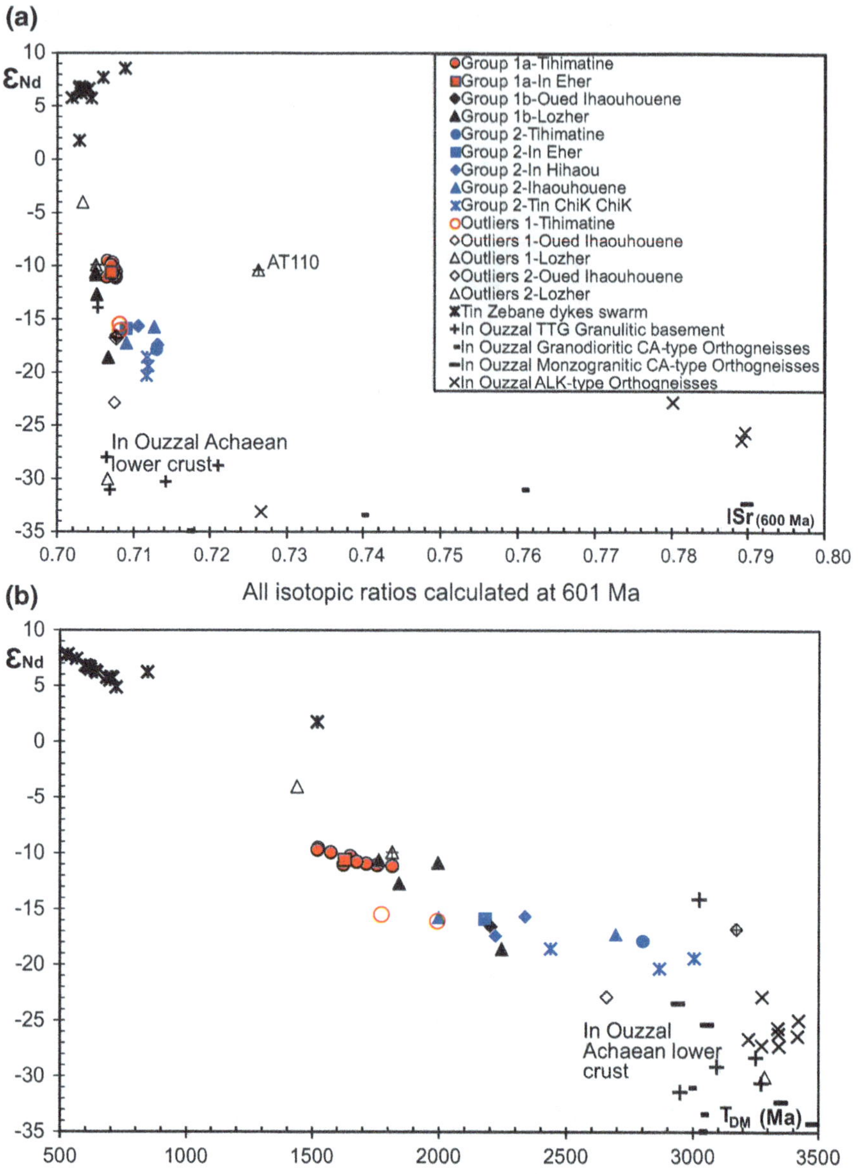

Fig. 11 **a** Epsilon Nd versus initial $^{87}Sr/^{86}Sr$ isotopic ratios, calculated at 601 Ma, the U–Pb zircon age of the Tihimatine pluton. **b** Epsilon Nd calculated at 601 Ma, the U–Pb zircon age of the Tihimatine pluton, versus Nd T_{DM} model ages, calculated following Nelson and DePaolo (1985). Tin Zebane dyke compositions are from Hadj Kaddour et al. (1998) and those of In Ouzzal basement gneiss samples are from Peucat et al. (1996)

Fig. 11a, b). Moreover, the alkali Group 2, whatever the pluton, presents higher I_{Sr} (Fig. 11a, already determined by their Rb–Sr isochron, Fig. 5) and older NdT_{DM} model ages (Fig. 11b). The NdT_{DM} model ages (1.4–1.8 Ga for Group 1a, 1.7–2.2 Ga for Group 1b, 2–3 Ga for Group 2) indicate an important participation of an Archean component, namely the In Ouzzal lithosphere, which displays 2.9–3.5 Ga NdT_{DM} model ages (Fig. 11b). The AT110 dyke, affected by late fluids, presents a much higher I_{Sr}.

The Sr and Nd isotopic composition of the In Ouzzal Pan-African granitoids (IOPAG) can be explained by the variability of the old In Ouzzal crust and by the variable proportion of the mantle and crustal melts in the mixing process that generated the IOPAG. This mixing can be modeled taking the G533 Tin Zebane gabbro (Hadj Kaddour et al. 1998) as the mantle pole and four In Ouzzal gneisses, representative of the main In Ouzzal lithologies (Peucat et al. 1996): samples Inz87 (TTG gneiss), Inz15 (granodioritic calc-alkaline gneiss), Inz41 (monzogranitic calc-alkaline gneiss) and Inz70 (alkaline orthogneiss).

Taking into account the isotopic compositions and Nd and Sr concentrations of these poles (Fig. 12), the Group 1a (Tihimatine) can be formed through the mixing of 75–80% of the mantle pole and 20–25% of the TTG-type gneiss, which is, by far, the most abundant lithology of the In Ouzzal terrane. If the granodioritic calc-alkaline-type gneiss is chosen as the crustal pole, 12–18% of this end-member is enough. Group 1b comprises more enriched samples and the TTG pole varies

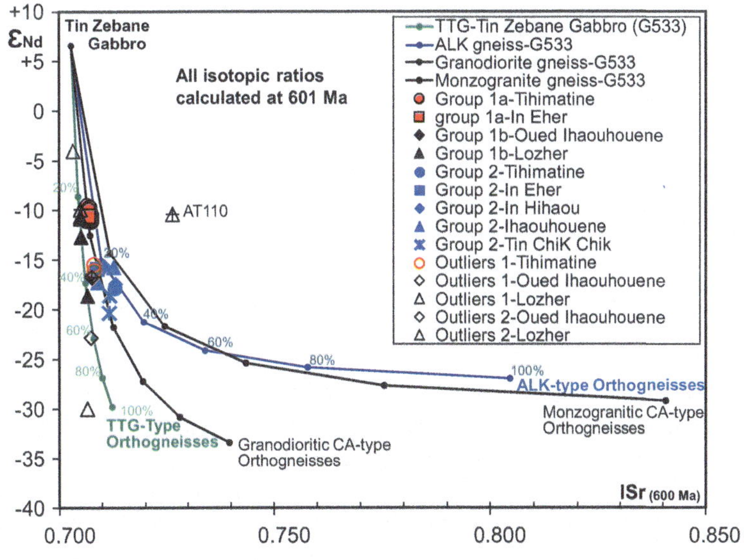

Fig. 12 Epsilon Nd versus initial $^{87}Sr/^{86}Sr$ isotopic ratios, calculated at 601 Ma and mixing curves between the Tin Zebane dyke swarm (Hadj Kaddour et al. 1998), representing the Pan-African asthenospheric mantle, and different gneisses from the In Ouzzal basement (Archean protoliths affected by an UHT Eburnean metamorphim, Peucat et al. 1996)

from 20 to 60%. Several outlier group samples are located on the same curve but with a large spread. The mafic enclave AT112 (48 wt% SiO_2) is close to the TTG-gneiss pole and the intermediate enclave AT106 (56 wt% SiO_2) needs only 15% of this TTG pole. These two enclaves belong to the Lozher pluton, located close to the East Ouzzalian shear zone that probably facilitated both the mantle melt to rise and blocks from the crust to be ripped off.

Group 2, alkali in composition and higher in I_{Sr}, cannot be modeled with the TTG pole. Rocks from this group need either the granodioritic CA-type, the monzogranitic-type or the alkalic-type pole as the crustal end-member. These three poles are much less important in volume but are more fusible than the TTG gneisses. This means that if one of these lithologies is available, it can become the preponderant end-member during crustal melting. Calculation shows that the monzogranitic CA-type pole is less suitable than the two other poles. Indeed, the I_{Sr} of the studied samples are too low when considering their ε_{Nd} (Fig. 12). Tin Chik-Chik pluton is closer to the mixing curve with the granodioritic CA-type pole, while Tihimatine TH30 and other samples are closer to the curve with the ALK-type pole (Fig. 12). Both poles could also participate together. Between 25 and 40% of one of these two poles is needed for Group 2. Considering the geochemistry of rocks from Group 2 and their clear difference with Group 1, we favor a mixing with the alkali gneisses.

5 Interpretation: Metacratonization of the Rigid Archean/Eburnean In Ouzzal Terrane Within the Tuareg Pan-African Orogen

The age of emplacement of the Tihimatine pluton (600 ± 5 Ma—LA-ICP-MS and 602 ± 4 Ma—SHRIMP) is in the range of the large high-K calc-alkaline granitic batholiths emplaced in the neighboring terranes (630–580 Ma; Caby et al. 1985; Bertrand et al. 1986; Bosch et al. 2016, when considering only U–Pb zircon ages). The emplacement of these batholiths was accompanied by high-temperature metamorphism and ductile deformation (Gravelle 1969; Bertrand and Davison 1981; Caby et al. 1985; Bertrand et al. 1986; Liégeois et al. 1987; Boissonnas 2008; Bosch et al. 2016). Tihimatine and the other In Ouzzal sub-circular plutons (Fig. 2) emplaced at shallow depth, in a 2 Ga old granulitic basement only affected by brittle tectonics, even if actually quite abundant (Deroin et al. 2014) and sparse greenschist facies metamorphism. This set apart, it is worth noting that the 2 Ga old basement preserved most of its Eburnean HT metamorphic paragenesis (Ouzegane et al. 2003). High-level sub-circular plutons emplaced later at ca. 580 Ma in the Tuareg shield outside the In Ouzzal terrane (Abdallah et al. 2007 and references therein). The 630–580 Ma period corresponds to the general tectonic escape of the Tuareg shield towards the north, in response to the collision with the large West African craton (Black et al. 1994; Liégeois et al. 2003). In that context, the In

Ouzzal terrane moved faster than adjacent terranes. The East Ouzzalian shear zone displays dextral strike-slip movements, while the West-Ouzzalian shear zone was sinistral (Fig. 2b; Caby 1970; Caby and Monie 2003). These shear zones are sub-vertical at least on a depth of 20 km; the situation in the lower crust is less clear but it seems they reach the Moho (Bouzid et al. 2008). In the LATEA metacraton, in Central Hoggar, major shear zones are sub-vertical at least to a depth of 100 km, i.e., they can be imaged in the lithospheric mantle (Bouzid et al. 2015). All these features suggest a rigid behavior of the In Ouzzal terrane during the Pan-African orogeny, i.e., a cratonic behavior implying a thick lithospheric mantle (Liégeois et al. 2013), in contrast to adjacent terranes but similarly to the West African craton (Fig. 13). A similar example of this would be the Tarim craton, which experienced little deformation during the Indo-Eurasian collision due to its thick lithosphere, in contrast to the widespread deformation experienced by the surrounding more ductile terranes (Deng et al. 2017). In this configuration, the Indian cratonic indentor is the equivalent of the West African craton. No event has been recorded in the In Ouzzal terrane between 1.8 and 0.6 Ga, leaving plenty of time to acquire a thick lithospheric mantle simply by cooling.

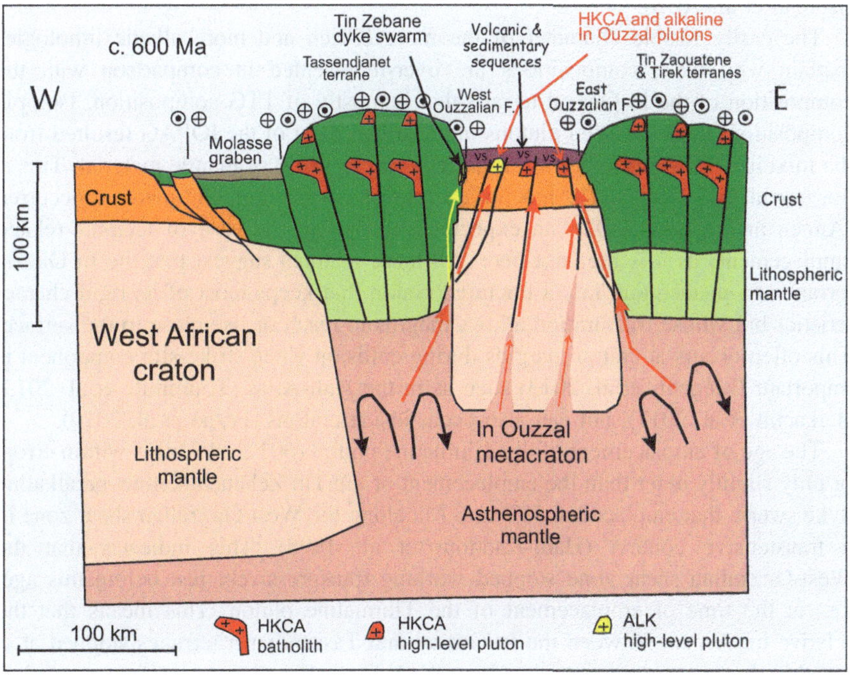

Fig. 13 Cross-section of western Hoggar at 600 Ma based on this study. See text for explanation

The studied In Ouzzal plutons are geochemically heterogeneous. They belong to two main trends, one being (high-K) calc-alkaline and one more restricted alkaline trend (the latter mainly as enclaves). However, not all have the geochemistry of TTG or show interaction with Cl–F fluids. Their Sr–Nd isotopic signatures are strongly enriched and variable. They imply a mixing process between a depleted mantle (represented by the nearby and sub-contemporaneous alkaline/peralkaline bimodal Tin Zebane dyke swarm; Hadj-Kaddour et al. 1998; Ait-Djafer et al. 2003) and at least three old crustal poles, with TTG, calc-alkaline and alkaline compositions, all being known in the In Ouzzal Archean/Eburnean basement. This allows proposing that the studied plutons were generated through the local melting of the In Ouzzal crust when hot mantle material rose up along the large shear zones bordering the In Ouzzal terrane, but also along the faults inside the terrane. Some of these faults are also of crustal scale and probably represent Eburnean structures separating different domains (Bouzid et al. 2008, 2015). This implies the existence of planar removal of the lithospheric mantle along these faults, allowing the asthenospheric mantle to rise (Fig. 13; Liégeois et al. 2003, 2005, 2013; Vander Auwera et al. 2003; Shang et al. 2007; Zhang et al. 2011; Ngalamo et al. 2017). Such structures have been delineated in LATEA metacraton, Central Hoggar (currently reactivated by the Europe–Africa collision generating volcanism, Liégeois et al. 2005) through electrical constraints from magnetotelluric data (Bouzid et al. 2015).

The easily fusible character of the more K-rich and more alkalic lithologies explain why these compositions are over-represented in comparison with the composition of the In Ouzzal terrane that is mostly of TTG composition. Isotopic compositions of the studied plutons indicate that most of the IOPAG resulted from the mixing of 20–40% old crustal material and 60–80% of mantle material. This is not unrealistic, especially if repetitive intrusion of mantle mafic material occurred (Annen and Sparks 2002), an expected situation in the case of tectonic-related emplacements as it is the case here. All these features suggest that the In Ouzzal terrane is a metacraton, i.e., a fractured craton that keeps most of its rigid characteristics but whose fracturation allows magma to reach or rise close to the surface. This often occurs at craton margins during collision when strike-slip component is important (Liégeois et al. 2013) such as in the Anti-Atlas (Toummite et al. 2013; Belkacim et al. 2017), but can also occur inside cratons (Fezaa et al. 2010).

The age of emplacement of the Tihimatine pluton (601 $\pm$ 4 Ma) is within errors or only slightly older than the emplacement of the Tin Zebane alkaline–peralkaline dyke swarm that emplaced at 592 $\pm$ 6 Ma along the West-Ouzzalian shear zone in a transtensive context (Hadj-Kaddour et al. 1998). This indicates that the West-Ouzzalian shear zone stopped working transpressively just before this age, i.e., at the time of emplacement of the Tihimatine pluton. This means that the relative movement between the In Ouzzal and Tassendjanet terranes stopped at c. 600 Ma. In contrast, movements along the E-Ouzzalian shear zone are recorded as late as 580 Ma (Ferkous and Monie 2002). We suggest that the absence of movement along its western boundary, whereas movements along its eastern boundary were still ongoing, induced a stress inside the In Ouzzal terrane that generated brittle fracturation. This in turn produced planar delamination, asthenospheric uprise, local partial melting of the In Ouzzal crust at depth and emplacement of the In Ouzzal Pan-African granitoids (Fig. 13). Only additional precise ages could determine how long these granitoids intruded the In Ouzzal terrane, but the 600–580 Ma age range appears likely. As demonstrated in this study, the difference of the alkalic versus calc-alkaline In Ouzzal plutons is related to different crustal sources, thus a difference in age between the two groups is possible but is not needed.

The metacratonic character of the In Ouzzal terrane, due to a thick lithospheric mantle, imposes that it was not thickened during the Pan-African period, remaining a low land during that period, as is the case currently for the cratonic Tarim basin north of Tibet and south of the Tien Shan range (Deng et al. 2017). The studied granitoid plutons, however, need some hundred meters of cover above them. This can be easily obtained with contemporaneous volcanic products, still observed at In Hihaou (Fig. 2b), and/or with sediments originating from the highlands encircling the In Ouzzal terrane (Fig. 13). This soft cover was removed probably during the Cenozoic uplift of the Tuareg shield, currently a swell with its Precambrian basement culminating at 2400 m with a mean elevation of 1200 m (Rougier et al. 2013), leaving the studied plutons as inselbergs (Fig. 3a).

6 Conclusion

- The In Ouzzal terrane is a ca. 2 Ga granulitic terrane with Archean protoliths bounded by two major structures, the East- and West-Ouzzalian shear zones, only slightly affected by the Pan-African orogeny.
- A series of Pan-African high-level sub-circular plutons intrude In Ouzzal; one of them, the Tihimatine pluton, is dated here at 601 ± 4 Ma by two independent methods (600 ± 5 Ma by LA-ICP-MS U–Pb zircon and 602 ± 4 Ma by SHRIMP). This age indicates that the In Ouzzal plutons emplaced under brittle conditions in the heart of the terrane undergoing regional tectonic escape, while elsewhere in the shield large batholiths emplaced under ductile conditions, prior to other Tuareg high-level plutons were emplaced (Abdallah et al. 2007; Paquette et al. 1998; Azzouni-Sekkal et al. 2003 and references therein).
- Major and trace element geochemistry and Sr–Nd isotopes indicate that the In Ouzzal granitoids resulted from the mixing of 60–80% mantle material represented by the mantle-origin Tin Zebane dyke swarm (Hadj Kaddour et al. 1998), with 20–40% of the heterogeneous and old 2 Ga In Ouzzal crust. The more fusible calc-alkaline and alkaline lithologies are over-represented as crustal end-members, whereas the dominant In Ouzzal TTG crust is represented only in some mafic enclaves. This leads to the varied geochemical affinities of the studied plutons, from medium-K calc-alkaline to alkaline.

- This localized melting and mixing process is related to a planar delamination along these faults allowing the uprise of asthenospheric mantle (Fig. 13), a typical metacratonic process (Liégeois et al. 2013).
- The Pan-African evolution of the In Ouzzal terrane demonstrates that a small but rigid terrane can subsist quite intact within a large orogen, as a result of tectonic escape due to a huge indentor such as the West African craton. The metacratonization at depth, where partial melting occurred, is probably of greater intensity than that observed currently at shallow depth level, but would likewise be localized along faults.
- The intrusion of the Tihimatine and similar In Ouzzal plutons is related to the metacratonization of the In Ouzzal rigid craton due to stress constraints within the general northern tectonic escape of the Tuareg terranes because of the collision with the West African craton (630–580 Ma). This is especially linked to the tearing of the In Ouzzal terrane due to the blocking of the movements 20 Ma earlier along the West In Ouzzal shear zone than along the East In Ouzzal shear zone. Additional zircon dating is needed to assess the period during which the In Ouzzal plutons emplaced, but we favor the 600–580 Ma time bracket.
- The In Ouzzal metacratonization can be compared with processes that occurred at the end of the Ediacaran along the margin of the Murzuq craton (Fezaa et al. 2010), in post-orogenic times in the LATEA metacraton (Taourirt granites; Azzouni-Sekkal et al. 2003) or in an intraplate setting in the Hoggar Mountains during the Cenozoic (silica-undersaturated volcanism; Liégeois et al. 2005; Ben El Khaznadji et al. 2017). Common features are an external compression on a rigid domain that eventually fractured, fissured, allowing mantle/crust magmas to rise and to emplace at shallow level, without lithospheric or crustal thickening and with a good preservation of the older lithologies (Liégeois et al. 2013).

Acknowledgements We warmly thank Jean-Yves Cottin and Abderrahmane Bendaoud for their thoughtful comments and suggestions that allowed us to ameliorate this paper.

Appendix: Analytical Techniques

LA-ICP-MS dating. Zircons were hand-picked in alcohol from the least magnetic concentrates (1° tilt at full amperage). Selected crystals were then embedded in epoxy resin, ground and polished to expose the internal structure. The sample mounts were later used for U–Th–Pb microanalyses using a Lambda Physik Compex 102 excimer laser generating 15 ns pulses of radiation at a wavelength of 193 nm. For analyses, the laser was coupled to a VG Plasmaquad II ICP-MS and analytical procedures followed those outlined by Bruguier et al. (2001) and described in earlier reports (e.g. Neves et al. 2006). Analyses where acquired during two analytical sessions where the spot size of the laser beam was 26 and 51 mm. Unknowns were bracketed by measurements of the G91500 zircon standard

(Wiedenbeck et al. 1995), which were used for mass bias and inter-element fractionation corrections. The calculated bias factors and their associated errors were then added in quadrature to the errors measured on each unknown. Accurate common Pb correction during laser ablation analyses is difficult to achieve, mainly because of the isobaric interference of ^{204}Hg with ^{204}Pb. The contribution of ^{204}Hg to ^{204}Pb was estimated by measuring the ^{202}Hg and assuming a ^{202}Hg/^{204}Hg natural isotopic composition of 0.2298. This allows monitoring of the common Pb content of the analyzed zircon domain, but corrections often resulted in spurious ages. Analyses yielding ^{204}Pb close to or above the limit of detection were thus rejected, and in Table 1 we report analyses uncorrected for common Pb. The ages have been calculated using the Isoplot/Ex, version 3.31 software (Ludwig 2003).

SHRIMP U–Pb zircon dating. The U–Pb dates have been obtained on the ion microprobe (Perth Consortium SHRIMP-II) from the Geochronological laboratory of the Curtin University of Technology in Perth, Western Australia. The zircons have been polished together with the zircon standards CZ3 and TEMORA-2 (U = 551 ppm, ca. 120 ppm, ages = 564, 418 Ma, respectively; Black et al. 2004; Pidgeon et al. 1994). The mounted zircons were analyzed under the following conditions: cycles of 7 scans, primary O^{2-} beam of $\sim$2 nA, spot size of $\sim$25 μm with a mass resolution of about 5000. The data have been reduced following the procedure described by Nelson (1997) using the software SQUID (Ludwig 2001). U/Pb ratios and U concentrations of the samples have been normalized to the TEMORA-2 and CZ3 standards, respectively. The ages have been calculated using the Isoplot/Ex, version 3.31 software (Ludwig 2003). The results are given in Table 2.

Sr and Nd isotopes have been acquired at the Royal Museum for Central Africa (RMCA):

After acid dissolution of the sample and Sr and/or Nd separation on ion-exchange resin, Sr isotopic compositions have been measured on Ta simple filament and Nd isotopic compositions on triple Ta–Re–Ta filament, both a VG Sector 54 mass spectrometer. Repeated measurements of Sr and Nd standards (4 standards for 16 samples) have shown that between-run error is better than 0.000015 (2σ). The NBS987 standard has given a value for ^{87}Sr/^{86}Sr of 0.710278 $\pm$ 0.000008 (2σ on the mean, normalized to ^{86}Sr/^{88}Sr = 0.1194) and the Rennes Nd standard a value for ^{143}Nd/^{144}Nd of 0.511970 $\pm$ 0.000008 (2σ on the mean, normalized to ^{146}Nd/^{144}Nd = 0.7219) during the course of this study. All measured ratios have been normalized to the recommended values of 0.710250 for NBS987 and 0.511963 for Nd Rennes standard (corresponding to a La Jolla value of 0.511858) based on the 4 standards measured on each turret together with 16 samples. Nd ages have been calculated following the Isoplot/Ex, version 3.31 software (Ludwig 2003). Used decay constant were 1.42×10^{-11} a^{-1} (^{87}Rb; Steiger and Jäger 1977) and 6.54×10^{-12} a^{-1} (^{147}Sm; Lugmair and Marti 1978). Sr and Nd isotopic ratios are listed in Table 3.

Whole-rock major and trace elements analyses have been acquired at the Royal Museum for Central Africa (RMCA): Si, Al, Ti, Fe, Ca and P have been analyzed by X-ray fluorescence, Mn, Mg, K and Na have been measured by atomic absorption after open acid digestion ($HF + HClO_4 + HNO_3$); Three different techniques have been used for the analysis of 29 trace elements: X-ray fluorescence on raw material for Rb and Sr; ICP-AES for Cu, Zn, Cr and Co; ICP-MS (VG PQ2+) for the other elements. For details, see the analytical appendix in Liégeois et al. 2003. The results are given in Table 4.

References

Abdallah N, Liégeois JP, De Waele B, Fezaa N, Ouabadi A (2007) The Temaguessine Fe-cordierite orbicular granite (central Hoggar, Algeria): U–Pb SHRIMP age, petrology, origin and geodynamical consequences for the late Pan-African magmatism of the Tuareg shield. J Afr Earth Sci 49:153–178

Abdelsalam M, Liégeois JP, Stern RJ (2002) The Saharan metacraton. J Afr Earth Sci 34:119–136

Aït-Djafer S, Ouzegane K, Liégeois JP, Kienast JR (2003) An example of post-collisional mafic magmatism: the gabbro-anorthosite complex from the Tin Zebane area (western Hoggar, Algeria). J Afr Earth Sci 37:313–330

Allègre CJA, Caby R (1968) Chronologie Rb/Sr du Précambrien de l'Ahaggar occidental. C R Som Soc Géol Fr 8:259–260

Annen C, Sparks RSJ (2002) Effects of repetitive emplacement of basaltic intrusions on thermal evolution and melt generation in the crust. Earth Planet Sci Lett 203:937–955

Azzouni-Sekkal A, Liégeois JP, Bechiri-Benmerzoug F, Belaidi-Zinet S, Bonin B (2003) The "Taourirt" magmatic province, a marker of the very end of the Pan-African orogeny in the Tuareg Shield: review of the available data and Sr–Nd isotope evidence. J Afr Earth Sci 37:331–350

Bechiri-Benmerzoug F, Liégeois JP, Bonin B, Azzouni-Sekkal A, Bechiri H, Kheloui R, Matukov DI, Sergeev SA (2011) The plutons from the Cryogenian Iskel composite oceanic island arc (Hoggar, Tuareg Shield, Algeria): U–Pb on zircon SHRIMP geochronology, geochemistry and geodynamical setting. In: Seventh Hutton symposium on granites and related rocks, Avila, Spain, 4–9 July 2011, p 17

Belkacim S, Gasquet D, Liégeois JP, Arai S, Ghalan H, Ahmed H, Ishida Y, Ikenne M (2017) The Ediacaran volcanic rocks and associated mafic dykes of the Ouarzazate Group (Anti-Atlas, Morocco): clinopyroxene composition, whole-rock geochemistry and Sr-Nd isotopes constraints from the Ouzellarh-Siroua salient (Tifnoute valley). J Afr Earth Sci. https://doi.org/10.1016/j.jafrearsci.2016.08.002 (in press)

Ben El Khaznadji R, Azzouni-Sekkal A, Benhallou A, Liégeois JP, Bonin B (2017) Neogene felsic volcanic rocks in the Hoggar province: volcanology, geochemistry and age of the Azrou trachyte-phonolite association (Algerian Sahara). J Afr Earth Sci 127:222–234

Bendaoud A, Ouzegane K, Godard G, Liégeois JP, Kienast JR, Bruguier O, Drareni A (2008) Geochronology and metamorphic P–T–X evolution of the Eburnean granulite-facies metapelites of Tidjenouine (central Hoggar, Algeria): witness of the LATEA metacratonic evolution. In: Ennih N, Liégeois J-P (eds) The boundaries of the west African craton, vol 297. Geological Society of London, Special Publication, pp 111–146

Berger J, Ouzegane K, Bendaoud A, Liégeois JP, Kiénast JR, Bruguier O, Caby R (2014) Continental subduction recorded by Neoproterozoic eclogite and garnet amphibolites from western Hoggar (Tassendjanet terrane, Tuareg Shield, Algeria). Precambr Res 247:139–158

Bernard Griffiths J, Fourcade S, Kienast JR, Peucat JJ, Martineau F, Rahmani A (1996) Geochemistry and isotope (Sr, Nd, O) study of Al–Mg granulites from the In Ouzzal archaen block (Hoggar Algeria). J Metamorph Geol 14:709–724

Bertrand JLM, Caby R (1978) Geodynamic evolution of the Pan-African orogenic belt: a new interpretation of the Hoggar shield (Algerian Sahara). Geol Rundsch 67:357–388

Bertrand JM, Davison I (1981) Pan-African granitoid emplacement in the Adrar des Iforas mobile belt (Mali): a Rb/Sr isotope study. Precambr Res 14:333–361

Bertrand JM, Meriem D, Lapique F, Michard A, Dautel D, Gravelle M (1986) Nouvelles données radiométriques sur l'âge de la tectonique pan-Africaine dans le rameau oriental de la chaîne pharusienne (région de Timgaouine, Hoggar, Algérie). C R Acad Sci Paris 302:437–440

Black R, Liégeois JP (1993) Cratons, mobile belts, alkaline rocks and continental lithospheric mantle: the Pan-African testimony. J Geol Soc London 150:89–98

Black R, Latouche L, Liégeois JP, Caby R, Bertrand JM (1994) Pan-African displaced terranes in the Tuareg shield (central Sahara). Geology 22:641–644

Black LP, Kamo SL, Allen CM, Davis DW, Aleinikoff JN, Valley JW, Mundil R, Campbell IH, Korsch RJ, Williams IS, Foudoulis C (2004) Improved $^{206}Pb/^{238}U$ microprobe geochronology by the monitoring of a trace-element-related matrix effect; SHRIMP, ID–TIMS, ELA–ICP–MS and oxygen isotope documentation for a series of zircon standards. Chem Geol 205:115–140

Boissonnas J (2008) Cisaillements ductiles et mise en place de plutons granitiques dans le nord de la chaîne panafricaine du Sahara central: le secteur de Tinnirt (Mouydir, Hoggar du NW, Algérie). Bull Serv Géol Natl 19:101–113

Bosch D, Bruguier O, Caby R, Buscail F, Hammor D (2016) Orogenic development of Adrar des Iforas (Touareg Shield, NE Mali): new geochemical and geochronological data and geodynamic implications. J Geodyn 96:104–130

Boullier A-M, Barbey P (1988) A polycyclic two stages corona-growth in the Iforas granulitic unit (Mali). J Metamorph Geol 6:235–254

Boullier AM, Davison I, Bertrand JML, Coward M (1978) L'unité granulitique des Iforas: une nappe de socle d'âge pan-Africain précoce. Bull Soc Géol Fr XX:877–882

Boullier AM, Liégeois JP, Black R, Fabre J, Sauvage M, Bertrand JM (1986) Late Pan-African tectonics marking the transition from subduction-related calc-alkaline magmatism to within-plate alkaline granitoids (Adrar des Iforas, Mali). Tectonophysics 132:233–246

Bouzid A, Akacem N, Hamoudi M, Ouzegane K, Abtout A, Kienast JR (2008) Magnetotelluric modeling of the deep geologic structure of In Ouzzal Granulitic Unit (western Hoggar). C R Geosci 340:711–722

Bouzid A, Bayou B, Liégeois J-P, Bourouis S, Bougchiche SS, Bendekken A, Abtout A, Boukhlouf W, Ouabadi A (2015) Lithospheric structure of the Atakor metacratonic volcanic swell (Hoggar, Tuareg Shield, southern Algeria): electrical constraints from magnetotelluric data. In: Foulger GR, Lustrino M, King SD (eds) The interdisciplinary earth: a volume in honor of Don L. Anderson, vol 514. Geological Society of America special paper, pp 239–255

Bruguier O, Telouk P, Cocherie A, Fouillac AM, Albarede F (2001) Evaluation of Pb-Pb and U-Pb laser ablation ICP-MS zircon dating using matrix-matched calibration samples with a frequency quadrupled (266 nm) Nd:YAG laser. Geostand Newsl 25:361–373

Caby R (1970) La chaine Pharusienne dans le nord-ouest de l'Ahaggar, Algerie. Sa place dans I'Orogenese du Precambrien superieur en Afrique. Ph.D. thesis, Universitt de Montpellier, p 336; and Alger (1983) Bull Dir Mines Géol 47:289

Caby R (2003) Terrane assembly and geodynamic evolution of central-western Hoggar: a synthesis. J Afr Earth Sci 37:133–159

Caby R, Andreopoulos-Renaud U (1985) Etude petrostructurale et geochronologie U/Pb sur zircon d'une metadiorite quartzique de la chaine pan-Africaine de l'Adrar des Iforas (Mali). Bull Soc Geol Fr 8:899–903

Caby R, Andreopoulos-Renaud U (1989) Age U-Pb à 620 Ma d'un pluton synorogénique de l'Adrar des Iforas (Mali). Consequences pour l'âge de la phase majeure de l'orogène pan-Africaine. C R Acad Sci 308:307–314

Caby R, Monié P (2003) Neoproterozoic subduction and differential exhumation of western Hoggar (southwest Algeria): new structural, petrological and geochronological evidence. J Afr Earth Sci 37:269–293

Caby R, Dostal J, Dupuy C (1977) Upper Proterozoic volcanic graywackes from north-western Hoggar (Algeria)—geology and geochemistry. Precambr Res 5:283–297

Caby R, Andreopoulos-Renaud U, Gravelle M (1982) Cadre géologique et géochronologie U/Pb sur zircon des batholites précoces dans le segment pan-Africain du Hoggar central (Algérie). Bull Soc Géol Fr 7:876–882

Caby R, Andreopoulos-Renaud U, Lancelot JR (1985) Les phases tardives de l'orogenèse pan-Africaine dans l'Adrar des Iforas oriental (Mali): lithostratigraphie des formations molassiques et géochronologie U/Pb sur zircon de deux massifs intrusifs. Precambr Res 28:187–199

Caby R, Andreopoulos-Renaud U, Pin C (1989) Late Proterozoic arc-continent and continent-continent collision in the Pan-African Trans-Saharan belt of Mali. Can J Earth Sci 26:1136–1146

Chikhaoui M, Dupuy C, Dostal J (1978) Geochemistry of late Proterozoic volcanic rocks from Tassendjanet area (N.W. Hoggar, Algeria). Contrib Mineral Petrol 66:157–164

Deng Y, Levandowski W, Kusky T (2017) Lithospheric density structure beneath the Tarim basin and surroundings, northwestern China, from the joint inversion of gravity and topography. Earth Planet Sci Lett 460:244–254

Deroin JP, Djemai S, Bendaoud A, Brahmi B, Ouzegane K, Kienast J-R (2014) Integrating geologic and satellite radar data for mapping dome-and-basin patterns in the in Ouzzal Terrane, western Hoggar, Algeria. J Afr Earth Sci 99:652–665

Ferkous K, Monie P (2002) Neoproterozoic shearing and auriferous hydrothermalism along the lithospheric N-S East In Ouzzal shear zone (west Hoggar, Algeria, north Africa). J Afr Earth Sci 35:399–415

Fezaa N, Liégeois JP, Abdallah N, Cherfouh EH, De Waele B, Bruguier O, Ouabadi A (2010) Late Ediacaran geological evolution (575–555 Ma) of the Djanet Terrane, eastern Hoggar, Algeria, evidence for a Murzukian intracontinental episode. Precambr Res 180:299–327

Frost BR, Barnes CG, Collins WJ, Arculus RJ, Ellis DJ, Frost CD (2001) A geochemical classification for granitic rocks. J Petrol 42:2033–2048

Gravelle M (1969) Recherches sur la géologie du socle Précambrien de l'Ahaggar centro-occidental dans la région de Silet Tibehaouine. Ph.D. thesis, Paris, 781 pp (unpublished)

Haddoum H, Choukroune P, Peucat JJ (1994) Structural evolution of the precambrian In Ouzzal massif, central Sahara, Algeria. Precambr Res 65:155–166

Hadj Kaddour Z, Liégeois JP, Demaiffe D, Caby R (1998) The alkaline-peralkaline granitic post-collisional Tin Zebane dyke swarm (Pan-African Tuareg shield, Algeria): prevalent mantle signature and late agpaitic differentiation. Lithos 45:223–243

Jahn BM, Caby R, Monié P (2001) The oldest UHP eclogites of the World: age of UHP metamorphism, nature of protoliths and tectonic implications. Chem Geol 178:143–158

Liégeois JP, Black R (1987) Alkaline magmatism subsequent to collision in the Pan-African belt of the Adrar des Iforas (Mali). In: Fitton JG, Upton BJG (eds) Alkaline igneous rocks, vol 30. Geological Society Special Publication, London, pp 381–401

Liégeois JP, Stern RJ (2010) Sr–Nd isotopes and geochemistry of granite-gneiss complexes from the Meatiq and Hafafit domes, eastern desert, Egypt: no evidence for pre-Neoproterozoic crust. J Afr Earth Sci 57:31–40

Liégeois JP, Bertrand JM, Black R (1987) The subduction- and collision-related Pan-African composite batholith of the Adrar des Iforas (Mali): a review. In: Kinnaird J, Bowden P (eds) African geology review. Wiley, London, pp 185–211 and Geol J 22(S2):185–211

Liégeois JP, Black R, Navez J, Latouche L (1994) Early and late Pan-African orogenies in the Aïr assembly of terranes (Tuareg Shield, Niger). Precambr Res 67:59–88

Liégeois JP, Navez J, Hertogen J, Black R (1998) Contrasting origin of post-collisional high-K calc-alkaline and shoshonitic versus alkaline and peralkaline granitoids. Lithos 45:1–28

Liégeois JP, Latouche L, Boughrara M, Navez J, Guiraud M (2003) The LATEA metacraton (central Hoggar, Tuareg shield, Algeria): behaviour of an old passive margin during the Pan-African orogeny. J Afr Earth Sci 37:161–190

Liégeois JP, Benhallou A, Azzouni-Sekkal A, Yahiaoui R, Bonin B (2005) The Hoggar swell and volcanism: reactivation of the precambrian Tuareg shield during Alpine convergence and west African Cenozoic volcanism. In: Foulger GR, Natland JH, Presnall DC, Anderson DL (eds) Plates, plumes and paradigms, vol 388. Geological Society of America special paper, pp 379–400

Liégeois JP, Abdelsalam MG, Ennih N, Ouabadi A (2013) Metacraton: nature, genesis and behavior. Gondwana Res 23:220–237

Ludwig KR (2001) Squid 1.02: a user's manual. Berkeley Geochronology Center. Special publication, 19 p

Ludwig KR (2003) User's manual for Isoplot/ 3, a geochronological toolkit for Microsoft excel. Berkeley Geochronology Center, Special Publication, 14, 71 p

Lugmair GW, Marti K (1978) Lunar initial 143Nd/144Nd: differential evolution of the lunar crust and mantle. Earth Planet Sci Lett 39:349–357

Marignac C, Semiani A, Fourcade S, Boiron MC, Joron JL, Kienast JR (1996) Metallogenesis of the late Pan-African gold-bearing East Ouzzal shear zone (Hoggar, Algeria). J Metamorph Geol 14:783–801

Mobbs PM (2004) The mineral industry of Algeria. In: U.S. geological survey minerals yearbook, pp 2.1–2.4

Nachit H, Rasafimhefa N, Stussi JM, Carron JP (1985) Composition chimique des biotites et typologie magmatique des granitoïdes. C R Acad Sci Paris 301(série II):813–818

Nelson DR (1997) Compilation of SHRIMP U-Pb zircon geochronology data, 1996, vol 2. Geological Survey of Western Australia, pp 1–198

Nelson BK, DePaolo DJ (1985) Rapid production of continental crust 1.7 to 1.9 by ago: Nd isotopic evidence from the basement of the north American midcontinent. Geol Soc Am Bull 96:746–754

Neves S, Bruguier O, Vauchez A, Bosch D, Rangel da Silva JM, Mariano G (2006) Timing of crust formation, deposition of supracrustal sequences, and Transamazonian and Brasiliano metamorphism in the East Pernambuco belt (central domain, Borborema Province, NE Brazil): implications for western Gondwana assembly. Precambr Res 149:197–216

Ngalamo JFG, Bisso D, Abdelsalam MG, Atekwana EA, Katumwehe AB, Ekodeck GE (2017) Geophysical imaging of metacratonizaton in the northern edge of the Congo craton in Cameroon. J Afr Earth Sci. http://dx.doi.org/10.1016/j.jafrearsci.2016.12.010 (in press)

Ouzegane K (1987) Les granulites Al-Mg et les carbonatites dans la série de l'In Ouzzal NW Hoggar, Algérie. Nature et évolution de la croûte continentale profonde pendant l'Archéen. Doctorate of Sciences, Paris, 433 p

Ouzegane K, Kienast JR, Bendaoud A, Drareni A (2003) A review of Archaean and Paleoproterozoic evolution of the In Ouzzal granulitic terrane (western Hoggar, Algeria). J Afr Earth Sci 37:207–227

Paquette JL, Caby R, Djouadi MT, Bouchez JL (1998) U–Pb dating of the end of Pan-African orogeny in the Tuareg Shield: the post-collisional syn-shear Tioueine pluton (western Hoggar, Algeria). Lithos 45:245–253

Peucat JJ, Capdevila R, Drareni A, Choukroune P, Fanning CM, Bernard-Griffiths J, Fourcade S (1996) Major and trace element geochemistry and isotope (Sr, Nd, Pb, O) systematic of an Archaean basement involved in 2.0 Ga very high-temperature (1000 °C) metamorphic event. In Ouzzal Massif, Hoggar, Algeria. J Metamorph Geol 14:667–692

Peucat JJ, Drareni A, Latouche L, Deloule E, Vidal P (2003) U–Pb zircon (TIMS and SIMS) and Sm–Nd whole rock geochronology of the Gour Oumalelen granulitic basement, Hoggar massif, Tuareg shield, Algeria. J Afr Earth Sci 37:229–239

Pidgeon RT, Furfaro D, Kennedy AK, Nemchin AA, Van Bronswijk W (1994) Calibration of zircon standards for the Curtin SHRIMP II. United States Geological Survey circular 1107, 251

Rougier S, Missenard Y, Gautheron C, Barbarand J, Zeyen H, Pinna R, Liégeois JP, Bonin B, Ouabadi A, Derder MEM, Frizon de Lamotte D (2013) Eocene exhumation of the Tuareg shield (Sahara, Africa). Geology 41:615–618

Shang CK, Satir M, Nsifa EN, Liegeois JP, Siebel W, Taubald H (2007) Archaean high-K granitoids produced by remelting of earlier Tonalite-Trondhjemite-Granodiorite (TTG) in the Sangmelima region of the Ntem complex of the Congo craton, southern Cameroon. Int J Earth Sci 96:817–841

Steiger RH, Jäger E (1977) Subcommission on geochronology: convention on the use of decay constants in geo- and cosmochronology. Earth Planet Sci Lett 36(3):359–362

Stern RJ, Ali KA, Liégeois JP, Johnson PR, Kozdroj W, Kattan FH (2010) Distribution and significance of pre-neoproterozoic zircons in juvenile neoproterozoic igneous rocks of the Arabian-Nubian shield. Am J Sci 310:791–811

Sun SS, McDonough WE (1989) Chemical and isotopic systematics of oceanic basalt: implications for mantle composition and processes. In: Saunders AD, Nrry MJ (eds) Magmatism in the oceanic basins, vol 42. Geological Society Special Publications, pp 313–345

Toummite A, Liégeois JP, Gasquet D, Bruguier O, Beraaouz EH, Ikenne M (2013) Field, geochemistry and Sr-Nd isotopes of the Pan-African granitoids from the Tifnoute valley (Sirwa, Anti-Atlas, Morocco): a post-collisional event in a metacratonic setting. Mineral Petrol 107:739–763

Vander Auwera J, Bogaerts M, Liégeois JP, Demaiffe D, Wilmart E, Bolle O, Duchesne JC (2003) Derivation of the 1.0–0.9 Ga ferro-potassic A-type granitoids of southern Norway by extreme differentiation from basic magmas. Precambr Res 124:107–148

Wiedenbeck M, Alle P, Corfu F, Griffin WL, Meier M, Oberli F, Von Quadt A, Rudi JC, Spiegel W (1995) Three natural zircon standards for U–Th–Pb Lu–Hf, trace element and REE analyses. Geostand Newsl 19:1–23

Zhang X, Mao Q, Zhang H, Zhai M, Yang Y, Hu Z (2011) Mafic and felsic magma interaction during the construction of high-K calc-alkaline plutons within a metacratonic passive margin: the early Permian Guyang batholith from the northern north China Craton. Lithos 125:569–591

An Overview of the Plutons Magnetic Fabric Studies in the Hoggar Shield: Evolution of the Major Shear Zones During the Pan-African

B. Henry, M. E. M. Derder, S. Maouche, O. Nouar, M. Amenna, B. Bayou and A. Ouabadi

Abstract The magnetic fabric obtained in the Hoggar shield on several plutons is related to various origins: simple flow, syn-deformation flow, stress field during late-magmatic stage and solid-state deformation. These results also evidenced the important role of hosting frame, acting as a more or less efficient "protection" against the effects on the magnetic fabric of the regional stress field. Combined studies, on neighboring sites of the main intrusion and of late-magmatic dykes crosscutting it, yield key-arguments about the acquisition age of this fabric. $P'(K_m)$ diagrams clearly highlight the strain gradient in plutons very close to shear zones. The fact that K_1 axis is mainly a mineral lineation, at least in three plutons, is evidenced by the determination of the magnetic zone axis. The plutons AMS is associated with different Pan-African stages in the Hoggar: gneissification of Eburnean plutons, syn-thrust pluton emplacement, pluton emplacement under regional shearing context. Accordingly, the main movements along the major shear zones, related to the oblique collision of the Hoggar shield with the West African Craton, are contemporaneous of the first stages. Shearing context during the following period points out that this collision had still active effects, probably associated with a progressive change in orientation of the continental convergence.

Keywords Shear zones · Plutons · Magnetic susceptibility · Anisotropy

B. Henry (✉)
Paléomagnétisme, Institut de Physique du Globe de Paris, Sorbonne Paris Cité, Univ. Paris Diderot and UMR 7154 CNRS, 4 Avenue de Neptune, Saint-Maur Cedex 94107, France
e-mail: henry@ipgp.fr

M. E. M. Derder · S. Maouche · O. Nouar · M. Amenna · B. Bayou
CRAAG, B.P. 63, Bouzaréah, Alger 16340, Algeria

A. Ouabadi
Laboratoire "Géodynamique, Géologie de l'Ingénieur et Planétologie", FSTGAT USTHB, BP 32, El-Alia Bab Ezzouar, Alger 16111, Algeria

A. Bendaoud et al. (eds.), *The Geology of the Arab World—An Overview*, Springer Geology, https://doi.org/10.1007/978-3-319-96794-3_4

1 Introduction

The study of shearing in the vicinity of major shear zones represents one of the tasks that have to be carried out in order to understand part of the structural evolution of orogenic belts. It is more and more often based on fabric analysis in plutons (e.g., Pitcher and Berger 1972; Marre 1982; Paterson et al. 1998; Schofield and D'Lemos 1998) emplaced under transpressive regimes (e.g., Acef et al. 2003; Neves et al. 2003). The visible effects of shearing in the rocks close to these shear zones can be explained by a combination of several key factors. One of these factors, often difficult to evaluate, is the mechanical behavior of the rocks, which depends in particular on their composition and petrofabric. The anisotropy of magnetic susceptibility (AMS) is a powerful tool (e.g. Tarling and Hrouda 1993; Borradaile and Henry 1997; Borradaile and Jackson 2004) to analyze emplacement conditions and deformation of intrusive bodies. This low time-consuming approach gives not only the orientation of foliation and lineation but also characterizes the intensity of the fabric and the relative importance of the foliation and lineation. The obtained magnetic fabric in plutons (e.g. King 1966; Bouchez 2000) is, according to the cases, mainly related to magma flow (e.g. Henry 1980; Hrouda and Lanza 1989), to stress field during magmatic to late-magmatic period (e.g. Pignotta and Benn 1999; Archanjo et al. 2002; Tomezzoli et al. 2003; Kratinová et al. 2007; Njanko et al. 2009; Henry et al. 2012; Adissin Glodji et al. 2014; Georgiev et al. 2014; Dawaï et al. 2017) or to solid-state deformation (e.g. Henry 1980). It could be composite, resulting from superimposed fabrics, whose characteristics can be distinguished only in favorable cases (e.g. Henry 1974, 1975). In fact, if partial weathering does not affect the orientation of the susceptibility principal axes (Krasa and Herrero-Bervera 2005), very strong weathering, strain or important mineralogical change can have more or less disturbed the initial magnetic fabric.

This paper presents a review of all the AMS studies that are presently published in plutons in the Hoggar shield. This shield is constituted by Archean/Paleoproterozoic and Neoproterozoic terranes. Its Pan-African evolution (Black et al. 1994; Liégeois et al. 1994, 1998, 2003) was first (900–680 Ma) characterized by the accretion of oceanic island arcs on previous cratons and microcratons. The following stage is represented by the regional northerly tectonic escape of the terranes linked to the oblique collision with the West African Craton. This stage is characterized by spectacular N–S shear zones and the intrusion of granitoid batholiths (615–580 Ma). Post-collisional and anorogenic high-level alkaline plutons are aligned around the same mega-shear zones (Lelubre 1952; Boissonnas 1974; Azzouni-Sekkal et al. 2003). Plutons studied using the AMS are related to different stages of the Hoggar evolution. Since they represent witnesses of the different tectonic phases, their study aimed at deciphering the complex interplay of the different successive tectonic events.

The Eburnean plutons are mostly characterized by high deformation, and their magnetic fabric studies have then a restricted interest because the structures can mostly be directly observed. That is not the case for the Pan-African plutons, which present various degree of deformation, even sometimes within the same massif. Most of the AMS-studied Pan-African plutons were chosen in areas close to the major shear zones, well known in the Hoggar shield (Table 1 and Fig. 1).

Most of these plutons were subject of other detailed (e.g., petrological, structural, geochemical, dating, etc.) studies, around the 4°50 shear zone (e.g., Boissonnas 1974; Azzouni-Sekkal and Boissonnas 1987, 1993; Djouadi et al. 1997; Paquette et al. 1998; Boulfelfel and Ouabadi 2000; Azzouni-Sekkal et al. 2003), in the Tamanrasset area (e.g. Cheilletz et al. 1992; Acef et al. 2003) or in eastern Hoggar (e.g. Zeghouane et al. 2008; Henry et al. 2009; Nouar et al. 2011). Magnetic fabric studies represent one of the components of the rich global knowledge of these plutons.

2 Sampling

For these studies, samples were taken using a portable gasoline-powered drill and oriented with magnetic and sun compasses. When possible 6 to more than 10 cores (except for Tesnou and Tioueine plutons, only two cores, but very neighboring sites) were collected in each sampling site covering several tens of square meters. For most plutons, numerous sites were distributed in different areas, in order to look for the variations of the structures within a same intrusion. Except for Tesnou and Tioueine plutons, additional specimens were cut from each core for rock magnetic analysis.

3 Magnetic Mineralogy

In the Tesnou and Tioueine plutons (Fig. 1 and Table 1; Djouadi and Bouchez 1992; Djouadi et al. 1997), the fabric is interpreted as mainly carried by biotite (± amphibole) because of low susceptibility values. For all other plutons, rock magnetism measurements were carried out. Thermomagnetic analyses of low field magnetic susceptibility $K(T)$ were performed on a KLY3 susceptibility bridge with CS2-3 and, for some plutons, with CSL attachments (AGICO, Brno). Hysteresis loops were obtained on small samples (3 cm^3) using a laboratory-made translation inductometer within an electromagnet capable of reaching 1.6 T.

For the Tifferkit and In Tounine (Fig. 1 and Table 1; Henry et al. 2006) samples, low susceptibility values shows that paramagnetic minerals (mainly biotite, muscovite) are their main magnetic fabric carriers. However, a small amount of magnetite (main decrease up to 580 °C) and hematite (final decrease up to 670 °C) visible on $K(T)$ curve (Fig. 2a) could have a minor effect. Hysteresis loop highlights a dominant paramagnetic behavior (Fig. 3a) but confirms (Fig. 3b) the presence of magnetite (partial saturation for low field) and hematite (open loop).

Table 1 AMS-studied plutons in the Hoggar shield; granite (γ), diorite (δ), granodiorite (γδ)

Plutons	Coordinates	Age (Ma)	Ref.	K_m (10^{-6} SI)	P' or P	T
Ain Kahla	27°31N, 3°24E	–	a	20,000 (18,100–21,700)	1.260 (1.197–1.319)	−0.19 (−0.41 to 0.16)
Tesnou	23°51N, 4°33E	506 ± 16[(1)]	b, c	Mostly >200 (20–6070)	Mostly >1.006 (1.001–1.020)	-
Tioueine	22°49N, 4°14E	523 ± 1[(2)] 520.5 ± 8.5[(1)]	c	Mostly >300 (130–5680)	Mostly >1.009 (1.001–1.015)	-
Tihaliouine	23°19N, 4°43E	–	d	1840 (24–10,700)	1.066 (1.007 and 1.174)	0.00 (−0.91 to 0.89)
Teg Orak	23°09N, 4°48E	519 ± 18[(1)]	d	2877 (8–9150)	1.074 (1.021–1.283)	−0.01 (−0.96 to 0.97)
Anfeg	22°47N, 5°43E	615[(3)]	e	10,547 and 25,375	1.042 and 1.196	−0.53 and 0.14
Tifferkit	22°44N, 5°31E	576[(3)]	e	114 (89–232)	1.021 (1.009–1.030)	0.31 (−0.44 to 0.84)
In Tounine	22°52N, 5°40E	539[(3)]; 502 ± 42[(1)]	e	85 (26–423)	1.014 (1.003–1.038)	0.25 (−0.67 to 0.90)
In Telloukh (γ)	19°53N, 5°56E	–	f	9100 (47–18,851)	1.073 (1.008–1.194)	0.19 (−0.82 to 0.86)
In Telloukh (δ)	19°53N, 5°56E	–	f	209 (46–347)	1.009 (1.003–1.017)	0.62 (0.44–0.71)
Alous-En-Tides	21°38N, 7°30E	570 ± 20?	g	1786 (57–6449)	1.134 (1.040–1.420)	0.37 (−0.32 to 0.89)
Arokam West	21°54N, 8°37E	1918 ± 5	h	8455 (141–29,488)	1.129 (1.046–1.302)	0.38 (0.09–0.73)
Arokam East	21°48N, 8°45E		h	3300 (1253–9324)	1.033 (1.013–1.054)	0.30 (0.03–0.78)
Oued Touffok	22°45N, 8°31E	793 ± 3	i	6796 (1970–9298)	1.160 (1.023–1.543)	0.05 (−0.72 to 0.76)
Yvonne (γ)	21°32N, 8°23E	595 ± 3	h	2112 (33–4190)	1.043 (1.035–1.066)	0.50 (0.02–0.77)
Yvonne (γδ)	21°35N, 8°26E		h	17,127 (15,269–18,238)	1.044 (1.018–1.089)	0.25 (−0.09 to 0.46)
Ohergehem	22°03N, 8°32E	594 ± 4	i	11,403 (111–30,708)	1.250 (1.021–1.556)	0.09 (−0.81 and 0.73)
Honag	23°33N, 8°10E	593 ± 17	i	6864 (102–11,116)	1.167 (1.060–1.414)	−0.04 (−0.77 to 0.96)
Hanane	21°57N, 8°38E	582 ± 3	h	7556 (1632–26,921)	1.106 (1.076–1.158)	0.46 (0.06–0.75)
Abdou	21°50N, 8°48E	<582	h	8154 (750–30,233)	1.085 (1.019–1.188)	0.20 (0.04–0.87)
Arigher	22°22N, 8°30E	554 ± 5	i	9757 (52–27,532)	1.148 (1.010–1.736)	0.22 (−0.94 and 0.95)

(continued)

Table 1 (continued)

Plutons	Coordinates	Age (Ma)	Ref.	K_m (10^{-6} SI)	P' or P	T
Oued Tiririne	23°40N, 8°41E	–	i	13,733 (754–27,841)	1.120 (1.013–1.246)	−0.02 (−0.71 to 0.71)
Kerkour	23°25N, 8°32E	–	i	7244 (3196–15,532)	1.202 (1.058–1.421)	0.01 (−0.73 to 0.66)
Tin Ghoras	23°41N, 8°30E	–	i	5162 (2538–6845)	1.070 (1.029–1.185)	−0.03 (−0.87 to 0.82)
Adjou	23°32N, 8°21E	–	i	8758 (1229–13,198)	1.056 (1.007–1.124)	0.09 (−0.64 to 0.71)

Ref.: (a) Guemache et al. (2009); (b) Djouadi and Bouchez (1992); (c) Djouadi et al. (1997); (d) Henry et al. (2008); (e) Henry et al. (2006); (f) Henry et al. (2007); (g) Henry et al. (2004); (h) Nouar et al. (2011); (i) Henry et al. (2009). References for age, when different from the AMS paper: (1) Azzouni-Sekkal et al. (2003); (2) Paquette et al. (1998); (3) Cheilletz et al. (1992); (4) Acef et al. (2003). Rb–Sr ages, obtained for the Taourirt plutons, have be considered with caution, these plutons having been subjected to fluid perturbation during an early Ordovician slight reactivation of shear zones (Azzouni-Sekkal et al. 2003)

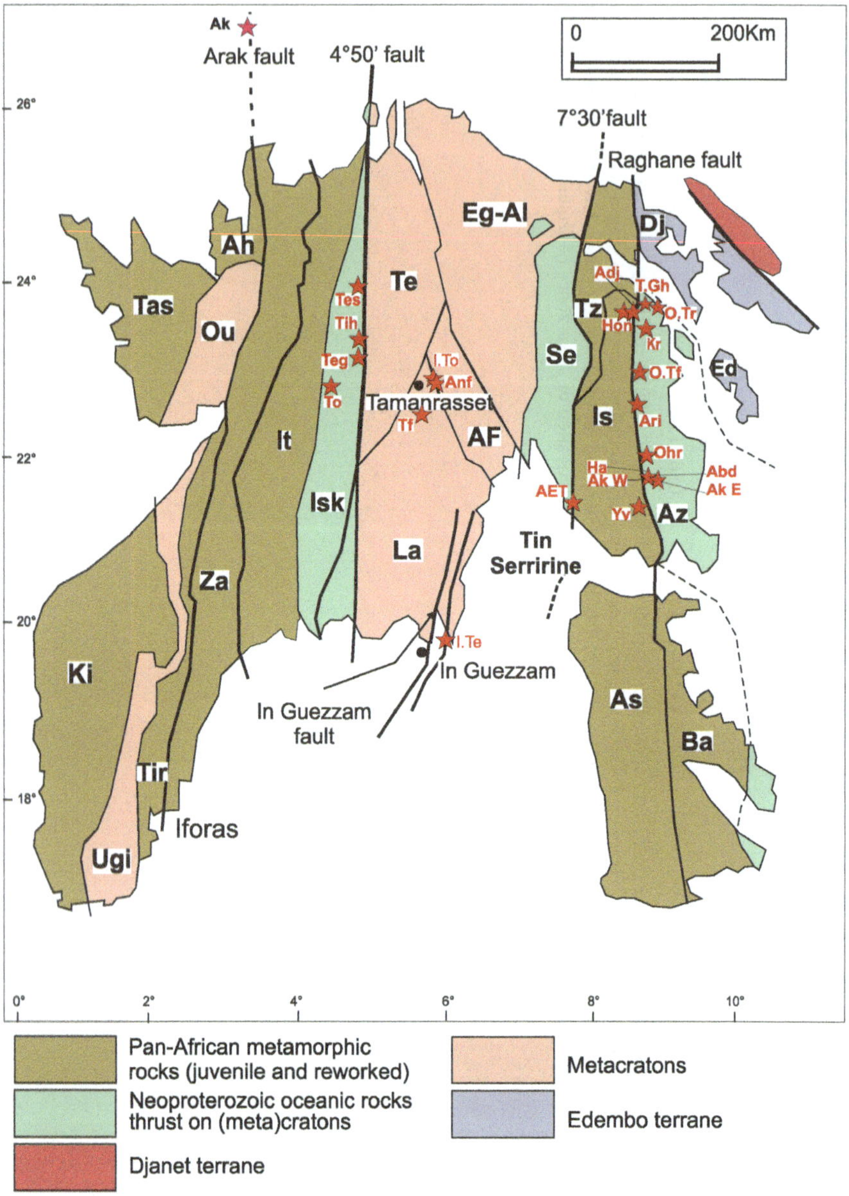

Fig. 1 Geological sketch map of the Hoggar shield showing the studied plutons (Table 1). *Terranes* Egere-Aleksod (Eg-Al), Azrou-n-Fad (AF), Tefedest (Te), Laouni (La), Iskel (Isk), In Teidini (It), Tin Zaouatene (Za), Tirek (Tir), Ahnet (Ah), In Ouzzal (Ou), Iforas granulitic unit (Ugi), Tassendjanet (Tas), Kidal (Ki), Assodé (As), Issalane (Is), Tazat (Tz), Aouzegueur (Az), Barghot (Ba), Djanet (Dj), Edembo (Ed). *Plutons* Ain Kahla (AK), Tihaliouine (Tih), Teg Orak (Teg), Tesnou (Tes), Alous-En-Tides (AET), In Telloukh (Ite), Tioueine (To), Tifferkit (Tf), In Tounine (Ito), Anfeg (Anf), Tin Ghoras (TGh), Oued Tiririne (OTr), Honag (Hon), Adjou (Adj), Kerkour (Kr), Oued Touffok (Otf), Arigher (Ari), Ohergehem (Ohr), Hanane (Ha), Arokam West (AkW), Arokam East (AkE), Abdou (Abd) and Yvonne (Yv)

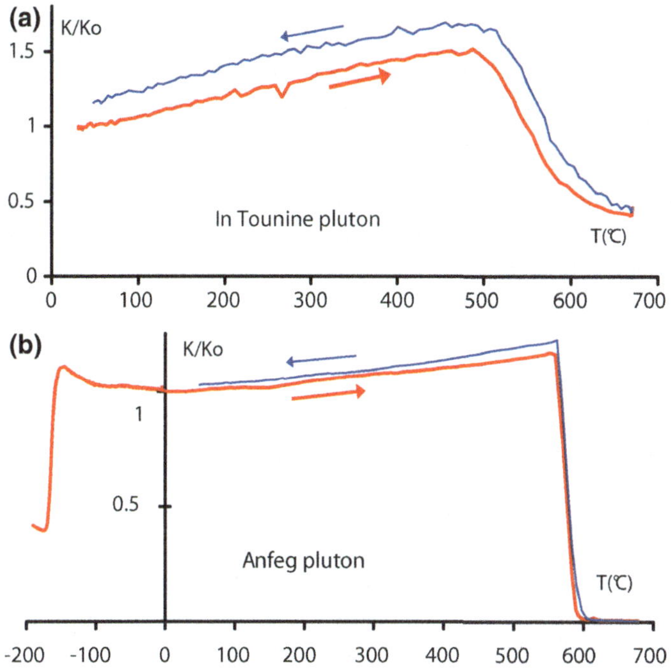

Fig. 2 Typical examples of normalized thermomagnetic curves (mean susceptibility K_m in low field as a function of the temperature T) of samples from the In Tounine (**a**) and Anfeg (**b**) plutons

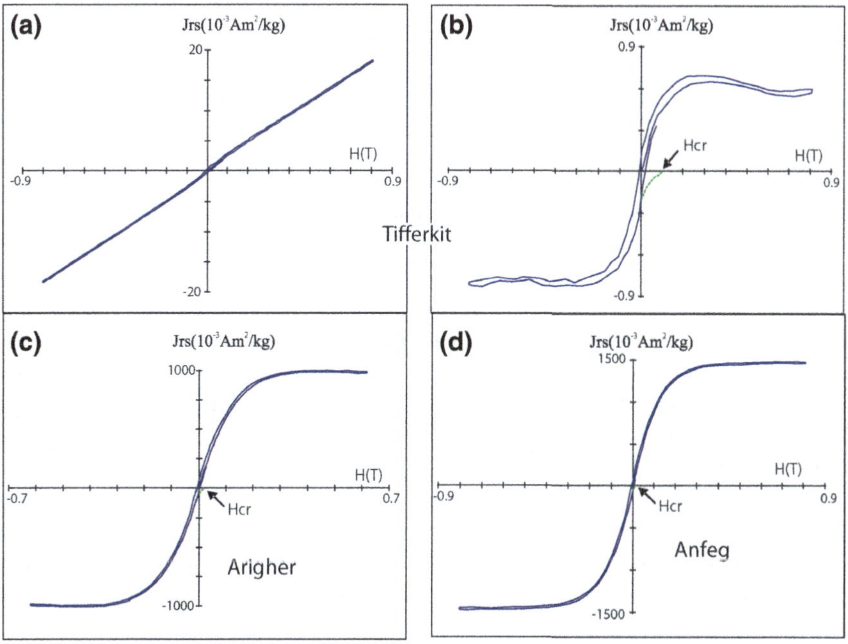

Fig. 3 Typical examples of hysteresis loops of samples from Tifferkit (**a**, **b**), Arigher (**c**) and Anfeg (**d**) plutons. Curves after correction of the paramagnetism for **b**, **c**, and **d**

In all the other plutons (Fig. 1 and Table 1), the samples give *K*(*T*) curves evidencing the presence of magnetite by a clear Verwey transition and a Curie temperature around 580 °C (Fig. 2b). Almost all the hysteresis loops are closed and with saturation for moderate field, confirming the presence of magnetite (Fig. 3c). Hysteresis parameters evidence large magnetite grains (multi-domain—MD—size). The magnetic fabric is then directly related to the shape of the magnetite grains. Significant weathering in part of the Anfeg, In Telloukh, and Arokam plutons, gave a partial oxidation of magnetite in hematite, evidenced by a weak opening of the hysteresis loop (Fig. 3d).

4 Magnetic Fabric

AMS in low field was measured using KLY Kappabridge (AGICO, Brno), giving the principal susceptibilities K_1, K_2, and K_3 ($K_1 \geq K_2 \geq K_3$) and the mean susceptibility K_m. The Jelinek (1981) parameters P' and T were used to describe the magnetic fabric: P' expresses the departure from a spherical AMS ellipsoid ($P' = 1$). The shape parameter T quantifies the shape of the magnetic ellipsoid, being linear when $1 \leq T < 0$ and planar when $0 \leq T \leq 1$ (obtained K_m, P' and T values in the different plutons are indicated in Table 1). The orientation of the AMS is characterized by the magnetic foliation, equivalent to the K_1–K_2 plane ($\perp$ to K_3) and the magnetic lineation, defined as the direction parallel to K_1. Data for a group of samples were analyzed using normalized tensor variability (Hext 1963; Jelinek 1978) and/or bivariate (Henry and Le Goff 1995) statistics.

4.1 Main Results Obtained in the Hoggar Shield (Fig. 1, Table 1 and References Herein)

The late Pan-African Aïn Kahla pluton, located close to the Arak fault has magnetic fabric expressing high deformation during the magmatic stage within an E–W regional stress field. Concerning the Taourirt plutons near the 4°50 shear zone, the dominant structural context of the Tesnou complex and of the eastern part of the Teg Orak pluton, reflected by its magnetic fabric, is a dextral shear strain related to this shear zone. For the western Teg Orak part, Tihaliouine and Tioueine plutons, located farther from the shear zone, the shearing effect is limited. Plutons of the Tamanrasset area (In Tounine and Tifferkit) have poorly defined paramagnetic fabric, rather suggesting strike-slip faults effects. Located close to the In Guezzam faults, the In Telloukh granite has fabric suggesting an emplacement mode likely related to N–S compression. Dioritic dykes crosscutting this granite have magnetic fabric, related on the contrary to a dextral N–S strike-slip shearing. The magnetic fabric of the Alous-En-Tides granite is clearly linked to the stress field associated with the 7°30 shear zone. All the other studied plutons are around the Raghane (8°30) shear zone and its satellites faults. The Eburnean Arokam West, Arokam East, and the old Oued Touffok pluton (793 Ma) plutons have deformational magnetic fabric related to the Pan-African orogeny. The Honag, Arigher, Kerkour plutons, and Yvonne granite present magnetic fabric related to Pan-African shearing tectonic context. That is also the case for the Adjou and Tin Ghoras pluton, though they sealed shear zones. The Oued Tiririne pluton has also magnetic fabric, which is not related to the magma flow. The fabric of the Ohergehem syntectonic pluton, and of the Yvonne granodiorite presents a magnetic lineation correlating an E–W shearing. The Hanane pluton has magnetic fabric indicating moderate effect of the regional stress field, probably because of pluton intrusion in a rigid basement. For the same reason, the Abdou pluton is the single one presenting magnetic fabric only related to magma flow.

4.2 Data Comparison

The dominant information concerning the orientation of the principal susceptibility axes is that, in most studied plutons, the magnetic lineation has a weak plunge toward a NNW–SSE to NNE–SSW direction, not far from the strike of the different neighboring shear zones.

The mean susceptibility K_m is high (6830×10^{-6} SI), only three plutons having paramagnetic fabric. K_m distribution within each site is, however, very large for some other plutons (Fig. 4a), highlighting a significant effect of the weathering.

Mean P' (1.103) is also relatively high for granitoids. In some plutons, large scattering of the P' values (Fig. 4a) is associated with the large distribution of K_m values, confirming a weathering effect. The highest P' corresponds to syntectonic emplacement (red dots on Fig. 4a) and the lowest ones to paramagnetic fabric (black dots). Green dots represent fabric related to solid-state deformation. Though plutons locally affected by visible deformation (violet dots) have relatively high P' values, the difference with the ones without visible deformation (blue dots) is not so clear. The effect of the mean susceptibility values on the P' ones, in fact, obscures a little bit the information. On the $P'(K_m)$ diagram (Fig. 4a), difference becomes clearer, all data corresponding to plutons without clearly visible structures being below a regression curve obtained for all the results.

The oblate shape is clearly dominant (mean T value: 0.18), but within most plutons, the T parameter presents high variation (Fig. 4b). A rough relationship between P' and T shows an increase of the prolateness with the "intensity" of the fabric. However, this variation is partly constrained by particular results, from the paramagnetic plutons, and from the syntectonic ones.

The fabric difference between plutons strongly depends on the nature of the host-rocks as exemplified by Fig. 5: Granitic host-rocks (γd2) around the Tihaliouine pluton and the western part of the Teg Orak pluton acted as a rigid block, protecting the intrusions from regional deformation, while basic plutonic and

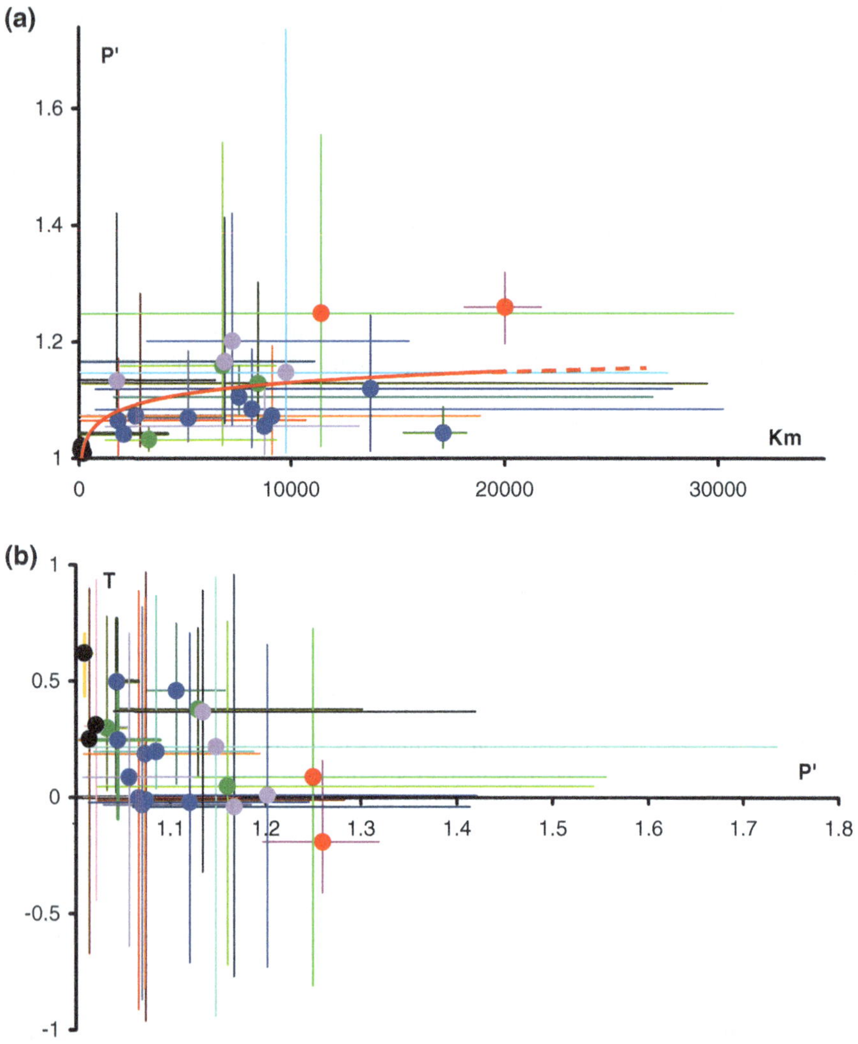

Fig. 4 **a** $P'(K_m)$ diagram for the different plutons with associated uncertainties (see Table 1); plutons with paramagnetic fabric (black dots), "ferrimagnetic" plutons with solid-state deformation (green dots), with syntectonic emplacement (red dots), with locally visible tardi-magmatic deformation (violet dots) and without visible structures (blue dots). Regression curve is presented only to highlight the main variation. **b** $P'T$ diagram for the different plutons with associated uncertainties (see Table 1 and (**a**))

metamorphic host-rocks (P2) around the eastern part of the Teg Orak pluton had a more plastic behavior and transmitted the regional strain to the intrusion. That also explains the "flow" fabric more or less preserved in some plutons (Abdou—Fig. 6a, Hanane, etc.). The nature of the basement has, therefore, a major role, particularly

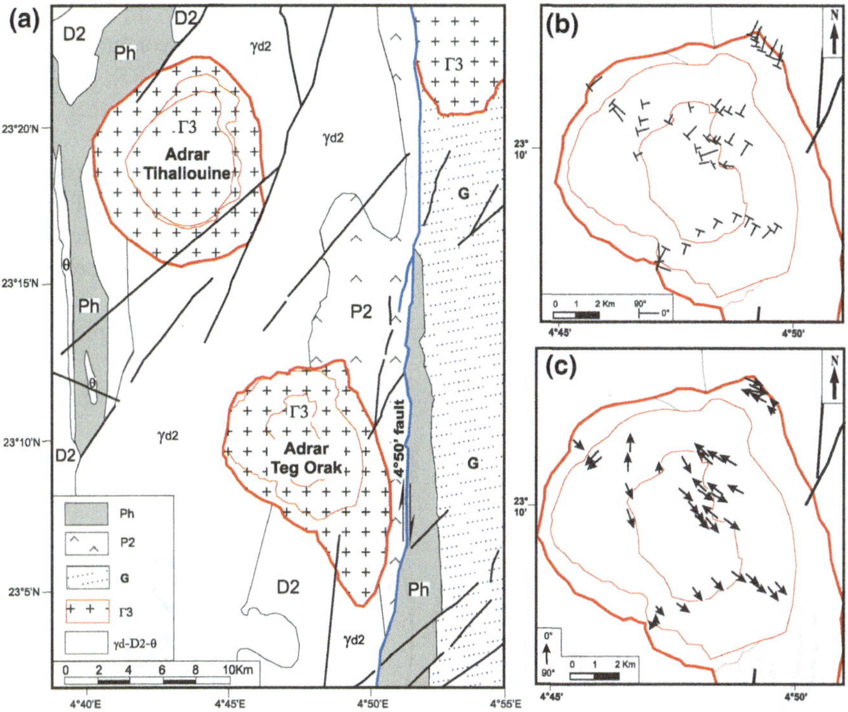

Fig. 5 Comparison of magnetic fabrics mainly related to flow (western part of Teg Orak pluton) and late-magmatic deformation (central and eastern Teg Orak parts): **a** geological setting (Geological map "Tin Felki" 2002) of the studied plutons (Γ3): Ph: volcanites (rhyolite, dacite, tuff); P2: metamorphic rocks; γd2-D2-θ: Afedafeda syntectonic granitoids (γd2: granodiorite and granite, D2: norite, gabbro and diorite, θ: lentil of ultramafite); G: Gneiss and migmatites of the LATEA metacraton. **b**, **c** Maps of Teg Orak pluton with the distribution of magnetic foliations and magnetic lineations, respectively. Figures reproduced from Henry et al. (2008), with permission of Springer

for "post-orogenic" plutons for which the regional stress field was moderate during emplacement.

5 Contributions of the Hoggar AMS Data

5.1 Magnetic Fabric "Tests"

These different studies highlighted the importance of complementary "tools" in the AMS interpretation, in addition to comparison with visible foliation and lineation.

- In favorable outcrops conditions (observation available on several different planar surfaces outcrops), preferential orientation of enclaves can be well-defined. Such an orientation is related to magma flow. In the studied plutons, only Abdou and Oued Tiririne intrusions presented favorable outcrops. For Abdou, that pointed AMS relative to magma flow (Fig. 6a), while the magnetic fabric of Oued Tiririne appeared clearly related to another origin (Fig. 6b).

- Comparison of fabric of late-magmatic dykes with that of the main intrusion is a very strong tool. In two plutons (Teg Orak, Tin Ghoras), the fabric of narrow dykes of various orientation and that of the "host" pluton appeared quite similar (Fig. 7), pointing a late-magmatic fabric acquisition. When the "frame" of the granite was acquired by the crystallization of the main minerals, the stress environment changed from the "hydrostatic" to the "anisotropic" type.

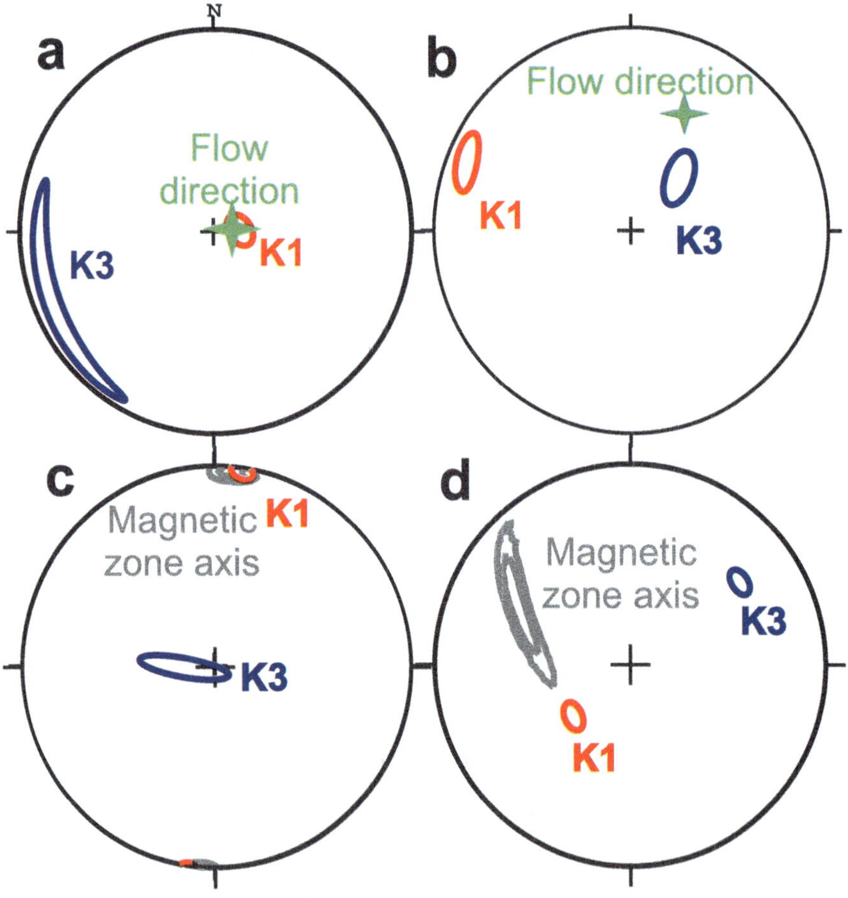

Fig. 6 Comparison of magnetic fabric (K_1 and K_3 confidence zones at 95%) with other known directions: **a**, **b** flow direction (star), from enclaves lengthening in Abdou (**a**) and Oued Tiririne (**b**) plutons. **c**, **d** Magnetic zone axis (confidence zones at 95 and 63%) in the Tin Ghoras (**c**) and Ohergehem (**d**) plutons (stereographic projection in the lower hemisphere)

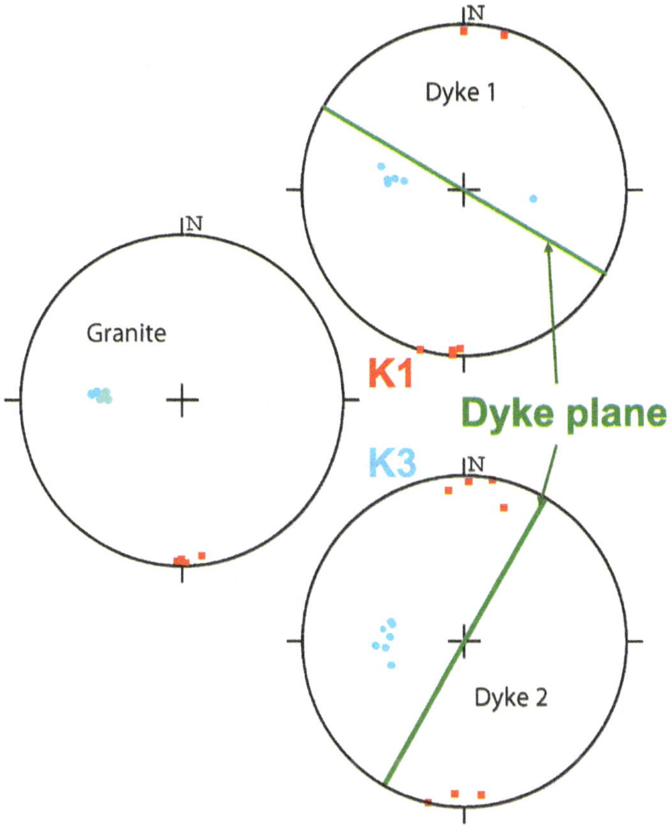

Fig. 7 Comparison of the magnetic fabric (K_1 open squares, K_3 full circles) in dykes (with an indication of their mean plane) and their host-granite: neighboring sites in the Tin Ghoras pluton (stereographic projection in the lower hemisphere)

The anisotropy of the magnetite, which crystallized mostly during this period, then reflects the regional stress conditions during the late-magmatic stage. This is confirmed by the homogeneity of the magnetic fabric in many plutons, particularly the magnetic lineation orientation that appeared independent from the plutons shape (Fig. 8).

- Comparison of fabric intensity (e.g., P' values) is only reliable for rocks with homogeneous characteristics. A diagram P' as a function the mean susceptibility K_m can evidence clear difference in deformation levels (Fig. 4a).
- The magnetic zone axis (Henry 1997) has mostly the same orientation as K_1 axis (Fig. 6c). When it is different from the K_1 axis, this K_1 axis is mainly related to a mineral lineation. That is the case for Ohergehem (Fig. 6d), Arigher and Honag data, evidencing superimposed elements in their magnetic fabric. The initial magnetic foliation was disturbed during a later event, then defining a zone axis different from the mineral lineation.

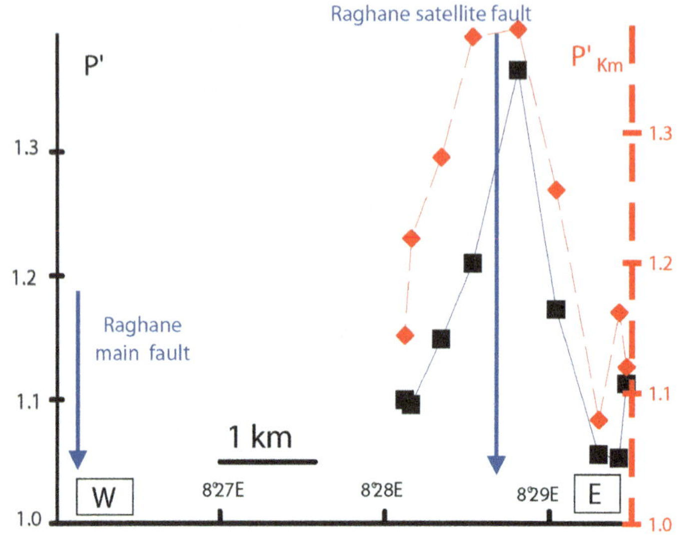

Fig. 8 Variation of the magnetic fabric intensity related to a major fault: E–W cross-section in the Oued Touffok pluton. Only to highlight and limit the effect of the mean susceptibility K_m variation on P' values, an arbitrary weighting was applied by the determination of P'_{K_m}: $P'_{K_m} = \left((P' - 1)/\sqrt{K\mathrm{m}}\right) + 1\right)$ with K_m in 10^{-2} SI (arbitrary scale for P'_{K_m})

5.2 The Different "Types" of Obtained Fabrics

Only Abdou pluton gave a flow fabric apparently without any effect of the regional stresses context. Upward flow direction is indicated by vertical K_1 axis and magnetic foliation is vertical and striking parallel to the pluton border (Fig. 6). The mean susceptibility is similar to that of the hosting rocks (Arokam orthogneisses) suggesting Abdou magma from melting of these orthogneisses, but none inherited zircons were found in the Abdou samples. The lack of effect of the regional stress field is clearly related to the hosting rocks, which acted as a rigid body protecting the intrusion from these stresses.

Syn-deformation flow fabric clearly characterizes the Ohergehem pluton (Fig. 6d), with apparent coincidence between magmatic and magnetic foliations. The Yvonne granodiorite fabric could have the same syn-deformation origin, though the lack of visible structures. That could be the case also for the Ain Kahla and In Telloukh granites. For these four plutons, magnetic lineation is dominantly oriented E–W.

Except In Tounine and Tifferkit plutons, the Pan-African plutons have similar characteristics, in particular, subhorizontal magnetic lineations striking around a N–S direction. Though some plutons (Adjou, Tin Ghoras) seal major shear structures, that points to a continuous regional shearing context. Some other structures, in particular, NNE–SSW were active during and after plutons emplacement (Honag,

Kerkour, Arigher). Magnetic foliation could be on the contrary still, at least partially, related to flow structure (e.g., Teg Orak, Tin Ghoras). As indicated for the Abdou pluton, an important factor for many Pan-African plutons in the Hoggar shield was the response of the hosting rocks to regional constraints. Some plutons (e.g., Tihaliouine, Tioueine, western part of Teg Orak) were partially "protected" from this stress field by "rigid" hosting rocks, while more basic plutonic and metamorphic host-rocks had a more plastic behavior and transmitted the regional strain (eastern part of Teg Orak). However, the most important information is that all the late Pan-African period was characterized by a continuity of the regional shearing context. The Eburnean orthogneiss (Arokam) and the relatively old Oued Touffok plutons are the only studied plutons that have a fabric related to a solid-state Pan-African deformation.

6 Conclusion: The Pan-African Events Evidenced by the Magnetic Fabric Data

All these methodological approaches, besides field observations as well as other key geochemical and dating laboratory data, contributed to a better knowledge of the Hoggar shield. The magnetic fabric of the oldest plutons is related to solid-state deformation on a cold body during Pan-African times. Some other plutons of the eastern Hoggar represent magmatism contemporaneous to the main northward movement of the Assodé-Issalane terrane along the western boundary of the Saharan metacraton (Liégeois et al. 1994). According to the ages of these plutons (Table 1), the end of this period should be between 598 and 580 Ma. The In Telloukh (central Hoggar) and Ain Kahla (western Hoggar) granites emplacement could be contemporaneous of this event, suggesting similar movements around In Guezzam and Arak faults. The later evolution is marked by the continuity of the shearing context, as evidenced in many plutons where the magnetic fabric is characterized by K_1 axes around a subhorizontal N–S direction.

Our different results contribute to better constrain the Pan-African geodynamical evolution of the Hoggar shield. The oblique collision of the Hoggar shield with the West African Craton played a key role during the main movements along shear zones, as evidenced by the syn-deformation flow fabric of some plutons. Regional shearing context during the following period shows that this collision was still active. The sealing by plutons of N–S main shear zones, though NNE–SSW secondary structures remained active, suggests a progressive change in orientation of the continental convergence. Such change could explain the displacement along the late NW–SE strike-slip sinistral fault. This NW–SE strike-slip affects both Raghane shear zone and Yvonne granite.

Acknowledgements We are very grateful to the civil and military authorities at Tamanrasset and Djanet, to the "Office de la Recherche Géologique et Minière" (ORGM) at Tamanrasset, to the "Office du Parc National de l'Ahaggar" (OPNA) and the "Office National du Parc Culturel du

Tassili N'ajjer"—ONPCTA for help during our field works. Thanks also to all people who helped us on the field, particularly H. Djellit, D. Belhai, A. Khaldi, A. Hemmi and M. Ayache. Special thanks also to all our drivers. Contributions of J. P. Liégeois and O. Bruguier were fundamental for the studies in the eastern Hoggar. J. P. Liégeois is also thanked for very constructive review.

References

Acef K, Liégeois JP, Ouabadi A, Latouche L (2003) The Anfeg post-collisional Pan-African high-K calc-alkaline batholith (Central Hoggar, Algeria), result of the LATEA microcontinent metacratonization. J Afr Earth Sci 37:295–311

Adissin Glodji L, Bascou J, Yessoufou S, Ménot RP, Villaros A (2014) Relationships between deformation and magmatism in the Pan-African Kandi Shear Zone: microstructural and AMS studies of Ediacaran granitoid intrusions in central Bénin (West Africa). J Afr Earth Sci 97:143–160. https://doi.org/10.1016/j.jafrearsci.2014.04.012

Archanjo CJ, Trindade, RIF, Bouchez JL, Ernesto M (2002) Granite fabrics and regional-scale strain partitioning in the Seridó belt (Borborema Province, NE Brazil). Tectonics 21. https://doi.org/10.1029/2000tc001269

Azzouni-Sekkal A, Boissonnas J (1987) Geochemistry of the Tioueine Pan-African granite complex (Hoggar, Algeria). Geol J 22:213–224

Azzouni-Sekkal A, Boissonnas J (1993) Une province magmatique de transition du calco-alcalin à l'alcalin: les granitoïdes panafricains à structure annulaire de la chaîne pharusienne du Hoggar (Algérie). Bull Soc Géol Fr 164:597–608

Azzouni-Sekkal A, Liégeois J-P, Bechiri-Benmerzoug F, Belaidi-Zinet S, Bonin B (2003) The "Taourirt" magmatic province, a marker of the closing stage of the Pan-African orogeny in the Tuareg shield: review of available data and Sr–Nd isotope evidence. J Afr Earth Sci 37:331–350

Black R, Latouche L, Liégeois JP, Bertrand JM (1994) Pan African displaced terranes in the Tuareg shield (central Sahara). Geology 22:641–644

Boissonnas J (1974) Les granites à structures concentriques et quelques autres granites tardifs de la chaîne pan-africaine en Ahaggar (Sahara central, Algérie). Thesis, Centre de Recherches sur les Zones Arides, Série Géologie, vol 16, 662 p

Borradaile GJ, Henry B (1997) Tectonic applications of magnetic susceptibility and its anisotropy. Earth Sci Rev 42:49–93

Borradaile GJ, Jackson M (2004) Anisotropy of magnetic susceptibility (AMS): magnetic fabric of deformed rocks. In: Martin-Hernandez F, Lünenburg CM, Aubourg C, Jackson M (eds) Magnetic fabric: methods and application. Geological Society London, Special Publication, vol 238, pp 299–360

Bouchez JL (2000) Anisotropie de susceptibilité et fabrique des granites. C R Acad Sci Paris Earth Planet Sci 330:1–14

Boulfelfel NE, Ouabadi A (2000) Le pluton granitique "Taourirt" pan-africain de Teg Orak (Hoggar Occidental): caractéristiques géochimiques. Bull Serv Géol Algérie 11:147–163

Cheilletz A, Bertrand JM, Charoy B, Moulahoum O, Bouabsa L, Farrar E, Zimmermannn JL, Dautel D, Archibald DA, Boullier AM (1992) Géochimie et géochronologie Rb–Sr, K–Ar et 40Ar/39Ar des complexes granitiques pan-africains de la région de Tamanrasset (Algérie): relation avec les minéralisations Sn-W associées et l'évolution tectonique du Hoggar central. Bull Soc Géol Fr 163:733–750

Dawaï D, Tchameni R, Bascou J, Awe Wangmene S, Fosso Tchunte PM, Bouchez B (2017) Microstructures and magnetic fabrics of the Ngaoundéré granite pluton (Cameroon): implications to the late-Pan-African evolution of Central Cameroon Shear Zone. J Afr Earth Sci 129:887–897

Djouadi MT, Bouchez JL (1992) Structure étrange du granite du Tesnou (Hoggar, Algérie). C R Acad Sci Paris 315 II:1231–1238

Djouadi MT, Gleizes G, Ferré E, Bouchez JL, Caby R, Lesquer A (1997) Oblique magmatic structures of two epizonal granite plutons, Hoggar, Algeria: late orogenic emplacement in a transcurrent orogen. Tectonophysics 279:351–374

Geological map "Tin Felki" (2002) Carte géologique de l'Algérie au 1/200 000 NF-31-XXIII. Service Géologique de l'Algérie, Office National de la Recherche Géologique et Minière

Georgiev N, Henry B, Jordanova N, Jordanova D, Naydenov K (2014) Emplacement and fabric-forming conditions of plutons from structural and magnetic fabric analysis: case study of the Plana massif (central Bulgaria). Tectonophysics 629:138–154. https://doi.org/10.1016/j.tecto.2014.02.018

Guemache MA, Henry B, Djellit H, Derder MEM (2009) Preliminary study of anisotropy of magnetic susceptibility of the Aïn Kahla Late Pan-African granites (north-west Hoggar, Algeria) and structural implications. Bull Serv Géol Algérie 20:189–203

Henry B (1974) Sur l'anisotropie de susceptibilité magnétique du granite récent de Novate (Italie du Nord). C R Acad Sci Paris 278C:11711174

Henry B (1975) Microtectonique et anisotropie de susceptibilité magnétique du massif tonalitique des Riesenferner Vedrette di Ries (frontière italo-autrichienne). Tectonophysics 27(155):165

Henry B (1980) Contribution à l'étude des propriétés magnétiques de roches magmatiques des Alpes: conséquences structurales, régionales et générales. Travaux du Laboratoire de Tectonophysique, Paris, CRE 80/07, 528 p

Henry B (1997) The magnetic zone axis: a new element of magnetic fabric for the interpretation of the magnetic lineation. Tectonophysics 271:325–329

Henry B, Le Goff M (1995) Application de l'extension bivariate de la statistique de Fisher aux données d'anisotropie de susceptibilité magnétique: intégration des incertitudes de mesure sur l'orientation des directions principales. C R Acad Sci Paris 320 II:1037–1042

Henry B, Djellit H, Bayou B, Derder MEM, Ouabadi A, Merahi MK, Baziz K, Khaldi A, Hemmi A (2004) Emplacement and fabric-forming conditions of the Alous-En-Tides granite, eastern border of the Tin Serririne/Tin Mersoï basin (Algeria): magnetic and visible fabrics analysis. J Struct Geol 26:1647–1657

Henry B, Derder MEM, Bayou B, Ouabadi A, Belhai D, Hemmi A (2006) Magnetic fabric of Late Panafrican plutons in the Tamanrasset area (Hoggar shield, Algeria) and structural implications. Afr Geosci Rev 13:41–52

Henry B, Bayou B, Derder MEM, Djellit H, Ouabadi A, Khaldi A, Hemmi A (2007) Late Panafrican evolution of the main Hoggar fault zones: implications of magnetic fabric study in the In Telloukh pluton (Tin Serririne basin, Algeria). J Afr Earth Sci 49:211–221. https://doi.org/10.1016/j.jafrearsci.2007.09.004

Henry B, Derder MEM, Bayou B, Guemache MA, Nouar O, Ouabadi A, Djellit H, Amenna M, Hemmi A (2008) Inhomogeneous shearing related with rocks composition: evidence from a major late-Panafrican shear zone in the Tuareg shield (Algeria). Swiss J Geosci 101:453–464. https://doi.org/10.1007/s00015-008-1262-4

Henry B, Liégeois JP, Nouar O, Derder MEM, Bayou B, Bruguier O, Ouabadi A, Belhai D, Amenna M, Hemmi A, Ayache M (2009) Repeated granitoid intrusions during the Neoproterozoic along the western boundary of the Saharan metacraton, Eastern Hoggar, Tuareg shield, Algeria: an AMS and U–Pb zircon age study. Tectonophysics 474:417–434. https://doi.org/10.1016/j.tecto.2009.04.022

Henry B, Naydenov K, Dimov D, Jordanova D, Jordanova N (2012) Relations between the emplacement and fabric-forming conditions of the Kapitan-Dimitrievo pluton and the Maritsa shear zone (Central Bulgaria): magnetic and visible fabrics analysis. Int J Earth Sci 101:747–749. https://doi.org/10.1007/s00531-010-0634-y

Hext G (1963) The estimation of second-order tensors, with related tests and designs. Biometrika 50:353

Hrouda F, Lanza R (1989) Magnetic fabric in the Biella and Traversella stocks (Periadriatic Line): implications for the mode of emplacement. Phys Earth Planet Inter 56:337–348

Jelinek V (1978) Statistical processing of magnetic susceptibility measured in groups of specimens. Stud Geophys Geod 22:50–62

Jelinek V (1981) Characterization of the magnetic fabric of rocks. Tectonophysics 79:63–67

King RF (1966) The magnetic fabric of some Irish granites. Geol J 5:43–66

Krasa D, Herrero-Bervera E (2005) Alteration induced changes of magnetic fabric as exemplified by dykes of the Koolau volcanic range. Earth Planet Sci Lett 240:445–453

Kratinová Z, Schulmann K, Edel JB, Ježek J, Schaltegger U (2007) Model of successive granite sheet emplacement in transtensional setting: integrated microstructural and anisotropy of magnetic susceptibility study. Tectonics 26:TC6003. https://doi.org/10.1029/2006tc002035

Lelubre M (1952) Recherches sur la géologie de l'Ahaggar central et occidental. Bull Serv Carte Géol Algérie 2:22

Liégeois JP, Black R, Navez J, Latouche L (1994) Early and late Pan-African orogenies in the Air assembly of terranes (Tuareg shield, Niger). Precambr Res 67:59–66

Liégeois JP, Navez J, Hertogen J, Black R (1998) Contrasting origin of post-collisional high-K calc-alkaline and shoshonitic versus alkaline and peralkaline granitoids. Lithos 45:1–28

Liégeois JP, Latouche L, Boughrara M, Navez J, Guiraud M (2003) The LATEA metacraton (Central Hoggar, Tuareg shield, Algeria): behaviour of an old passive margin during the Pan-African orogeny. J Afr Earth Sci 37:161–190

Marre J (1982) Analyse structurale des granitoïdes. BRGM Orléans, 80 p

Neves SP, Araújo AMB, Correia PB, Mariano G (2003) Magnetic fabrics in the cabanas granite (NE Brazil): interplay between emplacement and regional fabrics in a dextral transpressive regime. J Struct Geol 25:441–453

Njanko T, Nédélec A, Kwékam M, Siqueira R, Estaban L (2009) The emplacement and deformation of the Fomopea pluton: implications for the Pan-African history of Western Cameroon. J Struct Geol. https://doi.org/10.1016/j.jsg.2009.12.007

Nouar O, Henry B, Liégeois JP, Derder MEM, Bayou B, Bruguier O, Ouabadi A, Amenna M, Hemmi A, Ayache M (2011) Eburnean and Pan-African granitoids and the Raghane mega-shear zone evolution: image analysis, U–Pb zircon age and AMS study in the Arokam Ténéré (Tuareg shield, Algeria). J Afr Earth Sci 60:133–152. https://doi.org/10.1016/j.jafrearsci.2011.02.007

Paquette JL, Caby R, Djouadi MT, Bouchez JL (1998) U–Pb dating of the end of Pan-African orogeny in the Tuareg shield: the post-collisional syn-shear Tioueine pluton (Western Hoggar, Algeria). Lithos 45:245–253

Paterson SR, Fowler TK Jr, Schmidt K, Yoshinobu AS, Yuan ES, Miller RB (1998) Interpreting magmatic fabric patterns in plutons. Lithos 44:53–82

Pignotta GS, Benn K (1999) Magnetic fabric of the Barrington Passage pluton, Meguma Terrane, Nova Scotia: a two-stage fabric history of syntectonic emplacement. Tectonophysics 307: 75–92

Pitcher WS, Berger AR (1972) The geology of Donegal: a study of granite emplacement and unroofing. Wiley Interscience, New York

Schofield DI, D'Lemos RS (1998) Relationships between syn-tectonic granite fabrics and regional PTtd paths: an example from the Gander-Avalon boundary of NE Newfoundland. J Struct Geol 20:459–471

Tarling DH, Hrouda F (1993) The magnetic anisotropy of rocks. Kluwer Academic Publishers, 232 p

Tomezzoli RN, McDonald WD, Tickyj H (2003) Composite magnetic fabrics and S–C structure in granitic gneiss of Cerro de los Viejos, La Pampa province, Argentina. J Struct Geol 25: 159–169

Zeghouane H, Azzouni-Sekkal A, Liégeois JP (2008) Pétrologie et géochronologie des granitoïdes du massif d'Arirer, (Aouzegueur, Hoggar oriental, Algérie). Abstract. In: 22th Colloquium of African Geology, Tunis, Tunisia, p 259

Electrical Conductivity Constraints on the Geometry of the Western LATEA Boundary from a Magnetotelluric Data Acquired Near Tahalgha Volcanic District (Hoggar, Southern Algeria)

A. Bouzid, A. Bendekken, A. Deramchi, A. Abtout, N. Akacem, M. Djeddi and M. Hamoudi

Abstract The lithospheric structure of the Hoggar massif remains relatively unknown. The lack of high-resolution geophysical studies devoted to it is the source of controversial debates about its underlying structure and its geodynamics. This study targets the western edge of the LATEA microcontinent at the intersection of the 4°50′ sub-meridian major fault and the 4°35′ fault with the oued Amded NE–SW lineament. The study area also includes the northern flank of the Tahalgha Cenozoic Volcanic District. The analysis and modeling of magnetotelluric data collected at 12 sites forming an EW profile of 75 km length made it possible to build a resistivity model over a hundred km of depth. The magnetotelluric data reveal a heterogeneous upper crust made up of probably very compact and

A. Bouzid (✉) · A. Deramchi · A. Abtout
Division Géophysique de Subsurface, CRAAG, BP 63,
Route de l'Observatoire, Algiers 16340, Algeria
e-mail: a.bouzid@craag.dz

A. Deramchi
e-mail: aboubakarderamchi@gmail.com

A. Abtout
e-mail: a.abtout@craag.dz

A. Bendekken · N. Akacem
Unité de Recherche de Tamanrasset, CRAAG, BP 32,
Tamanrasset 11000, Algeria
e-mail: a_bendekken@yahoo.fr

N. Akacem
e-mail: akacemn@yahoo.com

A. Deramchi · M. Djeddi · M. Hamoudi
Département de Géophysique, FSTGAT, USTHB, BP 32,
Al Alia, Dar el Beida, Algiers 16123, Algeria
e-mail: dallidjeddi@gmail.com

M. Hamoudi
e-mail: hamoudi@ipgp.fr

A. Bendaoud et al. (eds.), *The Geology of the Arab World—An Overview*,
Springer Geology, https://doi.org/10.1007/978-3-319-96794-3_5

mechanically strong, electrically resistive structures corresponding to hypovolcanic granitoid and batholiths, and others more conductive probably more inhomogeneous, weak and more fractured corresponding to the Paleoproterozoic metamorphic basement. On the contrary, the lower crust and the lithospheric mantle down to a depth of about 100 km are fairly homogeneous. The 4°50′ mega-fault is rooted into the lithospheric mantle to a depth of about 70 km. This is corroborated by potential field data to at least the base of the crust. By comparison, the 4°35′ fault does not appear as important. The fault network highlighted by the magnetotelluric data has probably been used to transport melt from the asthenosphere up to the surface to give rise to the Tahalgha volcanic district. The fluids released or the precipitated mineralization are at the origin of the strong fall of the resistivity which gives a signature so peculiar to these faults.

Keywords Hoggar · LATEA · Tahalgha volcanic district · 4°50′ fault
Magnetotellurics

1 Introduction

The present study focuses on a geologically complex area located on the western edge of LATEA at the intersection of the major sub-meridian 4°50′ mega-fault with the NE–SW to ENE–WSW oued Amded lineament and the location of the Tahalgha volcanic district. Indeed, the geological structure underlying this area has been shaped by the different major geological events that have affected the African plate since the Precambrian. According to Liégeois et al. (2003), it corresponds to the closure during the Neoproterozoic of an oceanic domain followed by the accretion of island arcs against the LATEA microcontinent. This old suture zone appears today at the surface to be the sub-meridian 4°50′ mega-fault. In Pan-African times, during the formation of Gondwana, an intracontinental oblique convergence between the West African Craton and the Saharan Metacraton shaped the current structure of the Tuareg Shield and caused the dislocation of LATEA and the reactivation of the 4°50′ fault. This shear zone allowed a displacement of several hundreds of km on both sides. The opening of the South Atlantic had induced distensive movements within the African continent and the creation of numerous aborted rifts among which that located along the lineament of oued Amded. In the field, this lineament is represented by a corridor along which a large concentration of NE–SW faulting can be noticed. The magmatic districts of Tahalgha, Atakor/Manzaz, and Adrar N'Ajjer are also aligned along this lineament (Dautria and Lesquer 1989; Aït-Hamou and Dautria 1994). During the Cenozoic, the convergence between African and Eurasian plates resulted in a remote reactivation of the Tuareg shear zones and the advent of a tholeiitic volcanism in the center of the LATEA and alkaline volcanism towards the edges (Liégeois et al. 2005 and references therein). The volcanic district of Tahalgha is located at the periphery of

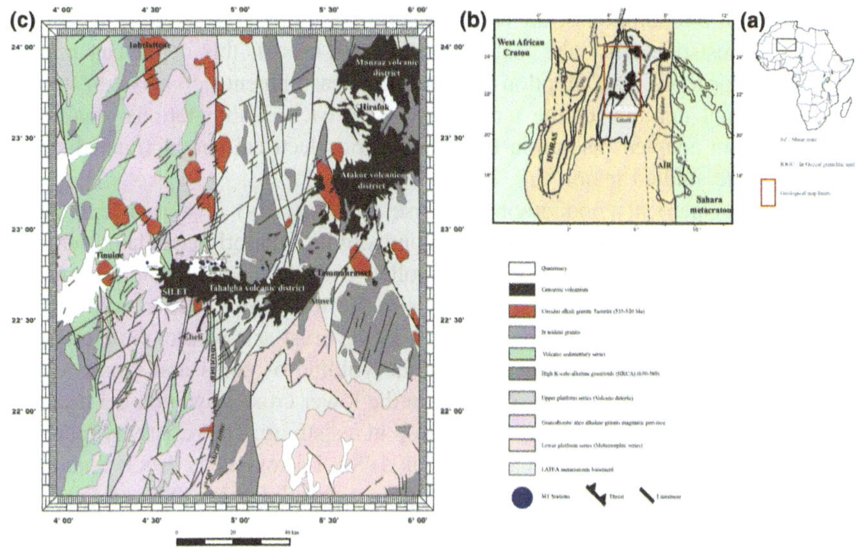

Fig. 1 The location of the Tuareg shield on the African continent map (in **a**), the location of the study region on a schematic map of the Tuareg shield (in **b**, after Black et al. 1994; Liégeois et al. 2003), and a simplified geological map of the study area (modified from Bertrand and Caby 1977; Zerrouk et al. 2016; in **c**)

LATEA, at the intersection of the fault 4°50′ and the lineament of oued Amded. It showed intense volcanic activity during the Neogene and the Quaternary (Fig. 1).

The Hoggar Shield has been well studied from surface geology, petrology, and xenoliths (Liégeois et al. 2003 and references therein; Dautria et al. 1988; Kourim et al. 2014). However, its deep structure (crustal and lithospheric) remains relatively unknown due to the paucity of geophysical studies (e.g., Bouzid et al. 2015). This is at the origin of controversies about its deep structure and its geodynamics. Global or continent-wide geophysical studies do not adequately resolve its underlying lithospheric structure, particularly beneath the volcanic districts and the mega-shear zones (Crough 1981; Lesquer et al. 1988; Pasyanos and Walter 2002; Sebaï et al. 2006; Gangopadhyay et al. 2007; Fishwick and Bastow 2011; Rougier et al. 2013). High-resolution geophysical studies are therefore of great importance in constraining the underlying geological structure of this region, thus bringing some elements into the debate on its deep structure.

Magnetotelluric (MT) is a passive geophysical method able to probe the crustal and upper mantle geological structure. From surface measurements of the natural telluric currents and the terrestrial magnetic field rapid fluctuations, the MT method allows to infer the distribution of the electrical conductivity in the lithosphere. Indeed, for a stratified Earth, and for a given frequency, the electrical resistivity of the lithosphere is proportional to the square of the ratio of the electric field component to the magnetic field perpendicular component (Tikhonov 1950; Cagniard 1953). The penetration depth of the electromagnetic energy is proportional to the square root of the duration of the recordings. For broadband measurements and for an old and resistive crust like that of the Hoggar massif, this depth can reach a hundred km or more. The adoption of a suitable measurement step between stations and an optimal duration of the measurements in each of the stations, enables to image with an adapted resolution the lithospheric electrical conductivity distribution under the studied region.

The bulk electrical conductivity (inverse of resistivity) of a rock can be an excellent geological marker. Indeed, the conductivity is poorly correlated with the nature of the rock matrix but highly controlled with the porosity, the presence of fluids, mineralization or melt as well as with the thermodynamic conditions that prevail in the crust and in the upper mantle. In the crust as a whole, the increase in pressure with depth reduces substantially the volume of pores and thus causes a strong increase in the resistivity of rocks. In the lower crust, however, the presence of fluids or graphite induces a significant drop in resistivity. The resistivity increases moderately in the lithospheric mantle (Haak 1980; Lastoviskova 1983; Shankland and Ander 1983; Jödicke 1992; Jones 1999; Jones and Ferguson 2001; Nover 2005; Jones 2013).

Furthermore, the lithospheric electrical resistivity may vary laterally for several reasons. Old suture zones may be areas of strong falls in resistivity due to sediments trapped during ocean closure or due to the presence of volcanic rocks formed during the collision (Gokarn et al. 2002; Rao et al. 2007, 2014). Similarly, the existence of faults and shear zones in the lithosphere which can act as drain for the circulation of fluids or zones of mineralization precipitation, or even zones of weakness allowing magma rising to the surface, are generally associated with a sharp drop in resistivity (see for example Jones et al. 1992; Ritter et al. 2003; Unsworth and Bedrosian 2004; Bouzid et al. 2008, 2015). The rise of melt through the lithospheric faults and their spreading under the Moho or under the upper crust, as well as the existence of crustal volcanic chambers and reservoirs, may also induce sharp falls in the electrical resistivity within the lithosphere (Wannamaker et al. 2008; Meqbel et al. 2014; Bouzid et al. 2015).

In the present study, magnetotelluric broadband data collected at 12 sites in the Silet/Abalessa area of the Hoggar region will be analyzed and modeled. A 2D lithospheric resistivity model will be proposed and will be interpreted in geological terms.

2 Geological and Geophysical Settings

The Tuareg Shield, situated in North Western Africa, is composed of three massifs, the Hoggar in Algeria, the Adrar des Iforas in Mali and the Air in Niger. Important sub-meridian lithospheric shear zones (4°50′ and 8°30′) characterize the structuration of the Hoggar as dislocated blocks (Caby 2003). Historically, this led to subdivide it into three parts, the Western Hoggar, the Central Polycyclic Hoggar and the Eastern Hoggar (Fig. 1). Later, the Tuareg Shield has been shown to be composed of 23 terranes (Black et al. 1994) that mostly are superposed on the previous subdivision. Especially, the terranes composing most of the Central Polycyclic Hoggar (Laouni, Azrou-n-fad, Tefedest, Egéré-Aleksod) have been shown to have a common history and have been grouped within the LATEA (acronym of the above terranes) metacraton (Liégeois et al. 2003), an Archeo-Eburnean basement partially reactivated during the Pan-African orogeny. The metacraton LATEA (Liégeois et al. 2003) had a complex history, it has been affected by an Eburnean highgrade metamorphism (amphibolite- to granulite-facies 2.1-1.9 Ga; Bendaoud et al. 2008). After that, LATEA has been dissected by shear zones along which high-K calc-alkaline granitoid batholiths has been implemented, generating local high-temperature amphibolite facies metamorphism (Bendaoud et al. 2008). During the Mesoproterozoic, LATEA was cratonized by acquiring a rigid lithospheric mantle (Black and Liégeois 1993; Abdelsalam et al. 2002; Liégeois et al. 2003) and behaved as a passive continental margin. The Neoproterozoic Iskel block represents the most eastern terrane of Western Hoggar. It is inlaid between the In Tedeini terrane to the west and the LATEA metacraton to the east and it is separated from the latter by the 4°50′ shear zone. Caby (2003) considers that Iskel corresponds to a Mesoproterozoic arc-type terrane, showing a low degree metamorphism (greenschist to amphibolite). It is characterized by Neoproterozoic wide magmatic-arc batholiths overlapping the early Neoproterozoic volcano sedimentary series. The presumed Paleoproterozoic granito-gneissic basement outcrop only in the Timgaouine area. Several west-dipping subduction episodes led to the establishment in the middle and in the upper crust of TTG batholiths and typical active margin volcano sedimentary series (Bechiri-Benmerzoug 2009). The 4°50′ is considered as a cryptic suture connecting different crustal blocs of LATEA microcontinent and Iskel terrane (Caby 2003) or an internal structure of LATEA, Iskel resting upon the LATEA basement (Azzouni-Sekkal et al. 2003; Liégeois et al. 2003, 2013). At the end of the Pan-African orogeny (during 580–540 Ma) and due to numerous stages of transtension, LATEA and Iskel have been intruded by circular alkali-calcic granitoids called "Taourirt" (Azzouni-Sekkal et al. 2003). Since the Oligocene period, the Tuareg shield has seen an intense volcanic activity due to an asthenospheric upwelling (Dautria and Girod 1991; Aït-Hamou et al. 2000) or to a reactivation of the shield by the distant Africa-Europe convergence (Liégeois et al. 2005). The volcanic activity is represented in several districts (Adrar N'Ajjer, North-Anahef, South Amadghor, Manzaz, Atakor, Eggéré, In Ezzane, and Tahalgha); some of them follow the North-East South-West oued Amded lineament. The Tahalgha district constitutes the largest volcanic area with 1800 km^2 and E–W lateral extension across tectonic blocks of LATEA and Iskel terrane. It is composed by extensive basaltic lava (Dautria 1988) issued by several strombolian stratovolcanoes whose volcanic activity started in the Miocene.

The current Hoggar massif is a Precambrian basement of a broad lithospheric swell morphology with a diameter of about 1000 km and an area of $\sim$300,000 km^2. This asymmetrical dome with a softer slope on the western side, has an altitude of 1000–1500 m and displays volcanic districts whose altitudes exceed 2000 m (Dautria and Lesquer 1989). Its highest point, Mount Tahat, is located in the Atakor volcanic district at an altitude of 2918 m (Rougier et al. 2013). Based on the modeling of the longitudinal profiles of three major Hoggar oueds, the crustal uplift rate of this massif was estimated by Roberts and White (2010). According to these authors, the uplift began at the Eocene (40–50 Ma) with an amount of 0.2–0.5 km, continued during the Miocene with an amplitude of 0.4–0.6 km. These results are consistent with those obtained from the analysis of the thermochronological data of apatite (U–Th)/He (Rougier et al. 2013). The Hoggar massif as a whole was eroded during the Eocene before the beginning of the magmatic activity at c. 35 Ma. The Hoggar basement that is now exposed on the surface would have been buried after the Early Cretaceous under a sedimentary cover of more than 1 km (Rougier et al. 2013).

In an attempt to study the nature and mantle structure at the origin of this important relief of Hoggar, several geophysical studies have been initiated: using seismic tomography, gravimetric, magnetism, heat flux, and magnetotelluric. Seismic tomography models obtained on a global or continental scale suffer from their low lateral resolution. These models cannot solve geological objects of less than few hundred kilometers in size, such as the different terranes constituting the Hoggar, suture zones or shear zones. However, these tomographic models show a slow and thin lithospheric mantle under the Hoggar relative to the West African Craton in the west, the Saharan metacraton to the east, and the Sahara platform to the north (Sebaï et al. 2006; Priestley et al. 2008; Begg et al. 2009; Fishwick and Bastow 2011; Liégeois et al. 2013). On a local scale, P-wave seismic data confirm that the Hoggar is generally characterized by a slow mantle. More particularly, the volcanic districts of Atakor and Tahalgha are underlain by zones of low seismic velocity in connection with the recent volcanic activity of these districts (Ayadi et al. 2000). On an even smaller scale, around Tamanrasset, Liu and Gao (2010) analyzed the receiver functions obtained at a single station (Geoscope seismic station of Tamanrasset). These authors revealed a Moho at about 34 km below the surface and spatial variations correlated with the rheological characteristics of the crust under the two adjacent terranes. The Tefedest to the west being very fractured contains volcanism, whereas Laouni to the east is less fractured so more rigid, is lacking. The Bouguer anomaly map of the African continent reveals the existence of a negative anomaly in the Hoggar massif (Pérez-Gussinyé et al. 2009). At the Hoggar scale, although the distribution of the data is not perfectly homogeneous and has many gaps, this anomaly appears to be localized in LATEA, more or less coinciding with Atakor, Manzaz, and Amadror. According to Lesquer et al. (1988), the Hoggar Bouguer anomaly map shows two types of trends: a trend corresponding to the long wavelengths can be correlated with the broad topographic swell, and the other of short wavelengths can be correlated with the structure of the Precambrian basement of the massif. The long wavelength negative anomaly was interpreted as the gravimetric response of a light structure underlying the crust (Lesquer et al. 1988). The effective elastic thickness of the African continent was estimated by Pérez-Gussinyé et al. (2009) from the coherency analysis between the continent topography and the gravimetric data. The Hoggar massif is characterized by a small effective elastic thickness probably due to a relatively thin lithosphere compared to the West African Craton, the Saharan Metacraton, and the Saharan Platform. Heat flow data show a relatively warm lithosphere in agreement with its Precambrian age with an average of 53 mW/m^2 (Lesquer et al. 1989). A strong anomaly is observed in the north in the Saharan sedimentary basins (Takherist and Lesquer 1989). But an important gap in the data between these two regions does not allow to locate the southern limit of the basins anomaly. However, a relatively important heat flux was observed near Tamanrasset town. It appears to be associated with the Cenozoic magmatic activity of the Atakor and Tahalgha volcanic districts (Lesquer et al. 1989). On the other hand, magnetotelluric data do not show a large-scale regional anomaly under the Hoggar (Bouzid et al. 2004). Lithospheric conductivity anomalies have been observed that may be associated with some geological faults (Bouzid et al. 2008) or with volcanic districts such as Atakor (Bouzid et al. 2015).

3 Data Collection and Analysis

3.1 Data Collection

The magnetotelluric data were collected at 12 sites during two field surveys. Three stations (53, 54 and 55) have been acquired in 2007 during a previously broad reconnaissance survey of Western and Central Hoggar at large scale with ~40 km interstation distance. The lateral resolution of the MT data was significantly improved by acquiring 9 new sites (1–9) in November 2010. The whole forms an east-west profile 75 km long, centered on the 4°50′ mega-fault near Abalessa town, located 70 km west of the city of Tamanrasset (Fig. 2). Another interesting aspect of the MT profile location is that it is located just on the northern flank of the Tahalgha volcanic district. To get a better lateral resolution at the upper crust depth, a 2 km measuring step was adopted near the surface trace of the 4°50′ fault. Then, it was increased to 4, 8, and 16 km towards the profile ends (Fig. 2). The horizontal telluric (electric) component and the magnetic vector time series have been acquired using the V5 system 2000 of Phoenix Geophysics. To attempt to reach a deep penetration across the lithosphere, each site has been occupied for about 20 h. The measuring sites are located far enough from any significant human activity and electromagnetic noise level was therefore very low, then no reference station was needed. Time series collected have been processed using a robust processing code based on Jones and Jödicke (1984) and provided by the instrument manufacturer. The magnetotelluric impedances and tippers obtained are of good quality and range between periods of 0.001 and 3000 s (Fig. 3). Furthermore, visual examination of apparent resistivity data of the tensor impedance main elements shows no large discrepancy to the very short periods. This offset would be indicative of the presence of static shift, then no such correction was applied to data.

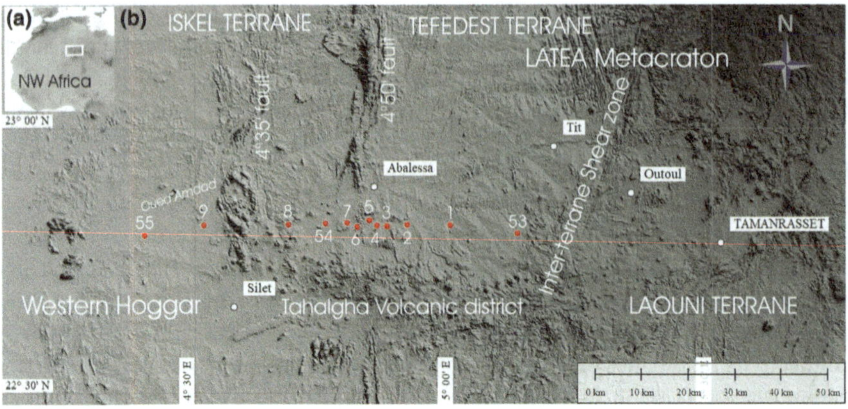

Fig. 2 Location map. **a** A North-West Africa topographic map (GTOPO30), the study area is pointed out by a rectangle. **b** A magnetotelluric sites (dots) location on SRTM3 map of the Silet-Abalessa area. The 4°50′ fault limiting Tefedest terrane from Iskel terrane as well as the 4°35′ fault and the volcanic district of Tahalgha are clearly visible on the topographic map

3.2 *Tipper/Induction Vector Data Analysis*

The tipper, also called magnetic transfer function, is defined as the ratio of the magnetic field vertical component to its horizontal component. It is theoretically zero for a layered earth but not zero for a 2D or 3D structure. Its magnitude exceeds generally the value of 0.1, sometimes reaches 0.3 at some sites, and is even equal to 0.5 for station 54 at short periods, indicating in these latter cases strong lateral changes in the conductivity distribution in the crust beneath the study area. This was expected if regarding the presence of faults and the vicinity of Tahalgha volcanic district as shown in the geological map (Fig. 1). The induction vector (with its real and imaginary parts) is another way to represent the magnetic transfer function. The real part of the induction vector, plotted in Parkinson's (1962) convention, points towards conductive areas that highly concentrate currents (Fig. 3). Figure 4 shows three maps representing the Parkinson induction vectors plotted for three periods corresponding to three penetration depths: the period of 1/8 s corresponds to upper crust penetration depth, that of 8 s to the lower crust and that of 512 s to the lithospheric upper mantle. For shorter periods, induction vectors are sensitive to local structures. They therefore indicate directions changing from site to another. They point towards two zones in the short period data (1/64–1/8 s): the 4° 50′ fault zone close to the center of the profile and another area situated west of station 54 that may be associated with the 4°35′ fault. Although in the latter case, the situation is less obvious due to the larger measuring step. Directions indicated by induction vectors at long periods ($T > 1$ s) are NNW to North and then NNE for longest periods revealing a regional conductive structure located north to NE of the MT profile (Fig. 4). This effect is due to the existence of a lithospheric zone of

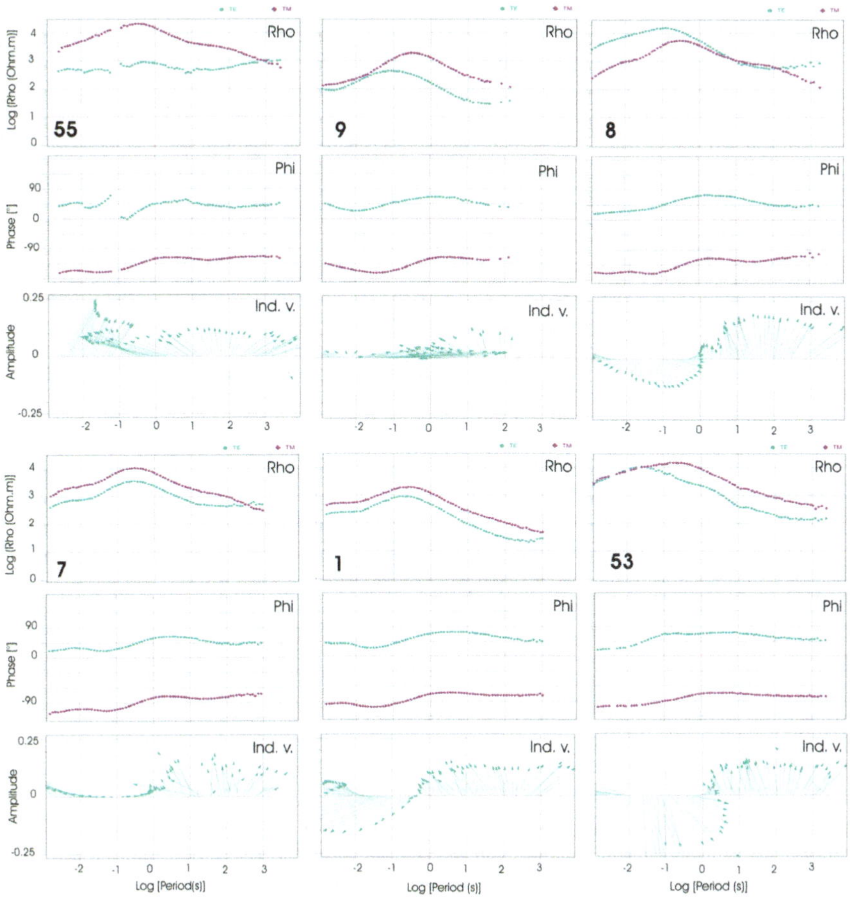

Fig. 3 TE and TM mode apparent resistivity (Rho) and phase (Phi) data, and real component of the induction vector (Ind. v.) plotted following Parkinson's (1962) convention corresponding to sites 55, 9, 8, 7, 1, and 53, respectively. TE and TM mode data are extracted using Groom and Bailey (1989) tensor decomposition for a fixed strike of N15°E

enhanced conductivity that is located out of the profile and which cannot be modeled by these MT sounding data.

3.3 *Impedance Tensor Analysis (Dimensionality, Strike)*

The magnetotelluric impedance tensor analysis consists in determining the geo-electric structure geometry (or dimensionality) and the direction of the regional structure elongation (strike) if the latter is two-dimensional (2D). This task could

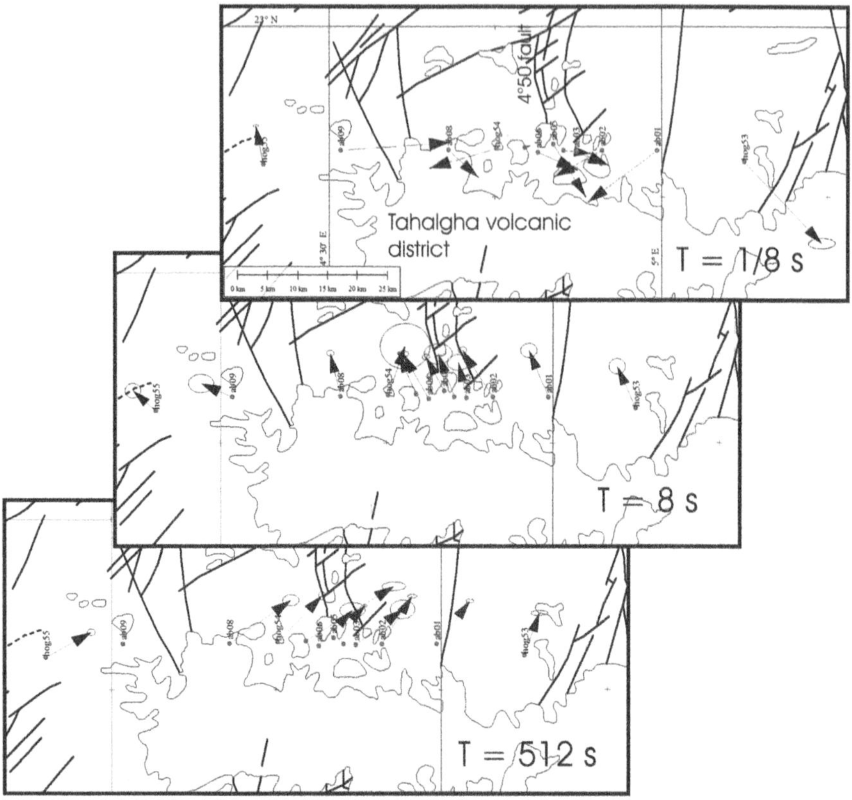

Fig. 4 Induction vectors (Parkinson 1962) maps plotted for three periods corresponding to three different investigation depths. In short and medium periods (T = 1/8 s), induction vectors point towards two specific area situated beneath the profile: the 4°50′ and the 4°35′ faults. At longer periods, lateral effects are situated north (T = 8 s) and NE (T = 512 s) from the MT profile

present some difficulties because the effect of near-surface small-scale inhomogeneities, considered as a geological noise by Bahr (1991), could distort the magnetotelluric response of the regional studied structure. The impedance tensor data analysis was performed according to the method of Bahr (1991) and then that of Caldwell et al. (2004). The Bahr's method consists in determining the values of four dimensionality indicators relative to respective empirical threshold values. The phase-sensitive skew of Bahr (1991) that is sensitive to the geometry of the regional structure only becomes strong (greater than 0.3 or 0.35) for a regional 3D structure. The Bahr skew, calculated frequency by frequency for all stations, remains mainly less than the threshold value of 0.3 (Fig. 5). This indicates that the regional structure under the magnetotelluric profile is roughly not 3D. On the contrary, the Swift (1967) skew, which can be strongly affected by the presence of distortion, exceeds the 0.06 threshold value (Marti et al. 2005) even if the regional structure is 2D or 1D. In the case of our data, this parameter is well above the threshold value

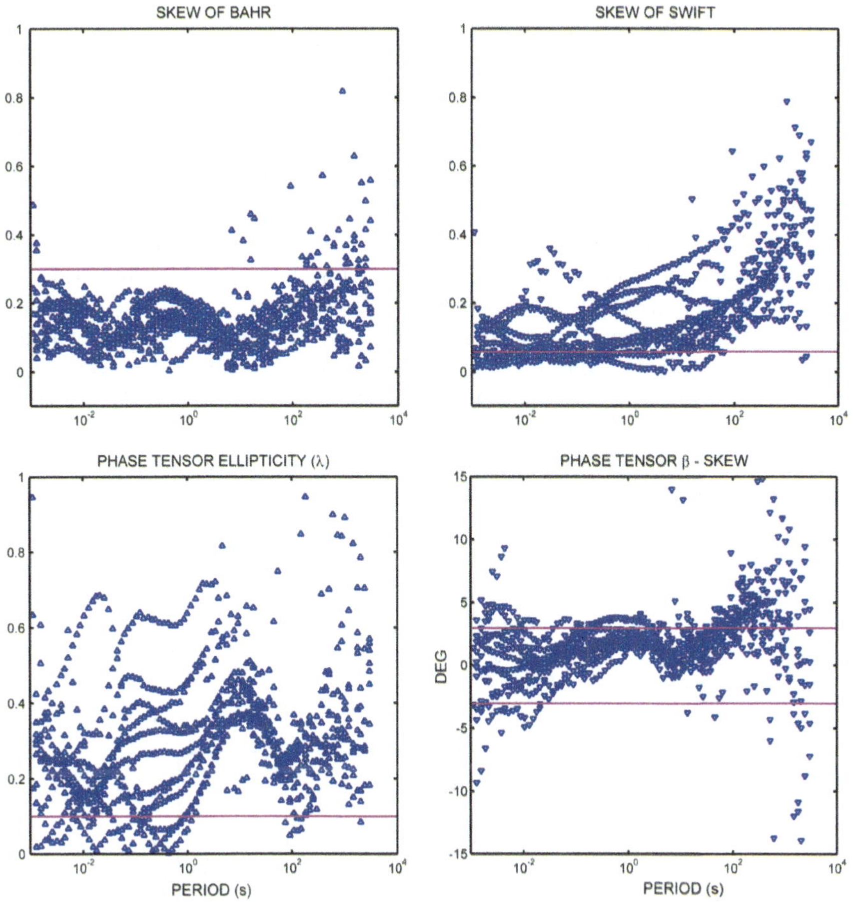

Fig. 5 Phase-sensitive skew (η, Bahr 1991) calculated frequency by frequency for the entire soundings. The line at 0.3 represents an empirical threshold that discriminates between a complex 3D structure ($\eta > 0.3$) and a simple 1D or 2D structure ($\eta < 0.3$). In the figure, some points above the empirical threshold represent a dispersion due to noise in data. The Swift Skew is sensitive to the effect of the galvanic distortion of superficial heterogeneities and may exceed the threshold of 0.06 even if the regional structure is 1D or 2D. Below, the ellipticity of the phase tensor indicates a non-1D structure. The β-skew indicates a roughly 2D regional structure

(Fig. 5). Having a weak Bahr skew concomitantly with strong Swift skew can be interpreted by the existence of a 1D or 2D regional structure to which are superimposed small superficial heterogeneities bellow the MT data resolution. To determine more accurately the regional structure geometry (i.e., 1D or 2D), it is necessary to call the other two Bahr's parameters, namely the rotationally invariant measure of two dimensionality (Σ) and the regional 1D indicator (μ). These two parameters were calculated systemically for all frequencies and for the entire MT stations. Their analysis site by site shows that the regional structure beneath the MT profile is 2D with a 3D superficial superimposed inhomogeneities, i.e., 3D/2D structure.

Bahr's analysis was corroborated by that of Caldwell et al. (2004), commonly referred to as the phase tensor method. The phase tensor is defined by the ratio of the imaginary component to the real component of the magnetotelluric impedance tensor (Caldwell et al. 2004; Bibby et al. 2005). Thus, it reflects the regional geoelectric structure without being sensitive to the presence of surface inhomogeneities. The ellipticity (λ) and β-skew parameters deduced from the phase tensor were calculated and plotted frequency by frequency and station by station (Fig. 5). On the whole, the ellipticity remains well above the threshold value (of 0.1) whereas β-skew remains low overall (<3°) except for long period data where it exceeds the threshold value. It is then reasonable to conclude that the magnetotelluric data reveal a regional 2D structure beneath the profile. The values of β-skew (low at short periods and stronger at long periods) can be interpreted by the response of some quasi-2D crustal geological structures which will be 2D at the medium periods but rather 3D at the long periods (Fig. 6). In addition, this effect will be more apparent on the TE mode data than on the TM mode data.

The regional structure being roughly 2D, it is necessary to determine the azimuth of its elongation, commonly known as the strike angle. To this end, we used three different approaches. The strike angle was calculated frequency by frequency for each survey using the least square method of Groom and Bailey (1989) then the

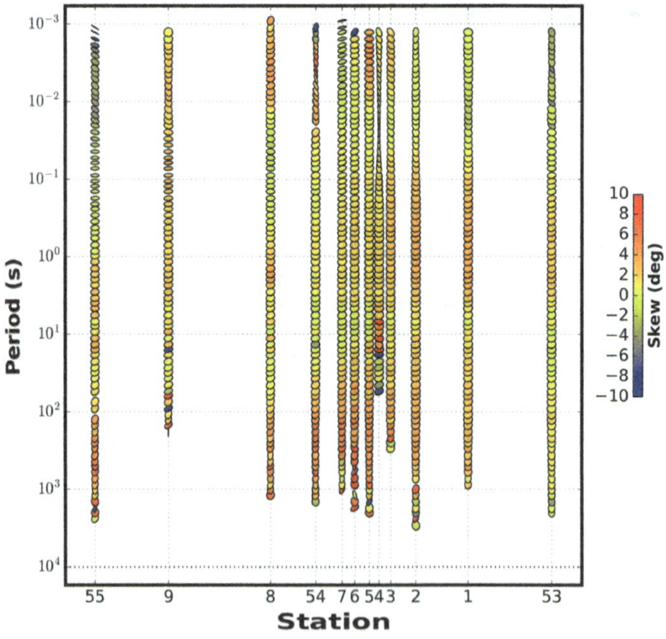

Fig. 6 Phase tensor section plotted using the toolbox of Krieger and Peacock (2014)

analytical formula of Bahr (1988, 1991), and lastly the phase tensor method of Caldwell et al. (2004).

For all three methods, the rose diagrams of the strike angles have been plotted decade by decade to highlight any changes in the strike according to the frequency. This revealed a change in strike at roughly 50 s of period (Fig. 7). Thus, the strike is mainly N15°W for short and medium periods (<50 s) corresponding to crustal depths, and N15°E for long periods (>50 s) corresponding to lithospheric upper mantle penetration (Fig. 7). A difference of 30° in the strike direction is thus shown by MT data between the crust and the mantle.

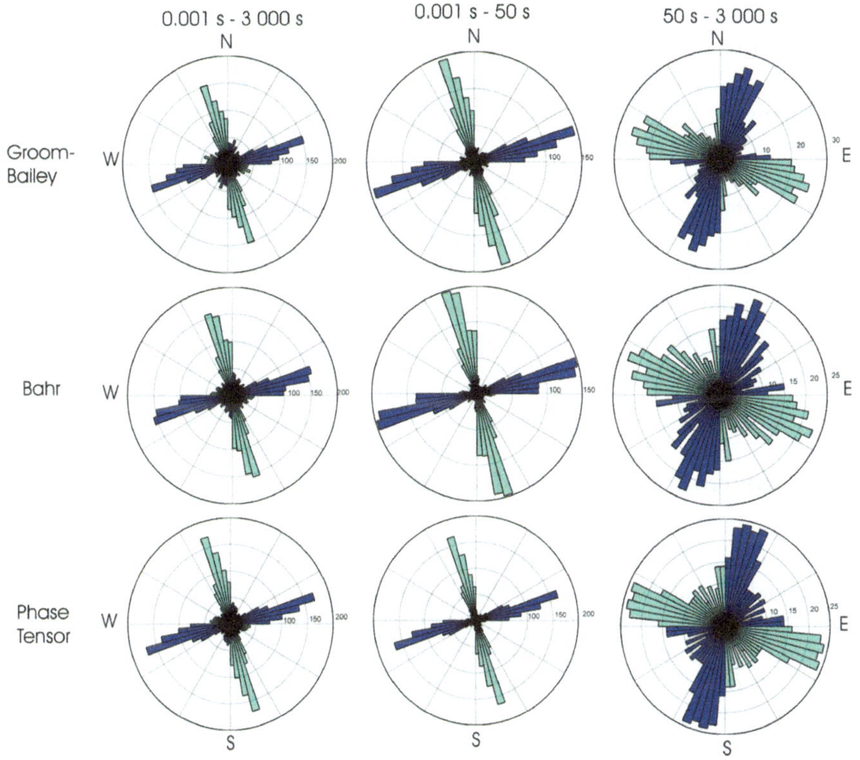

Fig. 7 Strike angles calculated frequency by frequency for all soundings using Groom and Bailey (1991) method (up), Bahr method (middle) and phase tensor method (bottom). The rose diagrams correspond to the entire periods (left column), short and medium periods (in the center) and long period data (right column)

4 Modeling

MT data analysis of all soundings shows a two-dimensional lithosphere structure striking N15°W for short and medium periods that are less than 50 s (crust) and N15°E for long period more than 50 s (lithospheric mantle). Moreover, the measuring site distribution along the profile is not regular. Sites are less spaced at the center (more or less than 2 km) allowing a better resolution of the firsts km of the crust, than at the ends of the profile where the interstation distance increases gradually until reaches 16 km. Taking into account the change in strike according to period and in the measurement step, it seems intuitively more ingenious to model the geoelectric structure under the profile at three different scales: at the central area (at the vicinity of 4°50′) with a shallow model, but higher spatial resolution, then at the crustal scale (strike of −15°) and at the lithospheric size (strike +15°).

Once the strike is fixed, the impedance tensor data of each sounding were decomposed using the Groom-Bailey method (1989). TE and TM modes data are then extracted. In all cases, the nonlinear conjugate gradient algorithm of Mackie et al. (1997) implemented in the Geotools package was used to invert the MT data. The algorithm consists in improving by successive iterations a starting model. At each iteration, a new model is calculated by minimizing a functional equal to the linear combination of a measure of the model simplicity (model roughness) and the fit between the model response and the observed data (root mean square or rms). The trade-off between the model roughness and the rms is made by a regularization parameter (τ).

4.1 Modeling Short and Medium Period Data

To model the crustal structure, the strike has been set to N15°W. Indeed, because of the redundancy in the data due to the short measuring step in the central area (only 2 km) and because of the presence of moderate noise in some stations especially to the long period data, three soundings were not used in the inversion, namely stations 4, 6 and 54. A set of 20 periods regularly distributed between 1/560 and 100 s were chosen to achieve the inversion. Several tests were conducted to find the best mesh that gives the lowest root mean square (rms). Then, a study was conducted to determine the optimum value of the regularization parameter (τ) that plays a role as a balance between data (rms) and the simplicity of the solution (model roughness). Thus, several inversions were performed by changing the τ value from 1, 2, 3, etc., to 100, while keeping the remainder unchanged. For each rms value, the model roughness is plotted with respect to the inverse of τ (Fig. 8). Three L-curves are therefore obtained in the case of the inversion of the TE mode data only, the TM mode data only and simultaneously both modes data (Hansen 1992). They provide three different optimum values of τ parameter, which are, respectively, 8, 10, and 8 (Fig. 8). At this stage, the inversion of apparent resistivity and phase data of the TE mode only and the TM mode only was carried out in two steps, by varying the noise floor on the data. This was taken as 50% on the apparent resistivity data and 2.87° on the phase data (equivalent to 5% on impedance). Then, in the second step, respectively of 10% and 2.87° (Fig. 9b). On the other hand, the simultaneous inversion of the data of both modes was carried out in two steps. In first step, the TM model previously obtained was used as starting model, with therefore the following noise floor values: 10% and 2.87° for respectively apparent resistivity and phase of the TM mode data, and 50% and 2.87° for those of the TE mode data. Then, in a second step, and for both modes, the error floor on apparent resistivity is fixed to 10% and phase to 2.87°. The 2D model was obtained with an rms of 3.67. A similar approach was followed to invert data of both modes but by taking TE 2D model previously calculated as initial model. The resultant model was obtained with an rms of 4.01.

The MT data show that the crustal structure is more conductive in the NNW direction (TE mode) than in the perpendicular direction (TM mode). As a result, the penetration depths obtained by the TM mode data are larger than those of the TE mode data. On the other hand, the MT sites located directly above conductive structures have a smaller investigation depth because they strongly attenuate the electromagnetic waves. This is clearly shown by the Niblett-Bostick transform. Indeed, for the TE mode data and for periods shortest than 50 s, the depth of investigation is about 40 km under stations 8, 2, and 53 while it does not exceed 15 km under the western part of the profile (under 55 and 9), and 18–25 km under the central part of the profile (5, 6 and 3) and under station 1. These three zones display strong drops in electrical resistivity. In contrast, the penetration depths are more important for the TM mode data. They exceed 40 km except under the center of the profile where they reach 30 km. Moreover, the TE mode data are more sensitive to the 3D effects of off-line structures than those of the TM mode. This explains the higher rms (4.39) in the first case and very acceptable one in the second (rms of 0.77). The geological map of the study area shows precisely the presence of changes in geological structure geometry from 10 to 15 km to the north and south of the MT profile. On the other hand, the model corresponding to the TE mode is more sensitive to thin conductive structures than that of the TM mode, which tends to smooth the structures in the model. This is most likely the case of the conductive body located beneath the central zone clearly indicated by the TE data but not in TM mode.

4.2 Modeling of the Central Area

To improve the spatial resolution of the resistivity model of the near surface crustal structure close to the 4°50′ fault area, a short profile of 7 stations and 16 km long has been selected. The interstation distance varies from 1.5 to 4 km. The strike was set to −15° (N15°W) then TE and TM mode data were extracted according to the GB method (Groom and Bailey 1989). Apparent resistivity and phase data corresponding to 19 periods ranging from 1/560 to 56 s (frequency of 0.018 Hz) were

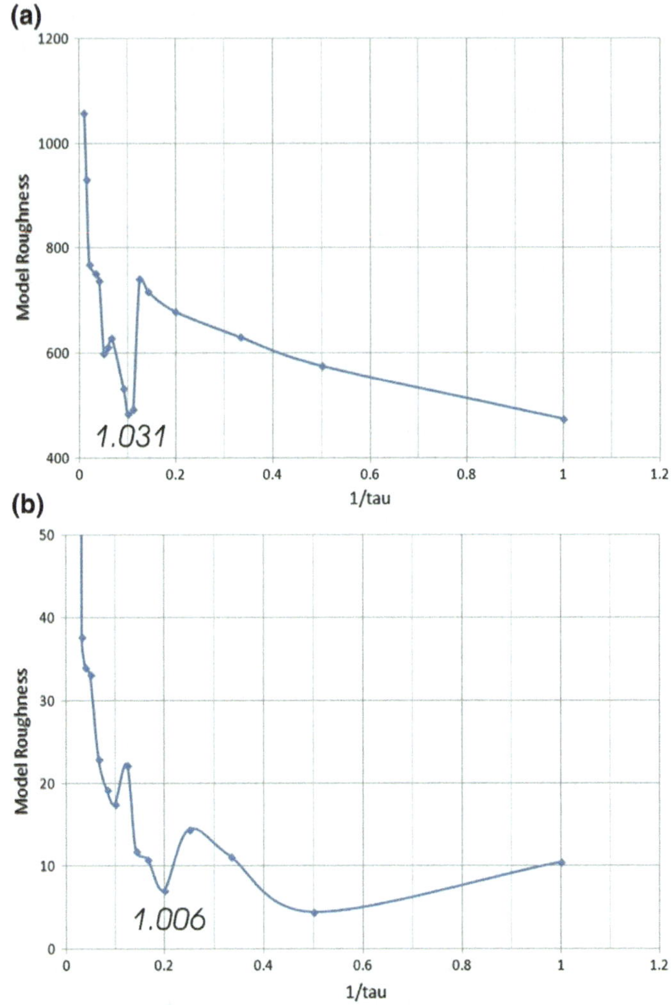

Fig. 8 L-curve representing the model roughness versus the reverse of regularization parameter (τ) for different root mean square (rms) for TM mode data 2D inversion corresponding to short and medium period data of models **a** and **b** in Fig. 9 (top in **a**), and to broadband data of model **c** in Fig. 9 (bottom in **b**). The corner of the curve which gives the optimum value of the regularization parameter corresponds to an rms of 1.031 in the case of the short and medium period data (**a**) and 1.006 in the case of the broadband data (**b**)

chosen to be inverted to estimate the crustal resistivity model. Models corresponding to TE mode only or TM mode only and both modes data were calculated in the same manner than that followed previously for the short and medium data of the previous section. TE model, TM model, TE+TM model (using the obtained TE model as starting model) and TM+TE model (using the obtained TM model as

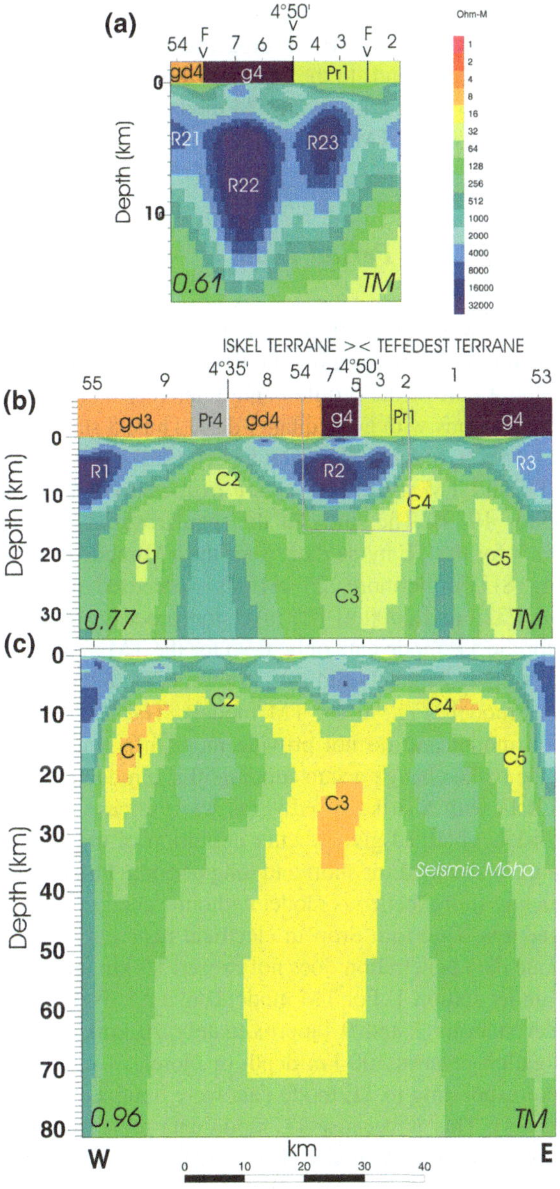

Fig. 9 Crustal 2D resistivity models obtained by inversion of apparent resistivity and phase of TM mode data corresponding to short period and central area, near the 4°50′ mega-fault (top in **a**), short and medium period data (in **b**) and broadband data (bottom in **c**). The description of the surface geology (colored bar on the top of the resistivity models of **a** and **b**) was taken from the Hoggar geological map at 1/1,000,000 of Bertrand and Caby (1977). The notations used are as follows: Pr1, for Paleoproterozoic LATEA metacraton basement; g4, for high-K calc-alkaline granitoids (HKCA, 630–580 Ma); Pr4, for Neoproterozoic volcano sedimentary series; gd3 and gd4, for Neoproterozoic granodiorite/calco alkaline granites magmatic province

starting model), were found with respectively the rms values of 4.72, 0.61, 3.85 and 5.47 (TM model is shown on Fig. 9a). TE model reveals mainly a strong conductive structure under site 6 that extends to the surface. However, this conductive structure appears with a much lower contrast on the TM model. It could then be a structure so thin that it is not resolved by TM mode data.

4.3 Modeling of Long Period Data

The same approach was followed for the inversion of the long period data. A significant penetration depth reaching the lithospheric mantle to about 100 km is obtained. The central part of the profile being relatively densely covered only was taken into account stations with good quality data. Thus, 8 stations were used in the inversion, namely, 55, 9, 8, 54, 5, 2, 1, and 53. The strike was set to N15°E and the Groom-Bailey TE and TM modes data were extracted from the measured tensor (Groom and Bailey 1989). The study of the regularization parameter τ gives the following values: 9 for the TE mode data, 5 for the TM mode data, and 7 for data of both modes (Fig. 8). Furthermore, 23 periods were used that are regularly distributed between 1/320 and 1000 s. The inversion process produces solutions with an rms of 5.82 for the TE mode data only, 0.96 for the TM mode data only, 4.93 for both modes data with TM model used as starting model, and 4.47 for that using TE model as starting model (Fig. 9c for TM model). The last two models are very similar to the TE model and do not provide more information especially for the deeper part. As the strike change occurs around 50 s, a modeling was performed for the long period data with a strike of −15° (as was the case for short and medium periods). The models obtained give a better definition of the crustal part (which is similar to the models obtained for short and long periods) but a very poor resolution for the deep structure (homogeneous model without a change in the resistivity) with however stronger rms. Due to a drop in electrical resistivity in some parts of the crust, the TE mode data penetration does not exceed 15 km under station 9, while it is about 60 km under station 1. For TM mode data, a relatively shallow penetration is obtained under stations 2 and 9 (approximately 70 km). Everywhere else, the depth of investigation reaches 100 km depth or more.

2D models corresponding to TE mode data were obtained with relatively strong rms (>4 in all cases). This is because TE mode data remains very sensitive to 3D effects and unlike those of TM mode. However, the latter give resistivity models with a very acceptable rms but having rather smooth structures. For this purpose, the resistivity models calculated from the TM mode data only have been selected for geological interpretation (Fig. 9).

5 Gravimetric and Aeromagnetic Data Constraints

The constraints provided by the gravimetric and magnetic data are complementary to those of the MT data. Indeed, the potential methods have a better lateral resolution than the MT, while the latter being a probing method better resolves the depths of the geological layers. Potential field data available in the study area will be analyzed and modeled. The results obtained will be integrated with those of magnetotellurics.

The gravimetric data processed in this study come from a compilation of data from several old gravimetric surveys to which new data are added. The new data were collected during a field survey carried out in 2005 along a profile from Tamanrasset to Silet. These new data, linked to the Algerian absolute measurement network, made it possible to homogenize the data of the different surveys (Bouyahiaoui et al. 2011). Their measurement accuracy is of the order of 0.5 mgal. The distribution of the measurement points is inhomogeneous, but a large part of the points is concentrated in the center of the map with an equidistance of about 2 km. The Bouguer anomaly was calculated on the basis of a correction density of 2.67 and a topographic correction at a distance of 110 km (Fig. 10). The gravimetric anomaly map (Fig. 10a) shows that anomalies with an amplitude greater than 40 mgals are elongated in an NS direction. The central area of the map is characterized by a negative anomaly almost wedged between positive anomalies, which suggests a mass deficit in the region centered on the 4°50′ fault. The vertical gradient map (Fig. 10b) enhances the high frequencies and highlights the gravimetric anomalous axes and contacts. It shows a succession of positive and negative anomalies, elongated in the NS direction, revealing the presence of faults within the crust. Thus, the 4°50′ fault signature can be highlighted (Fig. 10).

The magnetic data come from an aero-geophysical survey covering the entire Algerian territory and carried out in 1971 by the American company AERO-SERVICE CORPORATION (ASC). The direction of the flight plans is oriented in EW. The distance between the flight lines is 2 km. It is however 40 km for the tie lines. On a flight line, a measurement is taken every 50 m. The flight altitude is kept constant at 150 m (Bournas et al. 2003). Moreover, the reduced to the pole magnetic map (Fig. 11a) is dominated by a large positive anomaly covering the entire central part of the map. This strong anomaly superimposed rather well with the Tahalgha volcanic district corresponds to the magnetic signature of the lava field. It therefore partly hides the rest of the nearby magnetized geological formations. The vertical gradient map (Fig. 11b) makes it possible to reduce the effect of the lava and to highlight several magnetic discontinuities, as is the case of the positive magnetic anomaly of NS elongation and corresponding to 4°50′ fault.

The gravimetric and magnetic data along a profile that coincides with the MT stations were inverted jointly using GEOSOFT's GMSYS routine. The 2D model (Fig. 12) obtained shows that the lower crust thickens at the center of the profile under the 4°50′ fault to reach a depth of 43 km. Then, the Moho rises to a depth of 22 km to the west end and 30 km to the east end. The lower crust is bordered by

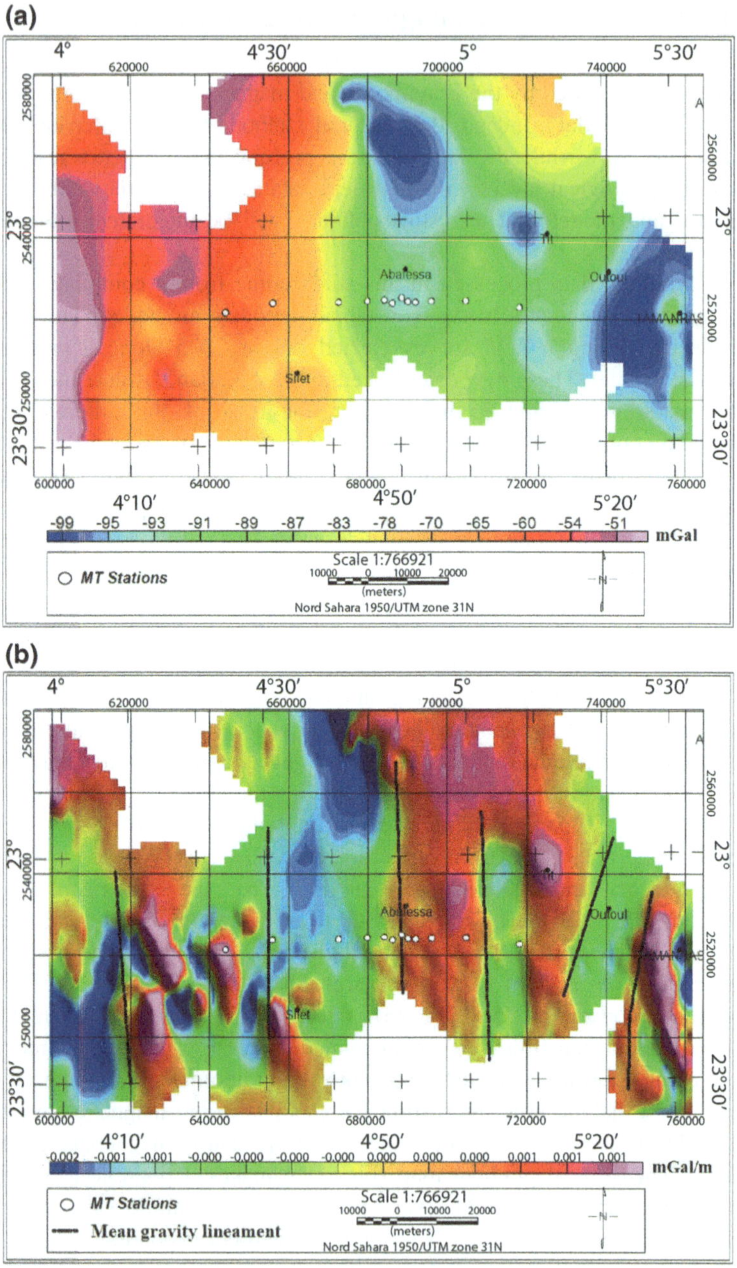

Fig. 10 Bouguer anomaly map around the study area (**a**) and that of its vertical derivative (**b**)

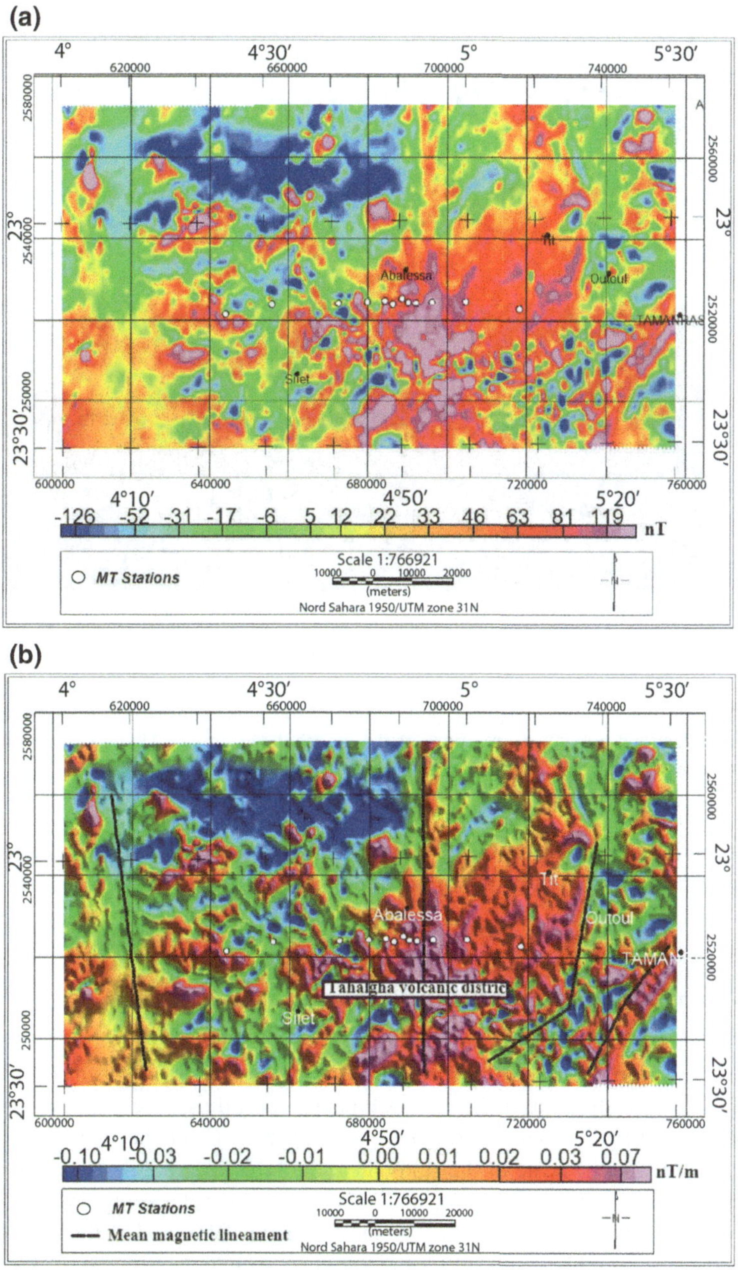

Fig. 11 Aeromagnetic map around the study area (**a**) and that of its vertical derivative (**b**)

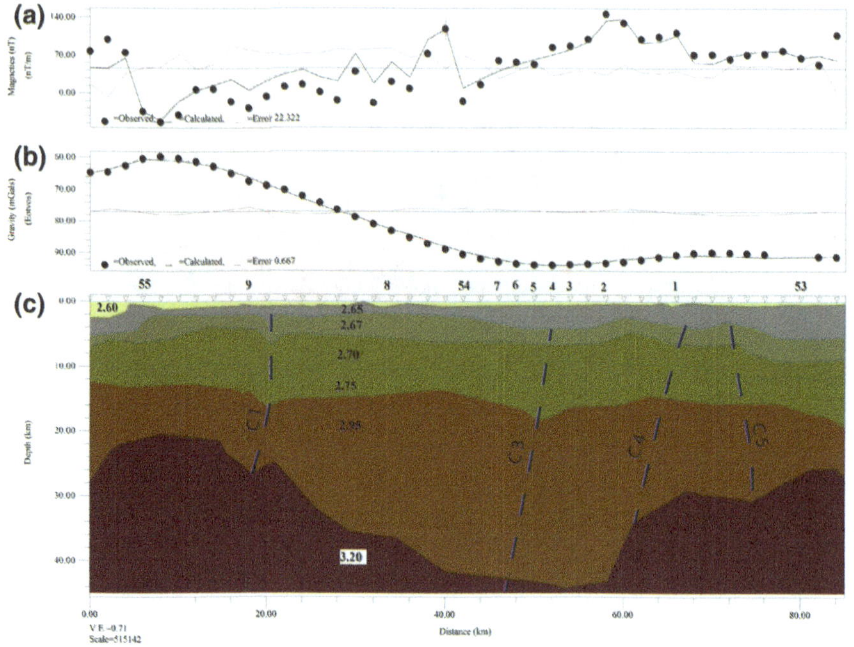

Fig. 12 Joint inversion of gravimetric and aeromagnetic data. **a** Misfit between observed magnetic data and magnetic response of the model shown in **c**. **b** Misfit between observed gravimetric data and gravimetric response of the model shown in **c**. **c** Model obtained by joint inversion of both data types

two discontinuities and is crossed by two faults. The upper crust of almost constant thickness reaches a depth of the order of 14 km.

6 Geological Interpretation

The magnetotelluric data acquired on the LATEA western boundary near Tahalgha volcanic district show a mainly two-dimensional lithosphere with a structure elongation in N15°W direction for the crust and N15°E for the underlying lithospheric mantle. At the scale of the study area, the N15°W strike is supported by the local surface trace directions of the major faults affecting the study area (4°50′, 4° 35′ faults). By contrast, the N15°E structural direction observed in the lithospheric mantle do not correlate with surface geological structures (Fig. 2). Taking into account the non-regular distribution of the MT sites, with a short measuring step in the center (up to 2 km) but wide towards the profile ends and the strike change with the depth, on the other hand, we calculated three resistivity models at three different scales (Fig. 9). A lithospheric model reaching an 80 km depth (Fig. 9c), then a crustal model (Fig. 9b) and lastly a high-resolution model of the central area around the 4°50′ fault (Fig. 9a). From depth to the surface, these three resistivity models show a fairly homogeneous lithospheric mantle with a resistivity of 100–200 Ωm. The lithospheric mantle contains a subvertical conductor rooted down to a depth of about 70 km and ascending to the base of the upper crust. This conductor, reaching a width of 10–20 km, perhaps less because it may appear more spread due to a loss in resolution with depth inherent to the magnetotelluric technique, is located directly below the 4°50′ fault (C3, Fig. 9). However, the model does not show any change in resistivity at the Moho depth, estimated at about 34 km below the surface by the analysis of the receiver functions obtained at the GEOSCOPE station at Tamanrasset (Fig. 9c). On the other hand, a very inhomogeneous upper crust is revealed. This modeled upper crust has an excellent correlation with the surface geology (Fig. 9b). From the crustal resistivity model, one can discern some crustal blocks corresponding to batholiths (R1, R2, and R3, Fig. 9b), characterized by a very resistive upper crust (>10,000 Ωm) and others corresponding to the Paleoproterozoic metamorphic basement, much less resistive (C2 and C4, Fig. 9b). In the vicinity of the 4°50′ fault, the central area of about 20 km lateral width is underlain by a relatively conductive middle and lower crust of about 100 Ωm (C3, Fig. 9b). But, at both profile ends, particularly under the metamorphic blocks of the Paleoproterozoic and late Neoproterozoic metamorphic basement, the crustal resistivity is relatively larger reaching a resistivity of about 500 Ωm (C2 and C4, Fig. 9b). These two portions of the crust are crossed by subvertical faults that are rooted in the crust.

The magnetotelluric profile spreads over the western edge of LATEA which is an old craton remobilized during the Pan-African Orogeny. At the scale of the studied area, the Tefedest terrane is characterized by a Polycyclic Paleoproterozoic gneissic basement (Pr1) that is intruded by (630–580 Ma) HKCA granitoids (g4) at the eastern edge of the profile. The two crustal blocks are separated by an intra-terrane sub-meridian shear zone of N10°E direction at the local scale (Fig. 1). The magnetotelluric model reveals a resistive (>10,000 Ωm) upper crust below Tefedest terrane of about 15 km thick beneath the Pan-African batholith of Tit (R3, Fig. 9b; see Fig. 2) to the east, and <10 km thick under the Paleoproterozoic metamorphic basement to the west (C4, Fig. 9b). The intra-terrane shear zone separating the two latter crustal portions, probably having played an important role in the HKCA batholith emplacement during the Pan-African Orogeny, is underlain by a subvertical conductor that reaches at least the base of the crust (C5, Fig. 9b). This conductive body could be interpreted as the electrical signature of a shear zone within the upper crust as shear zones often causes a decrease in the electrical resistivity due to an increase in porosity and to the presence fluids or mineralization in the pores.

The western part of the MT profile is located on the Iskel thrust terrane whose crust seems to be more inhomogeneous and complex with several conductive and resistive crustal portions as might be expected from the surface geology and the geologic history of this island arc (Fig. 9b). From the resistivity model, we can discern from east (near 4°50′ fault) to west, three crustal entities: a thick granitoid batholith probably the continuation of Anou-Eheli (640 Ma) batholiths (Bechiri-Benmerzoug 2009), rooted in the base of the upper crust (R2, Fig. 9b), a conductive block corresponding to a metagreywakes formation limited to the east by a 4°35′ sub-meridian fault (C2, Fig. 9b), and to the west a resistive block representing a thick Neoproterozoic batholiths (R1, Fig. 9b).

The mega-shear zone commonly known as 4°50′ fault limiting the Tefedest terrane (LATEA) to the east from the Iskel terrane (Western Hoggar) to the west, is clearly evidenced by the MT data (Fig. 9). It corresponds to a conductive sub-vertical body of 10 km width or more in the lower crust and upper mantle (C3, Fig. 9b and c) but only of few kms width near the surface where it joins the 4°50′ trace at the surface (Fig. 9a). This change in width with depth could be explained by the rheological behavior of the crust near the Abalessa area. Indeed, in the upper crust, the 4°50′ fault is sharp when it crosses the thick, resistive, and compact batholith, whose behavior is rather rigid (Fig. 9a). At the lithospheric scale, its signature is evident to a depth of several tens of kilometers. This fault appears to be connected with the conductive structures C2 and C4 (Fig. 9b and c) of the middle crust beneath the studied area. The loss in resolution of the method in the deep lithosphere does not make it highlighted to these depths (Fig. 9c). The model obtained by joint inversion of potential data (Fig. 12) is in agreement with the resistivity model (Fig. 9b). Unlike the case of 4°50′, the resistivity model shows that the fault of 4°35′, located inside the Iskel terrane, is not associated with any deep conductor. This is corroborated by the potential data model. Consequently, the 4°50′ appears as the major fault in the area, with a deep root in the lithospheric mantle.

The MT profile is located just on the northern flank and parallel to the Tahalgha volcanic district which extends in the E–W direction. Some stations are located even near or at some kilometers from volcanoes. The Pan-African structure, especially the various crustal faults and the fragile zones of the upper crust which correspond to conductive structures on the resistivity model, has certainly allowed the transport of melt from asthenosphere to surface. Fluids released by the magma and/or the precipitation of mineralization give rise to strong drops in the resistivity. In the absence of other constraints, the resistivity model cannot clearly state whether this signature corresponds to a Pan-African heritage, or to the Cenozoic reactivation. On another side, the relatively short length of the MT profile ($\sim$75 km) with respect to the width of the oued Amded lineament (see Fig. 1) does not make it possible to constrain the lithospheric structure of this important geological object.

7 Conclusions

The crustal and upper mantle structure of the Hoggar massif is the subject of hot scientific debates because it remains relatively unknown. Indeed, the lack of high-resolution geophysical studies in the region allows discrepancies between the various hypotheses formulated on the structure and deep nature of the Hoggar massif. In the present study, magnetotelluric was used to constrain the lithospheric geological structure of the Abalessa region, located on the LATEA western boundary. The study area is geologically complex because it is located at the intersection of the 4°50′ major fault with the oued Amded lineament, and on the northern flank of the Tahalgha volcanic district.

The collection of nine magnetotelluric sites added to three former sites belonging to a reconnaissance campaign allowed to model the crustal and upper mantle structure down to a hundred kilometers depth. Magnetotelluric data reveal crustal structures striking N15°W. This direction changes on 30° in the lithospheric mantle to be at N15°E. The resistivity model obtained by inversion of magnetotelluric data shows a rather inhomogeneous upper crust. The latter is made up of probably very compact and strong, electrically resistive structures corresponding to granitoids and batholiths, and others more conductive probably more inhomogeneous, weaker and more fractured corresponding to the Paleoproterozoic metamorphic basement. On the contrary, the lower crust and the lithospheric mantle down to a depth of about 100 km are fairly homogeneous.

The signature of some important faults is apparent at the crustal scale. Beneath the eastern side of the MT profile, the fault separating the Tit batholith (g4) to the east from the Proterozoic basement to the west can be discerned (Pr1, Fig. 9). In the profile on the western side, another crustal fault underlays the shallow batholith. The relatively short profile size makes it impossible to check whether these two faults are connected to others. The most important electrical signature is that of the 4°50′ fault which appears to extend into the lithospheric mantle to a depth of about 70 km. This is corroborated by the potential field data showing a root for the 4°50′ fault to at least the base of the crust. Near the surface and in the upper crust, this fault is relatively thin because it crosses a rigid medium. At depth, it becomes wider due to a lower ductile crust. On the other hand, the fault 4°35′ is not as deep. The fault network highlighted by the magnetotelluric model has probably been used to transport melt from the asthenosphere up to the surface to give rise to the Tahalgha volcanic district. The fluids released or the precipitated mineralization are at the origin of the strong fall of the resistivity which gives a so peculiar signature to these faults. The results obtained in this study will be more thorough during other magnetotelluric surveys planned soon in this area of the Hoggar massif.

Acknowledgements This study was conducted in the framework of the Algerian/French PHC Tassili cooperation program n° 09 MDU 787: « architecture et evolution du Bouclier Touareg: le role des grands accidents lithosphériques». The field campaign was carried out with the support of CRAAG. We thank M. Hani for his help during the fieldwork. The civil and military authorities of the Wilaya of Tamanrasset are thanked for their assistance in achieving the field campaign. The authors are grateful to A. Bendaoud, J.-L. Bodinier, C. Tibéri, A. Lesquer, and G. Vasseur for their fruitful discussions about this work. We warmly thank Prof. J.-P. Liégeois for his constructive criticism that has resulted in a substantial improvement of the manuscript.

References

Abdelsalam M, Liégeois J-P, Stern RJ (2002) The Saharan metacraton. J Afr Earth Sci 34:119–136. https://doi.org/10.1016/S0899-5362(02)00013-1

Aït-Hamou F, Dautria JM (1994) Le magmatisme cénozoïque du Hoggar: une synthèse des données disponibles. Mise au point sur l'hypothèse d'un point chaud. Bull Serv Géol Algérie 5:49–68

Aït-Hamou F, Dautria JM, Cantagrel JM, Dostal J, Briqueu L (2000) Nouvelles données géochronologiques et isotopiques sur le volcanisme cénozoïque de l'Ahaggar (Sahara algérien): des arguments en faveur d'un panache. C R Acad Sci Paris 330:829–836

Ayadi A, Dorbath C, Lesquer A, Bezzeghoud M (2000) Crustal and upper mantle velocity structure of the Hoggar swell (central Sahara, Algeria). Phys Earth Planet Inter 118:111–123. https://doi.org/10.1016/S0031-9201(99)00134-X

Azzouni-Sekkal A, Liégeois J-P, Bechiri-Benmerzoug F, Belaidi-Zinet S, Bonin B (2003) The 'Taourirt' magmatic province, a marker of the closing stage of the Pan-African orogeny in the Tuareg Shield: review of available data and Sr–Nd isotope evidence. J Afr Earth Sci 37:331–350

Bahr K (1988) Interpretation of the magnetotelluric impedance tensor: regional induction and local telluric distortion. J Geophys 62:119–127

Bahr K (1991) Geological noise in magnetotelluric data: a classification of distortion types. Phys Earth Planet Inter 66:24–38. https://doi.org/10.1016/0031-9201(91)90101-M

Bechiri-Benmerzoug F (2009) Pétrologie, géochimie isotopique et géochronologie des granitoïdes Pan-africains de type TTG de Silet: contribution à la connaissance de la structuration du bloc d'Iskel (Silet, Hoggar occidental). Doctoral dissertation, USTHB/FSTGAT, Algeria, 335 p

Begg GC, Griffin WL et al (2009) The lithospheric architecture of Africa: seismic tomography, mantle petrology and tectonic evolution. Geosphere 5:23–50

Bendaoud A, Ouzegane K, Godard G, Liégeois J-P, Kienast J-R, Bruguier O, Drareni A (2008) Geochronology and metamorphic P-T-X evolution of the Eburnean granulite-facies metapelites of Tidjenouine (Central Hoggar, Algeria): witness of the LATEA metacratonic evolution. In: Ennih N, Liégeois J-P (eds) The boundaries of the West African craton. Geological Society of London Special Publication, vol 297, pp 111–146. https://doi.org/10.1144/sp297.6

Bertrand JM, Caby R, SONAREM Geologists, Compilers (1977) Carte géologique du Hoggar (Algeria). SONAREM, Algiers, scale 1:1,000,000, 2 sheets

Bibby HM, Caldwell TG, Brown C (2005) Determinable and non-determinable parameters of galvanic distortion in magnetotellurics. Geophys J Int 163:915–930. https://doi.org/10.1111/j.1365-246X.2005.02779.x

Black R, Liégeois J-P (1993) Cratons, mobile belts, alkaline rocks and continental lithospheric mantle: the Pan-African testimony. J Geol Soc 150:89–98. https://doi.org/10.1144/gsjgs.150.1.0088

Black R, Latouche L, Liégeois JP, Caby R, Bertrand JM (1994) Pan-African displaced terranes in the Tuareg shield (central Sahara). Geology 22:641–644. https://doi.org/10.1130/0091-7613(1994)022%3c0641:padtit%3e2.3.co;2

Bournas N, Galdeano A, Hamoudi M, Baker H (2003) Interpretation of the aeromagnetic map of Eastern Hoggar (Algeria) using the Euler deconvolution, analytic signal and local wavenumber methods. J Afr Earth Sci 37:191–205

Bouyahiaoui B, Djeddi M, Abtout A, Boukerbout H, Akacem N (2011) Etude de la croûte Archéenne du môle In-Ouzzal (Hoggar Central) par la méthode gravimétrique: identification des sources par la transformée en ondelettes continue. Bull Serv Géol Nat 22(2):259–274

Bouzid A, Abtout A, Akacem N (2004) Electrical structure of the crust and upper mantle of the Hoggar shield from magnetotelluric data. In: Proceedings, 20th colloquium of African geology, Orléans, France, Abstracts, p 93

Bouzid A, Akacem N, Hamoudi M, Ouzegane K, Abtout A, Kienast J-R (2008) Modélisation magnétotellurique de la structure géologique profonde de l'unité granulitique de l'In Ouzzal (Hoggar occidental) [in French with an abridged English version]. C R Geosci 340:711–722. https://doi.org/10.1016/j.crte.2008.08.001

Bouzid A, Bayou B, Liégeois J-P, Bourouis S, Bougchiche SS, Bendekken A, Abtout A, Boukhlouf W, Ouabadi A (2015) Lithospheric structure of the Atakor metacratonic volcanic swell (Hoggar, Tuareg Shield, southern Algeria): electrical constraints from magnetotelluric data. In: Foulger GR, Lustrino M, King SD (eds) The interdisciplinary earth: a volume in honor of Don L. Anderson. Geological Society of America Special Paper 514 and American Geophysical Union Special Publication 71, pp 239–255. https://doi.org/10.1130/2015.2514(15)

Caby R (2003) Terrane assembly and geodynamic evolution of central-western Hoggar: a synthesis. J Afr Earth Sci 37:133–159

Cagniard L (1953) Principe de la méthode magnétotellurique, nouvelle méthode de prospection géophysique. Ann Geophys 9:95–125

Caldwell TG, Bibby HM, Brown C (2004) The magnetotelluric phase tensor. Geophys J Int 158:457–469. https://doi.org/10.1111/j.1365-246X.2004.x

Crough ST (1981) Free-air gravity over the Hoggar massif, northwest Africa: evidence for the alteration of the lithosphere. Tectonophysics 77:189–202. https://doi.org/10.1016/0040-1951(81)90262-6

Dautria J-M (1988) Relations entre les hétérogénéités du manteau supérieur et le magmatisme en domaine continental distensif. Exemple des basaltes alcalins du Hoggar (Sahara central, Algérie) et de leurs enclaves. Ph.D. thesis. Université de Montpellier 2, 421 p

Dautria JM, Girod M (1991) Relationships between Cainozoic magmatism and upper mantle heterogeneities as exemplified by the Hoggar volcanic area (Central Sahara, South Algeria). In: Kampunzu AB, Lubala RT (eds) Magmatism in extensional structural settings, Springer, Berlin, pp 250–269

Dautria JM, Lesquer A (1989) An example of the relationship between rift and dome: recent geodynamic evolution of the Hoggar swell and of its nearby regions (Central Sahara, Southern Algeria and Eastern Niger). Tectonophysics 163:45–61. https://doi.org/10.1016/0040-1951(89)90117-0

Dautria JM, Dostal J, Dupuy C, Liotard JM (1988) Geochemistry and petrogenesis of alkali basalts from Tahalra (Hoggar, Northwest Africa). Chem Geol 69:17–35. https://doi.org/10.1016/0009-2541(88)90155-6

Fishwick S, Bastow ID (2011) Towards a better understanding of African topography: a review of passive-source seismic studies of the African crust and upper mantle. In: van Hinsbergen DJJ, Buiter SJH, Torsvik TH, Gaina C, Webb SJ (eds) The formation and evolution of Africa: a synopsis of 3.8 Ga of earth history. Geological Society of London Special Publication, vol 357, pp 343–371. https://doi.org/10.1144/sp357.19

Gangopadhyay A, Pulliam J, Sen M-K (2007) Waveform modeling of teleseismic S, Sp, SsPmP, and shear-coupled PL waves for crust- and upper-mantle velocity structure beneath Africa. Geophys J Int 170:1210–1226. https://doi.org/10.1111/j.1365-246X.2007.03470.x

Gokarn SG, Gupta G, Rao CK, Selvaraj C (2002) Electrical structure across the Indus Tsangposuture and Shyok suture zones in NW Himalaya using magnetotelluric studies. Geophy Res Lett 29(8):1251. https://doi.org/10.1029/2001GL014325

Groom RW, Bailey RC (1989) Decomposition of magnetotelluric impedance tensor in the presence of local three-dimensional galvanic distortion. J Geophys Res 94:1913–1925. https://doi.org/10.1029/JB094iB02p01913

Groom RW, Bailey RC (1991) Analytic investigations of the effects of near-surface three-dimensional galvanic scatterers on MT tensor decompositions. Geophysics 56(4):496–518

Haak V (1980) Relations between electrical conductivity and petrological parameters of the crust and upper mantle. Geophys Surv 4:57–69. https://doi.org/10.1007/BF01452958

Hansen C (1992) Analysis of discrete ill-posed problems by means of the L-curve. SIAM Rev 34:561–580. https://doi.org/10.1137/1034115

Jödicke H (1992) Water and graphite in the earth's crust: an approach to interpretation of conductivity models. Surv Geophys 13:381–407. https://doi.org/10.1007/BF01903484

Jones AG (1999) Imaging the continental upper mantle using electromagnetic methods. Lithos 48:57–80. https://doi.org/10.1016/S0024-4937(99)00022-5

Jones AG (2013) Imaging and observing the electrical Moho. Tectonophysics 609:423–436. https://doi.org/10.1016/j.tecto.2013.02.025

Jones AG, Ferguson IJ (2001) The electric Moho. Nature 409:331–333. https://doi.org/10.1038/35053053

Jones AG, Jödicke H (1984) Magnetotelluric transfer function estimation improvement by a coherence-based rejection technique. In: Proceedings, 54th annual international society of exploration geophysicists meeting, Atlanta, Georgia, pp 51–55

Jones AG, Kurtz RD, Boerner DE, Craven JA, McNeice GW, Gough DI, DeLaurier JM, Ellis RG (1992) Electromagnetic constraints on strikeslip fault geometry: the Fraser River fault system. Geology 20:561–564. https://doi.org/10.1130/0091-7613(1992)020%3c0561:ecossf%3e2.3.co;2

Kourim F, Bodinier J-L, Alard O, Bendaoud A, Vauchez A, Dautria J-M (2014) Nature and evolution of the lithospheric mantle beneath the Hoggar swell (Algeria): a record from mantle Xenoliths. J Petrol 55(11):2249–2280. https://doi.org/10.1093/petrology/egu056

Krieger L, Peacock JR (2014) MTpy: a Python toolbox for magnetotellurics. Comput Geosci 72:167–175

Lastoviskova M (1983) Laboratory measurement of electrical properties of rocks and minerals. Geophys Surv 6:201–213. https://doi.org/10.1007/BF01454001

Lesquer A, Bourmatte A, Dautria JM (1988) Deep structure of the Hoggar domal uplift (Central Sahara, South Algeria) from gravity, thermal and petrological data. Tectonophysics 152:71–87. https://doi.org/10.1016/0040-1951(88)90030-3

Lesquer A, Bourmatte A, Ly S, Dautria JM (1989) First heat flow determination from the Central Sahara: relationship with the Pan-African belt and Hoggar domal uplift. J Afr Earth Sci 9:41–48. https://doi.org/10.1016/0899-5362(89)90006-7

Liégeois JP, Latouche L, Boughrara M, Navez J, Guiraud M (2003) The LATEA metacraton (Central Hoggar, Tuareg shield, Algeria): behaviour of an old passive margin during the Pan-African orogeny. J Afr Earth Sci 37:161–190. https://doi.org/10.1016/j.jafrearsci.2003.05.004

Liégeois J-P, Benhallou A, Azzouni-Sekkal A, Yahiaoui R, Bonin B (2005) The Hoggar swell and volcanism: reactivation of the Precambrian Tuareg shield during Alpine convergence and West African Cenozoic volcanism. In: Foulger GR, Natland JH, Presnall DC, Anderson DL (eds) Plates, plumes, and paradigms. Geological Society of America Special Paper 388, pp 379–400. https://doi.org/10.1130/0-8137-2388-4.379

Liégeois J-P, Abdelsalam MG, Ennih N, Ouabadi A (2013) Metacraton: nature, genesis and behavior. Gondwana Res 23:220–237. https://doi.org/10.1016/j.gr.2012.02.016

Liu HL, Gao SS (2010) Spatial variations of crustal characteristics beneath the Hoggar swell, Algeria, revealed by systematic analyses of receiver functions from a single seismic station. Geochem Geophys Geosyst 11:Q08011. https://doi.org/10.1029/2010GC003091

Mackie R, Rieven S, Rodi W (1997) User's manual and software documentation for two-dimensional inversion of magnetotelluric data. Earth Resources Laboratory Rpt. Massachusetts Institute of Technology, Cambridge, p 13

Marti A, Queralt P, Jones AG, Ledo J (2005) Improving Bahr's invariant parameters using the WAL approach. Geophys J Int 163:38–41. https://doi.org/10.1111/j.1365-246X.2005.02748.x

Meqbel NM, Egberta GD, Wannamaker PhE, Kelberta A, Schultz A (2014) Deep electrical resistivity structure of the northwestern U.S. derived from 3-D inversion of USArray magnetotelluric data. Earth Planet Sci Lett 402:290–304. https://doi.org/10.1016/j.epsl.2013.12.026

Nover G (2005) Electrical proprieties of crustal and mantle rocks: a review of laboratory measurements and their explanation. Surv Geophys 26:593–651. https://doi.org/10.1007/s10712-005-1759-6

Parkinson WD (1962) The influence of continents and oceans on geomagnetic variations. Geophys J Roy Astron Soc 6:441–449. https://doi.org/10.1111/j.1365-246X.1962.tb02992.x

Pasyanos ME, Walter WR (2002) Crust and upper-mantle structure of North Africa, Europe and the Middle East from inversion of surface waves. Geophys J Int 149:463–481. https://doi.org/10.1046/j.1365-246X.2002.01663.x

Pérez-Gussinyé M, Metois M, Fernàndez M, Vergés J, Fullea J, Lowry AR (2009) Effective elastic thickness of Africa and its relationship to other proxies for lithospheric structure and surface tectonics. Earth Planet Sci Lett 287(1–2):152–167

Priestley K, McKenzie D, Debayle E, Pilidou S (2008) The African upper mantle and its relationship to tectonics and surface geology. Geophys J Int 175:1108–1126

Rao CK, Jones GA, Moorkamp M (2007) The geometry of the Iapetus Suture zone in central Ireland deduced from a magnetotelluric study. Phy Earth Planet Inter 161:134–141. https://doi.org/10.1016/j.pepi.2007.01.008

Rao CK, Jones GA, Moorkamp M, Weckmann U (2014) Implications for the lithospheric geometry of the Iapetus suture beneath Ireland based on electrical resistivity models from deep-probing magnetotellurics. Geophy J Inter 198(2):737–759. https://doi.org/10.1093/gji/ggu136

Ritter O, Weckmann U, Vietor T, Haak V (2003) A magnetotelluric study of the Damara Belt in Namibia: 1. Regional scale conductivity anomalies. Phys Earth Planet Inter 138:71–90. https://doi.org/10.1016/S0031-9201(03)00078-5

Roberts GG, White N (2010) Estimating uplift rate histories from river profiles using African examples. J Geophys Res 115:B02406. https://doi.org/10.1029/2009JB006692

Rougier S, Missenard Y, Gautheron C, Barbarand J, Zeyen H, Pinna R, Liégeois J-P, Bonin B, Ouabadi A, Derder ME-M, Frison de Lamotte D (2013) Eocene exhumation of the Tuareg Shield (Sahara Desert, Africa). Geology 41:615–618. https://doi.org/10.1130/G33731.1

Sebaï A, Stutzmann E, Montagnera J-P, Sicilia D, Beuclerc E (2006) Anisotropic structure of the African upper mantle from Rayleigh and Love wave tomography. Phys Earth Planet Inter 155:48–62. https://doi.org/10.1016/j.pepi.2005.09.009

Shankland TL, Ander M (1983) Electrical conductivity, temperatures and fluids in the lower crust. J Geophys Res 88:9475–9484. https://doi.org/10.1029/JB088iB11p09475

Swift CM Jr. (1967) A magnetotelluric investigation of an electrical conductivity anomaly in the southwestern United States. Ph.D. thesis. Massachusetts Institute of Technology, Cambridge, p 223

Takherist D, Lesquer A (1989) Mise en évidence d'importantes variations régionales du flux de chaleur en Algérie. Can J Earth Sci 26(4):615–626. https://doi.org/10.1139/e89-053

Tikhonov AN (1950) Determination of electrical characteristics of the deep strata of the earth's crust. Dokl Akad Nauk SSSR 73:292–297

Unsworth M, Bedrosian PA (2004) On the geoelectric structure of major strike-slip faults and shear zones. Earth Planet Space 56:1177–1184

Wannamaker PhE, Hasterok DP, Johnston JM, Stodt JA, Hall DB, Sodergren TL, Pellerin L, Maris V, Doerner WM, Groenewold KA, Unsworth MJ (2008) Lithospheric dismemberment and magmatic processes of the Great Basin–Colorado Plateau transition, Utah, implied from magnetotellurics. Geochem Geophys Geosyst 9:Q05019. https://doi.org/10.1029/2007GC001886

Zerrouk S, Bendaoud A, Hamoudi M, Liégeois J-P, Boubekri H, Ben El Khaznadji R (2016) Mapping and discriminating the Pan-African granitoids in the Hoggar (southern Algeria) using Landsat 7 ETM+ data and airborne geophysics. J African Earth Sci 127:146–158. https://doi.org/10.1016/j.jafrearsci.2016.10.003

Regional Geology and Petroleum Systems of the Main Reservoirs and Source Rocks of North Africa and the Middle East

D. Lučić and W. Bosworth

Abstract Roughly 44% of the world's total known hydrocarbon resources are located in North Africa and the Middle East, from Algeria in the west to the Zagros region of Iran in the east. This includes more than 200 giant fields in the Middle East. North African basins, mainly in Algeria, Libya, and Egypt, contain 4% of the world's oil and gas reserves and nearly 40 giants. The most recent giant field resides in the offshore Levantine Basin. In this chapter, the region's geological history and petroleum systems are reviewed, including descriptions of the regional habitats and stratigraphy of the main reservoirs and source rocks. The Late Pre-Cambrian to Phanerozoic tectonostratigraphy is characterized by six major phases: (1) basement assembly and Infracambrian extension; (2) development of Cambrian to Carboniferous passive margins; (3) Late Carboniferous to Early Permian Hercynian orogeny; (4) post-Hercynian breakup of Gondwana; (5) collision with Eurasia; and (6) Red Sea rifting and the closing of Neotethys. For most of the Phanerozoic, the terranes of North Africa and Arabia were locked together at the northern rim of Gondwana. They shared a similar geodynamic history and generally similar climates. Their plate margins interacted over time with the water masses of three oceans: Proto-, Paleo-, and Neotethys. Throughout the Paleozoic, the region lay on a wide "ramp-like" passive margin facing northward toward Paleotethys. The main sediment source was the large hinterland to the south, with prevailing south to north and southeast to northwest-directed paleocurrents. Multiple marine transgressions across a low-relief continental platform were interrupted by at least four glacial events: Late Ordovician (Hirnantian), Silurian, Carboniferous, and Early Permian. Deposition of siliciclastics predominated throughout the Paleozoic whereas carbonates were much less common. The most important tectonic event was the mid-Carboniferous Hercynian composite-orogeny, which caused major uplift and erosion. The Hercynian represented the final closure of Paleotethys and produced

---

D. Lučić (✉)
Zagreb, Croatia
e-mail: dubravkolucic59@gmail.com

W. Bosworth
Apache Egypt Companies, Cairo, Egypt
e-mail: bill.bosworth@apacheegypt.com

A. Bendaoud et al. (eds.), *The Geology of the Arab World—An Overview*,
Springer Geology, https://doi.org/10.1007/978-3-319-96794-3_6

many fault blocks and arches that would later host many of the major hydrocarbon accumulations of North Africa and eastern Arabia. The Mesozoic–Cenozoic sedimentary sequence similarly consists of eustatically and tectonically controlled depositional cycles along the newly formed passive margin of Neotethys. Triassic to mid-Cretaceous facies along North Africa are almost everywhere shallow marine, nearshore, deltaic, and continental. Neotethys reached its maximum extent in the Late Cretaceous at which time carbonate sequences dominated. In addition to the Hercynian Orogeny, two other compressional events had major consequences on North Africa–Arabia petroleum systems. The closing of Neotethys began in the late Santonian ($\sim$84 Ma) and the convergence between Eurasia and Africa–Arabia sent pulses of compressional deformation across the plate. This corresponded to the first phase of the complex Alpine orogenic cycle and caused folding, basin inversion, and strike-slip faulting along the African–Arabian Neotethyan margin (the "Syrian Arc"), and thrusting and ophiolite obduction in Oman. Compression was renewed at the end of the Maastrichtian and continued into the early Paleocene, followed by even stronger effects in the Late Eocene. Eruption of the Afar plume at about 31 Ma marked the beginning of a new phase of continental rifting that had dramatic effects on all aspects of the geology of the region. The Gulf of Aden ruptured first in the Early Oligocene, followed by the southern Red Sea in the Late Oligocene. At the Oligocene–Miocene transition, the remainder of the Red Sea north to the Gulf of Suez underwent a regional dike event and accompanying extensional faulting. Initiation of the Gulf of Aqaba—Dead Sea transform plate boundary occurred in the Middle Miocene, completing formation of the independent Arabian plate. The Neotethys Ocean also ceased to exist in the Middle Miocene following a collision between Eurasia and the Arabian plate to form the Bitlis–Zagros suture and fold belt. The Arabian plate was progressively tilted to the northeast as a result of both uplift and rifting of Arabia from Africa, and structural loading of the northeast margin by the Zagros fold belt. Recent tectonic activity is mainly concentrated along the Maghrebian Alpine Belt, the offshore Nile Delta, the Red Sea–East African (or "Afro-Arabian") Rifts Province, the Aqaba–Dead Sea–Bekaa sinistral strike-slip fault zone, and some major intra-plate fault zones including the Guinean–Nubian, Aswan, and central Sinai lineaments. Our review of the petroleum systems of North Africa and Arabia is brief and includes only highlights of the hydrocarbon occurrences found across this broad and complex region. Based on the age of source rocks, it is possible to distinguish an Infracambrian Petroleum System, Palaeozoic-related Petroleum Systems, and linked Mesozoic–Cenozoic Petroleum Systems. The sedimentary fill contains numerous source rocks, some of them with exceptional quality and regional distribution, such as the Silurian "hot shale". Producing reservoirs are found in both siliciclastics and carbonates. The proximity and juxtaposition of source rocks with thick reservoirs minimized the need for complex oil migration pathways, but also facilitated hydrocarbon expulsion and migration over long distances. Huge amounts of evaporites and shales are present, providing excellent lateral and ultimate top seals. Hydrocarbons are trapped in literally all stratigraphic units from the fractured Neoproterozoic basement to the youngest Pliocene–Quaternary sediments.

Keywords North Africa · Middle East · Arabia · Petroleum systems
Tectonostratigraphy · Hydrocarbon reserves · Source rocks · Reservoir rocks

1 Introduction

The Arabian plate, and neighboring parts of Egypt, Turkey, Iraq, Iran, and the southern Caucasus, includes numerous active hydrocarbon seepages which have been recognized from ancient times. The gaseous seeps initially played a key role in some religions, and bitumen was also used extensively by ancient civilizations for different purposes. It was used as mortar or cement, for pavements, for ornaments and jewelery both in Mesopotamia and elsewhere, for fixing the blades of tools into their handles, for mummification in Egypt, for waterproofing pottery, and also as a military weapon known as "*Greek Fire*"—the exact composition of which remains unknown although it was hydrocarbon-based (Beydoun 1997).

The seeps and gas fires at the *Masjid Suleiman* locality in southwest Iran attracted some of the pioneers of the oil industry, and after several dry wells, finally on May 26, 1908 oil was found. From that point, a new era started in the Middle East. The Masjid Suleiman discovery had immense commercial implications but also geopolitical and cultural consequences. At about the same time, oil was found in Egypt first at *Gemsa* (1907) and then at the larger *Hurghada field* (1913). Soon other countries followed suit: *Kirkuk* in Iraq (1927), *Jebel Dukhan* in Bahrein (1932), *Dammam* in Saudi Arabia (1938), and *Oued Gueterini* in Algeria (1948). Oil started to run through vast pipeline networks draining the prolific Arabian reservoirs and feeding an increasingly demanding society worldwide. The region was changed forever.

Presently, oil and gas from the countries situated in North Africa and the Middle East, from Morocco in the west to the Zagros region of Iran in the east, contribute about 44% of the world's total known resource volume, with more than 90% of that quantity coming from the Middle East provinces within or adjacent to the Arabian plate (Fig. 1; Ahlbrandt et al. 2000, 2010, 2014). The geological processes are hidden in the subsurface and the very sensitive political, social, and cultural activities above ground all played key and sometimes contradictory roles during the long history of oil discovery and production in this region. In this review, only part of the story will be presented, focusing on critical aspects of North African and Arabian petroleum geology.

1.1 Why so Rich?

What makes this region so rich and so unique in its hydrocarbon reserves? To answer such an intriguing and complex question, many factors and geological processes need to be considered and evaluated. Many authors have considered this

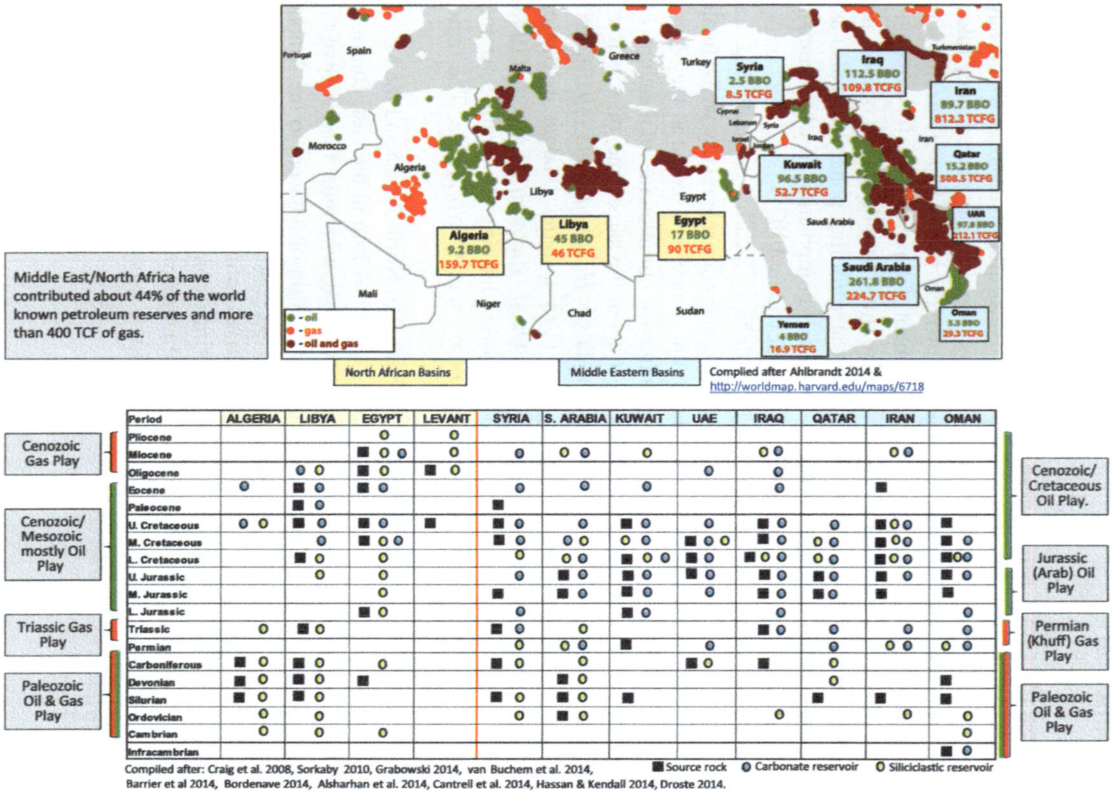

Fig. 1 Fields, reserves, and petroleum system summary

question. All of them in their own ways, using different methodologies, tried to explain this phenomenon and have provided useful insights.

Beydoun (1998) and Sorkhabi (2010, 2012) compiled lists summarizing and explaining the many positive coincidences that happened, through geological history, resulting in multiple working petroleum systems and enabling such incredible hydrocarbon richness. These include the following:

- Gondwana's continental passive margins had long and almost uninterrupted histories of sedimentation for ~500 Myr. Enormous thicknesses of sediments, including extensive porous and permeable limestone and sandstone reservoirs, were available for inclusion in future trap development.
- Sedimentation occurred largely between latitudes 30°N and 30°S, where the climate was warm with high organic activity, enabling repeated deposition of marine source rocks containing oil-prone kerogen types.
- This favorable low-latitude position was occupied during the mid-Cretaceous global warming, coupled with sea-level rise and plankton abundance.
- The proximity and juxtaposition of source rocks with thick reservoirs minimized the need for complex oil migration pathways but also facilitated hydrocarbon expulsion and migration over long distances.
- The presence of effective regional seals, both marine shales and evaporites, at the bottom, within, and on the top of the sedimentary package.
- The superposition of the Cenozoic Zagros foreland basin onto the older Tethyan shelf sediments.
- Large "whaleback" anticlines generated by the "gentle" Zagros compression (folding and thrusting) and salt diapirs provided abundant structural traps for oil accumulations.
- Tectonic fracturing of younger, stratigraphically higher carbonate reservoirs during the overprinting of the Zagros foreland basin in the Neogene to permit vertical migration into huge anticlines.

From a tectonic perspective, Burke et al. (2003) attributed much of Africa and Arabia's hydrocarbon development to four dominant events: (1) the Neoproterozoic assembly of Gondwana by multiple collisions (Pan-African orogenies) and the subsequent erosion of the resulting mountain belts to produce the region's extensive Cambro–Ordovician reservoir rocks; (2) the collisions of Baltica and Laurentia (~380 Ma) and Gondwana and Laurussia (~310 Ma; the "Hercynian" Orogeny) that produced numerous intra-plate rifts, later the site of hydrocarbon accumulations; (3) the Karoo plume (~183 Ma) which helped to pin Gondwana relative to the underlying mantle, established extensive new convection patterns and led to the prolific Mesozoic rift systems of the plate; and (4) the Afar plume (~31 Ma) which continued the arresting of Africa's movement and with other intra-plate swells provided the topography and sediment source that resulted in the great late Cenozoic hydrocarbon-prolific deltas that now ring the plate.

It is evident that in the terms of petroleum system elements, nothing is missing. The sedimentary fill contains numerous source rocks, some of them with exceptional regional distribution. Producing reservoirs can be found in both siliciclastics and carbonates. Huge amounts of evaporites and shales are present, providing excellent lateral and ultimate top seals. And, finally, there was a harmony of the stratigraphic and structural factors which were properly orchestrated within this extremely successful natural hydrocarbon-producing factory.

1.2 Previous Investigations

From historical times, when the first explorationists were riding camels, till today using state of the art technology, due to its petroleum richness the region was the subject of great interest by both academia and the petroleum industry. Petroleum systems and petroleum geology have been evaluated and assessed by numerous authors thereby defining its constitutive elements, both regionally and locally (e.g., Macgregor 1996; Boote et al. 1998; Macgregor and Moody 1998; Abu-Ali et al. 1999; Ghori et al. 2009; Dolson et al. 2014; Al-Johi and Al-Laboun 2015). "Total Petroleum Systems" (TPS) of different basins in North Africa and the Middle East were the subject of numerous reports prepared as a part of the U.S. Geological Survey World Petroleum Assessment 2000 (Ahlbrandt 1999, 2000a, b, c, 2002, 2010, 2014; Ahlbrandt et al. 2000; Klett 2000a, b, 2001; Fox and Ahlbrandt 2002; Pollastro 1999, 2000, 2003). Numerous published articles and studies are available detailing the controls exerted by plate motions and paleogeography on the geology of the African and Arabian plates (Schandelmeier and Reynolds 1997; Scotese et al. 1999; Stampfli 2000; Ziegler 2001; Stampfli and Borel 2002; Ruban et al. 2007; Johnson 2008; Torsvik and Cocks 2011; Berra and Angiolini 2014). Other processes, including tectonic events, associated deformation and uplift, basin formation, subsidence histories, eustasy, sediment supply, stratigraphic sequences, facies and depositional patterns of North Africa and Middle East, spanning the entire Phanerozoic, are the subjects of many publications (e.g., Brown et al. 1963; Biju-Duval et al. 1978; Klitzsch and Squyres 1990; Guiraud and Bosworth 1999; Konert et al. 2001; Sharland et al. 2001; Makhous 2001; Hallett 2002; Carr 2002, 2003; Badalini et al. 2002; Guiraud et al. 2005; Haq and Al-Qahtani 2005; Craig et al. 2008, 2010; Gerdes et al. 2010). Several comprehensive monographs and books have been written and are essential sources of data (Said 1990; Alsharhan and Nairn 1997; Tawadros 2001; Hallett 2002; Jassim and Goff 2006; Aqrawi et al. 2010; Hallett and Clark-Lowes 2016). However, the petroleum geology of North Africa and the Middle East have more often than not been discussed independently of one another. A recent exception is Marlow et al. (2014) who approached the petroleum systems of both regions together, from a plate-scale geologic perspective.

This chapter will briefly summarize key regional geological events, correlating common genetic processes and their role in simultaneously generating hydrocarbons in both North Africa and greater Arabia. One goal is to help underscore similarities that existed through geologic time across this broad region while it evolved as contiguous parts of Gondwana. Some basins containing giant fields will necessarily be considered in more detail than others. We use the Geologic Time Scale 2012 of Gradstein et al. (2012; ages in Ma rounded to whole numbers).

2 Geological History

During most of the Phanerozoic, the terranes of North Africa and Arabia were locked together at the northern rim of Gondwana—the "*motherland*"—sharing a similar geodynamic history. Their plate margins were interacting with the ingredients from the "*pra-soup*" water masses of three different oceans: Proto-, Paleo- and Neotethys. Hence, referral to it as the "Southern Tethyan Margin Region" (Kendall et al. 2014).

Depositional setting and sedimentary facies were similar for long periods of time throughout the entire area, enabling correlation of stratigraphic units and unconformities on a grand, regional scale. The average sedimentary thickness is about 6000 m, increasing up to 10,000 m northward toward the Tethyan Ocean basin (Kendall et al. 2014).

Throughout the Paleozoic, the region lay on a wide "ramp-like" passive margin facing northward toward Paleotethys (Fig. 9). The main sediment source was the large hinterland to the south, with prevailing south to north and southeast to northwest-directed paleocurrents (Craig et al. 2008). Multiple marine transgressions across a low-relief continental platform were interrupted by at least four glacial events: Late Ordovician (*Hirnantian*), Silurian, Carboniferous and Early Permian (Ghienne et al. 2007; Le Heron et al. 2009). This resulted in the deposition of predominantly siliciclastics and lesser carbonates. The facies are interpreted as estuarine, fluvial, deltaic, and glacial flanked by transitional-shallow marine, carbonate, and evaporite. The general stability of the platform was slightly and repeatedly interrupted during its geological history. The most important tectonic event was the mid-Carboniferous *Hercynian* orogeny, which caused major uplift and erosion (Guiraud et al. 2005; Faqira et al. 2009). The Hercynian orogeny represented the final closure of Paleotethys and produced many fault blocks and arches that later would host some of the major hydrocarbon accumulations of North Africa and eastern Arabia.

The Mesozoic–Cenozoic sedimentary sequence similarly consists of eustatically and tectonically controlled depositional cycles along the newly formed passive margin of Neotethys. Triassic to mid-Cretaceous facies along North Africa are almost everywhere shallow marine, nearshore, deltaic, and continental (Sestini 1984). Neotethys reached its maximum extent in the Late Cretaceous at which time carbonate sequences dominated (Guiraud et al. 2005; Gerdes et al. 2010).

In addition to the Hercynian Orogeny, two other compressional events had major consequences to North Africa–Arabia petroleum systems. These occurred in the Late Cretaceous and Late Eocene. The closing of Neotethys began in the late Santonian ($\sim$84 Ma) and the convergence between Eurasia and Africa–Arabia sent pulses of compressional deformation across the plate (Guiraud and Bosworth 1997; Guiraud 1998). The "*Santonian event*" caused folding, basin inversion, and strike-slip faulting along the African–Arabian Neotethyan margin (the "*Syrian Arc*" of Krenkel 1925) and thrusting and ophiolite obduction in Oman. The Santonian event corresponds to the first phase of the complex *Alpine orogenic cycle*.

Compression was renewed at the end of the Maastrichtian and continued into the early Paleocene, followed by even stronger effects in the Late Eocene. This Eocene "*Pyrenean-Atlasic*" event resulted in renewed folding, thrusting, and local metamorphism of the northern African–Arabian plate margin and rejuvenation of intra-plate fault zones (Guiraud et al. 1987, 2005).

At about 31 Ma, impingement of the Afar plume on the lithosphere of northeast Africa marked the beginning of a new phase of continental rifting that would have dramatic effects on all aspects of the geology of the region (Coleman 1993; Burke 1996). The Gulf of Aden ruptured first in the Early Oligocene, followed by the southern Red Sea in the Late Oligocene (Hughes et al. 1991; Hughes and Beydoun 1992). At the Oligocene–Miocene transition, the remainder of the Red Sea north to the Gulf of Suez underwent a regional dike event and accompanying extensional faulting (Sebai et al. 1991; Coleman 1993; Bosworth et al. 2005, 2015a; Bosworth and Stockli 2016).

Initiation of the Gulf of Aqaba—Dead Sea transform plate boundary occurred in the Middle Miocene, completing formation of the independent Arabian plate (Steckler et al. 1988; Bosworth and McClay 2001). The Neotethys Ocean also ceased to exist in the Middle Miocene following a collision between Eurasia and the Arabian plate to form the Bitlis–Zagros suture and fold belt (Hempton 1987; Bosworth et al. 2005). The previously dominant carbonate foredeep sedimentation was replaced by clastic deposition in a subaerial continental foreland basin setting. The Arabian plate was progressively tilted to the northeast as a result of both uplift and rifting of Arabia from Africa, and structural loading of the northeast margin by the Zagros fold belt (Sharland et al. 2001).

Recent tectonic activity is mainly concentrated along the Maghrebian Alpine Belt, the offshore Nile Delta, the Red Sea–East African (or "Afro-Arabian) Rifts Province, the Aqaba–Dead Sea–Bekaa sinistral strike-slip fault zone, and some major intra-plate fault zones including the Guinean–Nubian, Aswan, and central Sinai lineaments (Guiraud et al. 1985, 2005).

3 Tectonostratigraphy

The timing, intensity, spatial distribution, mechanisms, and driving forces of the tectonic events of North Africa–Arabia have been approached in differing fashions by many authors. Certain areas with higher hydrocarbon potential or geopolitical restrictions have been treated confidentially. Others of less economic value are limited largely to scientific researches. Inevitably this has resulted in unavoidable variations in details, nomenclature and granularity of the key tectonic and stratigraphic events, phases, and episodes. Boote et al. (1998), for example, defined only two Gondwana and Tethyan supercycles and Craig et al. (2008) recognized seven major tectonic phases. Other authors presented finer subdivisions. Badalini et al. (2002) addressed 6 tectonic events during the Paleozoic and 12 in the Mesozoic. Guiraud et al. (2005) divided the main rifting and orogenic phases based on the details of Pangea breakup, Alpine rifting, and Alpine foreland compression.

They defined 14 tectonic phases and unconformities. However, there is a general recognition that North Africa and Arabia shared some common tectonic events and similar depositional settings on the passive margins of Paleo- and Neotethys including (1) deformation was predominantly extensional during the post-Hercynian disintegration of northern Gondwana; (2) sedimentation was focused within fault-bounded half-grabens that formed along the southern margin of Neotethys; and (3) Alpine age compression then forced inversion of these half-grabens to form the well-known *Syrian Arc* and related structures.

Utilizing the concepts of sequence stratigraphy Sharland et al. (2001) identified 11 major surfaces bounding tectonostratigraphic megasequences (TMS) for the Arabian plate. This was generally accepted and widely used by later authors. Some of the surfaces are also applicable to North Africa as demonstrated by Kendall et al. (2014) while describing both areas from the plate-scale perspective. Discussion and rethinking of the post-Hercynian basin genesis were offered by Wood (2011, 2015). Our regional understanding will undoubtedly continue to improve with the results from future drilling campaigns and the continued application of new technologies.

We will discuss the tectonostratigraphy of North Africa and Arabia with a slightly modified and simplified scheme, consisting of six stages from the latest Neoproterozoic through the Phanerozoic:

1. **Basement assembly and Infracambrian extension**: Pan-African orogenies
2. **Cambrian to Carboniferous passive margin**: Extension–Compression and glaciation/deglaciation
3. **Late Carboniferous to Early Permian Hercynian Orogeny**: Final assembly of Pangea
4. **Post-Hercynian Breakup of Gondwana**: Permian–Mesozoic rifting
5. **Collision with Eurasia**: Mid-Cretaceous "Eo-Alpine" and Late Cretaceous—Cenozoic "Alpine" compression
6. **Red Sea Rifting and Closing of Neotethys**: Subduction of the NE margin of Gondwana

3.1 Basement Assembly and Infracambrian Extension

After being welded together by complex collisions during the Pan-African Orogenies, a number of Pre-Cambrian terranes and cratonic nuclei created the large, stable cratonic block of Gondwana (Stern 1985; 1994; Kröner and Stern 2004; Johnson and Kattan 2012). Arabia and North Africa occupied an intracratonic setting, having a contiguous passive margin bordering the Paleotethys Ocean (Sharland et al. 2001; Guiraud et al. 2005). Infracambrian extensional movements in northeast Africa and Arabia have generally been interpreted as the results of "escape tectonics" associated with the collision of West and East Gondwana leading to the formation of the Pan-African-East African-Antarctic Orogen. In Arabia, the forces associated with that collision resulted in the development of a 300–400 km wide northwest to southeast trending belt of transcurrent faults—the *Najd fault system*—with an overall sinistral displacement of some 300 km (Brown et al. 1963; Moor 1979; Stern 1985; Stoeser and Camp 1985; Stern and Johnson 2010). A related system of northeast–southwest trending faults is recognized further west on the North African side of the Red Sea rift and westwards below the *Kufra Basin* in southeast Libya (Craig et al. 2008). Infracambrian movement on these systems of sub-vertical faults generated numerous pull-apart basins of variable size in eastern Libya, Egypt, the Dead Sea region, and Arabia. Their detailed geometry and stratigraphic fill, however, is in many cases yet to be determined. Frequent rejuvenation of this fault network throughout the Phanerozoic played an important role during later regional and local tectonic evolution (Guiraud et al. 2005).

3.1.1 Habitats and Stratigraphy of the Main Infracambrian Reservoirs and Source Rocks

The basal Infracambrian succession generally consists of volcaniclastics (Fig. 2). The section overlying this is dominated by continental siliciclastics interbedded with shallow marine siliciclastics and carbonates, including some transitional marine deposits in the Salt Basins of Oman and in Saudi Arabia (lower part of TMS AP1; Sharland et al. 2001). In Oman, the Infracambrian succession is referred to the *Huqf Supergroup* which rests unconformably on top of a ~800 Ma igneous-metamorphic basement and is itself subdivided into three groups: the *Abu Mahara*, *Nafun* and *Ara Groups* (Cozzi et al. 2012). The oldest sedimentary rocks of the *Abu Mahara Group* were deposited during the Pan-African orogeny 800–712 Ma, and consist of diamictites, siliciclastics, and volcaniclastics, as well as pillow basalts related to intra-continental extension (Droste 2014). The siliciclastic-carbonate cycles of the *Nafun Group* were deposited during extensive regional subsidence and passive margin conditions. The siliciclastics were deposited in shallow to deep marine settings while the carbonates were dominated by stromatolites and reservoir-prone peloidal and oolitic grainstones that represent periods of extensive platform deposition. Deep-water-starved basins located between areas of platform sedimentation contained bituminous mudstones and black shales (Lüning et al. 2009). *Ara Group* deposition was marked by the development of the *South Oman, Fahud,* and *Ghaba* salt basins, with sharp lateral facies changes and deposition of thick stacked carbonate sequences on the basin margin changing into a "stringer" of carbonates within the salt basin, together with organic-rich shales (Cozzi et al. 2012).

3.2 Cambrian to Carboniferous Passive Margin

The structural evolution of the Paleozoic basins between the Infracambrian extensional phase and the late Carboniferous "Hercynian Orogeny" (upper part of TMS AP1 to AP4; Sharland et al. 2001) was not affected by major plate collisions or separations. Apart from the collision with Laurussia that commenced in the late Devonian with mild uplift in the platform interior, it was a relatively "calm period". Only local transpressional and transtensional reactivation processes occurred but these did contribute to the formation of numerous widespread unconformities. The Saharan Paleozoic basins were part of a large, interconnected North African shelf system that was in a "sag stage" following the Infracambrian extension (Burke et al. 2003; Craig et al. 2008). The region formed a wide "ramp-like" northward-facing passive margin of Gondwana, being blanketed by a regionally extensive succession of high-latitude, predominantly siliciclastic sequences.

Across North Africa, generalized sedimentary trends consisted of proximal fluvial-dominated deltas in the southeast prograding basinward, and more distal shallow marine and offshore marine sediments toward the northwest (Carr 2002). Major sediment sources were present to the south and deposition occurred in arid to glacial settings, reflecting the drift of the region from equatorial to high southern latitudes and back (Konert et al. 2001; Le Heron et al. 2010).

A prominent disturbance of these general sedimentary trends resulted from glaciation during the Late Ordovician (Le Heron and Craig 2008; Bell and Spaak 2007; Le Heron et al. 2010). The center of the ice sheet was located in central Africa. The legacy of the glaciation was incised relief, sometimes quite dramatic in magnitude, observed across North Africa and the Middle East. This had a strong influence on subsequent facies patterns and in particular later Silurian source rock deposition and distribution (Hirst et al. 2002; Ghienne et al. 2003; Le Heron et al. 2004; Moreau et al. 2007; Moreau 2011; Rubino et al. 2000; Ron Martin et al. 2016).

3.2.1 Habitats and Stratigraphy of the Main Cambrian Reservoirs

Many prolific reservoirs of the Saharan Platform lie within Cambro–Ordovician and Late Silurian–Ordovician sequences (Fikirine and Abdallah 1998). Cambrian sandstones were deposited over eroded basement in a variety of braided stream, deltaic, and marginal shallow marine environments. The Cambrian is productive in most fields of the Algerian and Libyan *Triassic province*, in the *Illizi* and *Timimoun* basins (Figs. 4, 5 and 6). The main producing horizons are assigned to the *Menkel* and *Hassaouna Formations* (Crossley and McDougall 1998). The distribution of permeability and intergranular or fracture porosity is irregular and controlled by sedimentological and diagenetic parameters. The sandstones consist of a succession of fining upward sequences capped by clay intervals. In most proximal areas, the basal deposits are pebbly, arkosic, cross-bedded sandstones associated with an extensive fluvial braid plain. In the region of the present-day Gulf of Suez, the metamorphic and granitoid pre-rift basement was extensively peneplained prior to deposition of the Cambrian clastics of the *Araba Formation* of the *Qebliat Group* (Hassan 1967; Said 1971; El Barkooky 1992).

In Saudi Arabia, the peneplained Pre-Cambrian crystalline rocks are unconformably overlain by the *Saq sandstone*, of probable Cambro–Ordovician age

(Fig. 8). The basal *Risha Member* consists of channelized, cross-bedded coarse sandstone deposited under high energy in a braided river system. The overlying *Sajir Member*, is a shallow marine littoral sandstone (McGillivray and Husseini 1992). On the southeastern margin of Saudi Arabia, the *Wajid sandstone* ranging in age from Cambrian to Devonian is a recent target for hydrocarbon exploration. This sandstone is composed of varicolored siliciclastics, mainly coarse- to fine-grained with granule conglomerate and siltstone interbeds, displaying diverse sedimentary primary structures (Wanas and Abdel-Maguid 2006; Keller et al. 2017).

3.2.2 Habitats and Stratigraphy of the Main Ordovician Reservoirs

Compared to the Cambrian, during the Ordovician a greater area is covered by marine facies. The Lower to Upper Ordovician continental or fluvio-deltaic shallow marine siliciclastic reservoirs contain both oil and/or gas (*Hamra Quartzite* and *Haouaz Formations* of Libya; Vos 1981; Ramos et al. 2006; Mohamed 2016). Across North Africa, the Upper Ordovician glacigene reservoirs (*Memouniat, Naqus, Zarqa*, and *Sarah Formations*) contain at least 5 billion barrels of oil equivalent in more than 50 fields (Fig. 7). Reservoirs were deposited in glacially influenced, generally shallow marine settings (Vaslet 1990; Mohammed et al. 2002; El-Hawat et al. 2003; Fello and Turner 2004; El-Ghali 2005; Craig et al. 2008; Le Heron et al. 2010). The presence of unconformities at the top and bottom of the latest Ordovician formations was attributed to the "Taconian" events (Guiraud and Bosworth 1999; Guiraud et al. 2005).

3.2.3 Habitats and Stratigraphy of the Main Silurian Reservoirs and Source Rocks

This period was of relatively short duration but became famous due to its enormous importance to North African–Arabian hydrocarbon systems (Figs. 3, 4, 5, 6 and 7). After the melting of the Late Ordovician ice cap, early Silurian sea-level rose by more than 100 m, leading to a major transgression that flooded the North African shelf and other parts of Gondwana. Anoxic water bottom conditions in the sediment-starved basin margins resulted in the preservation of organic-rich shales—the prolific Silurian "hot shale" (Lüning et al. 2000; Mustafa et al. 2014; Lüning and Fello 2006; Fello et al. 2006). These Silurian shales are proven sources for hydrocarbons throughout much of the margin of the Gondwana continent (Figs. 8 and 9), from Algeria in the west to Oman in the east (*Tanezzuft Formation* in Libya; *Mudawwara Formation* of Jordan; *Tanf* Formation of Syria, Iraq, and Turkey; *Sarchahan Formation* of Iran; *Qusaiba Formation* of Saudi Arabia, Bahrain, Qatar, and the UAE; and *Sahmah Formation* of Oman). Unfortunately, the time-equivalent strata in Egypt are comprised predominantly of siltstone with minor sandstone and shale (*Kohla Formation*; Keeley 1989; Tawadros 2001) and have no source potential. The upper Silurian regressive fluvio-deltaic shallow marine *Acacus*

Formation sandstones serve as reservoirs and contain both oil and/or gas in Libya (Fig. 7; Echikh 1998; Arduini et al. 2003). These same sands extend into the *Western Desert* of Egypt where their exploration is just beginning (*Basur Formation*; Keeley 1989). Subsidence of the north Gondwanan passive margin continued through the Silurian reflecting the development of the proto-Tethyan ocean between Gondwana, Armorica, and Avalonia.

3.2.4 Habitats and Stratigraphy of the Main Devonian Reservoirs and Source Rocks

Most workers consider the lowermost Devonian to be missing in the *Illizi Basin* of eastern Algeria and across much of Libya (Tawadros 2001). This is a consequence of the *Caledonian* or *Early Acadian/Ardennian* event (Boote et al. 1998; Guiraud and Bosworth 1999; Coward and Ries 2003; Guiraud et al. 2005). The Lower Devonian succession above the unconformity consists of continental sandstones with brief marine transgressions (Fig. 5). Lower Devonian sandstones deposited as offshore coastal bars in the northwestern Illizi Basin contain important hydrocarbon reservoirs (Klett 2000a, b; Wendt et al. 2009). The Lower Devonian is also the most productive horizon in the Algerian-Libyan *Ghadames-Hamra Basin* (Fig. 6) accounting for half the Ghadames reserves and about a third of those in the Hamra (Echikh 1998). The transition between the Early and Middle Devonian was marked by the *Mid- or Main-Acadian event* and entailed significant erosion (Guiraud et al. 2005). The Middle and Late Devonian consist of transgressive deposits and include the *Tadrart-Ouan Kasa* and *F5-F4 Formations*. Sand bars of the Late Devonian *F3* are important reservoirs in the Illizi Basin (Fig. 5), (Chaouchi et al. 1998). From the petroleum perspective, of importance is a regional Frasnian transgression resulting in the deposition of a widespread organic-rich shale, the "*Argile Radioactive*". This source rock extends across much of the North African Platform, and is estimated to be the origin of about 10% of all Paleozoic sourced hydrocarbons in North Africa (Boote et al. 1998; Lüning et al. 2003). The overlying late Frasnian to early Tournaisian succession is dominated by relatively thin, cyclic, platform deltaic sandstone deposits of the *Aouinet Ouenine (III and IV)-Ouenine-Shatti Formations* in Libya and the *F2 sequence* of the Illizi-Ghadames region (Vos 1981; Boote et al. 1998). These provide significant reservoirs in the *Illizi* and *Hamra* Basins. In the Western Desert of Egypt, the Pragian to Famennian section is assigned to the well-developed *Zeitoun Formation* (Keeley 1989). The formation is predominantly shale with occasional siltstone, sandstone, and limestone beds. Some of the shale units contain elevated organic carbon levels but these are not comparable in quality to the Argile Radioactive.

In contrast to North Africa, in the Middle East, the latest Silurian and Devonian periods are poorly represented in the rock record (Fig. 8). This is primarily due to Carboniferous Hercynian tectonism, uplifting and resulting erosion. Devonian deposits are partly preserved in Saudi Arabia where a large delta front developed (Saner et al. 2006). In the north, the delta is replaced by mixed marine siliciclastics and carbonates. The Devonian *Jauf Formation* is the main pre-Khuff reservoir in the *Northern Ghawar area*, where it hosts the recently discovered giant *Hawiyah gas-condensate field* (Saner et al. 2006). The *Jauf* consists of shallow marine sandstones that exhibit unusually high porosities considering the burial depths (Wender et al. 1998). In central Arabia, Syria and Iraq the Devonian consists of continental deposits, while marginal marine environments developed in Turkey and Oman (Konert et al. 2001).

3.2.5 Habitats and Stratigraphy of the Main Carboniferous Reservoirs

The Carboniferous is well developed in Libya and parts of Egypt. The Carboniferous stratigraphic nomenclature of Libya is very complex, partly because these units occur in outcrop around the rims of several basins which resulted in numerous local formation names. The basal Tournaisian to Visean section is widespread and most commonly referred to the *Mrar Formation* (Lelubre 1948; Burollet 1960). The lower part of this section can contain numerous, stacked coarsening upward shale-sandstone prograding deltaic sequences and is often capped by a stromatolitic limestone flooding unit (Tawadros 2001). The Mrar Formation is overlain by the upper Visean to Serpukhovian *Assedjefar Formation* and its lateral equivalents (Lelubre 1952). The lithologies of this interval vary dramatically both vertically and laterally and include continental sandstones, shales, marls, and limestones (Tawadros 2001). Both the Mrar and Assedjefar form significant reservoirs in the *Illizi* and *Ahnet Basins* and to a lesser extent in the *Ghadames-Hamra* region (Whitebread and Kelling 1982; Boote et al. 1998; Fröhlich et al. 2010). The Upper Carboniferous of Libya is generally assigned to the *Dembaba* and *Tiguentourine Formations* (De Lapparent and Lelubre 1948; Lelubre 1952; Burollet 1960). Lithologies include limestone, dolomitic marls and shales, feldspathic sandstones, stromatolites, and occasionally gypsum or anhydrite (Tawadros 2001). This marked the replacement of the earlier siliciclastic-dominated deltaic systems with a carbonate platform (Carr 2002). Some of the sandstone reservoir units in the deeper part of the Sirt Basin stratigraphy that were previously thought to be Cambro–Ordovician in age are now known to be Carboniferous (Wennekers et al. 1996).

Large parts of the Mississippian (Lower Carboniferous) are missing from the Western Desert of Egypt. This is likely a consequence of erosion during the Late Carboniferous glaciation and coeval Hercynian tectonism. Those Carboniferous strata that are preserved are largely massive, thick continental sandstones assigned to the *Desouqy Formation* (Keeley 1989). The Desouqy has recently produced oil in the *Faghur Basin* of the Western Desert (Bosworth et al. 2015b). In the Gulf of Suez, Carboniferous–Permian sandstones are important reservoirs and are variously referred to the *Abu Thora, Abu Durba, Rod el Hamal or Aheimer Formations* (Abdallah and El Adindani 1963; Alsharhan and Salah 1997; Kora 1998) or simply *Nubia B*.

As with the Devonian, Carboniferous deposits are largely missing in the Middle East, due to widespread uplift and erosion during the Hercynian Orogeny. In Syria,

Carboniferous sequences (*Doubayat Formation*) were deposited and preserved in a NE-trending *proto-Palmyra trough* (Al-Saad et al. 1992; Brew et al. 2001). *Markada Formation sandstones* are producing reservoirs in several gas fields in the Palmyrides (Al-Youssef and Ayed 1992; Lučić et al. 2007; Lučić 2008).

3.3 Hercynian Orogeny

The stability of the entire Gondwana platform was interrupted during the mid-Carboniferous by the *Hercynian Orogeny* (Fig. 8). The collision between Gondwana and Laurasia, and compression associated with the closure of Paleotethys took place along the line of the *Lawrence-Azores Shear Zone*. Following the disassembly of Pangaea in the late Paleozoic–early Mesozoic, the North African Platform became a broad Neotethyan-facing passive continental margin. The intensity of deformation decreases from west to east. The "*Hercynian*" unconformity is also strongly diachronous and often a composite of multiple events (Guiraud et al. 2005). In Algeria and westernmost Libya, the main unconformity is *Early Permian (Sakmarian)* in age and overlies *Autunian-Stephanian* clastic sediments. In eastern Libya and Egypt, there are two discrete unconformities, one, Late Carboniferous and the other, Early Permian in age (Craig et al. 2008). Further east in Arabia, *Westphalian* to *Artinskian* clastic sediments unconformably overlie older Carboniferous and pre-Carboniferous strata. Deformation included block tilting, domal uplift, and local folding associated with the movement of major N–S trending strike-slip faults (Guiraud and Bosworth 1999; Guiraud et al. 2005). During this orogeny, Gondwana was rotated through 90° in an anticlockwise direction (Konert et al. 2001), and central Arabia was uplifted, tilted down eastward, and deeply eroded.

The Hercynian Orogeny played an important role in the evolution of Paleozoic petroleum systems across North Africa and Arabia. Most of the major structural arches (Fig. 9) in central North Africa were active at this time, including: (1) the E-W oriented *Telemzane Arch/Dahar Dome* and *Charnian Uplift* along the northern edge of the Ghadames Basin; (2) the east–west to west-southwest—east-northeast trending *Qargaf Arch* separating the Ghadames and Murzuq basins in western Libya; and (3) the north–south to north-northeast—south-southwest oriented *Amguid-El Biod Arch* and *Hassi Messaoud Ridge* separating the Illizi-Ghadames and Oued Mya basins in western Algeria (Craig et al. 2008). Extensive erosion of lower Paleozoic rocks took place over the arches during the mid-Carboniferous and Early Permian events. These arches also affected the hydrocarbon migration from large areas, resulting in a number of supergiant oil and gas fields, such as *Hassi Messaoud* in Algeria. These structures were sufficiently large to preserve their integrity through two subsequent phases of uplift. The reservoir properties of some of the more deeply buried Cambrian and Ordovician sandstones were locally enhanced by weathering and fracturing during the uplift and erosion (Craig et al. 2008). In addition, reactivated Hercynian faults locally provided conduits for vertical hydrocarbon migration from source to reservoir.

In Arabia, the main effect of the deformation was the formation of broad NE-trending arches (Fig. 9) over which the Cambrian—Devonian section was deeply or entirely removed by erosion. These arches included: (1) the *Levant Arch* extending from Egypt along the Levant coast to Turkey; (2) the *Al-Batin Arch* extending from the Arabian Shield to Iran; and (3) the *Oman-Hadhramaut Arch* located along the southeast coast of Oman (Faqira et al. 2009). The main pre-Khuff oil and gas accumulations in the Middle East are concentrated along the flanks of these arches, as at the supergiant *Ghawar field* of Saudi Arabia (Faqira et al. 2009).

3.4 Post-Hercynian Breakup of Gondwana

From the Pennsylvanian (Late Carboniferous) to mid-Cretaceous times the North African-Arabian plate margin underwent multiple phases of rifting accompanying the oceanization of Neotethys. The newly formed passive margin subsided and enhanced the effects of eustatic marine transgressions (TMS AP5-AP8; Sharland et al. 2001). The region was subjected to north–south extension accompanied by related volcanism (flow basalts) and evidenced by the presence of numerous unconformities (Guiraud 1998; Guiraud and Bosworth 1999). Rifting initiated locally in the Late Carboniferous or Permian–Triassic, accelerated and spread during the Jurassic, and reached maximum subsidence rates in the Early Cretaceous (Şengör 1979; Stampfli et al. 2001; Garfunkel 2004).

Throughout most of the Jurassic, opening of the mid-Atlantic was accompanied by separation of North Africa and Europe along a major sinistral transcurrent shear zone. The eastern Mediterranean region opened during this period to create the new passive margin of Neotethys. Around the Jurassic–Cretaceous transition times, tectonic deformation that occurred in southeastern Europe, referred to the *Cimmerian* or *Berriasian* orogenic event, was registered along the northern Egyptian margin and the Levant (Guiraud 1998; Guiraud et al. 2005). Early Cretaceous extension reactivated preexisting basement structures and wrench zones throughout North Africa and disrupted the Paleozoic succession, creating new hydrocarbon traps and further modifying source rock maturity. This was followed by two major transgressions from Neotethys, first during the Aptian and subsequently in the late Albian/Cenomanian. Evaporite and then large neritic carbonate platforms developed around the North African–Arabian Neotethyan margins (Guiraud et al. 2005). Many of the peri-Tethyan basins were later folded, faulted, and inverted starting in the Santonian (Guiraud and Bosworth 1997). This had a major impact on the trapping mechanisms and hydrocarbon migration in many basins including in Algeria, Libya (Sirt), Egypt (eastern Western Desert), and Syria (Palmyrides).

Gondwana moved north, so its margin now straddled the equator with most of North Africa and the Arabian plate occupying an equatorial setting within 30°N and S latitude (Sharland et al. 2001). Late Permian, Triassic, and Jurassic sedimentation on the Arabian Plate was dominated by carbonate–evaporite deposition, whereas the Early Cretaceous was dominated by open marine-mixed siliciclastic-carbonate deposits.

3.4.1 Habitats and Stratigraphy of the Main Late Carboniferous–Permian Reservoirs and Source Rocks

The first extensive deposits following the Hercynian Orogeny are the Upper Carboniferous–Lower Permian clastics that rest with angular unconformity on the older Paleozoic rocks or basement (Fig. 8). They were partly deposited coevally with rift tectonics along the eastern and northern margins of the Arabian Plate (Konert et al. 2001). In North Africa, Early Permian rift basins similarly formed diachronously along the northernmost African margin, from Morocco to Egypt. These were infilled by thick series of continental (Morocco–Algeria), mixed (Egypt), or marine sediments (southern Tunisia–northwest Libya), whereas only thin and local continental series were deposited along the southern Sahara platform (Guiraud et al. 2005).

In Arabia, the syn-rift sedimentary succession contains important source rocks and reservoirs intercalated with shale and evaporate seals. Generally, the section consists of braided plain, channel fill, and eolian sandstones and siltstones that were deposited in semiarid conditions. The Lower Permian *Unayzah Formation* is the principal pre-Khuff oil and gas reservoir in Saudi Arabia, including the Ghawar area, and in Bahrain, Qatar, Kuwait and the U.A.E. The Unayzah consists of fluvial to marginal marine sandstones (McGillivray and Husseini 1992; Wender et al. 1998; Khan Tanoli et al. 2008). They are replaced basinward by shallow marine nearshore sediments in the Zagros Mountains.

Rifting accelerated throughout the Permian, increasing accommodation space in conjunction with the opening of Neotethys (Stampfli and Borel 2002). The base of the Arabian Upper Permian succession consists of nearshore continental to marine sandstones and shales, prograding from the west. Northward the continental deposits include coal deposits indicating wetter tropical environments. These were followed by the deposition of extensive carbonates and anhydrites over the entire Arabian shelf in shallow marine to tidal flat environments. The transgressive *Khuff Formation* carbonates mark a significant change in the pattern of sedimentation from dominantly siliciclastic to dominantly carbonate. Each depositional cycle began with a transgressive phase and ended with a regressive phase during which sedimentation occurred in the supratidal zone. Khuff deposition continued into the Triassic. Khuff fields located in Bahrain, Iran, Qatar, Saudi Arabia and the United Arab Emirates hold the world's largest nonassociated gas reserves (Vaslet et al. 2005). The primary reservoir lithology is microcrystalline dolomite.

3.4.2 Habitats and Stratigraphy of the Main Triassic Reservoirs and Source Rocks

Late Permian to Early Triassic transgression of the Neotethys over the northwesternmost African margin produced terrigenous to mixed marine sedimentation which unconformably overlies the Permian or older deposits. Sediments were initially deposited in topographic lows and Triassic rocks are not preserved over large areas of eastern Libya and Egypt, which were topographic highs until the Late Triassic or Early Jurassic (Carr 2003). Triassic fluvial sands and shales which transgressed southwards across the peneplained Hercynian unconformity form the most important reservoirs in the *"Triassic Basin"* of northeast Algeria and southern Tunisia (Fig. 4). These are the *"Triassic Argilo-greseux Inferieur" (TAGI)* and its lateral equivalent in the Hassi Touareg area. In the intra-continental domain, large fluviatile/lacustrine basins remained active, spreading sediments northward (Boote et al. 1998; Bibonne et al. 2015). Roughly equivalent fluvial sandstones are the main oil and gas reservoirs in the Algerian *Berkine* and *Oued Mya Basins*, including the giant gas field of *Hassi'R Mel* (Fig. 4). Similar Triassic sandstones also serve as reservoirs in the *Sirt Basin* (Craig et al. 2008). Faulting and local uplift of rift basin shoulders favored terrigenous sedimentation (Guiraud et al. 2005). The Late Triassic to Early Jurassic evaporites and shales in the northeast part of the Algerian Saharan Platform are up to 2 km thick and form a hydrocarbon cap rock for the Triassic reservoirs (Turner and Pelz 2017). In some cases, through the Hercynian unconformity, they also form the top seal for Paleozoic reservoirs.

The Arabian Plate persisted as a relatively peneplaned ENE sloping passive margin platform and Early Triassic transgressive–regressive cycles formed an alternating sequence of shallow marine sands, shales, and marly limestones (Ziegler 2001). A marginal marine platform developed over eastern Saudi Arabia during the Middle Triassic (Ziegler 2001) and then diminished in the Late Triassic. By the end Triassic, gentle tectonic activity was registered along the African–Arabian Neotethys margin, creating frequent unconformities due to the emersion of this large domain (Guiraud et al. 2005). In the Arabian Basin, east–west shortening (in the south) and northeast–southwest shortening (in the north) caused intermittent erosion of the Upper Triassic facies (Khalifa 2007). In contrast, subsident regions accumulated large amounts of carbonates and evaporates. The *Kurra Chine Formation* is one of the primary reservoirs in the central part of the margin especially in Syria and Kurdistan (Craig et al. 1994; Sadooni 1995; Hips and Árgyelán 2007; Lučić et al. 2002, Lučić 2008). Other Triassic reservoirs include the *Dolaa, Merga Mir, Mohilla, Mulussa, Geli Khana,* and *Jihl Formations* and the *Ma'in Group*. The shales of the *Kurra Chine* are proven source rocks in Syria and Iraq (Aboud et al. 2005; Leško et al. 2005; Al-Ameri et al. 2009; Lučić 2008; Vulama 2011). Early in the Carnian, a very pronounced relative drop in sea level was recorded throughout the circum-Mediterranean region, causing a *"salinity crisis"*. The resulting lowstand evaporite sequence was dominated by halite, continental sands, marls, and shales (Ziegler 2001; Marlow et al. 2011). Thick salt deposits acting as an efficient seal for the underlying reservoirs have been recorded in Syria (McBride et al. 1990; Best et al. 1993; Choquette et al. 1994; Brew et al. 2001; Lučić et al. 2003a, b, 2010), Lebanon (Nader and Swennen 2004), Israel (Eshet 1990a, b; Bialik et al. 2012), and Libya (Adloff et al. 1986).

3.4.3 Habitats and Stratigraphy of the Main Jurassic Reservoirs and Source Rocks

Jurassic extension continued during the separation of the North African and European plates and resulted in a broad zone of rift basins crosscutting the Pre-Cambrian basement and, in some areas, preexisting Paleozoic basins on the Neotethyan continental margin of North Africa (Guiraud et al. 2001, 2005). In northern Egypt (Fig. 12), half-grabens controlled the deposition of Jurassic source rocks and reservoirs (Keeley et al. 1990; Keeley and Wallis 1991; Moustafa 2008; Wescott et al. 2011; Bevan and Moustafa 2012; Bosworth et al. 2015b). General proximal-to-distal trends can be noted from eastern Algeria toward the west, in western Egypt from southwest to northeast, and in central Egypt from south to north (Carr 2003). Large rivers flowed north on both sides of the *Farafra-Bahariya Arch in* Egypt, supplying clastic sediments for the Jurassic *Khatatba Formation* (Rossi et al. 2001; Alsharhan and Abd El-Gawad 2008; Younes 2002, 2012; Shalaby et al. 2012; Bosworth et al. 2015b). The Khatatba Formation is comprised of fluvial siliciclastics and coals grading upwards into first nonmarine and then marine shales and limestones (Keeley and Wallis 1991; Bosworth et al. 2015b). The Khatatba sandstones are oil and gas reservoirs in many fields, and the overlying organic-rich nonmarine shales represent the most important source rock in the Western Desert (Hegazy and Tammam 1990; Shalaby et al. 2012; Abrams et al. 2016). Associated coals contribute both oil and gas. The coaly shales act also as seals, and thus the *Khatatba Formation* represents a complete hydrocarbon source-seal-reservoir system.

Maximum transgression of the sea occurred during the Jurassic-Cretaceous transition (end Tithonian) and at this time the sea covered almost the whole of northern Egypt. This resulted in the widespread deposition of the *Masajid Formation* (Keeley et al. 1990), generally limestone but locally dolomitic and in some cases an oil-producing reservoir. Similarly, the main reservoirs for the *Essaouira* and *Tarfaya Basins* in Morocco are carbonate buildups of Triassic to Late Jurassic age (Broughton and Trepaniér 1993). In Morocco and Tunisia, siliciclastic Jurassic reservoirs are less common but the Oxfordian–Tithonian *M'Rabtine sandstones* are important reservoirs at the *Essaouia field* in the *Gulf of Gabes* (Fig. 10; MacGregor and Moody 1998; Klett 2001).

On the Arabian Plate, narrow facies belts of marginal marine siliciclastics, shallow marine siliciclastics, and siliciclastic-free carbonate platform bordered the Arabian Shield (Fig. 14; Ziegler 2001). During the Early Jurassic Toarcian, a widespread westward marine transgression covered the Arabian shelf. The depositional environment was arid and varied from nondeposition and erosion in southern Arabia to a clastic platform in the west. A shallow marine, mostly carbonate, platform developed to the north and east. After a significant depositional hiatus in the Aalenian, Middle Jurassic (Bajocian–Bathonian) carbonates were deposited in a shallow, clear, warm-water shelf environment that occupied most of the Arabian Peninsula (Al-Husseini 1997). Locally, in the west and south, clastic input continued, while in the north, a deeper mixed siliciclastic-carbonate shelf existed where the *Gotnia Basin* developed. During the early Late Jurassic intra-shelf basins developed over most of Arabia. These basins were the depocenters of the main Jurassic source rocks as well as several important reservoirs. The Upper Jurassic contains about 25% of the total source rocks of the region (Ulmishek and Klemme 1990). The source rocks include transitional marine shales, deeper marine shales, and chalks generally assigned to the *Marrat, Dhruma, Sargelu* and *Najmah Formations* (Abdullah 2001; Kendall et al. 2014). The Tithonian was dominated by evaporites and shallow marine carbonates which correspond to the *Arab reservoirs* and *Arab, Hith,* and *Gotnia* evaporates (Swart et al. 2005; Eltom et al. 2013). The Jurassic formations of the western and southern Arabian Gulf constitute one of the most important petroleum systems of the world (Al-Husseini 1997; Ulmishek and Klemme 1990).

3.4.4 Habitats and Stratigraphy of the Main Early Cretaceous Reservoirs and Source Rocks

Late Jurassic rifting was followed in northern Egypt by an even stronger phase of extension in the *Valanginian-Hauterivian*. This resulted in a doubling or tripling of subsidence rates in most of the affected half-grabens (Bosworth et al. 2015b). Sedimentation was a mixture of fluvial, estuarine and shallow marine, transitioning to deeper marine nearer to the Neotethyan margin. These strata are assigned to the *Alam el Bueib Member* of the *Burg al Arab Formation* (Fig. 12; Norton 1967) and include some of the most productive siliciclastic reservoirs of the Western Desert (Egyptian General Petroleum Corporation 1992; note there is great diversity in the stratigraphic terminology used by industry in the Western Desert—we follow the original definitions). Tectonic quiescence followed in the *Barremian*, ending in a major transgressive event in the *Aptian*. The *Alamein carbonates* and *Dahab shales* were deposited at this time and these together form some of the most extensive top seals in the Western Desert. The Alamein itself, where dolomitic, is also an important local reservoir (Egyptian General Petroleum Corporation 1992; Dolson et al. 2014).

Much of the North African–Arabian margin was involved in a second Early Cretaceous rifting episode that occurred in the Aptian–Albian and affected the area from the *Anza graben* of Kenya, to the *South Sudan rifts*, the *Central African Shear Zone*, the *Benue trough* of Nigeria, the *Niger rifts* and north to the *Sirt basin* and *Western Desert* of Egypt (Browne and Fairhead 1983; Fairhead 1988; Schull 1988; Guiraud and Maurin 1992; Bosworth 1992, 1994; Guiraud et al. 2005). This rifting period was accompanied by volcanism (flow basalts) as observed from the Euphrates Trough, through the Palmyrides, to the Sirt region of Libya and into Tunisia (Fig. 12; Guiraud 1998; Wilson and Guiraud 1998; Guiraud and Bosworth 1999). Mainly, terrigenous fluviatile–lacustrine formations initially infilled rift basins as in the eastern Sirt *Hameimat trough* where the *"Sarir Formation"* fluvial sandstones are a significant reservoir (Lewis 1990; Gras 1998; Gras and Thusu 1998; Ambrose 2000; Hallett 2002; Abadi et al. 2008; El-Khatri et al. 2015). In the

Sirt Basin, deposition was accompanied by intensive basement block faulting from the Aptian–Albian onwards. In the Western Desert, the equivalent Aptian–Albian siliciclastic reservoirs are referred to the Burg el Arab *Kharita Member* and lower *Bahariya Formation* (Fig. 12; Said 1962; Norton 1967; Tawadros 2001).

In eastern Egypt, in the area of the present-day Gulf of Suez, another important *Aptian-Albian* reservoir, though not confined to penecontemporaneous rifted structures, is the *Malha Formation,* or "*Nubia A*" (Fig. 13; Abdallah et al. 1963; Alsharhan and Salah 1997). In the southern Gulf of Suez, the Malha and fractured Neoproterozoic basement are the only really significant pre-Miocene reservoirs as most of the Paleozoic sandstones are eroded (Bosworth 1995).

Sarir-Malha-Nubia deposition was followed by another major transgression in the Late Albian and culminating at the Early-Late Cretaceous boundary. This resulted again in the development of large marine gulfs in Egypt and Algeria. Carbonate platforms developed along the African–Arabian Neotethyan margin (Sadooni and Alsharhan 2003; Guiraud et al. 2005). From this point on into the Cenozoic much of North Africa was dominated by carbonate deposition and siliciclastics, when present, were typically marine shales.

During most of the Early Cretaceous, Arabia was surrounded to the north, east, and south by Neotethyan passive margins. Sediments were deposited on open platforms and within intra-shelf basins (Fig. 15; Ziegler 2001). The interior of the plate was rimed by proximal shallow marine and in places continental, lacustrine, and deltaic deposits. Distally toward the open marine environments, transitional and carbonate deposits predominated (Davies et al. 2002). Shallow-water platform limestones with fringing rudist reefs were present along the hinge line of the eastern Mediterranean shelf margin. Generally, the succession becomes pure limestone toward the late Cenomanian. The Early to mid-Cretaceous (Hauterivian to Cenomanian) rocks of the Arabian Plate contains some of the most important siliciclastic and carbonate hydrocarbon reservoirs in the world (Fig. 15; Davies et al. 2002). Siliciclastic reservoirs are the *Burgan, Nahr Umr*, and *Wasia* (*Safaniya-Khafji Members*) Formations of the northern Gulf, the *Zubair Formation* in Iraq and Kuwait, (Alsharhan et al. 2014a, b; Davis et al. 2002) and the *Qishn Formation* of Yemen. The best reservoir sandstones represent west to east trending estuarine channels cut into N-S oriented palaeoshorefaces. Rudist and associated carbonate reservoirs include the *Sarvak Formation* of Khuzestan Province, Iran; the *Shu'aiba, Upper Thamama* and *Mishrif* reservoirs of northeastern Saudi Arabia and the United Arab Emirates; and the *Natih* reservoirs of northern Oman (Davies et al. 2002). Cretaceous source rocks have been mostly deposited within the intra-shelf basins.

3.5 Collision with Eurasia

The Late Cretaceous times correspond to a crucial period for North Africa and Arabia (TMS AP 9–10; Sharland et al. 2001). Continued oceanic spreading of the Atlantic, Neotethyan and Indian Oceans was concomitant with the continuous northwest movement of the Arabian Plate. Subduction resulted along the northern margin of the rapidly closing Neotethys.

As discussed above, active Jurassic and Early Cretaceous rifting episodes over large domains led to the breakup of Western Gondwana and the opening of the South and Equatorial Atlantic Oceans. The inversion of many E–W trending basins followed in the mid- to Late Cretaceous, corresponding to the first stages of the formation of the *Alpine orogenic belt*. The first signs of shortening were observed in the Aptian, with an approximately north–south compressional event corresponding to the *Eo-Alpine (Austrian) Phase* described in the European Alps (Guiraud et al. 1987). This has been considered as a precursor to the Alpine Orogeny sensu stricto, and is recorded by a regionally important intra-Aptian unconformity evident in numerous African basins. Thrusting and nappe formation occurred in the northern Atlas of Morocco, passing into milder inversion in the northern Pelagian basin with additional minor effects in the Ghadames and related Triassic basins (Badalini et al. 2002).

The *Late Santonian* ($\sim$84 Ma) *event* corresponds to the first plate-scale compressional episode registered by Africa–Arabia during the *Alpine Cycle*, resulting in inversion or folding of most suitably oriented E–W to ENE–WSW trending basins (Guiraud et al. 1987; Guiraud and Bosworth 1997; Guiraud 1998). The African-Arabian plate rotated in an anticlockwise direction and this began the still ongoing collision with the Eurasian plate. Northeast Africa and Arabia were subducted beneath Eurasia along the Cretian–Cyprean–Bitlis–Zagros arc complexes. The final stage of the obduction of ophiolites was registered in Oman. Localized uplift occurred within the Sirt Basin, producing islands within an otherwise open to marginal marine setting, and uplift of *Gebel Akhdar* in Cyrenaica initiated.

One of the products of the compressional episode is a system of en-echelon, NE and ENE oriented fold-thrust structures that extends in an S-shaped "*Syrian Arc*" belt from the Western Desert *Farafra-Bahariya* Oases to *Abu Roash* near Cairo, through Sinai and the Levant and into Central Syria (Krenkel 1925; Said 1962; Ponikarov 1967a, b; Lovelock 1984; Quennell 1984; Moustafa and Khalil 1989; Abdel Aal et al. 1992; Chaimov et al. 1992, 1993; Barazangi et al. 1993; Best et al. 1990, 1993; Ayyad and Darwish 1996; Brew et al. 1999; Litak et al. 1997; Bosworth et al. 1999; Sawaf et al. 2001; Walley 2001; Abd El-Motaal and Kusky 2003). The Santonian event is one of the most important phases of deformation from the hydrocarbon system perspective. Many structural traps were formed or had significant growth during this period. Around the Maastrichtian–Paleocene boundary, a new compressional episode affected the northern African–Arabian Fold Belt (Guiraud and Bosworth 1997). The former Late Santonian folds were accentuated during this "*End Cretaceous event*" (Guiraud et al. 2005).

3.5.1 Habitats and Stratigraphy of the Main Late Cretaceous–Paleocene Reservoirs and Source Rocks

During the Cenomanian, regional subsidence related to Neotethyan rifting took place across the North African margin, and a marine transgression covered the entire northern African plate, culminating in the Turonian when a seaway linked Neotethys to the South Atlantic (Philip et al. 1993; Guiraud et al. 2005). During that period, also known as the Oceanic Anoxic Event (OAE) 2, organic-rich strata of black to grey laminated shales were deposited in intra-shelf rift basins and slopes across North Africa and in deep-sea basins of the adjacent oceans (Lüning et al. 2004; Zobaa et al. 2011; El-Sabbagh et al. 2011). These source rocks are part of the World's volumetrically most important petroleum generating stratigraphic sequence (Aptian-Turonian of Klemme and Ulmishek 1991).

In Libya, the collapse of the *Sirt arch* and the main rift phase began in the early Cenomanian, the base of which is marked by a regional unconformity (Fig. 11; El-Alami et al. 1989; Van der Meer and Cloetingh 1993, 1996; Thomas 1995; Anketell 1996; Baird et al. 1996; Wennekers et al. 1996; Guiraud and Bosworth 1999; Hallett 2002; Abadi et al. 2008; Hassan and Kendal 2014). Tectonism coincided with a southward transgression of Neotethys which progressively submerged the flanks of the Tunisian platform, the Sirt graben complex, basins of the Western Desert of Egypt, and the *Sinjar, Euphrates* and *Anah grabens* of Syria. Meanwhile, updip platform edges and crests of fault blocks were areas of either erosion, nondeposition, or high-energy deposition of predominantly biohermal carbonates. The main Upper Cretaceous clastic reservoirs are the Cenomanian *Bahi* and Coniacian–Santonian *Maragh Formations*. Carbonate reservoirs include the Campanian *Tagrifet* and Maastrichtian *Kalash* and *Waha* limestones and *Samah* dolostones (Hesendel et al. 1996; Hallett 2002). The three principal source rocks in the Sirt Province are the Upper Cretaceous *Rachmat, Sirte*, and *Etel Formations*, with the Campanian Sirte shale being the most important (Barr and Weegar 1972; Baird et al. 1996; Hallett 2002; Note that in Libya the basin is referred to as the "Sirt", but the formation was originally defined as the "Sirte").

In the Western Desert of Egypt, the Cenomanian began with deposition of interbedded sands and shales of the *Upper Bahariya Formation* (Said 1962; Darwish et al. 1994; Catuneanu et al. 2006). These constitute some of the basin's most important reservoir intervals because they are immediately capped by tight carbonate and shale units of the Late Cenomanian to Santonian *Abu Roash Formation* (Norton 1967; Aadland and Hassan 1972; Wasfi et al. 1988; Keeley and Massoud 1998). The "F" Member of the Abu Roash is an important source rock in the very deep *Abu Gharadig* basin and is the local manifestation of the Turonian OAE2.

In Eastern Egypt, the Late Cretaceous stratigraphy is very similar to that of the Western Desert (Fig. 13). Locally important siliciclastic reservoirs are found in the Cenomanian *Raha* and Coniacian–Santonian *Matulla Formations* (Ghorab 1961; Darwish et al. 1994). The Turonian *Wata Formation* is organically rich in the central and northern parts of the Gulf of Suez, but its significance is dwarfed by the Campanian *Duwi Formation* (Brown Limestone) that attains TOC of over 20%, though more typically 1.7–3.5% (Alsharhan 2003; El Diasty and Peters 2014).

The Paleocene section, like the Late Cretaceous, is part of the syn-rift megasequence in the Sirt basin. Although the rift-related paleotopography had been subdued by the end of the Cretaceous, old horst and graben blocks continued to undergo differential subsidence along with growth-type faulting (Wennekers et al. 1996; Guiraud and Bosworth 1999; Hallett 2002). The Paleocene consists almost entirely of carbonates and shales deposited in platform, platform-margin, pinnacle-reef, and basinal environments (Gumati 1992; Bezan 1996). Paleocene carbonate reservoirs are assigned to the *Upper Satal, Defa, Beda, Dahra, Sabil,* and *Zelten Formations* (Garea 1996; Garea and Braithwaite 1996; Spring and Hansen 1998; Pawellek 2009). The Paleocene accounts for approximately 12 BBO or over 25% of the oil reserves found in Libya (Baird et al. 1996; Hallett 2002). Across Egypt, the Paleocene was a time of shale and muddy carbonate deposition and is, therefore, not a reservoir objective. Further east in Israel and Syria localized folding and thrusting of the Syrian arc continued (Salel and Séguret 1994).

Due to mid-Turonian ophiolite obduction along the active eastern margin of the Arabian Plate, Late Cretaceous–Paleocene sediments were deposited within a compressive foreland basin setting (Searle and Cox 1999). A narrow NW-trending foredeep was formed west of the rising orogeny and the erosional products from the orogenic front were deposited as flysch sediments. Deeper marine shales of the *Shiranish* and *Sa'adi Formations* represent caprocks and the *Gurpi Formation* is a source rock. The interior of the Arabian shield was still occupied by a shallow-water carbonate platform (Ziegler 2001).

3.5.2 Habitats and Stratigraphy of the Main Early to Middle Eocene Reservoirs and Source Rocks

Relative tectonic quiescence continued through most of North Africa during the Early to Middle Eocene, though mild subsidence continued in the Sirt, Abu Gharadig and other major basins (Guiraud and Bosworth 1999; Guiraud et al. 2005). Active faulting was more pronounced in the *Gindi* and *Assiut basins* of the Nile River region. Carbonate platforms were largely maintained but a limited Ypresian regression led to *Hon Formation* evaporite deposition in the western and southern Sirt (Wennekers et al. 1996). Significant volcanic activity occurred at *Gebels Gharyan* and *Awaynat* (Busrewil and Wadsworth 1980; Flinn et al. 1991).

Eocene reservoirs are very important in the Pelagian basin offshore Tunisia and NW Libya. Early Eocene nummulite accumulations are common in the *El Garia* and *Jdeir Formations* (Loucks et al. 1998; Racey 2001; Racey et al. 2001; Anketell and Mriheel 2000; Klett 2001). Wet gas and minor oil within the El Garia Formation are sourced by organic-rich pelagic and hemipelagic biomicrites of the *Bou Dabbous Formation*, which is the deep-water lateral equivalent of the El Garia Formation.

Carbonate deposition also continued to dominate the Eocene of the Sirt basin with the most important reservoir units assigned to the *Gir, Gialo,* and *Augila Formations* (Barr and Weegar 1972; Tawadros 2001; Hallett 2002). At Gialo field, the reservoir includes banks of winnowed nummulites, each bank several meters thick, with the overall gross accumulation of nearly 500 m. Eocene reservoirs hold about 5 BBO in the Sirt basin (Baird et al. 1996; Hallett 2002).

Early to Middle Eocene limestone reservoirs are of local significance in Egypt. Minor oil and some gas are produced from the *Apollonia Formation* of the Western Desert Abu Gharadig basin and oil from the *Thebes* and *Darat Formations* of the Gulf of Suez (Snavely et al. 1979; Egyptian General Petroleum Corporation 1996; Bosworth et al. 2014). In both regions, secondary porosity predominates, generally resulting in good initial production rates with steep decline curves. The Thebes Formation is also a fair source rock with TOC typically 1.4–1.7% in the central part of the Gulf of Suez (Alsharhan 2003).

In Arabia, Early Eocene sediments were deposited within a "remnant" foredeep setting during rapid erosion and subsidence of the emergent ophiolite and sedimentary structures along the eastern margin of the plate (Ziegler 2001).

3.6 Red Sea Rifting and Closing of Neotethys

During this last stage (TMS AP11; Sharland et al. 2001), the geological evolution of North Africa and Arabia was dominated by strong tectonic and magmatic activities. The main events were the continued closing of the Neotethys Ocean resulting in the suturing of Eurasia and Arabia in the Middle Miocene and the opening of the Gulf of Aden and Red Sea starting in the Early Oligocene. These polyphased events generated deformation within the intra-plate domains and had significant impacts on the region's petroleum systems.

The eruption of the Afar plume at 31 Ma followed in rapid succession by the opening of Aden, the southern and then the northern Red Sea and Gulf of Suez were the start of the formation of the Arabian plate (McKenzie et al. 1970; Hempton 1987; Montenat et al. 1988; Hughes et al. 1991; Crossley et al. 1992; Ragab and El-Kaliouby 1992; Coleman 1993; Patton et al. 1994; Burke 1996; Bosworth et al. 2005). Red Sea rift initiation was marked by extensive basaltic dikes and local flows 24–22 Ma (Oligocene–Miocene transition; Sebai et al. 1991; Plaziat et al. 1998; Bosworth and Stockli 2016) followed by extensional faulting and concordant subsidence (Robson 1971; Garfunkel and Bartov 1977; Evans 1988; Steckler et al. 1988; Patton et al. 1994; Pivnik et al. 2003; Pejs et al. 2012). The final step was formation of the Gulf of Aqaba-Dead Sea transform fault boundary in the Middle Miocene (Bayer et al. 1988; Bosworth and McClay 2001; Bosworth et al. 2005). Coincident with the onset of Gulf of Aqaba movement connectivity of the Red Sea with the World Ocean became restricted and thick evaporite sections were deposited (Hassan and El-Dashlouty 1970; Fawzy and Abdel Aal 1986; Bunter and Abdel Magid 1989; Hughes and Beydoun 1992). In the Gulf of Suez, these would subsequently provide world-class top seals.

Another important Late Cenozoic event was the establishment of the Nile river drainage from the Oligocene onwards. During the Early Oligocene, the rise of African swells (Burke 1996) led to a change in river patterns across Africa and the systems which originally drained to the Niger switched northwards to create the current Nile system. This river not only cut a deep canyon on its way (Said 1981) but also via its delta delivered huge quantities of siliciclastics to the shores of the Mediterranean. The cone consists of 5.74 × 10^5 km^3 of compacted sediments (Macgregor 2011, 2012). Desiccation of the Mediterranean Sea during the Messinian caused enormous erosion and deposition of thick evaporites, later acting as a top seal (Roveri et al. 2016; Granado et al. 2016). "Grand Canyon scale" incisions occurred along many coastal areas and within the proto-Nile valley (Dolson et al. 2000). Large volumes of sediment were transported offshore through these canyons bypassing the shelf and into what would later be deep-water trends. As the Mediterranean finally refilled, subsequent flooding back-filled these valleys with fluvial and estuarine strata that are productive in many combination traps.

The collision between the African–Arabian and Eurasian plates episodically intensified during the Neogene and led to further inversion in northeast Libya and northern Tunisia, with strike-slip reactivation of older faults and the development of the Eastern Mediterranean Ridge (Craig et al. 2008). A new change in plate motions occurred around 22 Ma, as Africa then began to move northeastward. This change, synchronous with the opening of the Western Algerian–East Alboran ocean basin, generated compressional deformation during the Burdigalian along the northwestern African plate margin (Guiraud et al. 2005).

The Zagros fold and thrust belt is the deformational product of the Cretaceous to present-day convergence of the Arabian and Iranian (Eurasian) plates. Current seismicity and folding indicate that this deformation is still active (Jackson et al. 1981; Berberian 1995; ANSS 2018). The belt extends more than 2000 km from southern Turkey through northeastern Iraq to the Strait of Hormuz in southwestern Iran (Zainy et al. 2017). The belt is up to 300 km wide and is dominated by numerous large "whaleback" and smaller anticlines as well as localized thrust faults which trend NNW-SSE parallel to the general strike of the Zagros. 4–7 km of folded and faulted Paleozoic and Mesozoic section, resting on highly metamorphosed Proterozoic Pan-African basement, are overlain by 3–5 kilometers of Cenozoic siliciclastic and carbonate rocks (Ghazi and Moazzen 2015). This belt has also been affected by highly active, north–south trending strike-slip faults that cut across the region (Tillman et al. 1981; Hessami et al. 2001).

3.6.1 Habitats and Stratigraphy of the Main Neogene Reservoirs and Source Rocks

In North Africa, Neogene reservoirs and source rocks are best known from the Gulf of Suez, Nile Delta and partly from the Pelagian province, offshore Libya (Figs. 13, 17 and 18). The Gulf of Suez syn-rift succession consists of mainly siliciclastic deposits, including many reservoirs, source rocks and seals (Gawthrope et al. 1990; Patton et al. 1994; Bosworth and McClay 2001; Polis et al. 2005; Rohais et al. 2016). The main syn-rift phase of deposition is divided into the *Nukhul, Rudeis, and Kareem Formations* and all include reservoir quality sandstones (Saoudi and Khalil 1984; Patton et al. 1994; Wescott et al. 1996; Winn et al. 2001; Carr et al. 2003; Galil et al. 2012; Zaid 2013). Shales of the Rudeis and Kareem Formation contain significant amounts of the organic matter and act as generative source rocks, especially in the southern part of the Gulf.

In the Nile Delta, the stratigraphic succession can be subdivided into three sequences: pre-Miocene, Miocene, and post-Miocene (Fig. 18; Kellner et al. 2009; Dolson et al. 2000, 2001; Hanafy et al. 2016). The pre-Miocene sequence is topped by the Late Oligocene *Tineh Formation*, which includes lowstand overlapping sand-prone submarine fan facies and highstand prograding shale facies. The Miocene sequence is characterized by a thick shale-prone succession, deposited in an open marine environment and divided into four formations: the *Qantara, Sidi Salim, Wakar*, and *Rosetta Formations* (Farouk et al. 2014). The post-Miocene sequences include three formations: *Kafr El Sheikh* (early Late Pliocene), *El Wastani* (Late Pliocene), and *Mit Ghamr–Bilqas Formations* (Plio–Pleistocene to Recent) (Cross et al. 2009).

Similar depositional settings are present in the Levantine Basin (Fig. 19; Feinstein et al. 2002; Gardosh and Tannenbaum 2014; Gardosh et al. 2006, 2008a, b, 2009; Montadert et al. 2014; Calon et al. 2005; Homberg and Bachmann 2010; Roberts et al. 2010; Marlow et al. 2011; Hodgson 2012; Nader 2011, 2014). The main reservoirs are the *Tamar Sands* which were deposited in a series of submarine, turbidite fans of Chattian to Aquitanian age that accumulated at the base of the Levant slope and lower Pliocene *Yafo Sands* deposited in slope and basin-floor fans at the mouth of the Afiq and el-Arish Canyons. In both Tamar and Dalit fields, gas is trapped within large folds that were formed during the Late Miocene in a late stage of the Syrian Arc contractional deformation.

On the Arabian plate, dominant carbonate foredeep sedimentation was replaced in the Neogene by deposition in a subaerial continental foreland basin setting (Fig. 16). The Arabian Plate was progressively tilted to the northeast as a result of both uplift and rifting of Arabia from Africa, and structural loading of the northeast margin by the Zagros fold belt (Sharland et al. 2001). Up to 6000 m of sediments were deposited in this foredeep between the Oligocene and Present-day. Sedimentation was initially dominated by reefal carbonates of the *Kirkuk* and *Asmari Formations* which are some of the main reservoirs. The formations were deposited from the Early Oligocene to the Early Miocene. In addition to dolomitization, fracturing is one of the principal factors controlling reservoir quality and hydrocarbon production in SW Iran (McQuillan 1973, 1974).

4 Petroleum Systems

The "Petroleum System" is a very useful framework for describing the geological context of a prospect. It is not only a three-dimensional description but involves also the fourth dimension: time. The basic idea is that there are a number of independent processes leading to a hydrocarbon accumulation: generation, migration, entrapment, retention, and recovery (Magoon 1988; Magoon and Dow 1994). A very interesting, creative and stimulating task, and in geology, one of the most integrative fields of all. In practice, however, the definition of a petroleum system is not so simple and terminology is often used in differing manners. In addition to the petroleum systems definitions provided by the USGS, petroleum systems of North Africa and Arabia have been evaluated and described by many different authors. An excellent synthesis of North Africa petroleum systems was provided by MacGregor et al. (1998) and the petroleum systems of the "Southern Tethyan Margin Region" were recently compiled by Marlow et al. (2014). Here we will try to summarize at least the main elements that are most characteristic for each specific area or basin. Generally, due to the age of source rocks, it is possible to distinguish an Infracambrian Petroleum System, Palaeozoic-related Petroleum Systems, and Mesozoic-Cenozoic Petroleum Systems (Fig. 20). Hydrocarbons are trapped in literally all stratigraphic units from fractured Neoproterozoic basement to the youngest Pliocene–Quaternary sediments.

4.1 Infracambrian Petroleum System, Oman

Both production and hydrocarbon shows sourced from and reservoired in the uppermost Neoproterozoic–Cambrian successions occur worldwide. In northern Gondwana, the Infracambrian sedimentary fill developed during localized extension, and these basins have historically been of limited importance for hydrocarbon exploration. Deposition of marine organic-rich strata was generally restricted to the deeper parts of half-graben and pull-apart basins where anoxic environments developed under low-energy conditions (Lüning et al. 2009). In North Africa, hydrocarbon shows are documented in Morocco, and in the Taoudeni Basin which extends over southern and eastern Mauritania and into western Mali and SW Algeria. The widespread presence of stromatolitic carbonate units of potential reservoir facies in many parts of North Africa has been confirmed by new fieldwork in the Taoudenni Basin, in the Anti-Atlas region of Morocco and in the Al Kufrah Basin of Libya (Lottaroli et al. 2009). Hydrocarbons in the Taoudeni Basin are derived from source rocks in the Meso-Neoproterozoic *Atar Group*, which is composed of stromatolite-dominated carbonates and organic-rich basinal shales. The stromatolitic carbonates are dolomitized and contain solid hydrocarbons (Albert-Villaneuva et al. 2016). Following the assessment of the Abolag-1 well where both gas and liquids were tested, the "*Atar-Atar (!) Petroleum System*" was defined by Martin-Monge et al. (2016). The source rock and main reservoirs lie within the Atar Group and the petroleum recovered from the reservoir is correlative with the source rock.

To date, the commercial production of hydrocarbons from Neoproterozoic–Lower Paleozoic petroleum systems is known only from the Sultanate of Oman (Fig. 2; Pollastro 1999; Ghori et al. 2009; Droste 2014; Al-Johi and Al-Laboun 2015). The estimated reserves of 5.5 BBO and 30 TCFG ranks Oman among the relatively smaller producers in the Middle East (Droste 2014) and still stands out as the exception rather than the rule for Neoproterozoic plays. The petroleum systems

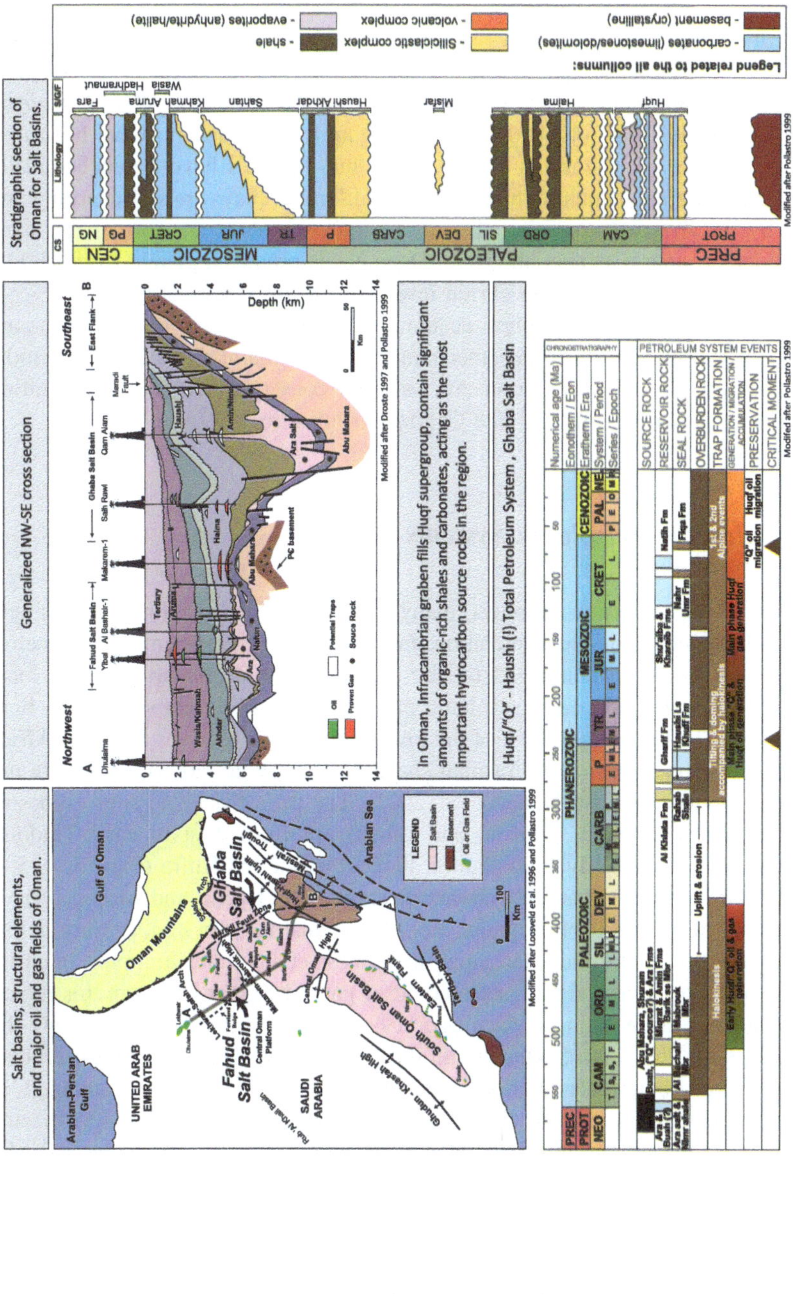

Fig. 2 Infracambrian petroleum system—Oman

in Oman are defined by (Pollastro 1999), who distinguished "*North Oman Huqf/ "Q"–Haushi(!) Total Petroleum System*" in Ghaba Salt Basin province, where hydrocarbons have been generated from the *Huqf* sources and producing from clastic reservoirs of the *Gharif* and *Al Khlata* Formations, *Haushi Group* (Middle Carboniferous to Late Permian), and "*North Oman Huqf–Shu'aiba(!) Total Petroleum System*" in the Fahud Salt Basin, where Cretaceous carbonates of the *Shu'aiba* and *Natih* Formation account for most of the production. In addition, Droste (2014) differentiated the Huqf *Petroleum System* from the "*Q*" *Petroleum System*. Both the Huqf oils, and the "Q" oil family is thought to have been also derived from the pre-Cambrian source rocks, but "Q" oils have a geochemically distinct and different signature than the Huqf oil family. Plays are self-charging and sealed by the *Ara Salt*, meaning that all the play elements are part of the same depositional system (Cozzi et al. 2012). Deep gas is produced mainly from Middle Cambrian to Lower Ordovician clastic reservoirs of the *Haima Supergroup*.

Traps in nearly all hydrocarbon accumulations of these petroleum systems are mainly structural. Trap-forming mechanisms were mainly related to periodic halokinesis of the thick Cambrian Ara salt and consequent folding and faulting during the Late Cretaceous and late Cenozoic (Pollastro 1999).

4.2 Paleozoic Petroleum Systems

The Paleozoic succession displays many similarities across North Africa and Arabia. However, petroleum generation, migration, and entrapment in each petroleum system has been controlled by the different tectonic histories of the local basins. Most of the Paleozoic petroleum originated from Silurian sources, in both regions (Fig. 3). In contrast, 10% of the total hydrocarbon reserves in North Africa are from the Upper Devonian Frasnian shales (Lüning et al. 2003), which are not present in the Middle East (Konert et al. 2001). The main Paleozoic-sourced basins in North Africa (Algeria, Tunisia, and Libya) are from west to east: Tindouf, Reggane, Ahnet, Oued Mya, Ghadames, Ilizi, Murzuk, and Kufra (Figs. 3, 4, 5, 6 and 7). The Paleozoic charged systems of southwest Algeria and Morocco have been mostly destroyed by Hercynian, Eo-Alpine/Austrian and mid-Tertiary deformation. Only the high-relief Hercynian anticlines retained their trapping integrity and still reservoir a very significant amount of gas. The other basins in this region now all appear to be extinct (Macgregor 1996). We will briefly present only the most important Petroleum Systems with the respective quantities of hydrocarbons.

The hydrocarbon reserves discovered to date in Algeria are contained in approximately 200 oil and gas fields, with most reserves (67%) coming from the provinces of Oued Mya and Hassi Messaoud, with two giant fields Hassi R'Mel (gas) and Hassi Messaoud (oil). These fields owe their enormous size to their position on large, regional structural arches. The Illizi Basin occupies third place with 14% of the reserves (Attar and Hammat 1995).

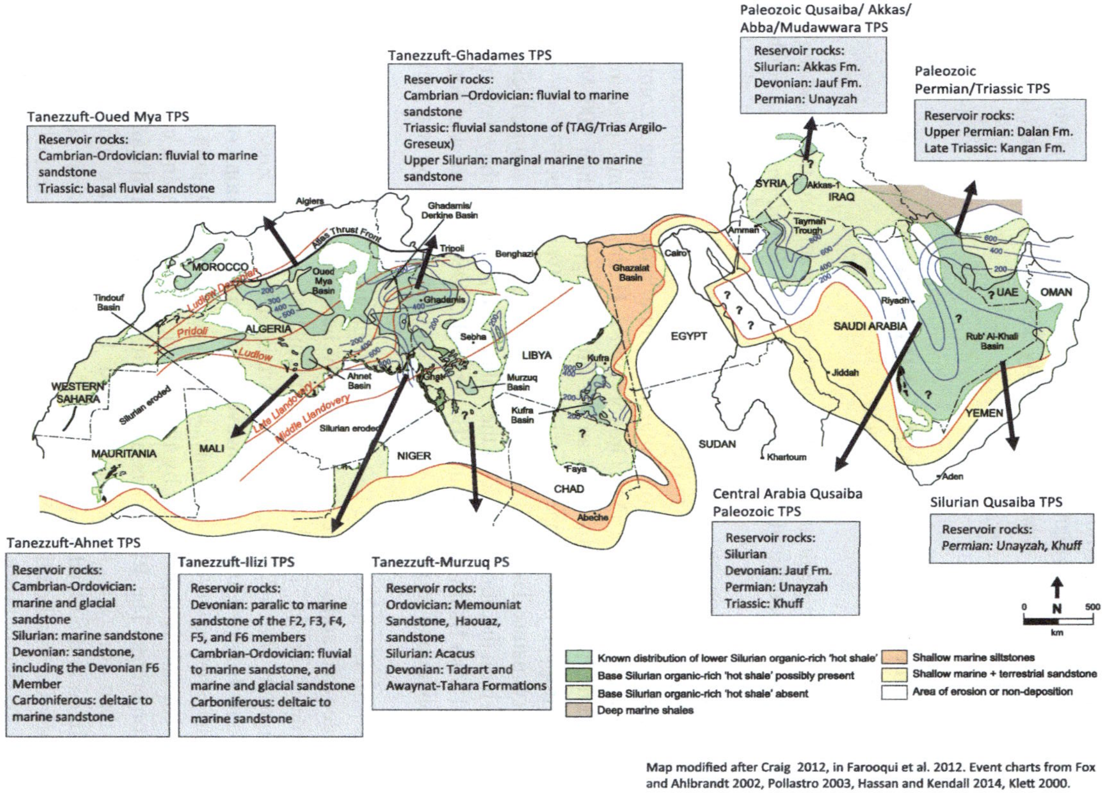

Fig. 3 Distribution of the Silurian source rock and related petroleum systems

Paleozoic-sourced petroleum systems on the Arabian Plate have been defined in Saudi Arabia, Bahrain, Qatar, Iraq, Syria, Jordan, and northern parts of the Iranian Gulf offshore along the Zagros Fold Belt (Fig. 8). The major Paleozoic petroleum system of Arabia is characterized by a proven Silurian (*Qusaiba Member, Qalibah Formation*) source rock, Devonian, Permian, and Carboniferous reservoirs, a laterally extensive, regional Permian seal (basal Khuff clastics and Khuff evaporites), and four-way closure Hercynian structures. The Khuff is the largest nonassociated gas reservoir in the world, with approximately 750 TCF of recoverable reserves (Konert et al. 2001).

4.2.1 North Oued Mya, Hassi R'Mel and Hassi Messaoud-El Biod Arch Petroleum Systems, Algeria

The *Hassi R'Mel* single giant gas field lies on the culmination of the broad regional *Talemzane Arch*, flanked by the *Benoud Trough-Atlas Foredeep* to the north and the northern *Oued Mya* depression to the southeast (Fig. 4; Boote et al. 1998). The field resides within the *Tanezzuft-Benoud Total Petroleum System* (Klett 2000a). The gas field developed in the pinch-out of Triassic fluviatile sands, *Trias Argilo-Greseux Inferieur* (*TAGI*) on a major structural arch, forming a combination of structural-stratigraphic trap (MacGregor 1998). Hydrocarbons, predominantly gas, (the accumulation also includes an oil rim with greater than 300 MMBOE reserves), are trapped below upper Triasic and Liassic evaporates (Fig. 4). The gas came from very mature Tanezzuft and possible Upper Devonian source rocks in the Benoud Trough on the northern side of the Talemzane Arch. Oil may have migrated from the Oued Mya Basin in the south (Boote et al. 1998). Closure was not formed until the Turonian and most of the earlier oil charge in this region may have been lost.

The *Hassi Messaoud* oil field lies on the culmination of a regional north–south trending Paleozoic arch, buried unconformably beneath basal Triassic silts, clays and evaporites (Fig. 4). There are five distinguished reservoir units in age from Upper Cambrian (*Ra* + *Ri*) to Tremadoc-Arenig (*Alternance Zone, El Atchane Sandstone* and *Hamra Quartzite*). However, the main production comes from the Ra, whereas other reservoirs are only locally productive (Djarnia and Fekirine 1998). The Ra unit is represented by fine to coarse sandstone and compact indurated conglomerates. The Ri consists of quartzite sandstones and has better lateral continuity than Ra. The field was charged by long-distance migration from *Tanezzuft* Shale in the Oued Mya depocenter, first through Triassic sands and then perhaps along the weathered unconformity surface toward the crest of the structure. The oil charge clearly post-dates the formation of closure in the late Jurassic–early Cretaceous and probably occurred during deposition of the thick pile of Late Cretaceous sediments over the Oued Mya depocentre (MacGregor 1998). The field was unaffected by mid-Cretaceous deformation and later Cenozoic uplift and erosion, with little or no subsequent loss of trap integrity (Boote et al. 1998).

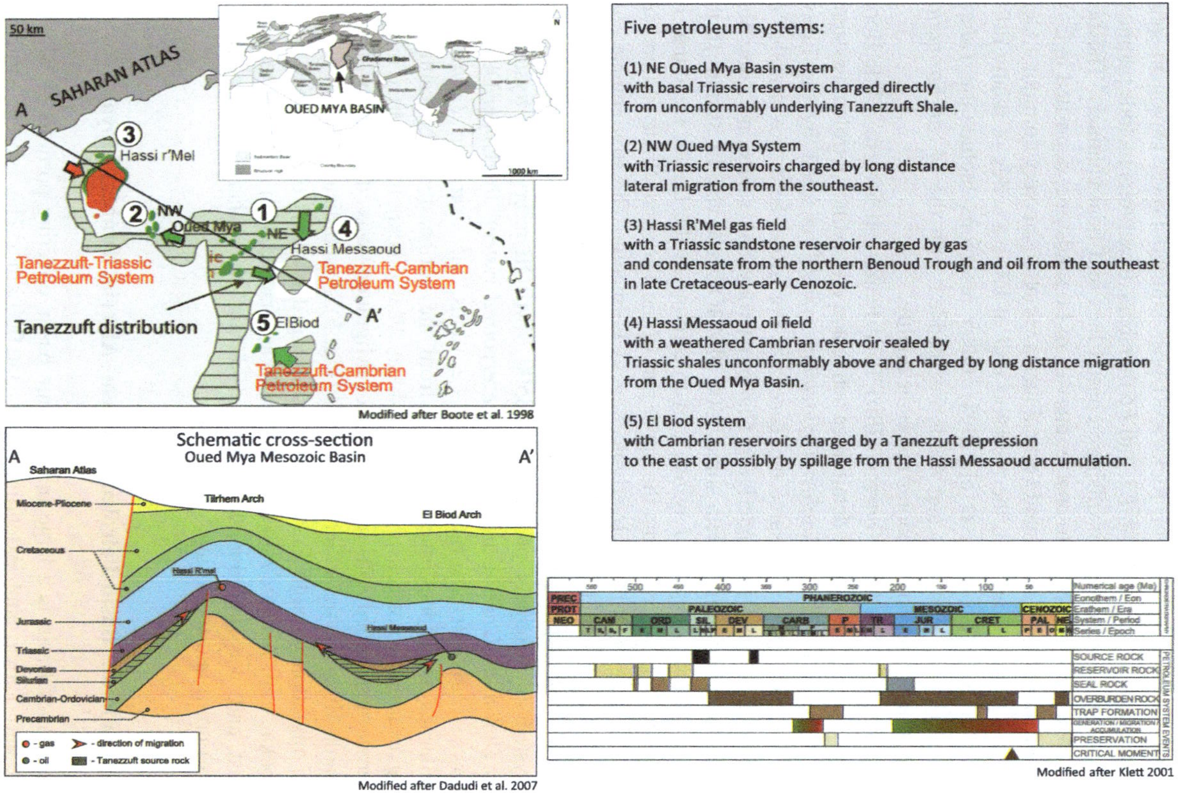

Fig. 4 North Oued Mya, Hassi R'Meland Hassi Messaoud-EI Biod Arch petroleum systems, Algeria

4.2.2 Illizi Basin Petroleum System, Algeria

Illizi Basin has a surface area of 108,424 km^2 (Fig. 5). This polycyclical basin has a Mesozoic sequence separated from the Paleozoic by the Hercynian unconformity.

The primary known reservoir rocks are Upper Silurian–Lower Devonian sandstones subdivided to *F2, F3, F4, F5,* and *F6* members. Reservoirs include paralic to marine sandstone deposited in offshore coastal bars and tidal channels. The best quality reservoirs correspond to coarse intervals of F6 sandstone, deposited in E–W trending offshore bars. Lateral and vertical facies variation are very common (Alem et al. 1998). In the Alrar field, the reservoir unit of F3 is a series of tidally influenced barrier bars (Chaouchi et al. 1998). Other reservoirs include the *Tahara Sandstone* (Upper Devonian) and sands interbedded with the *M'rar Formation* (Carboniferous) which have poorer reservoir quality.

Klett (2000b) defined the *Tanezzuft*-Illizi Total Petroleum System, while Boote et al. (1998) described in more detail three active and one extinct petroleum systems (Fig. 5). System I is the "*Cambro–Ordovician system*" charged by the Lower Tanezzuft. Besides being a source the Tanezzuft Formation here provides also an excellent regional seal for underlying Cambro–Ordovician sandstone reservoirs. The system is very prolific, with some 1500 MMBOE reserves. In System II, the "*Lower Devonian reservoired system*", hydrocarbons in the Lower Devonian F6 and F4 sands were sourced from the upper part of the Tanezzuft Shale and perhaps interbedded shales within the overlying *Acacus* Formation. The stratigraphic continuity of the F6 sands and their position directly above the *Acacus* sands encouraged long-distance lateral migration to the west and southwest. The petroleum system is extremely prolific, with 3500 MMBOE reserves. In System III, "*Upper Devonian–Lower Carboniferous reservoired system*" the Upper Devonian and Lower Carboniferous including the F2 sands, the overlying *Tahara* and Carboniferous sandstones, were charged from the northeastern depocentre and flanking areas to the northeast by Frasnian shales. Peak oil expulsion started in early Cretaceous and lasted till mid-Cenozoic. Migration was dominantly lateral, southwards along low-relief structural axes (Fig. 5). A significant vertical component of migration through Eo-Alpine/Austrian aged faults is on the flank of the Tihemboka Arch. A "*Late Paleozoic system*" was charged from a depocenter to the southeast and was subsequently dispersed during Hercynian (and Eo-Alpine/ Austrian) uplift and unroofing.

4.2.3 Ghadames Basin Petroleum System, Algeria, Tunisia, Libya

The Ghadames Basin is a large intracratonic basin, covering portions of Algeria, Tunisia and Libya, and extending over 350,000 km^2. The present structural framework of the Ghadames Basin was produced by tectonic movements related to the Taconic, Caledonian, Hercynian, and Austrian phases. The basin contains thick succession of Paleozoic and Mesozoic sediments (Fig. 6).

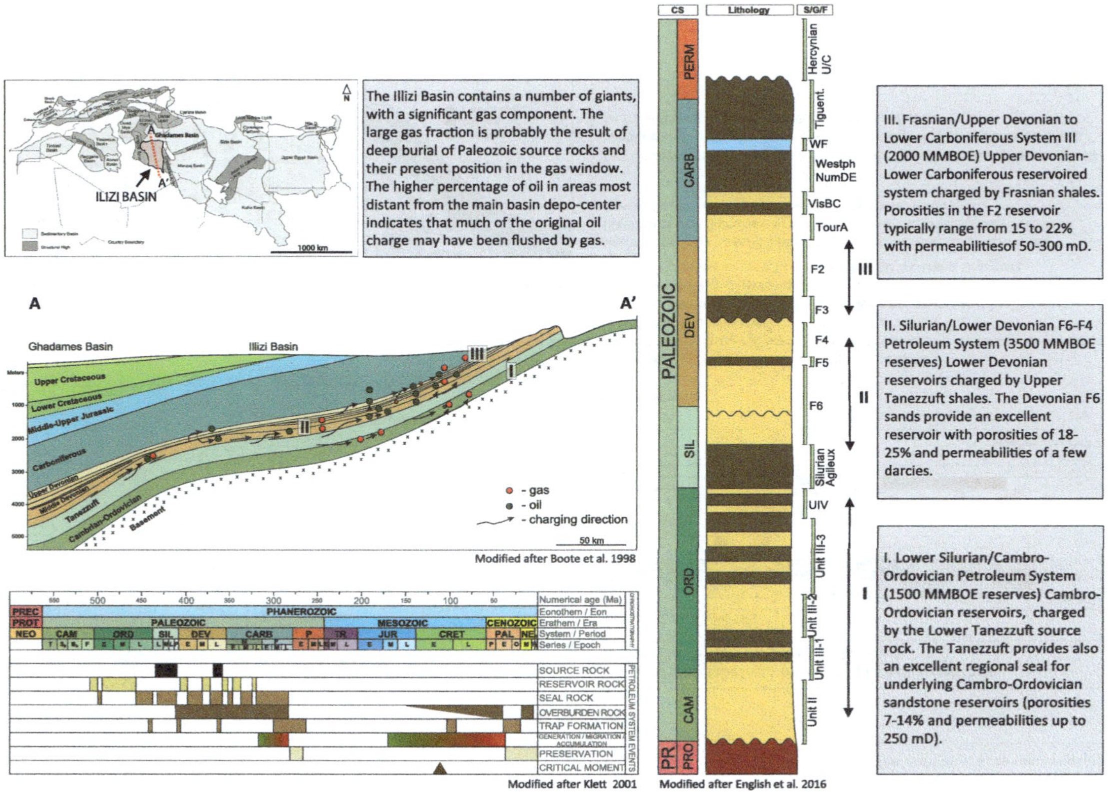

Fig. 5 Ilizi Basin petroleum system, Algeria

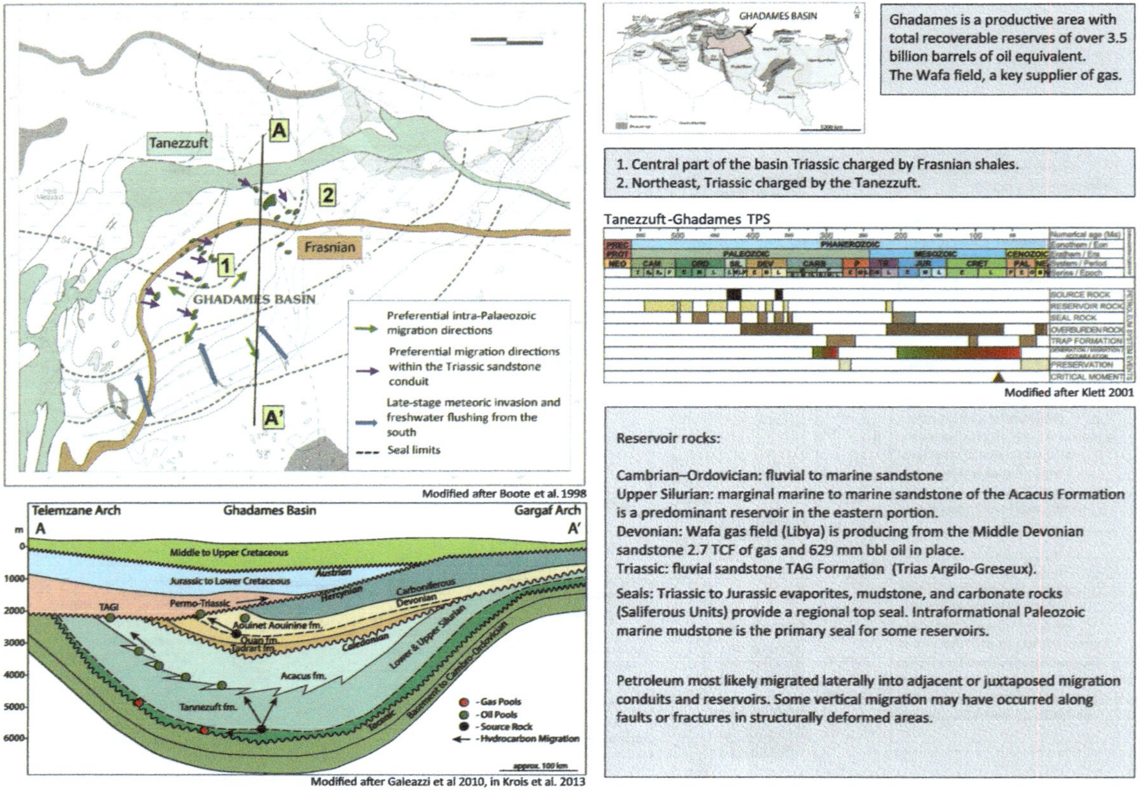

Fig. 6 Ghadames Basin petroleum system

Main source rocks are as elsewhere the Silurian *Tanezzuft Formation* and Devonian shale. The western part of the Ghadames Basin has the higher potential for gas due to the over-mature Silurian source and excellent potential for oil because of less mature Devonian source (Sikander 2000; Sikander et al. 2003). In the central and eastern part of the basin, the Silurian shale is in mature to over mature stage and can generate both oil and gas.

The main regional Silurian reservoir is the *Acacus Formation*, equivalent to the lower part of the *F6* in Algeria. Only in northwestern Libya, in a region where source-reservoir-seal relationships are optimized the Silurian *Lower Acacus* is the prolific reservoir (Echikh 1998). Twenty-two oilfields and three gas fields have been found in this region. Other Lower Devonian reservoirs, in Libya, are the *Tadrart* and *Ouan Kasa* Formations. These are equivalent to the upper part of the *F6* and the *F5* in Algeria. Middle and Upper Devonian reservoirs in Libya are within the *Aaouinet* and *Tahara* Formations, equivalent to the *F2* in Algeria. Minor production comes from the Ordovician *Memouniat* quartzitic sands. Ordovician and Devonian reservoirs have the best petrophysical properties on the southern basin flank but decrease in quality into the basin because of increased shaliness and diagenetic effects associated with deeper burial. The Lower Devonian is the most productive horizon in the Ghadames Basin in terms of both number of oil fields and total oil in place, with 40% of each. The Al Wafa gas field contains 2.7 TCFG and 629 MMBO in the Middle Devonian sandstone (Hassan and Kendal 2014). Roughly half of the Algerian reserves and about a third of the Libyan occur at this level (Echikh 1998). In Tunisia, the best reservoir is the Middle–Upper Triassic (Ladinian to Carnian) fluvio-deltaic sandstone of the *Kirchaou Formation*. The Kirchaou Formation unconformably overlies the deformed Palaeozoic section above the Hercynian unconformity. The formation has excellent reservoir properties with average porosity of 15–22% and permeability up to 1 Darcy (Aissaoui et al. 2016). Equivalent Triassic reservoirs in Algeria are within the *Argilo-Greseux Formation* (Echikh 1998).

Three distinct petroleum systems (Fig. 6) are defined in Libya by Dardour et al. (2004): (1) a "*NW Tanezzuft-Acacus System*" where the oil is accumulated within thinly interbedded, shelf-dominated facies of the Silurian *Acacus Formation*. The more moderate pre-Hercynian burial depth in this area, as opposed to that in the central depression to the west, resulted in preservations of generation potential of the Lower Silurian source rock into the Mesozoic–Cenozoic, with peak generation during the Late Cretaceous–Cenozoic. Hydrocarbons within the Upper Silurian sandstone reservoirs did not accumulate until the Cenozoic, after Alpine exhumation, which resulted in significant uplift and unroofing over the eastern margin (Underdown and Redfren 2007); (2) a central "*Tanezzuft-Tadrart/Acacus (Devonian, Carboniferous) System*" where oil and gas fields are charged by shorter distance laterally from the west; and (3) a southern "*Tanezzuft-Tadrart System* with hydrocarbons in Tadrat and Ouan Kasa sandstones, that originated by vertical migration through the sand-dominated Upper Silurian Acacus Formation. Accumulations are dominantly anticlinal structural traps, with reservoirs sealed by the overlying thick Middle–Upper Devonian shale (Underdown and Redfren 2007).

Similar petroleum systems have been defined in the Tunisian part of the Ghadames Basin: *Silurian Tannezuft—Djeffara* (Upper Ordovician), *Tannezuft—Acacus*, *Tannezuft—Kirchaou* (Triassic) and Devonian *Aouinette Ouinine Member III—Kirchaou* (Aissaoui et al. 2016). Tunisian daily oil production is approximately 43,000 BO, about half of which comes from the fields located in southern Tunisia and is produced from the Acacus Formation (Aissaoui et al. 2016).

4.2.4 Murzuq Basin Petroleum System, Libya

The Murzuq Basin is a large, symmetrical intracratonic basin that straddles Libya's borders with Algeria, Niger, and Chad (Fig. 7). The Murzuq Basin is bounded to the north by the Gargaf High, to the west by the Pre-Cambrian Hoggar massif and the north trending Tihemboka Uplift, and to the east by the Tibesti massif where Pre-Cambrian and Paleozoic rocks are cut by northeast-trending faults. It is separated from the Ghadamis Basin in the north by the Gargaf High and Atshan Saddle (Hassan and Kendal 2014; Hallett 2002). Major oil discoveries are restricted to the north of the central part of the basin. These include the "Elephant Field" in NC174, and further north is the A, B, C, H, and J-NC115 oil field complex. The system reservoirs about 600 MMBOE (Boote et al. 1998). The Ordovician *Memouniat Formation* is the main reservoir and Lower and Middle Ordovician *Haouaz Formation* is the second-most important reservoir.

The main source rock is the Silurian *Tanezzuft Formation*. Hydrocarbons generated in this area migrated northwards through the Ordovician to charge the NC 101 and NC 1115 fields (Boote et al. 1998). From the central part of the basin northwards, there is an increase in maturity of the "hot shale" from early mature-early oil window to condensate mature window. To the south and east, there has been less uplift (Fig. 7) and erosion so the *Tanezzuft* sequences are mostly within the oil window (Hall et al. 2010). Expulsion and migration probably occurred from the Jurassic to the end of the Cretaceous, although there may have been some generation during the Carboniferous (Hallett 2002). The hot shales were probably uplifted out of the hydrocarbon generating window during the Cenozoic (Lüning et al. 2000).

There is only one Petroleum system defined: the "*Tanezzuft-Ordovician Petroleum system*" (Boote et al. 1998). Efficient charging was facilitated by the proximity of mature source rocks with regionally continuous and locally fractured Ordovician migration conduits. The importance of direct contact between source and reservoir is underscored by the dominant role of the *Memouniat* fine-grained, generally tight but clean quartzitic sandstone that is not conducive to long-distance migration (Echickh and Sola 2000).

4.2.5 Arabian Plate, Silurian Petroleum Systems

Several different Paleozoic petroleum subsystems are described in the Middle East by the USGS (Fig. 8). In general, all subsystems are defined by the distribution of

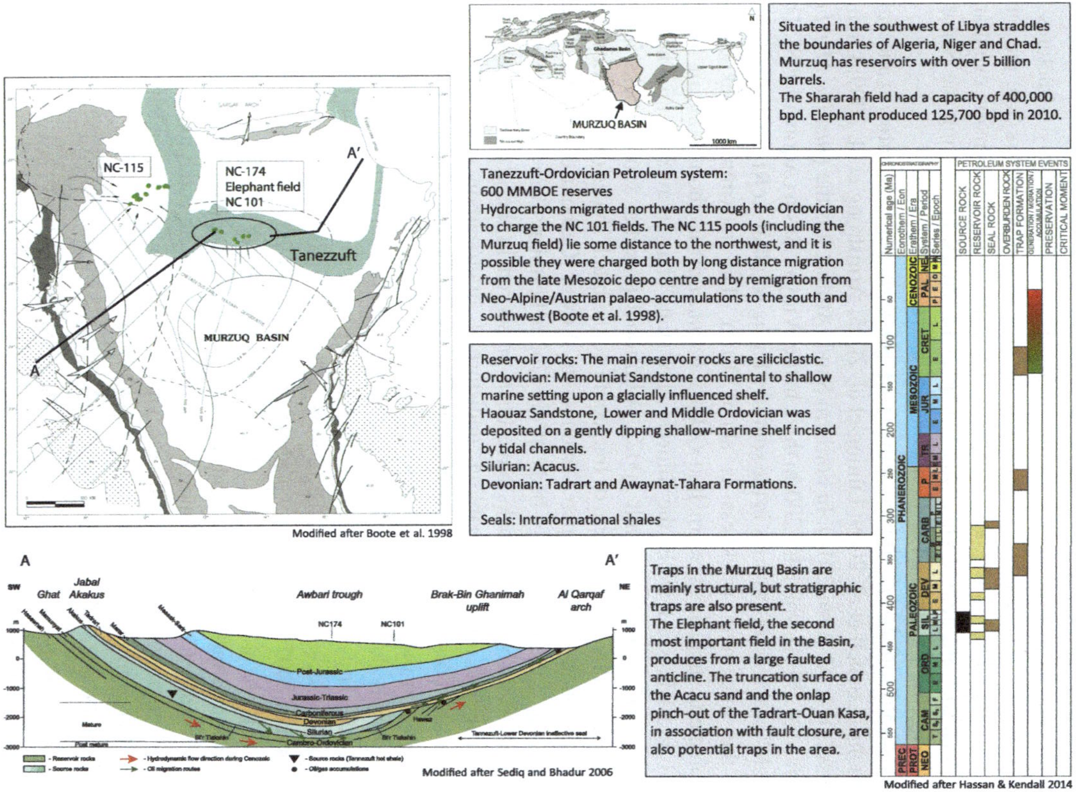

Fig. 7 Murzuq Basin petroleum system, Libya

the proven organic-rich "hot shale" of the Lower Silurian as the major source rock, Devonian, Permian, and Carboniferous reservoirs, a laterally extensive, regional Permian seal, and mostly, four-way closure Hercynian structures. Hydrocarbons found in these systems include nonassociated gas in Eastern Arabia and extra light oil in Central Arabia (Abu-Ali 2005). Major trap formation includes both reactivated basement fault blocks and structures related to movements of Infracambrian salt, and occurred in four main tectonic stages: Hercynian Orogeny (structures at Ghawar); Early Triassic Zagros rifting and opening of the Neotethys Sea; the First Alpine Orogeny during the Late Cretaceous, associated with the collision event and emplacement of the Semail Ophiolite Complex in Oman; and during the Second Alpine Orogeny, middle to late Cenozoic (Pollastro 2003).

In Syria and Jordan, the Silurian source rock is known as the *Mudawara* and *Tanf* Formations (Loydell et al. 2009; Lüning et al. 2005; Armstrong et al. 2009; Lüning and Kuss 2014; Al-Rawi 2014). In northern Saudi Arabia, it is the *Qusaiba Member* (Mustafa et al. 2014) which is 1200 m thick in the Widyan Basin (Aqrawi 1998). The "hot shale" is at least 60 m thick in the southern basin, south of Ar Riyad. In the Tabuk Basin, the maximum thickness is about 30 m. The maximum thickness of age-equivalent strata in western Iraq (*Akkas Formation*) is about 2000 m. In the Akkas-1 and Khiesia-1 wells, the Silurian "hot shale" has an average thickness of about 65 m (Fox and Ahlbrandt 2002). Additional potential source rocks in the western and southwestern deserts of Iraq could be the black shales of the Ordovician *Khabour Formation*, the Upper Devonian to Lower Carboniferous *Ora Shale Formation*, and the lower shaly beds of the Upper Permian *Chia Zairi Formation* (Aqrawi 1998). The source rocks have been preserved in the grabens during the Hercynian horst uplifting enabling undisturbed hydrocarbon generation. The Hercynian-age horsts and grabens pattern (Fig. 9) can be traced northwards from Saudi Arabia to the Saudi-Iraq border. The source rocks began generating hydrocarbons possibly during the late Paleozoic in the region of northern Arabia into Iraq, and during the Triassic in the region of the present-day Arabian Gulf. Peak generation and migration occurred through the Jurassic and ended in the Early Cretaceous (Fox and Ahlbrandt 2002). In Jordan, two-phase generation of hydrocarbons is interpreted for both the Risha and Wadi Sirhan areas, with the first during a pre-Hercynian phase in the Carboniferous and the second as a less intense generation phase during the Mesozoic to early Cenozoic (Lüning et al. 2005). In both Iraq and Saudi Arabia, the direction of migration is towards the west. The oil expulsion in northern Saudi Arabia started earlier, during the Devonian (Aqrawi 1998).

The Silurian Total Petroleum Systems recognized by the USGS are:

(1) a northern "*Paleozoic Qusaiba/Akkas/Abba/Mudawarra TPS*" covers the area in Saudi Arabia north of the Central Arabian Arch and Hercynian unconformity, the western and southwestern deserts of Iraq, and eastern portions of Syria and Jordan (Fox and Ahlbrandt 2002). Strata thicken to the north reaching up to more than 7000 m. In north-central Saudi Arabia, the primary Paleozoic reservoirs are Early Permian–Carboniferous *Unayzah* and *Ga'ara Formations*. The main Paleozoic reservoir rocks in the Western Desert of Iraq and East Jordan (Risha

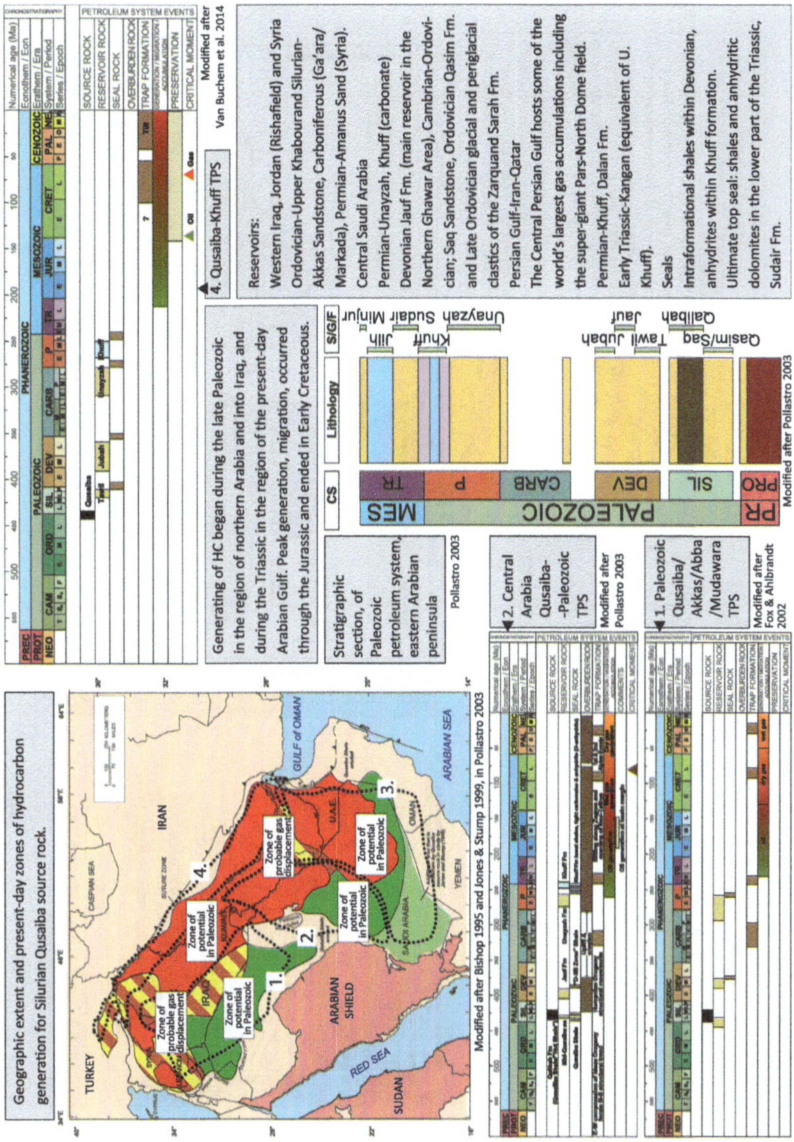

Fig. 8 Arabian Plate, Silurian petroleum systems

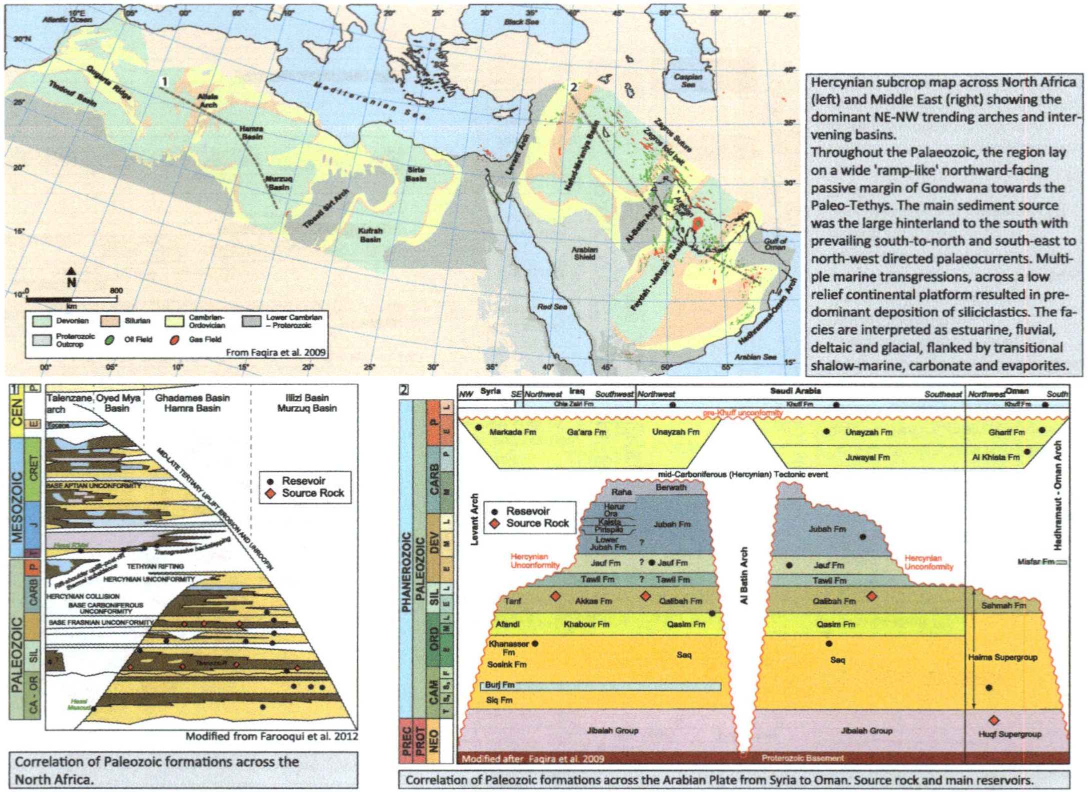

Fig. 9 North Africa and Middle East, Paleozoic summary

field) are sandstones of the Ordovician Upper *Khabour* and Silurian *Akkas Formations*. Additional reservoirs may be found in the Carboniferous *Ga'ara* sandstones and the carbonates of the Permian *Chia Zairi Formation* (Aqrawi 1998). In Syria, in the Central Palmyrides, gas production is coming from the Carboniferous channelized sandstones of the *Markada Formation*, while the Permian *Amanus Sand Formation* contains both oil and gas (Lučić et al. 2001, 2003, 2007; Lučić 2008). In Saudi Arabia, the migration route of hydrocarbons is from the structurally deeper mature source-rock areas in the overlying *Unayzah* reservoir. In Iraq, along the eastern flank of the Akkas field, hydrocarbons migrated downward from the Lower Silurian source into the Ordovician upper *Khabour* fractured sandstone reservoir (Fox and Ahlbrandt 2002).

(2) the "*Central Arabia Qusaiba-Paleozoic TPS*" (Pollastro 2003) encompasses an area mostly within Saudi Arabia, Bahrain, western Qatar, and northern portions of the Iranian Gulf along the Zagros Fold Belt. Primary reservoirs (Fig. 8) are sandstones of the fluvial to marginal marine Permian *Unayzah Formation*. The Devonian *Jauf Formation* is the main reservoir in the Northern Ghawar Area, where it hosts the giant Hawiyah gas-condensate field (Wender et al. 1998). In the Arabian Gulf area, the main reservoirs are basal transgressive marine sandstones and cyclic dolomitic shelf carbonates of the Late Permian *Khuff Formation*. Other reservoirs include clastics of the Cambrian–Ordovician *Saq Sandstone*, shallow marine sandstones of the Ordovician *Qasim Formation*, and Upper Ordovician glacial and periglacial clastics of the *Zarqu* and *Sarah* Formations. Structures are mainly domes over salt diapirs and salt-assisted or enhanced horst-block anticlines. The 'Usaylah-1' oil discovery in an upper *Unayzah* eolian dune facies confirms also stratigraphic-trap potential in Central Saudi Arabia. These stratigraphic traps include both updip pinch-out of reservoir sands and truncation traps below the regionally extensive Pre-Khuff Unconformity (Evans et al. 1997).

The regional seal in Saudi Arabia and Oman are Triassic shales of the *Sudair Formation*, which completely separate the Silurian hydrocarbon system from the overlying Mesozoic systems. Other seals are intraformational claystone and shale, or tight carbonates and evaporites.

Paleozoic oils derived from clastic *Qusaiba Member* in the "*Central Arabia Qusaiba-Paleozoic TPS*" (Pollastro 2003) are geochemically much different and more mature than those generated from younger Jurassic and Cretaceous carbonate source rocks. In the deepest part of the Central Arabian sub-basin, the source rock was thermally mature by the Late Jurassic, but hydrocarbon expulsion was not initiated until the Early Cretaceous. Oil and gas were expelled simultaneously, but gas expulsion continued for a longer time. Most of the gas was expelled during the Late Cretaceous. Expulsion of oil nearly ceased after the Late Cretaceous (Abu-Ali et al. 1999). In the Ghawar area, the basal *Qusaiba shales* began oil generation in the Triassic and proceeded through the wet gas-condensate window in the Cretaceous. The source rock entered the dry gas generation phase in the Early Cenozoic, and remains in the dry gas window at present (Wender et al. 1998).

(3) the "*Silurian Qusaiba TPS*" is recognized for the greater Rub 'al Khali Basin Province. In the northeastern part, the *Khuff* carbonates, mostly grainstones, are the primary reservoir. *Qusaiba* hot shale is in the dry gas generation phase. Hydrocarbons migrated to the east into the crestal part of the Qatar arch, forming the world's largest gas accumulations including the supergiant *South Pars-North Dome* field. Along the eastern flank of the Rub 'al Khali Basin, in Oman and Yemen, primary reservoirs are sandstones of the *Unayzah* and basal *Khuff Formations.*

(4) *the Paleozoic-Permian/Triassic TPS* is defined in Iran within the Zagros Fold Belt including the Northern Qatar Arch Extension (Ahlbrandt 2000a, b). This TPS includes the "*Paleozoic petroleum system*" defined in Iran by Bordenave (2008, 2014) and "*Paleozoic Qusaiba-Khuff Petroleum system*" defined in Qatar (Van Buchem et al. 2014). The gas in the Permo–Triassic reservoirs is assumed to be derived from "hot shale" *Sarchahan (Qusaiba) Formation.* Oil and gas generation commenced during Jurassic and Late Cretaceous time, respectively, and gas generation continued until the Neogene (Mohsenian et al. 2014). Present-day maturity of the source rock along the crest of Qatar High is within the gas window. The gas expulsion during geological history moved from west to east. Initially, maturation took place in Jurassic from the Northern Salt Basin which is north and west of the present Qatar Arch, later in Upper Cretaceous-Cenozoic the activity has been shifted to the Qatar High and towards the north as the plate was tilting while the Zagros foredeep was developing (Van Buchem et al. 2014). Huge volumes of oil, then of gas, accumulated in a few major regional highs and salt-related structures prior to the Zagros orogeny. Part of the gas was lost during Zagros folding as some of the anticlines were breached, and another portion, possibly associated with light oil, remigrated into unbreached Zagros anticlines (Bordenave 2008).
Reservoirs are Upper Permian (*Dalan Formation*) and Lower Triassic (*Kangan Formation*) dolomites, dolomitic limestones, and oolitic limestones (Moradpour et al. 2008; Esrafili-Dizaji and Rahimpour-Bonab 2013). The Early Triassic *Kangan Formation* is the main reservoir for natural gas in SW Iran and the northern Persian Gulf and is equivalent to the *Upper Khuff Formation* (Peyravi et al. 2010).

4.3 Mesozoic Petroleum Systems

The Mesozoic petroleum systems of North Africa formed in a suite of extensional basins developed along the Atlantic and Tethyan margins (MacGregor and Moody 1998). Hydrocarbons are unevenly distributed across the region, with one world-class petroleum province in the Sirt Basin of Libya, much smaller reserves defined in the Pelagian area offshore Tunisia and Libya, and several medium-sized hydrocarbon provinces in the Western Desert and Gulf of Suez in Egypt.

On the Arabian Plate, passive margin conditions during the Jurassic through Late Cretaceous produced a broad, stable shelf environment. In particular, Jurassic geologic conditions resulted in deposition of the ideal sequence of primary petroleum system elements: thick oil-prone source rocks in intra-shelf basins, extensive reservoir facies, and excellent seals (Ahlbrandt et al. 2000). Cretaceous rocks also comprise several petroleum systems along the eastern part of the Arabian subcontinent.

4.3.1 Pelagian Basin (Offshore Libya and Tunisia), Jurassic-Cretaceous Petroleum System

The "*Jurassic-Cretaceous Composite Total Petroleum System*" was defined by Klett (2001) for the predominantly offshore Pelagian basin of Tunisia and Libya (Fig. 10). Most of the discovered oil is in the Lower Cretaceous *Bireno Formation* (approximately 38 MMBO) and Jurassic (approximately 34 MMBO) reservoirs, whereas most of the discovered gas resides in Upper Cretaceous *Abiod* fractured reservoirs (approximately 1000 BCFG; Klett 2001).

The primary source rocks are mudstone of the Jurassic *Nara* Formation, Lower Cretaceous *M'Cherga* Formation, Albian *Lower Fahdene* Formation, and Cenomanian to Turonian *Bahloul* Formation. The Lower Fahdene is generally a fair to excellent oil and gas-prone source rock with TOC values that reach 10% (Mohamed et al. 2015). It is marginally mature to late mature and is rich in predominantly marine organic matter containing some terrestrial input. The Bahloul Formation exhibits high TOC content of 7–13% (Lüning et al. 2004). The organic matter is particularly well preserved and is type II. The M'Cherga Formations is a fair to good source rock characterized by TOC ranging from 0.9 to 4% also with predominantly marine organic matter (Mohamed et al. 2015). Most of the discovered oil in the Jurassic-Cretaceous Total Petroleum System is contained in anticlinal traps, whereas most of the discovered gas is in fault-related traps (Klett 2001).

4.3.2 Sirt Basin Petroleum System, Libya

The Sirt Basin is bordered in the northeast by the Cyrenaica Platform, in the west by the Ghadames Basin, in the southeast by the Al Kufrah Basin, in the southwest by the Murzuq Basin and in the north by the Mediterranean Sea (Fig. 11). It covers an area of approximately 400,000 km^2 onshore and extends offshore into the Gulf of Sirt. The Sirt Basin belongs to the World's top 20 petroleum provinces, having proven reserves of 43.1 billion BOE in 2002 (Ahlbrandt 2002). Early Mesozoic Neotethyan-related extension and volcanism was of limited effect and is represented by the Triassic *Amal Formation* (Hallett 2002; Guiraud et al. 2005). Broader structuration took place in the Aptian–Albian with deposition of the *Sarir* and *Faregh* sandstones and their lateral equivalents. This was most pronounced in the

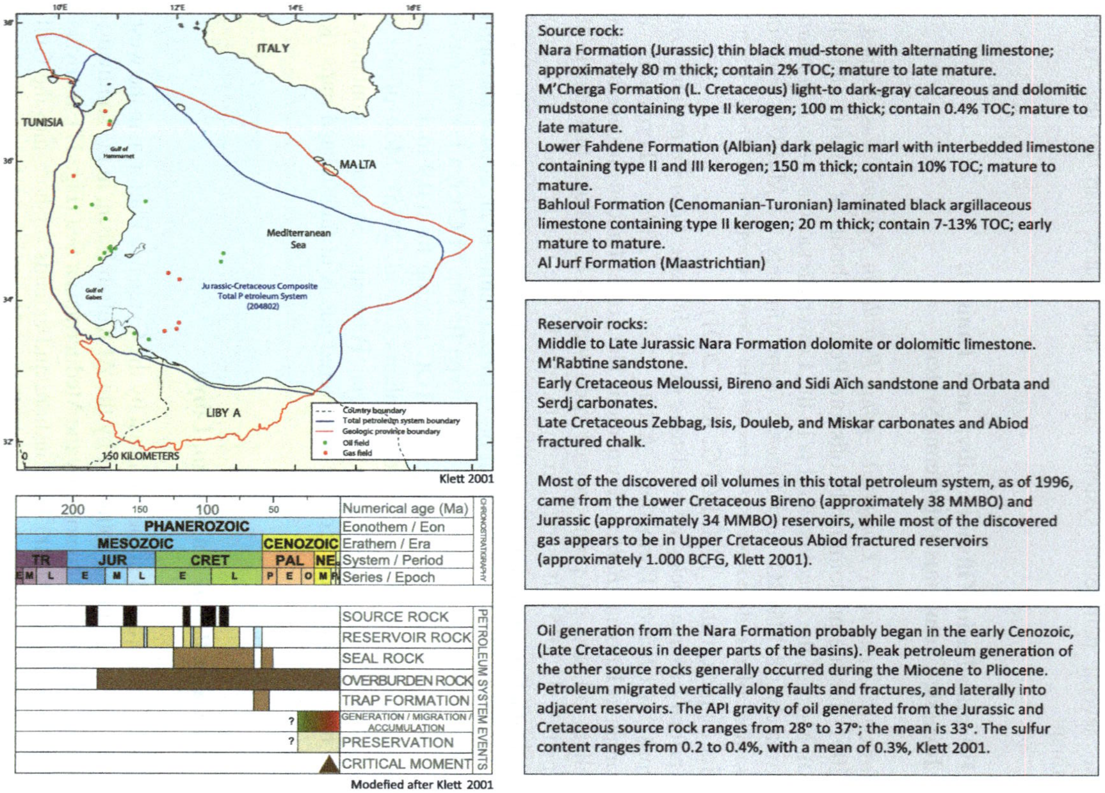

Fig. 10 Pelagian Basin (offshore Libya and Tunisia) Jurassic-Cretaceous petroleum system

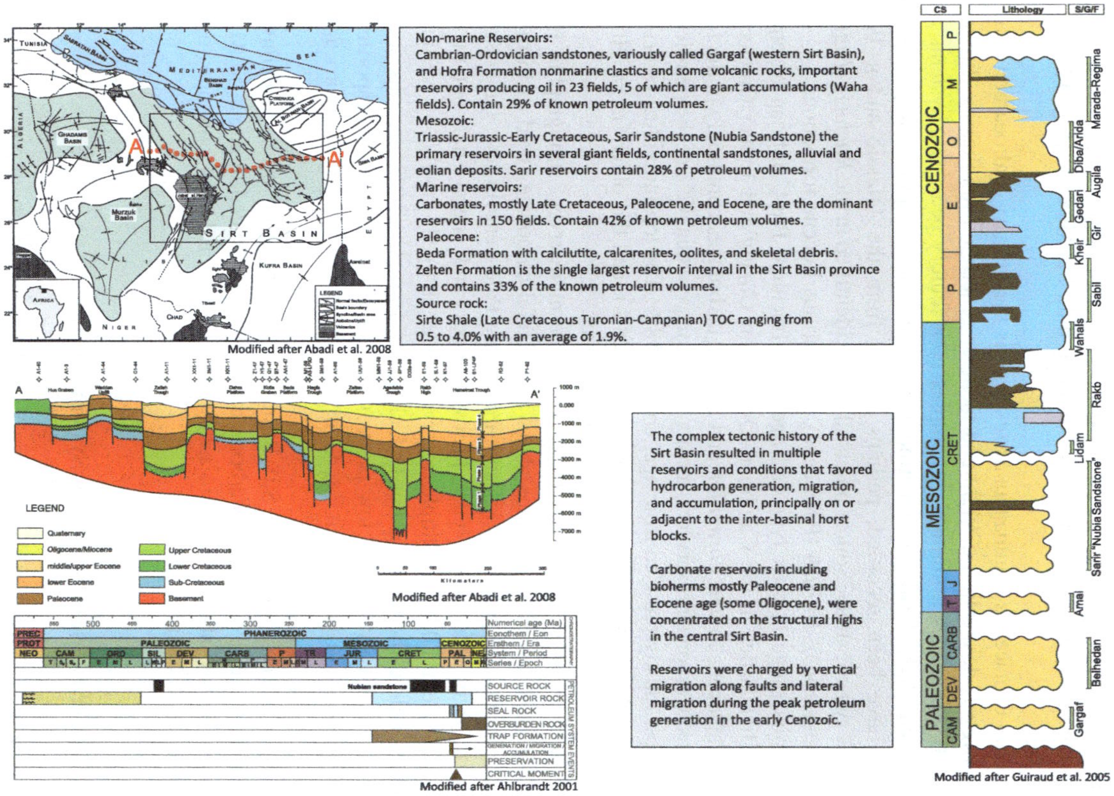

Fig. 11 Sirt Basin petroleum systems, Libya

east–west trending eastern Sirt basins. The main phases of extension and subsidence, however, are distinctly Late Cretaceous to Cenozoic in age and have been well-documented by many authors (El-Alami et al. 1989; Wennekers et al. 1996; Baird et al. 1996; Anketell 1996; Selley 1997; Salem 2000; Hallett 2002; Burwood et al. 2003; Hassan and Kendal 2014). The onshore Sirt Basin is for the most part not affected by Late Cretaceous and Cenozoic inversion, rather being marked by a thick succession of post-rift sediments extending until the Miocene (Fig. 11; Abadi et al. 2008).

The Late Cretaceous Campanian *Sirte Shale* is the primary hydrocarbon source rock. However, delineation of eight end-member generic oil families by Burwood et al. (2003) indicates a number of complex contributory petroleum systems and mixed-system hybrid oils. Nonmarine lacustrine source inputs with increased waxiness have been recorded, perhaps arising from the Triassic and/or Early Cretaceous syn-rift strata. Source potential is also identified in the *Rachmat-Tagrifet* (Upper Cretaceous), *Harash* (Paleocene), and various Eocene formations.

Reservoirs range in rock type and age from fractured Pre-Cambrian basement (rarely), siliciclastics in the Cambrian–Ordovician *Gargaf* and Lower Cretaceous *Sarir*, to Paleocene *Zelten Formation* and Eocene carbonates commonly in the form of bioherms. Carbonate reservoirs, mostly Cenozoic, contain 42% of the petroleum; siliciclastics, mostly pre-Cenozoic age, contain 58%. The top individual reservoirs are Zelten carbonates (33%), Sarir sandstones (28%), and the older Gargaf/Hofra/Amal sandstones (29%) percent (Ahlbrandt 2000c).

Structural traps contain about 68% of Sirt basin reserves, stratigraphic traps about 24%, and combination traps about 8%. Reefal carbonates tend to form the most important stratigraphic traps and are mostly Paleocene in age. Other stratigraphic traps include algal banks, facies changes, and lateral shale-out of calcarenites (Salem 2000). Multiple reservoirs are charged mostly by vertically migrating hydrocarbons along horst-block faults from Upper Cretaceous source rocks that occupy structurally low positions in the grabens. Post-rift evaporites in the middle Eocene provide an excellent seal for the Sirte-Zelten hydrocarbon system.

Timing of generation in the Sirt Basin is estimated from basin modeling as mainly Paleogene in age. Hydrocarbon generation commenced at about 50 Ma in the deeper basins, about 40 Ma in many other areas, and may continue to the Present-day (Ahlbrandt 2001).

4.3.3 Western Desert Petroleum Systems, Egypt

The Mesozoic rift basins of the northern Western Desert of Egypt cover an area of about 200,000 km^2 and include sedimentary sections locally in excess of 7000 m (Fig. 12; Hantar 1990; Sehim 1993; Moustafa 2008; Bosworth et al. 2008). These basins contain about 5.7 BBOE of proven reserves, and the Western Desert creaming curve shows no signs of leveling off (Dolson et al. 2014). Extension-driven subsidence is known to have occurred during the Jurassic but may

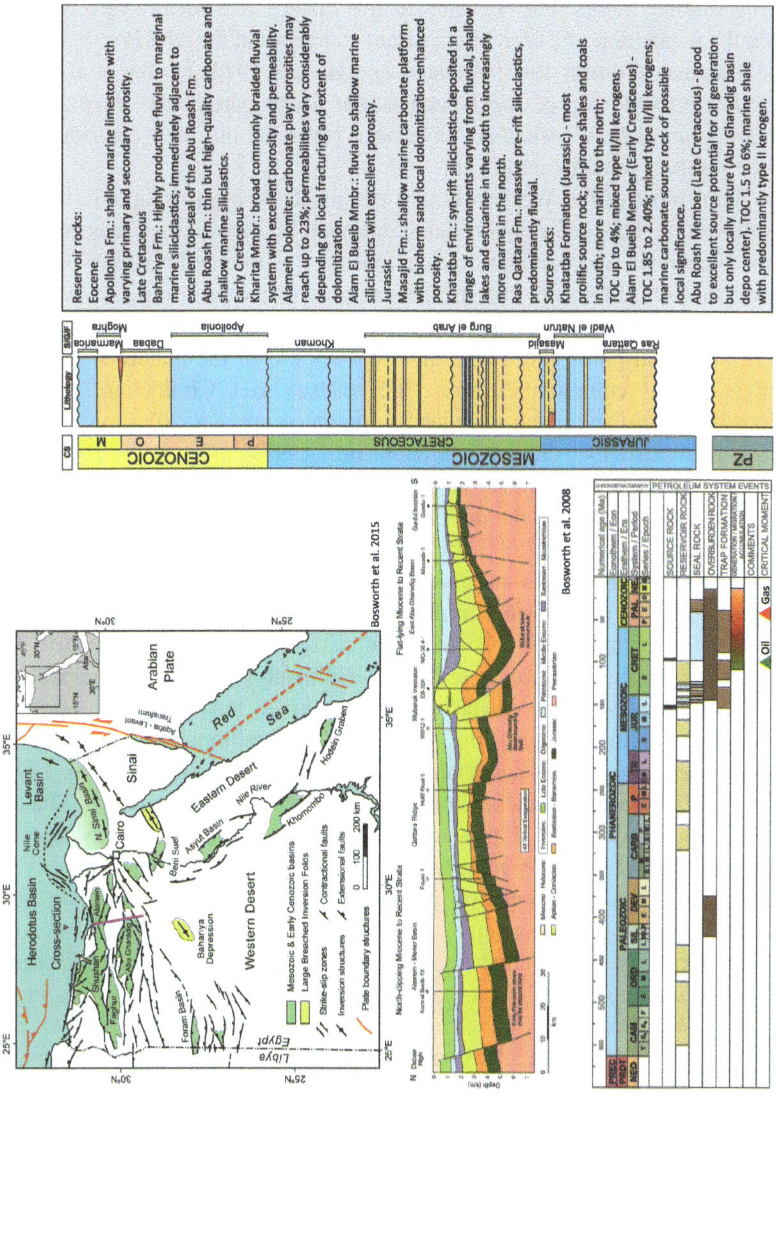

Fig. 12 Western Desert petroleum systems, Egypt (Jurassic-Cretaceous source rocks)

have initiated earlier during the Permo–Triassic opening of Neotethys (Şengör 1979; Stampfli et al. 2001; Garfunkel 2004). It was during this first phase of Jurassic rifting that the volumetrically most important source rock of the Western Desert, the *Khatatba Formation*, was deposited (Keeley et al. 1990; Keeley and Wallis 1991). Much later during the Cenomanian, most of northern Egypt was flooded by marine waters that persisted to the Eocene and locally beyond. This facilitated the deposition of a second important source rock, the *Abu Roash Member F*, in the Turonian (Norton 1967; Aadland and Hassan 1972; Keeley et al. 1990). Several other possible source rocks such as the lower part of the *Burg el Arab Formation* (*Alam el Bueib Member*) have been proposed but their contribution to the known reserves remains unproven.

Throughout most of the Western Desert extension during the Jurassic was oriented north–south. This deformation field continued into the Early Cretaceous, but during the Barremian and younger times extension was oriented NE–SW (Guiraud and Bosworth 1999; Moustafa 2008). The interplay of the resulting east–west and NW–SE striking extensional faults have a dominant role in controlling the development of structural plays and migration pathways out of the mature source rock depocenters. Local complexities arise due to the brief Cimmerian (Jurassic-Cretaceous transition) and much stronger Santonian compressional events (Fig. 12; Ayyad and Darwish 1996; Guiraud and Bosworth 1997).

The lithologies and depositional environments of the Khatatba Formation vary laterally from predominantly fluvial and estuarine in the south to progressively more marine in the north. The southern fluvial facies are often associated with both coals and organic-rich shales that tend to be oil-prone. Biomarkers indicate that these source rocks were deposited in nonmarine, well-mixed waters atypical of a stratified deep lacustrine setting (Abrams et al. 2012; Bosworth et al. 2015b). The geochemistry and core and electric log-based sedimentological data together suggest overbank swamps and small lakes transitioning to lagoons as the origin of these source rocks. In the north, the source rock depositional environments are marginal marine/deltaic shales and shallow marine carbonates (Keeley et al. 1990). Northern Western Desert Khatatba source rocks can be more gas-prone (Shalaby et al. 2012) but significant oil production in this region demonstrates that the distribution of organic facies is complex.

Because the Khatataba Formation is at the base of the syn-rift section, it is immediately adjacent to high-quality reservoir sands of the underlying Palezoics both stratigraphically and/or due to structural juxtaposition. Paleozoic sands sourced with Jurassic oil has become one of the most important play types in the southwestern Western Desert in recent years, principally in the *Faghur* and *Shushan* sub-basins (Fig. 12; Bosworth et al. 2015b). Sitting between the Paleozoics and the Khatatba in some areas is another reservoir interval that is now known to also be Jurassic in age but is pre-rift (pre-kinematic) in its tectonostratigraphic setting. Following Keeley (1994), we informally refer these sandstones to the *Ras Qattara*. They serve as the principal reservoir at the *Qasr* field which is the Western Desert's largest. This is a gas-condensate field with about 700 MMBOE reserves.

The Khatatba Formation itself also contains good quality siliciclastic reservoirs assigned to the *Safa Member* (Fig. 12; Keeley et al. 1990). The Khatatba Formation is, therefore, a complete petroleum system on its own. Top and lateral seal are provided by the Khatatba *Zahra Member* and the overlying *Masajid Formation* tight limestones. Leaving Qasr Field aside, the most important reservoir units sourced from the Jurassic are the main syn-rift (syn-kinematic) strata of the Early Cretaceous *Burg el Arab Formation* (*Alam el Bueib and Kharita Members*) and the overlying *Bahariya Formation* (Norton 1967). These are all siliciclastic reservoirs, ranging from fluvial to shallow marine. Intraformational shales of the Alam el Bueib and the Late Aptian *Dahab Member* provide top and lateral seal (Fig. 12). However, of historical note, the first field discovered in this Jurassic-Cretaceous petroleum system was *Alamein* in 1966, where the reservoir is Early Aptian dolostone (Egyptian General Petroleum Corporation 1992).

The Turonian Abu Roash F source rock was deposited under anoxic, shallow marine conditions (Barakat et al. 1987; Hegazy and Tammam 1990; Maky and El Sayed 2007; Bakr 2009). It is widespread across northern Egypt but is only buried deep enough to reach maturity in the central Western Desert deep *Abu Gharadig* basin and a few other localized areas. The Abu Roash F has predominantly produced oil and this is generally reservoired in siliciclastics of the Abu Roash Formation itself or fractured limestones of the Paleocene to Middle Eocene *Apollonia Formation*. Abu Roash shales and carbonates and overlying Campanian–Maastrichtian *Sudr Formation* chalks form the top and lateral seals for the Bahariya and Abu Roash reservoirs.

4.3.4 Gulf of Suez Petroleum System, Egypt

The Gulf of Suez Basin began its life about 23 Ma as a northern segment of the Red Sea rift, with similar structural timing and syn-rift stratigraphy (Tewfik and Ayyad 1984: Miller and Baraket 1988). The basin is about 325 km long and 90 km wide at the junction with the Red Sea, decreasing to about 50 km wide at the latitude of Suez City (Fig. 13). The basin covers an area of about 24,000 km^2 and contains approximately 11.5 BBO of proven reserves (Dolson et al. 2014). The average discovered oil per square kilometer is, therefore, ~480,000 barrels, or more than twice the richness of the much larger Sirt Basin.

The Gulf of Suez is comprised of three subbasins, each constituting a "mega-half-graben" that is internally broken into numerous, nested fault blocks (Fig. 13; Moustafa 1976; Angelier 1985; Bosworth 1985; Moretti and Colletta 1987; Colletta et al. 1988; Perry and Schamel 1990; Patton et al. 1994). Structural complexity increases toward the south and correspondingly the size of individual traps becomes smaller. The five largest fields—*Morgan, Belayim, July, Ramadan* and *October*—contain nearly 50% of the basin's total reserves and are all located in the central part of the Gulf. These and many other Gulf of Suez fields are characterized by good reservoir parameters, tall oil columns, high reserves per unit of surface area, good oil quality and shallow-water depths. The producing fields are

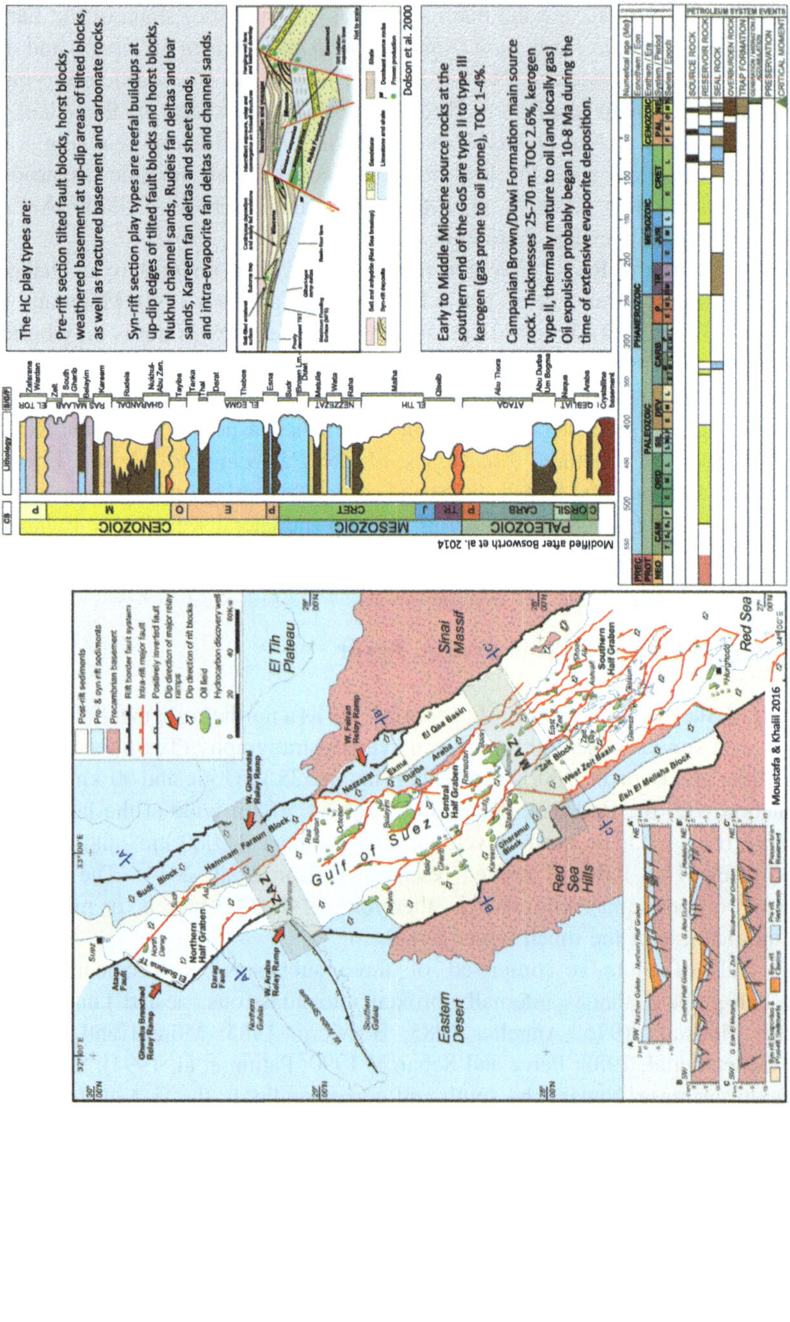

Fig. 13 Gulf of Suez petroleum systems, Egypt

predominantly rotated fault blocks, or combination structural-stratigraphic traps. About 60% of discovered reserves are held in Miocene strata deposited during rifting, and 90% of these are siliciclastics (Egyptian General Petroleum Corporation 1996). For the 40% of the oil found in pre-rift strata, only about 5% comes from carbonates. Superb top and lateral seals are provided by Early Miocene shales of the *Rudeis* and *Kareem Formations* and by thick evaporites of the Middle to Late Miocene *Belayim* and *South Gharib Formations* (Fig. 13; Evans 1988; Richardson and Arthur 1988; Patton et al. 1994; Tawadros 2001; Bosworth and McClay 2001).

Three source rock intervals are well-documented in the Gulf of Suez, and there are probably several other units contributing locally. All are marine strata, dominated by type II kerogen, but with significant differences. The richest interval is the Campanian black shale, often phosphatic, referred to the *Sudr* or *Duwi Formation* or more commonly by industry simply as the *Brown Limestone* (El Ayouty 1990; Alsharhan 2003). Lindquist (1998) referred to oils derived from this source rock as the Sudr-Nubia (!) total petroleum system. "Nubia" is a collective name used by Lindquist and many other authors for pre-Red Sea rifting, predominantly continental sandstones ranging in age from the Cambrian to the Early Cretaceous. It is an unfortunate name for this petroleum system because the Nubia reservoirs are actually less important volumetrically than the younger syn-rift reservoirs. The second major source rock was deposited within the rift basin during the Early to Middle Miocene main phase of extensionally driven subsidence and lies principally within the *Rudeis Formation* in Egypt and the *Maqna Group* in Saudi Arabia. Lindquist (1998) called this the Maqna (!) total petroleum system as the principal reservoir rocks are part of the same formations. The Maqna/Rudeis reservoirs are largely marine siliciclastics, and because they were syn-tectonic in origin their lateral and vertical facies distributions are extremely complex (Richardson and Arthur 1988; Patton et al. 1994; Plaziat et al. 1998; Bosworth et al. 1998; Pivnik et al. 2003). The Maqna (!) source rocks have stronger terrigenous input than the Sudr-Nubia (!), and hence a mixture of type II and III kerogens prevails.

The thickness of the Brown Limestone varies from about 25–70 m. The organic-rich facies are absent in the northern Gulf of Suez sub-basin and in the Red Sea it is present only in the very north. TOC is typically 1.7–3.5% (Alsharhan 2003) but values over 20% have been measured in Egyptian phosphate mines along the northern Red Sea margin. Thermal maturity (top of the oil window) is reached at about 2400 m along the axis of the central Gulf of Suez but not until up to 3900 m in cooler basin margin settings (Alsharhan and Salah 1995). The timing of hydrocarbon maturation and expulsion varies for each sub-region but generally commenced about 10 Ma and continues to the present-day (Mohamed et al. 2013). The Rudeis source rocks often reach several 100 m in thickness but their richness is much lower and typically 1.5–2.2% TOC (Alsharhan 2003). Due to their higher position in the stratigraphic section, the syn-rift source rocks did not typically reach maturity until about 4 Ma.

The third commonly cited Gulf of Suez source rock is the Eocene *Thebes Formation* with TOC generally 1.4–1.7% (Alsharhan 2003). In the northern and central Gulf of Suez, the Thebes is 250–300 m thick, but it decreases to zero on fault block crests in the southern Gulf due to erosion during the onset of rifting (Bosworth and McClay 2001). Synthesizing the distribution of the three principal documented source rocks shows that they are all only present in the central sub-basin, fortuitously where the largest structures are also available to be charged.

4.3.5 Jurassic Petroleum System on the Arabian Plate and Adjacent Areas

An estimated two-thirds of the World's ultimately recoverable oil is in the Arabian–Persian Gulf Region are predominately generated and produced from Jurassic carbonates (Fig. 14; Beydoun 1991; Cole et al. 1994a, b; Alsharhan and Nairn 1997). Widespread marine transgression deposited a thick sequence of shallow marine shelf Jurassic carbonates and platform evaporates on the passive margin to Neotethys to the northeast and north (Ziegler 2001). In Late Middle Jurassic differential subsidence and sea-level rise resulted in the formation of broad, intra-shelf sub-basins (Fig. 14). These intra-shelf sub-basins were depocenters for the main Jurassic source rocks, as well as several important reservoirs. The northern Gotnia Sub-basin, the central Arabian sub-basin, and southeastern South Arabian Gulf sub-basin define the three major Jurassic petroleum systems of the eastern Arabian subcontinent.

(1) the "*Arabian Sub-Basin Tuwaiq/Hanifa-Arab TPS*" (Pollastro 2003) is mostly defined within Saudi Arabia and includes the entire Greater Ghawar Uplift Province, western flank of the Qatar Arch Province, and northern Gulf offshore of southern Iran along the Zagros Fold Belt and Mesopotamian Foredeep Basin. The petroleum system includes *Ghawar* field, as well as several other super-giant Jurassic oil fields of the central Arabian Peninsula such as *Berri, Abqaiq, Abu Sa'fah, Khurais, Khursaniyah*, and *Dukhan*. Two organic-rich intervals of marine carbonates and marls of the Jurassic *Tuwaiq Mountain* and *Hanifa Formations*, generated a single superfamily of oils (Pollastro 2003). The *Tuwaiq Mountain* is 150 m thick in the central part of the Central Arabian sub-basin. TOC exceeds 13% and averages about 3.5%. The Hanifa source rock of central Saudi Arabia is less than 30 m thick and averages about 2.6% TOC. Kerogen in both source rocks is primarily Type II. The main phase of oil generation began in the Cretaceous about 75 Ma (Campanian) or later in the Eocene about 45–50 Ma (*Awali* field in Bahrein) for Tuwaiq Mountain/Hanifa source rocks (TOC > 1.0%). The source rocks are, presently, in the late stage of oil generation.
The primary reservoirs are commonly 10–30 m thick, cyclic shallow-water, carbonate grainstones of the Upper Jurassic *Arab Formation* (*Arab A, B, C*, and *D*). Carbonates are interbedded with organic-rich, muddy lime source rocks. Arab D is the primary reservoir with an average porosity of 25%, accounts for most of the Jurassic oil production forming many supergiant and giant fields (Swart et al. 2005; Sahin and Saner 2001; Eltom et al. 2013; Morad et al. 2012;

Al-Saad and Sadooni 2001; Al-Siddiqi and Dawe 1999). Secondary reservoirs are the porous carbonate–rock units within the *Hanifa* and *Tuwaiq Mountain Formations* (*Hadriya reservoir*) comprising large-scale upward-coarsening, shallowing-upward, carbonate–rock platform sequences.

Seal rocks for the major carbonate reservoirs are anhydrite beds of the *Arab* and *Hith Formations*. The Upper Jurassic Hith Formation consists of hundred meters of massive anhydrite.

Traps include anticlines related to basement block faulting, domes from tectonic salt pillows or other halokinetic deformation, combined structural traps such as salt-assisted fault block traps, and combined structural/stratigraphic (facies) traps (Alsharhan and Nairn 1997). Major trap formation and modification are a result of the First and Second Alpine Orogenic Events of Oman.

(2) the "*Jurassic Hanifa/Diyab-Arab TPS*" is designated for the Rub' al Khali Basin Province of Saudi Arabia, UAE, Qatar, eastern Oman, and northeastern Yemen. The organic-rich, argillaceous limestone of the Upper Jurassic *Hanifa Formation* (*Diyab Formation* in UAE) is the primary source rock. The regionally extensive, Upper Jurassic Arab Formation cyclic carbonate rocks are the primary reservoirs. Intraformational cyclic, Arab and overlying Hith Formation evaporites form the seals (Pollastro 2003).

(3) the "*Jurassic Gotnia/Barsarin/Sargelu/Najmah Total Petroleum System*" in the Widyan Basin–Interior Platform Province (Fox and Ahlbrandt 2002). Another important source rock on the Arabian Peninsula in the Gotnia Basin (Iraq) are organic-rich bituminous limestone and shales of the Jurassic *Sargelu* and *Naokelekan Formations* (Al-Ameri and Zuberge 2012; Grabowski 2014). The *Sargelu Formatio*n consists of marine shales, with up to 16% type II organic matter (Al-Ameri et al. 2014), deposited in a relatively deep, anoxic Jurassic depocenter. Bituminous shale or so-called coal horizon of the *Naokelekan Formation* was quarried for many decades by local communities in northern Iraq as a fuel supply. These organic-rich bituminous, papery shales, and bituminous limestones of the *Naokelekan* and *Sargelu Formations* are probably the richest source rocks in the world (Sadooni 1997). Organic-rich limestone and calcareous shale deposited in a marine environment under highly reducing conditions in the *Chia Gara Formation* (TOC > 18.5%, oil-prone Type II and mixed Type II–III kerogens) are the important prolific oil- and gas-prone source rocks in Kurdistan, Iraq (Mohialdeen et al. 2013). Maturation in Iraq began in the Late Cretaceous, around 90 Ma, in the Mesopotamian trough in southern Iraq and in the Zagros fold belt, but as late as Late Miocene at Kirkuk filed (Grabowski 2014). The northern Gotnia source basin was buried deeper during the Cretaceous than the Arabian platform and therefore passed into the oil window in the Early Cretaceous, some 35 million years before the Hanifa of the Arabian Basin. Petroleum generated from Jurassic source rocks in the Gotnia Basin migrated updip to the west and south into basement related, block-faulted structures that moved periodically since the Pre-Cambrian.

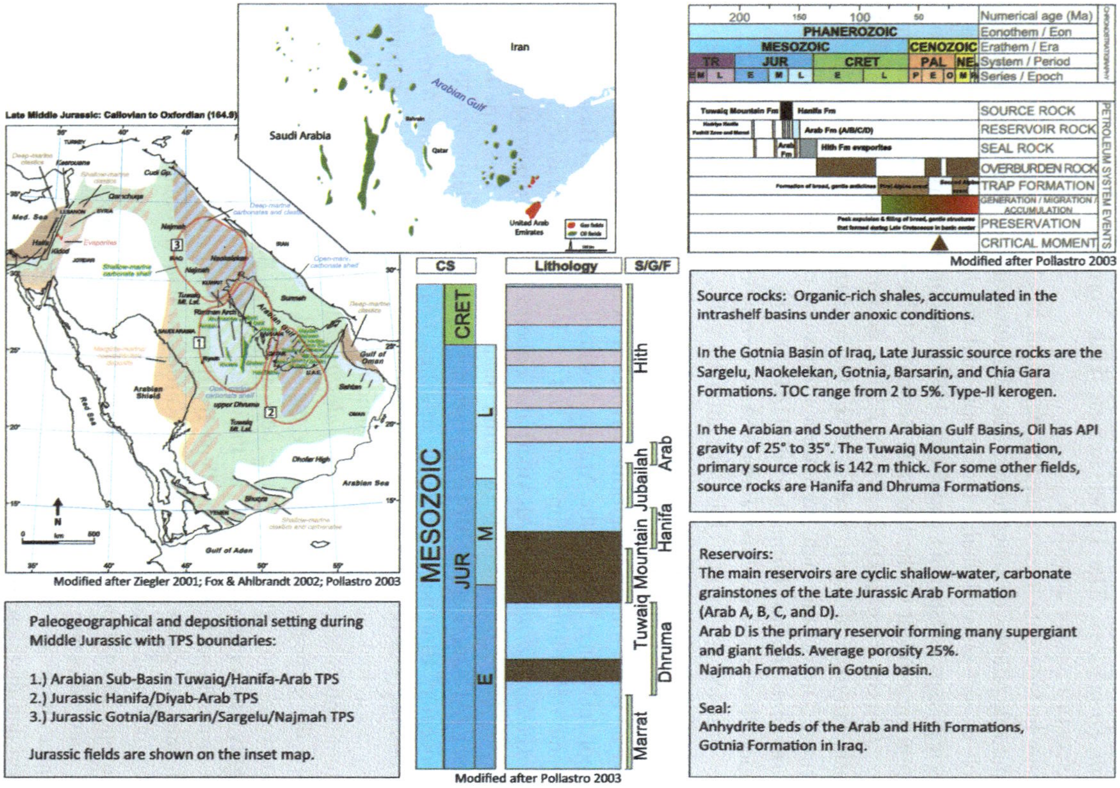

Fig. 14 Jurassic petroleum system on the Arabian Plate and adjacent areas

The *Najmah Limestone* is the primary reservoir and consists of oolitic limestone, dolomite, and anhydrite that were deposited in a shallow marine and transitional marine setting of lagoons and shoals, similar to the Arabian and Southern Arabian Gulf Basins (Sadooni 1997). Locally, oolitic and sandy, marly, argillaceous limestone reservoir facies of the *Sargelu* and Najmah Formations are laterally juxtaposed to argillaceous limestone marl and shale source-rock facies of the Sargelu.

The overlying *Gotnia Formation* consists of evaporite, anhydrite, oolitic limestone, and bituminous rocks deposited in an evaporite basin and lagoonal setting. The Gotnia is locally a minor oil reservoir. The anhydrite is a cap rock. Generally, trap formation took place through most of the Mesozoic, although Late Cretaceous and Miocene to recent are the peak periods of growth.

4.3.6 Cretaceous Petroleum System on the Arabian Plate and Adjacent Areas

Oil production from Cretaceous reservoirs is well known in Iraq, Kuwait, UAR, Saudi Arabia, and Iran (Grabowski 2014; Alsharhan et al. 2014a, b; Cantrell et al. 2014; Bordenave 2014). Cretaceous rocks on the Arabian Plate have been incorporated and described in several petroleum systems. In the Zagros fold belt and foreland they are included within the "*Zagros-Mesopotamian Cretaceous-Tertiary (Cenozoic) TPS*" (Ahlbrandt 1999). In the southern Arabian Gulf region and within the Rub' al Khali Basin they constitute the *Thamama/Wasia TPS* (Pollastro 2000). The *Zubair* (Barremian) and *Burgan* (Aptian) clastics are the main reservoirs in many fields in Kuwait and Iraq (Davies et al. 2002; Ehrenberg et al. 2008; Alsharhan et al. 2014). The Zubair Formation was deposited through deltaic, estuarine, and fluvial environments (Al-Mudhafar 2017). Burgan oil field, second largest in the world, and many surrounding giant oil fields in northern Arabia produce from Cretaceous deltaic reservoirs. The distribution of the Early Cretaceous oil fields is in coincidence and controlled by the reservoir (transitional to shallow marine shelf) and source rock (intra-shelf basins) setting (Fig. 15). Rudist carbonate buildups are common in both Lower Cretaceous carbonate reservoirs such as the *Shuaiba* (Aptian) and Upper Cretaceous *Mishrif Formations*. The Mishrif Formation is of Late Cenomanian age and is the most important oil reservoir in the Mesopotamian Basin, southern Iraq (Alsharhan and Nairn 1997; Aqrawi 1998; Al-Ameri et al. 2009), containing some 30% of Iraq's total oil reserves. It extends throughout the Mesopotamian Basin reaching thicknesses of 100–200 m in the Basrah District at subsurface depths of 2100–2400 m. The Mishrif Formation is dominated by a shallow-water, shelf carbonate sequence composed of bioclastic-detrital limestones, including algal, coral, and rudist bioherms with reservoir porosities exceeding 20% and permeabilities of 100 mD to one Darcy (Al-Ameri et al. 2009). This formation is widespread throughout the Arabian Peninsula. A variety of other reservoirs related to oolites and fracturing are known to occur in Cretaceous rocks.

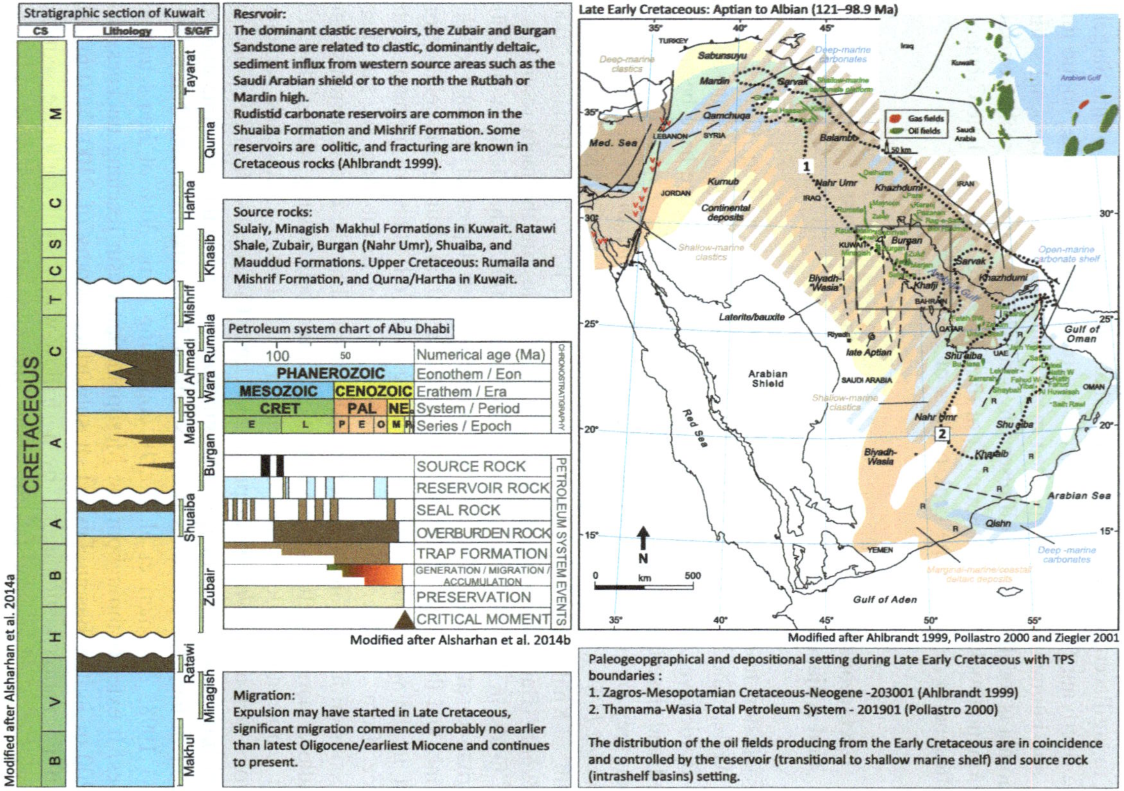

Fig. 15 Cretaceous petroleum system on the Arabian Plate and adjacent areas

Source rock intervals are dominated by Lower Cretaceous Shales (Abeed et al. 2011). The Berriasian *Sulaiy* and *Minagish Formations* are known source intervals in Kuwait and Iraq (Ahlbrandt et al. 2000). The Upper Jurassic–Lower Cretaceous *Sulaiy* and the Lower Cretaceous *Yamama Formations* have the best oil source potential with TOC values up to 7%, kerogen types II and III (Al-Ameri et al. 2009). The Hauterivian *Ratawi Shale* (0.06–1.9% TOC), kerogen type II and III, and the *Zubair Formation* (0.2–2.6% TOC) are important in the North Arabian Gulf region. Carbonate source rocks of the *Shu'aiba Formation* also provide hydrocarbons for much of the Arabian Gulf. The mid-Cretaceous Aptian *Burgan* (*Nahr Umr*), *Mauddud*, Cenomanian *Rumaila* and *Mishrif Formations* are also important source rocks in the Zagros Fold Belt and foreland basin areas. Peak generation occurred in the Middle Miocene time. This timing is coincident with Zagros collision and a thick accumulation of orogenic clastics in the Zagros foredeep (Ahlbrandt et al. 2000). The expulsion may have started in the Late Cretaceous, but significant migration was probably no earlier than latest Oligocene/earliest Miocene and continues to the present-day. The basinal facies of the *Shu'aiba Formation*, argillaceous layers in the Lower Cretaceous *Thamama Group*, and basinal facies of the middle Cretaceous *Shilaif* (*Khatiyah*) *Formation* of the *Wasia Group* are the primary source rocks within the *Thamama/Wasia TPS*, Rub' al Khali Basin Province (Ahlbrandt et al. 2000; Pollastro 2000). Thamama source rocks are presently mature for oil generation along the basin axis of the Rub 'al Khali basin. Shu'aiba source rocks started generating oil in the Eocene, with major expulsion from the Falaha syncline and Oman foreland basin commencing about 40 Ma. Both of these areas are, presently, in the gas generation window (Pollastro 2000).

There are several regional unconformities and shales that form significant seals. The *Nahr Umr Shale* (Aptian) is a major regional seal trapping major accumulations. Eocene and Miocene salt and evaporates in the Zagros fold belt are critical to block the vertical migration in the thrust belt structures of Iraq and Iran.

Giant fields, located in the 50,000 km^2 Dezful Embayment within the Zagros fold belt in Iran (Fig. 16), produce from the *Oligocene—Lower Miocene Asmari Formation* and from the Cenomanian *Sarvak limestone*, and contain some 400 BBO in-place, or 7% of the global oil reserves. Several of the fields are categorized as supergiants as they contain 10–50 BBO in-place, including Agha Jari, Ahwaz, Bibi Hakimeh, Gachsaran, Mansuri, Marun and Rag-e Safid (Bordenave 2002). Asmari Formation and its time-equivalents form one of the world's most important petroleum reservoirs which is estimated to contain more than 90% of the recoverable hydrocarbons of Iran and Iraq. It consists of approximately 400 m of cyclic platform limestones and dolostones with subordinate intervals of sandstone and shale (Ehrenberg et al. 2007). Lithofacies include bioclastic grainstones, ooidal and bioclastic, foraminiferal and intraclastic packstones, and mudstones. Multiple episodes of calcite cementation, dolomitization (Aqrawi et al. 2006) and fracturing (Mcquillan 1973) have affected these rocks to varying degrees and control porosity. Reservoir porosity is mostly dominated by microcrystalline pore spaces in muddy, dolomitized matrix and mouldic porosity in grainstone facies (Al-Aasm et al. 2009). The mid-Cretaceous *Sarvak Formation*, the second-most important reservoir unit in

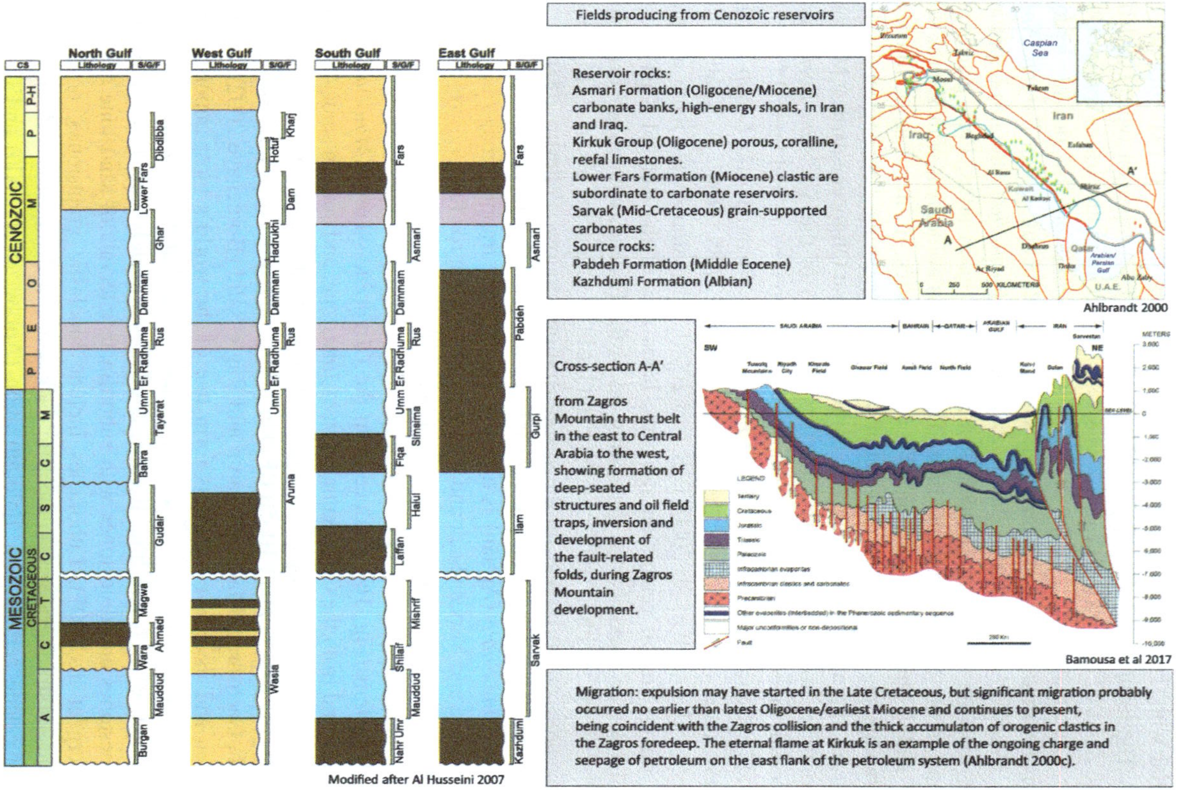

Fig. 16 Arabian Plate; Middle Cretaceous–Cenozoic petroleum system

Iran, is composed mainly of grain-supported carbonates (Beiranvand et al. 2007; Rahimpour-Bonab et al. 2013).

Two excellent source rocks are associated with these reservoirs, the *Albian Kazhdumi Formation*, and the Middle Eocene *Pabdeh Foramation* (Bordenave and Burwood 1990). The Kazhdumi was deposited in a depression corresponding to the present Dezful Embayment, and probably extending north-eastwards. Up to 300 m of dark grey marls and subordinate argillaceous limestone, containing a pelagic fauna of globigerina, globotruncana, and radiolaria, were deposited in strictly euxinic conditions. The organic matter is of algal origin. The Kazhdumi TOC ranges from 1 to 11% with an average of about 5% in the center of the depression (Bordenave 2002, 2014). The Pabdeh accumulated in a NNW–SSE trough, from Shiraz to Lurestan, and consists of 150–200 m of organic-rich marls that contain up to 11.5% TOC. Average TOC varies from 3% in Fars to 7.5% in Lurestan. The organic matter is mostly algal and terrestrial influences are visible in the vicinity of the NE shoreline. Modeling shows that Kazhdumi and Pabdeh reached the onset of oil expulsion at 1–10 Ma, after the beginning of Zagros folding. Migrations took place almost vertically and were facilitated by intense fracturing in high-relief anticlines (Bordenave 2002, 2014).

4.4 Cenozoic Petroleum Systems

Hydrocarbons sourced from Cenozoic source rocks are known from the Pelagian Basin, offshore Tunisia, and Libya, from the Nile Delta in both onshore and offshore settings in Egypt, and the Levantine Basin in the Eastern Mediterranean. In the Zagros, the Eocene source rock (Pabdeh Fm.) works together with the older Cretaceous source rocks charging Oligocene-Early Miocene reservoirs discussed above. Similarly, in the Gulf of Suez, the Early Miocene source intervals contribute to the same reservoirs receiving hydrocarbons from the underlying Late Cretaceous source rocks (Fig. 13).

4.4.1 Pelagian Basin (Offshore Libya and Tunisia) Bou Dabbous–Cenozoic Total Petroleum System

The Pelagian Province is generally located within the offshore shelf area of east-central Tunisia and northern Libya—the Gulfs of Hammamet and Gabes (Fig. 17). The western and southern boundaries are onshore.

The "Bou Dabbous–Tertiary (Cenozoic) Total Petroleum System" (Klett 2001) coincides with the potential extent of petroleum being generated by, and migrating from, Lower Eocene *Bou Dabbous* source rocks. The Bou Dabbous Formation is dark brown marl and mudstone containing type I and II kerogen and ranging in thickness from 50 to 300 m. TOC ranges from 0.4 to 4% and maturation is described as early mature to mature (Klett 2001; Mohamed et al. 2015). Petroleum

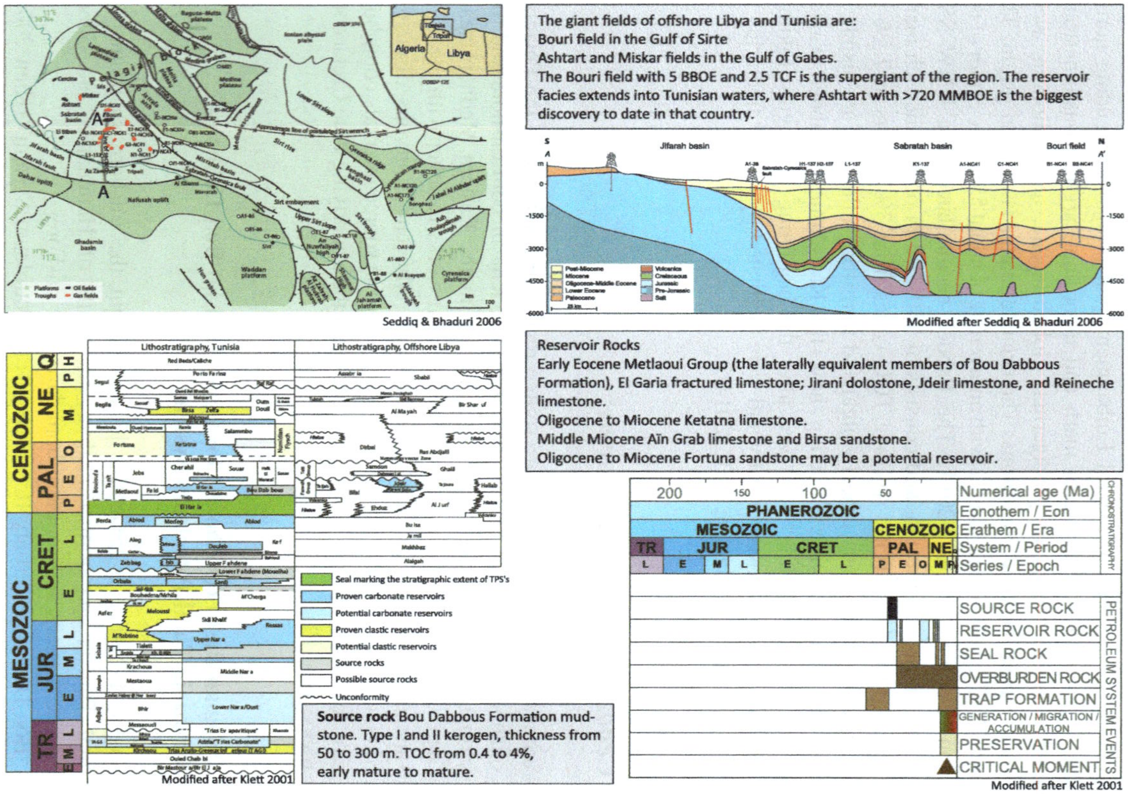

Fig. 17 Pelagian Basin (offshore Libya and Tunisia) Bou Dabbous-Cenozoic total petroleum system

migrated laterally into adjacent or juxtaposed reservoirs and vertically along faults or fractures.

Reservoir rocks include lateral equivalents of the Lower Eocene Bou Dabbous Formation, such as the, *El Garia fractured limestone*, *Jdeir* equivalent in Libya, and *Jirani dolostone* (Mriheel and Anketell 2000). The El Garia Formation was deposited on a shallow north to NE facing ramp composed of a belt of nummulitic wackestone–grainstone (Macaulay et al. 2001; Anketell and Mriheel 2000). Most hydrocarbon production is from the seaward side of the nummulite bank trend including Bourri, Ashtart, Sidi El Itayem, and Zarat fields. The giant Ashtart and Bourri fields are located on structural highs which protruded into the basin and were almost surrounded by Bou Dabbous source rocks; therefore, potential migration pathways are short (Racey 2001; Racey et al. 2001). Other reservoirs are Oligocene to Miocene *Ketatna limestone*; the Middle Miocene *Aïn Grab limestone*; and the Middle Miocene *Oum Douil sandstone* and laterally equivalent *Birsa* and *Mahmoud sandstones*. Sandstones vary from shoreface to shallow marine and typically exhibit excellent reservoir quality of 30–35% porosity and good permeability from 500–1100 mD. Sedimentary paleo-environment distribution varies from proximal deltaic/fluvial deposits in the northern part of the high central Birsa horst to a delta front and pro-delta coastal and shelf shoreface and shoreline channelized deposits in the surrounding borders of grabens (Bedir et al. 2016).

Known accumulations are in fault blocks, low-amplitude anticlines, high amplitude anticlines associated with reverse faults, wrench fault structures, and stratigraphic traps. Most of the traps formed before the Middle Miocene. Seals include Eocene and Miocene mudstone and carbonate rocks.

4.4.2 Nile Delta Petroleum System, Egypt

The Nile Delta begins some 30 km north of Cairo and is a triangular area covering approximately 200,000 km^2 where the River Nile divides into the eastern Damietta branch and the western Rosetta branch (Fig. 18). The lobate form of the Delta began its buildup during the Oligocene forming a very thick section of Late Cenozoic sediments indicating rapid and continuous sedimentation.

Exploration targets include Pliocene–Pleistocene deep-water channel and basin-floor turbidite sands in a variety of structural settings. The focus of pre-Messinian salt exploration is the delineation of distal turbidites within the Serravallian to Tortonian sequence and the identification of new reservoir sequences deposited on preexisting intra-basinal highs (Aal et al. 2001). The Miocene sedimentary succession reaches up to 3000 m of thickness in the central part of the Nile Delta Basin (Sestini 1984; Dolson et al. 2001). The *Sidi Salem Formation* (Early Tortonian) was deposited in a lower neritic to deep shelf environment. The *Qawasim Formation* (Tortonian—Early Messinian) consists predominantly of sandstones and interbeds of sandy claystones and siltstones. The *Abu Madi Formation* (Late Messinian) was formed by sedimentary infilling of a fluvial paleo-valley, the so-called "*Abu Madi paleo-valley*" (Dalla et al. 1997;

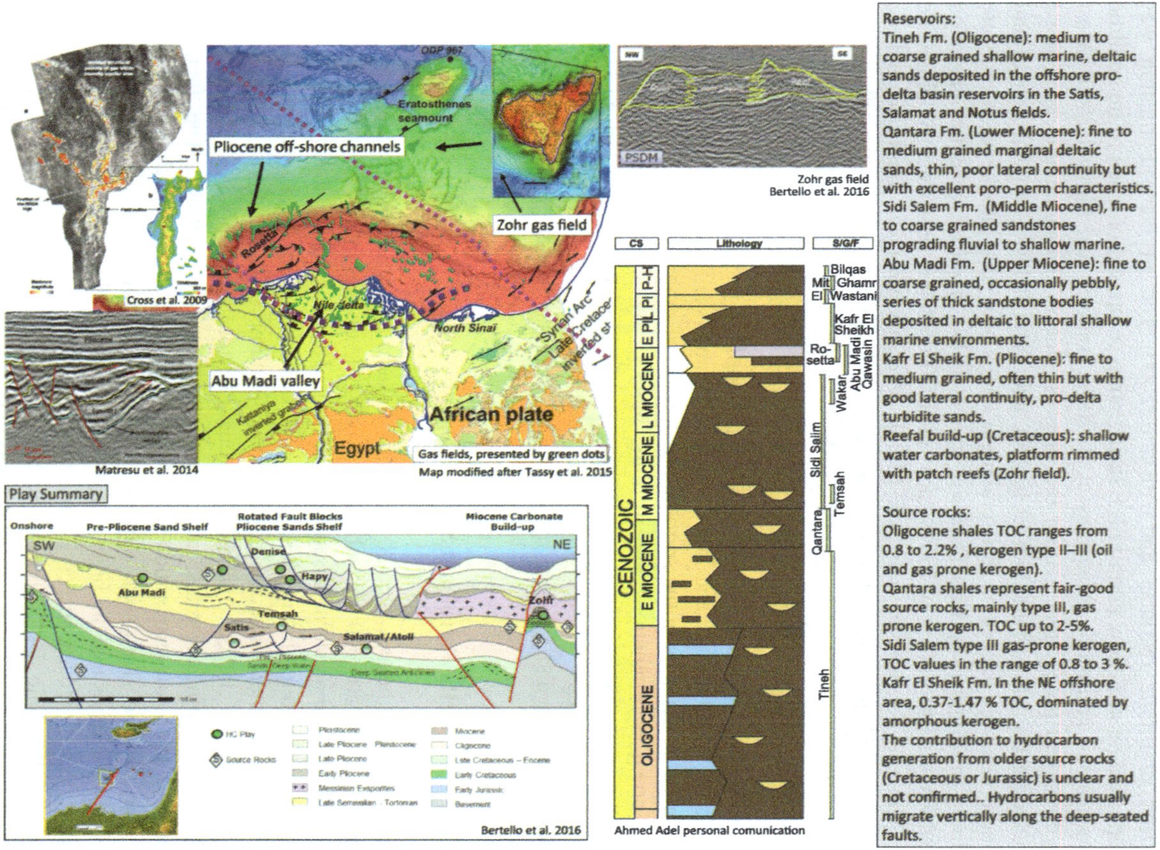

Fig. 18 Nile Delta petroleum system, Egypt

Matresu et al. 2014), and consists of thick sand layers, rarely conglomeratic, interbedded with shale which become thicker and more frequent in the upper part of the formation. From seismic interpretation, this valley is developed approximately from south to north for about 130 km and is at least 5 km wide. It has an average thickness of 300 m and is characterized by stacked fluvio-deltaic sandstones and shale's onlapping landward (southward) and to the valley flanks against the basal erosional surface (unconformity) cutting the Qawasim and Sidi Salim Formations (Matresu et al. 2014). The Messinian canyons were completely overstepped and infilled by 5.6 Ma, but large volumes of sediment from the Nile Valley continued to prograde into the Mediterranean, and the associated facies shifts have formed important reservoir fairways and traps. Pliocene deltaic sandstones form very significant reservoirs and have become the dominant "big play" (Dolson et al. 2001). To summarize, the continental shelf depositional environment, meandering fluvial channels, point bar and levees, shelf slope environment, progarding delta, slope fans, slope channels, sand lobes, and mouth bar sands (Miocene-Serravallian), basin-floor fan and deep-seated channels deposited as a thick turbidite sequences during Pliocene are the most important settings holding the most of the gas reserves in the Nile Delta province (Abu El-Ella 1990; Abdel Aal et al. 1994; Dolson et al. 2000; Wescott and Boucher 2000; Cross et al. 2009; Abd-Allah et al. 2012; Hanafy et al. 2016; Samuel et al. 2003; Kellner et al. 2009; Matresu et al. 2014; Farouk et al. 2014; Mokhtar et al. 2016; Leila and Moscariello 2017). Although the Nile Delta is predominantly a gas province, condensate and light oil have been found in a number of wells (Qantara, Tineh, Abu Qir fields). This indicates the possibility of other oil accumulations. Most of the production is coming from the Miocene and Pliocene deposits, but recently the deeper Oligocene channelized sediments have become more attractive targets (Dolson et al. 2002). The first deep discovery in the Oligocene was the Satis field with 1–3 TCFG and associated condensate (Munn et al. 2013; Dolson et al. 2014). Oligocene reservoirs consist of conglomeratic sandstone intercalated into marine shale.

The Nile Delta basin contains a thick sequence of potential hydrocarbon source rocks that generate essentially gas and condensate. The majority of analyzed Nile Delta shale samples show moderate organic carbon content (0.5-2% TOC). A few Jurassic and Cretaceous samples with high TOC have been described (Shaaban et al. 2006). In the NE offshore area, Pliocene *Kafr el Sheikh Formation* samples indicate 0.37–1.47% TOC, dominated by amorphous kerogen (Sharaf 2003). Within the Miocene intervals, the *Abu Madi, Qawasim, and Sidi Salem* have good organic carbon quantities in offshore wells (0.41–1.56% TOC), with the potential to generate mixed gas and oil. From the onshore, the Qawasim contains up to 2.2% TOC (Sharaf 2003). The main potential source rocks in the Abu Madi gas field are the Abu Madi Formation and Sidi Salem shales. Both contain mixed type II and III kerogens.

Basin modeling suggests that the Abu Madi Formation entered the oil window in the Middle to Late Miocene and the gas window in the Late Miocene to Early Pliocene, while the Sidi Salem Formation entered the oil window in the Early Miocene (Keshta et al. 2014). The Oligocene to Early Miocene shales of the

Qantara Formation also have source potential. The Oligocene fairway has been proven by good results and the presence of disseminated terrestrial organic matter within the Chattian turbidite system (Villinski 2013). In general, Nile Delta source rocks only reached mature stages relatively recently (Pleistocene–Holocene; Shaaban et al. 2006).

Biodegraded thermogenic gas and condensate co-occurs with dry microbial gas in the western Nile Delta (Vandre et al. 2007). This supports the interpretation that secondary microbial methane is formed by microbial degradation of thermogenic hydrocarbons (Vandre et al. 2007). The contribution of older Cretaceous or Jurassic source rocks to hydrocarbon generation is unclear and not fully confirmed. Hydrocarbons are interpreted to migrate vertically along deep-seated faults.

4.4.3 Tamar Petroleum System-Levantine Basin

The Levantine Basin is located in the easternmost Mediterranean (Fig. 19). It is bounded by the Cyprus Larnaca thrust zone to the north, the Eratosthenes Seamount to the northwest, the Nile Delta deep-sea fan to the southwest, and by the eastern Mediterranean coast. Rifting occurred in several phases in the early Mesozoic, after which passive margin conditions were established. Post-rift thermal subsidence and accompanying sedimentation continued for the next 100 Myr (Gvirtzman and Steinberg 2011). The basin is a proven hydrocarbon province, with numerous gas fields and discoveries, spanning from the highly explored Nile Delta of Egypt and offshore Israel in the south to the under-explored margins of Cyprus and Lebanon to the north and east. Technically recoverable gas resources discovered in the northern Levantine Basin are over 35 TCFG.

Many of the gas discoveries to the east are biogenic in origin, although a thermogenic source underlies the Nile Delta, as evidenced by minor oil discoveries and, especially in the pre-Pliocene, ubiquitous gas-condensate discoveries. Two petroleum systems are defined as offshore Israel (Gardosh and Tannebaum 2014). The "*Yafo Petroleum System*" includes several types of reservoir rocks, the most productive of which are siliciclastics of the *Yafo Sand Member* (YSM). The YSM reservoir is composed of poorly cemented quartzose sandstones with high porosity and permeability, deposited in slope and basin-floor fans at the mouth of the Afiq and el-Arish Canyons offshore Israel. The net thickness of the sandstones ranges from several tens of meters in the *Noa* field up to several hundred meters in the *Mari B* and *Nir* fields. This southeastern Levant basin petroleum system consists of dry gas at shallow depths of 300–2000 m. Traps are stratigraphic pinch-outs of sub-horizontal sand layers, sand mounds, and fold structures. Source rocks are organic-rich shale beds within the upper part of the Miocene *Saqiye Group*. The TOC values are from 0.7 to 1.65%, and the organic material is composed of mixed coaly and amorphous gas-prone, Type Ill and IV kerogens (Gardosh and Tannenbaum 2014). The shallow gas occurrences onshore and offshore are composed almost entirely of methane (> 99%) with an isotopic ratio that is characteristic of a microbial origin. The recoverable gas resources in the Mari B, Noa,

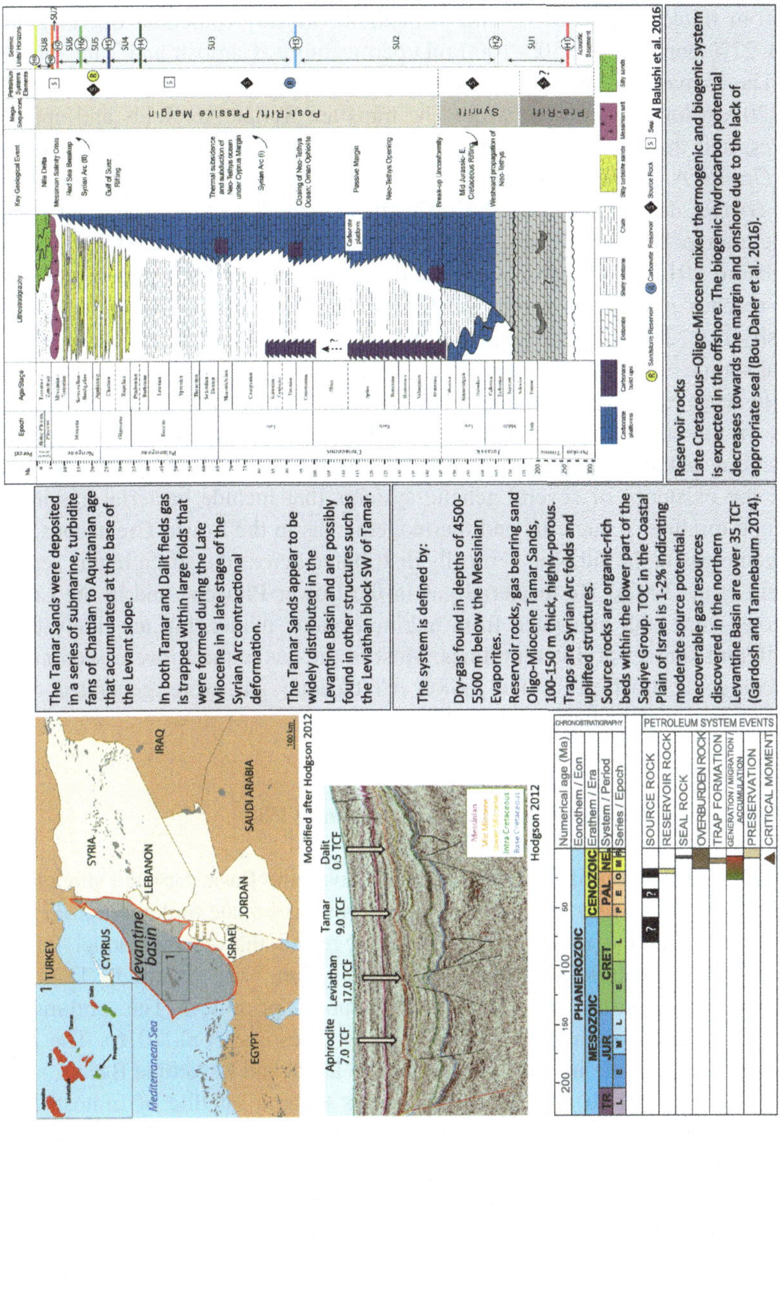

Fig. 19 Tamar petroleum system-Levantine Basin

Gaza Marine, Or, and Nir fields are estimated at 2.5 TCFG (Gardosh and Tannenbaum 2014).

The "*Tamar Petroleum System*" is the second offshore system and is characterized by dry gas found at depths of 4500–5500 m, below the Messinian evaporites, in reservoir rocks of the Oligo-Miocene *Tamar Sands*. The Tamar Sands are 100–150 m thick, composed of fine to medium, loosely cemented quartz grains with 20–27% porosity and 500–1000 mD permeability (Gardosh and Tannenbaum 2014). The Tamar Sand package is markedly continuous and covers an area that is about 120 km long and 60 km wide. The traps are Syrian Arc folds and uplifted structures, and the source rocks are organic-rich beds within the lower part of the *Saqiye Group*. A deeper, very prolific source rock is found in the Senonian *Mount Scopus Group*, deposited during the early rift reactivation phase. Campanian and Maastrichtian pelagic marls and chalks contain up to 20% TOC (Gardosh and Tannenbaum 2014). Thermal maturity modeling shows that the Upper Cretaceous section reaches maturation within the Levant Basin at depths greater than 4 km (Gardosh et al. 2008a, b). Senonian strata, therefore, should be considered as a potential source of oil and thermogenic gas in the deep part of the basin. It is not yet known to what extent these source rocks are participating in hydrocarbon generation, as oil-to-source correlations are inconclusive. Chemical and stable isotopic compositions obtained from offshore Middle Jurassic to Pliocene reservoir rocks indicate the existence of several genetic systems that include bacterial gas in the Pliocene, a mixture of bacterial and thermogenic gas in the Lower Cretaceous, and thermogenic gas in the Jurassic vertically distributed across the stratigraphic section (Feinstein et al. 2002). The bacterial gas in the Lower Pliocene and in the Lower Cretaceous were both generated from marine organic matter, but are likely to be from different sources at different times. Most of the gas in the Jurassic section and the thermogenic fraction in the Lower Cretaceous gas were generated from oil-prone kerogen (Feinstein et al. 2002).

4.4.4 Eastern Mediterranean "Zohr" Carbonate Play

Historically, exploration of the Nile Delta and Levantine Basin targeted siliciclastic plays. The 2015 discovery of a massive gas-bearing carbonate buildup at *Zohr field* far to the north of the modern Nile Delta suggested that a new model for the paleogeographic evolution of the Levantine Basin was required. This had far-reaching implications for regional exploration priorities. The preliminary resource estimate for Zohr is 30 TCFG in-place (Esestime et al. 2016). The Zohr structure is located at the intersection of the Nile Delta and Levantine Basins, south of the Eratosthenes High. The structure appears to be a satellite of Eratosthenes, forming a NE-SW elongated ridge with an area of about 190 km^2. It is 20 km long and the width varies from 5 km in the north to 15 km in the south. It is draped by a layer of evaporites reaching more than 1000 m thickness (Bertello et al. 2016). The initial discovery well encountered a biogenic gas column 630 m tall in a carbonate sequence of Cretaceous to Miocene age with excellent reservoir characteristics

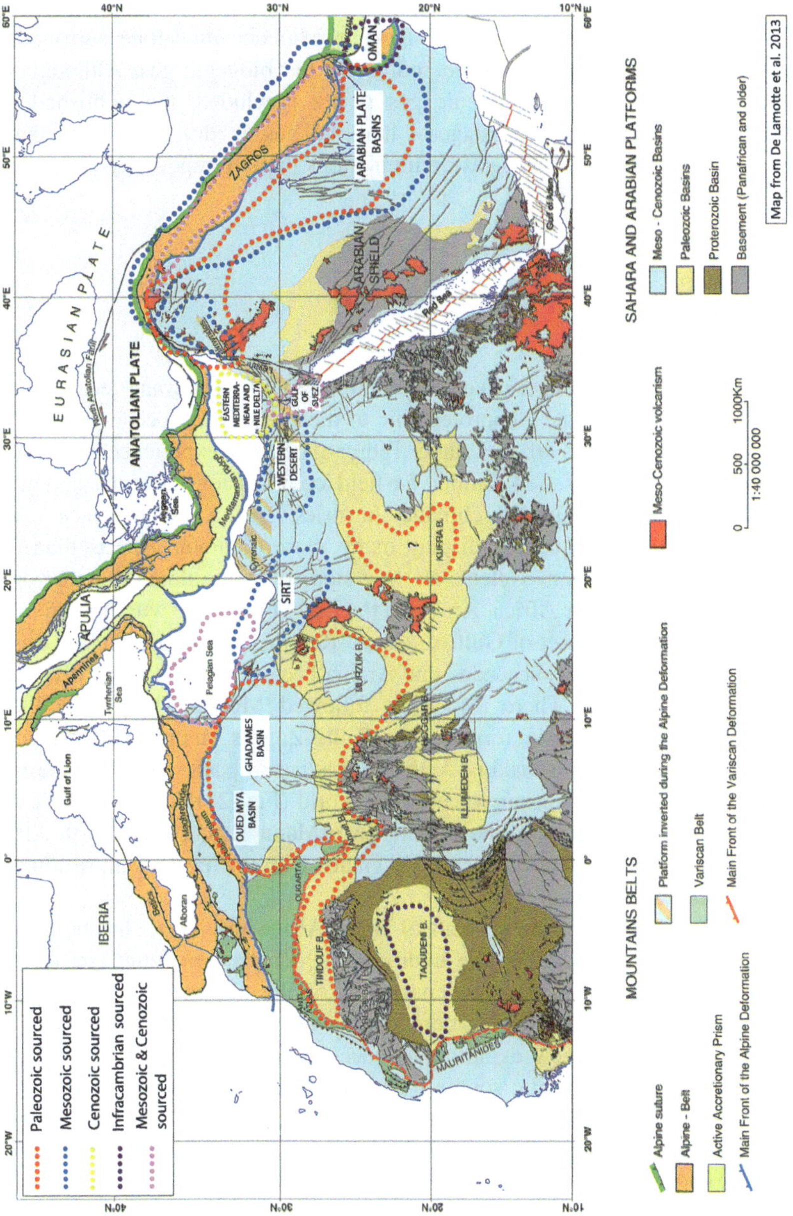

Fig. 20 Petroleum bearing Phanerozoic Basins and schematic play overview of North Africa and Arabia (with Oman and Taoudeni Infracambrian systems)

(> 400 m net pay; Giovannelli, personal communication, 2017). The reservoir facies are ascribed to a platform rimmed with patch reefs surrounding an internal lagoon that experienced significant tidal effects. The best reservoir intervals are characterized by the presence of rudists, coral fragments and molluscs (Giovannelli personal communication, 2017). The Oligocene–Miocene mudstone surrounding the carbonate buildup is the probable source for the biogenic gas. Although the potential involvement of thermogenic gas is not precluded, the published gas pressure gradient (0.26 g/cm^3) suggests that the gas is dry, mostly methane, and similar to the offshore Israel gas accumulations that are reportedly biogenic in origin (Esestime et al. 2016).

5 Conclusions

North Africa and Arabia experienced a common eastern Gondwana tectonostratigraphic history from the late Neoproterozoic to the end of the Eocene. They were adjacent to different parts of the same Tethyan oceans, experienced the same intra-plate rifting episodes, felt similar far-field compressional events, and were subjected to generally similar climatic histories. As a consequence, their Phanerozoic lithologic sequences display many more common aspects than differences, and their petroleum systems are best studied and explored together or in parallel. Times changed for Africa–Arabia in the Early Oligocene with the arrival of the Afar plume and the onset of Gulf of Aden—Red Sea rifting. Unlike the earlier intra-plate rifts, this system succeeded and Arabia became its own plate, diverging from the rest of Africa. There also soon followed Middle Miocene collision between Arabia with Eurasia forming the Bitlis–Zagros suture. This Neogene tectonism became critical in making Arabia and adjacent parts of the Middle East the most prolific hydrocarbon province in the World (Fig. 21). The end result was over 200 giant fields residing in the Middle East (Mann et al. 2003). North Africa did not fare as well but still tallied nearly 40 giants, with the most recent residing in the offshore Levantine basin.

Our review of the petroleum systems of North Africa and Arabia has been brief and includes only highlights of the hydrocarbon occurrences found across this broad and complex region. Our understanding of the geologic history, and its implications for hydrocarbon exploration continues to rapidly advance and evolve. This will result in many new exploration successes in a part of the Earth that is endowed with so many prolific source rocks, extensive reservoir-seal pairs, multiphase trap-formating deformation events, and burial histories conducive to the formation, migration, and preservation of oil and gas. In compiling our overview we have also attempted to provide a fairly extensive bibliography in the hope that this will prove useful to the readers of this volume. Future advances of our general understanding of the geology of North Africa and Arabia will benefit from new and more detailed integrative studies across this fascinating paleogeographic province.

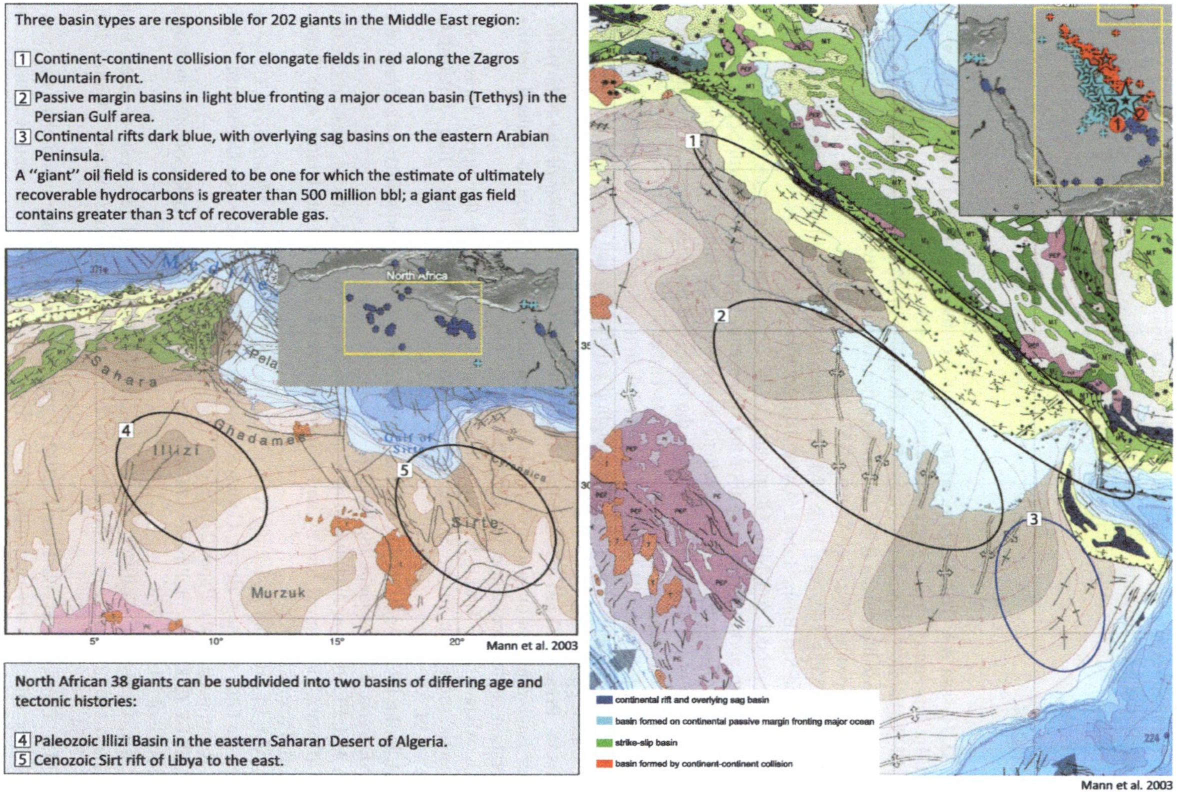

Fig. 21 Tectonic setting of the giant fields

Acknowledgements Gabor Tari provided a very helpful review of our manuscript and Ahmed El-Barkooky gave us much appreciated encouragement. Vlatko Brčić kindly assisted with the drafting of the figures. All are gratefully thanked for their efforts to improve this work. We have benefited from many fruitful discussions over the years from both our industry and academic colleagues. In particular, René Guiraud constantly emphasized the importance of correlating tectonostratigraphic events across the breadth of Gondwana. Errors of interpretation or presentation are solely the responsibility of the authors.

References

Aadland AJ, Hassan AA (1972) Hydrocarbon potential of the Abu Gharadig basin in the western desert. In: 8th Arab Petrol Congress, Algiers, vol 81, B3

Aal AA, El-Barkooky A, Gerrits M, Meyer HJ, Schwander M, Zaki H (2001) Tectonic evolution of the Eastern Mediterranean Basin and its significance for the hydrocarbon prospectivity of the Nile Delta deepwater area. GeoArabia 6:363–384

Abadi AM, van Wees J-D, van Dijk PM, Cloetingh SAPL (2008) Tectonics and subsidence evolution of the Sirt Basin, Libya. AAPG Bull 92:993–1027

Abd-Allah AMA, Abdel Aal MH, Ghandour A (2012) Structural characteristics and tectonic evolution of the northwestern margin of the Nile Delta, Egypt. J Afr Earth Sci 68:82–95

Abd El-Motaal E, Kusky TM (2003) Tectonic evolution of the intraplate S-shaped Syrian Arc fold-thrust belt of the middle east region in the context of plate tectonics. In: 3rd international conference on the geology of Africa, vol 2, December 2003, Assiut, Egypt, pp 139–157

Abdel Aal A, Day RA, Lelek JJ (1992) Structural evolution and styles of the northern Sinai, Egypt. In: 11th Egyptian general petroleum corporation exploration and production conference, Nov 1992, Cairo, pp 546–562

Abdel Aal A, Price RJ, Vaiti JD, Shrallow JA (1994) Tectonic evolution of the Nile Delta, its impact on sedimentation and hydrocarbon potential. In: 12th Egyptian general petroleum corporation conference, Cairo vol 1, pp 19–34

Abeed Q, Alkhafaji A, Littke R (2011) Source rock potential of the upper Jurassic-lower cretaceous succession in the Southern Mesopotamian basin, Southern Iraq. J Petrol Geol 34:117–134

Aboud M, Philip PR, Alen J (2005) Geochemical correlations of oils and source rocks from central and NE Syria. J Petrol Geol 28:203–218

Abdallah AM, El Adindani A (1963) Stratigraphy of the upper Paleozoic rocks, western side of the Gulf of Suez. Geological Survey and Mineral Research Department, United Arab Republic (Egypt), Paper 25, 18 p

Abdallah AM, El Adindani A, Fahmy N (1963) Stratigraphy of the lower Mesozoic rocks, western side of the Gulf of Suez, Egypt. Geological Survey and Mineral Research Department, United Arab Republic (Egypt), Paper 27, 23 p

Abdullah FHA (2001) A preliminary evaluation of Jurassic source rock potential in Kuwait. J Petrol Geol 24:361–378

Abrams MA, Greb MD, Collister JW, Thompson M (2016) Egypt far western desert basins petroleum charge system as defined by oil chemistry and unmixing analysis. Marine Petrol Geol 77:54–74

Abrams MA, Thompson M, Collister J (2012) Egypt western desert petroleum charge system as defined by oil chemistry and unmixing analysis. AAPG Annual convention abstracts long beach CA

Abu-Ali MA, Rudkiewicz J-LL, McGillivray JG, Behar F (1999) Paleozoic petroleum system of central Saudi Arabia. GeoArabia 4:321–336

Abu-Ali MA (2005) Organic petrology, maturation, thermal and burial history analysis, and hydrocarbon generation and migration modeling of the Saudi Arabian paleozoic petroleum systems. Ph.D. thesis, University of Aachen, 202 pp
Abu El-Ella R (1990) The Neogene-quaternary section in the Nile Delta, Egypt: geology and hydrocarbon potential. J Petrol Geol 13:329–340
Adloff MC, Doubinger J, Massa D, Vachard D (1986) Trias de Tripolitaine (Lybie). Nouvells donées biostratigraphiques et palynologiques. Revue de l'Institut Franc¸ ais du Pétrole 41:27–72
Ahlbrandt TS (1999) Zagros-Mesopotamian Cretaceous-Tertiary (203001). US Geol Surv World Petroleum Assessment 2000—Description and Results
Ahlbrandt TS (2000a) Paleozoic-Permian/Triassic total petroleum system 203002. US Geol Surv World Petroleum Assessment 2000—Description and Results
Ahlbrandt TS (2000b) Northern Qatar Arch Extension AU 20300201. US Geol Surv World Petroleum Assessment 2000—Description and Results
Ahlbrandt TS (2000c) Tertiary Reservoirs Assessment Unit 20300102. US Geol Surv World Petroleum Assessment 2000—Description and Results
Ahlbrandt TS (2001) The Sirte Basin Province of Libya—Sirte-Zelten total petroleum system. U. S. Geol Survey Bull 2202–F:1–29
Ahlbrandt TS (2002) Petroleum geology and total petroleum systems of the Widyan Basin and interior platform of Saudi Arabia and Iraq. US Geol Surv Bull 2202–E:1–26
Ahlbrandt TS (2010) The petroleum endowments of the total petroleum systems in the Middle East and North Africa Tethys. AAPG Search and Discovery Article #10244, 44 p
Ahlbrandt TS, Pollastro RM, Klett TR, Schenk CJ, Lindquist SJ, Fox JE (2000) Chapter R2. Region 2 assessment summary–Middle East and North Africa. In: US geological survey digital data series 60, World petroleum assessment 2000—description and results, pp 1–39
Ahlbrandt TS (2014) Petroleum systems and their endowments in the Middle East and North Africa portion of the Tethys. In: Marlow L, Kendall CCG, Yose LA (eds) Petroleum systems of the Tethyan region, AAPG Memoir, vol 106, pp 59–100
Aissaoui MN, Bedir M, Gabtni H (2016) Petroleum assessment of Berkine-Ghadames Basin, southern Tunisia. AAPG Bull 100:445–476
Albert-Villaneuva E, Permanyer A, Tritlla J, Levresse G, Salas R (2016) Solid hydrocarbons in Proterozoic dolomites, Taoudeni Basin, Mauritania. J Pet Geol 39:5–28
Al-Aasm IS, Ghazban F, Ranjbaran M (2009) Dolomitization and related fluid evolution in the Oligocene-Miocene Asmari formation, Gachsaran area, SW Iran: petrographic and isotopic evidence. J Petrol Geol 32:287–304
Al-Youssef W, Ayed H (1992) Evolution of upper Paleozoic sequences and the application of stratigraphy as a tool for hydrocarbons exploration in Syria. In: 11th Egyptian general petroleum corporation exploration and production conference, Cairo, Egypt, 7–10 Nov 1992
Al-Ameri TK, Zuberge J (2012) Middle and upper Jurassic hydrocarbon potential of the Zagross Fold Belt, North Iraq. Mar Petrol Geol 36:13–34
Al-Ameri TK, Al-Khafaji AJ, Zumberge J (2009) Petroleum system analysis of the Mishrif reservoir in the Ratawi, Zubair, North and South Rumaila oil fields, southern Iraq. GeoArabia 14:91–108
Al-Ameri TK, Najaf AA, Al-Khafaji AS, Zumberge J, Pitman J (2014) Hydrocarbon potential of the Sargelu formation, North Iraq. Arab J Geosci 7:987–1000
Al-Balushi AN, Neumaier M, Fraser AJ, Jackson CAL (2016) The impact of the Messinian salinity crisis on the petroleum system of the Eastern Mediterranean: a critical assessment using 2D petroleum system modelling. Petrol Geosci 22:357–379
Al-Husseini MI (1997) Jurassic sequence stratigraphy of the western and southern Arabian Gulf. GeoArabia 2:361–382
Al-Johi A, Al-Laboun A (2015) Neoproterozoic (infracambrian)—Paleozoic total petroleum systems of Arabian Peninsula; an overview. Arabian J Earth Sci 2:1–21
Al-Mudhafar WJ (2017) Geostatistical lithofacies modeling of the upper sandstone member/Zubair formation in south Rumaila oil field, Iraq. Arab J Geosci 10:153–167

Al-Rawi MM (2014) Petroleum systems in Jordan. GeoExpro 11:22–30
Al-Saad D, Sawaf T, Gebran A, Barazangi M, Best JA, Chaimov TA (1992) Crustal structure of central Syria: the intracontinental Palmyride Mountain belt. Tectonophysics 207:345–358
Al-Saad H, Sadooni FN (2001) A new depositional model and sequence stratigraphic interpretation for the upper Jurassic Arab "D" reservoir in Qatar. J Petrol Geol 24:243–264
Al-Siddiqi A, Dawe RA (1999) Qatar's oil and gas fields: a overview. J Petrol Geol 22:417–436
Al-Youssef W, Ayed H (1992) Evolution of upper Paleozoic sequences and the application of stratigraphy as a tool for hydrocarbons exploration in Syria. In: 11th Egyptian general petroleum corporation exploration and production conference, Cairo, Egypt 7–10 Nov 1992
Alem N, Assassi S, Benhebouche S, Kadi B (1998) Controls on hydrocarbon occurrence and productivity in the F6 reservoir, Tin Fouyé–Tabankort area, NW Illizi Basin. In: Macgregor DS, Moody RTJ, Clark-Lowes DD (eds) Petroleum geology of North Africa, Geol Soc London Spec Publ 132:175–186
Alsharhan AS (2003) Petroleum geology and potential hydrocarbon plays in the Gulf of Suez rift basin, Egypt. AAPG Bull 87:143–180
Alsharhan AS, Nairn AEM (1997) Sedimentary basins and petroleum geology of the Middle East: Elsevier Science BV, Amsterdam, 942 p
Alsharhan AS, Salah MG (1995) Geology and hydrocarbon habitat in rift setting: northern and central Gulf of Suez, Egypt. Bull Can Petrol Geol 43:156–176
Alsharhan AS, Salah MG (1997) Lithostratigraphy, sedimentology and hydrocarbon habitat of the Pre-Cenomanian Nubian Sandstone in the Gulf of Suez oil province, Egypt. GeoArabia 2:385–400
Alsharhan AS, Abd El-Gawad EA (2008) Geochemical characterization of potential Jurassic/Cretaceous source rocks in the Shushan Basin, north western desert, Egypt. J Pet Geol 31:191–212
Alsharhan AS, Strohmenger CJ, Abdullah FH, Al Sahlan G (2014a) Mesozoic stratigraphic evolution and hydrocarbon habitats of Kuwait. In: Marlow L, Kendall C, Yose L (eds) Petroleum systems of the tethyan region. AAPG Mem, vol 106, pp 541–611
Alsharhan AS, Strohmenger CJ, Al-Mansoori A (2014b) Mesozoic petroleum systems of Abu Dhabi, United Arab Emirates. In: Marlow L, Kendall CCG, Yose LA (eds) Petroleum systems of the Tethyan region. AAPG Mem, vol 106, pp 679–711
Ambrose G (2000) The geology and hydrocarbon habitat of the Sarir Sandstone, SE Sirt Basin, Libya. J Pet Geol 23:165–192
Angelier J (1985) Extension and rifting: the Zeit region, Gulf of Suez. J Struct Geol 7:605–612
Anketell JM (1996) Structural history of the Sirt Basin and its relationships to the Sabratah Basin and Cyrenaican Platform, Northern Libya. In: Salem MJ, El-Hawat AS, Sbeta AM (eds) The geology of Sirt Basin, vol III. Elsevier, Amsterdam, pp 57–88
Anketell JM, Mriheel IY (2000) Depositional environment and diagenesis of the Eocene Jdeir Formation, Gabes-Tripoli basin, western offshore, Libya. J Petrol Geol 23:425–447
ANSS (2018) Comprehensive earthquake catalog (ComCat) USGS earthquake hazards program. Available from http://earthquake.usgs.gov/earthquakes/search Search Earthquake Catalogue
Aqrawi AA (1998) Paleozoic stratigraphy and petroleum systems of the western and southwestern deserts of Iraq. GeoArabia 3:229–248
Aqrawi AA, Keramati M, Ehrenberg SN, Pickard N, Moallemi A, Svana T, Darke G, Dickson JAD, Oxtoby NH (2006) The origin of dolomite in the Asmari formation (Oligocene-Lower Miocene), Dezful Embayment, SW Iran. J Petrol Geol 29:381–402
Aqrawi AA, Goff JC, Horbury AD, Sadooni FN (2010) The petroleum geology of Iraq. Scientific Press, Bucks, p 424
Arduini M, Barassi M, Golfetto F, Ortenzi A, Serafini G, Tebaldi E, Trincianti E, Visentin C (2003) Silurian-Devonian sedimentary geology of the Libyan Ghadāmis Basin: example of an integrated approach to the Acacus Formation study. In: Salem MJ, Oun KM (eds) The geology of northwest Libya, vol II, Sedimentary basins of Libya, pp 183–210

Armstrong HA, Abbot GD, Turner BR, Makhlouf IM, Muhammad AB, Pedentchouk N, Peters H (2009) Black shale deposition in an upper Ordovician-Silurian permanently stratified, peri-glacial basin, southern Jordan. Palaeogeogr Palaeoclimatol Palaeoecol 273:368–377

Attar M, Hammat M (1995) Algeria's hydrocarbon potential. In: Algeria well evaluation conference, Schlumberger, vol IV, pp 1–17

Ayyad MH, Darwish M (1996) Syrian arc structures: a unifying model of inverted basins and hydrocarbon occurrences in North Egypt. In: Proceedings of 13th petroleum conference, exploration, vol 1. Egyptian General Petroleum Corporation, Cairo, pp 40–59

Badalini G, Redfern J, Carr ID (2002) A synthesis of current understanding of the structural evolution of North Africa. J Petrol Geol 25:249–258

Baird DW, Aburawi RM, Bailey NJL (1996) Geohistory and petroleum in the central Sirt Basin. In: Salem MJ, El-Hawat AS, Sbeta AM (eds) The geology of Sirt Basin. Elsevier, Amsterdam, vol III, pp 3–56

Bakr MMY (2009) Molecular organic geochemistry of crude oil from Shushan and Abu Gharadig Basins, Western Desert, Egypt. Earth Sci 20:97–125

Bamousa AO, Memesh AM, Dini SM, Al-Zahrani AA (2017) Tectonic style and structural features of Alpine-Himalayan Orogeny in Central Arabia. In: Roure F, Amin AA, Khomsi S, Al-Garni MAM (eds) Lithosphere dynamics and sedimentary basins of the Arabian plate and surrounding areas. Springer, Berlin, pp 173–185

Barakat MG, Darwish M, Abdel-Hamid ML (1987) Hydrocarbon source-rock evaluation of upper cretaceous (Abu Roash formation), east Abu-Gharadig area, North Western Desert, Egypt. MERC Ain Shams Univ Earth Sci Ser 1:120–150

Barazangi M, Serber D, Chaimov T, Best J, Litak R, Al-Saad D, Sawaf T (1993) Tectonic evolution of the northern Arabian plate in western Syria. In: Boschi E, Mantonani E, Morelli A (eds) Recent evolution and seismicity of the Mediterranean region. Kluwer Academic Publication, Dordrecht, pp 117–140

Barr FT, Weegar AA (1972) Stratigraphic nomenclature of the Sirte Basin, Libya. Petroleum Exploration Society of Libya, Tripoli, p 179

Barrier E, Machour l, Blaizot M (2014) Petroleum systems of Syria. In: Marlow L, Kendall CCG, Yose LA (eds) Petroleum systems of the Tethyan region. AAPG Mem 106:335–378

Bayer H-J, Hötzl H, Jado AR, Roscher B, Voggenreiter W (1988) Sedimentary and structural evolution of the northwest Arabian Red Sea margin. Tectonophysics 153:137–151

Bedir M, Arbi A, Khomsi S, Houatmia F, Aissaoui MN (2016) Seismic tectono-stratigraphy of fluvio-deltaic to deep marine Miocene siliciclastic hydrocarbon reservoir systems in the Gulf of Hammamet, northeastern Tunisia. Arab J Geosci 9(726):33

Beiranvand B, Ahmadi A, Sharafodin M (2007) Mapping and classifying flow units in the upper part of the Mid-Cretaceous Sarvak formation (Western Dezful embayment, SW Iran) based on a determination of reservoir rock types. J Petrol Geol 30:357–373

Bell A, Spaak P (2007) Gondwanan glacial events and their influence on petroleum systems in Arabia. AAPG Search and Discovery Article #30047:4 p

Berberian M (1995) Master "blind" thrust faults hidden under the Zagros folds: active basement tectonics and surface morphotectonics. Tectonophys 241:193–224

Berra F, Angiolini L (2014) The evolution of the Tethys region throughout the Phanerozoic: a brief tectonic reconstruction. In: Marlow L, Kendall C, Yose L (eds) Petroleum systems of the Tethyan region. AAPG Memoir 106:1–27

Bertello F, Harby H, Brandolese S (2016) Egypt: Zohr, an outstanding gas discovery in a new deep-water hydrocarbon play. In: 8th Mediterranean offshore conference and exhibition (MOC), 19–21 Apr 2016, Alexandria, 10 p

Best JA, Barazangi M, Al-Saad D, Sawaf T, Gebran A (1990) Bouguer gravity trends and crustal structure of the Palmyride Mountain belt and surrounding northern Arabian platform in Syria. Geology 18:1235–1239

Best JA, Barazangi M, Al-Saad D, Sawaf T, Gebran A (1993) Continental margin evolution of the northern Arabian platform in Syria. AAPG Bull 77:173–193

Bevan TG, Moustafa AR (2012) Inverted rift-basins of northern Egypt. In: Roberts DG, Bally AW (eds) Regional geology and tectonics: phanerozoic rift systems and sedimentary basins, 1B. Elsevier, Amsterdam, pp 483–507

Beydoun ZR (1991) Arabian plate hydrocarbon geology and potential—a plate tectonic approach. AAPG Stud Geol 33, 77 p

Beydoun ZR (1997) Prehistoric, ancient and mediaeval occurrences and uses of hydrocarbons in the greater Middle East region. J Petrol Geol 20:91–95

Beydoun ZR (1998) Arabian plate oil and gas: why so rich and so prolific? Episodes 21:74–81

Bezan AM (1996) The Palaeocene sequence in Sirt Basin. In: Salem MJ, Mouzughi AJ, Hammuda OS (eds) Geology of the Sirt Basin, vol 1. Elsevier, Amsterdam, pp 97–117

Bialik OM, Korngreen D, Benjamini C (2012) Lithofacies and cyclicity of Mohilla evaporite basins on the rifted margin of the Levant in the Late Triassic, Makhtesh Ramon, southern Israel. Sedimentology 59:2097–2124

Bibonne R, Duringer P, Raulin C, Leleu S, Ghienne JF, Bouaziz S, Rubino JL (2015) Sedimentology and stratigraphy of the early to middle triassic clastic formations of the Ghadames Basin (Tunisia-Libya). Paper Tu N110 13. In: 77th EAGE conference and exhibition, Madrid, 1–4 June 2015, 5 p

Biju-Duval B, Letouzey J, Montadert L (1978) Structure and evolution of the Mediterranean basins. In: Hsu KJ et al (eds) Initial reports of the Deep Sea Drilling Project, Leg 42, Pt 1, pp 951–984

Boote DRD, Clark-Lowes DD, Traut MW (1998) Palaeozoic petroleum systems of North Africa. In: MacGregor DS, Moody RTJ, Clark-Lowes DD (eds) Petroleum Geology of North Africa. Geol Soc Lond Spec Publ 132:7–68

Bordenave ML (2002) The Middle Cretaceous to Early Miocene petroleum system in the Zagros Domain of Iran, and its prospect evaluation. AAPG Annual Meeting March 10–13, 2002, extended abstract

Bordenave ML (2008) The origin of the Permo-Triassic gas accumulations in the Iranian Zagros foldbelt and contiguous offshore areas: a review of the Paleozoic petroleum system. J Petrol Geol 31:3–42

Bordenave ML (2014) Petroleum systems and distribution of the oil and gas fields in the Iranian part of the Tethyan Region. In: Marlow L, Kendall CCG, Yose LA (eds) Petroleum systems of the Tethyan region, AAPG Mem, vol 106, pp 505–540

Bordenave ML, Burwood R (1990) Source rock distribution and maturation in the Zagros orogenic belt: provenance of the Asmari and Bangestan reservoir oil accumulations. Org Geochem 16:369–387

Bosworth W (1985) Geometry of propagating continental rifts. Nature 316:625–627

Bosworth W (1992) Mesozoic and early Tertiary rift tectonics in East Africa. Tectonophys 209:115–137

Bosworth W (1994) A model for the three-dimensional evolution of continental rift basins, north-east Africa. Geol Rundsch 83:671–688

Bosworth W (1995) A high-strain rift model for the southern Gulf of Suez (Egypt). In: Lambiase JJ (ed) Hydrocarbon habitat in rift basins. Geol Soc London Spec Paper, vol 80, pp 75–112

Bosworth W, McClay K (2001) Structural and stratigraphic evolution of the Gulf of Suez rift, Egypt: A synthesis. In: Ziegler PA, Cavazza W, Robertson AHF, Crasquin-Soleau S (eds) Peri-Tethys Memoir 6: Peri-Tethyan Rift/Wrench Basins and passive margins. Mémoires du Muséum national d'Histoire naturelle de Paris, vol 186, pp 567–606

Bosworth W, Stockli DF (2016) Early magmatism in the greater Red Sea rift: timing and significance. Can J Earth Sci 53:1158–1176

Bosworth W, Crerello P, Winn RD Jr, Steinmetz J (1998) Structure, sedimentation, and basin dynamics during rifting of the Gulf of Suez. In: Purser BH, Bosence DWJ (eds) Sedimentation and tectonics of rift basins: Red Sea-Gulf of Aden. Chapman and Hall, London, pp 77–96

Bosworth W, Guiraud R, Kessler LG (1999) Late Cretaceous (±84 Ma) compressive deformation of the stable shelf of NE Africa (Egypt): far-field stress effects of the "Santonian event" and origin of the Syrian arc. Geology 27:633–636

Bosworth W, Huchon P, McClay K (2005) The Red Sea and Gulf of Aden Basins. J Afr Earth Sci 43:334–378

Bosworth W, El-Hawat AS, Helgeson DE, Burke K (2008) Cyrenaican "shock absorber" and associated inversion strain shadow in the collision zone of northeast Africa. Geology 36:695–698

Bosworth W, Khalil S, Clare A, Comisky J, Abdelal H, Reed T, Kokkoros G (2014) Integration of outcrop and subsurface data during the development of a naturally fractured Eocene carbonate reservoir at the East Ras Budran concession, Gulf of Suez, Egypt. In: Spence GH, Redfern J, Aguilera R, Bevan TG, Cosgrove JW, Couples GD, Daniel J-M (eds) Advances in the study of fractured reservoirs, Geol Soc Lond Spec Publ 374, pp 333–359

Bosworth W, Stockli DF, Helgeson DE (2015a) Integrated outcrop, 3D seismic, and geochronologic interpretation of Red Sea dike-related deformation in the Western Desert, Egypt—the role of the 23 Ma Cairo "mini-plume". J Afr Earth Sci 109:107–119

Bosworth W, Drummond M, Abrams M, Thompson M (2015b) Jurassic rift initiation source rock in the Western Desert, Egypt—relevance to exploration in other continental rift systems. In: Post PJ, Coleman JL Jr, Rosen NC, Brown DE, Roberts-Ashby T, Kahn P, Rowan M (eds) Petroleum systems in "rift" basins, 34th annual GCSSEPM foundation perkins-rosen research conference, 13–16 Dec 2015, Houston, CD-ROM, pp 615–650

Bou Daher S, Ducros M, Michel P, Hawie N, Nader FH, Littke R (2016) 3D thermal history and maturity modelling of the Levant Basin and its eastern margin, offshore–onshore Lebanon. Arab J Geosci 9:440

Brew G, Barazangi M, Al-Maleh AK, Sawaf T (2001) Tectonic and geologic evolution of Syria. GeoArabia 6:573–616

Brew G, Litak R, Barazangi M, Sawaf T (1999) Tectonic evolution of Northeast Syria: regional implications and hydrocarbon prospects. GeoArabia 4:289–318

Broughton P, Trepaniér A (1993) Hydrocarbon generation in the Essaouira Basin of Western Morocco. AAPG Bull 77:999–1015

Brown, GF, Schmidt, DL, Huffman Jr, AC (1963) Geology of the Arabian Peninsula, shield area of Western Saudi Arabia. U.S. geological survey professional paper 560-A, pp 1–188

Browne SE, Fairhead JD (1983) Gravity study of the Central African rift system: a model of continental disruption 1. The Nagoundere and Abu Gabra Rifts. Tectonophys 94:187–203

Bunter MAG, Abdel Magid AEM (1989) The Sudanese Red Sea: 1. New developments in stratigraphy and petroleum-geological evolution. J Petrol Geol 12:145–166

Burke K (1996) The African plate. South Afr J Geol 99:341–409

Burke K, MacGregor DS, Cameron NR (2003) Africa's petroleum systems: four tectonic 'Aces' in the past 600 million years. In: Arthur TJ, MacGregor DS, Cameron NR (eds) Petroleum geology of Africa: New themes and developing technologies. Geol Soc Lond Spec Publ 207:21–60

Burollet PF (1960) Léxique stratigraphique international, vol IV, Afrique, part IV, Libye. Cong. Géol. Internat. Commission de Stratig Rech Sci, Paris, 62 p

Burwood R, Redfern J, Cope M (2003) Geochemical evaluation of East Sirte Basin (Libya) petroleum systems and oil provenance. In: Arthur TJ, MacGregor DS, Cameron NR (eds) Petroleum geology of Africa: new themes and developing technologies. Geol Soc Lond Spec Publ 207:203–240

Busrewil MT, Wadsworth WJ (1980) The basanitic volcanoes of the Gharyan area, NW Libya. In: Salem MJ, Busrewil MT (eds) Second symposium on the geology of Libya, vol 3. Academic Press, London, pp 1095–1106

Calon TJ, Aksu AE, Hall J (2005) The Neogene evolution of the Outer Latakia Basin and its extension into the Eastern Mesaoria Basin (Cyprus), Eastern Mediterranean. Mar Geol 221:61–94

Cantrell DL, Nicholson PG, Hughes GW, Miller MA, Buhllar AG, Abdelbagi ST, Norton AK (2014) Tethyan petroleum systems of Saudi Arabia. In: Marlow L, Kendall CCG, Yose LA (eds) Petroleum systems of the Tethyan Region. AAPG Mem, vol 106, pp 613–639

Carr ID (2002) Second-order sequence stratigraphy of the Palaeozoic of North Africa. J Petrol Geol 25:259–280

Carr ID (2003) A sequence stratigraphic synthesis of the North African Mesozoic. J Petrol Geol 26:133–152

Carr ID, Gawthorpe RL, Christopher, Jackson AL, Sharp IR, Sadek A (2003) Sedimentology and sequence stratigraphy of early syn-rift tidal sediments: the nukhul formation, Suez Rift, Egypt. J Sediment Res 73(3):407–420

Catuneanu O, Khalifa MA, Wanas HA (2006) Sequence stratigraphy of the Lower Cenomanian Bahariya Formation, Bahariya Oasis, Western Desert, Egypt. Sedim Geol 190:121–137

Chaimov T, Barazangi M, Al-Saad D, Sawaf T, Gebran A (1992) Mesozoic and Cenozoic deformation inferred from seismic stratigraphy in the southwestern intracontinental Palmyride fold-thrust belt, Syria. GSA Bull 104:704–715

Chaimov T, Barazangi M, Al-Saad D, Sawaf T, Khaddouri T (1993) Seismic fabric and 3-D structure of the southwestern intracontinental Palmyride fold belt, Syria. AAPG Bull 77:2032–2047

Chaouchi R, Mall MS, Kechou F (1998) Sedimentological evolution of the Givetian-Eifelian (F3) sand bar of the West Alrar Field, Illizi Basin, Algeria. In: MacGregor DS, Moody RTJ, Clark-Lowes DD (eds) Petroleum geology of North Africa. Geol Soc Lond Spec Publ 132:187–200

Choquette PW, Craig DH, Cooper DM, Steinmetz JC (1994) Depositional history of a marine to evaporite succesion: Triassic of Central Syria. AAPG 78(13):121

Coleman RG (1993) Geologic evolution of the Red Sea. Oxford monographs on geology and geophysics, vol. 24. Oxford University Press, Oxford, 186 p

Cole GA, Abu-Ali MA, Aoudeh SM, Carrigan MJ, Chen HH, Colling EL, Gwathney WJ, Al-Hajii AA, Halpern HI, Jones PJ, Al-Sharidi SH, Tobey MH (1994a) Organic geochemistry of the Paleozoic petroleum system of Saudi Arabia. Energy Fuels 8:1425–1442

Cole GA, Carrigan HH, Colling EL, Halpern HI, Al-Khadhrawi MR, Jones PJ (1994b) Organic geochemistry of the Jurassic petroleum system in Eastern Saudi Arabia. In: Embry AF, Beauchamp B, Closs DJ (eds) Pangea, global environments and resources. Canadian Soc Petrol Geol Mem, vol 17, pp 413–438

Colletta B, Le Quellec P, Letouzy J, Moretti I (1988) Longitudinal evolution of the Suez rift structure (Egypt). Tectonophysics 153:221–233

Coward MP, Ries AC (2003) Tectonic development of North African basins. In: Arthur TJ, MacGregor DS, Cameron NR (eds) Petroleum geology of Africa: new themes and developing technologies. Geol Soc Lond Spec Publ 207:61–83

Cozzi A, Rea G, Craig J (2012) From global geology to hydrocarbon exploration: Ediacaran–Early Cambrian petroleum plays of India, Pakistan and Oman. In: Bhat, GM, Craig J, Thurow JW, Thusu B, Cozzi A (eds) Geology and hydrocarbon potential of Neoproterozoic–Cambrian Basins in Asia. Geol Soc Lond Spec Publ 366:131–162

Craig DH, Choquette PW, Cooper DM, Steinmetz JC (1994) High-resolution cyclostratigraphy of the Mid Triassic sucession: marine to penesaline lacrustine cycles, Palmyrid trough, Central Syria. AAPG 78(13):128

Craig J, Rizzi C, Said F, Thusu B, Lüning S, Asbali AI, Keely ML, Bell JF, Durham MJ, Eales MH, Beswetherick S, Hamblett C (2008) Structural styles and prospectivity in the Precambrian and Palaeozoic hydrocarbon systems of North Africa. Sedimentary basins of Libya, third symposium. In: Geology of East Libya, vol 4. Earth Science Society of Libya, pp 51–122

Craig J, Grigo D, Rebora A, Serafini G, Tebaldi E (2010) From Neoproterozoic to early Cenozoic; exploring the potential of older and deeper hydrocarbon plays across North Africa and the Middle East. In: Vining BA, Pickering SC (eds) Petroleum geology: from mature basins to new frontiers. Petroleum geology conference proceedings, vol 7, pp 673–705

Cross NE, Cunningham A, Cook RJ, Taha A, Esmaie E, El Swidan N (2009) Three-dimensional seismic geomorphology of a deep-water slope-channel system: the sequoia field, offshore west Nile Delta, Egypt. AAPG Bull 93:1063–1086

Crossley R, Watkins C, Raven M, Cripps D, Carnell A, Williams D (1992) The sedimentary evolution of the Red Sea and Gulf of Aden. J Petrol Geol 15:157–172

Crossley R, McDougall N (1998) Lower Palaeozoic reservoirs of North Africa. In: Macgregor DS, Moody RTJ, Clark-Lowes DD (eds) Petroleum geology of North Africa. Geol Soc Lond Spec Publ 132:157–166

Dalla S, Harby H, Serazzi M (1997) Hydrocarbon exploration in a complex incised valley fill: An example from the late Messinian Abu Madi Formation (Nile Delta Basin, Egypt). Leading Edge 1819–1824

Dardour AM, Boote DRD, Baird AW (2004) Stratigraphic controls on Paleozoic petroleum systems, Ghadames Basin, Libya. J Petrol Geol 27:141–162

Darwish M, Abu Khadrah AM, Abdel Hamid ML, Hamed TA (1994) Sedimentology, environmental conditions and hydrocarbon habitat of the Bahariya Formation, central Abu Gharadig basin, Western Desert, Egypt. In: Proceedings of 12th exploration and production conference, Exploration, vol 1. Egyptian General Petroleum Corporation, Cairo, pp 429–449

Davies RB, Casey DM, Horbury AD, Sharland PR, Simmons MD (2002) Early to mid-Cretaceous mixed carbonate-clastic shelfal systems: examples, issures and models from the Arabian Plate. GeoArabia 7:541–598

De Lamotte DF, Tavakoli-Shirazi S, Letrumy P, Averbuch O, Mouchot N, Raulin C, Leparmentier F, Blanpied C, Ringebach JC (2013) Evidence for Late Devonian vertical movements and extensional deformation in northern Africa and Arabia: integration in the geodynamics of the Devonian world. Tectonics 32:107–122

De Lapparent AF, Lelubre M (1948) Interpretation stratigraphique des series continentals entre Ohanet et Bourarhet (Sahara central). CR Acad Sci Séance du 22 novembre 1948, Paris

Djarnia MR, Fekirine B (1998) Sedimentological and diagenetic controls on Cambro-Ordovician reservoir quality in the southern Hassi Messaoud area (Saharan Platform, Algeria). In: Macgregor DS, Moody RTJ, Clark-Lowes DD (eds) Petroleum Geology of North Africa. Geol Soc Lond Spec Publ 132:167–175

Dolson JC, Shann MV, Matbouly SI, Hammouda H, Rashed RM (2000) Egypt in the twenty-first century: petroleum potential in offshore trends. GeoArabia 6:211–230

Dolson JC, Shann MV, Matbouly SI, Harwood C, Rashed R, Hammouda H (2001) The petroleum potential of Egypt. In: Downey MW, Threet JC, Morgan WA (eds) Petroleum provinces of the twenty-first century. AAPG Mem, vol 74, pp 453–482

Dolson JC, El-Barkooky AE, Wehr F, Gingerich P, Prochazka N, Shann M (2002) The Eocene and Oligocene paleo-geography of Whale Valley and the Fayoum Basins: implications for hydrocarbon exploration in the Nile delta and eco-tourism in the Greater Fayoum Basin. In: Cairo 2002 AAPG/EPEX/SEG/EGS/EAGE Field trip No.7 guide book, pp 1–79

Dolson JC, Atta M, Blanchard D, Sehim A, Villinski J, Loutit T, Romine K (2014) Egypt's future petroleum resources: a revised look into the 21st century. In: Marlow L, Kendall CCG, Yose LA (eds) Petroleum systems of the Tethyan region. AAPG Memoir, vol 106, pp 143–178

Droste H (2014) Petroleum geology of the Sultanate of Oman. In: Marlow L, Kendall C, Yose L (eds) Petroleum systems of the Tethyan region. AAPG Memoir, vol 106, pp 713–755

Echikh K (1998) Geology and hydrocarbon occurrences in the Ghadames Basin, Algeria, Tunisia, Libya. In: MacGregor DS, Moody RTJ, Clark-Lowes DD (eds) Petroleum geology of North Africa. Geol Soc Lond Spec Publ 132:109–129

Echickh K and Sola MA (2000) Geology and hydrocarbon occurrences in the Murzuq Basin, SW Libya. In: Sola MA and Worsley D (eds) Geological exploration in Murzuq Basin. Elsevier, Amsterdam, pp 175–222

Egyptian General Petroleum Corporation (1992) Western desert, oil and gas fields (a comprehensive overview). Al-Ahram Commercial Presses, Kalyoub, p 431

Egyptian General Petroleum Corporation (1996) Gulf of Suez oil fields (a comprehensive overview). Al-Ahram Commercial Presses, Kalyoub, p 736

Ehrenberg SN, Pickard NAH, Laursen GV, Monibi S, Mossadegh ZK, Svana TA, Aqrawi AAM, McArthur JM, Thirlwall MF (2007) Strontium isotope stratigraphy of the Asmari formation (Oligocene-Lower Miocene), SW Iran. J Petrol Geol 30:393–414
Ehrenberg SN, Aqrawi AA, Nadeau PH (2008) An overview of reservoir quality in producing Cretaceous strata of the Middle East. Petrol Geosci 14:107–128
El Ayouty MK (1990) Petroleum geology. In: Said R (ed) The geology of Egypt. Balkema, Rotterdam, pp 567–599
El Barkooky AN (1992) Stratigraphic framework of the Paleozoic in the Gulf of Suez Region, Egypt. In: 1st international conference on geology Arab World, Cairo Univ., Egypt (Abst)
El Diasty WSh, Peters KE (2014) Genetic classification of oil families in the Central and Southeren Gulf of Suez, Egypt. J Petrol Geol 37:105–126
El-Ghali MAK (2005) Depositional environments and sequence stratigraphy of paralic glacial, paraglacial and postglacial upper Ordovician siliciclastic deposits in the Murzuq Basin, SW Libya. Sed Geol 177:145–173
El-Alami M, Rahouma S, Butt AA (1989) Hydrocarbon habitat in the Sirte basin northern Libya. Pet Res Centre Tripoli 1:17–28
El-Hawat AS, Bezan A, Obeidi A, Barghathi H (2003) The upper Ordovician—lower Silurian succession of Western Libya: sequence stratigraphy and glacioeustatic-tectonic scenario. Geol Northwest Libya 2:65–78
El-Khatri F, El-Ghali MAK, Mansurbeg H, Morad S, Ogle N, Kalin RM (2015) Diagenetic alterations and reservoir quality evolution of Lower Cretaceous fluvial sandstones: Nubian Formation, Sirt Basin, North-Central Libya. J Pet Geol 38:217–240
El-Sabbagh A, Tantawy AA, Keller G, Khozyem H, Spangenberg J, Adatte T, Gertsch B (2011) Stratigraphy of the Cenomanian-Turonian Oceanic Anoxic vent OAE2 in shallow shelf sequences of NE Egypt. Cretac Res. https://doi.org/10.1016/j.cretres.2011.04.006
Eltom H, Abdullatif O, Makkawi M, Abdulraziq A (2013) Microporosity in the upper Jurassic Arab-D carbonate reservoir, Central Saudi Arabia: an outcrop analogue study. J Petrol Geol 36:281–297
English KL, Redfern J, Corcoran DV, English JM, Cherif RY (2016) Constraining burial history and petroleum charge in exhumed basins: New insights from the Illizi Basin, Algeria. AAPG Bull 100:623–655
Esestime P, Hewitt A, Hodgson N (2016) Zohr—a newborn carbonate play in the Levantine Basin, East-Mediterranean. First Break 34:87–93
Eshet Y (1990a) The palynostratigraphy of the Permian-Triassic boundary in Israel: two approaches to biostratigraphy. Israel J Earth-Sci 39:1–15
Eshet Y (1990b) Paleozoic-Mesozoic palynology of Israel. I. Palynological aspects of the Permo-Triassic succession in the subsurface of Israel. Geol Surv Israel Bull 81:1–57
Esrafili-Dizaji B, Rahimpour-Bonab H (2013) A review of Permo-Triassic reservoir rocks in the Zagros area, SW Iran: influence of the Qatar-Fars Arch. J Petrol Geol 36:257–279
Evans AL (1988) Neogene tectonic and stratigraphic events in the Gulf of Suez rift area, Egypt. Tectonophys 153:235–247
Evans DS, Bahabri BH, Al-Otaibi AM (1997) Stratigraphic trap in the Permian Unayzah Formation, Central Saudi Arabia. GeoArabia 2:259–278
Fairhead JD (1988) Late Mesozoic rifting in Africa. In: Manspeizer W (ed) Triassic-Jurassic rifting, developments in geotectonics, vol 22. Elsevier, Amsterdam, pp 821–831
Faqira M, Rademakers M, Afifi AM (2009) New insights into the Hercynian Orogeny, and their implications for the Paleozoic Hydrocarbon system in the Arabian Plate. GeoArabia 14:199–228
Farooqui MY, Farhoud K, Mahmoud D, El-Barkooky AN (2012) Petroleum potential of the interpreted Paleozoic geoseismic sequences in the South Diyur Block, Western Desert of Egypt. GeoArabia 17:133–176
Farouk S, Ziko A, Eweda SA, Said AE (2014) Subsurface Miocene sequence stratigraphic framework in the Nile Delta, Egypt. J Afr Earth Sci 91:89–109

Fawzy H, Abdel Aal A (1986) Regional study of Miocene evaporites and Pliocene—recent sediments in the Gulf of Suez. In: Proceedings of 7th exploration seminar, Mar 1984. Egyptian General Petroleum Corporation and Egypt Petroleum Exploration Society, Cairo, pp 49–74

Feinstein S, Aizenshtat Z, Miloslavski I, Gerling P, Slager J, McQuilken J (2002) Genetic characterization of gas shows in the east Mediterranean offshore of southwestern Israel. Organic Geochem 33:1401–1413

Fello NM, Turner BR (2004) Depositional environments of the upper Ordovician Mamuniyat formation, NW Murzuq Basin, Libya. In: Proceedings of 3rd International Symposium on Geophysics, Tanta, pp 166–182

Fello N, Lüning S, Štorch P, Redfern J (2006) Identification of early Llandovery (Silurian) anoxic palaeo-depressions at the western margin of the Murzuq Basin (southwest Libya), based on gamma-ray spectrometry in surface exposures. GeoArabia 11:101–118

Fikirine B, Abdallah H (1998) Paleozoic lithofacies correlatives and sequence stratigraphy of the Saharan Platform, Algeria. In: MacGregor DS, Moody RTJ, Clark-Lowes DD (eds) Petroleum geology of North Africa. Geol Soc Lond Spec Publ 132:97–108

Flinn D, Lagha S, Atherton MP, Cliff RA (1991) The rock-forming minerals of the Jabal Arknu and Jabal al Awaynat alkaline rign complexes of SE Libya. In: Salem MJ, Busrewil MT, Ben Ashour AM (eds) Third symposium on the geology of Libya, vol 7. Elsevier, Amsterdam, pp 2539–2558

Fox JE, Ahlbrandt TS (2002) Petroleum geology and total petroleum systems of the Widyan basin and interior platform of Saudi Arabia and Iraq. US Geol Surv Bull 2202–E, 26 p

Fröhlich S, Redfern J, Petitpierre L, Marshall JD, Power M, Grech PV (2010) Diagenetic evolution of incised channel sandstones: implications for reservoir characterisation of the Lower Carboniferous Marar formation, Ghadames Basin, western Libya. J Petrol Geol 33:3–18

Galil A, Hewaidy A, Farouk S, Ayyad HM (2012) Nukhul Formation in Wadi Baba, southwest Sinai Peninsula, Egypt. GeoArabia 17:103–120

Garea BB (1996) Environment of deposition and diagenesis of the Beda reservoir in Block NC 74F, SW Sirt Basin, Libya. In: Salem MJ, Mouzughi AJ, Hammuda OS (eds) Geology of the Sirt Basin, vol 1. Elsevier, Amsterdam, pp 115–133

Garea BB, Braithwaite CJR (1996) Geochemistry, isotopic composition and origin of the Beda dolomites, Block NC74F, SW Sirt Basin, Libya. J Petrol Geol 19:289–304

Gardosh M, Druckman Y, Buchbinder B, Rybakov M (2006) The Levant Basin offshore Israel: stratigraphy, structure, tectonic evolution and implications for hydrocarbon exploration. Geol Surv Israel, Report GSI/14/2006, 119 p

Gardosh M, Druckman Y, Buchbinder B, Rybakov M (2008a) The Levant Basin offshore Israel: stratigraphy, structure, tectonic evolution and implications for hydrocarbon exploration. Geol Surv Israel, Report GSI/4/2008, 118 p

Gardosh M, Druckman Y, Buchbinder B, Calvo R (2008b) The Oligo-Miocene deepwater system of the Levant Basin. Geol Surv Israel, Report GSI/33/2008, 73 p

Gardosh M, Druckman Y, Buchbinder B (2009) The late tertiary deep-water siliciclastic system of the levant margin—an emerging play offshore Israel. Search and Discovery Article #10211

Gardosh MA, Tannenbaum E (2014) The petroleum systems of Israel. In: Marlow L, Kendall CCG, Yose LA (eds) Petroleum systems of the Tethyan Region. AAPG Mem, vol 106, pp 179–216

Garfunkel Z (2004) Origin of the Eastern Mediterranean basin: a reevaluation. Tectonophys 391:11–34

Garfunkel Z, Bartov Y (1977) The tectonics of the Suez rift. Geol Surv Israel Bull 71:44

Gawthrope RL, Hurst JM, Sladen CP (1990) Evolution of miocene footwall-derived coarse—grained deltas, Gulf of Suez, Egypt, implications for exploration. AAPG Bull 74:1077–1086

Gerdes KD, Winefield P, Simmons MD, Van Oosterhout C (2010) The influence of basin architecture and eustacy on the evolution of Tethyan Mesozoic and Cenozoic carbonate sequences. In: Van Buchem FSP, Gerdes KD, Esteben M (eds) Mesozoic and cenozoic carbonate systems of the mediterranean and the middle east: stratigraphic and diagenetic reference models. Geol Spec Publ, vol 329, pp 9–41

Ghazi JM and Moazzen M (2015) Geodynamic evolution of the Sanandaj-Sirjan Zone, Zagros Orogen, Iran. Turkish J Earth Sci 24:513–528

Ghienne J-F, Deynoux M, Manatschal G, Rubino J-L (2003) Palaeovalleys and fault-controlled depocentres in the Late-Ordovician glacial record of the Murzuq Basin (central Libya). CR Geosci 335:1091–1100

Ghienne J-F, Le Heron DP, Moreau J, Denis M, Deynoux M (2007) The Late Ordovician glacial sedimentary system of the North Gondwana platform. In: Hambrey MJ, Christoffersen P, Glasser NF, Hubbard B (eds) Glacial sedimentary processes and products. Spec Publ Internat Assoc Sedimentologists, vol 29, pp 295–319

Ghorab MA (1961) Abnormal stratigraphic features in Ras Gharib oil field. In: 3rd Arab Petroleum Congress, Alexandria, 10 p

Ghori KAR, Craig J, Thusu B, Lüning S, Geiger M (2009) Global infracambrian petroleum systems: a review. In: Craig J, Thurow J, Thusu B, Whitham A, Abutarruma Y (eds) Global neoproterozoic petroleum systems: the emerging potential in North Africa. Geol Soc Lond Spec Publ 326:109–136

Grabowski GJ Jr (2014) Iraq. In: Marlow L, Kendall CCG, Yose LA (eds), Petroleum Systems of the Tethyan Region, AAPG Mem 106, pp 379–467

Gradstein FM, Ogg JG, Schmitz M, Ogg G (2012) The geologic time scale 2012, vols 1 and 2. Elsevier Science, Amsterdam, 1175 p

Granado P, Urgeles R, Sàbat F, Albert-Villanueva E, Roca E, Muñoz JA, Mazzuca N, Gambini R (2016) Geodynamical framework and hydrocarbon plays of a salt giant: the NW Mediterranean Basin. Petrol Geosci 22:309–321

Gras R (1998) Statistical analysis of syn-rift sediments: an example from the Sarir Sandstone, Messlah Field, Sirte Basin, Libya. J Petrol Geol 21:329–342

Gras R, Thusu B (1998) Trap architecture of the early cretaceous sarir sandstone in the eastern Sirt Basin, Libya. In: MacGregor DS, Moody RTJ, Clark-Lowes DD (eds) Petroleum geology of North Africa. Geol Soc Lond Spec Publ 132:317–334

Guiraud R (1998) Mesozoic rifting and basin inversion along the northern African Tethyan margin: an overview. In: Macgregor D., Moody RTJ, Clark-Lowes DD (eds) Petroleum geology of North Africa. Geol Soc Lond Spec Publ 132:217–229

Guiraud R, Bosworth W (1997) Senonian basin inversion and rejuvenation of rifting in Africa and Arabia: Synthesis and implications to plate-scale tectonics. Tectonophysics 282:39–82

Guiraud R, Bosworth W (1999) Phanerozoic geodynamic evolution of northeastern Africa and the northwestern Arabian platform. Tectonophysics 315:73–108

Guiraud R, Maurin JC (1992) Early cretaceous rifts of Western and Central Africa. An overview. Tectonophysics 213:153–168

Guiraud R, Issawi B, Bellion Y (1985) Les linéaments guinéo-nubiens: un trait structural majeur à l'échelle de la plaque africaine. CR Acad Sci Paris 300:17–20

Guiraud R, Bellion Y, Benkhelil J, Moreau C (1987) Post-Hercynian tectonics in Northern and Western Africa. In: Bowden P, Kinnaird JA (Eds) African geology reviews. Geological J 22:433–466

Guiraud R, Issawi B, Bosworth W (2001) Phanerozoic history of Egypt and surrounding areas. In: Ziegler PA, Cavazza W, Robertson AHF, Crasquin-Soleau S (eds) Peri-Tethys Memoir 6: Peri-Tethyan Rift/Wrench Basins and Passive Margins. Mémoires du Muséum national d'Histoire naturelle de Paris, vol 186, pp 469–509

Guiraud R, Bosworth W, Thierry J, Delplanque A (2005) Phanerozoic geological evolution of Northern and Central Africa: an overview. In: Catuneanu O, Guiraud R, Eriksson P, Thomas B, Shone R, Key R (eds) Phanerozoic evolution of Africa. J Afr Earth Sci 43:83–143

Gumati YD (1992) Lithostratigraphy of oil-bearing Tertiary bioherms in the Sirte Basin, Libya. J Petrol Geol 15:305–318

Gvirtzman Z, Steinberg J (2011) Inland jump of the Arabian northwest plate boundary from the Levant continental margin to the Dead Sea Transform. Geol Surv Israel, Report GSI/29/2011, pp 1–19

Hall PB, Bjoroy M, Ferriday IL, Ismail Y (2010) Libyan Murzuq basin source rocks. AAPG Search and Discovery Article #10271

Hallett D (2002) Petroleum geology of Libya. Elsevier Science BV, Amsterdam, p 503

Hallett D, Clark-Lowes D (2016) Petroleum geology of Libya. Elsevier BV, Amsterdam, p 392

Hanafy S, Nimmagadda SL, Mahmoud SE, Mabrouk WM, Farhood K (2016) Regional integrated interpretation of the hydrocarbon prospectivity of the Nile Delta, Offshore Egypt. Arab J Geosci 9:376, 18 p

Hantar G (1990) North western desert. In: Said R (ed) The geology of Egypt (Chap. 15). AA Balkema, Rotterdam, pp 293–319

Haq BU, Al-Qahtani AM (2005) Phanerozoic cycles of sea-level change on the Arabian Platform. GeoArabia 10:127–160

Hassan AA (1967) A new carboniferous occurrence in Abu Durba, Sinai, Egypt. In: 6th Arab Petroleum Congress, Baghdad, Paper 39 (B-3), 8 p

Hassan F, El-Dashlouty S (1970) Miocene evaporites of Gulf of Suez region and their significance. AAPG Bull 54:1686–1696

Hassan HS, Kendal CCG (2014) Hydrocarbon provinces of Libya: a petroleum system study. In: Marlow L, Kendall C, Yose L (eds) Petroleum systems of the Tethyan region. AAPG Mem, vol 106, pp 101–141

Hegazy A, Tammam TM (1990) Hydrocarbon source rock and maturity trend in the N.W. Desert, Egypt. In: Tenth Egyptian exploration and production conference, pp 138–167

Hempton MR (1987) Constraints on Arabian plate motion and extensional history of the Red Sea. Tectonics 6:687–705

Hessami K, Koyi HA, Talbot CJ (2001) The significance of strike-slip faulting in the basement of the Zagros fold and thrust belt. J Petrol Geol 24:5–28

Hesendel RGW, Cubitt GM, Borman P, Madi FM (1996) Lithofacies study of the Lidam and Maragh formations (late cretaceous) of the Masrab field and adjacent areas, Sirt Basin, Libya. In: Salem MJ, El-Hawat AS, Sbeta AM (eds) The geology of Sirt Basin II. Elsevier, Amsterdam, pp 191–110

Hips K, Árgyelán (2007) Controls on diagenesis of the Triassic Kurrachine Dolomite Syria. GeoArabia 12:41–64

Hirst JPP, Benbakir A, Payne DF, Westlake IR (2002) Tunnel valleys and density flow processes in the upper Ordovician glacial succession, Illizi basin, Algeria: influence on reservoir quality. J Petrol Geol 25:297–324

Hodgson N (2012) Play types and exploration potential offshore Lebanon. Lebanese international petroleum exploration forum, July 2012

Homberg C, Bachmann M (2010) Evolution of the Levant margin and western Arabia platform since the Mesozoic: introduction. In: Homberg C, Bachmann M (eds) Evolution of the levant margin and Western Arabia platform since the Mesozoic. Geol Soc Lond Spec Publ 341:1–8

Hughes GW, Beydoun ZR (1992) The Red Sea-Gulf of Aden: biostratigraphy, lithostratigraphy and palaeoenvironments. J Petrol Geol 15:135–156

Hughes GW, Varol O, Beydoun ZR (1991) Evidence for Middle Oligocene rifting of the Gulf of Aden and for Late Oligocene rifting of the southern Red Sea. Mar Petrol Geol 8:354–358

Jackson JA, Fitch TJ, McKenzie DP (1981) Active thrusting and the evolution of the Zagros fold belt. In: McClay KR, Price NJ (eds) Thrust and Nappe Tectonics. Geol Soc Lond Spec Publ 9:371–379

Jassim SZ, Goff JC (2006) Phanerozoic development of the northern Arabian Plate. In: Jassim SZ, Goff JC (eds) Geology of Iraq. Dolin, Prague and Moravian Museum, Brno, pp 32–44

Johnson CA (2008) Phanerozoic plate reconstructions of the Middle East: insights into the context of Arabian tectonics and sedimentation. In: SPE paper 118062, International petroleum exhibition and conference, Abu Dhabi, UAE, 3–6 Nov 2008, 16 p

Johnson PR, Kattan FH (2012) The geology of the Saudi Arabian Shield. Saudi Geological Survey, Jiddah, p 479

Keeley ML (1989) The Palaeozoic history of the Western Desert of Egypt. Basin Res 2:35–48

Keeley ML (1994) Phanerozoic evolution of the basins of Northern Egypt and adjacent areas. Geol Rundsch 83:728–742

Keeley ML, Massoud MS (1998) Tectonic controls on the petroleum gelogy of NE Africa. In: Macgregor DS, Moody RTJ, Clark-Lowes DD (eds) Petroleum geology of North Africa. Geol Soc Spec Publ 132:265–281

Keeley ML, Wallis RJ (1991) The Jurassic system in Northern Egypt: II. Depositional and tectonic regimes. J Pet Geol 14:49–64

Keeley M, Dungworth G, Floyd CS, Forbes GA, King C, Garva RM, Shaw D (1990) The Jurassic system in Northern Egypt: I. Regional stratigraphy and implications for hydrocarbon prospectivity. J Pet Geol 13:397–420

Keller M, Abu Amarah B, Al Ajmi HF (2017) The Wajid Sandstone around Aha/Khamis Mushayt and in its type area: lithostratigraphic architecture and correlations. Arab J Geosci 5:17 p

Kellner A, El-Khawaga H, Brink G, Brink-larsen S, Hesham M, Abu El-Saad H, Atef A, Young H, Finlayson B (2009) Depositional history of the West Nile Delta—upper Oligocene to Upper Pliocene. AAPG Search and Discovery Article #30092

Kendall CCG, Alsharhan AS, Marlow L (2014) Stratigraphy and depositional systems of the southern Tethyan region. In: Marlow L, Kendall CCG, Yose LA (eds) Petroleum systems of the Tethyan region. AAPG Memoir, vol 106, pp 29–57

Keshta S, Metwalli FI, Al-Arabi HS (2014) Analysis of petroleum system for exploration and risk reduction in Abu Madi/El Qar'a gas field, Nile Delta, Egypt. AAPG Search and Discovery Article #20253

Khalifa MA (2007) Triassic-Jurassic boundary of the southern margin of Tethys: implications of facies, tectonics and volcanism. In: Lucas SG, Spielmann JA (eds) The global triassic, New Mexico museum of natural history and science, Bulletin 41, pp 110–119

Khan Tanoli S, Husain R, Sajer AA (2008) Facies in the Unayzah formation and basal Khuff clastics in subsurface, northern Kuwait. GeoArabia 13:15–40

Klemme HD, Ulmishek GF (1991) Effective petroleum source rocks of the World: stratigraphic distribution and controlling depositional factors. AAPG Bull 75:1809–1851

Klett TR (2000a) Total petroleum systems of the Grand Erg/Ahnet Province, Algeria and Morocco—the Tanezzuft-Timimoun, Tanezzuft-Ahnet, Tanezzuft-Sbaa, Tanezzuft-Mouydir, Tanezzuft-Benoud, and Tanezzuft-Béchar/Abadla. US Geol Surv Bull 2202-B, 27 p

Klett TR (2000b) Total petroleum systems of the Illizi province, Algeria and Libya—Tanezzurt-Illizi. US Geol Surv Bull 2202-A, 15 p

Klett TR (2001) Total petroleum systems of the pelagian province, Tunisia, Libya, Italy, and Malta —The Bou Dabbous–Tertiary and Jurassic-Cretaceous Composite. US Geol Surv Bull 2202-D

Klitzsch EH, Squyres CH (1990) Paleozoic and Mesozoic geological history of northeastern Africa based upon new interpretation of Nubian strata. AAPG Bull 74:1203–1211

Konert G, Afifi AM, Al-Hajri SA, de Groot K, Al Naim AA, Droste HJ (2001) Paleozoic stratigraphy and hydrocarbon habitat of the Arabian Plate. In: Downey MW, Threet JC, Morgan WA (eds) Petroleum provinces of the twenty-first century. AAPG Memoir, vol 74, pp 483–515

Kora M (1998) The Permo-Carboniferous outcrops of the Gulf of Suez region, Egypt: stratigraphic classification and correlation. In: Crasquin-Soleau S, Izart A, Valset D, De Wever P (eds) Peri-Tethys: stratigraphic correlations 2, Geodiversitas, vol 20, pp 701–721

Krenkel E (1925) Geologie der Erde, Geologie Afrikas, vol I. Gebrüder Borntraeger, Berlin, p 461

Krois P, Adouani F, Pelz K, Cantarella G (2013) Oil and gas exploration in the Ghadames Basin of Southern Tunisia. In: 12th PESGB/HGS conference on African exploration and production, London, Extended abstract

Kröner A, Stern RJ (2004) Pan-African orogeny. Encyclopedia Geol 1. Elsevier, Amsterdam, 12 p

Le Heron DP, Craig J (2008) First order reconstructions of a Late Ordovician Saharan ice sheet. J Geol Soc London 165:19–29

Le Heron D, Sutcliffe O, Bourgig K, Craig J, Visentin C, Whittington R (2004) Sedimentary architecture of Upper Ordovician tunnel valleys, Gargaf Arch, Libya: implications for the genesis of a hydrocarbon reservoir. GeoArabia 9:137–160

Le Heron DP, Craig J, Etienne JL (2009) Ancient glaciations and hydrocarbon accumulations in North Africa and the Middle East. Earth-Sci Rev 93:47–76

Le Heron DP, Armstrong HA, Wilson C, Howard JP, Gindre L (2010) Glaciation and deglaciation of the Libyan Desert: the Late Ordovician record. Sed Geol 223:100–125

Leila M, Moscariello A (2017) Organic geochemistry of oil and natural gas in the west Dikirnis and El-Tamad fields, onshore Nile Delta, Egypt: interpretation of potential source rocks. J Petrol Geol 40:37–58

Lelubre M (1948) La Paléozoique du Fezzan sudoriental. Compte Rendus Société Géologique de France 18:79–81

Lelubre M (1952) Aperçu sur la géologie du Fezzan. Bull Serv Carte Géol Algérie 3:109–148

Leško J, Jasnić L, Krešić A, Tomašić D, Španić D, Troskot-Čorbić T (2005) Basin modelling and hydrocarbon potential of Northwestern part of the Hayan block, Syria. In: Proceedings 21/06, 3rd International Oil and Gas Conference, Zadar, Naftaplin

Lewis CJ (1990) Sarir field. AAPG treatise structural traps II: traps associated with tectonic faulting, pp 253–267

Litak RK, Barazangi M, Beauchamp W, Seber D, Brew G, Sawaf T, Al-Yousef W (1997) Mesozoic-Cenozoic evolution of the intraplate Euphrates Fault System, Syria: implications for regional tectonics. J Geol Soc London 154:653–666

Lindquist SJ (1998) The Red Sea basin province: Sudr-Nubia(!) and Maqna(!) petroleum systems. US Geol Surv Open-File Report 99-50-A, 21 p

Lottaroli F, Craig J, Thusu B (2009) Neoproterozoic–early Cambrian (Infracambrian) hydrocarbon prospectivity of North Africa: a synthesis. In: Craig J, Thurow J, Thusu B, Whitham A, Abutarruma Y (eds) Global Neoproterozoic petroleum systems: the emerging potential in North Africa. Geol Soc Lond Spec Publ 326:137–156

Loucks RG, Moody RTJ, Bellis JK, Brown AA (1998) Regional depositional setting and pore network systems of the El Garia Formation (Metlaoui Group, Lower Eocene), offshore Tunisia. In: Macgregor DS, Moody RTJ, Clark-Lowes DD (eds) Petroleum geology of North Africa. Geol Soc Spec Publ 132:355–374

Lovelock PER (1984) A review of the tectonics of the northern Middle East region. Geol Mag 121:577–587

Loydell DK, Butcher A, Fryda J, Lüning S, Fowler M (2009) Lower Silurian "hot shales" in Jordan: a new depositional model. J Petrol Geol 32:261–270

Lučić D (2008) Litostratigrafija karbonskih i trijaskih naftonosnika u području Hayan bloka, Palmiridi, Sirija. Lithostratigraphy of the Carboniferous and Triassic Reservoirs, Hayan Block, Palmyrides, Syria). PhD thesis, University of Zagreb, Croatian with English abstract, 197+ xxvii pp

Lučić D, Maretić S, Mesić I, Feher L (2001) Markada pješčenjaci – interpretacija sredine taloženja (Palmiridi, Sirija). Markada sandstones—depositional environment interpretation (Palmyra region, Syria). Naftaplin, izvanredni broj 3, International Oil and Gas Conference Zadar, Croatia, 2–5 Oct 2001, pp 321–333

Lučić D, Maretić S, Foršek G, Takač D, Mesić I (2002) Mesozoic carbonate facies and reservoirs in central palmyrides, Syria. poster and abstract AAPG Conf. Cairo

Lučić D, Foršek G, Ivković Ž, Maretić S, Marić-Dureković Ž, Balaz-Boromisa E, Bubnić J, Koch G, Velić J (2003a) Depositional environment and reservoir properties of Kurrachine Dolomite and Markada Formation, in central palmyrides, Syria. In: Vlahović I (ed) Abstract book, 22nd IAS Meeting of Sedimentology Opatija, Croatia, p 110

Lučić D, Koch G, Foršek G, Balaž-Boromisa E, Krizmanić K, Španić D, Mesić I, Maretić S (2003b) SW Tethyan triassic evaporite facies; Paleogeographical correlation of deep-wells from off-shore Croatia and on-shore central Syria. AAPG Int. Conf. Barcelona Abstract, AAPG Bull. 87

Lučić D, Ayed H, Ivkovic Z, Takac D, Forsek G, Bubnić J (2007) The Carboniferous sedimentary succession, lithofacies and hydrocarbon potential in Central Palmyrides. In: AAPG European Region, energy conference and exhibition, Athens, Greece (poster and abstract)

Lučić D, Ivković Z, Foršek G, Takač D, Bubnić J, Koch G (2010) Depositional sequences and palynology of Triassic carbonate-evaporite platform deposits in the Palmyrides, Syria. In: Van Buchem FSP, Gerdes KD, Esteban M (eds) Mesozoic and Cenozoic carbonate systems of the mediterranean and the middle east: Stratigraphic and diagenetic reference models. Geol Soc London Spec Publ 329:43–63

Lüning S, Fello N (2006) Silurian 'Hot Shales' in the Murzuq and Al Kufrah Basins (S Libya): improved predictability of source rock distribution based on gamma-ray spectrometry in surface exposures. Geology of East Libya

Lüning S, Kuss J (2014) Petroleum geology of Jordan. In: Marlow L, Kendall CCG, Yose LA (eds) Petroleum systems of the Tethyan region. AAPG Memoir, vol 106, pp 217–239

Lüning S, Loydell DK, Sutcliffe O, Ait Salem A, Zanella E, Craig J, Harper DAT (2000) Silurian—lower Devonian black shales in Morocco: which are the organically richest horizons? J Petrol Geol 23:293–311

Lüning S, Adamson K, Craig J (2003) Frasnian organic-rich shales in North Africa: regional distribution and depositional model. In: Arthur TJ, MacGregor DS, Cameron N R (eds) Petroleum geology of Africa: new themes and developing technologies. Geol Soc London Spec Publ 207:165–184

Lüning S, Shahin YM, Loydell D, Al-Rabi HT, Masri A, Tarawneh B, Kolonic S (2005) Anatomy of a world-class source rock: distribution and depositional model of Silurian organic-rich shales in Jordan and implications for hydrocarbon potential. AAPG Bulletin 89:1397–1427

Lüning S, Schulze F, Marzouk AM, Kuss J, Gharaibeg A, Kolonic S (2004) Uranium-enriched horizons refine the stratigraphic framework of Cenomanian-Turonian (Late Cretaceous) strata in central Jordan and northern Tunisia. Z dt Geol Ges 155:49–60

Lüning S, Kolonic S, Geiger M, Thsu B, Bell JS, Craig J (2009) Infracambrian hydrocarbon source rock potential and petroleum prospectivity of NW Africa. Geol Soc Lond Spec Publ 326:157–180

Macaulay CI, Beckett D, Braithwaite K, Bliefnick D, Philps B (2001) Constraints on diagenesis and reservoir quality in the fractured Hasdrubal field, offshore Tunisia. J Petrol Geol 24:55–78

Macgregor D (2011) Rift shoulder source to prodelta sink: the Cenozoic development of the Nile drainage system. AAPG Search and Discovery Article #50506, 16 p

Macgregor DS (1996) The hydrocarbon systems of North Africa. Mar Petrol Geol 13:329–340

Macgregor DS (1998) Giant fields, petroleum systems and exploration maturity of Algeria. In: Macgregor DS, Moody RTJ, Clark-Lowes DD (eds) Petroleum geology of North Africa. Geological Society, London, Special Publication 132:79–96

Macgregor DS (2012) The development of the Nile drainage system: integration of onshore and offshore evidence. Petrol Geosci 18:417–431

Macgregor DS, Moody RTJ (1998) Mesozoic and Cenozoic petroleum systems of North Africa. In: MacGregor DS, Moody RTJ, Clark-Lowes DD (eds) Petroleum geology of North Africa. Geol Soc London Spec Publ 132:201–216

Macgregor DS, Moody RTJ, Clark-Lowes DD (1998) Petroleum geology of North Africa. Geol Soc Spec Publ 132:1–443

Magoon LB (1988) The petroleum system—a classification scheme for research, exploration and resource assessment. In: Magoon LB (ed) Petroleum systems of the United States. US Geol Surv Bull 1870, 15 p

Magoon LB, Dow WG (eds) (1994) The petroleum system—from source to trap. American AAPG Memoir, vol 60, 655 p

Makhous M (2001) The formation of hydrocarbon deposits in the North African basins; geological and geochemical conditions. Lecture Notes in Earth Sci 89, 341 p

Maky ABF, El Sayed SM (2007) Source rock characteristics of some upper Cretaceous rock units, Abu Gharadig basin, north Western Desert, Egypt, and relevance to petroleum generation. Sedimentol Egypt 15:121–138

Mann P, Gahan L, Gordon MB (2003) Tectonic setting of the World's giant oil and gas fields. In: Halbouty MT (ed) Giant oil and gas fields of the decade 1990–1999. AAPG Mem, vol 78, pp 15–105

Marlow L, Kornpihl K, Kendall CGStC (2011) 2-D basin modeling study of petroleum systems in the Levantine Basin, Eastern Mediterranean. GeoArabia 16:17–42

Marlow L, Kendall CCG, Yose LA (eds) (2014) Petroleum systems of the Tethyan region. AAPG Mem 106, pp 1–780

Matresu J, Nassar M, El Belasy A, Talaat A (2014) New insights on Miocene exploration potential in Nile Delta-Egypt. In: 6th Mediterranean offshore conference and exhibition (MOC), Alexandria, 18 p

Martin-Monge AM, Baudino R, Garifo-Ferreira LM, Tocco R, Badali M, Ochoa M, Haryono S, Soriano S, El Hafizi N, Hernan-Gomez J, Chaco B, Brissoni I, Grammatico G, Varade R, Abdallah H (2016) An unusual proterozoic petroleum play in Western Africa: the Atar Group carbonates (Taoudeni Basin, Mauritania). In: Sabato Ceraldi T, Hodgkinson RA, Backe G (eds) Petroleum geoscience of the West Africa Margin. Geological Society, London, Special Publications, p 438

McBride JH, Barazangi M, Best J, Al-Saad D, Sawaf T, Al-Otri M, Gebran A (1990) Seismic reflection structure of intracratonic Palmyride fold-thrust belt and surrounding Arabian Platform, Syria. AAPG Bull 74:238–259

McGillivray JG, Husseini MI (1992) The Paleozoic petroleum geology of central Arabia. AAPG Bull 76:1473–1490

McKenzie DP, Davies D, Molnar P (1970) Plate tectonics of the Red Sea and east Africa. Nature 226:243–248

McQuillan H (1973) Small-scale fracture density in asmari formation of southwest Iran and its relation to bed thickness and structural setting. AAPG Bulletin 57:2367–2385

McQuillan H (1974) Fracture patterns on Kuh-e Asmari anticline, Southwest Iran. AAPG Bulletin 58:236–246

Miller PM, Barakat H (1988) Geology of the Safaga Concession, northern Red Sea, Egypt. Tectonophysics 153:123–136

Mohamed AB, Saidi M, Jrad MB, Soussi M, Riahi S, Bouazizi I, Saadani B (2015) The organic geochemistry of source rocks, bitumens, asphalts and oil stains from Northern Tunisia. In: 77th EAGE conference and Exhibition, IFEMA Madrid, Spain, extended abstract

Mohamed AK (2016) Reservoir quality of Hawaz formation. J oil field, Concession NC186, NW Murzuq basin, SW Libya. Arab J Geosci 9, 17 p

Mohammed K, Corbett PWM, Bowen D, Gardiner AR, Buckman J (2002) Solution seams in the Mamuniyat formation, El-Sharara—a field, SW Libya: impact on reservoir preformance. J Petrol Geol 25:281–296

Mohamed NS, Shahin AN, El Kammar AM (2013) Hydrocarbon generating basins and migration pathways in the Gulf of Suez, Egypt. Life Sci J 10:229–235

Mohialdeen IMJ, Hakimi MH, Al-Beyati FM (2013) Biomarker characteristics of certain crude oils and the oil-source rock correlation for the Kurdistan oilfields. Northern Iraq. Arab J Geosci

Mohsenian E, Fathi-Mobarakabad A, Sachsenhofer RF, Asadi-Eskandar A (2014) 3D basin modelling in the Central Persian Gulf, offshore Iran. J Petrol Geol 37:55–70

Mokhtar N, Saad M, Selim S (2016) Reservoir architecture of deep marine slope channel, Scarab field, offshore Nile Delta, Egypt: application of reservoir characterization. Egypt J Petrol

Montadert L, Nicolaides S, Semb PH, Lie O (2014) Petroleum systems offshore cyprus. In: Marlow L, Kendall C, Yose L (eds) Petroleum systems of the Tethyan region. AAPG Mem, vol 106, pp 301–334

Montenat C, Ott D'estevou P, Purser B, Burollet P, Jarrige J, Sperber F, Philobbos E, Plaziat J-C, Prat P, Richert J, Roussel N, Theiriet J (1988) Tectonic and sedimentary evolution of the Gulf of Suez and the northern western Red Sea. Tectonophys 153:166–177

Moor JM (1979) Tectonics of the Najd transcurrent fault system, Saudi Arabia. J Geol Soc Lond 136:441–454

Morad S, Al-Aasm IS, Nader FH, Ceriani A, Gasparrini M, Mansburg H (2012) Impact of diagenesis on the spatial and temporal distribution of reservoir quality in the Jurassic Arab D and C members, offshore Abu Dhabi oilfield, United Arab Emirates. GeoArabia 17:17–56

Moradpour M, Zamani Z, Moallemi SA (2008) Controls on reservoir quality in the Lower Triassic Kangan formation Southern Persian Gulf. J Petrol Geol 31:367–386

Moreau J (2011) The late Ordovician deglaciation sequence of the SW Murzuq Basin (Libya). Basin Res 23:449–477

Moreau J, Degermann L, Ghienne JF, Rubino JL (2007) Large-scale physiography of the Murzuq basin shelf during hirnantian ice-sheet final retreat and silurian transgression. Paper A27, 3rd North African/Mediterranean petroleum and geosciences conference and exhibition Tripoli, Libya, 26–28 Feb 2007, 5 p

Moretti I, Colletta B (1987) Spatial and temporal evolution of the Suez Rift subsidence. J Geodyn 7:151–168

Moustafa AM (1976) Block faulting of the Gulf of Suez. Presented at the 5th Egyptian general petroleum corporation exploration Seminar, Cairo, unpublished report, 19 pp

Moustafa AR (2008) Mesozoic-Cenozoic basin evolution in the northern Western Desert of Egypt. In: Salem M, El-Arnauti A, Saleh, A (eds) Geology of East Libya, vol 3, pp 29–46

Moustafa AR, Khalil MH (1989) North Sinai structures and tectonic evolution. Middle East Res Center, Earth Sci Ser 3:215–231

Moustafa AR, Khalil SM (2016) Control of extensional transfer zones on syntectonic and post-tectonic sedimentation: implication for hydrocarbon exploration. J Geol Soc 174:318–335

Mriheel IY, Anketell (2000) Dolomitization of the Early Eocene Jirani dolomite formation, Gabes-Tripoli Basin, western offshore, Libya. J Petrol Geol 23:449–474

Munn C, Wild L, Cowper D (2013) Unlocking the deep HPHT Oligocene fairway in the Nile Delta and the 20K technology promise. In: AAPG Conference, Barcelona 2013, Abstract

Mustafa A, Sahin A, Abouelresh MO, Hariri M (2014) Geological characteristics of the Lower Silurian Qusaiba Shale, Rub Al-Khali Basin, Saudi Arabia. Paper P10, 2nd EAGE/SPE/AAPG Shale Gas Workshop 21–24 September 2014, Dubai, UAE, 5 p

Nader FH, Swennen R (2004) The hydrocarbon potential of Lebanon: new insights from regional correlations and studies of Jurassic dolomitization. J Petrol Geol 27:253–275

Nader FH (2011) The petroleum prospectivity of Lebanon. J Petrol Geol 34:135–156

Nader FH (2014) Insights into teh petroleum prospectivity of Lebanon. In: Marlow L, Kendall CCG, Yose LA (eds) Petroleum systems of the Tethyan region. AAPG Memoir, vol 106, pp 241–278

Norton P (1967) Rock stratigraphic nomenclature of the Western Desert, Egypt. Gulf of Suez Petroleum Company internal report, 18 p

Patton TL, Moustafa AR, Nelson RA, Abdine SA (1994) Tectonic evolution and structural setting of the Suez Rift. In: Landon SM (ed) Interior rift basins, AAPG Memoir 59, pp. 7–55

Pawellek T (2009) From thin sections to discovery: the upper Satal in the central Sirte Basin (north-central Libya)—a success story. Geotectonic Res 96:85–98

Pejs JAMM, Bevan TG, Piombino TG (2012) The Gulf of Suez rift basin. Phanerozoic rift systems and sedimentary basins. Elsevier BV, Amsterdam, pp 165–194

Perry SK, Schamel S (1990) The role of low-angle normal faulting and isostatic response in the evolution of the Suez rift, Egypt. Tectonophys 174:159–173

Philip J, Babinot JF, Tronchetti G, Fourcade E, Guiraud R, Bellion Y, Herbin JP, Combes PJ, Cornée JJ, Dercourt J, Ricou LE (1993) Late Cenomanian (94–92 Ma). In: Dercourt J, Ricou LE, Vrielynck B (eds) Tethys palaeoenvironmental maps, explanatory notes. Gauthier-Villars, Paris, pp 153–178

Peyravi M, Kamali MR, Kalani M (2010) Depositional environments and sequence stratigraphy of the early Triassic Kangan formation in the northern part of the Persian Gulf: implications for reservoir characteristics. J Petrol Geol 33:371–386

Pivnik D, Ramzy M, Steer BL, Thorseth J, El Sisi Z, Gaafar I, Garing JD, Tucker RS (2003) Episodic growth of normal faults as recorded by syntectonic sediments, July oil field, Suez rift, Egypt. AAPG Bull 87:1015–1030

Plaziat J-C, Montenat C, Barrier P, Janin M-C, Orszag-Sperber F, Philobbos E (1998) Stratigraphy of the Egyptian syn-rift deposits: correlations between axial and peripheral sequences of the north-western Red Sea and Gulf of Suez and their relations with tectonics and eustacy. In: Purser BH, Bosence DWJ (eds) Sedimentation and tectonics of rift basins: Red Sea-Gulf of Aden. Chapman and Hall, London, pp 211–222

Polis SR, Angelich MT, Beeman CR, Maze WB, Reynolds DJ, Steinhauff DM, Tudoran A, Wood MV (2005) Preferential deposition and preservation of structurally-controlled synrift reservoirs: northeast red sea and Gulf of Suez. GeoArabia 10:97–124

Pollastro RM (1999) Ghaba salt basin province and Fahud salt basin province, Oman-geological overview and total petroleum systems. US Geol Surv Bull 2167, 41 p

Pollastro RM (2000) Cretaceous Thamama/Wasia total petroleum system (201901). US Geol Surv World Petroleum Assessment 2000 – Description and Results

Pollastro RM (2003) Total petroleum systems of the Paleozoic and Jurassic, greater Ghawar uplift and adjoining provinces of central Saudi Arabia and northern Arabian-Persian Gulf. US Geol Surv Bull 2202-H, 100 p

Ponikarov VP (1967a) The geology of Syria, explanatory notes on the geological map of Syria, scale 1:500,000 part I: stratigraphy, igneous rocks and tectonics. Ministry of Industry, Syrian Arab Republic, Damascus, p 229

Ponikarov VP (1967b) The geology of Syria, explanatory notes on the geological map of Syria, scale 1:500,000 part II: mineral deposits and underground-water resources. Ministry of Industry, Syrian Arab Republic, Damascus, p 70

Quennell AM (1984) The western Arabia rift system. In: Dixon JE, Robertson AHF (eds) The geological evolution of the eastern Mediterranean. Geol Soc Lond Spec Publ 17:775–788

Racey A (2001) A review of Eocene nummulite accumulations: structure, formation and reservoir potential. J Petrol Geol 24:79–100

Racey A, Bailey HW, Beckett D, Gallagher LT, Hampton MJ, McQuilken J (2001) The petroleum geology of the Early Eocene El Garia Formation, Hasdrubal field, offshore Tunisia. J Petrol Geol 24:29–53

Ragab AI, El-Kaliouby BA (1992) Geodynamic of the Gulf of Suez-Red Sea rifting and origin of within plate magmatism. J Afr Earth Sci 14:351–360

Rahimpour-Bonab H, Mehrabi H, Navidtalab A, Omidvar M, Enayati-Bidgoli AH, Sonei R, Sajjadi F, Amiri-Bakhtyar H, Arzani N, Izadi-Mazidi E (2013) Palaeo-exposure surfaces in Cenomanian-Santonian carbonate reservoirs in the Dezful embayment, SW Iran. J Petrol Geol 36:335–362

Ramos E, Marzo M, de Gibert JM, Tawengi KS, Khoja AA, Bolatti ND (2006) Stratigraphy and sedimentology of the Middle Ordovician Hawaz formation (Murzuq Basin, Libya). AAPG Bull 90:1309–1336

Richardson M, Arthur MA (1988) The Gulf of Suez–northern Red Sea Neogene rift: a quantitative basin analysis. Mar Petrol Geol 5:247–270

Roberts G, Harmer C, Peace D (2010) Plays and prospectivity offshore Lebanon, Syria and Cyprus: new insights from depth-imaged seismic data. Search and Discovery Article #10262

Robson DA (1971) The structure of the Gulf of Suez (Clysmic) rift, with special reference to the eastern side. J Geol Soc 127:247–276

Rohais S, Barrois A, Colletta B, Moretti I (2016) Pre-salt to salt stratigraphic architecture in a rift basin: insights from a basin-scale study of the Gulf of Suez (Egypt). Arab J Geosci 9:317–346

Ron Martin MR, Erquiaga M, Blake B, Buitrago J, Reveron J, Obregon F, Cobos C, González Muñoz JM (2016) Seismic expression of Intra-Ordovician unconformities in Murzuq Basin (Libya) through pre-stack inversion and modelling. In: Paper Th LHR3 11, 78th EAGE conference and exhibition, Vienna, Austria, 30 May–2 June 2016, 5 p

Rossi C, Marfil R, Ramseyer K, Permanyer A (2001) Facies-related diagenesis and multiphase siderite cementation and dissolution in the reservoir sandstones of the Khatatba Formation, Egypt's Western Desert. J Sedim Res 71:459–472

Roveri M, Gennari R, Lugli S, Manzi V, Minelli N, Reghizzi M, Riva A, Rossi ME, B. Schreiber BC (2016) The Messinian salinity crisis: open problems and possible implications for Mediterranean petroleum systems. Petrol Geosci 22:283–290
Ruban DA, Al-Husseini MI, Iwasaki Y (2007) Review of middle east Paleozoic plate tectonics. GeoArabia 12:35–56
Rubino J-L, Anfray R, Blanpied C, Ghienne J-F, Manatschal G (2000) Meander belt complex within the Lower Mamuniyat Formation in western Al Qarqaf area, Libya. In: Geology of Northwest Libya pp 3–18
Sadooni FN (1995) Petroleum prospects of upper Triassic carbonates in northern Iraq. J Petrol Geol 18:171–190
Sadooni FN (1997) Stratigraphy and petroleum prospects of upper Jurassic carbonates in Iraq. Petrol Geosci 3:233–243
Sadooni FN, Alsharhan AS (2003) Stratigraphy, microfacies, and petroleum potential of the Mauddud formation (Albian-Cenomanian) in the Arabian Gulf basin. AAPG Bull 87:1653–1680
Sahin A, Saner S (2001) Statistical distribution and correlations of petrophysical parameters in the Arab-D reservoir, Abqiq oilfield, Eastern Saudi Arabia. J Petrol Geol 24:101–114
Said R (1962) The geology of Egypt. Elsevier, Amsterdam, p 377
Said R (1971) Explanatory notes to accompany geological map of Egypt. Geol Surv Egypt, Cairo, Paper 56, 123 p
Said R (1981) The geological evolution of the river Nile. Springer, New York, p 151
Said R (ed) (1990) The geology of egypt. A.A. Balkema, Rotterdam, 734 p
Saoudi A, Khalil B (1984) Distribution and hydrocarbon potential of Nukhul sediments in the Gulf of Suez. In: Egyptian general petroleum corporation. 7th EGPC Exploration Seminar, Cairo, pp 75–96
Salel JF, Séguret M (1994) Late cretaceous to palaeogene thin-skinned tectonics of the palmyrides belt (Syria). Tectonophys 234:265–290
Salem AAB (2000) Hydrocarbon prospectivity to the north-west of the Assumood field, Sirt Basin, Libya. Durham theses, Durham University, 113 pp
Samuel A, Kneller B, Raslan S, Sharp A, Parsons C (2003) Prolific deep-marine slope channels of the Nile Delta, Egypt. AAPG Bull 87:541–560
Saner S, Hassan HM, Al-Ramadan KA, Abdulghani WM (2006) Mineralogical, pore and petrophysical characteristics of the Devonian Jauf sandstone reservoir, Hawiya field, eastern Saudi Arabia. J Petrol Geol 29:257–272
Sawaf T, Brew G, Litak R, Barazangi M (2001) Geological evolution of the intraplate palmyride basin and Euphrates fault system, Syria. In: Ziegler PA, Cavazza W, Robertson AHF, Crasquin-Soleau S (eds) Peri-Tethys Memoir 6: Peri-Tethyan rift/wrench basins and passive margins. Mémoires du Muséum national d'Histoire naturelle, vol 186, pp 441–467
Schandelmeier H, Reynolds PO (1997) Palaeogeographic–palaeotectonic atlas of North-Eastern Africa, Arabia and Adjacent Areas. Balkema, Rotterdam, pp 1–160, 17 plates
Schull TJ (1988) Rift basins of interior Sudan: Petroleum exploration and discovery. AAPG Bull 72:1128–1142
Scotese CR, Boucot AJ, McKerrow WS (1999) Gondwanan palæogeography and palæoclimatology. J Afr Earth Sci 28:99–114
Searle M, Cox J (1999) Tectonic setting, origin and obduction of the Oman ophiolite. Geol Soc Amer Bull 111:104–122
Sebai A, Zumbo V, Féraud G, Bertrand H, Hussain AG, Giannérini G, Campredon R (1991) $^{40}Ar/^{39}Ar$ dating of alkaline and tholeiitic magmatism of Saudi Arabia related to the early Red Sea Rifting. Earth Planet Sci Lett 104:473–487
Seddiq H, Bhaduri A (2006) The petroleum geology of Libya, a synthesis. In: Schlumberger reservoir optimization conference book, pp 20–44
Sehim A (1993) Cretaceous tectonics in Egypt. Egypt J Geol 37:335–372
Selley RC (1997) The Sirte Basin of Libya. In: Selley RC (ed) Sedimentary basins of the world, vol 3: African Basins (Chap. 3). Elsevier, Amsterdam, pp 27–37

Şengör AMC (1979) Mid-Mesozoic closure of Permo-Triassic Tethys and its implications. Nature 279:590–593

Sestini G (1984) Tectonic and sedimentary history of the NE African margin (Egypt–Libya). In: Dixon JE, Robertson RHF (eds) The geological evolution of the eastern Mediterranean. Geol Soc Lond Spec Publ 17:161–175

Sikander AH (2000) Structural development, geology and hydrocarbon potential of the Ghadamis and Murzuq Basins—an overview. The geology of Northwest Libya, pp 281–326

Sikander AH, Basu S, Rasul SM (2003) Geochemical source-maturation and volumetric evaluation of lower Palaeozoic source rocks in the West Libyan Basins pp 3–53

Shaaban F, Lutz R, Littke R, Bueker C, Odisho K (2006) Source–rock evaluation and basin modelling in NE Egypt (NE Nile Delta and northern Sinai). J Petrol Geol 29:103–124

Shalaby MR, Hakimi MH, Abdullah WH (2012) Organic geochemical characteristics and interpreted depositional environment of the Khatatba formation, northern Western Desert, Egypt. AAPG Bull 96:2019–2036

Sharaf LM (2003) Source rock evaluation and geochemistry of condensates and natural gases, offshore Nile Delta, Egypt. J Petrol Geol 26:189–209

Sharland PR, Archer R, Casey DM, Davies, Roger B, Hall, Stephen H, Heward AP, Horbury AD, Simmons MD (2001) Arabian Plate sequence stratigraphy. GeoArabia Spec Publ 2, 371 p

Snavely PD, Garrison RE, Meguid AAA (1979) Stratigraphy and regional depostional history of the Thebes formation (lower Eocene), Egypt. Ann Geol Surv Egypt 9:344–362

Sorkhabi R (2010) Why so much oil in the Middle East? GEO ExPro 7:20–26

Sorkhabi R (2012) The Zagros uplift. GEO ExPro, Febr 2012, pp 18–24

Spring D, Hansen OP (1998) The influence of platform morphology and sea level on the development of a carbonate sequence: the Harash Formation, Eastern Sirt Basin, Libya. In: MacGregor DS, Moody RTJ, Clark-Lowes DD (eds) Petroleum geology of North Africa, Geol Soc London Spec Publ, vol 132, pp 335–353

Stampfli GM (2000) Tethyan oceans. In: Bozkurt E, Winchester JA, Piper JDA (eds) Tectonics and magmatism in Turkey and the surrounding area. Geol Soc London Spec Publ 173:1–23

Stampfli GM, Borel GD (2002) A plate tectonic model for the Paleozoic and Mesozoic constrained by dynamic plate boundaries and restored synthetic oceanic isochrons. Earth Planet Sci Lett 196:17–33

Stampfli GM, Mosar J, Favre P, Pillevuit A, VannayJC (2001) Permo-Mesozoic evolution of the western Tethys realm: the Neo-Tethys east Mediterranean Basin connection. In: Ziegler PA, Cavazza W, Robertson AHF, Crasquin-Soleau S (eds) Peri-Tethys Memoir 6: Peri-Tethyan Rift/Wrench Basins and Passive Margins: Mémoires du Muséum National d'Histoire Naturelle de Paris, vol 186, pp 51–108

Steckler M, Berthelot F, Lyberis N, Le Pichon X (1988) Subsidence in the Gulf of Suez: implications for rifting and plate kinematics. Tectonophys 153:249–270

Stern RJ (1985) The Najd fault system, Saudi Arabia and Egypt: a late precambrian rift-related transform system? Tectonics 4:497–511

Stern RJ (1994) Arc assembly and continental collision in the Neoproterozoic East African Orogen: implications for the consolidation of Gondwanaland. Ann Rev Earth Planet Sci 22:319–351

Stern RJ, Johnson P (2010) Continental lithosphere of the Arabian Plate: a geologic, petrologic, and geophysical synthesis. Earth Sci Rev 101:29–67

Stoeser DB, Camp VE (1985) Pan-African microplate accretion of the Arabian shield. Geol Soc Am Bull 96:817–826

Swart PK, Cantrell DL, Westphal H, Handford CR, Kendall CG (2005) Origin of dolomite in the Arab-D reservoir from the Ghawar field, Saudi Arabia: evidence from petrographic and geochemical constraints. J Sediment Res 75:476–491

Tassy A, Crouzy E, Gorini C, Rubino JL, Bouroullec JL, Sapin F (2015) Egyptian Tethyan margin in the Mesozoic: evolution of a mixed carbonate-siliciclastic shelf edge (from Western Desert to Sinai). Mar Petrol Geol 68:565–581

Tawadros E (2001) Geology of Egypt and Libya. AA Balkema, Rotterdam, p 468

Tewfik N, Ayyad M (1984) Petroleum exploration in the Red Sea shelf of Egypt. In: Proceedings of the 6th exploration seminar, March, vol 1. Egyptian general petroleum corporation and Egypt petroleum exploration society, Cairo, pp. 159–180

Thomas D (1995) Exploration limited since '70 s in Libya' Sirte basin. Oil Gas J 13:99–104

Tillman JD, Poosti A, Rosselo S, Eckert A (1981) Structural evolution of Sanandaj-Sirjan ranges near Esfahan, Iran. AAPG Bull 674–687

Torsvik TH, Cocks RM (2011) The Palaeozoic palaeogeography of central Gondwana. In: Van Hinsbergen DJJ, Buiter SJH, Torsvik TH, Gaina C, Webb SJ (eds) The formation and evolution of Africa: a synopsis of 3.8 Ga of earth history. Geol Soc London Spec Publ 357:137–166

Turner P, Pelz K (2017) Development of an upper Triassic-lower Jurassic evaporite basin on the Saharan Platform, North Africa (Chap. 27). In: Soto JI, Flinch JF, Tari G (eds) Permo-Triassic salt provinces of Europe, North Africa and the Atlantic Margins. Elsevier, Amsterdam, pp 581–599

Underdown R, Redfren J (2007) The importance of constraining regional exhumation in basin modelling: a hydrocarbon maturation history of the Ghadames Basin, North Africa. Petrol Geosci 13:253–270

Ulmishek GF, Klemme HD (1990) Depositional controls, distribution, and effectiveness of world's petroleum source rocks. US Geol Surv Bull 1931, 59 pp

Van Buchem FSP, Svendsen N, Hoch E, Pedersen-Tatalovic, Habib K (2014) Depositional history and petroleum habitat of Qatar. In: Marlow L, Kendall C, Yose L (eds) Petroleum systems of the Tethyan region. AAPG Mem, vol 106, pp 641–677

Van der Meer F, Cloetingh S (1993) Intraplate stresses and the subsidence history of the Sirte Basin (Libya). Tectonophysics 226:37–58

Van der Meer F, Cloetingh S (1996) Intraplate stresses and the subsidence history of the Sirt Basin. In: Salem MJ, Mouzughi AJ, Hammuda OS (eds) The geology of Sirt Basin: first symposium on the sedimentary basins of Libya, Tripoli, 10–13 Oct 1993, vol 1. Elsevier, Amsterdam, pp 211–230

Vandre C, Cramer B, Gerling P, Winsemann J (2007) Natural gas formation in the western Nile delta (Eastern Mediterranean): thermogenic versus microbial. Organic Geochem 38:523–539

Vaslet D, Le Nindre Y-M, Vachard D, Broutin J, Crasquin-Soleau S, Berthelin M, Gaillot J, Halawani M, Al-Husseini M (2005) The Permian-Triassic Khuff Formation of central Saudi Arabia. GeoArabia 10:77–134

Vaslet D (1990) Upper Ordovician glacial deposits in Saudi Arabia. Episodes 13:147–161

Villinski J (2013) Unusual but effective petroleum system-disseminated terrestrial organic matter of the upper Oligocene as the primary source rock offshore Nile Delta, Egypt. In: AAPG Conference, Barcelona 2013, Abstract

Vos RG (1981) Sedimentology of and Ordovician fan delta complex, western Libya. Sedim Geol 29:153–170

Vulama I (2011) Source rock generative potential and volumetric characteristics of the Kurrachine dolomite formation, Hayan Block, central Syrian Palmyrides. Geol Croat 64(3):259–272

Walley C (2001) The Lebanon passive margin and the evolution of the Levantine Neo-Tethys. In: Ziegler PA, Cavazza W, Robertson AHF, Crasquin-Soleau S (eds) Peri-Tethys Memoir 6: Peri-Tethyan rift/wrench basins and passive margins. Mémoires du Muséum national d'Histoire naturelle, vol 186, pp 407–439

Wanas HA, Abdel-Maguid NM (2006) Petrography and geochemistry of the Cambro-Ordovician Wajid Sandstone, southwest Saudi Arabia: implications for provenance and tectonic setting. J Asian Earth Sci 27:416–429

Wasfi S, Hassouba M, Shafy AA, Mokhtar A, Azazi G, Badawi A, Sakr S, El Sisi Z (1988) Abu Roash "C" sandstone environmental model for hydrocarbon exploration. In: Proceedings of the eighth exploration conference, 17–23 Nov 1986, Cairo, vol 2. Egyptian General Petroleum Corporation, pp 97–125

Wescott WA, Nummedal D, Krebs WN, Karamat SA (1996) Depositional facies of the early synrift strata on the Sinai margin of the Gulf of Suez. In: Proceedings of 13th Egyptian general petroleum corporation conference, Cairo, Oct 1996, Exploration, vol 1, pp 297–312
Winn RD, Crevello PD, Bosworth W (2001) Lower Miocene Nukhul formation, Gebel el Zeit, Egypt: model for structural control on early synrift strata and reservoirs, Gulf of Suez. Bull Am Assoc Pet Geol 85(10):1871–1890
Wender LE, Bryant JW, Dickens MF, Neville AS, Al-Moqbel AM (1998) Paleozoic (Pre-Khuff) hydrocarbon geology of the Ghawar Area, Eastern Saudi Arabia. GeoArabia 3:273–302
Wendt J, Kaufmann B, Zdzislaw B (2009) Devonian stratigraphy and depositional environments in the southern Illizi Basin (Algerian Sahara). J Afr Earth Sci 54:85–96
Wennekers JHN, Wallace FK, Abugares YI (1996) The geology and hydrocarbons of the Sirt Basin: a synopsis. In: Salem MJ, Mouzughi AJ, Hammuda OS (eds) The geology of the Sirt Basin. Elsevier, Amsterdam, vol 1, pp 3–56
Wescott WA, Boucher PJ (2000) Imaging submarine channels in the western Nile Delta and interpreting their paleohydraulic characteristics from 3-D seismic. Leading Edge 580–591
Wescott WA, Atta M, Blanchard DC, Cole RM, Georgeson ST, Miller DA, O'Hayer WW, Wilson AD, Dolson JC, Sehim A (2011) Jurassic rift architecture in the northeastern Western Desert, Egypt. AAPG Search and Discovery Article #10379, 5 p
Whitebread T, Kelling G (1982) Mrar formation of Western Libya—evolution of an early carboniferous delta system. AAPG Bull 66:1091–1107
Wilson M, Guiraud R (1998) Late permian to recent magmatic activity on the African–Arabian margin of Tethys. In: MacGregor DS, Moody RTJ, Clark-Lowes DD (eds) Petroleum geology of North Africa. Geol Soc London Spec Publ 132:231–263
Wood B (2011) Whole lithospheric folding as a mechanism of basin formation and tectonic implications for Gondwana evolution: evidence from the Palmyride Trough, Syria. AAPG Search and Discovery Article #30176, 29 p
Wood BGM (2015) Rethinking post-Hercynian basin development: eastern Mediterranean region. GeoArabia 20:175–224
Younes MA (2002) Alamein basin hydrocarbon potential of the jurassic-cretaceous source rocks, North western desert, Egypt. Oil Gas Eur Mag 3:22–28
Younes MA (2012) Hydrocarbon potentials in the northern Western Desert of Egypt. In: Younes M (ed) Crude oil exploration in the world, pp 23–46
Zaid SM (2013) Provenance, diagenesis, tectonic setting and reservoir quality of the sandstones of the kareem formation, Gulf of Suez, Egypt. J Afr Earth Sci 85:31–52
Zainy MT, Al-Ansari, Bauer TE, Ask M (2017) The tectonic and structural classifications of the western part of the Zagros fold and thrust belt, North Iraq, review and discussion. J Earth Sci Geotech Eng 7:71–89
Ziegler MA (2001) Late Permian to Holocene paleofacies evolution of the Arabian plate and its hydrocarbon occurrences. GeoArabia 6:445–503
Zobaa MK, Oboh-Ikuenobe FE, Ibrahim MI (2011) The Cenomanian/Turonian oceanic anoxic event in the Razzak Field, north Western Desert, Egypt: source rock potential and paleoenvironmental association. Marine Petrol Geol 28:2011–1482

Paleomagnetism of the Western Saharan Basins: An Overview

M. E. M. Derder, B. Henry, S. Maouche, N. E. Merabet, M. Amenna and B. Bayou

Abstract Numerous paleomagnetic studies were performed in the western Saharan basins, particularly during the last decades. Primary magnetization of the sedimentary formations older than Bashkirian appeared as totally overprinted. By contrast, 23 new coherent paleomagnetic poles, mainly from Bashkirian to Autunian age and from Middle Triassic to Lias age, were determined. These new data greatly improved the Apparent Polar Wander Path (APWP) for Africa, and consequently for the whole Gondwana, especially for the Upper Carboniferous. The corresponding paleoreconstruction strongly argued for an A2 Pangea during this last period. By its comparison with paleomagnetic data from undated geological units, this new APWP provided dating of these units. Paleomagnetic data highlighted also the existence of a post-Liassic regional tectonic event having affected the Paleozoic cover in the Sahara platform. Finally, several magnetic overprints, pointed out in these studies, are of chemical origin, with likely a significant role of ground-fluids. Indeed, fluids migration phenomena often favored chemical changes and remagnetization process. Upper Carboniferous, Permian and Upper Cretaceous–Cenozoic overprinting ages were thus probably linked to regional geochemical events that occurred in the Saharan Platform.

Keywords Paleomagnetism · Sahara · Basins · Geodynamic · Tectonics
Geochemistry

M. E. M. Derder (✉) · S. Maouche · N. E. Merabet · M. Amenna · B. Bayou
CRAAG, B.P. 63, 16340 Bouzaréah, Algiers, Algeria
e-mail: m.e.m.derder@gmail.com

B. Henry
Paléomagnétisme, Institut de Physique du Globe de Paris, Sorbonne Paris Cité, Univ. Paris Diderot and UMR 7154 CNRS, 4 avenue de Neptune, 94107 Saint-Maur cedex, France

A. Bendaoud et al. (eds.), *The Geology of the Arab World—An Overview*, Springer Geology, https://doi.org/10.1007/978-3-319-96794-3_7

1 Introduction

The geodynamical evolution of the Hoggar shield since the Precambrian is highly linked to that of the surrounding Saharan basins. It has conditioned the formation and the evolution of these basins during the Phanerozoic (Fabre 2005 and references herein). The paleomagnetism can be considered as one of the best tools for studying such evolution. Since post-Hercynian tectonic events in the Saharan Platform are still considered as a matter of debate (e.g., Haddoum et al. 2001), it could point out such movements, which should be related to reactivation of the deep mega-shear zones of the Hoggar (Bertrand and Caby 1978; Black et al. 1994; Liégeois et al. 1994). Then, it could specify the Phanerozoic evolution of the Hoggar basement.

On one hand, it is well known that the paleomagnetism is also a powerful tool to determine drift of the main tectonic plates and to provide paleocontinental reconstructions (e.g., Van der Voo 1993; McElhinny et al. 2003; Derder et al. 2006; Torsvik et al. 2012; Henry et al. 2017). Thus, it could give a global vision of the Paleozoic drift patterns of whole Gondwana and of the African plate after the opening of the Atlantic Ocean. The migration of these continents could also have an impact on the Hoggar tectonic evolution (e.g., Liégeois et al. 2005).

At the end of the 1980's, very few paleomagnetic data were available from the Hoggar surroundings (Morel et al. 1981; Daly and Irving 1983; Kent et al. 1984; Aifa et al. 1990), and the Apparent Polar Wander Path (APWP) of the African plate was poorly defined, especially for the Paleozoic times. The Gondwana APWP was also doubtful for this period, leading to the existence of very different APWPs for this supercontinent (e.g., Smith and Hallam 1970; Van der Voo and French 1974; Morel and Irving 1981; Smith et al. 1981; Bachtadse and Briden 1991; Schmidt and Clark 2000). From a geodynamical point of view, the convergence model of Gondwana and Laurussia plates during the Upper Paleozoic interval remained hypothetical. The evolution pattern of the Pangea supercontinent from Carboniferous to Triassic was still controverted (e.g., Torcq et al. 1997).

To improve, by a better constraint of APWP, the knowledge of the geodynamical evolution of the Pangea, Gondwana, and Africa plates, numerous paleomagnetic studies were performed in the stable Saharan Platform during the last decades. Many areas were thus investigated, i.e., Tindouf, Bechar–Abadla, Mezarif, Timimoun, Reggane, Ahnet-Mouydir, Illizi, Murzuq, Tin Serririne, and Taoudeni basins surrounding the Hoggar shield (see Fig. 1; Tables 1 and 2). In this chapter, we present an overview of the paleomagnetic studies conducted in these basins of the Saharan Platform and highlight their different geodynamical, structural, geochemical, and chronological implications.

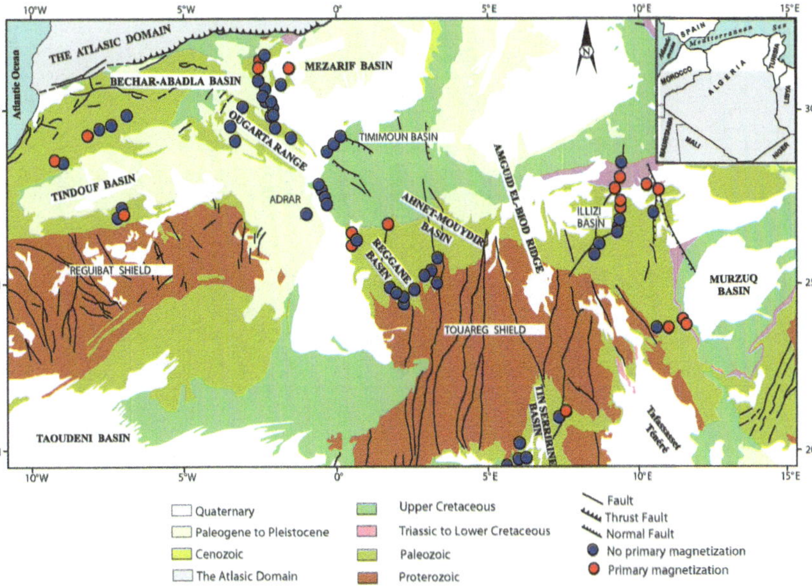

Fig. 1 Geological map of the studied Saharan basins; paleomagnetic sites yielding primary magnetization are in red (full) dots, whereas those giving remagnetizations or unstable remanences are in blue (open) dots. Locations for two studies (Kent et al. 1984; Boudzoumou et al. 2011), evidencing remagnetizations, far on the western and southern borders of the Taoudeni basin are out of the limits of the map

2 Geological Setting

The Saharan Platform is a Precambrian basement unconformably overlain by transgressive Phanerozoic thick deposits. Various tectonic events delineated sedimentary basins having their own more or less complete stratigraphic sedimentary column (Fabre 2005; Haddoum et al. 2001). The stratigraphic correlations between the basins highlight lateral facies changes and discontinuities. These basins contain mainly Paleozoic deposits, Mesozoic and Cenozoic sediments being scarce (Conrad and Le Mosquet 1984). The Paleozoic sediments reach thicknesses of over 8000 m in the Tindouf Basin, 6000 m in Reggane one and up to 8000 m in the Bechar Basin. From lithological point of view, the Cambrian formations consist of sandstones, quartzites, and conglomerates, deposited over the infra-Cambrian crystalline basement. The overlying formations up to the Upper Carboniferous are represented by various facies (shales, sandstones, limestones, etc), mainly marine and with some sedimentary gaps. The Stephano–Autunian formations essentially consist of red beds. There is a large pre-Mesozoic sedimentation gap. The Triassic is represented by sandy shales and lacustrian–continental deposits unconformably overlaying the Paleozoic formations. When present, the Jurassic formations are present in the form of marine and lacustrian to continental deposits. The transgressive

Table 1 Paleomagnetic data from Saharan Platform associated with primary magnetization

Basin	Rock unit	Site	Age	*N* NS	Reliability	Paleomagnetic pole	References
Tindouf	Djebel Reouiana	28.9°N, 8.0°W	Namurian	**11** 12		28.4°S, 56.9°E, $K = 642$, $A_{95} = 1.7°$	Merabet et al. (1999)
	Merkala	28.5°N, 8.5°W	Low. Stephanian Autunian	**9** 9		32.4°S, 56.6°E, $K = 399$ $A_{95} = 2.3°$	Henry et al. (1999)
	Dolerites dykes	27.1°N, 7.0°W	198.9 ± 1.8 Ma (Chabou 2008)	**10** 17		69.8°S, 62.1°E, $K = 490$, $A_{95} = 2.0°$	Boussada et al. (2015)
Bechar–Abadla	Ben Zireg	31.2°N, 1.8°W Ben Zireg	Famenian	**3** 3	Fold test	19.2°N, 19.8°E, $A_{95} = 3.7°$	Aifa et al. (1990)
	Low unit of Abadla	31°N, 2.7°W	Autunian	**11** 12		29.1°S, 57.8°E, $K = 462$, $A_{95} = 2.0°$	Merabet et al. (1998)
	Up. unit of Abadla	31°N, 2.7°W	*Autunian*	**13** 13		29.0°S, 60.0°E, $A_{95} = 5°$	Morel et al. (1981)
	Up. unit of Abadla	31°N, 2.7°W	*Autunian*	**1** 1	Coherence with Morel et al. (1981)	29.2°S, 60.0°E,	Merabet et al. (1998)
Mezarif	Nekheila	31.4°N, 1.5°W	Autunian	**7** 7	Fold test with Abadla	29.3°S, 56.4°E, $K = 322$, $A_{95} = 3.40$	Merabet et al. (2000, 2005)
Reggane	Aïn Ech Chebbi	26.5°N, 0.3°W Aïn Chebbi	Low. Moscovian	**4** 4		22.9°S, 51.8°E, $K = 123$, $A_{95} = 6.6°$	Daly and Irving (1983)
	Aïn Ech Chebbi	26.5°N, 0.3°W Aïn Chebbi	Low. Serpukhovian–Low. Moscovian	**10** 18	Fold test + reversal test	26.5°S, 44.7°E, $K = 383$, $A_{95} = 4.7°$	Derder et al. (2001a)
	Dolerites	26.5°N, 0.3°W Aïn Chebbi	195.0 ± 1.6 Ma (Chabou et al. 2007)	**20** 24	Fold test “synfolding”	57.6°N, 254.3°E,	Smith et al. (2006)

(continued)

Table 1 (continued)

Basin	Rock unit	Site	Age	*N* NS	Reliability	Paleomagnetic pole	References
Ahnet	Hassi Bachir	26.6°N, 1.0°E Hassi Bachir	Up. Namurian–Low. Moscovian	**7** 7		26.8°S, 56.6°E, $K = 204$, $A_{95} = 3.7°$	Daly and Irving (1983)
	Hassi Bachir	26.6°N, 1.8°E Hassi Bachir	Up. Namurian–Low. Moscovian	**15** 28	Fold test	32.8°S, 55.7°E, $K = 328$, $A_{95} = 2.0°$	Derder et al. (2009)
Illizi	Up. Oubarakat Low. El Adeb Larache	27.5′N, 8.8″E; El Adeb Larache	Bashkirian	**7** 24		28.2°S, 55.5°E, $K = 207$, $A_{95} = 3.4°$	Derder et al. (2001b)
	El Adeb Larache	27.45°N, 8.9°E El Adeb Larache	Moscovian	**10** 18		28.7°S, 55.8°E, $K = 235$, $A_{95} = 2.9°$	Henry et al. (1992)
	El Adeb Larache	27.4°N, 9.5°E Edjeleh	Moscovian	**6** 16	Fold test	28.3°S, 58.9°E, $K = 157$, $A_{95} = 4.2°$	Derder et al. (2001c)
	Low. Tiguentourine	7.7°N, 9.0°E La Reculée	Stephano–Autunian	**10** 19		35.3°S, 60.3°E, $K = 195$, $A_{95} = 3.2°$	Derder et al. (1994)
	Low. Zarzaitine	27.9°N, 9.3°E La Reculée	*Carnian–Rhaetian*	**8** 14	*Reversal test*	70.9°S, 55.2°E, $K = 478$, $A_{95} = 2.3°$	Kies et al. (1995)
	Mid. Zarzaitine	28°N, 9.7°E Zarzaïtine	Liassic	**14** 21		71.8°S, 54.9°E, $K = 91$, $A_{95} = 3.9°$	Derder et al. (2001d)

(continued)

Table 1 (continued)

Basin	Rock unit	Site	Age	*N* NS	Reliability	Paleomagnetic pole	References
Murzuq	Up. Dembaba	23.5°N, 11.9°E In Ezzane	Lower Moscovian	**12** 12	Fold test	25.2°S, 59.9°E, $K = 55$, $A_{95} = 5.4°$	Amenna et al. (2014)
	Zarzaïtine	24.2°N, 11.5°E Anai	*Up. Permian*	**11** 14	*	43.8°S, 70.9°E, $K = 80$, $A_{95} = 4.5°$	Henry et al. (2014)
	Gabbro sill	24°0′N, 10°4′E Arrikine	*Low. Devonian*	**10** 13	Contact test	42.8°S, 22.9°E, $K = 88$, $A_{95} = 4.7°$	Derder et al. (2016)
Tin Serririne	dolerites sills and dykes	21.1°N, 7.38°E 20.8°N, 6.46°E	347.6 ± 16.2 Ma (Djellit et al. 2006)	**12** 33		18.8°S, 31.2°E, $K = 29$, $A_{95} = 7.5°$	Derder et al. (2006)

Italics in "Age" column correspond to age from paleomagnetic dating. *N* (number of sites yielding primary magnetization), NS (number of studied sites), Symbol "*" indicates reliable variation of the paleomagnetic direction, related to stratigraphical level

Table 2 Paleomagnetic data from Saharan platform (a) associated with "published" remagnetizations data, (symbol "?" means that age remains doubtful or undetermined, possibly related to composite magnetizations), or (b) with "unpublished" remagnetizations or unstable magnetization data (Merabet Nacer Eddine, Pers. com. for all basins except Tin Serririne; and Derder Mohamed El Messaoud, Pers. com for Tin Serririne Basin)

Basin	Rock unit	Coordinates	Age of main remagnetizations	References
(a)				
Tindouf	Djebel Reouiana	28.9°N, 8.0°W	Permian	Merabet et al. (1999)
	Merkala	28.5°N, 8.5°W	Permian	Henry et al. (1999)
	Dolerites dykes	27.1°N, 7.0°W	Cenozoic	Boussada et al. (2015)
Bechar–Abadla	Famenian	31.9°N, 0.8°E	Tournaisian + Permian	Aifa et al. (1990); Aifa (1993)
	Lower Devonian	29.7°N, 2.1°W		
Ougarta	Magmatic complexes	29.8°N, 2.7°E	Tournaisian + Visean	Lamali et al. (2013)
		30.2°N, 3.2°W		
		29.2°N, 1.2°W		
Timimoun	Up. Visean	29.3°N, 02° E	Cenozoic	Kherroubi (2003)
Ahnet	Givetian Hazzel Matti	24.9°N, 2.2°E	Cenozoic + ?	Smith et al. (1994)
	Famennian–Tournaisian	26.8°N, 0.4°E		Bayou et al. (2000)
	Frasnian	26.7°N, 0.7°E	Cenozoic	
			Liassic Dogger?	
			Carboniferous + ?	
	Givetian	26.7°N, 1.0°E	Cenozoic	
			Carboniferous? + ?	
	Emsian	26.7°N, 0.9°E	Cenozoic	
			Liassic Dogger?	
			Carboniferous? + ?	
	Hassi Bachir	26.7°N, 1.8°E	Permian	Daly and Irving (1983)
			Mesozoic	Derder et al. (2009)

(continued)

Table 2 (continued)

Basin	Rock unit	Coordinates	Age of main remagnetizations	References
Illizi	Albian sandstone and clay	28.7°N, 9.2°E	Cenozoic	Henry et al. (2004b)
	Liassic limestone and clay	27.9°N, 9.3°E		
	Up. Triassic-Rhaetian sandstone	27.8°N, 9.3°E		
	Stephano–Autunian clays	27.8°N, 9.0°E		
	Moscovian limestone and clay	27.6°N, 9.8°E		
	Bashkirian limestone and clay	27.3°N, 8.8°E		
	Namurian limestone	27.2°N, 8.7°E		
	Visean sandstone	27.0°N, 8.7°E		
	Tournaisian sandstone	26.8°N, 8.8°E		
	Tournaisian red beds	26.7°N, 8.9°E		
	Strunian sandstone	26.6°N, 9.0°E		
	Strunian shelled limestone	26.4°E, 8.5°E		
	Givetian limestone	26.3°N, 8.5°E		
	Emsian sandstone	26.4°N, 8.4°E		
	Silurian sandstone	26.2°N, 9.1°E		
	Low. Devonian	26.3°N, 8.2°E		Lamali et al. (2014)
	Silurian	25.7°N, 7.9°E		
Murzuq	Low. Silurian	24.0°N, 10.5°S	Cenozoic	Amenna et al. (2017)
	Up. Ordovician			

(continued)

Table 2 (continued)

Basin	Rock unit	Coordinates	Age of main remagnetizations	References
Tin Serririne	Givetian	20.0°N, 6.0°E	Cenozoic	Amenna (2009)
			Jurassic?	Bayou et al. (2004).
			Carboniferous?	
	Emsian		Cenozoic	Amenna (2009)
	Low. Devonian	20.0°N, 6.0°E	Liassic?	
		21.1°N, 7.4°E	Carboniferous?	Bayou et al. (2004)
	Cambrian Ignimbrites	19.9°N, 6.0°E	Jurassic? + ?	
Taoudeni	Devonian Gneiguira	18.0°N, 12.3 W	?	Kent et al. (1984)
	Up. Proterozoic–Cambrian Mejeira	17.9°N, 12.3°W	?	
	Neoproterozoic	11.2°N, 4.3°W	?	Boudzoumou et al. (2011)

Basin	Rock unit	Coordinates	Age of main remagnetizations
(b)			
Tindouf	Westphalian	28.5°N, 8.5°W	Permian
	Up. Visean	28.6°N, 8.6°W	
	Low. Visean Strunian Silurian	29.1°N, 7.3°W	Cenozoic
	Strunian Tournaisian Visean	26.7°N, 7.5°W	Cenozoic
Bechar Abadla	Bashkirian Namurian Visean	31.6°N, 2.2°W	Cenozoic
			Permian
	Bashkirian Moscovian	31.6°N, 2.4°W	Cenozoic
	Moscovian	31.0°N, 2.7°W	
	Visean Namurian	30.9°N, 2.0°W	
Ougarta	Up. Visean	30.4°N, 2.3°W	Cenozoic
	Up. Visean	30.4°N, 2.3°W	Cenozoic
	Strunian	30.0°N, 2.1°W	
	Famenian	30.2°N, 2.2°W	Permian
	Emsian	29.9°N, 2.1°W	Jurassic
Timimoun	Up. Visean Tournaisian	29.4°N, 0.2°E	Cenozoic

(continued)

Table 2 (continued)

Basin	Rock unit	Coordinates	Age of main remagnetizations
	Strunian		
Adrar	"Gothlandian" Low. Devonian Mid. Devonian	29.0°N, 0.3°W	Cenozoic Carboniferous + Permian Permian + Jurassic
	Visean Namurian	27.2°N, 0.1°W	Permian
Illizi	Cambro–Ordovician	24.5°N, 9.5°E	Cenozoic
Tin Serririne	Silurian	19.7°N, 5.8°E	Cenozoic
		20.9°N, 7.4°E	
	Ordovician	19.9°N, 5.8°E	

Upper Cretaceous shaly sandstone and carbonate sequence are found over the Saharan Platform. The Cenozoic formations are represented by clastic continental sediments from Oligocene to Pliocene and Quaternary.

The Bechar Basin is bounded to the north by the High Atlas and to the south by the Ougarta mountain range. Tindouf and Reggane basins are asymmetrically located on the N and NE of the Reguibat shield. The basement has been encountered in some deep wells in the Illizi and Ahnet area. This basement, structured as several crustal blocks by major faults under the Paleozoic sedimentary cover, would be (Freulon 1964; Beuf et al. 1971; Fabre 2005) the follwoing:

- The Western Neoproterozoic of the Hoggar, under the Ahnet-Timimoun basins.
- The Central Hoggar, Polycyclic Paleoproterozoic, under the Amguid-El Biod Ridge.
- The Eastern Hoggar, Precambrian, under the Illizi Bassin.
- The West African Craton under the Tindouf and Reggane basins.

In the Saharan Platform, basin inversion, uplift and reactivation of regional to local structures are documented as mainly related to the Hercynian collisional event (Haddoum et al. 2001). Recent observations highlight evidence of permanent mobility of this platform (Nedjari et al. 2011). Many tectonic events during different periods are linked to the basin's location, particularly for those of the western ones. During the Hercynian period, the configuration, evidenced by new structural elements, shows the platform as a foreland regarding the Variscan chain, which was edified and later eroded. The Saharan Platform is structured into synclyses. In the northwestern Sahara, the Bechar–Abadla basin is considered as atypical because of extreme mobility due to location close to the suture zone of the Variscan orogeny. This basin represents an appropriate model of foredeep basin (Nedjari et al. 2011).

It is highly subsident with very thick sedimentary filling. The deformations within the platform have been explained by distal effects of stresses generated by collision and mechanical coupling between the Gondwana and Laurussia plates. Basin exhumation and compressional reactivation of structures also occurred during the Tertiary collision between the African and European plates (Liégeois et al. 2005; Galeazzi et al. 2010).

3 Sampling and Analysis Procedure

Depending on outcropping conditions and on facies features, different stratigraphic levels of the studied geological formations were mostly sampled. To have significant and reliable data, at least seven sites were generally selected (Table 1). In favorable cases, more than 300 cores distributed on up to 33 independent sites have been drilled in the same rock unit. In most cases, the sampling was made by a portable gasoline-powered drill. Large hand samples (precisely oriented using a plaster cap) were also collected when the facies was too friable and cores were drilled in the laboratory. All samples were oriented with magnetic and sun compasses. One to three specimens of standard size (cylinders of 11 cm^3) were cut from each core, allowing demagnetization treatments and additional rock magnetic studies to be performed. Prior to any demagnetization analysis, the specimens were put in a zero field for at least 1 month, in order to reduce possible viscous magnetization.

The Remanent Magnetization was mostly measured using JR4 or JR5 spinner magnetometer (AGICO, Brno, Czech Republic). In each study, whatever the kind of the rocks (sedimentary or volcanic) both thermal and Alternating Field (AF) demagnetization were performed on pilot specimens, and, when needed, by combined AF-thermal procedure. In order to correctly isolate and identify the magnetization components, numerous demagnetization steps were used (10 °C, and until 5 °C increment in high temperatures when necessary).

The results of demagnetization analysis were carried out using classical methods: they were presented on orthogonal vector plots (As and Zijderveld 1958; Zijderveld 1967). The remaining vectors after each step and the difference vectors removed between two consecutive demagnetization steps were plotted on equal area projections. When applicable, the remagnetization circles methods (Halls 1976, 1978; McFadden and McElhinny 1988), were also used. The direction of the magnetization components was calculated by principal component analysis (Kirschvink 1980). Fisher (1953) statistics were used to determine the mean characteristic directions. When appropriate, progressive unfolding was performed, allowing fold test (e.g., McElhinny 1964; McFadden 1990; Tauxe and Watson 1994). "Synfolding" magnetizations were analyzed using small circles method (Shipunov 1997; Henry et al. 2004a; Waldhör and Appel 2006). When applicable, reversal (McFadden and McElhinny 1990) and contact tests were also performed.

4 Rock magnetism

Rock magnetism analyses were carried out on all the studied geological formations. Representative samples from each formation were chosen to determine their magnetic mineralogy from different magnetic approaches, based on thermal demagnetization curves, thermomagnetic *K*(*T*) curves (magnetic susceptibility as a function of temperature), and hysteresis loops. Mean susceptibility and *K*(*T*) curves were determined on AGICO KLY2 (or KLY3), CSL and CS2 (or CS3) equipments and hysteresis loops on a laboratory-made translation inductometer for small samples (about 3 cm^3) within an electromagnet.

The results showed that, generally, for sedimentary formations, the main magnetic mineral carriers were hematite (Fig. 2a, b), magnetite (Fig. 3a, b) or a mixture of these two minerals (Fig. 4a, b). For the magmatic formations, rock experiments result suggested that Ti-poor titanomagnetites of pseudo-single-domain grain size (Derder et al. 2016), or a mixing of Ti-rich titanomaghemite and magnetite are the main carriers (Derder et al. 2006).

5 Paleomagnetic Results

During the demagnetization process, the analysis of the natural remanent magnetization gave different kinds of evolution on the Zijderveld plot, after elimination of a viscous component. The first one (i) is illustrated by a stable magnetic direction characteristic of a single component (Fig. 5a). The second type (ii) shows the

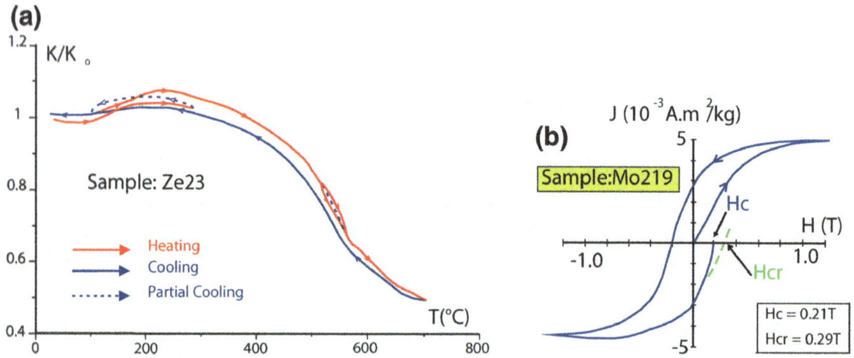

Fig. 2 **a** Typical thermomagnetic (variation of the normalized magnetic susceptibility K/K_0 during a cycle of progressive heating and cooling in air in low magnetic field) curve for ZE23 sample pointing out presence of hematite (from Fig. 3a of Derder et al. 2001d, modified). **b** Hysteresis loop for Mo219 sample (H: magnetic field; Hc: coercive force; Hcr: remanent coercive force; all in Tesla; J: Magnetization) suggesting presence of hematite (from Fig. 2b of Derder et al. 2001c, modified)

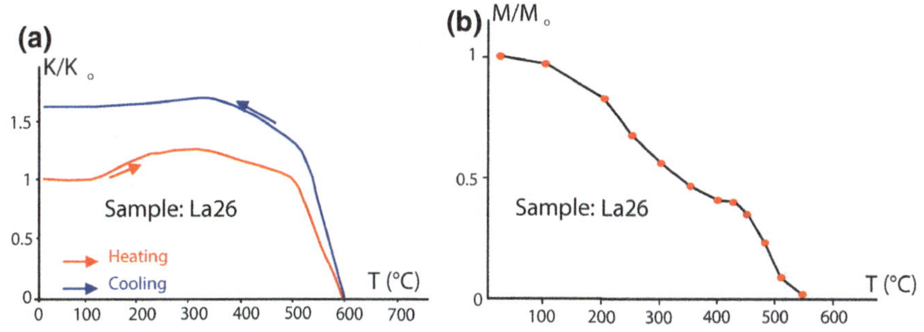

Fig. 3 **a** Thermomagnetic curve (see Fig. 2a caption) for La26 sample pointing out presence of magnetite (from Fig. 3a of Henry et al. 1992, modified). **b** Typical thermal demagnetization curve for La26 sample pointing out existence of magnetite (from Fig. 3b of Henry et al. 1992, modified)

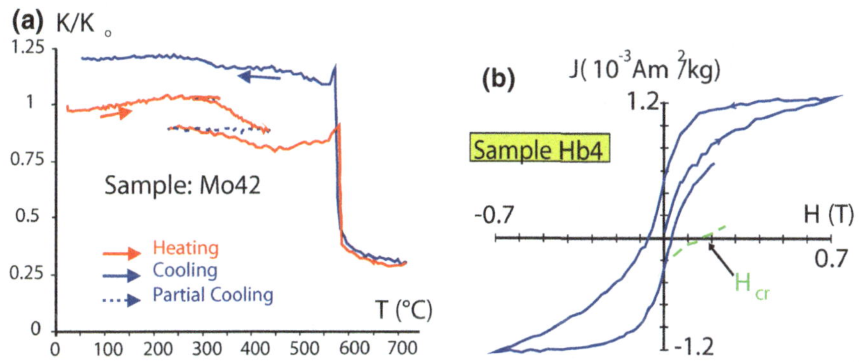

Fig. 4 **a** Typical thermomagnetic curve (see Fig. 2a caption) for Mo42 sample showing, in addition to an "Hopkinson" peak related to magnetite, the presence of hematite (from Fig. 3 of Derder et al. 2001c, modified). **b** Hysteresis loop (see Fig. 2b caption) for Hb4 sample showing wasp-waisted shape suggesting existence of magnetite and hematite (from Fig. 3b of Derder et al. 2009, modified)

isolation of several components distributed over the unblocking temperature spectra (Fig. 5b). The third one (iii) points out magnetic directions evolving along a great circle and reaching a "stable endpoint" direction for the highest demagnetization steps (Fig. 5c). The fourth one (iv) is characterized by directions evolving along a great circle (superimposition of unblocking spectra for at least two components), but no stable component was reached even for the highest demagnetization steps (Fig. 5d). Finally, the last kind (v) is characterized by an erratic behavior of the magnetization.

According to the sites, the percentage of usable data (i, ii, iii, and sometimes iv) is very variable from one study to another (e.g. Table 1). Reliability criteria (Van der Voo 1990; Henry et al. 2017) and paleomagnetic tests yielded separation of

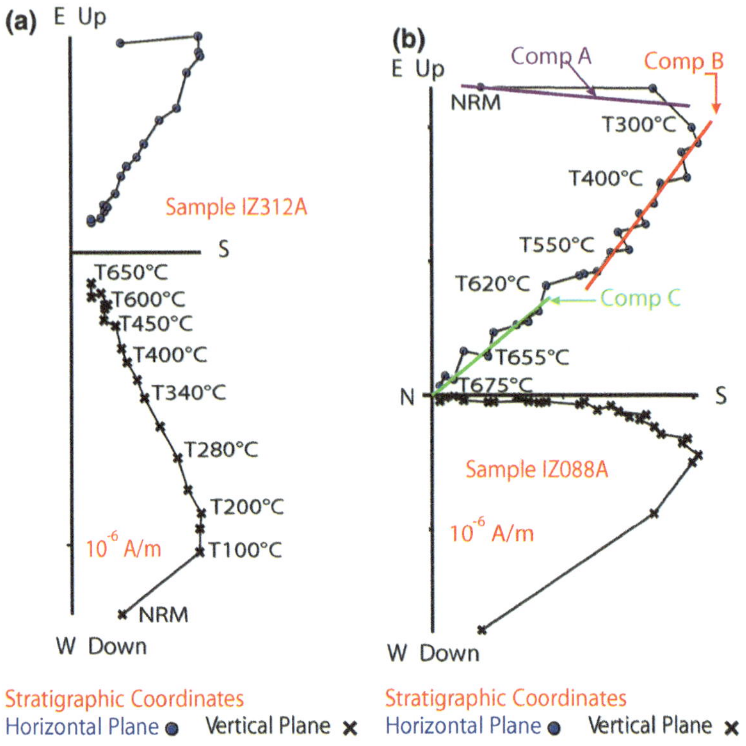

Fig. 5 Orthogonal vector plots (filled circles: horizontal plane, crosses: vertical plane), in stratigraphic coordinates for **a** samples IZ312A (from Fig. 6a of Amenna et al. 2014, modified); **b** sample: IZ088A (from Fig. 6a of Henry et al. 2014, modified); **c** sample D332 (from Fig. 6 of Derder et al. 2006, modified) and **d** sample TG019A (from Fig. 6e of Derder et al. 1994, modified)

primary (Fig. 1; Table 1) and secondary (Table 2) magnetizations. Used tests (see Table 1) were fold test (Aifa et al. 1990; Derder et al. 2001a, c, 2009; Merabet et al. 2005; Smith et al. 2006; Amenna et al. 2014) (Fig. 6), reversal test (Derder et al. 2001a) (Fig. 7) and contact test (Derder et al. 2016) (Fig. 8). Significant variability of the paleomagnetic direction pointed out in different stratigraphic levels within a same formation was also used as a criterion to highlight primary magnetization (e.g., Henry et al. 2014). Several examples (e.g., Bayou et al. 2000; Bouabdallah et al. 2003) of data of type *i* appeared to be composite (i.e., resulting from superimposition of different magnetization components with similar blocking characteristic spectra). In some favorable cases, separation of these components was possible (Merabet et al. 1999; Henry et al. 1999; Derder et al. 2001d). In all, 93 geological formations or intrusions were studied, giving 23 new paleomagnetic poles related to primary magnetizations.

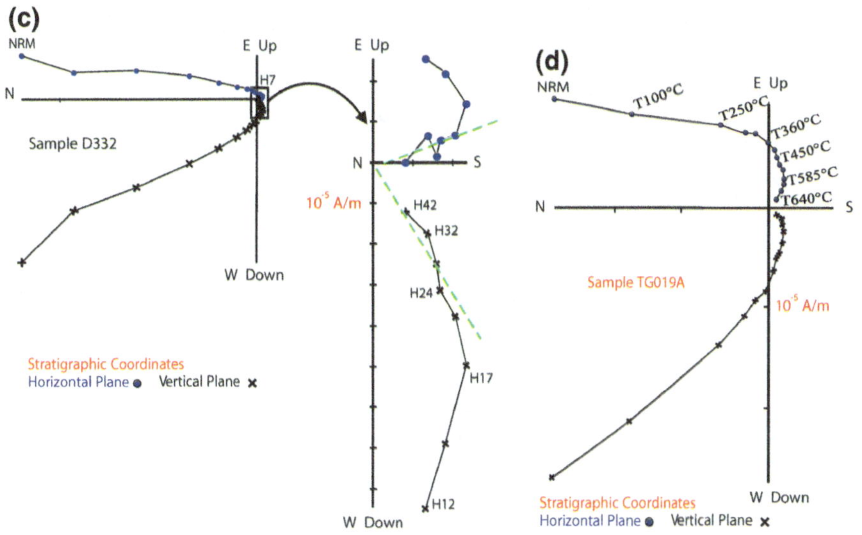

Fig. 5 (continued)

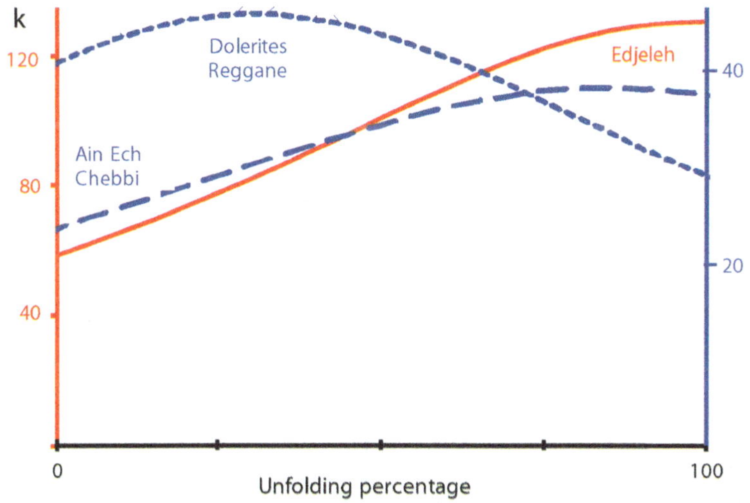

Fig. 6 Variation of *k* parameter (Fisher 1953) during progressive unfolding for the Moscovian Formation at Edjeleh in Illizi basin (Derder et al. 2001c), Jurassic dolerites in Reggane basin (Smith et al. 2006) and lower Serpukhovian—lower Moscovian Ain Ech Chebbi formation in Ahnet basin (Derder et al. 2009)

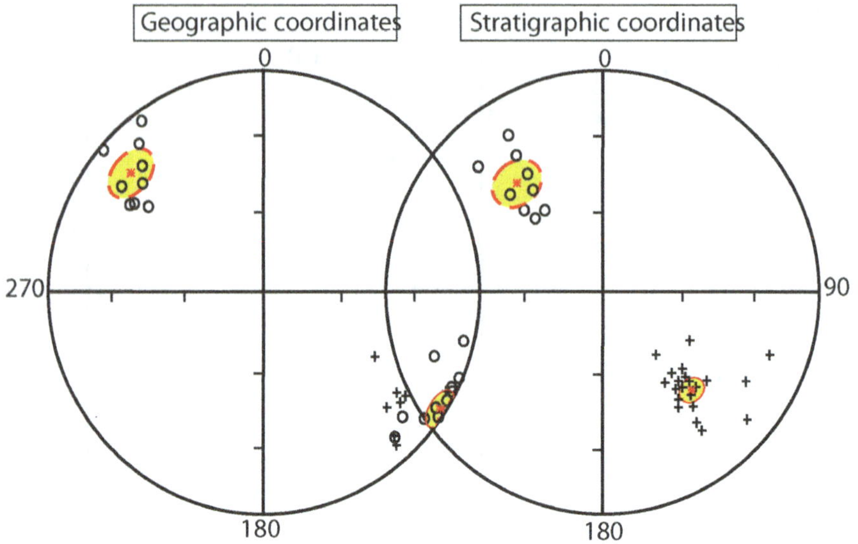

Fig. 7 Equal area plot (crosses: positive inclinations, open circles: negative inclinations) of paleomagnetic directions of middle–upper Carboniferous formations from Reggane basin. The reversal test (McFadden and McElhinny 1990) is positive after dip correction, with angular difference γ (5.0°) between mean directions for normal and reversed data lower than the critical value γc (9.2°), but this test is negative (γ = 17.8°; γc = 10.7°) before dip correction (from Fig. 9a of Derder et al. 2001a, modified)

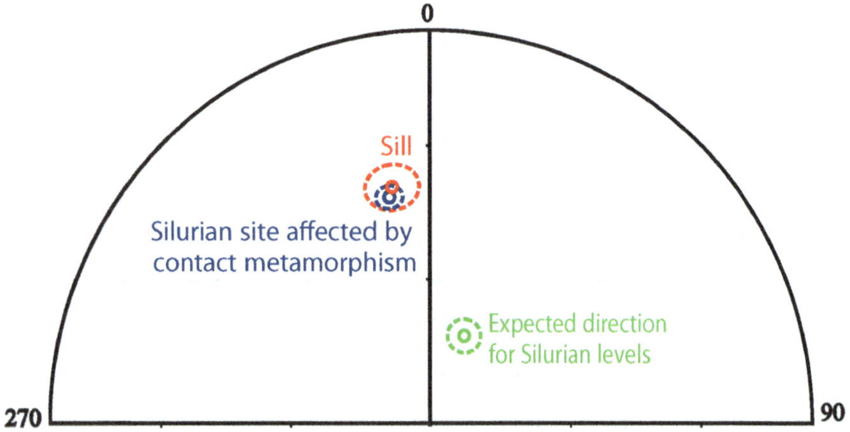

Fig. 8 Equal area plot (open circles: negative inclinations) of paleomagnetic directions obtained from the sill, after dip correction, from the sedimentary Silurian site affected by contact metamorphism and that of the expected direction of the Silurian levels (calculated from the APWP for 430 Ma), showing positive contact test

6 Discussion

6.1 *Apparent Polar Wander Path (APWP) for the Gondwana and Its Implications*

Unfortunately, all studies from dated geological units older than the Bashkirian did not yield primary magnetization, except for the Lower Carboniferous dolerites of the Tin Serririne basin (Derder et al. 2006). In addition, because of the lack of existing geological formations during several periods, the paleomagnetic poles obtained in dated formations correspond thus to two age windows, from the Bashkirian to the Autunian (19 poles) and from the Middle Triassic to the Lias (4 poles). For the first window, the poles were obtained in different basins, widespread between the far western Tindouf and far eastern Murzuq ones, i.e., along a distance of about 2000 km. That gives an important significance to each of the data, which represent independent areas of the Saharan Platform. These poles were integrated in the classical paleomagnetic pole selection of McElhinny et al. (2003), Derder et al. (2006) and Torsvik et al. (2012) to determine (Le Goff et al. 1992) an improved APWP for the Gondwana (Amenna 2015).

However, a detailed comparison of the African paleomagnetic poles obtained in the Saharan basins with the Gondwana APWP highlighted an incoherency in the used selection. This incoherency appeared as due to the selected poles for South America (Tomezzoli et al. 2013). In fact, poles positions obtained from areas close to the eastern border of the Andean Cordillera are different from those obtained from the cratonic eastern part of this continent. This discrepancy pointed out tectonic disturbances not previously suspected in this border (Henry et al. 2017). The pole selection for the Gondwana, and therefore the Gondwana APWP, were then reevaluated and improved (Fig. 9).

Mainly due to the reliable paleomagnetic data from the Saharan basins, this APWP is very precise for the Upper Carboniferous, yielding a well-defined location of the Gondwana at this period (Fig. 10). By contrast, for the Laurussia, the limited number of available paleomagnetic data yields large uncertainty in its APWP (Domeier et al. 2012) and then of its location in the reconstruction. Because of the uncertainty in longitude in such reconstructions, other relative positions of the two supercontinents remain possible, and this reconstruction strongly argues for an A2 Pangea (Van der Voo and French 1974), as presented on the Fig. 10.

This APWP yielded also dating by comparison with the paleomagnetic data obtained in undated geological units (Fig. 11).

- K/Ar age of the dolerites studied in the Tin Serririne basin (Djellit et al. 2006) has been confirmed by this method (Derder et al. 2006).
- The Zarzaïtine Formation, well dated in the Illizi basin (Lehman 1971; Bourquin et al. 2010; Aït Ouali et al. 2011), was studied in the Murzuq basin at Anaï area. Because of lack of paleontological arguments in this last basin, the age of this formation, at the base of the post-Hercynian deposits, was of particular interest.

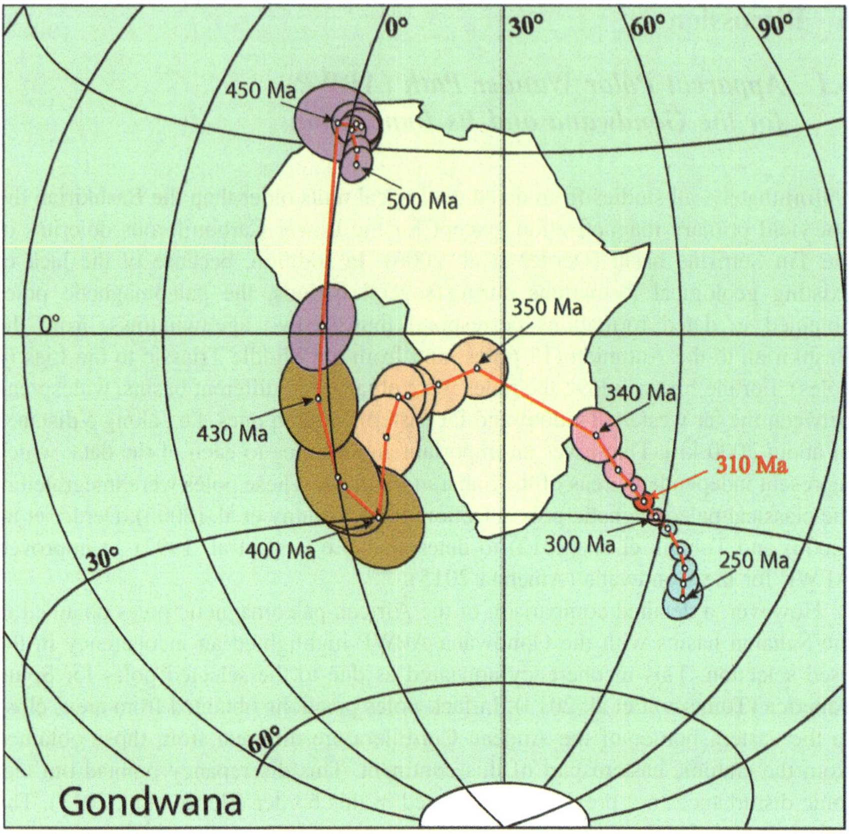

Fig. 9 Gondwana APWP (from Henry et al. 2017) for the period 500–250 Ma, with associated uncertainty zone at 95% A_{95} (Fisher 1953)

The obtained Late Permian age evidenced a very large diachronism (40 Ma) within the Zarzaïtine Formation (Henry et al. 2014). This strong diachronism, as well as the local character of the Stephano–Autunian Tiguentourine deposits in newly formed basins in Libya (Hallett 2002), clearly indicates that the post-Hercynian structural evolution of the Saharan Platform included vertical movements, which gave differential uplifts. The latter was at the origin of erosion, hiatus or sediments deposition according to areas and likely according to time for a same area.

- The Lower Devonian age (415–400 Ma—Derder et al. 2016) of the very large sill discovered in the Murzuq basin shows that this magmatic event corresponds also to that (407 ± 8 Ma—Moreau et al. 1994) of the Aïr intrusives (Liégeois et al. 1994) and could be also at the origin of sand injections in different borders of the Murzuq basin (Moreau et al. 2012). In Niger, large sills in the same stratigraphical position (Menchikoff 1962) are probably of the same age. These

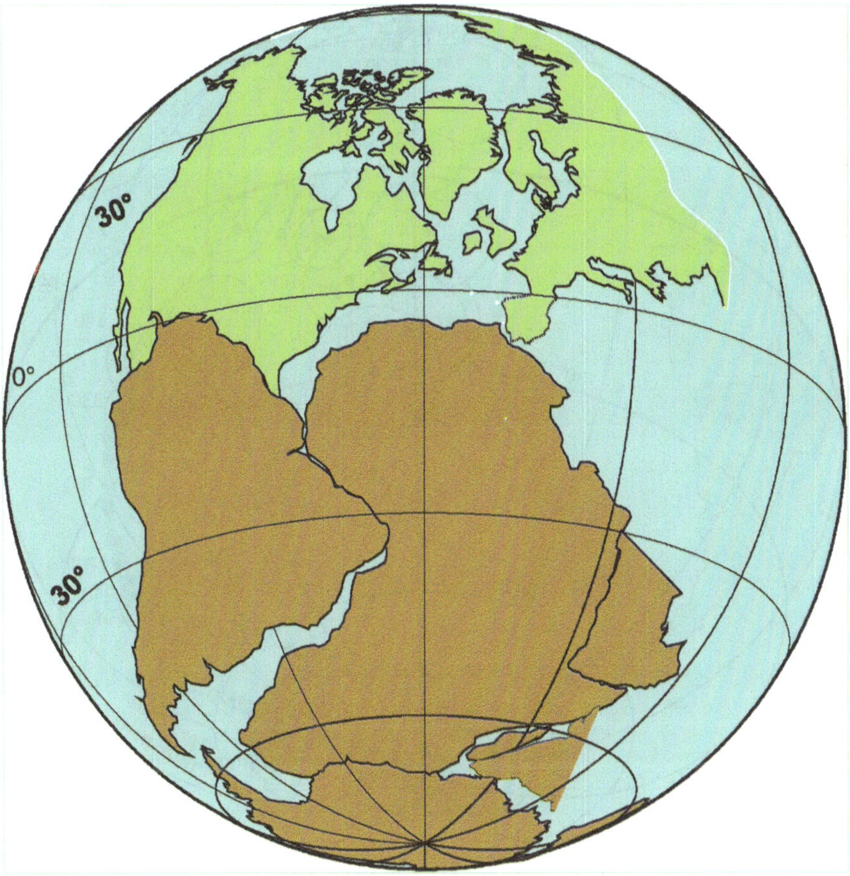

Fig. 10 Paleogeographic reconstruction for the Moscovian (310 Ma) using the data in north west African coordinates of Laurussia (Domeier et al. 2012) and Gondwana (Henry et al. 2017) to restore the landmasses independently. This reconstruction is in favor a Pangea A2 type, suggesting that such a reconstruction had existed since the Upper Carboniferous

different results imply a regional magmatic event affecting a very large area of the Saharan Platform. On the other hand, the loop of the Gondwana APWP (Fig. 9) between the Late Ordovician (pole 450 Ma) and Late Devonian (370 Ma) was not generally deemed as very reliable. The Aïr (Hargraves et al. 1987) and Arrikine (Derder et al. 2016) coherent mean paleomagnetic datum is "unique" and supports the existence of this loop, which has to be considered for the future paleocontinental reconstructions for this period.

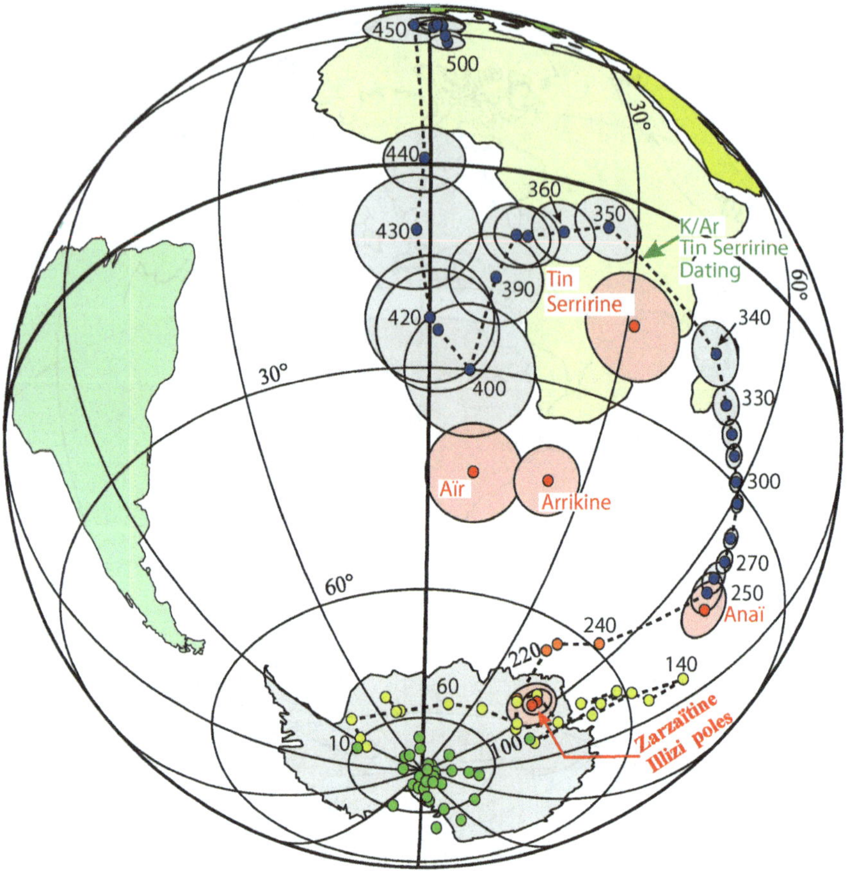

Fig. 11 Comparison of the Zarzaitine Illizi (Kies et al. 1995; Derder et al. 2001d), Anaï (Henry et al. 2014), Arrikine (Derder et al. 2016), Aïr (Hargraves et al. 1987) and Tin Serririne (Derder et al. 2006) paleomagnetic poles with the Gondwana and Africa APWP (500–250 Ma—Henry et al. 2017; 240–210 Ma—Domeier et al. 2012 and 200–0 Ma—Besse and Courtillot 2002). K/Ar age of the Tin Serririne dolerites (Djellit et al. 2006). "Cenozoic" remagnetizations poles from the Saharan basins are in green dots

6.2 Structural Implications

The study, in the Reggane basin, of a Jurassic sill emplaced within folded Paleozoic series, highlighted that actually part of this folding occurred after dolerites intrusion (Smith et al. 2006). The dip of the Paleozoic formations during intrusion has been determined by a small circles approach (see Henry et al. 2004a and references herein). It was interesting to notice that this dip had in some sites higher values than

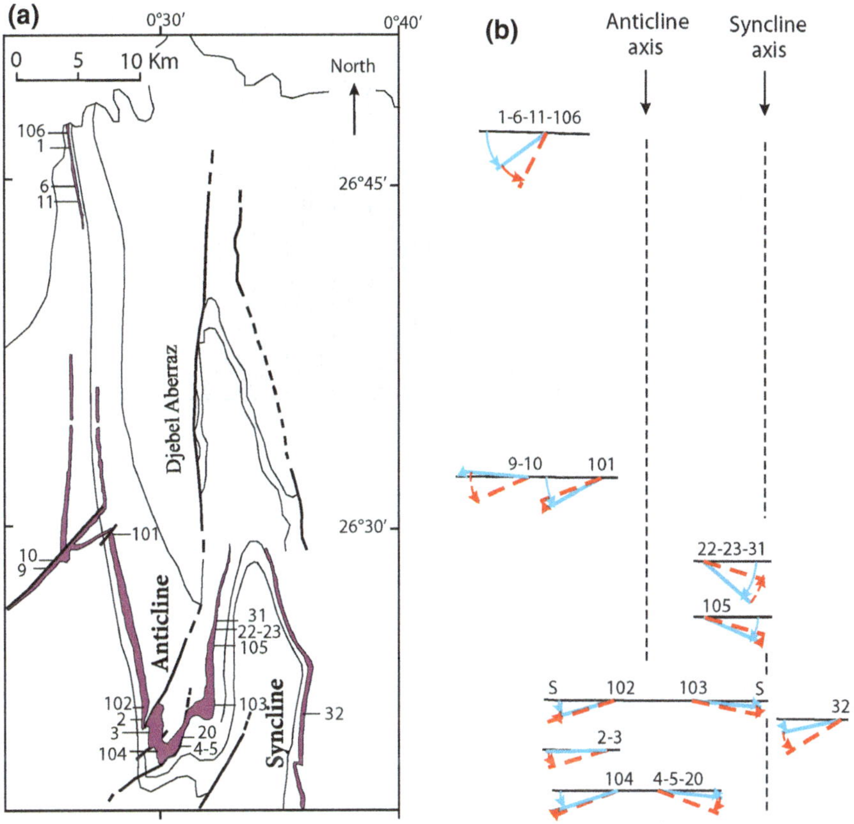

Fig. 12 **a** Sketch map of the Jurassic Reggane sill (pink color) area with the sites location and **b** dip values related to the two tilting events having occurred, respectively before (Hercynian dips, blue arrows) and after (red arrows) the dolerite emplacement for the different sites (from Fig. 11 of Smith et al. 2006)

the present dip (Fig. 12), implying locally tilting in opposite direction during the Hercynian and post-intrusion foldings. This proves the validity of the assumption of post-Hercynian tectonics in the Saharan Platform (Conrad 1972, 1981).

At Hassi Bachir area, Daly and Irving (1983) evidenced superimposition of paleomagnetic components. A new analysis (Derder et al. 2009) allowed isolation of the primary magnetization C and of another B component (possibly composite—see above). A statistical approach (Tauxe and Watson 1994), based on 10,000 bootstrap resamplings for progressive unfolding (Fig. 13), indicates that the B component is "syntectonic" (i.e., for paleomagnetists acquired either after the beginning of the first deformation and before the end of the last one, or during the folding in case of single tectonic event). This means that B component acquisition is related to the Hercynian folding or to existence of a second folding event, as shown with the Reggane dolerites.

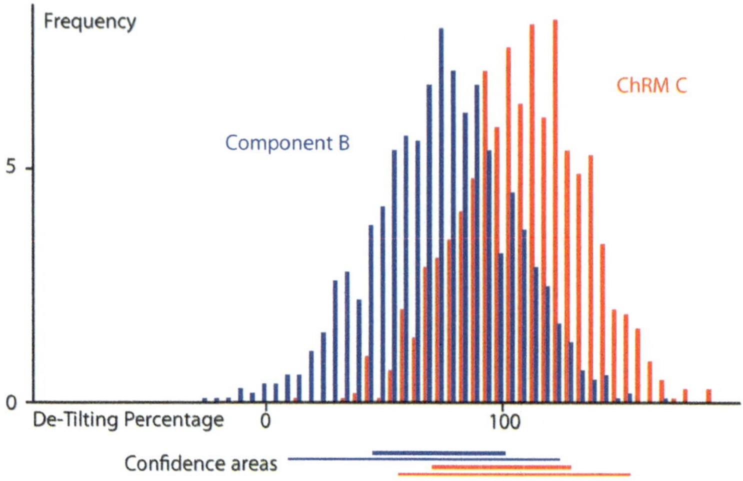

Fig. 13 Frequency (in percentage) of optimal untilting value (by window of 5°) obtained by the bootstrap method of Tauxe and Watson (1994) for B component (blue) and ChRM C (red), with confidence areas at 95% (thin line) and 63% (thick line), (from Fig. 7 of Derder et al. 2009)

Paleomagnetic data had also structural important implications in the "Ougarta" range, by pointing out two different tectonic phases affecting the magmatic complexes (Lamali et al. 2013).

6.3 *Geochemical Implications*

Remagnetizations phenomena, either partial or total, have been evidenced in many basins (e.g. Aifa 1993; Henry et al. 2004b). Only one Silurian site, metamorphized during the Arrikine sill intrusion (Murzuq basin), shows magnetic overprint related to heating (Derder et al. 2016). For the other studied geological units, the thickness of the overlying series at the time of remagnetization was insufficient to produce significant heating effects due to simple burial. In many areas, presence of high temperatures components of normal and reversed polarities during demagnetization process shows that remagnetizations are not related to effects of the recent field (Viscous Remanent Magnetization). Chemical phenomena represent, therefore, the main origin of the magnetic overprints of these different geological formations.

These overprints were acquired during two principal periods: the Permian and Cenozoic (Fig. 11). These two ages correspond in the Saharan Platform to erosion and continental environment. However, the presence of both magnetic polarities for the most recent remagnetizations in the different geological formations of the Illizi and Murzuq basins (Henry et al. 2004b; Lamali et al. 2014; Amenna et al. 2017) shows that they were not due to a simple superficial surface weathering. In addition, in the Illizi basin, a relation between the magnetic polarity and subhorizontal stratigraphic levels (giving a pseudo-magnetostratigraphy) suggests a significant role of ground-fluids (Henry et al. 2004b). Both periods of remagnetization also correspond to differential uplifts in the Saharan Platform after the Hercynian main tectonics (Henry et al. 2014) or to the Cenozoic Hoggar uplift (Rougier et al. 2013). Fluids migration resulting from such vertical movements often favored chemical changes and remagnetization process (e.g., Oliver 1986; McCabe and Elmore 1989; Symons et al. 1996; Rouvier et al. 2001).

Paleomagnetic data yielded another geochemical implication. As mentioned above, the primary magnetization was not obtained in the studied sedimentary sites older than the Bashkirian of the Saharan basins. Indeed, all these rocks did not give stable remanent magnetization or were totally remagnetized. This Bashkirian age precisely corresponds to that obtained in the Arrikine sill by K/Ar (325.6 ± 7.7 Ma —Derder et al. 2016), age attributed to rejuvenation related to cryptocirculations of fluids, suggesting a significant regional geochemical event at this period.

7 Conclusion

More than a quarter century of paleomagnetic studies in Saharan basins surrounding the Hoggar Shield have largely contributed to a better knowledge of the Saharan Platform. These investigations demonstrated that the paleomagnetism is a strong tool for different geological and geodynamical purposes, being even in some cases the unique possible approach. Some applications, such as dating or tectonic analyses, used successfully in Algeria, could be applied in other countries of the Saharan Platform, thus opening other perspectives. As pointed out in this overview, analyses of rocks remagnetization and of their acquisition processes could have major geochemical implications. The remagnetizations, previously neglected by most paleomagnetists, could then become a new geochemical indicator. Finally, the example of the structural implications of the improvement of the South American APWP, mentioned in this overview, underlines that for tectonic purposes, it will be interesting to extend paleomagnetic investigations to the craton border zones, as around the south Atlasic flexure, which separates the Alpine domain from the Saharan craton in North Africa.

Acknowledgements We are very grateful to the Algerian research Ministry MESRS, to the French Foreign Office and to the French CNRS (Programs DPRS (DGRU)-CNRS, CMEP and PICS yielded financial supports). Special thanks also to the SONATRACH, ORGM, OPNA (Tamanrasset), ONPCT (Djanet) and to all the civil and military authorities everywhere in the Algerian Sahara for their constant and important help in the field. We are very grateful to Pr Mohamed Hamoudi for his detailed and constructive review.

References

Aifa T (1993) Different styles of remagnetization in Devonian sediments from the north-western Sahara (Algeria). Geophys J Int 115:529–537

Aifa T, Feinberg H, Pozzi JP (1990) Devonian/Carboniferous paleopoles for Africa. Consequences for Hercynian geodynamics. Tectonophysics 179:288–304

Aït Ouali R, Nedjari A, Taquet P, Bitam L, Tayeb-Cherif L, Bouras R (2011) Le Zarzaïtine inférieur (In Amenas, Sahara algérien): Derniers développements dans une série du Trias pro-parte. Mém Serv Géol Natl 17:6–26

Amenna M (2009) Paléomagnétisme de formations sédimentaires paléozoïques du bassin de Tin Seririne et du bassin de l'Ahnet (Sud-Est et Nord-Ouest du Hoggar). Master thesis, USTHB University, Algiers

Amenna M (2015) Applications du paléomagnétisme dans la bordure occidentale du bassin de Murzuq en Algérie. PhD thesis, U. S. T. H. B. University, Algiers

Amenna M, Derder MEM, Henry B, Bayou B, Maouche S, Bouabdallah H, Ouabadi A, Beddiaf M, Ayache M (2014) Improved Moscovian part of the Gondwana APWP for paleocontinental reconstructions, obtained from a first paleomagnetic pole, age-constrained by a fold test, from In Ezzane area in the Murzuq basin (Algeria, Stable Africa). J Afr Earth Sci 99:342–352. https://doi.org/10.1016/j.jafrearsci.2013.12.006

Amenna M, Derder MEM, Henry B, Maouche S, Bayou B, Ouabadi A, Bestandji R, Bouabdallah H, Ayache M, Beddiaf M (2017) Chemical remagnetization acquisition process: case study of the widespread Cenozoic remagnetization of the Saharan basins. Arab J Geosci 10:379. https://doi.org/10.1007/s12517-017-3165-z

As JA, Zijderveld JDA (1958) Magnetic cleaning of rocks in paleomagnetic research. Geophys J Roy Astron Soc 1:308–319

Bachtadse V, Briden JC (1991) Paleomagnetism of Devonian ring complexes from the Bayuda desert, Sudan—new constraints on the apparent polar wander path for Gondwanaland. Geophys J Int 104(3):635–646

Bayou B, Smith B, Derder MEM, Yelles Chaouche K, Henry B, Djellit H (2000) Paleomagnetic investigations in Devonian rocks from the central Sahara, Algeria. In: Final proceeding of the first international symposium on geophysics, Tanta University, Egypt, 8–9 Sept 1998, pp 119–128

Bayou B, Derder MEM, Henry B, Amenna M, Djellit H, Ouabadi A, Khaldi A, Baaziz K, Hemmi A (2004) Paléomagnétisme des Roches sédimentaires du Dévonien et volcaniques du Cambrien du bassin de Tin Serririne. In: Final proceeding of the international colloque: « Structure et Évolution de la Bordure Sud du Hoggar Central » C.R.A.A.G. and I.P.G.P., Algiers, 27 Nov 2004

Bertrand JML, Caby R (1978) Geodynamic evolution of the Pan-African orogenic belt: a new interpretation of the Hoggar shield (Algerian Sahara). Geol Rundsch 67:357–388

Besse J, Courtillot V (2002) Apparent and true polar wander and the geometry of the geomagnetic field over the last 200 Myr. J Geophys Res 107:2300. https://doi.org/10.1029/2000JB000050

Beuf S, Biju-Duval B, de Charpal O, Rognon P, Gariel O, Benacef A (1971) Les grès du paléozoique inferieur au Sahara; Sédimentation et discontinuités, évolution structurale d'un craton. Editions technip, Paris, pp 464–644

Black R, Latouche L, Liégeois JP, Caby R, Bertrand JM (1994) Pan-African displaced terranes in the Tuareg shield (central Sahara). Geology 22:641–644

Bouabdallah H, Henry B, Merabet NE, Maouche S (2003) Juxtaposed and superimposed magnetizations in Carboniferous formations of the Tindouf basin. Bull Serv Cart Géol Algér 14:53–64

Boudzoumou F, Vandamme D, Affaton P, Gattacceca J (2011) Neoproterozoic paleomagnetic poles in the Taoudeni basin (west Africa). C R Geosci 343:284–294

Bourquin S, Eschard R, Hamouche B (2010) High-resolution sequence stratigraphy of upper Triassic succession (Carnian–Rhaetian) of the Zarzaitine outcrops (Algeria): a model of fluvio-lacustrine deposites. J Afr Earth Sci 58:365–386

Boussada MA, Merabet N, Mahdjoub Y, Maouche S, Chabou MC, Aifa T, Ayache M (2015) Paleomagnetic and magnetostructural study of the Gara Djebilet Jurassic magmatic formations (Tindouf Basin, Southwest Algeria). In: EGU General Assembly 2015, Vienne, Austria. Geophysical research abstracts, 17, pp EGU2015–2953, 2015

Chabou MC (2008) Datation $^{39}Ar/^{40}Ar$ et Géochimie de la Province Magmatique de l'Atlantique Central dans le Sud-Ouest algérien. PhD thesis, Ecole Nationale Supérieur Polytechnique, Algiers

Chabou MC, Sebai A, Féraud G, Bertrand H (2007) Datation $^{40}Ar/^{39}Ar$ de la province magmatique de l'Atlantique central dans le Sud-Ouest algérien. C R Geosci 339:970–978

Conrad J (1972) Distension jurassique et tectonique éocrétacée sur le Nord-Ouest de la plate-forme africaine (Bassin de Reggane, Sahara central). C R Acad Sci Paris 274:2423–2426

Conrad J (1981) La part des déformations posthercyniennes et de la néotectonique dans la structuration du Sahara central algérien, un domaine relativement mobile de la plate-forme africaine. C R Acad Sci Paris 292:1053–1056

Conrad G, Le Mosquet Y (1984) Du craton vers sa marge : évolution sédimentaire et structurale du bassin Ahnet-Timimoun-Béchar (Sahara algérien) au cours du Carbonifère; données paleoclimatiques. Bull Soc Géol Fr XXVI(6):987–994

Daly L, Irving E (1983) Paléomagnétisme des roches carbonifères du Sahara central: analyse des aimantations juxtaposées: Configurations de la Pangée. Ann Geophys 1:207–216

Derder MEM, Henry B, Mérabet N, Daly L (1994) Paleomagnetism of the Stephano-Autunian lower Tiguentourine formation from stable saharan craton (Algeria). Geophys J Int 116:12–22

Derder MEM, Smith B, Henry B, Bayou B, Yelles Chaouche A, Djellit H, Ait Ouali A, Gandriche H (2001a) Juxtaposed and superimposed paleomagnetic components from the folded middle Carboniferous sediments in the Reggane basin (Saharan craton, Algeria). Tectonophysics 332:403–422

Derder MEM, Henry B, Merabet N, Amenna M, Bourouis S (2001b) Upper Carboniferous paleomagnetic pole from stable Saharan craton and the Gondwana reconstructions. J Afr Earth Sci 32:491–502

Derder MEM, Henry B, Bayou B, Amenna M, Djellit H (2001c) New Moscovian paleomagnetic pole from the Edjeleh fold (Saharan craton, Algeria). Geophys J Int 147:345–355

Derder MEM, Henry B, Merabet N, Bayou B, Amenna M (2001d) Paleomagnetism of the Liassic member of the Zarzaitine formation (stable Saharan craton, Illizi basin, Algeria). Ann Geofis 44:995–1010

Derder MEM, Henry B, Bayou B, Ouabadi A, Bellon H, Djellit H, Khaldi A, Amenna M, Baziz K, Hemmi A, Guemache MA (2006) New African lower Carboniferous paleomagnetic pole from intrusive rocks of the Tin Serririne basin (southern border of the Hoggar, Algeria). Tectonophysics 418:189–203

Derder MEM, Henry B, Amenna M, Bayou B, Djellit H, Guemache MA, Hemmi A (2009) New structural implications for central Sahara (Algeria) from revisited upper Carboniferous "Hassi Bachir" formation: paleomagnetic constraints. Tectonophysics 463:69–76. https://doi.org/10.1016/j.tecto.2008.09.012

Derder MEM, Maouche S, Liégeois JP, Henry B, Amenna M, Ouabadi A, Bellon H, Bruguier O, Bayou B, Bestandji R, Nouar O, Bouabdallah H, Ayache M, Beddiaf M (2016) Discovery of a Devonian mafic magmatism on the western border of the Murzuq basin (Saharan metacraton): paleomagnetic dating and geodynamical implications. J Afr Earth Sci 115:159–176

Djellit H, Bellon H, Ouabadi A, Derder MEM, Henry B, Bayou B, Khaldi A, Baziz K, Merahi MK (2006) Age $^{40}K/^{40}Ar$, Carbonifère inférieur, du volcanisme basique filonien du synclinal paléozoïque de Tin Sirririne, Sud-Est Hoggar (Algérie). C R Geosci 338:624–631

Domeier M, Van der Voo R, Torsvik TH (2012) Paleomagnetism and Pangea: the road to reconciliation. Tectonophysics 514–517:14–43

Fabre J (2005) Géologie du Sahara occidental et central. Tervuren Afr Geosci Collect 108:572p

Fisher RA (1953) Dispersion on a sphere. Proc Roy Soc Lond A 217:295–305
Freulon JM (1964) Etude géologique des séries primaires du Sahara central, Publ. du centre de Recherche Sahariennes (CNRS). Serv Géol Algér 3:1–198
Galeazzi S, Point O, Haddadi N, Mather J, Druesne D (2010) Regional geology and petroleum systems of the Illizi-Berkine area of the Algerian Saharan Platform: an overview. Mar Petrol Geol 27:143–178
Haddoum H, Guiraud R, Moussine-Pouchkine A (2001) Hercynian compressional deformations of the Ahnet-Mouydir basin, Algerian Saharan platform: far-field stress effects of the late palaeozoic orogeny. Terra Nova 13:220–226
Hallett D (2002) Petroleum geology of Libya. Elsevier editors
Halls HC (1976) A least squares method to find a remanence direction from converging remagnetization circles. Geophys J Roy Astron Soc 45:297–304
Halls HC (1978) The use of converging remagnetization circles in paleomagnetism. Phys Earth Planet Inter 16:1–11
Hargraves RB, Dawson EM, Houten FB, (1987) Palaeomagnetism and age of mid-palaeozoic ring complexes in Niger, West Africa, and tectonic implications. Geophys Roy Astron l Soc 90:705–729. Doi:https://doi.org/10.1111/j.1365-246x.1987.tb00750.x
Henry B, Merabet N, Yelles Chaouche A, Derder MEM, Daly L (1992) Geodynamical implications of a Moscovian palaeomagnetic pole from the stable Sahara craton (Illizi basin, Algeria). Tectonophysics 201:83–96
Henry B, Merabet NE, Bouabdallah H, Maouche S (1999) Nouveau pôle paléomagnétique stéphanien inférieur pour le craton saharien (Formation de Merkala, bassin de Tindouf, Algérie). C R Acad Sci Paris 329(IIa):161–166
Henry B, Rouvier H, Le Goff M (2004a) Using syntectonic remagnetizations for fold geometry and vertical axis rotation: example of the Cévennes border (France). Geophys J Int 157:1061–1070
Henry B, Merabet N, Derder MEM, Bayou B (2004b) Chemical remagnetizations in the Illizi basin (Saharan craton, Algeria). Geophys J Int 156:200–212
Henry B, Derder MEM, Amenna M, Maouche S, Bayou B, Ouabadi A, Bouabdallah H, Ayache M, Beddiaf M, Bestandji R (2014) Paleomagnetic dating of continental geological formations: strong diachronism evidenced in the Saharan platform and geodynamical implications. J Afr Earth Sci 99:353–362. https://doi.org/10.1016/j.jafrearsci.2014.02.010
Henry B, Derder MEM, Amenna M, Maouche S, Bayou B (2017) Better constrained selection of the paleozoic west Gondwana (South America) paleomagnetic poles for the APWPs determination. Stud Geophys Geod 61. https://doi.org/10.1007/s11200-016-1036-9
Kent DV, Dia O, Sougy JMA (1984) Paleomagnetism of lower-middle Devonian and upper Proterozoic-Cambrian(?) rocks from Mejeria (Mauritania, West Africa), In: Van der Voo R, Scotese CR, Bonhommet N (eds) Plate reconstruction from Paleozoic paleomagnetism. AGU Geodynamics series, vol 12, pp 99–115
Kherroubi A (2003) Paléomagnétisme des formations paléozoïques des bassins Ouest Sahariens. Master thesis, USTHB Houari Boumediene University, Algiers
Kies B, Henry B, Merabet N, Derder MEM, Daly L (1995) A new late Triassic-Liasic paleomagnetic pole from superimposed and juxtaposed magnetizations in the Saharan craton. Geophys J Int 120:433–444
Kirschvink JL (1980) The least-squares line and plane analysis of palaeomagnetic data. Geophys J Roy Astron Soc 62:699–718
Lamali A, Merabet N, Henry B, Maouche S, Graine-Tazerout K, Mekkaoui A, Ayache M (2013) Polyphased geodynamical evolution of the Ougarta (Algeria) magmatic complexes evidenced by paleomagnetic and AMS studies. Tectonophysics 588:82–89
Lamali A, Merabet N, Henry B, Maouche S, Hamoudi M, Ayache M (2014) Réaimantations énigmatiques des formations du Silurien et du Dévonien inférieur du Tassili N-Ajjer (Bassin d'Illizi, Algérie). Bull Serv Géol Natl Algér 25:181–199
Le Goff M, Henry B, Daly L (1992) Practical method for drawing VGP paths. Phys Earth Planet Int 70:201–204

Lehman JP (1971) Nouveaux vertébrés du Trias de la série du « Zarzaïtine ». Ann Paléont Vertébr Paris 57:71–93

Liégeois JP, Black R, Navez J, Latouche L (1994) Early and late Pan-African orogenies in the Aïr assembly of terranes (Tuareg shield, Niger). Precambr Res 67:59–66

Liégeois J-P, Benhallou A, Azzouni-Sekkal A, Yahiaoui R, Bonin B (2005) The Hoggar swell and volcanism: reactivation of the precambrian Tuareg shield during Alpine convergence and West African Cenozoic volcanism. In: Foulger GR, Natland JH, Presnall DC, Anderson DL (eds) Plates, plumes, and paradigms, vol 388. Geological Society of America special paper, pp 379–400

McCabe C, Elmore RD (1989) The occurrence and origin of late Paleozoic remagnetization in the sedimentary rocks of North America. Rev Geophys 27:471–494

McElhinny MW (1964) Statistical significance of the fold test in palaeomagnetism. Geophys J Roy Astron Soc 8:338–340

McElhinny MW, Powell C McA, Pisarevsky SA (2003) Paleozoic terranes of eastern Australia and the drift history of Gondwana. Tectonophysics 362:41–65

McFadden PL (1990) A new fold test for palaeomagnetic studies. Geophys J Int 103:163–169

McFadden PL, McElhinny MW (1988) The combined analysis of remagnetization circles and direct observations in palaeomagnetism. Earth Planet Sci Lett 87:161–172

McFadden PL, McElhinny MW (1990) Classification of the reversal test in palaeomagnetism. Geophys J Int 103:725–729

Menchikoff N (1962) Carte géologique du Nord-Ouest de l'Afrique: Sahara central. CNRS

Merabet N, Bouabdallah H, Henry B (1998) Paleomagnetism of the lower Permian redbeds of the Abadla basin (Algeria). Tectonophysics 293:127–136

Merabet N, Henry B, Bouabdallah H, Maouche S (1999) Paleomagnetism of the Djebel Reouina Namurian formation (Tindouf basin, Algeria). Stud Geophys Geod 43:376–389

Merabet N, Kherroubi A, Maouche S, Henry B (2000) Paleomagnetism of the Permian series of the Mezarif basin (Algeria). Geol Carpath 51:179

Merabet N, Henry B, Kherroubi A, Maouche S (2005) Autunian age constrained by fold tests for paleomagnetic pole from Mezarif and Abadla basins (Algeria). J Afr Earth Sci 43:556–566

Moreau C, Demaiffe D, Bellion Y, Boullier AM (1994) A tectonic model for the location of Paleozoic ring-complexes in Aïr (Niger, West Africa). Tectonophysics 234:129–146

Moreau J, Ghienne JF, Hurst A (2012) Kilometre-scale sand injectites in the intracratonic Murzuq basin (south-west Libya): an igneous trigger? Sedimentology 59:1321–1344

Morel P, Irving E (1981) Paleomagnetism and the evolution of Pangea. J Geophys Res 86 (B3):1858–1872

Morel P, Irving E, Daly L, Moussine-Pouchkine A (1981) Paleomagnetic results from Permian rocks of the northern Saharan craton and motions of the Moroccan Meseta and Pangea. Earth Planet Sci Lett 55:65–74

Nedjari A, Aït Ouali R, Debaghi F, Hamdidouche R, Benhamouche A, Amrouche F, Messamri K (2011) La Géologie Saharienne Revisitée (1980–2009). Mém Serv Géol Natl 17:85–176

Oliver J (1986) Fluids expelled tectonically from orogenic belts: their role in hydrocarbon migration and other geologic phenomena. Geology 14:99–102

Rougier S, Missenard Y, Gautheron C, Barbarand J, Zeyen H, Pinna R, Liégeois JP, Bonin B, Ouabadi A, Derder MEM, Frizon de Lamotte D (2013) Eocene exhumation of the Tuareg Shield (Sahara Desert, Africa). Geology 41:615–618. https://doi.org/10.1130/G33731.1

Rouvier H, Henry B, Macquar JC, Leach D, Le Goff M, Thibiéroz J, Lewchuk MT (2001) Réaimantation régionale éocène, migration de fluides et minéralisations dans la bordure cévenole. Bull Soc Géol Fr 172:503–516

Schmidt PW, Clark DA (2000) Paleomagnetism apparent polar wander path and paleolatitude. In: Veevers JJ (ed) Billion year earth history of Australia and neighbours in Gondwanaland. GEOMOG Press, Department of Earth and Planetary Sciences, Macquarie University, Sydney, pp 12–17

Shipunov SV (1997) Synfolding magnetization: detection testing and geological applications. Geophys J Int 130:405–410

Smith AG, Hallam A (1970) The fit of the southern continents. Nature 225(5228):139–144
Smith AG, Hurley AM, Briden JC (1981) Phanerozoic paleocontinental world maps. Cambridge Univ. Press, London, UK
Smith B, Moussine-Pouchkine A, Ait Kaci A (1994) Paleomagnetic investigation of middle Devonian limestones of Algeria and the Gondwana reconstruction. Geophys J Int 119:166–186
Smith B, Derder MEM, Henry B, Bayou B, Amenna M, Djellit H, Yelles AK, Garces M, Beamud E, Callot JP, Eschard R, Chambers A, Aifa T, Ait Ouali R, Gandriche H (2006) Relative importance of the Hercynian and post-Jurassic tectonic phases in the Saharan platform: a palaeomagnetic study of Jurassic sills in the Reggane basin (Algeria). Geophys J Int 167:380–396
Symons DTA, Sangster DF, Leach D (1996) Paleomagnetic dating of Mississippi valley-type Pb-Zn-Ba deposits. In: Sangster DF (ed) Carbonate hosted Pb/Zn deposits, special vol 4. Society of Economic Geologists
Tauxe L, Watson GS (1994) The fold test: an eigen analysis approach. Earth Planet Sci Lett 122:331–341
Tomezzoli RN, Vizán H, Tickyj H, Woroszylo ME (2013) Revisión de la posición del polo paleomagnético de Sierra Chica en la curva de desplazamiento polar aparente del Gondwana. Latinmag Lett 3(Special Issue, OB25):1–8. http://www.geofisica.unam.mx/LatinmagLetters/LL13-03-SP/LLv3spIndex.html
Torcq F, Besse J, Vaslet D, Marcoux J, Ricou LE, Halawani M, Basahel M (1997) Paleomagnetic results from Saudi Arabia and the Permo-Triassic Pangea configuration. Earth Planet Sci Lett 148:553–567, 1997
Torsvik TH, Van der Voo R, Preeden U, Mac Niocaill C, Steinberger B, Doubrovine PV, van Hinsbergen DJJ, Domeier M, Gaina C, Tohver E, Meert JG, McCausland PJA, Cocks LRM (2012) Phanerozoic polar wander, paleogeography and dynamics. Earth Sci Rev 114:325–368. https://doi.org/10.1016/j.earscirev.2012.06.007
Van der Voo R (1990) The reliability of paleomagnetic data. Tectonophysics 184:1–9
Van der Voo R (1993) Paleomagnetism of the Atlantic, Tethys and Iapetus oceans. Cambridge University Press, 441 pp
Van der Voo R, French RB (1974) Apparent polar wandering for the Atlantic-bordering continents: late Carboniferous to Eocene. Earth Sci Rev 10:99–119
Waldhör M, Appel E (2006) Intersections of remanence small circles: new tools to improve data processing and interpretation in palaeomagnetism. Geophys J Int 166:33–45
Zijderveld JDA (1967) Demagnetization of rocks: analysis of results. In: Collision DW,Creer KM, Runcorn SK (eds) Methods in paleomagnetism, pp 254–286

Archaeoseismology in Algeria: Observed Damages Related to Probable Past Earthquakes on Archaeological Remains on Roman Sites (Tel Atlas of Algeria)

K. Roumane and A. Ayadi

Abstract For the period before 1365, the catalogue of historical earthquakes in Algeria remains sparse. A number of earthquakes have been identified in archived documents, and yet others can be inferred from their damage to archaeological structures. In this study, we focus on the Roman period (BC 146–429), the Vandal and Byzantine period (AD 429–533) in the region of the seismically active Tell Atlas. The Tell Atlas of Algeria retains numerous archaeological records of former earthquakes. At the Roman sites of Lambaesis (Lambèse), Thamugadi (Timgad) Thibilis (Salaoua Announa) or Thevest (Tebessa), we interpret damage to monuments as having been caused by strong shaking, ground subsidence, and landslides effects. In this study, we aim at contributing towards archaeoseismology in Algeria by presenting examples of observed damage and disorders on several Roman sites.

Keywords Archaeoseismology · Ancient earthquakes · Antiquity
Roman sites · Thamugadi · Lambaesis · Cuicul · Theveste · Tipasa

1 Introduction

The seismicity of Algeria is attributed to its location within the Africa-Eurasia plate boundary. The Tell Atlas of Algeria has experienced numerous strong and devastating shallow earthquakes, 5–20 km depth (Ayadi and Bezzeghoud 2015). The goal of this paper is to analyze the various Algerian archaeological sites, located on seismogenic zones through the identification and examination of the features observed on structures that are probably related to the effects of earthquake. The selected sites (Fig. 1) were chosen based on their proximity to seismogenic areas

K. Roumane (✉)
Institute of Archaeology, University of Algiers, Algiers, Algeria
e-mail: roumanekahina@gmail.com

A. Ayadi
Center of Research in Astronomy, Astrophysics and Geophysics,
CRAAG, Algiers, Algeria

A. Bendaoud et al. (eds.), *The Geology of the Arab World—An Overview*,
Springer Geology, https://doi.org/10.1007/978-3-319-96794-3_8

Fig. 1 Location of the Roman sites along the eastern part of the Tell Atlas (Northern Algeria)

that experienced strong seismic events and frequently guided by ancient texts that reported destruction and extensive damage following an earthquake in these sites (Ferdi and Harbi 2013; Roumane 2016). For this, we studied several cases and we attempted to compare them to other studies such that of Sintubin and Stewart (2008), Stiros (1996) and Rodriguez-Pascua et al. (2011), that reported typological identification of disorders on archaeological sites.

2 The Algerian Archaeological Sites Located in Earthquake Prone Areas

The Tell Atlas in Northern Algeria is a wide zone of tectonic deformation induced by the convergence between the Africa and Eurasia plates. Most of the seismicity is located in this extended zone bounded to the south by the south Atlasic faults system and to the north by the Mediterranean Sea. This area experienced several strong earthquakes (Ayadi and Bezzeghoud 2015). Roman archaeological sites in the Tellian Atlas (Fig. 1) are numerous and most of them are located near seismogenic zones regarding the distribution of the seismicity along the Tell Atlas (Fig. 2). In this paper, we investigated the following sites:

- Tipasa (Tipasa) is located on the Algerian coast west of Algiers Capital city. According to Lancel (2005), its location was strategic and considered as a

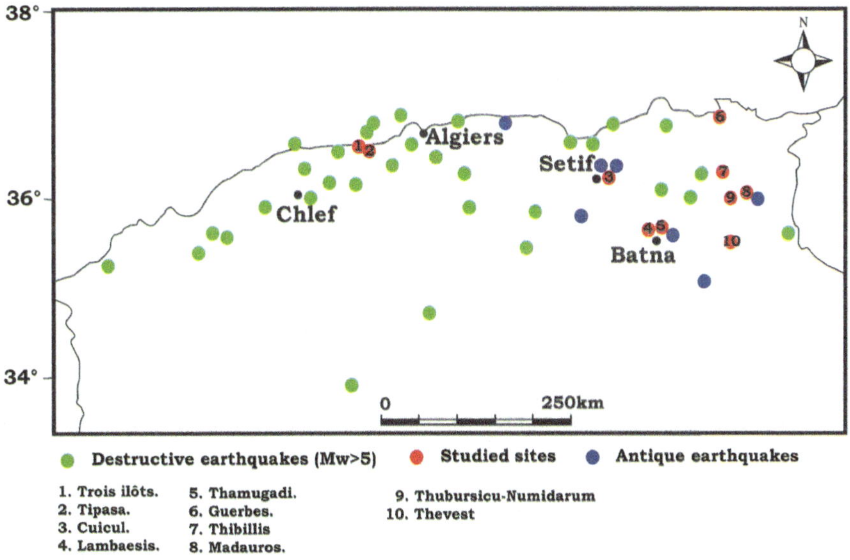

Fig. 2 Significant earthquakes in the Tell Atlas of Algeria from Maouche et al. (2010). Green circles: significant earthquakes with magnitude > 5.0, Red circles: Archaeological sites in this study. Blue circles: other archaeological sites during the Antic period

trading center and a stopover for Phoenician sailors on their route to the west (Gsell 1894). Baradez (1952) specified that the city was built in the fifth century BC (Phoenician counter) and became a Roman city in AD 39. Tipasa comprises several archaeological complexes and many of them were not restored.

- Cuicul (Djemila), an ancient Roman city which was built far from the coastline on the high plateau in a mountainous area on a rocky site at about 900 m elevation, 40 km east of Setif (known as Sitifis during the Roman empire). According to Février (1978), who referred to Cagnat (1923), the city was founded between AD 96 and 97 during the reign of Nerva. Other authors, such as Gascou (1972), proposed the beginning of the reign of Trajan (AD 98) as a period during which Cuicul was created. The remains of the archaeological structures of Cuicul have been restored in the beginning of the twentieth century.
- Thamugadi (Timgad) is located in the Aurès region. Its original name was Colonia Marciana Trajana Thamugadi (Courtois 1951) and it was built during the reign of Trajan (AD 100) as a Roman veteran's city by the legate Lucius Munatius Gallus. Thamugadi was believed to be a fortification defending the occupied area from southern threats.
- Lambaesis (Tazoult) is a military camp which was built in AD 81 by the Third Augustus legion. An independent source reports it to have attained City status during a period of enlargement by the emperor Trajan in AD 100, and its military camp to have been inspected by Emperor Hadrian in AD 128.
- Thibilis (Sellaoua-Announa) is a Numidian city which became an important Roman settlement. Administratively, the town was ruled by the Roman colony of Cirta, 57 km to the northwest. Thibilis became an autonomous municipality, probably between AD 260 and 268.
- Madauros (M'Daourouch) was the home of the famous writer *Lucius Apuleius* who reported that the city was located in the border of Numidia and Getulia (Gsell 1901, 1914, 1922). Madauros was built during the Flavian dynasty between AD 69 and 96.
- Thubursicu-Numidarum (Khemissa) in Guelma province, northeastern Algeria is one of the important Roman cities in Africa as suggested by its extensive ruins. According to Gsell (1914), about 600 epitaphs were found reporting the importance of the city which was founded during the reign of Trajan (AD 100) when it became a Municipium (Municipium Ulpium Traianum Augustum Thubursicu) and Colonia (Roman colony) around AD 270.
- Thevest (Tébessa) was founded during the reign of Vespasian at the end of the first century (AD 69–79) and became the principal residence of the Third legion of Augustus. Subsequently, Theveste became the richest city in Africa after Carthage (Ballu 1893).
- Mons (Mopth) in eastern Algeria between Cuicul and Sitifis is a Roman settlement built on a rock hill founded under the reign of Nerva (AD 96–98) (Gsell 1901).

- Guerbes is a town in eastern Algeria on the coastline which consists of a large rectangular enclosure, typical of many sites that served as overnight shelters for passing travelers (Gsell 1901).

3 Seismicity of Algeria and Earthquakes of the Ancient Times

In earlier earthquake catalogues, Ayadi and Bezzeghoud (2015) and Harbi et al. (2015) identified several damaging earthquakes in the Tell Atlas (Algeria), over a period starting from the fourteenth century to the present (Table 1). The most important ones are those of Algiers (1365, I_0 X, 1716, I_0 IX), Blida (1825) and the El Asnam of October 10th 1980 with Ms7.3. Testimonials in ancient documents and epigraphs prior to 1365 are scarce (Ferdi and Harbi 2013). This is probably due to the destruction of historical records during numerous invasions. According to an epighraphic inscription, Ad Maiores (Henchir Besseriani now, South of Tébessa) (Lepelley 1981; Laporte and Dupuis 2009; Laporte 2016) (Table 2) experienced in

Table 1 Table of significant earthquake that have occurred in Algeria since 1365 (Ayadi and Bezzeghoud 2015)

Locality	Date	Intensity	Magnitude	Observations
Algiers	3/01/1365	Strong	Strong	Algiers completely destroyed
Algiers	10/03/1673	Strong	Strong	71 aftershocks
Mediterranean	3/02/1716	X	7.5	Several houses destroyed
Algiers	29/11/1722	Strong	Strong	Damage 75 km SW Algiers
Oran	9/10/1790	IX–X	6.5–7.5	Felt in Malta
Mascara	–03/1819	IX	6.5	Little damage to Oran
Blida	02/03/1825	X–XI	7.5	Destruction of Blida
Jijel	22/08/1856	X	7.5	Generated a tsunami
Kherba	09/03/1858	IX	6.5	Damage in the plain of Mitidja
Mitidja	02/01/1867	X–XI	7.5	Mouzaïa destroyed
Biskra	16/11/1869	IX	6.5	
Gouraya	15/01/1891	X	7.5	Damage in Algiers. Felt in Saida and Djelfa
Sour El Ghozlane	24/06/1910	X	6.4	
A. El Hassan	25/08/1922	IX–X	5.1	Damage reported in Cherchell
El Attaf	07/09/1934	IX	5.0	Landslides triggered
Bejaia	12/02/1946	VIII–IX	5.6	
Chlef	09/09/1954	X–XI	6.7	
M'Sila	21/02/1960	VIII	5.6	
M'Sila	01/01/1965	VIII	5.6	

(continued)

Table 1 (continued)

Mansourah	24/11/1973	VII	5.1	
Chlef	10/10/1980	IX	7.3	
Constantine	27/10/1985	VIII	5.9	
El Affroun	31/10/1988	VII	5.4	
Djebel Chenoua	29/10/1989	VIII	6.0	

Table 2 Table of the major earthquakes in Algeria during the Antiquity period according to ancient texts

Date	Locality	References
Antic period	Rusucurru	Robert (1891), Laporte (2016)
267 AD	Lambaesis (Praetorium)	Bull. Soc. Archéologique du midi de la France (24 June 1902); Lepelley (1981)
267 AD	Lambaesis	Ballu (1893), Ferdi and Harbi (2013)
267 AD	Ad Maiores and Thamugadi	Ballu (1897), Lepelley (1981, 1984), Laporte and Dupuis (2009)
355 AD	Thubursicu Numidarum	Lepelley (1981, 1984), Ferdi and Harbi (2013)
365 AD	Cuicul	Albertini (1949), Rebuffat (1980)
419 AD	Sitifis	Guidoboni (1994); Augustin Sermon 19.6 in Ferdi and Harbi (2013)

AD 267 an earthquake which caused damage and was responsible of the collapse of an arch at night. In his book on the ruins of antique Thamugadi, Ballu (1897) related damaging earthquake which was felt in Thamugadi in AD 267. Finally, the earthquake that occurred on 21 July 365 affected much of the Mediterranean, including the Roman city of Cuicul (Djemila) (Rebuffat 1980; Di Vita 1990) (Table 2). Lepelley (1984), by reading several epigraphic texts, has drawn up a picture in which the term “ruina” (in Latin), which most often means falls of rubble, is highlighted in Table 3. This table describes various damage observed in some archaeological sites of Algeria. The AD 365 earthquake was recently studied by Stiros (2001, 2010) and conclude that the earthquake was not related to a single event but a series of at least three earthquakes in Cyprus, Crete, and Sicily/Libya giving the impression of a “universal” earthquake.

4 Damage Identification on Archaeological Sites

Deformation associated with earthquakes in the vicinity of an archaeological site can be classified into two categories according to the effects they induced. We can have direct and indirect effects. The study of damage on archaeological sites related

Table 3 Inventory of destructions observed on sites in North Africa (after Lepelley 1984)

	Date	Cities name (in Latin)	Place	Type of damage (in Latin)	Type of damage (in English)
1	Before AD 305	Thubursicu Bure (Proc.) – *C.* 25998	Square	ruinam manans	Risk of collapse
2	AD 324–326	Lepcis Magna (Tr.) *I.R.T.* 467	Basilica	ruina … deformata … [cum] diunio ictu confla-Graret incendio	Collapse… degradation as well as destruction by fire
3	AD 324–326	Lepcis Magna (Tr.) *I.R.T.* 468	Gantry	in ruinam [la] bemque conuersa	Removal of trace of destruction
4	AD 361–362	Thubursicu Numidarum (Proc.) – *I.L. Alg.*, I, 1247 et 1274	Statues transferred from a place in ruins	signum Traiani de ruinis ablatum; de ruinis asigno titulisque translatis	Removal of stigmas by cleaning and transferring statues to another location
5	AD 364	Madauros (Proc.) *I.L. Alg.*, I, 2101	Baths	[tot re] tro annis ruinarum labe deformes	A few years ago, the bath was cleared of any sign of destruction
6	AD 364–367	Lambaesis (Num.) *C.* 2656	Fountain	r[u]inis [obrutam ?]	Cover the ruins
7	AD 364–467	Mascula (Num.) *A.E.* 1911, 217	Baths	ruinarum deformitas	The ugliness of the ruins
8	AD 368–370	Abbir Maius (Proc.) *A.E.* 1975, 873	Baths	soliaris ruina conlapsus	Became ruin
9	AD 378	Sabrtha (Tr.) *I.R.T.* 103	Baths	post ruinam	After ruin
10	AD 383–392	Sitifis (Maur. S.) *C.* 8480 (*I.L. S.*, 5596)	Public Bakery	ruinis imminentibus destitutis	The ruins were left
11	Lower-Empire (AD 284–533)	Thamugadi (Num.) *B.C.T.H.*, 1907, p. 262	House	ruinis tamdiu informibus tristem, felicius quam condita est restituit	As long as these ruins are in desolation, glory to who will restore them.
12	Lower-Empire (AD 284–533)	Thibilis (Num.) *A.E.*, 1969-70, 691; *I.L. Alg.*, II, 2, 4724	Fountain	inca(s)um funditus superante ruina	Weak foundations, defeated by the ruins

to earthquakes is a recent approach, several studies such those of Stiros (1995, 1996), Stiros and Pytharouli (2014), Monaco and Tortorici (2004), and Stiros et al. (2006), highlighted the importance of archaeological, geological, and seismological investigations to retrieve the impact of ancient earthquakes on archaeological structures.

Direct effects are directly observed on structures and the indirect effects are due to other geological phenomena such as landslides, folding, liquefaction, or subsidence. According to Schaub et al. (2009), it is necessary to determine, the type of material and its use by ancient civilizations on archaeological sites (masonry, type of mortar, type of column, etc.). Thus, knowledge of the local architectural framework allows good discrimination between buildings that potentially record the first effects, associated to the ground shaking from those due to the secondary effects related to other phenomena.

5 Seismic and Non-seismic Damage on Archaeological Structures

In archaeoseismological research, it is important to discuss some ambiguous cases when damage could not be identified whether related to tectonic activity or caused by other geological or anthropic origin. In this context, we met two significant cases. We observed the first on the site of Tipasa (70 km west of Algiers) at the base of the new temple where a displacement of blocks was not due to a seismic effect but to the development of the roots of tree, which generated a vertical displacement of about 12 cm (Fig. 3). The second case was noticed in Cuicul (Djemila), and consists of a large inclination that was observed on walls of the Septimius Severe (Septime Severe) Temple. This inclination was attributed to a poor restoration of the Temple. Old pictures show the original form of the ruins, and compared with its present form it is easy to attribute the deformation to bad restoration (Fig. 4). In the Oum El Kanatir site in Jordan, Marco (2008) observed a large break in an old basin that recent investigation concludes to the effect of the landslide. To avoid erroneous conclusions on the origin of the damage, care should be given to the various deformations observed on ruins especially those which have been subject to restoration.

6 Observed Damage on Archaeological Sites

To retrieve the cause of the damage on archaeological structures and identify its relation to a seismic action, we have been inspired by the approach used by several studies. We may mention as examples particular fractures, falls of columns,

Fig. 3 Displacement of blocks of steps in a Roman Temple, Tipasa. **a** View of the Temple, **b** zoom on the tree roots, origin of the displacement of the block. **c** The block uplift was measured to about 12 cm

extrusions and intrusions of masonry blocks, as well as permanent ground deformation.

The damages caused by an earthquake on buildings are various. We can have disorders on walls, fractures, blocks, and keystones displacement and folded steps. All these observations were noted on structures in the sites we investigated.

a. Deformation on walls

In all the sites investigated, we observed folded walls. When walls are relatively subject to large distortion and displacements, they may indicate that these effects were caused by earthquakes. The best example of these deformations is located in the upper town of Tazoult (Lambaesis) where a distortion was observed in the propylaea of the Asclepieum (religious and prestigious complex, organized around the center of the city). A wall of the structure was rotated with a large tilt component. The investigated structure is of rectangular shape, but in its eastern part, the wall is no longer continuous and exhibit an outward rotation (Fig. 5a) which has been measured at more than 1 m (Fig. 5b). It should be noted that this rotation was made possible by the existence of an inflection point where the torsion of the wall begins (Fig. 5c). This torsion at the center (point of rotation) is measured at 8 cm

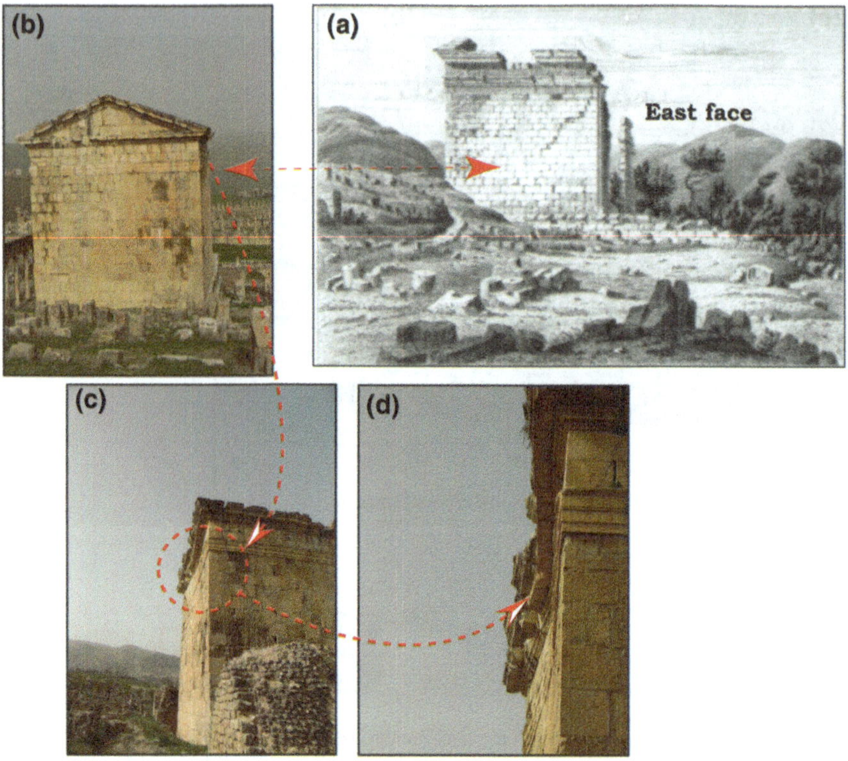

Fig. 4 Bad repairing of the Septime Severe Temple in Cuicul **a** old picture in black and white, before repairing, **b** recent picture after repairing, **c**, **d** view and zoom on the deformed wall

(Fig. 5d). This rotation induced a considerable inclination of the structure towards the interior of the building (Fig. 5e) visible, also from inside the structure (Fig. 5f). The resulting inclination and rotation (Fig. 5g) resulted in an extrusion blocks (13 cm) on a wall perpendicular to this structure (Fig. 5h). Even if they are of minor importance, inclined walls were observed in many Algerian archaeological sites, notably in Thubursicu Numidarum and Cuicul at the thermal baths with 10° inclination (Fig. 6). The walls under lateral seismic load could be tilted with various inclination and this is a good indicator of seismic action. Rodriguez-Pascua et al. (2011) indicate that in ancient cities, the inclination of walls, in general, may be a consequence of a significant horizontal movement. Moreover, Korjenkov and Kaiser (2003) and Korjenkov et al. (2003) specify that the types of deformation observed on walls are dependent on the orientation of the walls. We should note that these effects are better seen on the brick walls than on walls with large and heavy stones (opus quadratum) which relatively have more rigidity for being not tilted by earthquake shaking.

b. Fractures

Active faults could reach the surface (strong earthquakes) and generates traces on ground, damaging the constructions with different degrees. The ground shaking may also cause fractures on structures due to transient movements. But this is not always the case because in most cases, fractures could have other origin. The best examples were noticed in Thamugadi where several forms of fractures are observed. Fractures were observed on an oil roller mill near the Lambaesis gate with more than one meter in length (Fig. 7a, b), and with E-W orientation (Roumane 2016).

c. Displaced blocks

The displacement of large blocks in archaeological structures is attributed to damage due to seismic shaking according to many authors, such as Sintubin and Stewart (2008), Hinzen et al. (2010) and Hinzen and Yerli (2010). Yerli et al. (2010) links these displacements to the proximity of a fault. For Bilham et al. (2010), who worked on a Hindu temple, the displacement of large blocks is related to a seismic action (intensity VII to IX, MSK scale).

For this type of deformation, several observations were made on all sites. In the site of Thamugadi, in the lower part of the capitol (Fig. 8a), there is a movement of a block with a slight inward orientation (Fig. 8b, c), quite large of about 15 cm (Fig. 8d). Karakhanyan et al. (2010), described the same deformation at the level of a temple in Luxor in Egypt. In the site of Thibilis, the same disorder was recorded at the Christian basilica at the lower part of the eastern section. The investigation on this typical case shows block extrusion to outside (Fig. 9a, b), with an offset of about 14 cm (Roumane and Ayadi 2016). The same pathology was found at Thevest on the byzantine rampart, not far from the arc of Caracalla (symbol of the Metropolis). The observed deformation shows a masonry block extrusion (Fig. 10a), in the lower part of the rampart. The two extruded blocks do not exhibit any rotation, but the displacement is variable between 13 and 15 cm (Fig. 10b). The significant example was observed in the Lambaesis mausoleum. The observed deformation is represented by a lateral displacement of blocks and horizontal rotation also reported by Gsell (1893) (Fig. 11a) in the upper part of the structure (Fig. 11b, c) and in the lower part (Fig. 11d). Similox-Tohon et al. (2006) reported similar deformation in the upper part of a Roman mausoleum at Pinara (Southwest Turkey).

d. Dropped keystones and lintels

Keys stones of arches are common in old structures usually in doors, large windows, bridges, and other corridors. In many cases, the deformations recorded in keys and arches suggest a displacement of the vault key downwards, which may be according to Marco (2008), and Silva et al. (2009) an indicator of seismic origin. Kamai and Hatsor (2007) point out that the sliding of the lower lintels is also an earthquake shaking result. There is a relation between the vertical sliding of the key

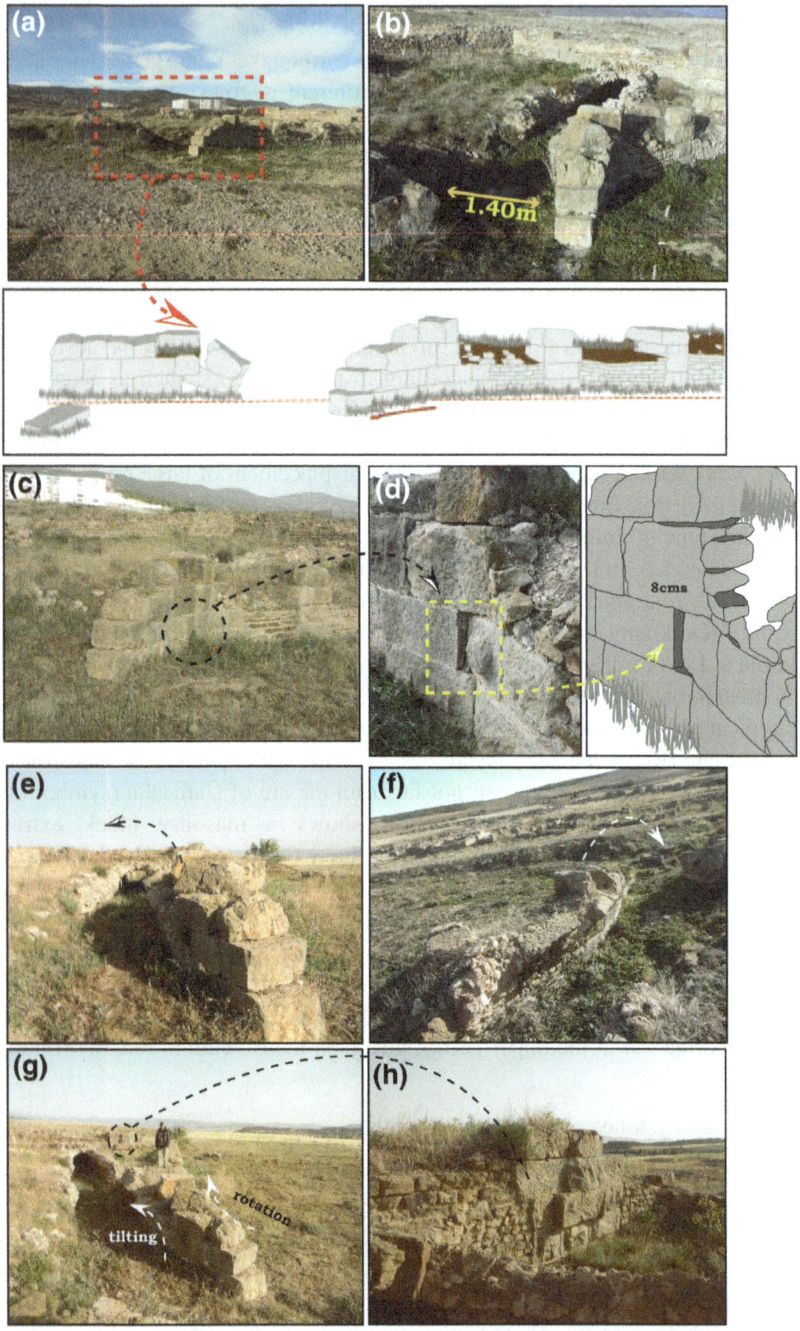

Fig. 5 Displaced and rotated wall at Lambaesis site. **a** Rotated wall. **b** Displacement of about 1.4 m. **c**, **d** Point of rotation. **e** Tilting of the wall towards the interior of the building. **f** Rear view of the tilting. **g**, **h** Rotation and tilting and extrusion of blocks in the background

Fig. 6 Displaced and rotated wall at Cuicul site. **a** Thermal baths. **b**, **c** Main street

Fig. 7 Fractures observed at Thamugadi site. **a**, **b** Basis of oil mills

Fig. 8 Displaced blocks at Thamugadi site. **a** In the Capitole. **b** Zoom on the displacement measured at 15 cm

Fig. 9 Extrusion of blocks in the Christian basilica at Thibillis site

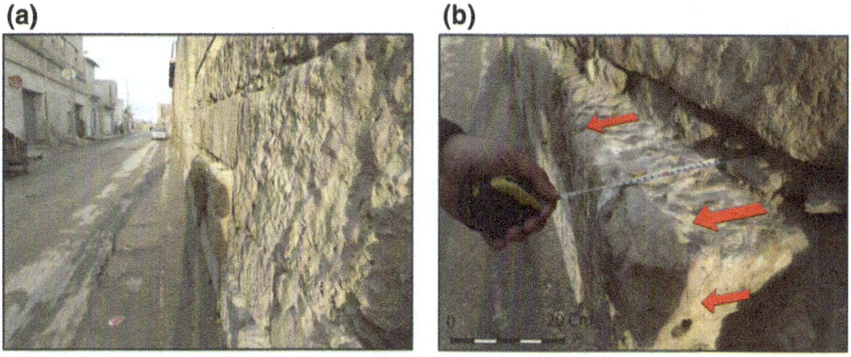

Fig. 10 Extrusion of blocks at Thevest **a** on a wall of Caracalla Temple, **b** zoom on the displacement observed

to the bottom and the direction of the seismic shaking. Thubursicu Numidarum city recorded an important example of this pathology. Three kilometers east of this site, we investigated an aqueduct in perfect conservation and its foundations have not suffered any deformation. The central part of the vault key of the Roman aqueduct moved down by about 10 cm (Fig. 12). On the site of Madauros, collapsed lintels were studied at the thermal baths on the entrance facing the slave tunnel. In the central part, attempts of restoration can be seen on the facade, allowing the building to remain in place (Fig. 13). These lintels are degraded in the left part, while in the right part, the offset is clearly visible when we measured it, to about 5 cm.

e. Folded steps

Soft sedimentary rocks, especially clays, react differently compared with other limestones and sandstones (competent rocks). The recording of the deformations and the reaction of these elements are different. The soft sedimentary rocks exhibit sort of plastic deformation, on the other hand, competent rocks will be displaced or fractured depending on the seismic loading. The deformations observed on the pavements and paving stones seem most often to be related to the local instability of the soil. Stairs are a good indicator of seismic disorder. Blocks of rock forming the steps could be folded or displaced. We observed such deformation on a western view of the arch of Trajan at Thamugadi, and the steps on the path to the arch of Trajan were deformed. This deformation is shown by large undulations of the different steps (Fig. 14). Slight deformation of the ground due to various non-seismic geotechnical effects can produce extension on arches and keystone drops or even collapse (Stiros 1996).

Fig. 11 Lateral displacement of blocks at Lambaesis site. **a** Observations made on picture from Gsell (1893). **b**, **c** Actual view of the upper level. **d** Lateral displacement in the basal part

Fig. 12 Collapsed keystones at roman aqueduct in Thibursicu Numidarum site

Fig. 13 Collapsed lintels near the thermal baths at Madauros site

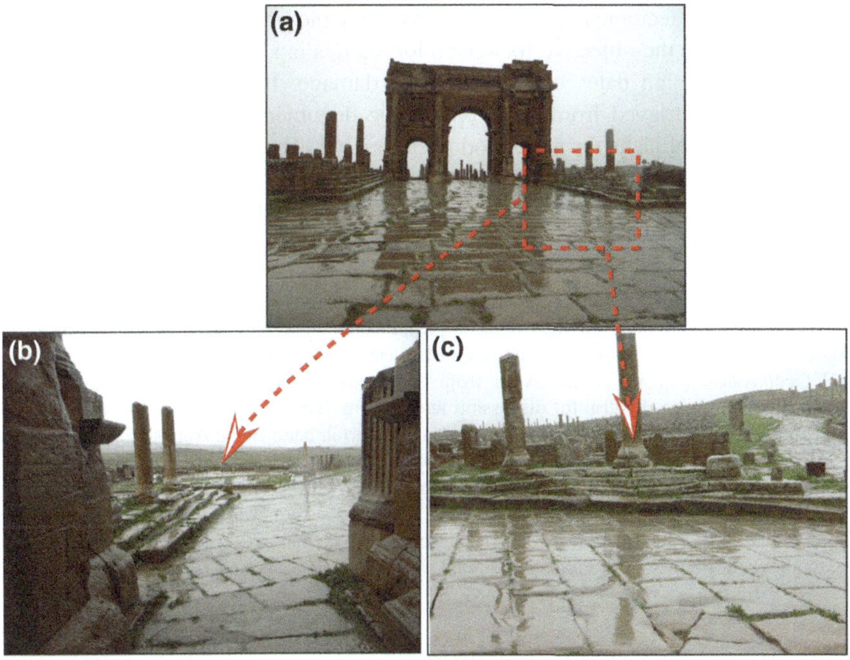

Fig. 14 Folded steps on stairs on the path to the Arche of Trajan at Thamugadi site

7 Conclusion

The aforementioned deformations and disorders in this study were carefully analyzed following observations that have been made by other authors on other sites. We have compared our results to that obtained by Stiros (1996), Sintubin and Stewart (2008) and Rodriguez-Pascua et al. (2011). During our study, we have visited ten sites of the Roman period in Algeria and most of them that are located within the Tell Atlas considered as a seismic active zone in northern Algeria. We tried to identify the action of probable past strong earthquakes considering the damage observed on archaeological structures. Attention was given to deformation that is not of seismic origin but of bad repairing procedures or other source of action. The bad repairing process was observed on several sites as on the Septime Severe Temple and Arch of Caracalla in Cuicul which are the best examples. The Septime Severe Temple was badly repaired as shown by the old and recent pictures of the structure, and the Arch of Caracalla was disassembled for transportation to France (during the French colonization in the beginning of the twentieth century) and reassembled on site, unfortunately with some deformation and disorders on it. We focused in this paper on damage and disorders that are related to probable earthquakes. This work should be completed by a detailed tectonic investigation in the archaeological sites and their surroundings to identify active faults. Much remains to be done on the dating of damage and the seismic events associated in relation with the tectonic context where the archaeological site is located. This will be useful and be the objective for seismologists dealing with historical seismology who need to assign dates on the observed damage. In Algeria, except for few seismic event retrieved from some epigraphic documents, the seismic catalogues seriously suffer of lacks for the period prior to 1365. The recognition of earthquake effects on archaeological sites should lead us to pay more attention to this heritage to better save them by engaging a serious policy of preservation of the archaeological heritage.

Acknowledgements The authors would like to thank colleagues from various institutions, Salim Drici, and Mustapha Filah, from the Archaeological Institute, University Abu Al Kacem Saadallah, Algiers, Said Maouche, and Farida Ousadou from Center of Research in Astronomy Astrophysics and Geophysics, Algiers, Kamel Amri from the Earth Sciences Faculty, University Houari Boumedienne, Algiers for fruitful discussion and support. The authors would like also to thank particularly Ghilles Rabai, Mohamed Beresse and Abderrahmane Bellal from the Earth Sciences Faculty, University Houari Boumedienne, Algiers for their help during the preparation of the figures. The authors would like to thank the anonymous reviewers for their fruitful comments and suggestions. This work was conducted in the framework of the MEDYNA FP7-PEOPLE-2013-IRSES project, WP-1: Present-day Kinematics and seismic hazards, funded by the Seventh Framework European Program FP-7.

References

Albertini E (1949) Une nouvelle basilique civile à Cuicul (Djemila). Comptes rendus des séances de l'Académie des Inscriptions et Belles-Lettres, vol 87(3)

Ayadi A, Bezzeghoud M (2015) Seismicity of Algeria from 1365 to 2013: maximum observed intensity map (MOI2014). Seismol Res Lett

Ballu A (1893) Monuments antiques de l'Algérie: Tébessa, Lambèse, Timgad.: Conférence faite au palais du Trocadéro le 11 décembre 1893 […]

Ballu A (1897) Les ruines de Timgad (antique Thamugadi). Eds. Leroux, Paris

Baradez J (1952) TIPASA, ville antique de Maurétanie. Service des Antiquités de l'Algérie, p 80, 52 ill., I carte

Bilham R, Bali BS, Bhat MI, Hough S (2010) Historical earthquakes in Srinagar, Kashmir: clues from the Shiva temple at Pandrethan. Geol Soc Am Spec Pap 471:107–117

Bulletin archéologique du Midi de la France. Deuxième série du 19/11/1901 au 07/07/1903.p 456

Cagnat R (1923) L'Asclepieium de Lámbese (Numidie). Atti della Pontificia Accademia romana di Archeologia, . s. Ili, Memorie, voi. I, Rome 1923, pp 81–88

Courtois C (1951) Timgad, Antique Thamugadi, Edition du Gouvernement Général de l'Algérie. Direction de l'Intérieur et des Beaux-Arts Service des Antiquités. (Alger, Imprimerie Officielle, 1951, 104 p. romain, pp 425–494, CERF, 134, Rome

Di Vita A (1990) Sismi, Urbanistica e cronologia assoluta. Terremoti e urbanistica nelle citta di Tripolitania fra il I secolo A.C. ed il IV D.C. Collection de l'Ecole Française de Rome 134:425–494

Ferdi S, Harbi A (2013) Roman literary and epigraphic sources for the study of historical seismicity in Algeria circa 42–420 AD. J Seismol. https://doi.org/10.1007/s10950-013-9400-5

Février A (1978) Djemila. Direction des arts, musées, monuments historiques et antiquités. Alger, p 118

Gascou J (1972) La politique municipale de l'Empire romain en Afrique proconsulaire, de Trajan à Septime-Sévère. Ecole française de Rome

Gsell S (1893) Recherches archéologiques en Algérie. 4e partie Inscriptions de Khamissa, de M'daourouch et de Ksiba-Mraou. Paris, 434 p

Gsell S (1894) Tipasa, ville de la Maurétanie Césarienne. Mélanges d'archéologie et d'histoire T. 14(1894):291–450

Gsell S (1901) Les monuments antiques de l'Algérie. Services des monuments historiques de l'Algérie., 2 tomes (pp 366–464)

Gsell S (1914) Khamissa, Mdaourouch, Announa (avec la collaboration de Ch. A. Joly). 1ère partie, Khamissa, 1er fascicule, Alger et Paris, p 114

Gsell S (1922) Khamissa, Mdaourouch, Announa. Services des monuments historiques de l'Algérie, p 510

Guidoboni E (1994) Catalogue of ancient earthquakes in the Mediterranean area up to the 10th century. Instituto Nazionale di Geofisica Press, p 504

Harbi A, Sebaï A, Benmedjber M, Ousadou F, Rouchiche Y, Grigahcene A, Aini D, Maouche S , Ayadi A (2015) The Algerian homogenized macroseismic database (267–1989): a deeper insight into the Algerian historical seismicity. Seismol Res Lett

Hinzen KG, Yerli B (2010) The lycian Sarcophagus of Arttumpara, Pinara (Turkey)—testing seismogenic and anthropogenic damage scenarios. Bull Seismol Soc Am 100(6):3148–3164

Hinzen KG, Fleischer C, Reamer SK, Schreiber S, Schütte S, Yerli B (2010) Quantitative methods in archaeoseismology. Quatern Int. https://doi.org/10.1016/j.quaint.2010.11.006

Kamai R, Hatzor YH (2007) Numerical analysis of block displacements in ancient masonry structures: a new method to estimate historic ground motions. Int J Numer Anal Meth Geomech 32:1321–1340

Karakhanyan A, Avagyan A, Sourouzian H (2010) Archaeoseismological studies at the temple of Amenhotep III, Luxor, Egypt. Geolog Soc Am Spec Pap 471:199–222

Korjenkov AM, Kaiser D (2003) Historical-macroseismic study of the town church in Wittstock, Northern Germany. In: Proceedings, 11th FIG symposium on deformation measurements, Santorini, Greece

Korjenkov AM, Baipakov C, Chang Y, Peshkov-Savelieva T (2003) Traces of ancient earthquakes in medieval cities along the Silk Road, northern Tien Shan and Dzhungaria. Turk J Earth Sci 12:241–261

Lancel S (2005) Les hommes de l'Antiquité face aux séismes. In: Comptes-rendus des séances de l'Académie des Inscriptions et Belles-Lettres, 149e année, N. 4 (pp 1281–1289)

Laporte J-P, Dupuis X (2009) De Negrines Maiores à Negrine. Antiquités africaines 45:51–102

Laporte J-P (2016) Peuplement et catastrophes. In: Le peuplement du Maghreb. Actes du troisième colloque international SoussseTunisie)

Lepelley C (1981) Les cités de l'Afrique romaine au Bas empire. études augustiniennes. t 2, p 213

Lepelley C (1984) L'Afrique du Nord et le séisme universel du 21 Juillet 365. MEFRA, pp 463–491

Maouche S, Meghraoui M, Mohrange S, Bellabes S, Bouhaddad Y (2010) Active coastal thrusting and folding, and uplift rate of the Sahel Anticline and Zemmouri earthquake area (Tell Atlas, Algeria). Tectonophysics 509(1):69–80

Marco S (2008) Recognition of earthquake-related damage in archaeological sites: examples from the Dead Sea fault zone. Tectonophysics 453:148–156

Monaco C, Tortici L (2004) Faulting and effects of earthquakes on Minoan archaeological sites in Crete (Greece). Tectonophysics 382:103–16

Rebuffat R (1980) Cuicul le 21 juillet 365. Ant Afr 15:309–328

Robert G (1891) Voyage à travers l'Algérie: Notes et croquis. Edtions E, Dentu, Paris

Rodriguez-Pascua MA, Perez-Lopez R, Gines-Robles JM, Silva PG, Garduno-Monroy VH, Reicherter K (2011) A comprehensive classification of earthquake archaeological effects (EAE) in archaeoseismology: application to ancient remains of Roman and Mesoamerican cultures. Quatern Int 242:20–30

Roumane K (2016) Identification des dommages induits par les tremblements de terre sur les structures archéologiques. Contribution à l'Archéosismologie en Algérie. Thèse Doctorat, Université, Alger, 195 p

Roumane K, Ayadi A (2016) Impact of earthquakes on archaeological sites in Algeria. In: The first ArabGU international conference Algiers, 17/18 Feb 2016

Schaub A, K Reicherter, C Grützner, T Fernández-Steeger (2009) Evidence for a medieval earthquake in the Aachen area (Germany), revealed by structural damage in the cathedral. Abstracts volume of the 1st INQUA-IGCP 567 international workshop on earthquake archaeology and palaeoseismology, 7th–13th Sept. In: Pérez-López R, Grützner C, Lario J, Reicherter K, Silva PG (eds) Archaeoseismology and palaeoseismology in the Alpine-Himalayan collisional zone, p 132. Baelo Claudia, Spain

Silva PG, Reicherter K, Grutzner C, Bardaji T, Goy JL, Zazo C, Becker-Heidmann P (2009) Surface and subsurface palaeoseismic records at the ancient Roman city of Baelo Claudia and the Bolonia Bay area, Cádiz (south Spain). Geol Soc London 316:93–121

Similox-Tohon D, Sintubin M, Fernández M, Muchez P, Waelkens M (2006) Identification of an active fault by a multidisciplinary study at the archaeological site of Sagalassos (SW Turkey). Tectonophysics 420:371–87

Sintubin M, Stewart IS (2008) A logical methodology for archaeoseismology: a proof of concept at the archaeological site of Sagalassos, Southwest Turkey. Bull Seismol Soc Am 98(5):2209–2230

Stiros SC (1995) Archaeological evidence of antiseismic construction in antiquity. Ann Geophys XXXVIII(N° 5–6):725–736

Stiros S (2001) The AD 365 Crete earthquake and possible seismic clustering during the 4–6th centuries AD in the Eastern Mediterranean: a review of historical and archaeological data. J Struct Geol 23:545–562

Stiros S (2010) The 8.5+ magnitude, AD365 earthquake in Crete: coastal uplift, topography changes, archaeological and historical signature. Quatern Int 216:54–63

Stiros SC, Pytharouli SI (2014) Archaeological evidence for a destructive earthquake in Patras. J Seismol, Greece. https://doi.org/10.1007/s10950-014-9437-0

Stiros SC (1996) Identification of earthquakes from archaeological data: methodology, criteria and Limitations. Fitch Laboratory Occasional Paper 7. In: Stiros, S, Jones R (eds) Archaeoseismology. British School at Athens, pp 129–152

Stiros SC, Papageorgiou S, Kontogianni V, Psimoulis P (2006) Church repair swarms and earthquakes in Rhodes Island. Greece J Seismol 10:527–537

Yerli B, Ten-Veen J, Sintubin M, Karabacak V, Yalciner CC, Altunel E (2010) Assessment of seismically induced damage using LIDAR: the ancient city of Pınara (SW Turkey) as a case study. Geol Soc Am Spec Pap 471:157–170

A Glimpse at the History of Seismology in Algeria

Assia Harbi, Amal Sebaï and Mohamed Salah Boughacha

Abstract Seismology has a long tradition stretching back over three centuries in Algeria since the country is an earthquake-prone area. This paper presents a quick overview on the development of seismology in Algeria from the first written records of seismic events in the fourteenth century up to the monitoring of earthquakes and modern, and historical seismological studies nowadays. We particularly focus on the important milestones on the way of progress of seismology in Algeria and present a summary on what has been achieved so far in seismology research and education.

Keywords Algeria · Seismology · History · Progress · Research
Education

1 Introduction

Algeria is among the most seismic zones in North Africa. The seismicity of Algeria, which is essentially concentrated in the North of the country (Fig. 1) in the Atlas Mountains (Tell and Sahara Atlas ranges), is due to the 4–6 mm/year NW–SE Africa-Eurasia plate convergence (Nocquet and Calais 2004). Algeria experienced moderate-sized and strong earthquakes in the past. Figure 2 illustrates the most damaging and destructive earthquakes that occurred in the region during its seismic

A. Harbi (✉) · A. Sebaï
Centre de Recherche en Astronomie, Astrophysique et Géophysique,
BP 63, Bouzaréah, 16340 Algiers, Algeria
e-mail: harbi.assia@gmail.com

A. Harbi
Simons Fellow at Abdus Salam, International Centre for Theoretical Physics,
Strada Costiera, 11, 34151 Trieste, Italy

M. S. Boughacha
Université des Sciences et de la Technologie Houari Boumediène,
BP 32, El Alia, 16111 Bab Ezzouar, Algiers, Algeria

A. Bendaoud et al. (eds.), *The Geology of the Arab World—An Overview*,
Springer Geology, https://doi.org/10.1007/978-3-319-96794-3_9

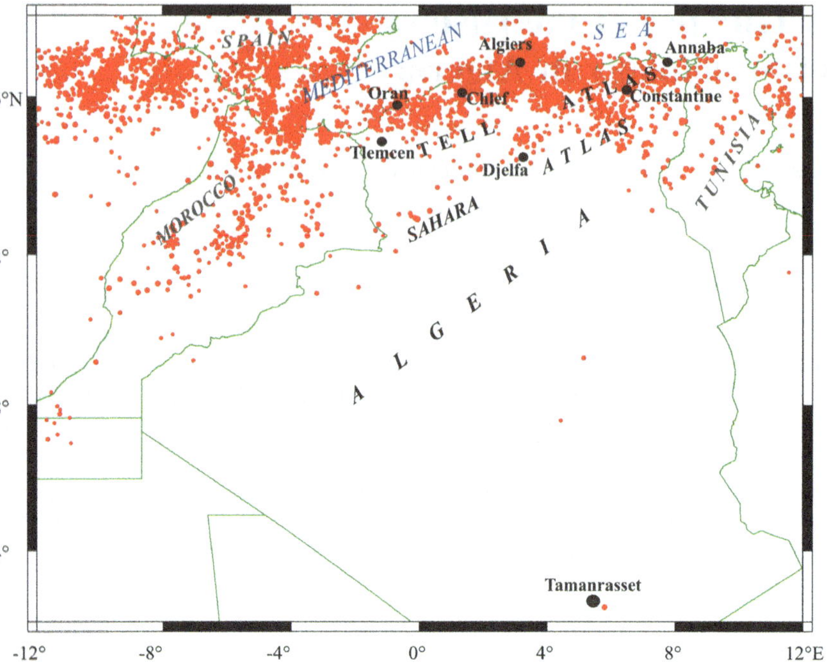

Fig. 1 The spatial distribution of earthquakes from 1970 to 2016 ($M \geq 3$). Data are from International Seismological Centre, On-line Bulletin, http://www.isc.ac.uk, Internatl. Seismol. Cent., Thatcham, United Kingdom, 2014

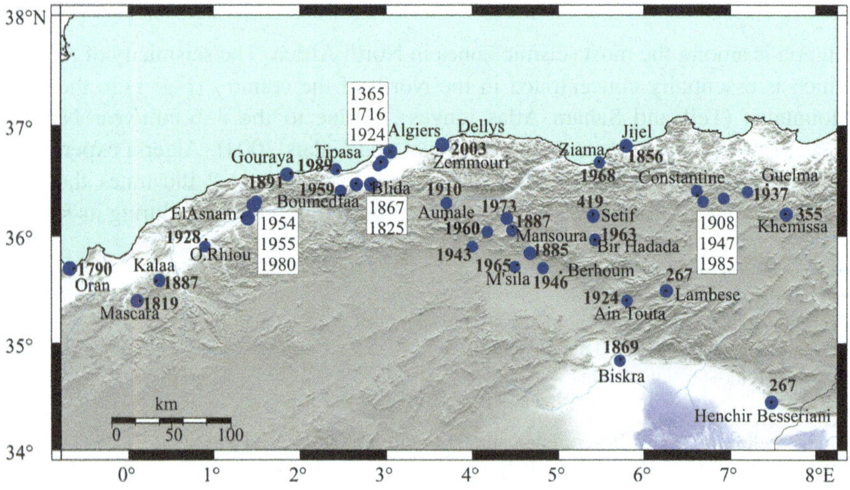

Fig. 2 Spatial distribution of damaging and destructive earthquakes in Algeria ($I_0 \geq$ VIII EMS). Data are from Harbi et al. (2015)

history. It is worthwhile noting that light to minor earthquakes occurred and continue to take place until recently in an area remote from the plate boundary, in the Hoggar shield, near Tamanrasset (Grandjean et al. 1960; Boughacha 2005; Boughacha et al. 2006; Harbi et al. 2010; Bourouis et al. 2013). The origin of this seismicity, whether or not it is from intraplate tectonics, is still subject of debate.

The perception of earthquakes is well known by the Algerian population from north to south. Therefore, the Algeria intellectuals and scientists took an interest in understanding and studying the earthquakes from time immemorial. In this paper, a chronological approach will be adopted in order to give a quick overview on what was done so far in terms of seismology in Algeria (Table 1). The objective of this paper is not to review the studies of Algerian earthquakes, but to present a summary of some important milestones on the way of progress of seismology in Algeria. The reader will notice that this discipline almost flourishes after each damaging to destructive event, which gives a fresh start and a boost to the earthquake science in Algeria and leads to the installation/acquisition of new high-performance equipment and to sound results. We do not pretend in this article to give a definitive and comprehensive account of the history of seismology in Algeria, but just a glimpse drawn from our experience as scientists mainly involved in historical seismology research. No new investigations were carried out to recount the history of seismology in Algeria. In fact, when seeking information on past earthquakes, we have often found very interesting details, which are more related to the history of seismology than to historical seismology itself. We have gone from surprise to surprise in discovering that Algeria, a colonized, then a developing country was not so far from what was done in terms of progress of seismology in its geographical area. The work of earthquake scientists of Algeria has significantly contributed to a better understanding of seismicity and active tectonics in Mediterranean and North Africa.

2 Pre-French Period (Before 1830)

The historical Algerian earthquake catalogue (Ambraseys and Vogt 1988; Mokrane et al. 1994; Boughacha 2005; Harbi et al. 2007a, 2010, 2015) does not span a period of time as long as the period covered by earthquake catalogues worldwide, e.g., 815 BC (the Iberian Peninsula, Roca et al. 2004), 1365 BC (Syria, Sbeinati et al. 2005), and 23rd century BC (China, Wang 2004). In fact, we know very little about Algerian earthquakes earlier than 1825. The town of Dellys (Fig. 2), which experienced disastrous effects following the 2003 Zemmouri earthquake, was alleged to have been destroyed in 42 A.D. (see Harbi et al. 2007b for more details), but until now we have found no original sources confirming the occurrence of that earthquake. As already reported in Ferdi and Harbi (2014), earthquakes were relatively frequent in Algeria in the first century. According to Beaujeu (1973), the first Algerian scientist who discussed earthquakes and tsunami is Apulée de Madaure (125–170 A.D.) in his cosmographic book "De mundo". Little information about the earthquakes that occurred in Algeria during ancient times was found

Table 1 Summary of the main events of the history of seismology in Algeria between 1830 and present

Before 1830	1845	1847	1856	1858	1864	1867	1891
First "seismic code" or preventive measures in reconstructing the city after the 1716 earthquake were issued by the Algiers Governor, Dey Ali Chaouche	First compilation of earthquakes of Algeria	The Perrey earthquake lists	First descriptions of the effects of an earthquake, first picture on the damage caused by a seismic event	Creation of the Algiers Observatory	The Earthquakes in the Algerian scientific periodicals	A step toward teamwork, from a simple pendulum in 1856 to a seismograph, first mapping of geological effects	Very first questionnaires

1906	1907–1910	1916	1919	1922	1924	1925	1931
First seismicity map	Creation of the service of seismology, first seismic stations, use of reinforced concrete	First intensity estimate?	Annals of the Institute of Earth Physics of Strasbourg	First macroseismic and instrumental study and first attempt of the identification of the causative fault	First isoseismal map drawn by a contemporaneous author	First map of the earthquakes frequency	The Algerian meteorological service became the "Institut de Météorologie et de Physique du Globe d'Algérie" (IMPGA)

1935–1949	1950	1955–1958	1960	1962	1963–1968
Installation of the following seismic stations: 1935: Oued Fodda 1948: Tamanrasset 1949: Algiers	First mapping of the causative fault of an earthquake, first determination of epicenter of past earthquakes and first seismicity map of Algeria	Installation of the following seismic stations: 1955: Relizane 1958: Setif	First analysis of the seismicity of the Hoggar shield	Independence of Algeria	Research activity of the IMPGA under the umbrella of Franco-Algerian agreements

1960–1970	1970–1979	1980	1985–1989	1990s	2000–Present
Seismology in Algeria faced tough times	Revival of seismology in Algeria	Creation of CNAAG, renaissance of seismology in Algeria, El Asnam earthquake: the largest event ever known in Algeria	Creation of the department of geophysics at USTHB First group of geophysics researchers at CRAAG Inception of the Department of seismological studies and survey at CRAAG at the end of 1990	Installation of the first modern seismological network: RSTA First tomography study, First earthquake catalogues in independent Algeria Inception of NAGET in 2000 at ICTP	ADSN network has been developed on the former network RSTA, then upgraded Development of the REGAT GPS Network Inception of SAG in 2007 at USTHB

RSTA Réseau Sismologique Telemetré Algérien; *ADSN* Algarian Digital Seismic Network; *NAGET* North-African group for earthquake and tsunami studies; *ICTP* The Abdus Salam International Center for Theoretical Physics; *SAG* Société Algérienne de Géophysique; *USTHB* Université des Sciences et de la Technologie Houari Boumediène; *CRAAG* Centre de Recherche en Astronomie, Astrophysique et Géophysique; *CNAAG* Centre National d'Astronomie, d'Astrophysique et de Géophysique in epigraphic and literary sources such as the 267 Lambaesis earthquake and the 419 Sitifis earthquake (Ferdi and Harbi 2014).

The Algerian scholars of the Middle Ages and those from the Maghreb countries, and Andalusia who traveled to Algeria (called Central Maghreb in medieval times) used to report the earthquakes, particularly in their chronicles or when they dealt with historiography. The Andalusian scholar Ibn El-Haj Ennoumeyri (1313–1367) who traveled to Algeria and worked for some years in Bejaia, reported in his travel notes, which are collected in a book presented by Ibn Chakroun (1990), that when he was in Miliana, he read a letter of the judge of this city to the Sublime Porte (Istanbul during the Ottoman Empire) informing that three shocks occurred in 1344 in Miliana and were followed by hail that damaged some houses. In his book "History of the Algiers Pashas", Ibn Al Mofti Hussein Ibn Rajeb Chaouche (1095H–1144H ~ 1683/1684–1731/1732 A.D.) referred to the religious scholar El Brechki who experienced the 1365 Algiers tsunamigenic earthquake, and related the effects of this earthquake and the sea wave that inundated Bab El Oued following this seismic event (see the original information on http://naget.ictp.it/PUBLICATIONS/resources/AMD.pdf, last accessed in October 2017). Until recently (Harbi et al. 2015), this event was the first, which was reported in the Algerian earthquake catalogue (Ambraseys and Vogt 1988; Mokrane et al. 1994). Ibn Al Mofti also reported the effects of the earthquakes that occurred in Algiers (1585 and 1716), Dellys (1631), and Médéa (1632). It is important to recall that the first "seismic code" or preventive measures in reconstructing the city after the 1716 earthquake were issued by the Algiers Governor, Dey Ali Chaouche. It was recommended to build the houses in such a way that the balconies lie against one another over the streets, with an additional timber propping (in cedar or eucalyptus) in order to strengthen the supports, as we may see nowadays in the Algiers Casbah (Rothé 1970; Abdessemed-Foufa and Benouar 2010).

In a book on astronomy and astrology published in 1192 H (1778/1779 A.D.), Mohamed Ibn Ali Ech-Chellati, better known as Ibn Ali Cherif, related the destructive effects of an earthquake, which occurred in his region, Chellata in 1767 (Harbi et al. 2015). In the Western world, one claim that the 1755 Lisbon earthquake served as milestone on the road of progress of seismology, which is true. This somewhat applies for Algeria and the change of Algerian scholars' interest in earthquakes is particularly perceptible in the work of Ahmed *Ibn Sahnoun Al Rachidi* (1791). For the first time, one has not only a summary of the effects of the earthquake, as the previous scholars used to do, but also a detailed description of these effects and how an earthquake is perceived at the time of the author. Ibn Sahnoun Al Rachidi experienced the 1790 Oran earthquake whose disastrous effects on the population were at the origin of the departure of the Spanish who were colonizing Oran from the sixteenth century. Al Rachidi reported a famous poem of Abou Ras on the effects of the 1790 earthquake and provided us with information on the definition of an earthquake by three categories of persons of his time in Central Maghreb (Algeria now): (1) ordinary people, (2) scientists, and (3) clerics (mainly Sufis). All these points of view were detailed by Ghalem (1998) and deserve to be known. In the diary of Al Hadj Ahmed Cherif Al-Zahar (1754–1830) presented by Al-Madani (1974), one finds some details on the effects of the 1802 Kolea and 1825 Blida destructive earthquakes, and especially on the emergency relief measures taken by the Agha (civil or military officer in the Ottoman empire) toward the population. For all authors, the earthquake is just an event that they experienced since the first objective of their respective work is merely historiographical.

Over a century before the French colonization, French consuls and clerics systematically reported on the earthquakes felt in Algiers and surroundings. This was the case of the events felt in Algiers between 1716 and 1811 (more details in Sebaï and Bernard 2008). It seems that the British consuls also were used to send accounts on the Algeria earthquakes since we found in the archives two letters describing the effects of the 1825 Blida earthquake (Harbi et al. 2017). As already reported in our previous works, press reports largely contributed to the survival of macroseismic information of Algeria earthquakes. The first press report available to us comes from "La Gazette de France" of February 20, 1723 and concerned the November 29, 1722 Algiers damaging earthquake (Harbi et al. 2015). British newspapers also reported information on Algeria earthquakes such as "The London Chronicle" (1763, 1766) and "The Times" (1790, 1825).

3 French Period (1830–1962)

3.1 1845: The First Compilation of Earthquakes of Algeria

It was long believed that the first compiler of the earthquakes of Algeria is Perrey (1847). The macroseismic survey that we recently carried out (Harbi et al. 2015) has shown that the first list of earthquake appeared in 1845 and was compiled by Dr. Finot who was physician in chief of the Blida hospital. This list contains the significant earthquakes that occurred in Blida and surroundings from 1760 to 1840 and seems to be based on Arabic documents since the dates are converted from the Hegira calendar by the author. However, the Finot's list contains less than 10 events.

3.2 1847: The Perrey Earthquake Lists

Perrey is one of the most famous compilers of historical earthquakes in Europe and Mediterranean. He published annual lists of earthquakes that occurred throughout the world from 1844 to 1871. In 1847, he published a note on the earthquakes of Algeria and North Africa and continued reporting earthquakes of Algeria in his different lists until 1869. The work produced by Perrey, while providing useful data on earthquakes of those periods, did not involve studies of historical earthquakes. The second descriptive earthquake catalogue of Algeria was published in 1892 by Chesneau. The description of these catalogues and the successive catalogues compiled for Algeria, and briefly presented hereafter, is outside the scope of this paper.

3.3 1856: First Comprehensive Descriptions of the Effects of an Earthquake, Early Attempt to Record the Ground Motion, and First Picture on the Damage Caused by a Seismic Event in Algeria

The occurrence of a destructive seismic event in Algeria has always sparked the interest in earthquakes and boosted the earthquake studies. We will see throughout this paper the progress made in seismology for better knowledge of the seismicity, understanding of seismotectonics, and assessment of earthquake hazard in Algeria.

Two destructive seismic events struck Djidjelli and surrounding areas on August 21 and 22, 1856, with an estimated intensity VIII, IX (EMS), respectively. These shocks, which triggered tsunami, are the first well-documented historical earthquakes in Algeria (Harbi et al. 2011), thanks to the wealth of information produced in the wake of these earthquakes. At the time it was the largest and most intense investigation of an earthquake in Algeria. Among all the available materials that we found with regard to these events, two detailed and invaluable reports drew our attention. These reports were made by military officers: Schoenagel (1856) and De Sénarmont (1857). Both authors reported on the underground noise heard during the shaking, the ground and hydrological earthquake effects, the tsunami effects, the damage to buildings, and the number of aftershocks and how they were felt. This was the first time when one estimates the height of the sea wave triggered by an earthquake in Algeria (see Harbi et al. 2011 for more details). Schoenagel (1856) detailed the damage to private houses and public buildings at Djidjelli by providing us with drawings, which highlight the damage to some specific buildings. He also recommended new ways of improving local constructions procedures following these earthquakes. De Sénarmont (1857) provided a clear description of all the effects at each site of the felt area and the direction of the ground motion on these sites. He reported the effects on objects, while Schoenagel reported the effects on animals. De Sénarmont wrote an additional section on the absence of any meteorological phenomenon before, during, and after the earthquake. The correlation between the occurrence of earthquakes and the atmospheric disturbance was frequently made at that time as we will see hereafter. According to De Sénarmont, a pendulum was installed in the morning of August 22 (before the earthquake) by the Engineer Maevus at Constantine. Little is known of its operation except that it recorded a ground motion in an NE–SW direction. The reports of Schoenagel and De Sénarmont were the first such accounts of destructive earthquakes and illustrate the level of the understanding about seismic phenomena at that time in Algeria. It is true that none of the authors estimated the intensity of the events; however, the way with which they synthesized the information makes it easy for any analyst to estimate and draw an intensity map for these earthquakes. 1856 was also the first year in which an engraving showing the damage caused by an earthquake (Djidjelli 1856) is published in Algeria (Ambraseys 1982).

3.4 1858: Creation of the Algiers Observatory

The Algiers Observatory was created on November 26, 1858, under the name "stations of astronomical observations." It was under the aegis of the Ministry of Education until July 6, 1861 when it moved to the General Government of Algeria, and then it returned to the Ministry of Education on December 26, 1873. The mission of the institution consisted in making meteorological, magnetic, and astronomical observations (see http://www.obs-hp.fr/dictionnaire/observatoires.pdf). The meteorological observations included the earthquakes too. Previously located at Algiers at its inception in 1858 (Bulard 1873) then at Kouba in 1881 (Trépied 1884), the Algiers observatory moved to its present location at Bouzaréah in the outskirts of Algiers in 1885.

3.5 1864: The Earthquakes in the Algerian Scientific Periodicals

The scientific periodical "*La Gazette Médicale de l'Algérie*", which was founded in 1856, started reporting the earthquakes that occurred in Algeria in June 1864 after the publication of a decree of the Marshal-Governor General of Algeria, dated February 23, 1864, prescribing the centralization of all the meteorological observations, which were daily collected in the meteorological stations throughout the country, at Algiers Observatory (see La Gazette Médicale of March 1864). From that year onward, the "Meteorological Bulletin of the Algiers Observatory" was regularly published in the Algiers Press (see Le Moniteur de l'Algérie of July 26, 1864). Other Algerian scientific periodicals took the example of "La Gazette Médicale" and published information on the earthquakes of Algeria whenever they occur (for more details, see Harbi et al. 2015).

3.6 1867: A Step Toward Teamwork, from a Simple Pendulum in 1856 to a Seismograph, First Mapping of Geological Effects,...

The year 1867 for the first time saw the beginning of a team effort by scientists while the previous earthquakes have been undertaken almost exclusively by military officers. This happened after the destructive Mouzaia-El Affroun earthquake, which occurred on January 2, with an estimated intensity IX EMS (Harbi et al. 2017), when some meetings gathering experts were organized to present and discuss the observations that the experts made. The bulletin of the Algerian society of climatology, physical, and natural sciences published in 1867 revealed (on page 294) that Cochard presented detailed explanations on the recording provided by the seismograph of the naval dockyard of Algiers after the earthquake. This was the first time when the word "seismograph" is used. After that presentation, the Society instructed Cochard, Vatonne, and Marès to prepare as complete a summary of the earthquake effects as possible. We are told, on page 297, that Cochard, the Rapporteur of a committee working on the Mouzaia-El Affroun earthquake, presented the required details and informed the Algerian society of climatology and physical sciences about the improvement that he made in the seismograph. Vatonne presented a map on the ground fissures caused by the earthquake and Marès accounted on the boreholes carried out at Oued El Alleug by the Mines service of Algiers province 1 year before (Vatonne 1866). On the other hand, Marès drew the attention of the Society of climatology, physical, and natural sciences to the increase of the water flow in August or September 1866 in the Mitidja basin. Surprisingly, it is the bulletin of the "Société de Médecine d'Alger", which published a comprehensive report on the effects of the earthquake. The structure of this report is quite similar to the structure of the report by *De Sénarmont* (1857) on the 1856 Djidjelli earthquakes (see above), but with some differences. This report inter alia has addressed the following issues: (1) Physical effects, which mainly concern the damage caused to the buildings at several sites of the epicentral area; (2) meteorological phenomena with an attempt to provide the temperature at the time of the occurrence of an earthquake; (3) physiological and pathological effects; and (4) number of aftershocks, the extent of the affected area, duration, direction, and origin time (see the original information in Harbi et al. 2017). We also note the availability of several engravings and photographs on the damage caused by the earthquake at Mouzaïa, El Affroun, and Blida in contrast to the 1856 Djidjelli earthquake for which one has only one picture. These illustrations, which document damage, were very recently used as a research tool in the retrospective construction of the macroseismic field of the 1867 Mouzaïa-El Affroun earthquake (Harbi et al. 2017). It is worthwhile noting that from 1867 onward, the directors of the observatory of Algiers started to author the "Meteorological Bulletin of the Algiers Observatory". In 1867, the Director Charles Bulard (from December 30, 1858 to 1880) sent the exact origin time (hour, minute, second) of the Mouzaia-El Affroun earthquake to the national press. Bulard used to correlate the occurrence of the earthquake with the change of temperature or the atmospheric disturbances as he often wrote in most of his reports (for the period 1870–1880) to the national press and sometimes to the scientific periodical "*Comptes Rendus des Séances de l'Académie des Sciences*".

3.7 *1891: Very First Questionnaires*

On January 15, 1891, a destructive earthquake struck the locality of Gouraya and its surrounding villages, with an estimated intensity IX EMS (Maouche et al. 2008). It was the first time when a senior geologist (Pomel) paid attention to an earthquake in Algeria. Pomel (1891) described in details the earthquake effects in almost the same way as previous authors for the 1856 and 1867 earthquakes. He informed us that the first shock was recorded by the seismograph. He observed the first coastal uplift (30 cm) documented for an Algerian earthquake and set up the first questionnaires in order to get details on the earthquake effects at many sites and delineate the perceptibility area. Pomel said that these questionnaires, sent to the officials of the primary schools, allowed him to collect 380 useful information. Unfortunately, we could not find any of these questionnaires in Algerian archives and therefore we ignore the information that they requested.

3.8 *1906: First Seismicity Map of Algeria and Its Adjacent Areas*

The first seismicity map available for the Maghreb region including Algeria was drawn by Montessus de Ballore (1906). One may see from this map (Fig. 3) that Algeria was already more seismically active than its adjacent areas.

3.9 *1907–1910: Creation of the Service of Seismology, First Seismic Stations, Use of Reinforced Concrete…*

We mentioned above that the successive directors of the Algiers observatory contributed to the circulation of information on Algeria earthquakes. This was the case of its second director Charles Bulard as previously cited. However, we could find any work or communication in the field of seismology left by his immediate successor who is Charles Trépied (1880–1907). François Gonnessiat, Director of the Algiers observatory from 1907 to 1931 gave an impulse to earthquake science and created the service of seismology under the umbrella of the "*Service météorologique de l'Algérie*". Gonnessiat who was an astronomer compiled a list of the earthquakes that occurred in Algeria from 1871 to 1881 on the basis of the meteorological bulletin that Bullard sent to the newspapers "Akhbar" and "Le Moniteur de l'Algérie".

Instrumental recordings of earthquakes became possible in the late 1800s with the development of the seismograph as one has seen for the 1867 Mouzaia-El Affroun earthquake. A description of this seismograph is available in Cochard (1867). We learnt from the archives (Harbi et al. 2015, 2017) that this seismograph installed at the naval dockyard of Algiers recorded the earthquakes of Mouzaia on September 20, 1869, Boufarik on October 16, 1873, Cherchell on March 28, 1874,

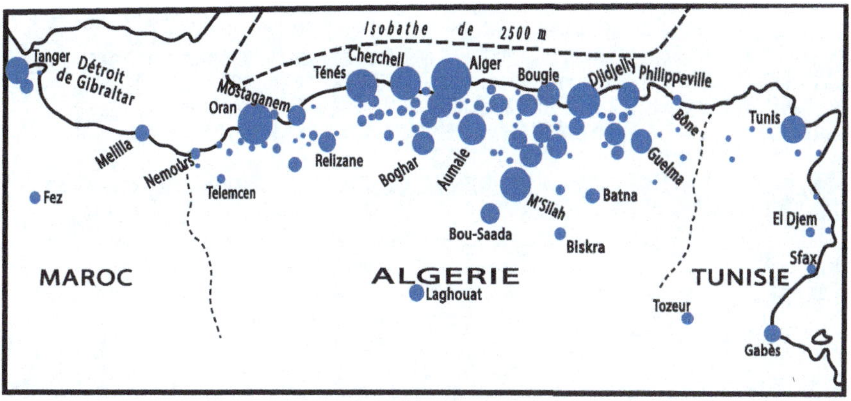

Fig. 3 Seismicity map of Algeria by Montessus de Ballore (1906)

and Marengo on October 22, 1882. According to Chesneau (1892), this seismograph was still operating in 1892. The only information provided by this instrument is the direction of the ground motion. One may notice from the figure drawn by Cochard that one was still at the preliminary stage of trial and error (see Fig. 4 in Harbi et al. 2017). Other seismometers were installed across the country as reported in the "Annales des sciences physiques et naturelles, d'agriculture et d'industrie" following the November 22, 1872 Mostaganem earthquake. We know from the "Echo d'oran" that the May 21, 1883 Oran earthquake was recorded by a seismometer installed at the Direction of Artillery of Oran. The astronomer Gonessiat, a tireless worker as described in the press reports whenever a strong earthquake occurs, was the pioneer in seismology at the Algiers observatory, and it really is to him that we owe the first seismic station installed at Bouzaréah in 1910. We read in the February 27, 1909 annual report of the Commission inspection of the Algiers observatory that the meteorological service is led by Gonnessiat proposing to carry out research in Earth physics. The 1911 report of the same commission informs us that a Bosh-Mainka seismograph with a 450 kg mass was installed in the basement of the library, which was specially designated to include it. In the 1914 report, one learns that the seismograph operated without interruption under the control of Mr. Maubert and that the director Gonnessiat was providing seismograms readings that he circulated to ten European stations and printed out the seismograms in the "Bulletins mensuels du Bureau Central Météorologique" of the meteorological service. Actually, the meteorological observations were daily collected at the Algiers observatory (also called Bouzaréah observatory) and sent to the meteorological service, which was located at Algiers City Hall and linked to the observatory by an active telegraph (Boutequin 1911). The first seismic event reported in the earthquake catalogue as event recorded by the Bouzaréah station is the February 25, 1911 Aumale earthquake (Hée 1925). The first known seismogram available to us

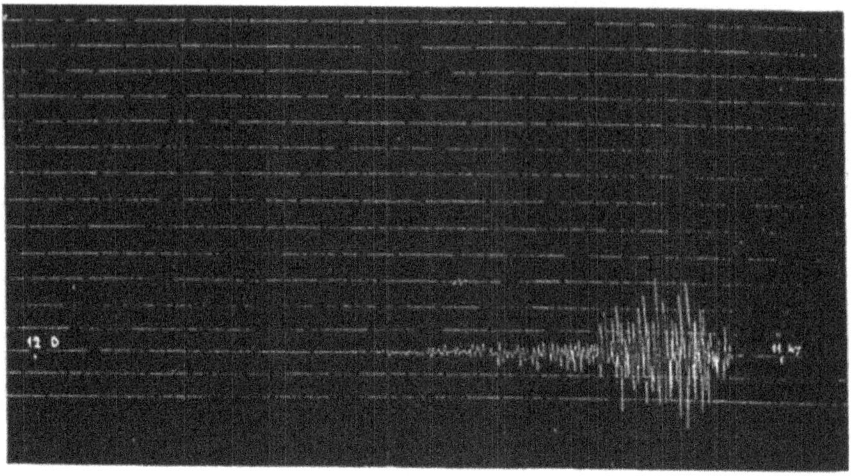

Fig. 4 Seismogram of the August 25, 1922 Cavaignac earthquake

was recorded on August 25, 1922 during the Cavaignac earthquake and sent to the national press by Gonnessiat (Fig. 4). However, we know from the annual reports of the Commission inspection of the Algiers observatory that the Bouzaréah seismic station operated on a regular basis and constantly communicated with the "Bureau Central" of Strasbourg. This station was gradually supplemented by other stations distributed in northern Algeria to constitute the first seismic network of the country (see hereafter).

We think that the 1906 San Francisco earthquake boosted the interest in seismology over the world at that time. In addition to the creation of the service of seismology by Gonessiat, we found in the literature (L'Afrique du Nord Illustrée of June 19, 1909) an article entitled "Earthquakes", dealing with the necessity to reduce the disastrous effects of the earthquakes by using steel-reinforced concrete in constructing buildings. A picture of the first villa built with reinforced concrete in Algiers (Fig. 5) and a detailed description on the mode of construction are provided in that newspaper (see an extract in the original language in Appendix).

3.10 1916: First Intensity Estimate?

We know that the first macroseismic scale which is the Egen scale was set up in 1828 according to Rothé (1925). The first attribution of an intensity estimate to a seismic event in Algeria, as communicated by the meteorological service to the national press, dates back to 1916 following the October 18 Algiers earthquake, for which an intensity VI on the Rossi scale is allocated (Echo d'Alger of October 19, 1916). However, one will see hereafter that the assessment of intensity started much earlier.

Fig. 5 First building constructed with steel-reinforced concrete at Telemly (Algiers) in 1909 (L'Afrique du Nord illustré)

3.11 1919: Annals of the Institute of Earth Physics of Strasbourg

The year 1919 saw the appearance of "***Annales de l'Institut de Physique du Globe de Strasbourg***" (***AIPGS***), which are authored for Algeria by *Hée* from 1919 to 1939. Hée mainly based her work on the bulletins of Algiers observatory and its meteorological service. As mentioned above, the Algerian meteorological service was linked to the "Bureau Central" of Strasbourg, which belongs to the institute of earth physics. Hée's lists of earthquakes are invaluable and comprise brief descriptions of the shocks, the date, origin time, intensity, and the name of the sites where each event was felt. This is therefore the first attempt at anything like a parametric catalogue of Algerian earthquakes. However, we note that except for some destructive events for which a detailed report is available, there is no information justifying the estimation of any intensity and there is seldom any attempt to determine the position of the epicenter. Despite these shortcomings, the annals of the institute of earth physics of Strasbourg remain invaluable. In addition to her contribution to these annals, Hée compiled three earthquake catalogues for Algeria. The first two catalogues by Hée (1925, 1933) start in 1911 to 1918, then merged with contemporary observations from 1919 to 1932 (already included in the AIPGS by Hée). The third catalogue by Hée (1950) went back to 1850, running forward to 1911. The materials used by Hée for this investigation are mainly the bulletins of the Algiers observatory, the bulletins of the Algerian meteorological service, and the press reports. We mentioned above that the first intensity estimate dates back to 1916. It seems that the observers of the Algerian meteorological service started assessing the intensity earlier since we found in Hée (1950) an intensity estimate for the August 23, 1904 Relizane earthquake. The main drawback of the Hée's catalogues is the difficulty to attribute a location to the seismic event. Hée divided the Algerian territory into 19 quadrangles. There is any attempt to discover the relation between shocks felt at the same time at different places. One may find an earthquake that is reported at several quadrangles whenever it was felt at sites belonging to the different quadrangles. This seems to be the primary reason for which many of the subsequent authors ignored her invaluable work (Harbi et al. 2010).

3.12 1922: First Macroseismic and Instrumental Study and First Attempt of the Identification of the Causative Fault

The first damaging earthquake in history to be studied comprehensively both macroseismically and instrumentally by a contemporaneous author is the August 25, 1922 Cavaignac earthquake (**AIPGS by Hée 1922**). Based on press reports and accounts of anonymous observers of the Algiers observatory, Hée described the effects of that earthquake and allocated maximum intensity to 12 sites, without drawing an intensity map. Hée determined the instrumental epicenter using the seismograms of the following stations: Algiers, Cartuja-Granada, Barcelone, San Fernando, Coïmbra, Rome, Zurich, Strasbourg, and Helwan. It is also following the 1922 Cavaignac earthquake that Brives and Dalloni (1922) attempted to show the probable fault that generated the seismic event. However, the authors just described the geological structure of the affected region and made some correlations without presenting any illustration and/or evidence of the causative fault. They reported the hydrological effects caused by the earthquake, but nothing about the geological effects that one can often observe after the occurrence of a damaging earthquake (M 5.9, Aoudia and Meghraoui 1995), as it was the case for the aforementioned earthquakes. It seems that the study performed by Brives and Dalloni (1922) is based on their own knowledge of the geological structure of the region, not on fieldwork carried out after the occurrence of the earthquake.

3.13 1924 Questionnaires and First Isoseismal Map Drawn by a Contemporaneous Author

In a recent paper (Harbi et al. 2015), we said that the oldest questionnaire archived in the Center of Research in Astrophysics, Astronomy and Geophysics (CRAAG) dates back to 1954. It seems that the questionnaires, which were probably set up by the seismological service of the Algiers observatory or by the French institution to which that service was also affiliated, were used by Hée to perform the macroseismic study of the November 5, 1924 Douéra earthquake. This is also the first time where an isoseismal map was drawn and published (***AIPGS by Hée 1924***). Grandjean (1954) who used this kind of data sources in her work indicated that the dispatching of questionnaires by the IMPGA has spread after the Second World War. However, none of the questionnaires of the pre-1954 period exists in Algeria and we do not know if they are similar to the questionnaires available to us, and which were set up to be used according to the Rossi–Forel scale (see details in Harbi et al. 2017).

3.14 1925: First Map on the Earthquakes Frequency

In 1925, an anonymous author mapped the frequency of earthquakes of Algeria and Tunisia. This map showed the distribution in Algeria and Tunisia of areas with high, medium, and low earthquake frequencies (Ayadi and Bezzeghoud 2015).

3.15 1931: The Algerian Meteorological Service Became the "Institut de Météorologie et de Physique du Globe d'Algérie" (IMPGA)

The Institute of Meteorology and Earth Physics of Algeria was operating under the Algiers Faculty of Sciences until the end of the 1970s. Its main tasks, at its inception, were to carry out research in meteorology and geomagnetism while seismology remained at the Bouzaréah observatory, which was also affiliated to the Algiers Faculty of Science. It seems that seismology was undertaken at the Bouzaréah observatory till 1962. An astronomer who worked at the observatory from 1950 to 1962 informs us that the header of the official documents was labeled with three letters O.A.S, which correspond to "Observatoire d'Astronomie et de Sismologie" (Milet 1999). The Algiers observatory and IMPGA were liaising with the scientific agencies of North Africa and France, and were affiliated to the Council of the institutes and observatories of Earth Physics in France (Hubert 1932). Following the installation of the "Bureau Central International de Séismologie" at the institute of earth physics of Strasbourg (IPGS) in 1921 (Rothé 1983), the studies in earth physics including seismology were developed in the French colonies and the IMPGA was based on the model of IPGS. Joanny Lagrula, a geophysicist, was appointed as a new director of the Algiers observatory (1931–1938). The same year, M. Vesselovski was appointed as head of the seismological service and was in charge of the analysis of the seismograms. Lagrula wrote a document entitled "*Rapport sur les travaux et la vie scientifique de l'Observatoire d'Alger*" (report on the scientific works and life at the Algiers observatory, Lagrula 1931). Regarding seismology, Lagrula reported what follows: "*Ludies Canovas and Feirar prepare the paper with black smoke (noir de fumée) and every month, M. Reiss prints a table of the measures read by M. Vesselovsky. Copies of these tables are regularly sent to the seismological observatories, particularly to the institute of earth physics of Strasbourg, which sends from its part its three-monthly bulletin. This bulletin is helpful for the final check of our seismograms, the original of which are often sent to the institute of Strasbourg upon their request*" (translated from French).

3.16 1934: First Picture of the Seismograph of Bouzaréah

On September 7, 1934, a moderate damaging earthquake (M 5.1) struck Carnot in the Cheliff basin (Benouar 1994). As used by most of the directors of the Algiers observatory, Lagrula sent to the national press the information on this earthquake. Lagrula also sent to the newspaper "l'Echo d'Alger" a photograph of the Bosh-Mainka seismograph, installed at the Bouzaréah observatory (Fig. 6).

3.17 1935: Oued Fodda Seismic Station

After the installation of the first seismic station at Bouzaréah, a second one was installed in the Cheliff basin after the 1934 Carnot earthquake to monitor the induced seismicity near the Oued Fodda dam. This station operated until 1982 (see its characteristics in Bezzeghoud et al. 1994).

Fig. 6 The Bosch-Mainka seismograph of the Bouzaréah observatory (photograph taken in 1934)

3.18 1948: Tamanrasset Seismic Station

The Tamanrasset station which is still in operation was installed in 1948 (Bezzeghoud et al. 1994) and was dedicated to teleseismic recording, thanks to its strategic position at a very quiet site. The data of this station have been already merged with data of other stations to study the anisotropy of northeast Africa (Hadiouche and Jobert 1988) and the anisotropic structure of the African upper mantle (Sebaï et al. 2006). The data of the Tamanrasset station were also recently used in studying the seismic activity in the Hoggar shield (Bourouis et al. 2013).

3.19 1949: The Algiers Seismic Station

A second seismic station (after that of Bouzaréah in the outskirts of Algiers) was installed in 1949 in the premises of IMPGA at Algiers University in the city center and operated until 1982. The Algiers station was equipped with two Coulomb-Grenet vertical seismographs, two horizontal seismographs and a Willmore vertical seismograph (Rothé 1970).

3.20 1950: First Mapping of the Causative Fault of an Earthquake, First Determination of Epicenter of Past Earthquakes and First Seismicity Map of Algeria

During the period March 13–18, 1949 upon the request of the General Government of Algeria and the "Centre National de la Recherche Scientifique (CNRS)", Rothé (1950) investigated the M 4.9 Kherrata February 1949 earthquake, visiting and mapping the geology of the Babor mountains. Rothé benefited in this work from the expertise of the senior geologist Dalloni (see above for the 1922 seismic event) and the questionnaires of IMPGA already analyzed by Seltzer. This investigation was published in 1950 in Algeria in the bulletin of the "Service de la Carte Géologique de l'Algérie" along with two additional studies: (1) an overview on the seismicity of the Babor region and (2) a catalogue of the destructive earthquakes of Algeria (1716–1949) for which Rothé determined the epicenter in most of cases, and a seismicity map showing the distribution of the destructive earthquakes of Algeria. Grandjean who worked at the IMPGA until 1963 continued the work of Rothé (1950) and published in 1954 a quite similar catalogue for the earthquakes that occurred from 1940 to 1950. For many years until the 2000s, these catalogues remained the definitive account of historical Algerian earthquakes.

3.21 1955: Installation of the Relizane Seismic Station

This station operated until 1961 (Bezzeghoud et al. 1994). We know from the newspaper "l'Echo d'Oran" of December 19, 1959 that it recorded the December 12, 1959 Oran earthquake (I_0 VII, Ms 4.5)

3.22 1958: Installation of the Setif Seismic Station

This station was equipped with a Coulomb-Grenet vertical short-period seismograph (Rothé 1970; Bezzeghoud et al. 1994).

There was also a seismic station located at Beni Abbes in the Sahara but we do not know when it was installed. Rothé (1970) informs us that the data of this station were handled by the French "Centre National de la Recherche Scientifique" (CNRS) and were sent to Algiers, then transmitted to Paris. We are also told that the Beni Abbes and the Tamanrasset stations were installed by the French for nuclear tests in the Hoggar and Sahara.

3.23 1960: First Analysis of the Seismicity of the Hoggar Shield

As mentioned above, the seismicity of Algeria is confined to the north of the country. To the south in the Hoggar shield (Tamanrasset region, Fig. 1), a seismic activity is also observed as already reported in Grandjean et al. (1960). Grandjean et al. revealed that 48 earthquakes (including nine aftershocks) were recorded between January 1, 1949 and December 31, 1958 by the seismological station of the Tamanrasset observatory (Fig. 7). About 70% of the recorded earthquakes have been located in the Silet area (SW of Tamanrasset) and would be linked to a recent volcanism according to Bordet (1952).

4 Post-independence Period (1963–Today)

4.1 1963–1968: Research Activity of the IMPGA Under the Umbrella of Franco-Algerian Agreements

After the war of national liberation which lasted from 1954 to 1962, research activities in Algeria were carried out within the framework of an agreement (1963–1968) between Algeria and France. In this context, a council of the scientific research (CRS) was created. A second agreement signed for the period 1968–1971

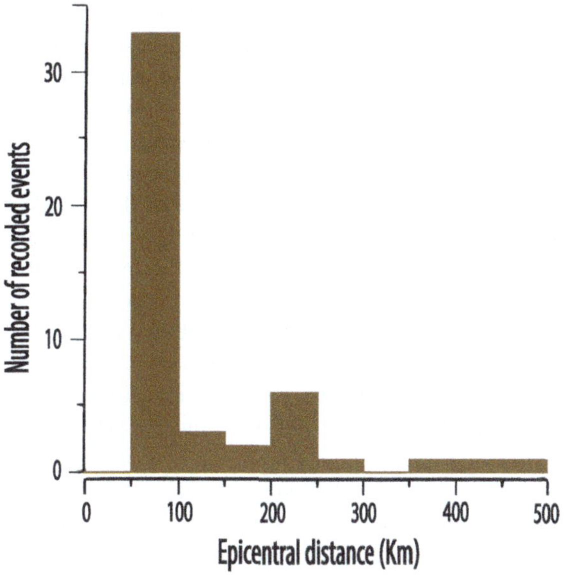

Fig. 7 Histogram of the number of seismic events recorded in the Hoggar by the Tamanrasset station between January 1, 1949 and December 31, 1958. 67% of the earthquakes are located 50–100 km from Tamanrasset (Boughacha 2005, data are from Grandjean et al. 1960)

led to the creation of the "Organisme de Coopération Scientifique" (OCS). During the period 1963–1971, CRS, then OCS handled the research activity of the Algerian institute of earth physics (Khelfaoui 2001). The seismological service of the IMPGA has continued its routine tasks regarding the determination of magnitude and epicenter of the earthquakes, which occurred during this period in Algeria. However, we noticed that the upturn of research activity has been somewhat slow at the beginning. For example, no attention was paid by the contemporaneous seismologists to the September 9, 1963 Bir Hadada destructive earthquake (M 5.7, Mokrane et al. 1994). This event was neglected in the subsequent catalogues and studies until the end of the 2000s when Harbi and Maouche (2009) reconstructed the macroseismic field of the Bir Hadada earthquake and estimated the epicentral intensity at VIII-IX EMS. It was not until 1966 that a first earthquake study appeared. This was related to the study of the January 1, 1965 M'sila earthquake (M 5.5, Grandjean et al. 1966). In this study, the authors attempted to use a multidisciplinary approach to analyze the 1965 event including macroseismology, tectonics, and gravimetric anomalies of the M'sila region. It was the first time when gravimetric data are integrated into a seismological study in Algeria.

4.2 *1969–1970: Seismology in Algeria Faced Tough Times*

This is what we learned from the mission report of Rothé (1970). Under the UNESCO/SC/1769/69, Rothé paid a visit to the seismological service of the

IMPGA the December 5–9, 1969 to get inquired about the status of seismology in Algeria. He regrettably concluded that both macroseismic and microseismic surveys, which were before performed on a regular basis stopped, due to recruitment difficulties that IMPGA was facing at that time. We can confirm, as researchers working in the field of historical seismology and macroseismology in Algeria, that IMPGA stopped to disseminate questionnaires just before the independence in 1961, 1962, then in 1969 and 1970 after the end of the Franco-Algerian agreement. Benhallou who was physicist at IMPGA and affiliated to Algiers University where he had been teaching since 1961 (Bezzeghoud 2012) informed Rothé about the status of the seismic network that included the seismic stations mentioned above, most of them were operating in 1969. We are told that the seismological service planned to develop the seismic network and to acquire additional seismic stations, aiming at surveying the natural seismicity in northern Algeria and the induced seismicity in the vicinity of the dams of Oued Fodda, Meffrouch (Tlemcen region), Zerdaza (Skikda region), and near Biskra. During the mission of Rothé who was accompanied by Benhallou in the field, there were discussions about the necessity to (1) publish the macroseismic observations from 1951 onward (the last publication concerned the earthquakes prior to 1950 studied by Grandjean 1954), (2) update the map of maximum observed intensities, which was already prepared by Grandjean at IMPGA, and (3) develop the seismic network. These three important tasks constituted the first objective of the future head of the seismological service at IMPGA and future founder of the present Centre of Research of Astronomy, Astrophysics and Geophysics (CRAAG), Hadj Benhallou who met practical impediments in achieving results. We will see hereafter that the occurrence of destructive and damaging earthquakes in 1980, then in 1989 greatly contributed in realizing Benhallou's ambitions.

4.3 1970–1979: Start of the Revival of Seismology in Algeria

In 1970, the ministry of high education and scientific research (MESRS) is created in Algeria. In 1971, the "Conseil Provisoire de la Recherche Scientifique" (CPRS) is created under the MESRS aegis and was replaced in 1973 by the "Office National de la Recherche Scientifique" (ONRS). Therefore, the IMPGA was affiliated to CPRS (Khelfaoui 2001), then to ONRS in December 1974. In the meanwhile, Benhallou was struggling for the revival of seismology in Algeria. Actually, after the Algeria independence, there was a drastic shortage of financial, material, and human resources. In addition, most of seismic data were generally moved to France. We noticed for example that even for the questionnaires left in Algeria and which cover the period 1954–1969, the questionnaires which are related to the strongest events, such as the 1954 Orléansville (M 6.7) and the 1959 Boumedfaa (M 5.6) earthquakes are lacking. Through thick and thin, Benhallou finally succeeded to publish the macroseismic catalogue of the Algerian earthquakes from 1951 to 1970. For the first time, this publication was edited by IMPGA (Benhallou et al. 1971). In

1972, Bockel (1972) who was a French consultant working at IMPGA published a second article first edited by the same institute, on the structure of the earth crust in Algeria using the seismic waves and determined a velocity model for Algeria. Roussel (1974) who was working at IMPGA and Algiers University presented a short note on the 1973 Mansoura earthquake. The resumption of the publications regarding Algeria seismicity materialized the revival of the IMPGA activities. Later on, Girardin et al. (1977) based on these last important papers (Benhallou et al. 1971; Bockel 1972) to relocate 80 earthquakes of the period 1950–1970 using P-wave arrival time of near-field stations (Algeria, Morocco, Spain, and Italy) and determine focal solutions for five events of the same period. This was the first time when focal mechanisms are calculated including Algerian data.

Geophysics with its various branches, such as seismology, geomagnetism, gravimetry, seismics, etc., started to be taught at the Department of Physics at Algiers University by the 1970s, while petroleum geophysics was taught at the "Institut National des Hydcrocarbures" since 1964 and at "Institut Algérien du Pétrole" since 1971, both at Boumerdes.

4.4 1980: Creation of CNAAG, El Asnam Earthquake the Largest Earthquake Ever Known in Algeria, Renaissance of Seismology in Algeria

In April 1980, the "Centre National d'Astronomie, Astrophysique et Géophysique" is created. Hadj Benhallou, the CNAAG founder, was appointed director of this center (1980–1998), which inherited from the Algiers Observatory and IMPGA. The headquarters of CNAAG, which is under the Ministry of High Education and Scientific Research, are based in the Bouzaréah Observatory. Macroseismic and microseismic surveys were continued by the technicians of CNAAG under the Benhallou supervision. Benhallou can be stated to have been the first Algerian seismologist who was involved in conducting microseismic and macroseismic surveys of recent events, in historical seismology, and in developing the local instrumental network. He was helped in this by the occurrence of the largest seismic event that Algeria had ever known, the 1980 El Asnam earthquake (M 7.5) and by the young researchers that Benhallou later on recruited at CRAAG (see hereafter).

Since the El Asnam earthquake, Algerian seismology has received a tremendous impetus. The CNAAG acquired the first portable network (MEQ 800), several studies of El Asnam earthquake were carried out in the framework of the Algero-French bilateral cooperation and many earth scientists from over the world visited the epicentral area and investigated the earthquake. These studies covered many subtopics of seismology and active tectonics including paleoseismological investigations and geodesy measurements, which were carried out for the first time in Algeria. This resulted in the publication of a collection of scientific papers (e.g., Ouyed et al. 1981, 1983; Deschamps et al. 1982; Yileding et al. 1981; Ruegg et al. 1982; Philip and Meghraoui 1983; King and Yielding 1984; Meghraoui et al. 1986,

1988a, b; Dimitrov et al. 1987, 1991 etc.) and the education of some postgraduate students who formed the first nucleus of Algerian seismologists. Up to now, the 1980 El Asnam earthquake, which can fairly be considered as a textbook case, continues capturing the scientists' interest who are still working on many aspects related to this event (e.g., Roger et al. 2011; Bellalem et al. 2015).

4.5 1985–1989: Creation of the Department of Geophysics at USTHB, First Group of Geophysics Researchers at CRAAG

In February 1985, the CNAAG became a research center under the name "Centre de Recherche en Astronomie, Astrophysique et Géophysique" CRAAG. Another research center, with a role focusing on earthquake engineering, was created under the aegis of the Ministry of housing and urban planning in April 1985, the "Centre de Recherche Appliquée en Génie Sismique" (CGS). At the Algiers University which is the so-called "Université des Sciences et de la Technologie Houari Boumediène" (USTHB), a group of geophysicists created the Department of geophysics after receiving their degree of first post-graduation (in Algeria, for long time the first post-graduation degree corresponded to the so-called doctorate of third academic cycle or magister, whereas the second post-graduation corresponded to the doctorate or Ph.D.) and started to teach geophysics (fundamental and applied geophysics) to the first class of engineers in geophysics from 1986 onward. In 1987 after the 1985 Constantine earthquake (M 6.0), Benhallou, the director of CRAAG succeeded to get, from the government, a budget that allowed him to recruit the first core of geophysics researchers who had just defended their doctoral thesis in France.[1] A special tribute should be paid to Hadj Benhallou who spared no efforts in laying the foundations of a research center dealing with fundamental geophysics in Algeria. Benhallou took on the task of traveling to Paris (France) in order to meet six Ph.D. students who were preparing their defense thesis, and to convince them to return home and to helping him in driving geophysics forward at CRAAG. In the meanwhile he continued, until his death, to supervise the successive Ph.D. students in geophysics, particularly in seismology at the USTHB.

During this period, the earthquake studies were generally carried out by the CRAAG researchers or in the framework of local and international cooperation mainly between CRAAG, USTHB, and the traditional collaborators of the Algerian seismologists based in France. This was the case with the 1985 Constantine and Chenoua Mount 1989 earthquakes (M 6.0) (e.g., Bounif et al. 1987; Deschamps et al. 1991; Meghraoui 1991). The MEQ 800 portable visual recorders acquired

[1]The first nucleus of earth scientists at CRAAG was constituted by Abdeslam Abtout, Mourad Bezzeghoud, Mohamed El Messaoud Derder, Nacer-Eddine Merabet, Mustapha Meghraoui, and Abdelkarim Yelles-Chaouche.

after El Asnam earthquake were used in the investigation of the 1985 and 1989 earthquakes. In 1989, the first class of the engineers in geophysics of USTHB received their graduation degrees.

4.6 1990–2003: Inception of the Department of Seismological Studies and Survey, Installation of the First Modern Seismological Network, First Tomography Study, First Earthquake Catalogues in Independent Algeria

After the 1989 earthquake, which occurred in October at Chenoua mount near the Capital Algiers (Meghraoui 1991; Bounif et al. 2003), the installation of a modern seismological network was needed more than ever to increase understanding of the seismicity and seismotectonics of Algeria. CRAAG moved to the Ministry of Interior in April 1990. Right after, CRAAG received funding to install the first modern telemetered seismic network RSTA (Réseau Sismologique Télémétré Algérien). The funding for the acquisition of the Telemetred seismological network was done from Arab Fund for Economic and Social Development (AFESD). The Kinemetrics seismological network was shipped to Algiers and installed between 1989 and 1990 (personnal communication from Abdelhakim Ayadi, head of the seismic network from 1990 to 1998). In the meanwhile, things were beginning to take off for the ancient seismological service, which grew up and became "Department of seismological studies and survey", which was constituted by important laboratories involved in modern and historical seismology, active deformation (geodetic measurements), seismotectonics, and paleoseismology. This department was created and headed by Mourad Bezzeghoud. The young researchers who just returned from France and joined CRAAG had started recruiting the engineers in geophysics and geology who were trained at the Algiers University (USTHB) and get them involved in several studies covering several aspects of fundamental geophysics. As mentioned above, the widespread interest that the 1980, 1985, and 1989 earthquakes generated was accompanied by a realization for the need to continue the work on the 1980 El Asnam earthquake regarding paleoseismicity and post-seismic deformation (Bezzeghoud et al. 1995; Meghraoui and Doumaz 1996; Lammali et al. 1997), to take interest in offshore seismicity and tectonics (Yelles-Chaouche 1991; Yelles-Chaouche et al. 1996), to get back to historical earthquakes through a multidisciplinary approach for a better understanding of the seismotectonics of the Tell Atlas (Aoudia and Meghraoui 1995), and above all to deploy a modern seismological network in the country.

The CRAAG which acquired and installed a modern Algerian Telemetred Seismological Network constituted of 32 short-period stations had the opportunity to study the Rouina (central Algeria) earthquake of January 19, 1992 (M 5.2), which was the first moderate seismic events recorded by this first telemetred network and studied by Bezzeghoud et al. (1994). Unfortunately, from 1992 to the beginning of the 2000s, the country experienced a bad security situation, which disrupted seismological research in Algeria. It was difficult to maintain the seismic stations distributed in isolated sites in Northern Algeria, most of them stopped, and few were lost or destroyed. This has not prevented the CRAAG researchers and their collaborators of USTHB to do what they could for keeping the flame of seismology alive in Algeria. As such, papers were devoted to the earthquakes that occurred during this decade of unstable security situation, which are the earthquakes of Mascara (1994, M 5.9, Benouar et al. 1994; Bezzeghoud and Buforn 1999; Ayadi et al. 2002), Algiers (1996, M 5.7, Maouche et al. 1998; Harbi et al. 2004), and Ain Temouchent (1999, M 5.8, Yelles-Chaouche et al. 2004).

As the political situation was getting increasingly unstable in Northern Algeria in the 1990s, the seismologists of CRAAG led an important project in southern Algeria, in the Hoggar shield, which is known as one of the most important swells in Africa. The objective was to explore the structure of the crust and upper mantle beneath the Hoggar swell and the Sahara basins through a teleseismic field experiment, which was carried out using a portable network of 33 seismic stations installed along a 700-km-long NNW–SSE profile running from Tamanrasset to In-Salah at north. This was the first time that teleseismic tomography is performed in Algeria with the collaboration of a French team from the CNRS and the University of Strasbourg (Ayadi et al. 2000).

In 1994, CRAAG published the first parametric earthquake catalogue of the whole Algeria (1365–1992) (Mokrane et al. 1994). This catalogue, which is the continuation of the earthquake catalogue of Algeria by Benhallou (1985), was compiled by young scientists under the direction of Mourad Bezzeghoud and Hadj Benhallou. The earthquakes included in this catalogue are mainly studied by means of questionnaires and instrumental data of IMPGA and CRAAG, and a macroseismic atlas is also provided. The same year, Benouar who was the PhD student of Ambraseys published a parametric catalogue for the seismicity that occurred in the Maghreb countries between 1900 and 1990 (Benouar 1994). Benouar reviewed the most important destructive events of Algeria through primary and secondary sources found in National and regional archives and relocated the instrumental earthquakes using the procedures of the international seismological center (ISC). Both catalogues (Mokrane et al. 1994; Benouar 1994) are still considered important guides for the knowledge of the seismic history of Algeria. In 1996, based on the first CRAAG earthquake catalogue, Bezzeghoud et al. (1996) mapped the maximum observed intensities of Algeria (1365–1989). This map was first published as a report for the Algerian National Insurance and Reinsurance Company (CAAR) and then as a scientific paper (Bezzeghoud et al. 1996). This map was updated until 2014 by Ayadi and Bezzeghoud (2015).

In 1998, Benhallou left the CRAAG and was then appointed dean of the "Faculté des Sciences de la Terre, de la Géographie et de l'Aménagement du Territoire" (FSTGAT, USTHB). Benhallou was replaced at CRAAG by Abdelkarim Yelles-Chaouche (1998–present) who, at that time, was the head of the Geophysics Department at CRAAG. Yelles-Chaouche published the second

CRAAG earthquake catalogue (Yelles et al. 2002), which is the continuation of the catalogue by Mokrane et al. (1994), running forward to 2001, and particularly focused, with the technical staff at CRAAG that he supervised, on maintaining and restoring the seismic network. These efforts allowed increasing the number of operating seismic stations from 4 to 28 in 2000 and led to studying the Ain Temouchent earthquake of December 22, 1999 using Algerian data (Yelles-Chaouche et al. 2004). The second step toward the improvement of the Algerian seismic network consisted in its digitization and the acquisition of new seismic stations, among which mobile stations. These new acquisitions allowed performing comprehensive seismological studies of the second largest Algeria earthquake that occurred during the instrumental era (see the next section). The study of the Algerian margin for a better understanding of offshore seismicity and tectonics was one of the main objectives of Yelles-Chaouche who initiated in 2000 the MARADJA project with French partners (Domzig et al. 2006).

4.7 2003–Present: Seismology in Algeria, a Second Wind, New Motivations and Projects After the 2003 Zemmouri-Boumerdes Earthquake

It is commonly known that one learns more whenever a damaging seismic event occurs anywhere around the world. The progress in seismology owes much to the occurrence of earthquakes and we have seen throughout this article how true it is for Algeria. The security situation somewhat improved in Algeria when on May 21, 2003, a second strongest earthquake with magnitude 6.8 (after the 1980 El Asnam earthquake M 7.5) struck the region of Boumerdes, 50 km east of the Capital Algiers (Ayadi et al. 2003; Yelles-Chaouche et al. 2003). This seismic event, which had disastrous effects (Harbi et al. 2007b) mobilized the seismologists in Algeria, particularly the CRAAG newly trained scientists, and at international level (particularly from France, Turkey, Japan, Portugal, and USA). This was not just because of its size and its proximity to the Capital, but particularly to the tsunami that it triggered and the large coastal uplift of marine terraces that it induced, and which implied an important continental deformation related to an SE dipping and 55-km-long thrust fault (Meghraoui et al. 2004). This earthquake has been a subject of several studies covering many topics such as relocation of the main shock using the double difference model and the analysis of the aftershocks or the accelerometric record, teleseismic inversion, study of the coseismic and post-seismic deformation from GPS measurements, modeling the fault from geodetic and accelerograms, tsunami modeling, tomography, interferometric aperture radar, combination of gravity data and aftershocks sequence, stress transfer, earthquake engineering, macroseismology, etc. (Bounif et al. 2004; Delouis et al. 2004; Yelles et al. 2004; Semmane et al. 2005; Alasset et al. 2006; Laouami et al. 2006; Harbi et al. 2007b; Ayadi et al. 2008; Mahsas et al. 2008; Maouche et al. 2008;

Belabbès et al. 2009; Sahal et al. 2009; Ouyed et al. 2011; Lin et al. 2011; Maouche et al. 2011; Cetin et al. 2012; Heidarzadeh and Satake 2013; Santos et al. 2014; Heddar et al. 2016; Kherroubi et al. 2017, etc.).

The 2003 earthquake did not just mobilize the scientists, but also funding of the Algeria Government. It was a good opportunity for the new director of the CRAAG, Yelles-Chaouche to upgrade the seismological network since the network installed in 1990 was in a bad state due to the difficulty of maintaining the seismic stations during the period of insecurity. The present ADSN network (Algerian Digital Seismic Network) has been developed on the former network RSTA, then updated. It consists of 20 broadband, 2 very broadband, and 47 short-period stations, and 21 accelerometers (for more details, see Yelles-Chaouche et al. 2013a). Refinements of the ADSN are continuously made toward an effective earthquake early-warning system (EEWS). From 2000 onward, precise instrumental data on earthquakes began to accumulate, and seismology has developed from the qualitative toward the quantitative side. Thanks to these improvements in recording continuously and steadily the events, the Algerian seismologists (mainly from CRAAG, CGS, and USTHB) started to systematically study recent earthquakes, such as the 2006 Laalam earthquake (M 5.2, Beldjoudi et al. 2009; Guemache et al. 2009; Bouhadad et al. 2010), the 2006 Tadjenna earthquake (M 5.0, Beldjoudi et al. 2011), the 2010 Béni Ilman earthquake (M 5.2, Yelles-Chaouche et al. 2013b; Abacha et al. 2014; Beldjoudi et al. 2016), the 2012 Béni Haoua earthquake (M 4.9, Abbès et al. 2016), the 2014 Bordj-Menaïel earthquake (M 4.1, Semmane et al. 2015), the 2013 Hammam Melouane earthquake (M 5.0, Yelles et al. 2017), the 2014 Algiers earthquake (M 5.3, Benfedda et al. 2017), the 2014 Mihoub earthquake (4.3, Semmane et al. 2017), etc.

As known, the 2003 Zemmouri earthquake triggered a tsunami (Alasset et al. 2006) that generated interest to Algerian scientists in exploring past tsunami (Maouche et al. 2008; Yelles-Chaouche et al. 2009; Harbi et al. 2011), tsunami deposits (Maouche et al. 2009), and studying tsunami risk (Amir et al. 2012, 2013, 2015; Amir and Theilen-Willige 2017) in Algeria. The study of tectonic activity of the Algerian margin was undertaken before and after the 2003 Zemmouri earthquake by two Algero-French collaborative projects, MARADJA and SPIRAL projects initiated by Yelles-Chaouche (https://spiral.oca.eu/), and resulted in a slew of papers (Yelles et al. 2009; Kherroubi et al. 2009; Bouyahiaoui et al. 2015; Hamai et al. 2015, etc.). The study of induced seismicity is also a topic that interested the Algerian seismologists and the first study was performed near the Beni Haroun dam (Semmane et al. 2012). The configuration of seismic stations and high quality of the collected data enable Algerian seismologists in using modern techniques in seismology at national, regional, and global scales (Radi et al. 2015, 2017; Haned et al. 2016; Kariche et al. 2017). The GPS technology is also currently used in Algeria by CRAAG and the GPS network REGAT (Réseau Géodésique de l'Atlas) with 60 GPS stations is collocated with the ADSN (Yelles-Chaouche et al. 2010). First results concerned the velocity field around the Ain Smara fault that was reactivated during the 1985 Constantine earthquake (Bellik et al. 2014). Reviewing the instrumental earthquakes for which a consistent database of seismic signals related to aftershocks is available is also a project carried out at CRAAG. Using new procedures, Ousadou et al. (2013) relocated the 1985 Constantine earthquake, and then they constructed a focal mechanisms database for the aftershocks of the 1985 Constantine, 1989 Chenoua Mount, 2003 Zemmouri, and 2004 Al-Hoceima (Morocco) earthquakes. This allowed them to calculate by inverting focal solutions the stress field and its variation along the northern part of the Maghreb area (Ousadou et al. 2014). This was the first time when focal mechanisms are calculated using Algerian data exclusively.

The development of quantitative seismology in Algeria did not mean that the qualitative seismology has been abandoned. To the contrary, the occurrence of the 2003 Zemmouri earthquake in a region not known before to have experienced such a destructive event had reinforced the belief that the best way to predict damage of future earthquakes is to explore past earthquakes of Algeria. In fact for many things in life, only the past is happening over and over again. It can be inferred from the above that the time range of reliable instrumental data is far too short when assessing seismic hazard for Algeria. Therefore, historical seismicity represents a wealth of potential information on long-term seismicity, which in its turn contributes to understanding seismotectonics and assessing seismic hazard. After 2003, there was a reawakening of interest in historical earthquake studies at CRAAG. Mining historical earthquake information with the aim to find out primary information sources is a time consuming and sometimes frustrating task, but very challenging for the historical seismology group created at the end of the 2010s at CRAAG. The group members are diligently constructing the Algerian earthquake archive. The macroseismic database still in development and available online reproduces verbatim historical accounts with references that permit verification and further work (see Harbi et al. 2015). The objective of the group is twofolds: (1) to improve the knowledge on the seismicity of Algeria and (2) to provide a credible input for seismic hazard assessment. Considering Ambraseys as a pioneer of the modern vision of historical seismology, the group is following the modus operandi which Ambraseys recommended in handling historical data and analyzing macroseismic information (Ambraseys et al. 1983; Ambraseys 2001). This cogent approach allowed obtaining efficient and reliable results. Nevertheless, the experience showed that the study of the historical earthquakes by the local scientists could give more sound results because they are more grounded in field reality. The historical earthquake research is becoming a matter of increasing returns in Algeria since it allowed us to go back further in time regarding the seismic history of Algeria (Ferdi and Harbi 2014). The acquisition of historical materials from a large variety of sources (Harbi et al. 2015) allowed intensity to be estimated/re-estimated for many events and isoseismal maps to be drawn/re-drawn, without forgetting to mention the newly discovered earthquakes, numbering today about 300 events only for eastern and a part of central Algeria (Sebaï and Bernard 2008; Harbi et al. 2015, 2017). As work continues for the rest of the country, this number will certainly be revised upward.

Besides the historical seismology group, seven other teams were created at CRAAG in quantitative seismology, earthquake warning, active deformation, seismotectonics, and seismic hazards and risk. The department of seismological studies and survey, which was created in the 90s, was developed and splitted into two divisions with eight research teams. After the period of late Hadj Benhallou who started almost from the scratch with very few resources to lay the foundations for the development of seismology in independent Algeria came the period of Abdelkarim Yelles-Chaouche with a modern strategy to strengthen and extend these foundations toward all the aspects related to earthquake science including the mitigation of seismic hazard and risk (for more details, see Yelles-Chaouche 2015). Both directors gained effective and whole-hearted support from the scientists and technicians involved in seismology, and the number of seismologists in Algeria is currently steadily increasing.

As mentioned above, efforts were made since the 1970s toward training and capacity building in geophysics. For now, a few dozen students defended their PhD thesis in seismology at USTHB. For long time, geophysics and/or earth physics were taught at Algiers and Boumerdes only. In 2005, an academic program devoted to earth physics including seismology has been initiated, with the help of CRAAG researchers, at Setif University (300 km east of Algiers), then at Khemis Miliana University (120 km west of Algiers) since 2015. Hence, one expects more scientists involved in seismology than before. The syllabus regarding seismology and related topics is quite the same in Setif and Khemis Miliana Universities and is inspired from the syllabus of USTHB (Algiers). It consists for the graduate students of plate tectonics and associated seismicity, faults, elasticity, propagation, seismometry, and source mechanism, while postgraduate students are involved in seismotectonic studies, interferometry, fault interaction, earthquake catalogues, aftershocks analysis, inversion, and very recently, active deformation (GPS), and historical seismology.

International conferences and symposia on seismology and related fields provided useful forums to discuss and spread the results obtained by Algerian earthquake scientists. CRAAG started this tradition during the Benhallou period in 1990 to commemorate the 1980 El Asnam earthquake and has continued in 1997 after the 1996 Ain Benian earthquake, and during the Yelles-Chaouche period in 2000 (on the Mediterranean seismicity), and on the occasion of the commemoration of the most significant Algeria earthquakes, which are the 1980 El Asnam earthquake (in 2010), the 2003 Zemmouri-Boumerdes earthquake (in 2013), and the 1985 Constantine earthquake (in 2015).

We shall not forget the continuous efforts of the regional science network *The North-African Group for Earthquake and Tsunami studies* (NAGET) and the Algerian academic association *Société Algérienne de Géophysique* (SAG) in disseminating the seismological information at regional and national level, and toward seismology in general. NAGET (http://naget.ictp.it), which was founded at the Abdus Salam International Centre for theoretical physics (ICTP, Trieste, Italy) in 2000, fosters and coordinates advanced research in earthquake science in North Africa through bilateral and/or multilateral projects by putting emphasize on interconnections within the group and maintaining joint links with the international scientific community. SAG (http://www.sag.dz), which was founded at USTHB in

2007, hosts the commission of seismology and natural risks, which aims at improving knowledge of the seismicity, seismic, and natural hazards of the country for a better risk mitigation.

5 Outlook

In this chapter, we wanted to share with the reader what we have learnt about the history of seismology in Algeria (see the timeline in Table 1) in our capacity of Algerian actors in the field of seismology. However, we believe that additional search in the archives of the institutions, which inherited from the Algiers observatory and the former IMPGA and, which are the CRAAG (Algiers) and especially the IPGS (Strasbourg), could provide substantial data related to many aspects of seismology in Algeria, such as the early seismograms (before 1922), the first stations and how they operated, the first questionnaires (before 1954), etc. Further archive search will allow filling some gaps and enriching the history of seismology presented here with respect to Algeria. This will be also a very good opportunity to perform a quantitative analysis of early seismograph recordings. This issue matters to us, but unfortunately most of early seismograms are lacking in Algeria and could be available at IPGS as we may infer from the present study (see above).

Several studies and projects lie ahead in terms of seismology in Algeria in addition to the projects, which had already been launched. A due contribution of the historical earthquakes to the seismic hazard assessment implies the estimation of their magnitudes on the basis of a robust method using generally intensity attenuation. The current work on historical seismicity of Algeria (Harbi et al. 2015) may increase the sample of the macroseismic data points on the basis of which a new intensity attenuation law could be derived. The calibration to the instrumental earthquakes may give good results especially if the intensity of all the earthquakes is objectively estimated according to a unique macroseismic scale. In the same way, the maximum calculated intensities (MCI) already mapped by Boughacha et al. (2004) could be updated. Algerian seismologists are currently taking a keen interest in induced seismicity and projects will be developed soon in this topic. A definite plan for a network in the Hoggar shield exists at CRAAG and USTHB to study and understand the mechanism of the seismicity of the region (Boughacha et al. 2006; Yelles-Chaouche 2015).

The reader certainly noticed that the history of seismology in Algeria is intertwined with the history of CRAAG and its father, the seismological service of the Algiers observatory and IMPGA to which the seismologists of the Algiers University were associated. After the Zemmouri–Boumerdes earthquake, the "Centre de Recherche Appliquée en Génie Sismique" (CGS) also developed seismological and accelerometric networks and its researchers published many valuable papers in earthquake engineering, which is, with seismic hazard and risk, outside the scope of the present paper. However, from last year onward, the CGS researchers started to perform earthquake studies (Abbès et al. 2016; Benfedda et al.

2017). The community of Algerian seismologists is growing up and this augurs well for the future of seismology in Algeria.

Last but not least, Algeria being an earthquake-prone country, raising awareness of living in a seismic country must be our first goal to reach. Therefore, it would be beneficial for the progress of seismology in the country and very challenging to attract students to earthquake science. The best way to achieve this, in addition to the awareness-raising campaigns on earthquakes that CRAAG takes in charge toward Algerian schools, is to introduce students to seismology through installing seismometers in schools. Our experience in teaching seismology to students showed us that they are more motivated when one allows them to have a close look to a seismic signal or damage caused by an earthquake that occurred in Algeria, or when we involve them in practical work, such as picking the P and S waves or training them in observational seismology (macroseismology and seismometry). They easily interact with real data. This concept produced successful results in the United Kingdom, USA, and Australia. Such a project permits connecting the school with research, higher education, and the world of the profession (monitoring). Algerian seismologists should draw on the experience of Australian seismologists for example (see Balfour et al. 2014 for more details) and, as such, will contribute to the risk assessment and its mitigation through education.

Acknowledgements The work presented in this paper was conducted as a part of the CRAAG project E007/08. This work benefited from thoughtful comments and suggestions from the main actors of the development of seismology in Algeria during different recent periods of time. Special thanks are due to Yelles-Chaouche Abdelkarim, director of CRAAG (1998–present); Mourad Bezzeghoud, head of the Seismology and Survey Department at CRAAG in 1989–1996 (presently at Évora University, Portugal); Ayadi Abdelhakim, head of the Algeria seismic network in 1990–1998 (CRAAG, Algeria). Sincere thanks are also due to Djamel Mati for his helpful information on the IMPGA where he worked in the 1970s before moving to CNAAG, then CRAAG, and to Toufik Abdelatif, Astrophysicist at CRAAG for the useful documents that he kindly provided. The authors wish to pay tribute to the first Algerian seismologist who is the late Professor Hadj Benhallou (1937–2011), for his continued guidance and support when they were young scientists, and for his dedication to the development of research in fundamental geophysics in Algeria.

Appendix: Extract of the Text on the New Mode of Construction of Buildings in Order to Protect the Buildings from Earthquakes, Published in French in Afrique du Nord Illustrée of June 19, 1909

Le principe qui a présidé à l'établissement du plan a été de proscrire tous les éléments constructifs n'ayant pas de stabilité propre, comme la voûte formant plancher qui, par ses poussées, jette les murs dehors au premier frisson du sol. La carcasse de cette villa se compose essentiellement de douze colonnes en béton armé, entrecroisées entre elles, à la base, par une série de nervures et, à chaque tiers de leur hauteur, par les planchers en béton armé qui forment avec elles une ossature indéformable. Chaque panneau extérieur de la «cage» a été cloisonné par deux muretles de 10 centimètres d'épaisseur en moellons de machefer fabriqués sur place. Entre ces deux murettes, il a été ménagé un vide de 20 centimètres, en sorte que ce matelas d'air intercepte, d'une façon parfaite et bien mieux que ne le ferait un mur de 50 centimètres d'épaisseur, le bruit, le froid et la chaleur. Les cloisons intérieures sont également en machefer. Qu'une secousse sismique survienne. Le pis qu'il puisse arriver, c'est la projection de droite et de gauche des murettes et cloisons; mais l'ossature, les colonnes et planchers ne broncheront pas, et le propriétaire non seulement n'encourera aucun danger, mais, six jours après, il aura reconstruit sa maison.

References

Abacha I, Koulakov I, Semmane F, Yelles-Chaouche A (2014) Seismic tomography of the area of the 2010 Beni-Ilmane earthquake sequence, north-central Algeria. http://www.springerplus.com/content/3/1/650. Accessed 14 Feb 2017

Abbès K, Dorbath L, Dorbath C, Djeddi M, Ousadou F, Maouche S, Benkaci N, Slimani A, Larbes S, Bouziane D (2016) The Beni Haoua, Algeria, Mw 4.9 earthquake: source parameters, engineering, and seismotectonic implications. J Seismol. https://doi.org/10.1007/s10950-015-9549-1

Abdessemed-Foufa a, Benouar D (2010) Investigation of the 1716 Algiers (Algeria) earthquake from historical sources: effect, damages, and vulnerability. Int J Archit Herit 3:270–293

Alasset PJ, Hébert H, Maouche S, Calbini V, Meghraoui M (2006) The tsunami induced by the 2003 Zemmouri earthquake (Mw = 6.9, Algeria): modelling and results. Geophys J Int. https://doi.org/10.1111/j.1365-246x.2006.02912.x

Al-Madani AT (1974) مذكرات الحاج أحمد الشريف الزهار (Al Haj Ahmed Sharif-al-Zahar diary). Société Nationale d'Edition et de Diffusion, Algiers, Algeria

Al-Rachidi A (1791).الثغر الجماني في ابتسام الثغر الوهراني (The Jamanian mouth in the Oranian smile). Publications of the Ministry of Education and Religious Affairs, heritage series, Algiers, Algeria

Ambraseys NN (1982) The seismicity of North Africa: the earthquake of 1856 at Jijeli, Algeria. Boll Geofis Teorica Appl XXIV(93):31–37

Ambraseys NN (2001) The earthquake of 1509 in the Sea of Marmara, Turkey (Revisited). Bull Seismol Soc Am 91(6):1397–1416

Ambraseys NN, Vogt J (1988) Material for the investigation of the seismicity of the region of Algiers. Eur Earthq Eng 3:16–29

Ambraseys N, Banda E, Irving J, Mallard D, Melville C, Morse T, Muir-Wood R, Muñoz D, Serva L, Shilston D, Surinach E, Vogt J (1983) Notes on historical seismicity. Bull Seismol Soc Am 73(6):1917–1920

Amir L, Theilen-Willige B (2017) Coastal risk and water flow analysis in eastern algeria (Western Mediterranean). Univ J Geosci 5(4):99–111

Amir L, Cisternas A, Vigneresse JL, Dudley W, Mc Adoo GB (2012) Algeria's vulnerability to tsunamis from near-field seismic sources. J Tsunami Soc Int Sci Tsunami Hazards 31(1):82–98

Amir L, Cisternas A, Dudley W, Mc Adoo BG, Pararas-Carayannis G (2013) A new tsunami risk scale for warning systems—application to the bay of Algiers in Algeria, west Mediterranean sea. J Tsunami Soc Int Sci Tsunami Hazards 32(2):116–130

Amir L, Dahlab AA, Douaifia MB, Theilen-Willige B (2015) Coastal risk in the Algiers region (Algeria): insights from tsunami velocities, seismic ground motion and remote sensing. J Tsunami Soc Int Sci Tsunami Hazards 34(3):173–188

Aoudia A, Meghraoui M (1995) Seismotectonics in the Tell Atlas of Algeria: the Cavaignac (Abou El Hassan) earthquake of 25.08.1922 (Ms = 5.9). Tectonophysics 248:263–276

Ayadi A, Bezzeghoud M (2015) Seismicity of Algeria from 1365 to 2013: maximum observed intensity map (MOI2014). Seismol Res Lett 86(1):236–244

Ayadi A, Dorbath C, Lesquer A, Bezzeghoud M (2000) Crustal and upper mantle velocity structure of the Hoggar swell (Central Sahara, Algeria). Phys Earth Planet Inter 118:111–123

Ayadi A, Ousadou-Ayadi F, Bourouis S, Benhallou H (2002) Seismotectonics and seismic quietness of the Oranie region (Western Algeria): The Mascara earthquake of August 18th 1994, Mw = 5.7, Ms = 6.0. J Seismolog 6:13–23

Ayadi A, Maouche S, Harbi A, Meghraoui M, Beldjoudi H, Ousadou F, Mahsas A, Benouar D, Heddar A, Rouchiche Y, Kherroubi A, Frogneux M, Lammali K, Benhamouda F, Sebaï A, Bourouis S, Alasset PJ, Aoudia A, Cakir Z, Merahi M, Nouar O, Yelles A, Bellik A, Briole P, Charade O, Thouvenot F, Semane F, Ferkoul A, Deramchi A, Haned SA (2003) Strong Algerian earthquake strikes near capital city. EOS Trans Am Geophys Union 84(50):561–568

Ayadi A, Dorbath C, Ousadou F, Maouche S, Chikh M, Bounif MA, Meghraoui M (2008) Zemmouri earthquake rupture zone (Mw 6.8, Algeria): aftershocks sequence relocation and 3D velocity model. J Geophys Res 113:B09301. https://doi.org/10.1029/2007JB005257

Balfour NJ, Salmon M, Sambridge M (2014) The Australian seismometers in schools network: education, outreach, research, and monitoring. Seismol Res Lett 85(5):1063–1068

Beaujeu J (1973) Apulée, opuscules philosophiques (du Dieu, de Socrate, Platon et sa doctrine du monde) et fragments. Edition les Belles lettres

Belabbès S, Wicks C, Cakir Z, Meghraoui M (2009) Rupture parameters of the 2003 Zemmouri (Mw 6.8), Algeria, earthquake from joint inversion of interferometric synthetic aperture radar, coastal uplift, and GPS. J Geophys Res 114:B03406. https://doi.org/10.1029/2008JB005912

Beldjoudi H, Guemache MA, Kherroubi A, Semmane F, Yelles-Chaouche AK, Djellit H, Amrani A, Haned A (2009) The Lâalam (Béjaïa, North-East Algeria) moderate earthquake (Mw = 5.2) on March 20, 2006. Pure Appl Geophys. https://doi.org/10.1007/s00024-009-0462-9

Beldjoudi H, Delouis B, Heddar A, Nouar OB, Yelles-Chaouche A (2011) The Tadjena Earthquake (Mw = 5.0) of December 16, 2006 in the Cheliff Region (Northern Algeria): waveform modelling, regional stresses, and relation with the Boukadir Fault. Pure Appl Geophys. https://doi.org/10.1007/s00024-011-0337-8

Beldjoudi H, Delouis B, Djellit H, Yelles-Chaouche A, Gharbi S, Abacha I (2016) The Beni-Ilmane (Algeria) seismic sequence of May 2010: Seismic sources and stress tensor calculations. Tectonophysics. https://doi.org/10.1016/j.tecto.2015.12.021

Bellalem F, Bounif MA, Koulakov I (2015) P and S waves tomographic analysis of the area of El Asnam 1980 Ms 7.3 earthquake (Algeria) from its aftershock sequence. J Seismolog 19 (1):253–264

Bellik A, Lammali K, Yelles A, Mahsas A, Terki T (2014) The velocity field around Ain Smara fault (Constantin region) by GPS. In: Abstracts of the second European conference on earthquake engineering and seismology, Istanbul, 24–29 Aug 2014

Benfedda A, Abbes K, Bouziane D, Bouhadad Y, Slimani A, Larbes S, Haddouche D, Bezzeghoud M (2017) The August 1st, 2014 (Mw 5.3) moderate earthquake: evidence for an active thrust fault in the Bay of Algiers (Algeria). Pure Appl. Geophys. https://doi.org/10.1007/s00024-017-1481-6

Benhallou H (1985) Les Catastrophes Séismiques de la Région d'Echéliff dans le contexte de la Séismicité Historique de l'Algérie. Thesis, University of Algiers, Algeria, p 294

Benhallou H, Ferrer A, Roussel J (1971) Catalogue des séismes algériens de 1951 à 1970. Publication of the Institut de Météorologie et de Physique du Globe de l'Algérie (IMPGA), Algiers University, p 198

Benouar D (1994) Materials for the investigation of the seismicity of Algeria and adjacent regions during the twentieth century. Ann Geofis XXXVII(4):459–860

Benouar D, Aoudia A, Maouche S, Meghraoui M (1994) The 18 August 1994 Mascara (Algeria) earthquake-a quick-look report. Terra Nova 6:634–637

Bezzeghoud (2012) Hadj Benhallou 1937-2011. J Seismol 16:373

Bezzeghoud M, Buforn E (1999) Source parameters of the 1992 Melilla (Spain, Mw = 4.8), 1994 Al Hoceima (Morocco, Mw = 5.8) and Mascara (Algeria, Mw = 5.7) earthquakes and seismotectonic implications. Bull Seismol Soc Am 89:359–372

Bezzeghoud M, Ayadi A, Sebaï A, Benhallou H (1994) Seismogenic zone survey by the Algerian Telemetred Seismological Network, case study of Rouina earthquake, 19 January 1992. Phys Earth Planet Inter 84:235–246

Bezzeghoud M, Dimitrov D, Ruegg JC, Lammali K (1995) Faulting mechanism of the El Asnam (Algeria) 1954 and 1980 earthquakes from modelling of vertical movements. Tectonophysics 249:249–266

Bezzeghoud M, Ayadi A, Sebaï A, Aït Messaoud M, Mokrane A, Benhallou H (1996) Seismicity of Algeria between 1365 and 1989: map of maximum observed intensities (MOI). Avances en Geofisica y Geodesia, Ministerio de Obras Publicas, Transportes y Media Ambiente, vol I. Pub. De IGN, Madrid, 107–114

Bockel M (1972) Structure de la croûte terrestre en Algérie d'après les ondes sismiques. Publication of the Institut de Météorologie et de Physique du Globe de l'Algérie (IMPGA), Algiers University, pp 330–358

Bordet P (1952) Les appareils volcaniques récents du Hoggar. In: Abstracts of the XIXe Congrès Géologique International, Monographies Régionales,1ère série n° 11

Boughacha MS (2005) Analyse de la sismicité de l'Algérie. Application à l'établissement des cartes des I.M.O et CFF. Doctorate thesis USTHB, Algiers, p 289

Boughacha MS, Ouyed M, Ayadi A, Benhallou H (2004) Seismicity and seismic hazard mapping of northern Algeria: map of maximum calculated intensities (MCI). J Seismolog 8:1–10

Boughacha MS, Ouyed M, Benhallou H (2006) Perspective d'étude de la sismicité du Hoggar par la caractérisation des sources. In: Abstracts of the 4th international geological correlation program IGCP 485. Cratons, metacratons and mobile belts: keys from the West African craton boundaries Eburnian versus Pan-African signature, magmatic, tectonic and metallogenic implications. Algiers-Tamanrasset (Algeria), 2–9 Dec 2006

Bouhadad Y, Benhamouche A, Bourenane H, Ait Ouali A, Chikh M, Guessoum N (2010) The Laalam (Algeria) damaging landslide triggered by a moderate earthquake (Mw = 5.2) (2010). Nat Hazards 54(2):261–272

Bounif A, Haessler H, Meghraoui M (1987) The Constantine (Northeast Algeria) earthquake of October 27, 1985: surface ruptures and aftershock study. Earth Planet Sc Lett. 85:451–460

Bounif A, Bezzeghoud M, Dorbath L, Legrand D, Deschamps A, Rivera L, Benhallou H (2003) Seismic source study of the 1989, October 29, Chenoua (Algeria) earthquake from aftershocks, broad-band and strong ground motion records. Ann Geophys 46(4):625–646

Bounif A, Dorbath C, Ayadi A, Meghraoui M, Beldjoudi H, Laouami N, Frogneux M, Slimani A, Alasset PJ, Kharroubi A, Oussadou F, Chikh M, Harbi A, Larbes S, Maouche S (2004) The 21 May, 2003 (Mw 6.8) Zemmouri (Algeria) earthquake relocation and aftershock sequence analysis. Geophys Res Lett 31:L19606. https://doi.org/10.1029/2004GL020586

Bourouis S, Sebaï A, Yelles-Chaouche A, N Akacem (2013) The recent intraplate seismicity of the Hoggar massif, Southern Algeria. In: Abstracts of the large Mediterranean Earthquakes: from past to present, Algiers, 21–23 May 2013

Boutequin M (1911) Ciel et Terre, vol 32

Bouyahiaoui B, Sage F, Abtout A, Klingelhoefer F, Yelles-Chaouche K, Schnurle P, Marok A, Deverchere J, Arab M, Galve A, Collot JY (2015) Crustal structure of the eastern Algerian continental margin and adjacent deep basin: implications for late Cenozoic geodynamic evolution of the western Mediterranean. Geophys J Int 201:1912–1938

Brives A, Dalloni M (1922) Le tremblement de terre du 25 août 1922 et la structure géologique de la région de Ténès-Cavaignac (Algérie) Comptes Rendus Des Séances de l'Académie des Sciences 175(14):527–529

Bulard Ch (1873) Observatoire d'Alger – 1ère partie – Panorama météorologique du climat d'Alger – Observations météorologiques

Cetin E, Meghraoui M, Cakir Z, Akoglu AM, Mimouni O, Chebbah M (2012) Seven years of postseismic deformation following the 2003 Mw = 6.8 Zemmouri earthquake (Algeria) from InSAR time series. Geophys Res Lett 39:L10307. https://doi.org/10.1029/2012GL051344

Chesneau M (1892) Note sur les tremblements de terre en Algérie. Ann Mines 9(I):5–46

Cochard (1867) Lettre sur le tremblement de terre d'Alger adressée à M. Wolf et comuniquée par M. Le Verrier. In: Comptes rendus des séances de l'Académie des Sciences 64:34

De Sénarmont H (1857) Analyse des documents recueillis sur les tremblements de terre ressentis en Algérie du 21 aout au 15 octobre 1856. CRAS 44:586–594

Delouis B, Vallée M, Meghraoui M, Calais E, Maouche S, Lammali K, Mahsas A, Briole P, Benhamouda F, Yelles A (2004) Slip distribution of the 2003 Boumerdes-Zemmouri earthquake, Algeria, from teleseismic, GPS, and coastal uplift data. Geophys Res Lett 31: L18607. https://doi.org/10.1029/2004GL020687

Deschamp A, Bezzeghoud M, Bounif A (1991) Seismological study of the Constantine (Algeria) earthquake (27 October 1985). Publication IGN Série Monografia 8:163–173

Deschamps A, Gaudemer Y, Cisternas A (1982) The El Asnam, Algeria earthquake of 10 October 1980: multiple-source mechanism determined from long-period records. Bull Seismol Soc Am 72(4):1111–1128

Dimitrov D, Benhallou H, Khalfoun A, Lammali K (1987) Mesures geodesiques liees aux deformations de l'écorce terrestre dans la region d'Ech Cheliff (ex El Asnam), Algérie du Nord. C R Acad Sci 305(11):1365–1368

Dimitrov D, Lammali K, Khalfoun A, Benhallou H (1991) Post-seismic movements from 1981 to 1988 on the El Asnam thrust fault (Algeria). Publication IGN Série Monografia 8:241–254

Domzig A, Yelles K, Le Roy C, Déverchère J, Bouillin JP, Bracène R, Mercier de Lépinay B, Le Roy P, Calais E, Kherroubi A, Gaullier V, Savoye B, Pauc H (2006) Searching for the Africa-Eurasia Miocene boundary offshore western Algeria (MARADJA'03 cruise). C. R. Geoscience 338(2006):80–91

Ferdi S, Harbi A (2014) Roman literary and epigraphic sources for the study of historical seismicity in Algeria circa 42–420 AD. J Seismol 18:277–287

Finot PX (1845) Note extraite des mémoires de Médecine, Chirurgie et Pharmacie Militaire, t. LVI

Ghalem M (1998) Earthquakes in traditional Algerian historiography: between history and memory. Insaniyat, 47–67. https://doi.org/10.4000/insaniyat.11586

Girardin N, Hatzfeld D, Guiraud R (1977) La sismicité du Nord de l'Algérie. C R somm Soc géol Fr 2:95–100

Grandjean A (1954) Séismes d'Algérie de 1940 à 1950 inclus. Annales de l'Institut de Physique du Globe de Strasbourg, VII, 83

Grandjean A, Pinon R, Bucher A (1960) Les séismes du Hoggar. Note interne, Institutde Physique du Globe d'Algérie, p 4

Grandjean A, Guiraud R, Polveche J (1966) Le séisme de M'sila. Bull Serv Géol Algérie N 33

Guemache MA, Machane D, Beldjoudi H, Gharbi S, Djadia L, Benahmed S, Ymmel Hayet (2009) On a damaging earthquake-induced landslide in the Algerian Alps: the March 20, 2006 Laâlam landslide (Babors chain, northeast Algeria), triggered by the Kherrata earthquake (Mw = 5.3). Nat Hazards 54(2):273–288

Hadiouche O, Jobert N (1988) Evidence for anisotropy in north east Africa, from geographical and azimuthal distribution of Rayleigh wave velocities, and average upper mantle structure. Geophys Res Lett 15(4):365–368

Hamai L, Petit C, Abtout A, Yelles-Chaouche A, Deverchere J (2015) Flexural behaviour of the north Algerian margin and tectonic implications. Geophys J Int 201:1426–1436

Haned A, Stutzman E, Schimmel M, Kiselev S, Davaille A, Yelles-Chaouche A (2016) Global tomography using seismic hum. Geophys J Int 204:1222–1236

Harbi A, Amal S, Yasmina R, Said M, Farida O, Khadidja A, Dalila AB, Manel B (2017) Reappraisal of the seismicity of the southern edge of the Mitidja basin (Blida region, North-Central Algeria). Seismol Res Lett 88(4). https://doi.org/10.1785/0220160217

Harbi A, Maouche S (2009) Les principaux séismes du Nord-Est de l'Algérie. Mém Serv Géol Natl 16:81

Harbi A, Maouche S, Ayadi A, Benouar D, Panza GF, Benhallou H (2004) Seismicity and tectonic structures in the site of Algiers and its surroundings: a step towards microzonation. Pure appl Geophys 161:949–967

Harbi A, Maouche S, Vaccari F, Aoudia A, Oussadou F, Panza GF, Benouar D (2007a) Seismicity, seismic input and site effects in the Sahel-Algiers Region (North Algeria). Soil Dyn Earthq Eng 27(5):427–447

Harbi A, Maouche S, Oussadou F, Rouchiche Y, Yelles-Chaouche A, Merahi M, Heddar A, Nouar O, Kherroubi A, Beldjoudi H, Ayadi A, Benouar D (2007b) Macroseismic study of the Zemmouri earthquake of 21 May 2003 (Mw 6.8, Algeria). Earthq spectra 23(2):315–332

Harbi A, Peresan A, Panza GF (2010) Seismicity of Eastern Algeria: a revised and extended earthquake catalogue. Nat Hazards. https://doi.org/10.1007/s11069-009-9497-6

Harbi A, Meghraoui M, Maouche S (2011) The Djidjelli (Algeria) earthquakes of 21 and 22 August (I0 IX, VIII) and related tsunami effects Revisited. J Seismolog 15(1):105–129

Harbi A, Sebaï A, Benmedjber M, Ousadou F, Rouchiche Y, Grigahcene A, Aïni D, Bourouis S, Maouche S, Ayadi A (2015) The Algerian homogenized macroseismic database (267–1989): a deeper insight into the Algerian historical seismicity. Seismol Res Lett 86(6):1705–1716

Heddar A, Beldjoudi H, Authemayou C, Si Bachir R, Yelles-Chaouche A, Boudiaf A (2016) Use of the ESI-2007 scale to evaluate the 2003 Boumerdès earthquake (North Algeria). Ann Geophys 59(2):S0211. https://doi.org/10.4401/ag-6926

Hée A (1919, ….., 1939) Les séismes de l'Algérie. Annuaires puis Annales de l'Institut de Physique du Globe de Strasbourg, 2e partie, Séismologie, t, I, …., IV

Hée A (1925) La fréquence des tremblements de Terre en Algérie 1911-1924. Monogr Bur Cent Séismol Int B(2):111–154

Hée A (1933) La fréquence des tremblements de terre en Algérie, 1911–1932Monogr Bur Cent Séismol Int, p 99

Hée A (1950) Catalogue des séismes algériens de 1850 à 1911. Annales de l'Institut de Physique du Globe de Strasbourg, 41–49

Heidarzadeh M, Satake K (2013) The 21 May 2003 tsunami in the Western Mediterranean Sea: statistical and wavelet analyses. Pure appl Geophys 170:1449–1462

Hubert H (1932) La Physique du globe dans les territoires d'Outre-Mer. Actes et Comptes Rendus de l'Assoiation Colonies-Sciences, N° 81, Paris

Ibn Ali Al Sharif (1778) معالم الاستبصار (Foresight landmarks). Ancient manuscripts available at Algiers Library, Algeria

Ibn Al Mofti H (no date)تاريخ بشوات الجزائ وعلمائها (History of the Algiers Pashas). Manuscript available at Beit Al Hikma, Algiers

Ibn Chakroun M (1990)فيض العُباب و افاضة قداح الآداب في الحركة السّعيدة إلى قسنطينة و الزاب لابن الحاج النميري (Notes of travel from Constantine to the Zibans by Ibn El Haj Ennoumeyri). Dar Al-Gharb Al-Islami, Beyrouth, p 536

Kariche J, Meghraoui M, Ayadi A, Boughacha MS (2017) Stress change and fault interaction from a two century-long earthquake sequence in the central tell atlas. BSSA, Algeria. https://doi.org/10.1785/0120170041

Khelfaoui H (2001) La science en Algérie. In: Waast Roland, Gaillard Jacque (eds) La Science en Afrique à l'aube du 21ème siècle. Commission Européenne, DG, XII, Paris

Kherroubi A, Déverchère J, Yelles A, Mercier de Lépinay B, Domzig A, Cattaneo A, Bracène R, Gaullier V, Graindorge D (2009) Recent and active deformation pattern off the easternmost Algerian margin, Western Mediterranean Sea: new evidence for contractional tectonic reactivation. Mar Geol 261:17–32

Kherroubi A, Yelles-Chaouche A, Koulakov I, Deverchere J, Beldjoudi H, Haned A, Semmane F, Aidi C (2017) Full aftershock sequence of the Mw 6.9 2003 Boumerdes Earthquake, Algeria: space–time, distribution, local tomography and seismotectonic implications. Pure Appl Geophys. https://doi.org/10.1007/s00024-017-1571-5

King G, Yielding G (1984) The evolution of a thrust fault system: process of rupture initiation, propagation and termination in the 1980 El Asnam (Algeria) earthquake. Geophys J R Soc 77:915–933

Lammali K, Bezzeghoud M, Oussadou F, Dimitrov D, Benhallou H (1997) Postseismic deformation at El Asnam (Algeria) in the seismotectonic context of northwestern Algeria. Geophys J Int 129:597–612

Laouami N, Slimani A, Bouhadad Y, Chatelain JL, Nour A (2006) Evidence for fault-related directionality and localized site effects from strong motion recordings of the 2003 Boumerdes (Algeria) earthquake: consequences on damage distribution and the Algerian seismic code. Soil Dyn Earthq Eng 26:991–1003

Lin J, Stein R, Meghraoui M, Toda S, Ayadi A, Dorbath C, Belabbes S (2011) Stress transfer among en echelon and opposing thrusts and tear faults: triggering caused by the 2003 Mw = 6.9 Zemmouri, Algeria, earthquake. J Geophys Res 116:B03305. https://doi.org/10.1029/2010JB007654

Mahsas A, Lammali K, Yelles K, Calais E, Freed AM, Briole P (2008) Shallow afterslip following the 2003 May 21, Mw = 6.9 Boumerdes earthquake, Algeria. Geophys J Int 172:155–166

Maouche S, Benouar D, Harbi A, Benhallou H (1998) The Algiers (Algeria) earthquake of 4 September 1996. Int J Earthq Eng Eng Seismol XII(1):51–55

Maouche S, Harbi A, Meghraoui M (2008) Attenuation of intensity for the Zemmouri earthquake of 21 May 2003 (Mw 6.8): insights for the seismic hazard and historical earthquake sources in northern Algeria. In: Fréchet J, Meghraoui M, Stuchi M (eds) Historical seismology, interdisciplinary studies of past an recent earthquakes. Springer, New York, pp 327–350

Maouche S, Morhange C, Meghraoui M (2009) Large boulder accumulation on the Algerian coast evidence tsunami events in the western Mediterranean. Mar Geol 262(2009):96–104

Maouche S, Meghraoui M, Morhange C, Belabbès S, Bouhadad Y, Haddoum H (2011) Active coastal thrusting and folding, and uplift rate of the Sahel anticline and Zemmouri earthquake area (Tell Atlas, Algeria). Tectonophisics 509:69–80

Meghraoui M (1991) Blind Reverse Faulting System Associated with the Mont Chenoua-Tipaza Earthquake of 29 October 1989 (North-Central Algeria). Terra Nova 3:84–93

Meghraoui M, Doumaz F (1996) Earthquake-induced flooding and paleoseismicity of the El Asnam, Algeria, fault-related fold. J Geophys Res 101(B8):17617–17644

Meghraoui M, Cisternas A, Philip H (1986) Seismotectonics of the lower Cheliff basin: structural background of the El Asnam (Algeria) Earthquake. Tectonics 5(6):809–836

Meghraoui M, Philip H, Albarede F, Cisternas A (1988a) Trench investigations through the trace of the 1980 El Asnam thrust fault: evidence for paleoseismicity. Bull Seismol Soc Am 78 (2):979–999

Meghraoui M, Jaegy R, Lammali K, Albarede F (1988b) Late Holocene earthquake sequences on the El Asnam (Algeria) thrust fault. Earth Planet Sci Lett 90:187–203

Meghraoui M, Maouche S, Chemaa B, Cakir Z, Aoudia A, Harbi A, Alasset PJ, A Ayadi, Bouhadad Y, Benhamouda F (2004) Coastal uplift and thrust faulting associated with the Mw = 6.8 Zemmouri (Algeria) earthquake of 21 May, 2003. Geophys Res Lett 31:L19605 https://doi.org/10.1029/2004gl020466

Milet B.(1999) L'Observatoire de la Bouzaréah. L'Algérianiste n°86, Les Presses Littéraires, Montpellier, France

Mokrane A, Ait Messaoud A, Sebai A, Menia N, Ayadi A, Bezzeghoud M, Benhallou H (1994). Les séismes en Algérie de 1365 à 1992. Publication du Centre de Recherche en Astronomie, Astrophysique et Géophysique, Département: Etudes et Surveillance Sismique, ESS, C.R.A.A.G, Alger-Bouzaréah, p 277

Montessus de Ballore F (1906) Les Tremblements de terre. Librairie armand colin, Paris, France

Nocquet JM, Calais E (2004) Crustal velocity field of Western Europe from permanent GPS array solutions, 1996–2001. Geophys J Int 154:72–88

Ousadou F, Dorbath L, Dorbath C, Bounif MA, Benhallou H (2013) The Constantine (Algeria) seismic sequence of 27 October 1985: a new rupture model from aftershock relocation, focal mechanisms, and stress tensors. J Seismol 17:207–222

Ousadou F, Dorbath L, Ayadi A, Dorbath C, Gharbi S (2014) Stress field variations along the Maghreb region derived from inversion of major seismic crisis fault plane solutions. Tectonophysics. https://doi.org/10.1016/j.tecto.2014.06.017

Ouyed M, Meghraoui M, Cisternas A, Deschamps A, Dorel J, Fréchet J, Gaulon R, Hatzfeld D, Philip H (1981) Seismotectonic of the El Asnam earthquake. Nature 292(5818):26–31

Ouyed M, Yielding G, Hatzfeld D, King GCP (1983) An aftershock study of the El Asnam (Algeria) earthquake of 1980 10. Geophys J R Astr Soc 73:605–639

Ouyed M, Idres M, Bourmatte A, Boughacha MS, Samai S, Yelles A, Haned A, Aidi C (2011) Attempt to identify seismic sources in the eastern Mitidja basin using gravity data and aftershock sequence of the Boumerdes (May 21, 2003, Algeria) earthquake. J Seismol 15:173–189

Perrey A (1847) Note sur les tremblements de terre en Algérie et dans l'Afrique septentrionale. Mémoires de l'Académie des Sciences, Arts et Belles Lettres de Dijon, années 1845-1846, pp 299–323

Philip H, Meghraoui M (1983) Structural analysis and interpretation of the surface deformation of the El Asnam earthquake of October 10, 1980. Tectonics 2:17–49

Pomel MA (1891) Les tremblements de terre du 15 et 16 Janvier en Algérie. Compt Rendus Séances l'Acad Sci CXII(13):643–646

Radi Z, Yelles-Chaouche A, Bokelmann G (2015) Seismic anisotropy of northeastern Algeria from shear-wave splitting analysis. Phys Earth Planet. https://doi.org/10.1016/j.pepi.2015.08.002

Radi Z, Yelles-Chaouche A, Corchete V, Guettouche S (2017) Crust and upper mantle shear wave structure of Northeast Algeria from Rayleigh wave dispersion analysis. Phys Earth Planet Inter 270(2017):84–89

Roca A, Izquierdo A, Sousa-Oliveira C, Martinez-Solares JM (2004) An outline of earthquake catalogues, databases and studies of historical seismicity in the Iberian Peninsula. Ann Geophys 47(2/3):561–570

Roger J, Hébert H, Ruegg JC, Briole P (2011) The El Asnam 1980 October 10 inland earthquake: a new hypothesis of tsunami generation. Geophys J Int. https://doi.org/10.1111/j.1365-246x.2011.05003.x

Rothé E (1925) Les échelles employées en séismologie. In: Compte-rendu de la 48e session de l'Association française pour l'avancement des sciences, congrès, Liège 1924, 365-371

Rothé JP (1950) Les séismes de Kherrata et la séismicité de l'Algérie. Bull Serv Cart Geol Algérie, 4 (3)

Rothé JP (1970) Mission d'information séismologique, novembre-décembre 1869, Moyen-Orient et Afrique du Nord, 1759/BMS.RD/SCE, Paris

Rothé JP (1983) Sismicité de la France entre 1971 et 1977. Publication of Bureau Central Sismologique Français, Strasbourg

Roussel J (1974) Note sur le séisme de Mansourah (Biban) du 24 novembre 1973. Bull Soc Hist nat Af Nord Alger, 65 (1, 2)

Ruegg JC, Kasser M, Tarantola A, Lepine JC, Chouikrat B (1982). Deformation associated with the El Asnam earthquake of 10 October 1980: Geodetic determination of vertical and horizontal movements. Bull Seismol Soc Am 72(6):2227–2224

Sahal A, Roger J, Allgeyer S, Lemaire B, Hébert H, Schindelé F, Lavigne F (2009) The tsunami triggered by the 21 Mays 2003 Boumerdès-Zemmouri (Algeria) earthquake: field investigations on the French Mediterranean coast and tsunami modelling. Nat Hazards Earth Syst Sci 9:1823–1834

Santos R, Caldeira B, Bezzeghoud M, Borges JF (2014) The rupture process and location of the 2003 Zemmouri–Boumerdes Earthquake (Mw 6.8) inferred from seismic and geodetic data. Pure Appl Geophys. https://doi.org/10.1007/s00024-014-0978-5

Sbeinati MR, Drawcheh R, Mouty M (2005) The historical earthquakes of Syria: an analysis of large and moderate earthquakes from 1365 BC to 1900 AD. Ann Geophys 48(3):347–435

Schoenagel (1856) Effets des tremblements de terre sur les constructions en maçonnerie. Document du génie militaire de la direction de Constantine, place de Djidjelli

Sebaï A, Bernard P (2008) Contribution à la connaissance de la sismicité d'Alger et de ses alentours au XVIIIe siècle, extraite des archives françaises. C R Geosci 340:495–512

Sebai A, Stutzmann E, Mantagner JP, Sicilia D, Beucler E (2006) Anisotropic structure of the African upper mantle from Rayleigh and Love wave tomography. Phys Earth Planet Inter 155:48–62

Semmane F, Campillo M, Cotton F (2005) Fault location and source process of the Boumerdes, Algeria, earthquake inferred from geodetic and strong motion data. Geophys Res Lett 32: L01305. https://doi.org/10.1029/2004GL021268

Semmane F, Abacha I, Yelles-Chaouche AK, Haned A, Beldjoudi H, Amrani A (2012) The earthquake swarm of December 2007 in the Mila region of northeastern Algeria. Nat Hazards 64:1855–1871

Semmane F, Benabdeloued BYN, Beldjoudi H, Yelles-Chaouche AK (2015) The 22 February 2014 Mw 4.1 Bordj-Menaïel Earthquake, Near Boumerdes-Zemmouri, North-Central Algeria. SRL https://doi.org/10.1785/0220140196

Semmane F, Benabdeloued BYN, Heddar A, Khelif MF (2017) The 2014 Mihoub earthquake (Mw4.3), northern Algeria: empirical Green's function analysis of the mainshock and the largest aftershock. J Seismol. https://doi.org/10.1007/s10950-017-9671-3

Trépied Ch (1884) Notice sur l'observation d'Alger. Bull Ast 1(1):214–216

Vatonne M (1866) Notice sur les sondages par le service des mines de la province d'Alger sur le territoire du village de l'Oued El Alleug, plaine de la Mitidja. Annales des Mines, série 6, vol. 9

Wang J (2004) Historical earthquake investigation and research in China. Ann Geophys 47(2/3)

Yelles AK, Deramchi A, Ferkoul A, Aoulaiche K (2002) Les séismes en Algérie de 1992 à 2001. C.R.A.AG Publication, Alger, p 216

Yelles AK, Lammali K, Mahsas A, Calais E, Briole P (2004) Coseismic deformation of the May 21, 2003, Mw = 6.8 Boumerdes earthquake, Algeria, from GPS measurements. Geophys Res Lett 31:L13610. https://doi.org/10.1029/2004GL019884

Yelles AK, Domzig A, Déverchère J, Bracène R, Mercier de Lépinay B, Strzerzynski P, Bertrand G, Boudiaf A, Winter T, Kherroubi A, Le Roy P, Djellit H (2009) Plio-Quaternary reactivation of the Neogene margin off NW Algiers, Algeria: The Khayr al Din bank. Tectonophysics. https://doi.org/10.1016/j.tecto.2008.11.030

Yelles-Chaouche AK (1991) Coastal Algerian earthquakes. A potential risk of tsunamis in Western Mediterranean? Preliminary investigations. Sci Tsunami Hazards 9(1):47–54

Yelles-Chaouche AK (2015) Prés d'un siecle de sismologie en Algerie. Journée thématique SAG, USTHB, 20 October

Yelles-Chaouche AK, Harbi A, Djelitt H (1996) Tectonique récente de la marge Est-Algérienne. Publication In Annales de l'équipement, Actes du VIème Séminaire Maghrébin de Génie Parasismique, publication semestrielle, vol VI, N° 2 99:127–144

Yelles-Chaouche AK, Djellit H, Hamdache M (2003) The Boumerdes-Algiers (Algeria) earthquake of May, 21st, 2003 (Mw 6.8). CSEM Lett 20:1–3

Yelles-Chaouche AK, Djellit H, Beldjoudi H, Bezzeghoud M, Buforn E (2004) The Ain Temouchent (Algeria) Earthquake of December 22nd, 1999 Pure Appl Geophys 161:607–621

Yelles-Chaouche AK, Roger J, Déverchère J, Bracène R, Domzig A, Hébert H, Kherroubi A (2009) The 1856 tsunami of Djidjelli (Eastern Algeria): seismotectonics, modeling and hazard implications for the Algerian coast. Pure appl Geophys 166:283–300

Yelles-Chaouche AK, Lammali K, Mahsas A (2010) The REGAT (Réseau Géodésique de l'Atlas) Network in northern Algeria. Wegener 15th Assembly, Istanbul, Sept 15–17

Yelles-Chaouche AK, Allili T, Alili A, Messemen W, Beldjoudi H, Semmane F, Kherroubi A, Djellit H, Larbes Y, Haned S, Deramchi A, Amrani A, Chouiref A, Chaoui F, Khellaf K, Nait Sidi Said C (2013a) The new Algerian Digital Seismic Network (ADSN): towards an earthquake early-warning system. Adv Geosci 36:31–38

Yelles-Chaouche AK, Abacha I, Semmane F, Beldjoudi H, Djellit H (2013b) The Beni-Ilmane (North-Central Algeria) Earthquake Sequence of May 2010. Pure Appl Geophys. https://doi.org/10.1007/s00024-013-0709-3

Yelles-Chaouche AK, Haned A, Aidi C, Beldjoudi H, Kherroubi A, Semmane F, Benabdeloued BYN, Larbes Y, Alili A, Khelif MF, Belheouane A (2017) The Mw 5.0 Hammam Melouane Earthquake (North Central Algeria) of 17 July 2013 in the Context of the Tellian Atlas Seismicity. Pure Appl Geophys https://doi.org/10.1007/s00024-017-1492-3

Yielding G, Jackson JA, King GCP, Sinval H, Vita-Finzi C, Wood RM (1981) Relations between surface déformation, fault geometry, seismicity, and rupture characteristics during the El Asnam (Algeria) earthquake of 10 October 1980 (1984). Earth Planet Sci Lett 56:287–304

Active Tectonics and Seismic Hazard in the Tell Atlas (Northern Algeria): A Review

Said Maouche, Youcef Bouhadad, Assia Harbi, Yasmina Rouchiche, Farida Ousadou and Abdelhakim Ayadi

Abstract The Tell Atlas of Algeria, which experienced several destructive earthquakes in the past, is among the most seismic active zones in the western Mediterranean. The seismicity is not randomly distributed but directly related to active geological structures, which mainly correspond to faulted folds. The comprehensive studies of the El Asnam and Zemmouri faults allowed identifying similar structures distributed all over the Tell Atlas, which generated moderate earthquakes. The available paleoseismic data attest that the recurrence of strong earthquakes ($M > 7.0$) is about 300–500 years while seismicity data suggest 25–30 years for moderate earthquakes. This paper presents a review of active tectonics and seismic hazard in the Tell Atlas.

Keywords Active tectonics · Earthquakes · Seismic hazard · Tell Atlas of Algeria

1 Introduction

Active tectonics in northern Algeria is relatively well understood thanks to the numerous studies which were carried out whenever a destructive earthquake occurs in the country (e.g., 1980 El Asnam (Mw 7.3), 1994 Mascara (Mw 5.9), 1996 Algiers (Ml 5.7), and 2003 Zemmouri (Mw 6.8) earthquakes) (King and Vita-Finzi 1981; King and Yielding 1984; Philip and Meghraoui 1983; Bounif et al. 1987,

---

S. Maouche (✉) · A. Harbi · Y. Rouchiche · F. Ousadou · A. Ayadi
Centre de Recherche en Astronomie, Astrophysique et Géophysique, BP 63, 16340 Bouzaréah, Algiers, Algeria
e-mail: said.maouche@gmail.com

Y. Bouhadad
Centre Nationale de Recherche Appliquée en Génie Parasismique, rue Kaddour Rahim prolongée, Hussein Dey, Algiers, Algeria

A. Harbi
Simons Fellow at Abdus Salam, International Centre for Theoretical Physics, Strada Costiera, 11, 34151 Trieste, Italy

A. Bendaoud et al. (eds.), *The Geology of the Arab World—An Overview*,
Springer Geology, https://doi.org/10.1007/978-3-319-96794-3_10

2004; Meghraoui 1991; Benouar et al. 1994; Bouhadad et al. 2003; Harbi et al. 2004; Yelles-Chaouche et al. 2006; Ayadi et al. 2008; Meghraoui et al. 2004; Belabbès et al. 2009). The seismicity level of the thrusts and folds belt in Northern Algeria is relatively low compared to the other earthquake prone areas in the Mediterranean basin, such as Italy and Greece. However, the Tell Atlas of Algeria is marked by the occurrence of strong to major earthquakes (Algiers 1365 and 1716, Oran 1790, Blida 1825, Djidjelli 1856 (6.5 < Ms < 7.0), Aumale 1910 (Ms 6.6), Orléansville 1954 (Ms 6.7), El Asnam 1980 (Mw 7.3), Constantine 1985 (Ms 6.0), Mt Chenoua–Tipaza 1989 (Ms 6.0), Mascara 1994 (Mw 5.9), Ain Témouchent 1999 (Mw 5.7), Béni Ourtilène 2000 (Mw 5.7), Zemmouri 2003 (Mw 6.8)) (Benhallou 1985; Mokrane et al.1994; Benouar 1994; Benouar et al. 1994; Bezzeghoud and Buforn 1999; Ayadi et al. 2002; Yelles-Chaouche et al. 2004; Maouche et al. 2008; Harbi et al. 2004, 2011, 2015; Belabbès et al. 2008; Ayadi and Bezzeghoud 2015). Despite the wealth of studies performed in the Tell Atlas, the discrimination between active and non-active faults in Algeria remains a complex issue. Neotectonic studies of the last decades provided a considerable amount of data on young geological structures, but the history of seismically active regions was not sufficiently investigated. Therefore, Pleistocene and Holocene behaviors of active zones are still unclear (Bouhadad et al. 2008; Maouche et al. 2011).

Detailed studies on the correlation between strong seismic events and geological features in the Tell Atlas of Algeria have raised two main points:

(1) Moderate-size earthquakes ($4.5 < M < 6.5$) are frequent and recurrent in the same seismogenic zones (Benouar 1994; Mokrane et al. 1994; Mourabit et al. 2014; Harbi et al. 2015);
(2) Seismically active regions correspond to zones of highly deformed young deposits and prominent Quaternary geological structures (Meghraoui 1988; Meghraoui et al. 1986; Maouche et al. 2011).

Compressional tectonics in the continental domain in northern Algeria are guided by the convergent movement of Africa towards Eurasia (Meghraoui and Pondrelli 2012 and references therein) characterized by geological structures such as folds, thrusts and overthrusts, and nappes that prevail in this part of the orogenic Alpine belt. Asymmetrical folds associated to thrust faults, uplifted terraces, and flexural slip folding, which are mainly located in intermountains Neogene and Quaternary basins, constitute the main structural features of the Tell Atlas. The identification of active zones capable of producing large or moderate seismic events could be analyzed through new neotectonic evidence. Similar geological structures to that of El Asnam and their relation with strong historical seismic events have been identified along the Tell Atlas (Meghraoui 1988; Aoudia and Meghraoui 1995; Meghraoui et al. 1996; Harbi et al. 1999).

In this paper, we present a review of the seismotectonics of the Tell Atlas of Algeria in the light of new neotectonic observations. This investigation is carried out in terms of a comprehensive literature review including seismotectonics, active tectonics, and earthquake distribution. Our objective is to contribute to a better understanding of the active tectonics in the Tell Atlas towards an efficient seismic hazard assessment. This is in fact one of the main challenges to mitigate the seismic risk in Northern Algeria.

2 Tectonic Framework of the Tell Atlas

The Tell Atlas of Algeria is an east–west trending fold and thrust belt located in the northern part of the African continent. This geological domain belongs to the southern branch of the Alpine chain surrounding the western Mediterranean basin (Fig. 1a). The tectonic structures of this region are the result of nearly N–S compressional movements involving the African and Eurasian plates from the Cainozoic to present. Neogene post-nappes basins, which correspond to EW-elongated intermountain structures, are characterized by compressional deformations that extended during the Quaternary. These neotectonic features correspond to E–W- to NE–SW-striking folds and related reverse faults that may affect very young Quaternary deposits. The Cheliff, Mitidja, Soummam, Hodna and Constantine intermountains sedimentary basins represent the main structural features in the Tell Atlas (Fig. 1b) concentrating most of the seismic activity of the Tell Atlas of Algeria. These east–west elongated basins show recent compressional tectonic structures, which are characterized by NE–SW to EW trending fold systems. These basins are filled with Neogene and Quaternary deposits with a thickness reaching 5000 m in the middle of the basins at most (i.e., 4000 m in the Hodna basin, Boudiaf 1996). The thickness of deposits gets more thin on the edges where they unconformably cover pre-Neogene flysch and epimetamorphic bedrocks (Perrodon 1957; Thomas 1985; Meghraoui et al. 1986).

2.1 *Neotectonics*

In Oran area in the vicinity of the Habra plain (Fig. 1b), the Murdjadjo structure consists of an asymmetric fold of about 32 km, which is trending N050 (Thomas 1985; Bouhadad 2001; Meghraoui et al. 1996). Field observations reveal the existence of a 60°NW dipping fault, which affects the southeastern flank of the Murdjadjo anticline. This fault, which is emphasized by the offset of the stream pattern, shows a strike lateral slip component. Marine terraces that one may observe along the Oran coast show a uniform uplift rate which could be compared to the coseismic rate determined in the central coastal area (Maouche et al. 2011). Based on geomorphic evidence (triangular facets, diversion and offset of drainage network, anticline direction, etc.), we can see that the Murdjadjo fault is segmented into two segments (Bouhadad 2001). In the southern part of the Habra alluvial basin, which corresponds to the western continuation of the Cheliff basin, the Beni

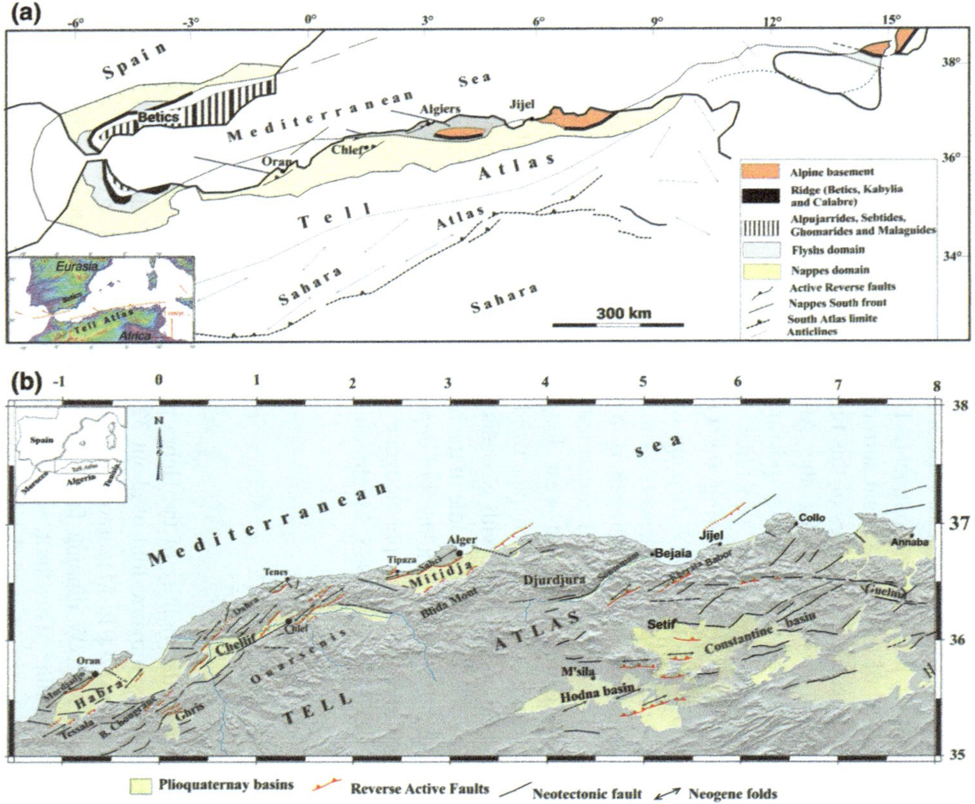

Fig. 1 **a** Geological setting of northern Africa (Durand Delga 1969), **b** neotectonic map showing active and potentially active structures in the Tell Atlas of Algeria (Vila 1980; Meghraoui 1988; Meghraoui et al. 1996, 2004; Harbi et al. 1999; Maouche et al. 2011)

Choughrane mounts located north of Mascara region are made of Cretaceous nappes that are unconformably covered by folded Neogene and Quaternary deposits (Meghraoui et al. 1996; Benouar et al. 1994). Further south, the Beni Choughrane mount is separated from the Ghriss alluvial basin by 40-km-long thrust (trending NE–SW and dipping NW) that shows highly deformed Quaternary deposits. The Beni Choughrane mounts are bounded to the north by a NE–SW and SE dipping thrust fault affecting young deposits. The Ain Temouchent area is considered of great interest from tectonic point of view. It was affected by a Mw 5.7 earthquake on December 22, 1999 associated with a reverse fault trending NE–SW affecting Neogene structures as it was highlighted by the interferometry (Belabbès et al. 2008).

In the Cheliff basin, the active tectonic features are NE–SW folds, which affect recent Quaternary deposits. These folds systematically exhibit an asymmetric shape with a transport direction towards the SE and a faulted southeastern flank (Philip and Meghraoui 1983; Meghraoui 1988). The folds represented by hills on the landscape are juxtaposed with flat alluvial plains, with a maximum topographic offset of about 300 m. On the northern edges of the Quaternary subbasins, the NE–SW Quaternary folds appear to be oblique with regard to the elongated basin and are often terminating by E–W flexural structures. The El Asnam and Tenes-Abou El Hassan structures are limited to the SE by reverse active faults (Philip and Meghraoui 1983; Aoudia and Meghraoui 1995). Paleoseismic investigations carried out by Meghraoui et al. (1988a, b), Meghraoui and Doumaz (1996) on trenches on the El Asnam fault evidenced a minimum uplift rate of 0.26 mm $year^{-1}$ during the Holocene. The observations of Meghraoui and Doumaz (1996) led to identifying and dating several strong events prior to the El Asnam 1980 earthquake. The El Asnam seismogenic zone is considered as the most active area of the Tell Atlas of Algeria since it experienced two strong events within 25 years only (1954, Ms 6.7 and 1980, Mw 7.3).

To the east of the Cheliff basin, there is another important Quaternary basin, the Mitidja basin (Fig. 1b). The thickness of the Neogene post-nappes and Quaternary deposits inside the Mitidja basin is around 3500 m (Gleangeaud et al. 1952). From a morphological point of view, the basin is limited to the north by the 70-km-long, N070-trending active Sahel fold. This anticline is an asymmetrical structure with a transport direction to the SE suggesting the presence of a blind thrust fault on its southern limb (Maouche et al. 2011). The moderate-sized 1989 Mt Chenoua–Tipaza earthquake (Ms 6.0), which occurred on the western branch of the Sahel folds system, corroborates previous studies that evidenced the hypothetical existence of a non-emergent thrust fault (Meghraoui 1991; Harbi et al. 2004). The northern flank of this active fold is covered by a step-like morphology of uplifted Pleistocene and Holocene marine terraces yielding a long-term uplift rate of 0.2 mm $year^{-1}$ while on the southern flank, the average topographic offset of 240 m is marked by a flat topography of the Mitidja alluvial plain (Maouche et al. 2011). To the south, the Mitidja Basin is characterized by the Blida active faults system which consists of an ENE–WSW trending and right-stepping reverse faults overthrusting the Neogene post-nappes units (Meghraoui 1988; Guemache 2010;

Maouche et al. 2011; Maouche and Harbi 2017). The present-day morphology of the Blida Mountains with an altitude greater than 1500 m contrasts with the Mitidja 60-m-high Quaternary flat. In the vicinity of Blida, Bouchelouh et al. (2017) highlighted the Miocene roof at 900 m depth which can suggest an important uplift rate of the southern border of the Mitidja Basin. This fault system extends to the ENE reaching the coastline, and the offshore fault continuation of this system was reactivated during the 2003 Zemmouri earthquake (Mw 6.8; Meghraoui et al. 2004; Ayadi et al. 2008; Maouche et al. 2011).

Along the coastal zone, geomorphic features were used to recognize witness of coastal tectonics. Fieldwork undertaken in the central coast of Algeria by Maouche et al. (2011) evidenced differential uplift of marine terraces along the coast. Several marine terraces and notches level were interpreted as geomorphic markers of recent coastal tectonic activity.

Regarding the Soummam region, Bouhadad (2015) suggested a 0.14 mm year^{-1} uplift rate using alluvial and tilted terraces related to the Djemila fault. The seismic potential of this fault could be comparable to the Kherrata fault. In the vicinity of Kherrata, after the earthquake of February 17, 1949, Rothé (1950) observed surface ruptures and mapped the fault relating to an anticline trending N070 E (Fig. 1b).

Further east, the faults affecting Plio-Quaternary deposits are well identified in the Constantine, Hodna, and Guelma basins (Guiraud 1973; Vila 1980; Meghraoui 1988; Bounif et al. 1987; Harbi et al. 1999; Maouche et al. 2013). In the M'sila region, Plio-Quaternary anticlines were observed in the Hodna basin. The folds are associated with reverse active faults which limit the southeast limb of the anticlines. In the Constantine region, Ain Smara fault and Sigus fault, which are oriented in the NE–SW and E–W direction, respectively, provide neotectonic activity indicators (Vila 1980). The surface ruptures which were generated by the October 27, 1985 Constantine earthquake may correspond to the northeastern continuation of the Ain Smara fault (Bounif et al. 1987, 2003; Ousadou et al. 2013). The Guelma basin is a seismogenic zone. It is a "pull-apart" structure located between east–west segmented dextral strike–slip faults (Meghraoui 1988; Meghraoui et al. 1996; Maouche et al. 2013). This basin is rather particular because pull-apart basins are not very common in Algeria. According to Maouche et al. (2013), the western border of the Guelma basin is represented by the Hammam Debbagh–Roknia NW–SE fault which has neotectonic signification. Fieldwork carried out in the Hammam Debbagh area revealed normal fault structures visible in the recent travertine deposits (Maouche et al. 2013). Moreover, juxtaposed alluvial terraces which are present in this zone attest on the vertical movement related to the recent activity on NW–SE to NNW–SSE fault (Maouche et al. 2013). In the Guelma basin, we also observe the Bouchegouf fault and the Hammam N'baïlis fault which affected the Quaternary sediments and are implicated in hydrothermal springs (Vila 1980; Harbi et al. 1999). One of the major issue in the Tell Atlas is related to the neotectonic structures which are sometimes considered as active faults (Fig. 2a, b). To overcome these ambiguities and inconsistencies, detailed studies of the recent Quaternary by using isotopic dating are the fundamental tool.

Fig. 2 Surface faulting showing **a** neotectonic movement (coastal neotectonic fault in western Oran), **b** reverse active fault (Oued Fodda segment, 1.5 m vertical movement during 1980, *M* 7.3 earthquake)

Several studies published in the framework of MARADJA and SPIRAL projects suggest that offshore Algeria displays active tectonic structures. Using detailed bathymetry and seismic reflection data, Domzig et al. (2006) discussed the offshore tectonic and identified large Neogene reverse faults and folds which are still active. Ratzov et al. (2015) discussed the synchroneity of turbidites deposits group that supports a regional trigger, interpreted as coseismic. Plio-Quaternary diffuse deformation related to the compressional reactivation of the central Algerian Neogene margin was discussed by Leprêtre et al. (2013).

2.2 Seismicity

The Tell Atlas displays strong and moderate-size seismic events with significant thrust or reverse focal mechanisms in its central part (Figs. 3 and 4). These earthquakes are associated with the main two Quaternary basins, the El Asnam and Mitidja basins. The 1980 El Asnam earthquake (Mw = 7.3) was generated by a 36-km-long thrust fault associated with an NE–SW trending broken-fold (King and Vita-Finzi 1981; Philip and Meghraoui 1983). The focal solution of the mainshock shows an NW dipping nodal plane corresponding to the main fault plane which is in good agreement with the observed surface deformations, the en-echelon-faults, and bookshelves tectonic pattern proposed by Meghraoui et al. (1996), Meghraoui and Pondrelli (2012), Derder et al. (2013). Similar fold structures to that of El Asnam

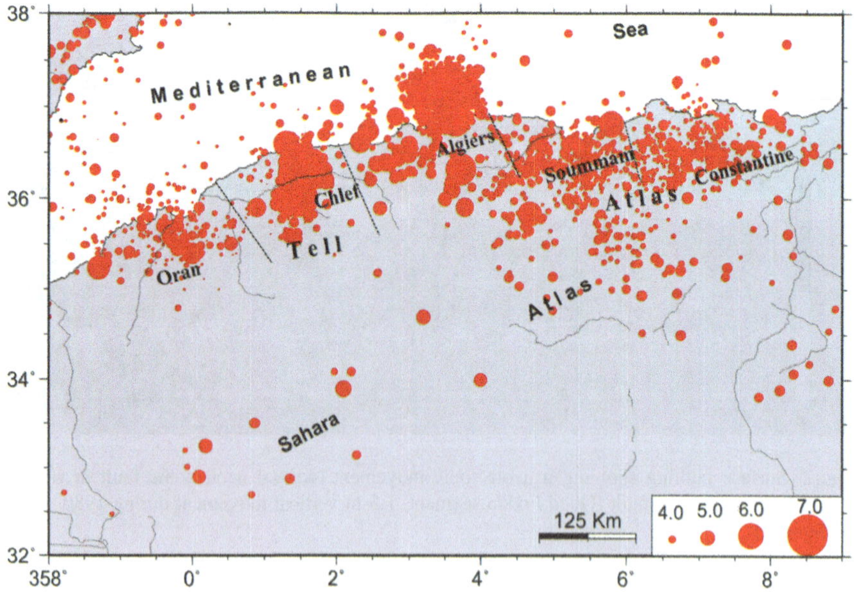

Fig. 3 The spatial distribution of earthquakes ($M \geq 3$) in the Tell Atlas

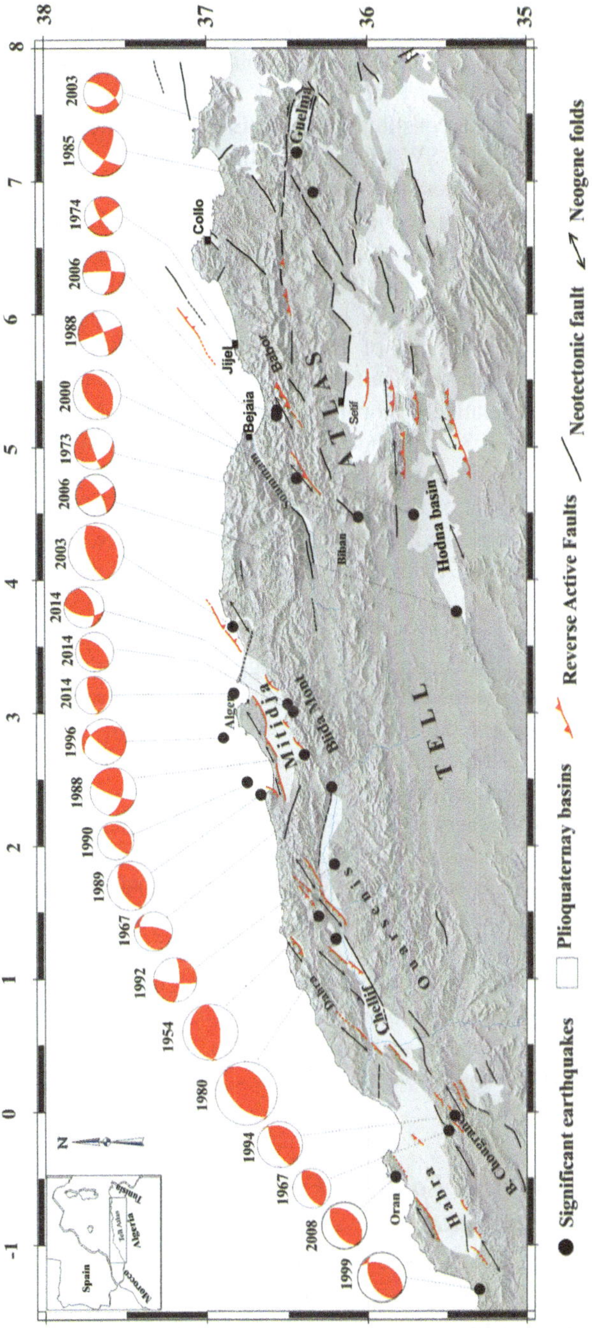

Fig. 4 Focal mechanism of significant earthquakes in the Tell Atlas (Maouche 2010 updated)

can be observed in different sites, such as the Tenes-Abou El Hassan fault (Aoudia and Meghraoui 1995), the Sahel fold and associated fault system, and south fault thrust system of the Mitidja basin (Meghraoui 1991; Harbi et al. 2004; Ayadi et al. 2008; Maouche et al. 2011). These compressional active structures, which generated the earthquakes of Tenes (1922, *M* 6.0), Chenoua–Tipaza (1989, Ms 6.0), Zemmouri (2003, Mw 6.8), constitute with the El Asnam anticline the most active zones within a series of NE–SW to E–W trending folds and associated faulting.

Figure 3, which illustrates the earthquake distribution ($M > 3$) in Algeria (Benouar 1994; Ayadi and Bezzeghoud 2015; Harbi et al. 2015), shows that seismic activity is generally scattered all along the Tell Atlas. The El Asnam and Mitidja basins exhibit a dense seismic activity because of 1954, 1980, 1989 and 2003 seismic sequences. The seismicity of El Asnam region, which extends in a NNE–SSW direction and over an area of 150 km long by 50 km wide and which is associated with the main active faults, is the consequence of a high level energy strain release during a relatively short period of time (with regard to the geological records). Based on the 1980 aftershocks, Yielding et al. (1989) showed that more than 50% of seismic events are located at depth ranging from 3.5 to 7 km. In the Mitidja basin, the analysis of the 1989 and 2003 aftershock distribution at depth corroborates what was observed during the El Asnam seismic sequence (Meghraoui 1991; Bounif et al. 2003; Ayadi et al. 2008). Using the different investigations carried out in the Tell Atlas (Cisternas et al. 1982; King and Yielding 1984; Yielding 1985; Yielding et al. 1989; Meghraoui 1991), we can say that in the study area, the strong or moderate earthquakes are triggered at a depth ranging between 8 and 14 km. Therefore, the mean value of depth distribution given by Yielding et al. (1989) does not represent the initiation levels of significant earthquakes but the activity in superficial sedimentary layers. Hence, immediately after the mainshock, coseismic ruptures propagate towards the surface showing a maximum concentration of events at an average of 5–7 km. Prehistoric evidence of seismic activity remains in geological records, such as paleoliquefaction features, boulders, and coastal uplifts (Bouhadad et al. 2008; Maouche et al. 2009; Benhamouche et al. 2014). Trenches analysis along the El Asnam fault zone allowed inferring the recurrence time of major earthquake ($M \geq 7.0$) of about 300–500 years during periods of high seismic activity (Meghraoui 1988; Meghraoui and Doumaz 1996). The earthquake catalogues (Benouar 1994; Harbi et al. 2010, 2015; Ayadi and Bezzeghoud 2015) give an average recurrence time of 20–25 years for magnitudes ranging between *M* 5.0 and *M* 6.0 (Fig. 5).

2.3 Active Tectonics: Folding, Faulting, and Associated Strong Earthquakes

The seismogenic signature of thrust faults is primarily related to the growth of associated anticline structures. The degree of activity of a fault is represented by the folding process visible on the hanging wall. Therefore, we think that compressional

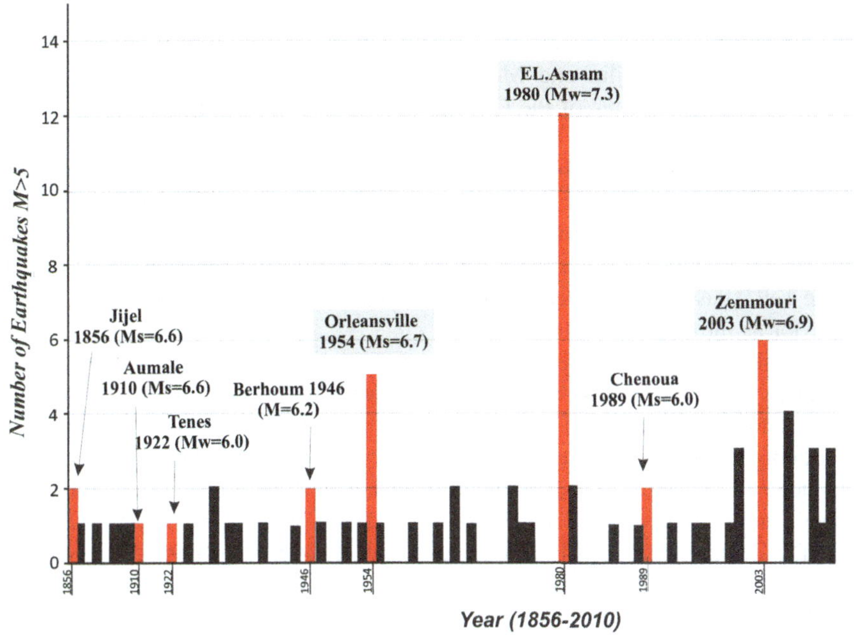

Fig. 5 Histogram of the number of earthquakes versus time (from 1850 to 2010 for magnitude $M > 5.0$)

movements at depth in the continental lithosphere have a direct impact on surface features. On a larger scale, active folding systems of the Tell Atlas are distributed over a narrow band, parallel to the coastline, which coincides with the seismicity distribution (Figs. 3 and 4). Surface faulting is however very scarce in this zone, except for the 1980 El Asnam earthquake where thrust faults have a complex pattern of surface deformation. In most of the seismic zones (Mourabit et al. 2014), the identification of active thrust or reverse faults is a real problem because they are generally hidden by folding or geological inherited structures. Outstanding examples of such seismogenic broken-folds are presented by Stein and Yeats (1989). This intricacy in identifying the seismogenic fault, particularly before the occurrence of strong or moderate earthquakes, is related to the folding which can mask the movement on the fault. The interaction between folds and faults is a complex mechanical system, which can be observed in the field at different levels of structural maturation. This is not specific to Algeria only since several plates boundary zones that comprise active folds, experienced destructive large earthquakes; some of them with surface faulting as in Iran (Tabas and Golshan 1979, Ms 7.5), California (San Fernando 1971, Ms 6.6) and Armenia (Spitak 1988, Ms 6.9), and some others without surface faulting as in California (Coalinga 1983, Ms 6.5; Whittier-Narrows, 1987, Ms 5.9). However, all these earthquakes revealed a consistency between the surface deformation and faulting at depth as illustrated by thrust focal solutions and aftershocks distribution.

Coseismic deformations during the 1980 El Asnam earthquake showed simultaneous vertical and left-lateral movements, which reached 1.30 m in the central segment (Fig. 2b) (Philip and Meghraoui 1983; Meghraoui et al. 1986). Both neotectonic investigations of the Cheliff and Mitidja basin and focal mechanisms solutions agree with a main NNW–SSE shortening component (Fig. 4) (Ousadou et al. 2014). Active folding arrangements in the Tell Atlas are consistent with a EW en-echelon fold distribution striking NE–SW (Fig. 4). Knowing that other active folds were the site of moderate or strong earthquakes (Tenes-Abou El Hassan 1922, Ms 6.0; Mt Chenoua–Tipaza 1989, Ms 6.0), we may assume that Quaternary folds of the Tell Atlas and associated blind or emergent thrust faults have similar dynamic behavior as the El Asnam structure.

Crustal shortening in the Tell Atlas of Algeria is partly accommodated by seismicity (Fig. 4). Field observations of active folds in the Cheliff and Mitidja basins and the modeling of the fold and thrust structures indicate a maximum shortening rate of 2.2 mm year^{-1} in the Tell Atlas (Meghraoui and Doumaz 1996; Maouche et al. 2011). The coseismic displacements observed during the 1922 Tenes-Abou El Hassen, Ms 6.0; Orléansville 1954, Ms 6.7; El Asnam 1980, Ms 7.3, and Zemmouri 2003, Mw = 6.8 show that about 50% of this convergence is due to the seismic deformation (Meghraoui and Doumaz 1996; Aoudia and Meghraoui 1995; Maouche et al. 2011). The aseismic deformation may be related to the folding process. This value is comparable to the estimated ratio of seismic/aseismic E–W extensional deformation of 2.6 mm year^{-1} calculated in the Italian Apennines by Westaway (1990). Nocquet and Calais (2004) and Serpelloni et al. (2007) using geodetic measurements yielded convergent rates of this region ranging between 4 and 6 mm year^{-1}, which are comparable to the 2.1 mm year^{-1}proposed shortening rate in the Tell Atlas using paleoseismology and active tectonics (Meghraoui and Doumaz 1996).

Broken-folds that affect Neogene and Quaternary formations in the Tell Atlas (Algeria) are potential zones of future moderate or strong earthquakes. Such seismogenic structures became the subject of great concerns recently for the seismic hazards in various part of the world (Armenia, Zagros, Transverse Ranges of California).

3 Seismic Hazard

A better knowledge of active tectonics in any region constitute the first step towards an efficient assessment of seismic hazard. Paleoseismological investigations are of great interest to retrieving the chronology and the recurrence of strong earthquakes in any area of study. This information will contribute in the assessment of seismic hazard by using a deterministic approach instead of the probabilistic modeling generally based on incomplete earthquake catalogue.

3.1 Background

Seismic hazard analysis consists in calculating the expected seismic shaking for a given site depending upon its nature (rock or firm site). This shaking can be expressed in terms of ground motion parameter, such as acceleration, velocity, intensity or spectral acceleration, which could be generated by potentially active geological structures located at a given distance around the considered site. Two different approaches are usually used throughout the world, namely the probabilistic approach and the deterministic approach (Reiter 1991). The probabilistic seismic hazard approach was proposed by Cornell (1968) and developed by several authors (McGuire 1978; Bender and Perkins 1987; Geomatrix Consultants 1993; Bommer and Abrahamson 2006 and references therein). In the probabilistic approach, the seismic hazard is defined as ground motion with an annual probability of exceedance, which is calculated from a mathematical model and based on the statistical relationships of earthquake and ground motion. It is assumed that the occurrence of an earthquake follows a Poissonian process. The deterministic approach is based on realistic parameters obtained during field investigations (fault parameters, geotechnical parameters, radiation pattern of the source, etc.). In the deterministic approach, seismic hazard is defined as the maximum ground motion from a single earthquake or from a set of earthquakes, including maximum credible earthquake, calculated considering the available physical knowledge on earthquake sources and wave propagation processes by means of deterministic models (Mourabit et al. 2014; Wang 2011; Panza et al. 2012).

Both methods require four (04) basic steps:

(1) Definition of a seismic source model that shows the potential source zone which can contribute to the seismic hazard in the considered area. The seismic source model may contain the well identified active fault.
(2) Seismic source characterization including the determination of seismotectonic context, seismicity (Gutenberg and Richter 1954), and geometrical parameters for each considered source.
(3) Choice and/or development of attenuation relationships that define the decrease of the seismic parameter (intensity or acceleration) from the site to the considered site. Several attenuation laws have been developed throughout the world based on strong motion records obtained during moderate-sized and strong earthquakes.
(4) Seismic hazard calculation in a given site, which can be performed for a given area by generalizing the calculation within a grid to obtain seismic hazard maps. The deterministic approach is often used for engineering purpose on critical facilities, while for seismic building code and land planning purposes, the probabilistic approach is privileged. The advantage of the probabilistic approach is in the capability to manage a great amount of earth science data in the framework of a logic tree model (NRC 1988).

3.2 Seismic Hazard in the Tell Atlas

During the last decades, several seismic hazards studies were performed in northern Algeria. The first seismic hazard map for engineering purpose was an intensity map prepared for the regulation rules AS55 (i.e., anti-seismic regulation decided in 1955 following the 9 September 1954 Orléasnville (Chlef now) earthquake, Ms 6.5). The first modern seismic hazard study of northern Algeria was performed by Mortgat and Shah (1978) with the aim to elaborate the first Algerian seismic building code "RPA 1981". This work was based on a probabilistic approach and resulted in several thematic maps in terms of acceleration, velocity, and displacement for different return periods (100, 200, 500, and 1000 years). This seismic code was updated twenty years later by Jimenez et al. (1999). The 10 October 1980 El Asnam earthquake (Ms 7.3) helped the earth scientists to better understand the seismotectonic of northern Algeria and realize that seismic hazard was previously underestimated. Indeed, this earthquake showed that earthquakes in northern Algeria are not randomly distributed but associated to clear active geological structures (Ouyed et al. 1981; Meghraoui et al. 1986; Meghraoui 1988, 1991; Aoudia and Meghraoui 1995; Meghraoui and Doumaz 1996; Bouhadad 2001; Bouhadad et al. 2003; Maouche et al. 2011). From that point onward, several studies were performed for the assessment of seismic hazard in Northern Algeria. These studies were based on various approaches: (i) on the seismic potential of geological structures (WCC 1984; Geomatrix and CGS 1998; Bouhadad 2000; Bouhadad and Laouami 2002), (ii) on earthquake catalogues (Benouar et al. 1996; Naili and Benouar 2000), (iii) on deterministic methods through the computation of synthetic seismograms (Aoudia et al. 2000; Harbi et al. 2007; Mourabit et al. 2014), (iv) on observed or calculated intensities (Bezzeghoud et al. 1996; Boughacha et al. 2004; Ayadi and Bezzeghoud 2015). The thematic maps obtained from all these studies constitute very useful tools for seismic hazard assessment.

4 Discussion and Conclusion

Despite the human lives and economic losses caused by the 10 October 1980 El Asnam earthquake, it can be considered as a starter of a new era for seismotectonic and active tectonic studies in the Tell Atlas of Algeria. The El Asnam earthquake, which was generated by a faulted fold clearly revealed by extensive surface breaks, allowed understanding and studying a typical active and seismogenic geological structure in compression context. Active faults and seismic hazard assessment in the tectonically active Tell Atlas are crucial because of the increasing urbanization and population growth in the fault adjacent areas. The Cheliff basin with its 1922, 1934, 1954, and 1980 damaging earthquakes represents a particular and interesting case of population growth in the foothills of active fault. Thanks to the availability of new datasets and our own observations, we cannot rule out the possibility of similar

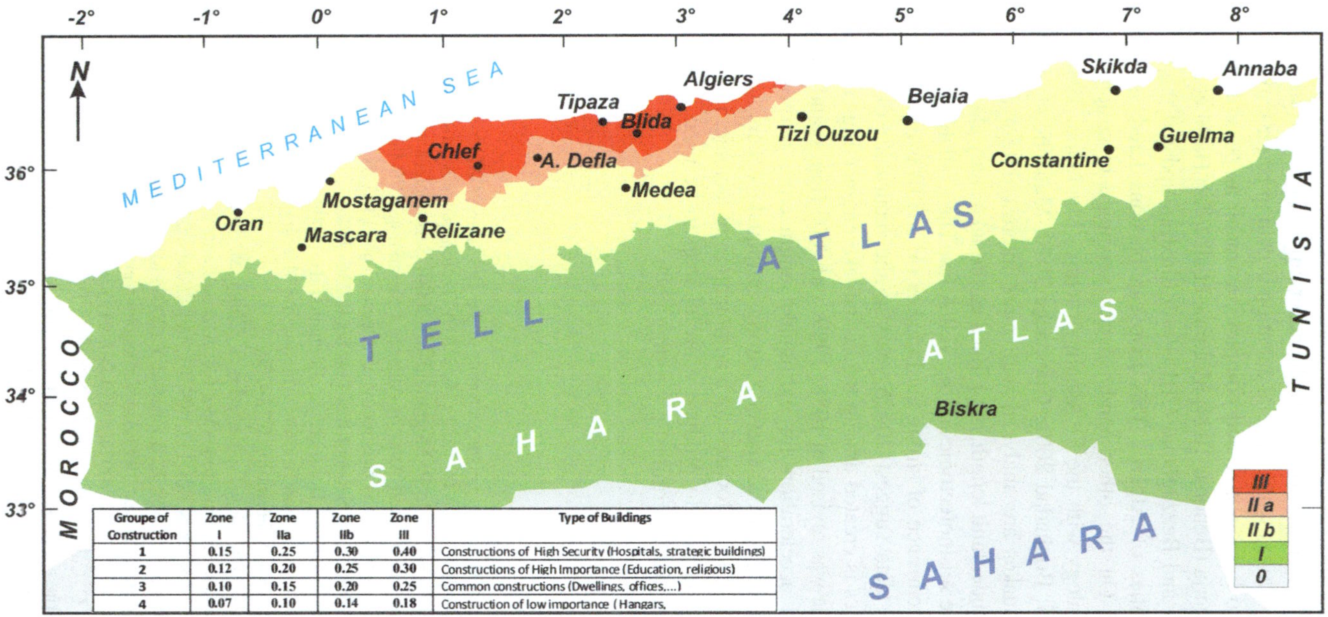

Groupe of Construction	Zone I	Zone IIa	Zone IIb	Zone III	Type of Buildings
1	0.15	0.25	0.30	0.40	Constructions of High Security (Hospitals, strategic buildings)
2	0.12	0.20	0.25	0.30	Constructions of High Importance (Education, religious)
3	0.10	0.15	0.20	0.25	Common constructions (Dwellings, offices,...)
4	0.07	0.10	0.14	0.18	Construction of low importance (Hangars,

Fig. 6 Seismic hazard map according to the Algerian seismic code in term of hazard (III high, IIa, IIb moderate, I low, 0 negligible) with corresponding peak ground acceleration in % of g, given in the table inserted on the figure (RPA-1999 version 2003), Algeria

earthquakes visiting the area in the future. The need to understanding seismic hazard is continuously increasing. The recent earthquakes may result in the reactivation of other neighboring active faults according to Coulomb stress transfer (Lin and Stein 2004). This is why a special attention is paid to better know the Tell Atlas seismicity and to homogenize and complete, as much as possible the earthquake catalogue of Algeria (Benouar 1994; Mokrane et al. 1994; Harbi et al. 2010, 2015). Recently, Ayadi and Bezzeghoud (2015) published an updated map of maximum observed intensities of Algeria. This map can be considered as a new step of seismic hazard. On the other hand, several seismogenic structures have been identified by analogy with the El Asnam structure (Meghraoui 1988; Aoudia and Meghraoui 1995; Bouhadad 2001; Guemache 2010; Maouche et al. 2011). Other moderate earthquakes are analyzed with more interest (return period, maximum expected magnitude, fault length, etc.). Each earthquake study allows improving the knowledge of the seismotectonics of the Tell Atlas. The available paleoseismic data show that the recurrence of strong earthquakes ($M > 7.0$) is about 300–500 years while seismicity data suggest 20–25 years as the recurrence of moderate earthquakes. Dynamics is created among the earthquake engineering community where the Algerian code was revised in 1988, 1999, and 2003 (Fig. 6 and the related table of peak ground acceleration for earthquake engineering purposes). Instead of revising the building code whenever a strong earthquake occurs as it is the case of Algeria, efforts should be focused on a combination of tectonic, intensity, and peak ground acceleration data to produce a useful map to be used in seismic hazard studies, land use, and deterministic building code.

References

Aoudia A, Meghraoui M (1995) Seismotectonics in the Tell Atlas of Algeria: the Cavaignac (Abou El Hassan) earthquake of 25.08.1922 (Ms = 5.9). Tectonophysics 248:263–276

Aoudia A, Vaccari F, Suhadolc P, Meghraoui M (2000) Seismogenic potential and earthquake hazard assessment in the Tell Atlas of Algeria. J Seismolog 4:79–98

Ayadi A, Bezzeghoud M (2015) Seismicity of Algeria from 1365 to 2013: maximum observed intensity map (MOI2014). Seismol Res Lett 86(1):236–244

Ayadi A, Ousadou-Ayadi F, Bourouis S, Benhallou H (2002) Seismotectonics and seismic quietness of the Oranie region (western Algeria): the Mascara earthquake of August 18th 1994, Mw = 5.7, Ms = 6.0. J Seismolog 6:13–23

Ayadi A, Dorbath C, Ousadou F, Maouche S, Chikh M, Bounif MA, Meghraoui M (2008) Zemmouri earthquake rupture zone (Mw 6.8, Algeria): aftershocks sequence relocation and 3D velocity model. J Geophys Res 113:B09301. https://doi.org/10.1029/2007JB005257

Belabbès S, Meghraoui M, Çakir Z, Bouhadad Y (2008) InSAR analysis of a blind thrust rupture and related active folding: the 1999 Ain Temouchent earthquake (Mw 5.7, Algeria) case study. J Seismol. https://doi.org/10.1007/s10950-008-9135-x

Belabbès S, Wicks C, Cakir Z, Meghraoui M (2009) Rupture parameters of the 2003 Zemmouri (Mw 6.8), Algeria, earthquake from joint inversion of interferometric synthetic aperture radar, coastal uplift, and GPS. J Geophys Res 114:B03406. https://doi.org/10.1029/2008JB005912

Bender B, Perkins DM (1987) SEISRISK III: a computer program for seismic hazard estimation. U.S. Geological Survey Bulletin 1772, 48 pp

Benhallou H (1985) Les Catastrophes Séismiques de la Région d'Echéliff dans le contexte de la Séismicité Historique de l'Algérie. Thesis, University of Algiers, Algeria, 294 pp

Benhamouche A, Nedjari A, Bouhadad Y, Machane D, Oubaiche E, Sidi Said N (2014) Field evidence of seismites in quaternary deposits of the Jijel (eastern Algeria) coastal region. J Seismol. https://doi.org/10.1007/s10950-013-9384-1

Benouar D (1994) Materials for the investigation of the seismicity of Algeria and adjacent regions during the twentieth century. Ann Geofis XXXVII(4):459–860

Benouar D, Aoudia A, Maouche S, Meghraoui M (1994) The 18 August 1994 Mascara (Algeria) earthquake-a quick-look report. Terra Nova 6:634–637

Benouar D, Molas GL, Yamazaki F (1996) Earthquake hazard mapping in the Maghreb countries. Earthq Eng Struct Dyn 25:1151–1164

Bezzeghoud M, Buforn E (1999) Source parameters of the 1992 Melilla (Spain, Mw = 4.8), 1994 Al Hoceima (Morocco, Mw = 5.8) and Mascara (Algeria, Mw = 5.7) earthquakes and seismotectonic implications. Bull Seismol Soc Am 89:359–372

Bezzeghoud M, Ayadi A, Sebaï A, Aït Messaoud M, Mokrane A, Benhallou H (1996) Seismicity of Algeria between 1365 and 1989: map of maximum observed intensities (MOI). Avances en Geofisica y Geodesia, Ministerio de Obras Publicas, Transportes y Media Ambiente, vol I. Pub. De IGN, Madrid, pp 107–114

Bommer JJ, Abrahamson NA (2006) Why do modern probabilistic seismic-hazard analyses often lead to increased hazard estimates? BSSA 96(6):1967–1977

Bouchelouh A, Bensalem R, Zaourar N, Machane D, Moulouel H, Oubaiche E (2017) The miocene roof mapping using microtremor recording and electrical survey method in Blida city, Algeria. Pure Appl Geophys. https://doi.org/10.1007/s00024-017-1684-x

Boudiaf A (1996) Etude sismotectonique de la region d'Alger et de la Kabylie (Algerie): Utilisation des modeles numeriques de terrain (MNT) et de teledetection pour la reconnaissance des structures tectoniques actives: contribution a l'evaluation de l'alea sismique. These de Doctorat, Universite de Montpellier II, France

Boughacha MS, Ouyed M, Ayadi A, Benhallou H (2004) Seismicity and seismic hazard mapping of northern Algeria: map of maximum calculated intensities (MCI). J Seismolog 8:1–10

Bouhadad Y (2000) Probabilistic seismic hazard assessment in the Beni Chougrane, western Algeria, region. Earthq Eng J (EEE) XIV(3):82–87

Bouhadad Y (2001) The Murdjadjo, western Algeria, fault-related fold: implication for seismic hazard. J Seismolog 5:541–558

Bouhadad Y (2015) Geomorphic evidence of active tectonics: the case of Djemila fault (eastern Algeria). Eng Geol Soc Territ 6:125–127. https://doi.org/10.1007/978-3-319-09060-3_22

Bouhadad Y, Laouami N (2002) Earthquake hazard assessment in the Oran region (northwest Algeria). J Nat Hazard 26(3):227–243

Bouhadad Y, Nour A, Laouami N, Belhai D (2003) The Tachaouaft-Beni-ourtilane fault and seismotectonic aspects of the Babors region (eastern Algeria). J Seismolog 7:79–88

Bouhadad Y, Benhamouche A, Maouche S, Belhai D (2008) Evidence for quaternary liquefaction-induced features in the epicentral area of the 21 May 2003 Zemmouri earthquake (Algeria, Mw = 6.8). J Seismolog 13(1):161–172

Bounif A, Haessler H, Meghraoui M (1987) The Constantine (northeast Algeria) earthquake of October 27, 1985: surface ruptures and aftershock study. Earth Planet Sci Lett 85:451–460

Bounif A, Bezzeghoud M, Dorbath L, Legrand D, Deschamps A, Rivera L, Benhallou H (2003) Seismic source study of the 1989, October 29, Chenoua (Algeria) earthquake from aftershocks, broad-band and strong ground motion records. Ann Geophys 46(4):625–646

Bounif A, Dorbath C, Ayadi A, Meghraoui M, Beldjoudi H, Laouami N, Frogneux M, Slimani A, Alasset PJ, Kharroubi A, Oussadou F, Chikh M, Harbi A, Larbes S, Maouche S (2004) The 21 May, 2003 (Mw 6.8) Zemmouri (Algeria) earthquake relocation and aftershock sequence analysis. Geophys Res Lett 31:L19606. https://doi.org/10.1029/2004GL020586

Cisternas A, Dorel J, Gaulon R (1982) Models of the complex sources of El Asnam earthquake. Bull Seismol Soc Am 72(6):2245–2266

Cornell CA (1968) Engineering seismic risk analysis. Bull Seismol Soc Am 58:1583–1606

Derder MEM, Henry B, Maouche S, Bayou B, Amenna M, Besse J, Bessedik M, Belhai D, Ayacheb M (2013) Transpressive tectonics along a major E-W crustal structure on the Algerian continental margin: blocks rotations revealed by a paleomagnetic analysis. Tectonophysics 593:183–192

Domzig A, Yelles K, Le Roy C, Déverchère J, Bouillin JP, Bracène R, Mercier de Lépinay B, Le Roya P, Calais E, Kherroubi A, Gaullier V, Savoye B, Pauch H (2006) Searching for the Africa–Eurasia Miocene boundary offshore western Algeria (MARADJA'03 cruise). C R Geosci 338(1–2):80–91. https://doi.org/10.1016/j.crte.2005.11.009

Durand Delga M (1969) Essai sur la structure du NE de la Berberie. Bull Serv Carte geol Algér 39:89–181

Geomatrix, CGS (1998) Probabilistic seismic hazard assessment in the Algiers region. Technical report, p 68

Geomatrix Consultants (1993) Probabilistic seismic hazard analysis computer program: a user manual

Gleangeaud L, Ayme A, Caire A, Mattauer M, Muraour P (1952) Histoire geologique de la province d'alger. Monog region, 1re serie, Algerie n 25

Guemache M (2010) Évolution géodynamique ges bassins sismogènes de l'Algérois (Algérie): Approche pluridisciplinaire (méthodes géologiques et géophysiques). Thesis, FSTGAT/ USTHB, Algiers, Algeria

Guiraud R (1973) Evolution post-triasique de l'avant-pays de la chaîne alpine en Algérie d'après l'étude du bassin du Hodna et des régions voisines. Publ Serv Géol Algér Alger Bull 47:270

Gutenberg B, Richter CF (1954). Seismicity of the earth and associated phenomena. Princeton University Press, p 310

Harbi A, Maouche S, Ayadi A (1999) Neotectonic and associate seismicity in the eastern Tellian Atlas of Algeria. J Seismolog 3(1):95–104

Harbi A, Maouche S, Ayadi A, Benouar D, Panza GF, Benhallou H (2004) Seismicity and tectonic structures in the site of Algiers and its surroundings: a step towards microzonation. Pure Appl Geophys 161:949–967

Harbi A, Maouche S, Vaccari F, Aoudia A, Oussadou F, Panza GF, Benouar D (2007) Seismicity, seismic input and site effects in the Sahel-Algiers region (north Algeria). Soil Dyn Earthq Eng 27(5):427–447

Harbi A, Peresan A, Panza GF (2010) Seismicity of eastern Algeria: a revised and extended earthquake catalogue. Nat Hazards. https://doi.org/10.1007/s11069-009-9497-6

Harbi A, Meghraoui M, Maouche S (2011) The Djidjelli (Algeria) earthquakes of 21 and 22 August (I0 IX, VIII) and related tsunami effects revisited. J Seismolog 15(1):105–129

Harbi A, Sebaï A, Benmedjber M, Ousadou F, Rouchiche Y, Grigahcene A, Aïni D, Bourouis S, Maouche S, Ayadi A (2015) The Algerian homogenized macroseismic database (267–1989): a deeper insight into the Algerian historical seismicity. Seismol Res Lett 86(6):1705–1716

Jiménez MJ, Garcia-Fernandez M, GSHAP Ibero-Maghreb Working Group (1999) Seismic hazard assessment in the Ibero-Maghreb region. Ann Geofis 42:1057–1066

King GCP, Vita-Finzi C (1981) Active folding in the Algerian earthquake of 10 October 1980. Nature 292:22–26

King GCP, Yielding G (1984) The evolution of a thrust fault system: processes of rupture initiation, propagation and termination in the 1980 El Asnam (Algeria) earthquake. Geophys J Int 77(3):915–933

Leprêtre A, Klingelhoefer F, Graindorge D, Schnurle P, Beslier MO, Yelles K, Déverchère J, Bracene R (2013) Multiphased tectonic evolution of the central Algerian margin from combined wide-angle and reflection seismic data off Tipaza, Algeria. J Geophys Res Solid Earth 118:3899–3916

Lin J, Stein RS (2004) Stress triggering in thrust and subduction earthquakes, and stress interaction between the southern San Andreas and nearby thrust and strike slip faults. J Geophys Res 109 (B02303). https://doi.org/10.1029/2003jb002607

Maouche S (2010) Tectonique active et géodynamique le long de l'atlas Tellien: Etude des soulèvements côtiers. Thèsis, USTHB, Algiers, Algeria, p 299

Maouche S, Harbi A (2017) The active faults of the Mitidja basin (north central Algeria): what does the seismic history of the region tell us? Submitted to Euro-Mediterr J Environ Integr

Maouche S, Harbi A, Meghraoui M (2008) Attenuation of intensity for the Zemmouri earthquake of 21 May 2003 (Mw 6.8): insights for the seismic hazard and historical earthquake sources in northern Algeria. In: Fréchet J, Meghraoui M, Stuchi M (eds) Historical seismology, interdisciplinary studies of past an recent earthquakes. Springer, New York, pp 327–350

Maouche S, Morhange C, Meghraoui M (2009) Large boulder accumulation on the Algerian coast evidence tsunami events in the western Mediterranean. Mar Geol 262(2009):96–104

Maouche S, Meghraoui M, Morhange C, Belabbès S, Bouhadad Y, Haddoum H (2011) Active coastal thrusting and folding, and uplift rate of the Sahel anticline and Zemmouri earthquake area (Tell Atlas, Algeria). Tectonophisics 509:69–80

Maouche S, Abtout A, Merabet N, Aïfa T, Lamali A, Bouyahyaoui B, Boughchiche S, Ayache M (2013) Tectonic and hydrothermal activities in Debbagh, Guelma basin (Algeria). J Geol Res. https://doi.org/10.1155/2013/409475

McGuire RK (1978) FRISK: computer program for seismic risk analysis using faults as earthquake sources. U.S. geological survey open file report 78-1007

Meghraoui M (1988) Géologie des zones sismiques de l'Algérie du Nord: Paléosismologie, tectonique active et Synthèse sismotectonique. Thèsis, p 356

Meghraoui M (1991) Blind reverse faulting system associated with the Mont Chenoua-Tipaza earthquake of 29 October 1989 (north-central Algeria). Terra Nova 3:84–93

Meghraoui M, Doumaz F (1996) Earthquake-induced flooding and paleoseismicity of the El Asnam, Algeria, fault-related fold. J Geophys Res 101(B8):17617–17644

Meghraoui M, Pondrelli S (2012) Active faulting and transpression tectonics along the plate boundary in north Africa. Ann Geophys. https://doi.org/10.4401/ag.4970

Meghraoui M, Cisternas A, Philip H (1986) Seismotectonics of the lower Chellif basin: structural background of the El Asnam (Algeria) earthquake. Tectonics 5:809–836

Meghraoui M, Philip H, Albarede F, Cisternas A (1988a) Trench investigations through the trace of the 1980 El Asnam thrust fault: evidence for paleoseismicity. Bull Seismol Soc Am 78 (2):979–999

Meghraoui M, Jaegy R, Lammali K, Albarede F (1988b) Late Holocene earthquake sequences on the El Asnam (Algeria) thrust fault. Earth Planet Sci Lett 90:187–203

Meghraoui M, Morel JL, Andrieux J, Dahmani M (1996) Tectonique plio-quaternaire de la chaîne tello-rifaine et de la mer d'Alboran. Une zone complexe de convergence continent–continent. Bull Soc Géol Fr 167(1):141–157

Meghraoui M, Maouche S, Chemaa B, Cakir Z, Aoudia A, Harbi A, Alasset PJ, A Ayadi, Bouhadad Y, Benhamouda F (2004) Coastal uplift and thrust faulting associated with the Mw = 6.8 Zemmouri (Algeria) earthquake of 21 May, 2003. Geophys Res Lett 31, L19605. https://doi.org/10.1029/2004gl020466

Mokrane A, Ait Messaoud A, Sebai A, Menia N, Ayadi A, Bezzeghoud M, Benhallou H (1994) Les séismes en Algérie de 1365 à 1992. Publication du Centre de Recherche en Astronomie, Astrophysique et Géophysique, Département: Etudes et Surveillance Sismique, ESS, C.R.A.A. G, Alger-Bouzaréah, p 277

Mortgat CP, Shah HC (1978) Seismic hazard analysis of Algeria. Technical report, the John A Blume Earthquake Engineering Center, California, p 97

Mourabit T, Abou Elenean KM, Ayadi A, Benouar D, Ben Suleman A, Bezzeghoud M, Cheddadi A, Chourak M, ElGabry MN, Harbi A, Hfaiedh M, Hussein HM, Kacem J, Ksentini A, Jabour N, Magrin A, Maouche S, Meghraoui M, Ousadou F, Panza GF, Peresan A, Romdhane N, Vaccari F, Zuccolo E (2014) Neo-deterministic seismimic hazard assessment in north-Africa. J Seismolog 18(2):301–318. https://doi.org/10.1007/s10950-013-9375-2

Naili M, Benouar D (2000) Influence of attenuation models on seismic hazard assessment in Algiers (capital of Algeria). Eur Earthq Eng 1:40–47

Nocquet JM, Calais E (2004) Crustal velocity field of western Europe from permanent GPS array solutions, 1996–2001. Geophys J Int 154:72–88

NRC (National Research Concil) (1988) Probabilistic seismic hazard analysis. National academy press, Washington, DC, p 96

Ousadou F, Dorbath L, Dorbath C, Bounif MA, Benhallou H (2013) The Constantine (Algeria) seismic sequence of 27 October 1985: a new rupture model from aftershock relocation, focal mechanisms, and stress tensors. J Seismol 17:207–222

Ousadou F, Dorbath L, Ayadi A, Dorbath C, Gharbi S (2014) Stress field variations along the Maghreb region derived from inversion of major seismic crisis fault plane solutions. Tectonophysics. https://doi.org/10.1016/j.tecto.2014.06.017

Ouyed M, Meghraoui M, Cisternas A, Deschamps A, Dorel J, Fréchet J, Gaulon R, Hatzfeld D, Philip H (1981) Seismotectonic of the El Asnam earthquake. Nature 292(5818):26–31

Panza GF, La Mura C, Peresan A, Romanelli F, Vaccari F (2012) Seismic hazard scenarios as preventive tools for a disaster resilient society. Adv Geophys 53:93–165. https://doi.org/10.1016/B978-0-12-380938-4.00003-3 ISSN: 0065-2687

Perrodon A (1957) Étude géologique des basins néogènes sublittoraux de l'Algérie Occidentale, vol 12. Publications du Service de la Carte Géologique de l'Algérie, p 328

Philip H, Meghraoui M (1983) Structural analysis and interpretation of the surface deformation of the El Asnam earthquake of October 10, 1980. Tectonics 2:17–49

Ratzov G, Cattaneo A, Babonneau N, Déverchère J, Yelles K, Bracene R, Courboulex F (2015) Holocene turbidites record earthquake supercycles at a slow-rate plate boundary. Geology. https://doi.org/10.1130/G36170.1

Reiter L (1991) Earthquake hazard analysis. Issues and insights. Columbia University Press, p 253

Rothé JP (1950) Les séismes de Kherrata et la séismicité de l'Algérie. Bull Serv Cart Geol Algérie, 4(3)

RPA (1981, 1999, 2003) Réglement parasismique algérien

Serpelloni E, Vannucci G, Pondrelli S, Argnani A, Casula G, Anzidei M, Baldi P, Gasperini P (2007) Kinematics of the western Africa–Eurasia plate boundary from focal mechanism and GPS data. Geophys J Int 169:1180–1200

Stein RS, Yeats RS (1989) Hidden earthquakes. Sci Am 260(6):48–57

Thomas G (1985) Géodynamique des bassins intramontagneux: Le bassin du bas Chéliff occidental (Algérie) durant le Mio-plioquaternaire. Thèsis Pau University, France, p 594

Vila JM (1980) La chaine alpine d'Algérie orientale et des confins algéro-tunisiens. Thèsis, Pierre et Marie Curie University, Paris VI, p 665

Wang Z (2011) Seismic hazard assessment: issues and alternatives. Pure Appl Geophys 168:11–25. https://doi.org/10.1007/s00024-010-0148-3

Westaway R (1990) Present-day kinematics of the plate boundary zone between Africa and Europe, from the Azores to the Aegean. Earth Planet Sci Lett 96:393–406

Woodward Clyde consultants (WCC) (1984) Microzonation sismique de la région de Ech-chelif. Technical report V1, p 145

Yelles-Chaouche AK, Djellit H, Beldjoudi H, Bezzeghoud M, Buforn E (2004) The Ain Temouchent (Algeria) earthquake of December 22nd, 1999. Pure Appl Geophys 161:607–621

Yelles-Chaouche AK, Boudiaf A, Djellit H, Bracene A (2006) La tectonique active de la région nord-algérienne. C R Geoscience 338:126–139

Yielding G (1985) Control of rupture by fault geometry during the El Asnam (Algeria) earthquake. Geophys J R Astr Soc 81:641–670

Yielding G, Ouyed M, King GCP, Hatzfeld D (1989) Active tectonics of the Algerian Atlas mountains-evidence from aftershocks of the 1980 El Asnam earthquake. Geophys J Int 99:761–788

Seismicity of the Algerian Tell Atlas and the Impacts of Major Earthquakes

Farida Ousadou and Mourad Bezzeghoud

Abstract The seismicity of the Tell Atlas, which extends from the Algerian margin to the South Atlasic fault system, is related to the dynamics of Quaternary basins under an oblique NW–SE convergent stress regime, including the basins of Mleta and L'Habra in the west, Cheliff and Mitidja in the centre, and Soummam, Hodna and Guelma in the east. This seismicity is characterized by moderate to low magnitudes with strong events occurring generally once a decade. Over the last six decades, several moderate, strong and major events occurred that were associated with extensive and severe damage, such as those of El Asnam (1954, M_s 6.7; 1980, M_s 7.3), Constantine (1985, M_s 6.0), Tipasa–Chenoua (1989, M_s 6.0), Mascara (1994, M_s 6.0), Ain Temouchent (1999, M_s 5.8), Beni Ouartilane (2001, M_s 5.6), Zemmouri—Boumerdes (2003, M_w 6.8) and Laalam (2005, M_s 5.8), in addition to numerous large historical seismic events, including those that occurred in Algiers (1365 and 1716, I_o = X), Oran (1790, I_o = X), Mascara (1819, I_o = X), Djidjelli (1856, I_o = VIII) and M'sila (1885, I_o = IX). This chapter presents a review of the seismicity of North Algeria and a detailed analysis of the main earthquakes that have occurred in the Tell Atlas since 1980. Finally, the impacts of several significant earthquakes that occurred during the period between 1364 and 2015 are presented and discussed in terms of seismic energy.

Keywords Seismicity · Focal mechanisms · Seismic energy
Impact of earthquakes · North Algeria

F. Ousadou
Centre de Recherche en Astronomie, Astrophysique et Géophysique, BP 63, 16340 Bouzaréah, Algiers, Algeria

M. Bezzeghoud (✉)
Departamento de Física, Escola de Ciências e Tecnologia (ECT), Instituto de Ciências da Terra (IIFA), University of Évora, Colégio Luis Antonio Verney, Romão Ramalho, 59, 7002-554 Évora, Portugal
e-mail: mourad@uevora.pt

A. Bendaoud et al. (eds.), *The Geology of the Arab World—An Overview*, Springer Geology, https://doi.org/10.1007/978-3-319-96794-3_11

1 Introduction

The western part of the Eurasia–Nubia plate boundary extends from the Mid-Atlantic Ridge in the west to Tunisia in the east, and includes the Iberian Peninsula and the northern part of the Maghreb region. The interaction between Iberia and Nubia results in a complex region located in the western part of the Eurasian–African plate boundary that has various tectonic features (Buforn et al. 2004; Borges et al. 2007, 2008; Bezzeghoud et al. 2014). This region corresponds to the transition from an oceanic boundary (between the Azores islands and the Gorringe Bank) to a continental boundary where Iberia and Nubia collide.

According to several authors, (Buforn et al. 2004; Borges et al. 2007; Bezzeghoud et al. 2014), in the Ibero-Maghrebian region between the Gulf of Cadiz and Algeria, the Eurasian–Nubian plate boundary corresponds to a well-defined narrow band of seismicity, where large earthquakes occur in association with N–S to NNW–SSE horizontal compression due to the convergence of Eurasia and Nubia (Fig. 1). In this region, earthquakes are concentrated between 2°W and 4°E, and $M > 6.0$ earthquakes have occurred on 9 September 1954 (M_s 6.5, El Asnam, formerly Orléansville, Algeria); 10 October 1980 (M_s 7.3, El Asnam, formerly Chlef, Algeria); 27 October 1985 (M_w 5.9, Constantine, Algeria); 29 October 1989 (M_w 6.0 Tipasa–Chenoua, Algeria); 18 August 1994 (M_w 5.7, Mascara, Algeria); 22 December 1999 (M_w 5.7, Ain Temouchent, Algeria); 10 November 2000 (M_s 5.5, Beni Ouartilane, Algeria); 21 May 2003 (M_w 6.8, Zemmouri—Boumerdes, Algeria) (Ayadi and Bezzeghoud 2015); 24 February 2004 (M_w 6.4, Al Hoceima, Morocco) (Buforn et al. 2004); 20 March 2006, (M_w 5.2, Laalam, Algeria) (Bouhadad et al. 2010) and 2016 south Alboran earthquake (M_w 6.4) (Buforn et al. 2017).

Intermediate-depth seismicity (60–150 km) in the Alboran Sea region extends, in the N–S direction, in a very narrow vertical band of 50 km wide. This intermediate-depth seismicity is distributed in E–W direction and limited by a narrow band less than 20 km wide that broadens as we move to the Strait of Gibraltar. This seismicity could also be associated with the convergence process of the Eurasia–Africa plates (Buforn et al. 2004). This intermediate-depth seismic activity on the eastern side of the Strait of Gibraltar may be explained by the existence of a seismogenic block in the upper mantle that has approximate dimensions of 200 km long, 150 km deep and 50 km wide (Buforn et al. 2004). In this region, the material is relatively rigid, and the stresses are released by larger earthquakes. The presence of very deep earthquakes (650 km) under southern Spain is a further sign of the complexity of this area. In the Alboran Sea region, the material is more fragmented with a large number of small faults, and the stresses are released by frequent small-to-moderate earthquakes. Consequently, the plate boundary is more diffuse and corresponds to a wider area that includes the Betics, the Alboran Sea and the Rif Cordilleras (Buforn et al. 2004), where deformation is manifested by the continuous occurrence of small earthquakes and only occasionally by moderate to strong earthquakes with magnitudes greater than six as shown by the historical seismicity (Martínez Solares and Mezcua 2002; El Mrabet

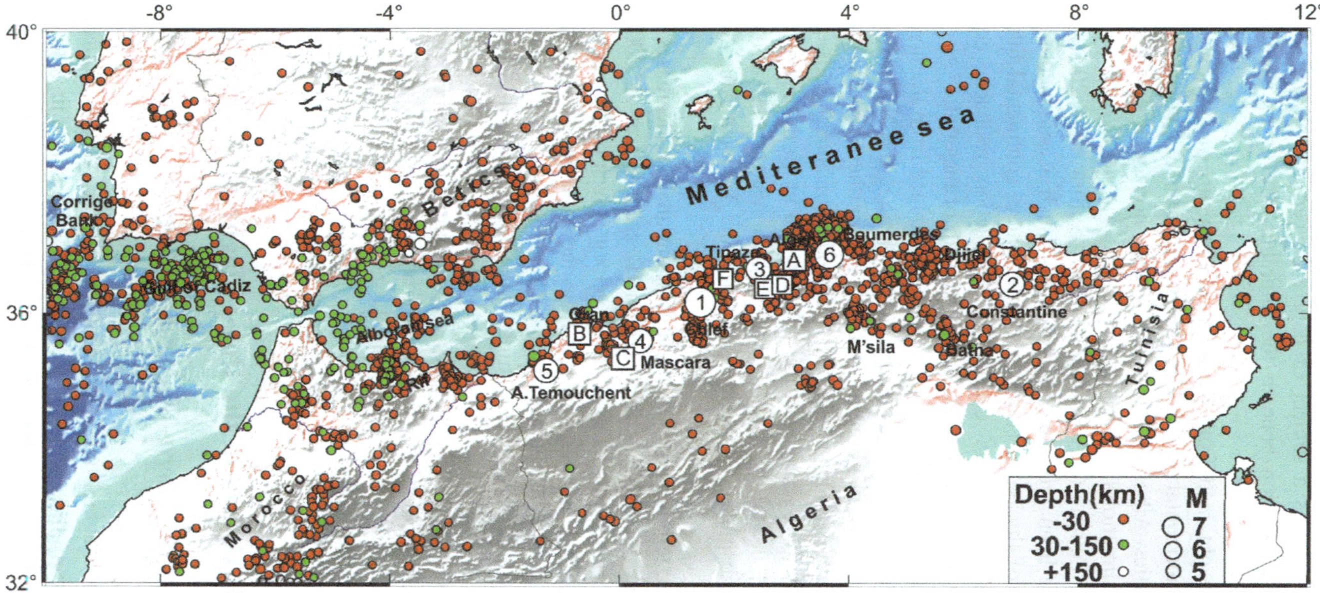

Fig. 1 Seismicity ($M \geq 4.5$) of the Ibero-Maghrebian region for 1973–2015. Seismic data were taken from the NEIC database. White squares are the major historical earthquakes (A: 02/01/1365, B: 09/10/1790, C: –/03/1819, D: 02/03/1825, E: 02/01/1867, F: 15/01/1891). White circles are the main earthquakes discussed in this study (1: 1980 El Asnam EQ, 2: 1985 Constantine EQ, 3: 1989 Tipaza Chenoua EQ). The topography and bathymetry data have been extracted from the GEBCO (General Bathymetric Chart of the Oceans: http://www.gebco.net). Figures 1, 2, 3, 4, 5 and 6 are plotted with GMT software (Wessel and Smith 1991)

2005; Bahrouni et al. 2013; Ayadi and Bezzeghoud 2015). This intermediate and deep seismicity recorded beneath the Gibraltar Arc, the westernmost Alboran Sea and southern Spain remains a subject of debate concerning whether it may be attributed to active subduction beneath the Gibraltar Arc or to another origin (Gutscher et al. 2012).

The stress regime obtained from the focal mechanisms of shallow events is compatible with the horizontal N–S to NW–SE convergence of Eurasia and Africa. However, in the Betics–Alboran area, there is also horizontal extension in an approximately E–W direction (De Vicente et al. 2008; Van der Woerd et al. 2014). According to several authors (e.g. Houseman 1996; Buforn et al. 2004) different tectonic models have been proposed for this region such as subduction process, extensional collapse of thickened continental lithosphere, continental lithospheric delamination, back-arc extension caused by subduction rollback, convective thinning, or subduction and breaking of a vertical slab of material.

Thus, for the Central Maghreb, the stress regime obtained from the inversion of focal mechanisms reflects a 'pure' convergence dynamic in the NW–SE direction and for both sides (the Rif in northern Morocco and northern Tunisia) reflects a strike–slip regime. Both sides are affected by the geodynamics of the adjacent areas that are influenced by the dynamics of the Alboran Sea in northwestern Morocco (Ousadou et al. 2014) and by lateral slab migration/segmentation and deep dynamics such as lithosphere–mantle interactions in the Tunisian Atlas (Soumaya et al. 2015).

The convergence rate determined by seismic strain analysis decreases from the Azores Plateau to the Ibero-Maghrebian region (Bezzeghoud et al. 2008, 2014; Buforn 2008). The corresponding values of slip velocity given by Bezzeghoud et al. (2014) range between 1.4 and 6.7 mm/year from the triple junction to the north-western part of Algeria and the Tell Mountains, where the velocity is approximately 3.7 mm/year. These values may underestimate the geological deformation and do not include the energy released by aseismic processes (e.g. folding, thickening, plastic deformation and slow aseismic slip). Additionally, the slip velocity may be considered to be instantaneous and independent from that derived from the geodetic data because the earthquake cycle is much shorter than the history of available earthquakes (Borges et al. 2007).

2 Seismicity of Algeria in the Ibero-Maghrebian Context

2.1 Seismicity of the Ibero-Maghrebian Region

The seismicity of the Maghreb region is located along the African–Eurasian plate boundary starting from the Azores triple junction and continuing to Tunisia, crossing the Strait of Gibraltar, Morocco and northern Algeria. It is related to the closure of Quaternary basins under an oblique NW–SE convergent stress regime.

Most of the activity is linked to fault-related folds striking NE–SW and is distributed along the main tectonic features. Because the plate boundary is not defined by a single discontinuity, as found in the western region of Gibraltar, but is more diffuse, the earthquakes are distributed over a wide zone. Most of the seismicity is shallow, located between depths of 5 and 20 km (Fig. 1). Nevertheless, intermediate and deep seismic events have been observed beneath the Gibraltar Arc, the westernmost Alboran Sea and southern Spain. The seismicity distribution in the Maghreb displays a clear arrangement for the activity since 1973, with an E–W trend along the African–Eurasian plate boundary (Fig. 1). Morocco and Algeria appear as more active areas than Tunisia, and the seismicity is mainly concentrated in the Rif, the High Atlas of Morocco, the Algerian Tell and Middle Atlas and the Tunisian Atlas, where the seismicity spreads along a band extending south to the Saharan Atlas (Fig. 1). The easternmost Maghrebian region, the Tunisian Atlas area, has not experienced strong earthquakes; most of the seismicity is of low magnitude located in the Northwestern Atlas, Southern Atlas and Eastern Pelagic platform (Bahrouni et al. 2013, 2014). However, Tunisia is still a key region because it is located at the junction between the eastern and western Mediterranean domains and straddles the orogenic province and the stable platform (Bouaziz et al. 2002).

In the Ibero-Maghrebian region, the NW–SE oblique convergence between Africa and Eurasia plates is consistent with the earthquake focal mechanisms that show mostly reverse faulting on NE–SW trending often combined with a strike–slip component. Most of the focal mechanisms in this area exhibit strike–slip faulting in northern Morocco, reverse faulting in northwestern Algeria and strike–slip mechanisms in northeastern Algeria and Tunisia (Fig. 2) (Ayadi et al. 2002; Bezzeghoud and Buforn 1999; Harbi et al. 1999; Buforn et al. 2004; Bahrouni et al. 2014). This agrees with earlier studies that give a NW–SE convergence direction for Africa and Eurasia with translation of the African plate to the east.

2.2 *Seismicity of Algeria*

The seismicity of the Tell Atlas, which extends from the Algerian margin to the South Atlasic fault system, is related to the dynamics of Quaternary basins under an oblique NW–SE convergent stress regime, including the basins of Mleta and L'Habra in the west, Cheliff and Mitidja in the centre, and Soummam, Hodna and Guelma in the east. Most of the activity is linked to fault-related folds striking NE–SW and is distributed along the main tectonic features. This seismicity is characterized by moderate to low magnitude with strong events occurring generally once a decade. Over the last six decades, several strong events occurred and were associated with extensive and severe damage. The seismic catalogues show several moderate, strong and major events offshore or inland, such as those listed in the introduction in addition to numerous large historical events such as those that occurred in Algiers (1365 and 1716, I_0 = X), Oran (1790, I_0 = X), Mascara (1819,

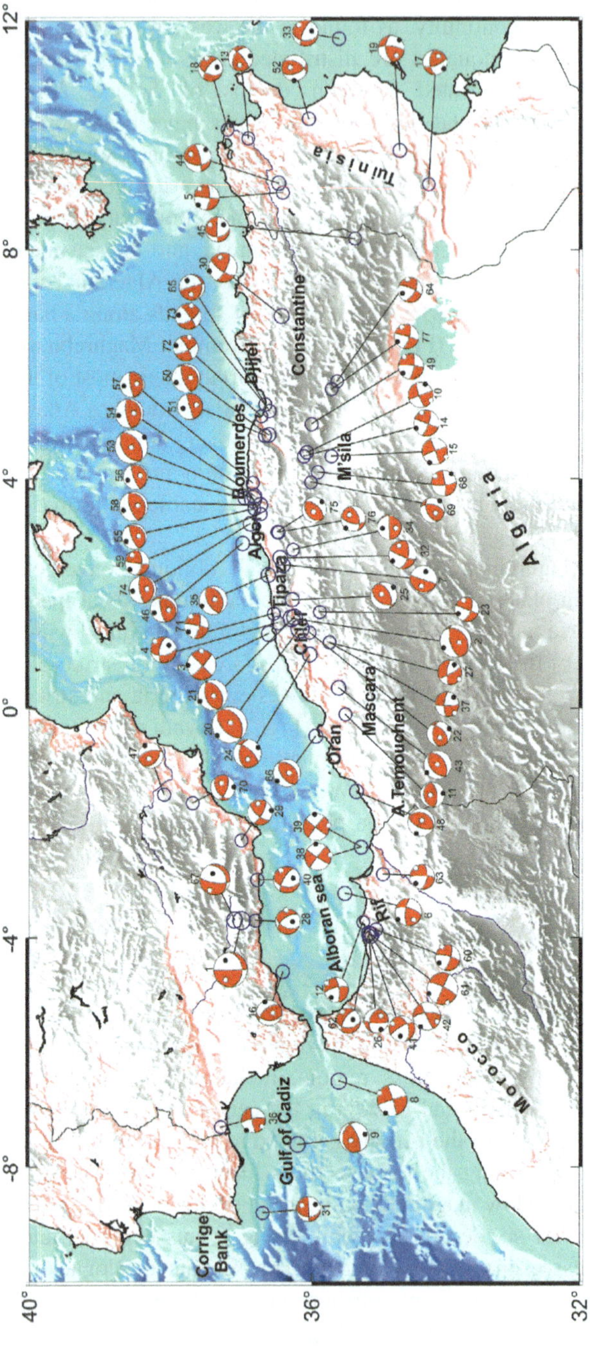

Fig. 2 Focal mechanisms for $M \geq 5.0$ shallow earthquakes (depth $\leq$ 40 km) in the Ibero-Maghrebian region for the period 1950–2015. Size is proportional to magnitude. Solution parameters for focal mechanisms are listed in Table 2

I_o = X), Djidjelli (1856, I_o = VIII) and M'sila (1885, I_o = IX) (Benouar 1994; Mokrane et al. 1994; Harbi et al. 2010; 2015; Ayadi and Bezzeghoud 2015).

The focal mechanisms in Algeria are dominated by reverse faulting for earthquakes located in the western part of the country from 2°W to 5°E, which are associated with the strong and moderate earthquakes that occurred in the area (Fig. 2). Meanwhile, the focal mechanisms in the eastern part are dominated by strike–slip faulting for earthquakes located from 5°E to 8°E (Fig. 2). Most of these are associated with identified active Quaternary faults and folds (Bouhadad et al. 2003; Ayadi et al. 2002, 2008; Meghraoui 1988; Bounif et al. 1987; Maouche et al. 2011; Abbes et al. 2016).

3 Significant Earthquakes in Algeria Since 1980

Among all the earthquakes that have occurred in Algeria, we present only those with magnitudes greater than or equal to 5.7 and caused serious damage (Table 1) for the period 1980–2016. These earthquakes include the following: (i) the El Asnam earthquake of 10 October 1980, with M_w 7.1; (ii) the Constantine earthquake of 27 October 1985, with M_w 5.9; (iii) the Tipasa–Chenoua earthquake of 29 October 1989, with M_w 6.0; (iv) the Mascara earthquake of 22 August 1994, with M_w 5.7; (v) the Ain Temouchent earthquake of 22 December 1999, with M_w 5.7 and (vi) the Zemmouri—Boumerdes earthquake of 21 May 2003, with M_w 6.8.

3.1 *The El Asnam Earthquake of 10 October 1980*

The strongest event that occurred in northern Africa is that of El Asnam on 10 October 1980, at 12 h 25 min (TU) with M_s 7.3 (Table 1). This event has been

Table 1 Damage caused by the most significant earthquakes occurred in Algeria since 1980 discussed in the text. The damage is given by Mr. Chergui Abdelkader and Mrs. Bradai Kheira from the 'Direction Générale de la Protection Civile' (Algiers, Algeria)

Date	Location	M_W	Damage
10/10/1980	El Asnam (now known as Chlef)	7.1	2633 killed, 8369 injured and 30,022,000 inhabitants homeless
27/10/1985	Constantine	5.9	10 killed and 300 injured
29/10/1989	Tipasa–Chenoua	6.0	22 killed, 250 injured and 50,000 inhabitants homeless
18/08/1994	Mascara	5.7	171 killed, 270 injured and 10,000 inhabitants homeless
22/12/1999	Ain Témouchent	5.7	28 killed, 174 injured and 25,000 inhabitants homeless
21/05/2003	Zemmouri—Boumerdes	6.8	2278 killed and 11,452 injured

Table 2 Source parameters of focal mechanisms plotted in Fig. 2

Date	LAT	LON	Depth	Strike	DIP	Rake	M	$n°$	References
20/03/1954	37	−3.7	640	179	88	−122	7.0	1	Buforn et al. (1991)
09/09/1954	36.28	1.57	7	253	61	104	6.5	2	Espinoza and Lopez (1984)
10/09/1954	36.6	1.3	8	44	90	−8	6.0	3	Dewey (1990)
05/06/1955	36.4	1.6	13	172	56	−32	5.2	4	Shirokova (1967)
20/02/1957	36.4	9	10	185	87	18	5.2	5	Hfaiedh et al. (1985), Soumaya et al. (2015)
23/08/1959	35.51	−3.23	20	276	70	153	5.5	6	Bezzeghoud and Buforn (1999)
07/11/1959	36.4	2.5	7	203	10	−5	5.6	7	Henares et al. (2003)
05/12/1960	35.6	−6.5	15	73	86	−178	6.2	8	Buforn et al. (2004)
15/03/1964	36.2	−7.6	12	276	24	117	6.1	9	Buforn et al. (2004)
01/01/1965	35.7	4.4	13	340	90	12	5.1	10	Hatzfeld (1978)
13/07/1967	35.5	−0.1	5	260	30	87	5.1	11	Buforn et al. (2004)
17/04/1968	35.24	−3.73	22	83	70	−162	5.0	12	Buforn et al. (2004)
01/12/1970	36.9	9.95	15	49	74	50	5.1	13	Hfaiedh et al. (1985), Soumaya et al. (2015)
24/11/1973	36.1	4.4	15	200	89.9	14	5.2	14	Hatzfeld (1978)
24/11/1973	36.06	4.47	8	70	86	−176	5.3	15	Hatzfeld (1978)
07/08/1975	36.41	−4.59	28	186	42	138	5.2	16	Henares et al. (2003)
08/02/1978	34.27	9.15	30	266	54	26	5.0	17	Mezcua and Martinez Solares (1983)
09/04/1979	37.2	10.1	12	305	67	−140	5.0	18	Gueddiche et al. (1992), Soumaya et al. (2015)
08/12/1979	34.7	9.74	33	7	72	35	5.4	19	Ben Ayed and Zargouni (1990)
10/10/1980	36.16	1.39	12	225	54	83	7.3	20	Ouyed et al. (1981)
10/10/1980	36.24	1.59	10	58	43	81	6.1	21	Harvard CMT
08/11/1980	36.02	1.32	5	270	45	126	5.0	22	Bezzeghoud et al. (2012)
05/12/1980	35.87	1.68	5	112	61	−179	5.0	23	Bezzeghoud et al. (2012)
07/12/1980	36.02	0.94	5	277	40	140	5.8	24	Bezzeghoud and Buforn (1999)

(continued)

Table 2 (continued)

Date	LAT	LON	Depth	Strike	DIP	Rake	M	$n°$	References
01/02/1981	36.27	1.9	11	210	43	64	5.5	25	Bezzeghoud and Buforn (1999)
07/04/1981	35.12	−3.98	4	182	75	132	5.0	26	Bezzeghoud et al. (2012, 2014)
15/11/1982	35.73	1.15	7	274	70	−169	5.0	27	Bezzeghoud et al. (2012, 2014)
24/06/1984	36.8	−3.7	5	201	48	−46	5.0	28	Bezzeghoud and Buforn (1999)
13/09/1984	37	−2.3	9	121	73	156	5.1	29	Buforn et al. (2004)
27/10/1985	36.42	6.85	10	213	71	20	5.9	30	Harvard CMT
20/10/1986	36.7	−8.8	37	180	37	3	5.0	31	Bezzeghoud et al. (2012, 2014)
31/10/1988	36.44	2.63	13	103	55	167	5.7	32	Bezzeghoud and Buforn (1999)
06/01/1989	35.59	11.69	11	183	41	−26	5.2	33	RCMT, Soumaya et al. (2015)
12/02/1989	36.26	2.77	7	10	11	10.6	5.0	34	NEIC
29/10/1989	36.61	2.33	13	242	55	87	6.0	35	Bounif et al. (2003)
20/12/1989	37.3	−7.3	23	351	77	10	5.0	36	Buforn et al. (2004)
15/11/1992	35.73	1.15	7	274	70	−169	5.0	37	Harvard CMT
23/04/1993	35.27	−2.42	6	308	86	4	5.4	38	Bezzeghoud and Buforn (1999)
23/05/1993	35.27	−2.42	6	308	86	4	5.4	39	Bezzeghoud et al. (2012, 2014)
23/12/1993	36.77	−2.99	8	300	70	−130	5.4	40	Buforn et al. (2004)
26/05/1994	35.14	−3.92	7	330	77	−45	5.3	41	Bezzeghoud and Buforn (1999)
26/05/1994	35.16	−3.92	8	355	69	2.5	5.7	42	Bezzeghoud and Buforn (1999)
18/08/1994	35.6	0.36	4	58	45	95	5.7	43	Bezzeghoud and Buforn (1999)
17/09/1994	36.46	9.17	10	181	47	20	5.6	44	Bahrouni et al. (2014), Soumaya et al. (2015)
22/09/1995	35.36	8.2	10	173	41	−8	5.0	45	Harvard CMT
04/09/1996	36.98	2.88	14	260	70	108	5.5	46	Stich et al. (2003)
02/02/1999	38.1	−1.5	1	125	39	56	5.1	47	Borges et al. (2001)
22/12/1999	35.34	−1.45	6	25	31	92	5.7	48	Bezzeghoud et al. (2014)

(continued)

Table 2 (continued)

Date	LAT	LON	Depth	Strike	DIP	Rake	M	$n°$	References
18/08/2000	35.99	4.96	10	99	71	−171	5.2	49	Harvard CMT
10/11/2000	36.60	4.773	6	251	70	97	5.7	50	Stich et al. (2003)
16/11/2000	36.65	4.759	4	254	87	88	5.2	51	Stich et al. (2003)
24/06/2002	36.03	10.29	15	28	48	128	5.2	52	Harvard CMT
21/05/2003	36.83	3.65	10	237	43	92	6.8	53	Santos et al. (2014)
21/05/2003	36.81	3.702	18	72	20	82	5.9	54	Braunmiller and Bernardi (2005)
21/05/2003	36.81	3.485	15	41	21	65	5.1	55	Braunmiller and Bernardi (2005)
22/05/2003	36.96	3.67	9	57	14	78	5.1	56	Braunmiller and Bernardi (2005)
22/05/2003	36.83	3.934	12	91	14	101	5.1	57	Braunmiller and Bernardi (2005)
27/05/2003	36.72	3.547	18	63	21	83	5.9	58	Braunmiller and Bernardi (2005)
29/05/2003	36.72	3.39	12	187	87	32	5.0	59	Braunmiller and Bernardi (2005)
02/02/2004	35.05	−3.86	10	108	73	−167	5.2	60	IAG
24/02/2004	35.13	−3.955	7	25	80	1	6.4	61	Ven der Woerd et al. (2014)
07/03/2004	35.16	−3.96	12	178	43	−43	5.1	62	Harvard CMT
04/12/2004	34.95	−2.89	12	98	61	−168	5.1	63	Harvard CMT
09/03/2005	35.62	5.707	12	22	77	−33	5.0	64	ZUR_RMT
20/03/2006	36.65	5.302	5.95	344	85	113	5.2	65	IGN
06/06/2008	35.93	−0.48	12	44	36	81	5.5	66	Harvard CMT
11/04/2010	37.1	−3.69	616.5	63	28	−39	6.3	67	GCMT
14/05/2010	35.9	4.12	12	174	77	5	5.2	68	Harvard CMT
16/05/2010	36	3.94	12	245	50	128	5.0	69	Harvard CMT
11/05/2011	37.7	−1.65	12	234	45	43	5.1	70	RCMT
25/04/2012	36.53	1.65	10	103	71	−174	5.0	71	Harvard CMT
28/11/2012	36.71	5.1	12	152	73	−170	5.0	72	RCMT

(continued)

Table 2 (continued)

Date	LAT	LON	Depth	Strike	DIP	Rake	*M*	*n*°	References
19/05/2013	36.59	5.19	15	149	63	−170	5.3	73	Harvard CMT
01/08/2014	36.87	3.22	10.4	101	19	109	5.3	74	Benfedda et al. (2017)
23/12/2014	36.46	3.072	10	42	58	85.9	5.1	75	CGS
26/12/2014	36.45	3.075	12	72.1	53.2	138.5	5.1	76	CGS
21/03/2015	35.7	5.57	13	107	71	−178	5.0	77	CGS

studied by numerous researchers (Deschamps et al. 1982; Ouyed et al. 1981, 1983; Philip and Meghraoui 1983; Yielding et al. 1981, 1989; Ousadou et al. 2014). The event was relocated by the International Seismological Centre (ISC) at 36.16°N–1.40°E with a fixed depth of 10 km based on 514 station records. The Harvard CMT seismic moment is 5.07×10^{19} N m (M_w 7.1). The mainshock mechanism constructed using *P*-wave polarities at teleseismic distances shows reverse faulting along a plane striking N45°E and dipping N54° with an 83° slip angle (Ouyed et al. 1981; Deschamps et al. 1982), which is similar to the Harvard CMT solution (azimuth 50°N, dip 61°, slip 81°). The *P* and SH wave modelling show that the source consisted of two sub-events separated by 4 s (Yielding et al. 1981) (Fig. 3). This event is associated with 40-km-long surface ruptures that have been clearly mapped (King and Vita-Finzi 1981; Philip and Meghraoui 1983). Three segments are identified on the rupture zone as shown by the aftershock distribution and surface fault traces. The aftershock zone extends for a length of approximately 40 km in the NE–SW direction (Fig. 3), ranging between 5 and 14 km in depth, as given by the relocation using HypoDD by Ousadou et al. (2014). The distribution exhibits three segments, and the NE edge is leading in approximately the N–S direction.

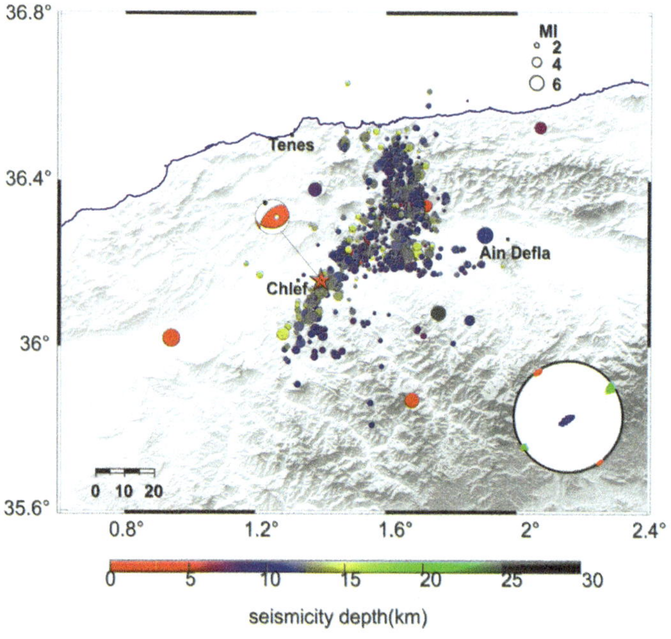

Fig. 3 The focal mechanism (Deschamps et al. 1982) and the corresponding aftershock sequence (relocated by Ousadou et al. 2014) following the mainshock of the El Asnam earthquake of 10 October 1980. Red star: mainshock after ISC relocation (2005). The regional stress tensor and principal stress (Ousadou et al. 2014) are shown in the lower right corner (Sh_{max} in red and Sh_{min} in blue colours). The scale bar units are in km

The focal mechanisms of the 1980 El Asnam aftershocks are well constructed using the *P* arrivals (Ouyed et al. 1983; Yielding et al. 1989). Most of the focal solutions for the aftershocks show reverse faulting along the El Asnam rupture zone, with an important strike–slip component both northeast and southwest of the fault. The aftershock focal mechanisms are inverted to calculate the stress field for the Cheliff basin yielding a reverse tensor with maximum and minimum horizontal stress Sh_{max} and Sh_{min} striking in the N325° and N235° directions, respectively (Ousadou et al. 2014). This is in good agreement with the directions of the *P* and *T* axes of the mainshock (Fig. 3).

The Cheliff basin experienced the 9 September 1954 Orléansville earthquake, which had a magnitude of M_s 6.7, maximum intensity of I_o X (Mercalli scale) and macroseismic depth of approximately 9 km (Rothé 1955; Rothé et al. 1977). According to Heezen and Ewing (1955), breaks in submarine cables were caused by the motion of a mass of sediments detached from the continental slope by the mainshock of 1954 and transformed into a strong turbidity current. All of the focal mechanisms established from the *P* arrivals show reverse solutions with a fault plane dipping to the NW (Shirokova 1967; McKenzie 1972; Espinosa and Lopez 1984), which is not much different from the El Asnam one. Referring to the work of Dewey (1990), the 1954 earthquake was relocated by the Joint Epicenter Determination method, taking the 1980 earthquake near the central segment of the El Asnam fault as the reference. As a result, Dewey proposes a model of the 1954 fault, associated with the 1980 thrust fault. This model agrees with the dislocation model proposed by Bezzeghoud et al. (1995) deduced from geodetic measurements.

3.2 *The Constantine Earthquake of 27 October 1985*

The eastern part of Algeria experienced a moderate-sized earthquake (M_w 5.9) on 27 October 1985 at 18 h 34 min (TU) (Table 1). It was the strongest earthquake recorded in the Tell Atlas in the 5 years following the 1980 El Asnam earthquake. The mainshock has been relocated and showed a remarkable jump of the epicentre from outside the cloud of aftershocks (Bounif et al. 1987) to inside it (Ousadou et al. 2013). The focal solution using *P* and SH waveform inversions shows a left-lateral strike–slip mechanism on a fault striking N217° and dipping 84° with a rake of 19°, a depth fixed at 9 km and a seismic moment of 5.2×10^{17} N m (Deschamps et al. 1991).

The aftershocks have been located (Bounif and Dorbath 1998) and, once more, relocated in a 3D model using TomoDD (Ousadou et al. 2013). The aftershock relocation by TomoDD is better defined and less scattered than that of the 1980 El Asnam earthquake (Fig. 4). It extends for a length of approximately 30 km, but only the 14-km-long central part shows an alignment in the N210°/215° direction, which corresponds to one of the two nodal planes given by Harvard CMT (2005) and Deschamps et al. (1991). This segment extending from 5 to 15 km down has been subsequently interpreted as the source of the mainshock rupture, and the

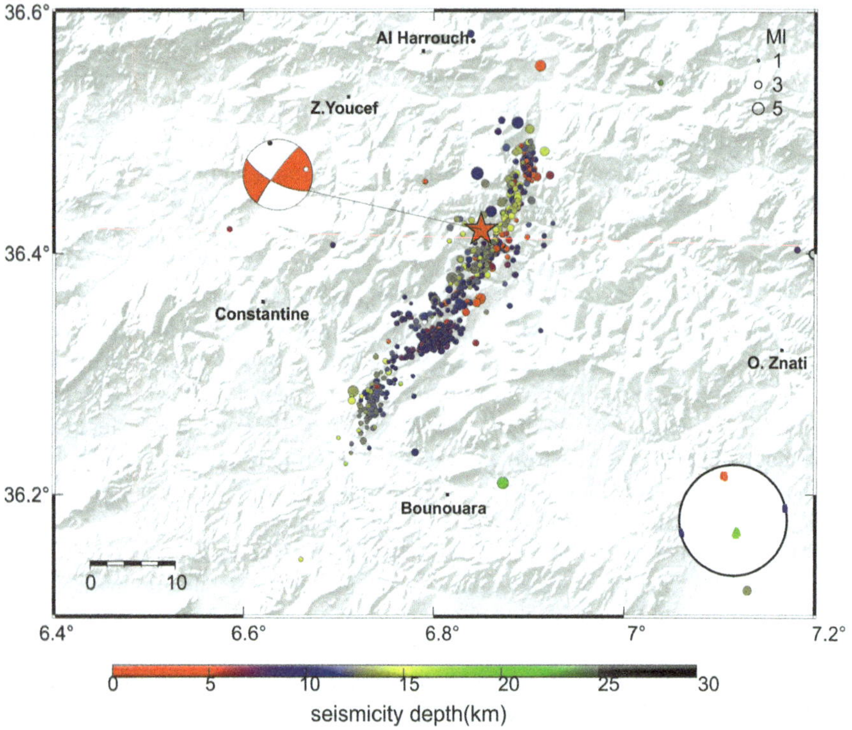

Fig. 4 The focal mechanism (Deschamps et al. 1991) and the corresponding aftershock sequence (relocated by Ousadou et al. 2013) following the mainshock of the Constantine earthquake of 27 October 1985. Red star: main shock after relocation (Ousadou et al. 2014). The regional stress tensor and principal stress (Ousadou et al. 2014) are shown in the lower right corner (Sh_{max} in red and Sh_{min} in green colours). The scale bar units are in km

4.5-km-long surface breaks, organized in an *en échelon* system, are secondary ruptures due to ground shaking around the epicentral area (Ousadou et al. 2013).

The aftershock focal mechanisms have been manually constructed, and the majority of them exhibit strike–slip faulting with some reverse or normal components, located between 10- and 15-km depths and spread all along the seismic zone (Ousadou et al. 2013). The inversion of aftershock focal mechanisms yields a strike–slip tectonic regime with quasi-horizontal Sh_{max} and Sh_{min} striking in the N347° and N257° directions, respectively. This is in accordance with the tectonic observations made in the Constantine basin, which is defined by the left-lateral strike–slip character of the NE–SW trending Ain Smara fault (Fig. 4). The Sh_{max} and Sh_{min} orientations correspond to the *P* and *T* axes of the mainshock mechanism (Ousadou et al. 2014).

3.3 The Tipasa–Chenoua Earthquake of 29 October 1989

The moderate earthquake that occurred on 29 October 1989 with magnitude M_s 5.9 struck the northwestern part of the Mitidja basin in the locality of Tipasa–Chenoua at 19 h 09 min (TU) (Table 1). The mainshock was located by NEIC and relocated by Bounif et al. (2003) (Fig. 5). The *P* and SH body wave modelling reveals almost pure reverse faulting along a plane striking 246°N and dipping 56° with an 86° slip angle and a seismic moment of 8.2 × 10^{17} N m, M_w 6.0 (Bounif et al. 2003) (Fig. 5).

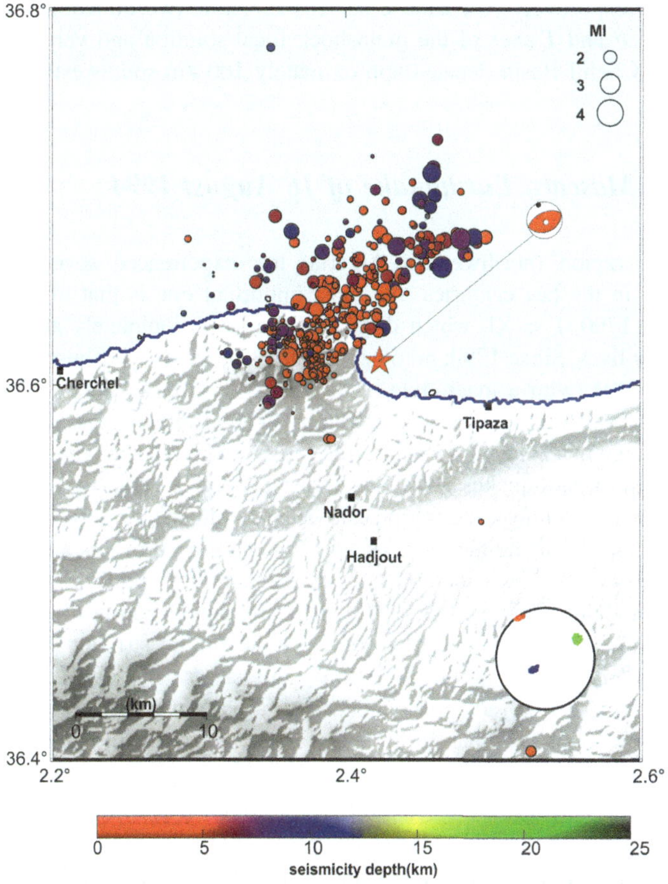

Fig. 5 The focal mechanism (Bounif et al. 2003) and the corresponding aftershock sequence (relocated by Ousadou et al. 2014) following the mainshock of the Tipasa–Chenoua earthquake of 29 October 1989. Red star: mainshock after relocation (Bounif et al. 2003). The regional stress tensor and principal stress (Ousadou et al. 2014) are shown in the lower right corner (Sh_{max} in red and Sh_{min} in blue colours)

The epicentral distribution of the aftershocks depicts an elongated area with a general SW–NE direction approximately 15 km long and 7 km wide. Most of these aftershocks lie offshore, and there is no clear segmentation of the aftershock zone (Fig. 5).

The aftershock focal solutions reconstructed manually (Ousadou et al. 2014) show predominantly pure reverse faulting with a weak left-lateral strike–slip component on a plane dipping to the NW and striking NE–SW. This is in agreement with the geometry of the aftershock area and the focal mechanism of the mainshock (Fig. 5).

The set of aftershock focal mechanisms has been inverted to calculate a compressive stress tensor with well-defined horizontal Sh_{max} and Sh_{min} trending N326° and N239°, respectively (Ousadou et al. 2014). These two directions are close to those of the *P* and *T* axes of the mainshock focal solution and very close to that obtained in Cheliff Basin, located approximately 100 km southwestward.

3.4 The Mascara Earthquake of 18 August 1994

The Oranie region (northwestern Algeria) has experienced several significant earthquakes in the last centuries; the most important one is that of Oran city on February 9, 1790, I_o = XI, which destroyed the town completely and caused the loss of many lives. Since 1790, no other event has been as disastrous, except that of 18 August 1994, with a magnitude of M_w 5.7 (Ayadi et al. 2002), which struck Mascara Province (Algeria) at 01 h 13 min GMT (Table 1). The Mascara earthquake is located in a zone with low seismic activity but with a maximum intensity I_o of IX from past earthquakes (Ayadi and Bezzeghoud 2015). However, since the beginning of this century, the region has been dominated by seismic quiescence, and no event with magnitude larger than 5.5 has occurred in this area.

The focal solutions computed by Harvard, Bezzeghoud and Buforn (1999) using waveform modelling and Thio et al. (1999) using single station inversion (hybrid method), show a reverse mechanism on a single rupture striking in the NE–SW direction at a shallow depth (4.5 km) and a scalar seismic moment of 3.3×10^{17} N m. These focal mechanisms seem to be in good agreement with the tectonic observations (Thomas 1985) and with the macroseismic map (Ayadi et al. 2002).

3.5 The Ain Temouchent Earthquake of 22 December 1999

The Ain Temouchent region situated in the northwestern part of Algeria is characterized by low-to-moderate seismic activity in comparison with the Oran and Mascara regions. Nevertheless, the region was shaken on 22 December 1999, by an earthquake of magnitude M_w 5.7 (Table 1). The focal mechanism of the mainshock has been established using broadband data at regional and teleseismic distances (Yelles-Chaouche et al. 2004), and the scalar seismic moment estimated from waveform modelling is 4.7×10^{17} N m. Unfortunately, there is no aftershock study for this earthquake.

Indeed, throughout history, no important earthquakes have been mentioned in the seismic catalogues for this region (Fig. 1) (Roussel 1973; Benhallou 1985; Mokrane et al. 1994; Benouar 1994; Ayadi and Bezzeghoud 2015). These catalogues all indicate a low level of seismic activity with the occurrence of earthquakes having magnitudes less than 5.5. The Ain Temouchent earthquake is the largest seismic event ever recorded in the region with a maximum observed intensity of VII (MSK scale).

3.6 The Zemmouri—Boumerdes Earthquake of 21 May 2003

The easternmost edge of the Mitidja Basin experienced a strong earthquake of M_w 6.8 on 21 May 2003, at 18 h 44 min (TU), located to the east of the city of Algiers; this was the largest seismic event felt since that of 3 February 1716, which had I_o = X (Rothé 1950; Harbi et al. 2015) (Table 1). The Zemmouri—Boumerdes earthquake occurred along the complex thrust and fold system of the Tell Atlas in the northeastern continuation of the Blida Mountain front and the related Mitidja Quaternary Basin.

Bounif et al. (2004) relocated the mainshock hypocenter, using three major aftershocks, with results that show a shift of this hypocenter from offshore to the coastline at 8–10 km depth. This coastal epicentre suggests a rupture along a previously unidentified offshore fault. According to Ayadi et al. (2003), this fault is an offshore continuation of the south Mitidja fault system. This is also supported by field observations of coastal uplift marked by a continuous white band of some sort of deposit attributed to a period of uplift, which is visible in exposed cliff faces (Meghraoui et al. 2004). Geodetic measurements of the 50-km-long coast show an average uplift of approximately 0.55 m along the shoreline with a maximum of 0.75 m east of Boumerdes and a minimum close to 0 near Cap Djenet (Meghraoui et al. 2004).

According to the study on teleseismic waveform inversion performed by Delouis et al. (2004), the Zemmouri—Boumerdes event was associated with an SE-dipping thrust fault mechanism with a seismic moment of 2.86×10^{19} N m (M_w 6.9), implying a 50-km-long fault rupture that should appear on the sea bottom at 6–12 km offshore from the coast (Fig. 6). Santos et al. (2015) investigate the rupture process to shed light on the location and geometry of the seismic fault and slip distribution using methodology based on teleseismic data, uplift measurements and synthetic aperture radar data. The methodology used by Santos et al. (2015) processes the available seismic and geodetic data to evaluate the two complementary planes of the focal mechanism. The location and geometry of the model proposed by Santos et al. (2015) are in agreement with the aftershock relocation proposed by Ayadi et al. (2008), where one aftershock cloud corresponded to a fault that dipped to the SE.

The event was followed by a large sequence of aftershocks. More than 900 aftershocks that were relocated with TomoDD extend NE–SW along an approximately 60-km-long zone crossing the coastline and exhibit at least three seismic clouds and a well-defined SE-dipping main fault geometry (Ayadi et al. 2008). The distribution of seismic events presents a clear contrast between a dense SW zone and a NE zone with scattered aftershocks (Fig. 6).

The aftershock focal mechanisms were determined by inverted seismic broadband (Braunmiller and Bernardi 2005) and constructed using *P*-wave onsets (Ayadi et al. 2008; Ousadou et al. 2014). The majority of the solutions are reverse mechanisms, similar to the mainshock and two of the three largest aftershocks. However, at the eastern extremity of the seismic cloud, some events show reverse faulting on planes striking approximately N–S. Some strike–slip and normal faulting solutions are found, especially at the westernmost end of the aftershock cloud (Ousadou et al. 2014).

The aftershock focal mechanism database was inverted to calculate the stress tensor for the eastern part of the Mitidja basin, which is a compressive stress triaxial tensor with well-defined horizontal Sh_{max} and Sh_{min} in the N340° and N251° directions, respectively (Ousadou et al. 2014).

4 The Impact of Earthquakes in Algeria

Earthquakes are the expression of the dispersion of energy through seismic waves. The quantification of this seismic energy led seismologists to define different magnitudes. Several energy–magnitude empirical relations have been determined, the most suitable ones being those proposed by Gutenberg and Richter (1956), which express, in this study, the energy in joule from the magnitudes m_b, and M_s given by the formulas $\log E = 2.4m_b - 1.2$ and $\log E = 1.5M_s + 4.8$, respectively, as well as the relationship established by Kanamori (1977) giving the energy in joules from the moment magnitude M_w by the same formula that for M_s. However, to calculate the energy released from the historical earthquakes (1365–1900) we use Mokrane et al. (1994) empirical relation which is the magnitude versus intensity: $M_m = 0.97I_o - 2.24$ for $\text{IX} < I_o \leq \text{XI}$.

Based on the catalogue of Ayadi and Bezzeghoud (2015) and using the formulas above, we estimated the seismic energy released by the 23 earthquakes with $M > 5.5$ that occurred in Algeria during the period between 1900 and 2015 (Fig. 1). The results show that 69% of the total energy (Fig. 7) comes from the El Asnam earthquake of 10 October 1980 (indicated by a red star in Fig. 4), if we consider that its magnitude was M_s 7.3. The El Asnam earthquake is the seismic event that dominates all the seismicity of the region for this period (1900–2015). A more recent earthquake that occurred in the Zemmouri—Boumerdes region on 21 May 2003 with a magnitude M_w 6.8 (Fig. 6) represents only 12% of the total energy (Fig. 7) released by the earthquakes occurring in the same period (1900–2015). Meanwhile, the other five recent earthquakes discussed in the previous sections released very low amounts of energy (1%, Fig. 7) when compared with the total seismic energy released by the El Asnam earthquake of 10 October 1980.

However, for the total period (1365–2015) listed in the catalogue of Ayadi and Bezzeghoud (2015), the seismic energy released by the El Asnam earthquake of 10 October 1980 and the Zemmouri—Boumerdes earthquake of 21 May 2003, at 8 and 1%, respectively, is much less significant when compared to that released (88%) (Fig. 8) by the six major historical earthquakes (2 January 1365; 9 October 1790; March 1819; 2 March 1825; 2 January 1867 and 1 January 1891) (Fig. 1). The other seven recent earthquakes discussed in the previous sections released an insignificant amount of energy (<1%, Fig. 8) when compared with the total seismic energy released by all the seismic events recorded in the period 1365–2015.

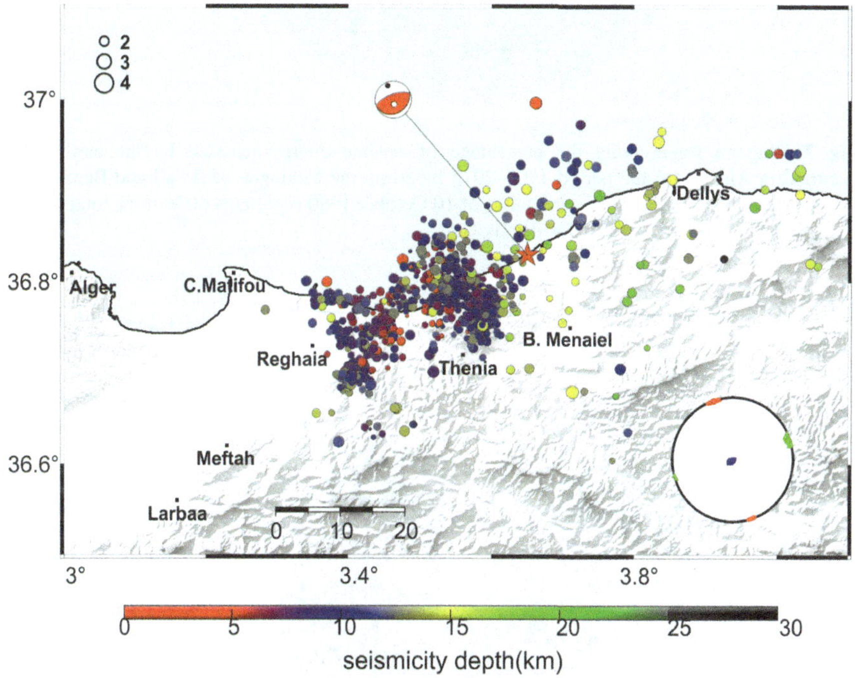

Fig. 6 The focal mechanism (Braunmiller 2005) and the corresponding aftershock sequence (relocated by Ayadi et al. 2008) following the mainshock of the Zemmouri—Boumerdes earthquake of 21 May 2003. Red star: mainshock after relocation (Bounif et al. 2004). The regional stress tensor and principal stress (Ousadou et al. 2014) are shown in the lower right corner (Sh_{max} in red and Sh_{min} in blue colours). The scale bar units are in km

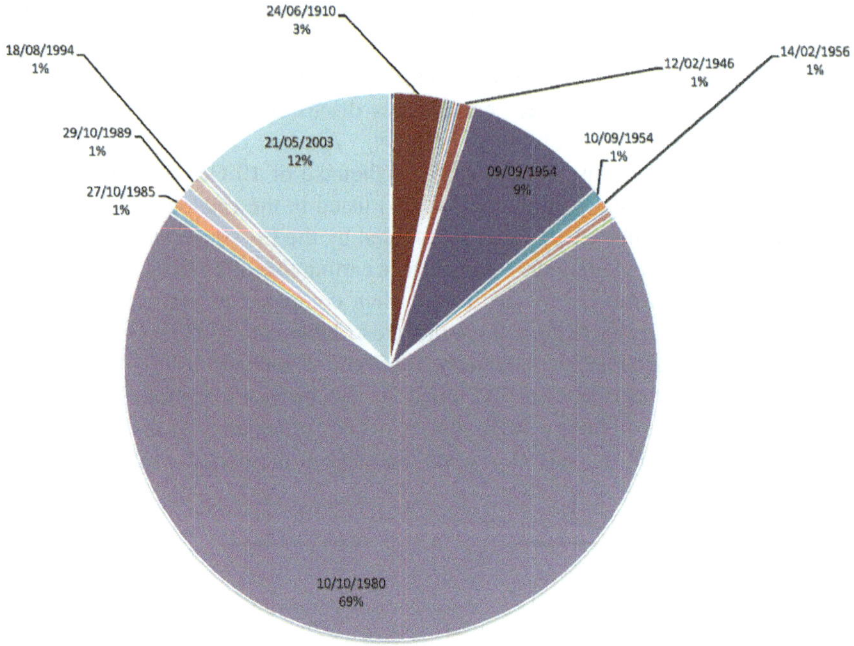

Fig. 7 Diagram representing the percentage of seismic energy released by the earthquakes occurred in Algeria for the period 1900–2015 based on the catalogue of Ayadi and Bezzeghoud (2015). Note that the El Asnam earthquake of 10 October 1980 represents 69% of the total seismic energy for this period. See text for details

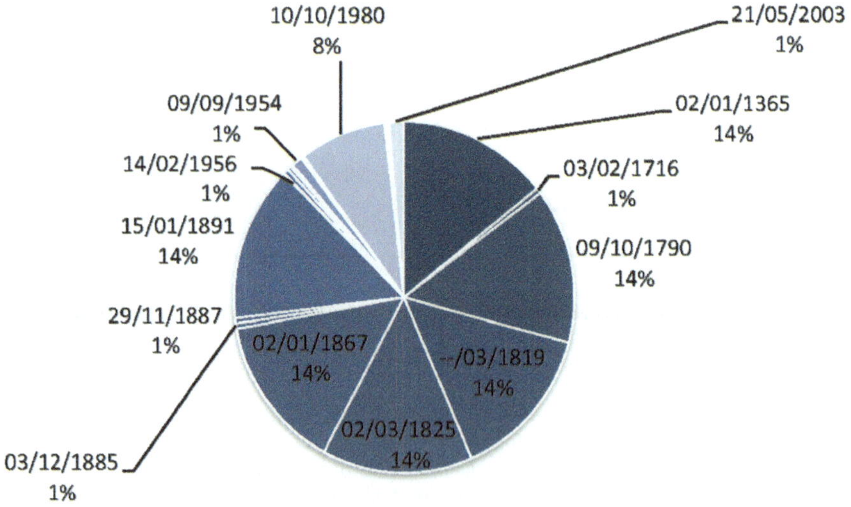

Fig. 8 Diagram representing the percentage of seismic energy released by the earthquakes occurring in Algeria for the period 1365–2015 based on the catalogue of Ayadi and Bezzeghoud (2015). Note that the six major historical earthquakes (02/01/1365; 09/10/1790; 03/1819; 02/03/1825; 02/01/1867 and 15/01/1891) (see Fig. 1) represent 84% of the total seismic energy for this period. See text for details

5 Discussion and Conclusion

Generally, northern Africa has experienced moderate to strong earthquakes. However, between 1920 and the present, three earthquakes with magnitudes of approximately 8.0 occurred in the oceanic region (M_w 8.2, 25 November 1941; M_s 8.0, 28 February 1969; M_w 7.9, 26 May 1975), and four earthquakes with magnitudes of approximately 7.0 (M_w 7.1, 8 May 1939, Santa María Island; M_w 7.1, 1 January 1980, Azores; M_s 7.1, 20 May 1931, Azores–Gibraltar fracture zone; M_w 7.3, 10 October 1980, El Asnam, Algeria) occurred along the western part of the Eurasian–Nubian plate boundary (Bezzeghoud et al. 2014). In general, the major earthquakes ($M \geq 7.0$) have occurred within the oceanic region, except for the El Asnam (Algeria) earthquake. The deadliest earthquakes that occurred in the study area since 1920, measured in numbers of victims, were those recorded in El Asnam, north Algeria (1954 and 1980), Azores, Portugal (1980 and 1998), Al Hoceima, north Morocco (1994 and 2004) and Zemmouri—Boumerdes, north Algeria (2003) (Benouar 1994; Bezzeghoud et al. 1996, 2014; Bezzeghoud and Buforn 1999; Ayadi et al. 2003).

The Ibero-Maghrebian region is vulnerable due to the shallow character of its seismicity generated by the convergence of the Eurasian and Nubian tectonic plates, to the specific mechanical properties of the crust and to local site effects (Bezzeghoud and Buforn 1999; Mourabit et al. 2014). To lead to a better assessment of the hazards and risks in and around urban areas, we need to know the detailed behaviour of the fault rupture processes that cause the earthquakes. Next, to produce reliable seismic hazard maps and mitigate the destructive impact of the earthquakes, the regional seismic hazards must be assessed using different approaches (e.g. deterministic and probabilistic) based on available historical and instrumental seismicity data (Ayadi and Bezzeghoud 2015; Harbi et al. 2015), earthquake sources and seismotectonic zoning studies and structural models associated with attenuation laws (Mourabit et al. 2014).

In summary, Algeria presents a seismotectonically complex area in which earthquakes responsible for high social and economic impacts have occurred and will continue to occur in the future. An Earthquake Early Warning System (EEWS) for this zone could be very useful in mitigating the damage caused by large earthquakes (Buforn et al. 2015). The complexity of the Ibero-Maghrebian region results in the need to address in depth the problem of the design of the early warning network to develop an appropriate EEWS. One of the issues that is particularly important is whether a single alert system could effectively cover the entire region or whether more than one system would be required instead, which takes into consideration the different seismic behaviours found in the areas under study. A preliminary feasibility study for an EEWS in the Gulf of Cadiz was performed as part of the ALERT-ES project coordinated by the Complutense University of Madrid with the participation of the Royal Institute and Naval Observatory of San Fernando, Cádiz, the Geological Institute of Catalonia in Barcelona and the

University of Évora. The extension of this system to the entire Ibero-Maghrebian region is now in progress with the ALERTE-RIM project, which began in 2015 with the same participants.

Data and resources

Data are available from the literature listed in the references section and from (1) the GCMT, Global Centroid Moment Tensors: http://www.globalcmt.org; (2) Harvard CMT (2005) Centroid Moment Tensor Project: http://www.globalcmt.org/CMTsearch.html; (3) IAG, Instituto Andaluz de Geofisica: http://www.iagpds.ugr.es; (4) IGN (2004), Instituto Geográfico Nacional, Spanish seismic network: http://www.ign.es/ign/en/IGN/home.jsp; (5) ISC (2008), International Seismological Centre: http://www.isc.ac.uk, Internat. Seis. Cent., Thatcham, United Kingdom; (6) NEIC, National Earthquake Information Center: http://neic.usgs.gov; (7) RCMT, Regional Centroid Moment Tensors: http://www.bo.ingv.it/RCMT; and (8) ZUR_RMT, Swiss Seismological Service (SED): http://www.seismo.ethz.ch/en/. Topography and bathymetry data have been extracted from the GEBCO (General Bathymetric Chart of the Oceans: http://www.gebco.net).

Acknowledgements This study was conducted within the scope of the MEDYNA FP7-PEOPLE-2013-IRSES project, WP-1: Present-day Kinematics and seismic hazards, funded by the Seventh Framework European Programme. The authors acknowledge the support provided by the Instituto de Ciências da Terra da Universidade de Évora (Portugal) and the Centre de Recherche en Astronomie, Astrophysique et Géophysique (Algiers, Algeria), under contract with the Portuguese Science and Technology Foundation (FCT, Portugal), PEst-OE/CTE/UI0078/2011. We would like to thank the Editors, Prof. Abderrahmane Bendaoud and Prof. Mohamed Hamoudi (Algiers University, USTHB), for their help.

References

Abbes K, Dorbath L, Dorbath C, Djeddi M, Ousadou F, Maouche S, Benkaci N, Slimani A, Larbes L, Bouziane D (2016) The Beni Haoua, Algeria, Mw 4.9 earthquake: source parameters, engineering, and seismotectonic implications. J Seismolog 20(2):655–667. https://doi.org/10.1007/s10950-015-9549-1

Ayadi A, Bezzeghoud M (2015) Seismicity of Algeria from 1365 to 2013: maximum observed intensity map (MOI2014). Seismol Res Lett 86(1). https://doi.org/10.1785/0220140075

Ayadi A, Ousadou-Ayadi F, Bourouis S, Benhallou H (2002) Seismotectonic and seismic quietness of the Oranie region (western Algeria); the Mascara earthquake of August 18th 1994, Mw = 5.7, Ms = 6.0. J Seismolog 6(1):13–23

Ayadi A, Maouche S, Harbi A, Meghraoui M, Beldjoudi H, Oussadou F, Mahsas A, Benouar D, Heddar A, Rouchiche Y, Kherroubi A, Frogneux M, Lammali K, BEnhamouda F, Sebai A, Bourouis S, Alasset PJ, Aoudia A, Cakir Z, Merahi M, Nouar O, Yelles A, Bellik A, Briole P, Charade O, Thouvenot F, Semane F, Ferkoul A, Deramchi A, Haned SA (2003) Strong Algerian earthquake strikes near capital city. Eos Trans Am Geophys Union 84:561–568

Ayadi A, Dorbath C, Ousadou F, Maouche S, Chikh M, Bounif MA, Meghraoui M (2008) Zemmouri earthquake rupture zone (Mw 6.8, Algeria): aftershocks sequence relocation and 3D velocity model. J Geophys Res Solid Earth 113(B9):B09301. https://doi.org/10.1029/2007JB005257

Bahrouni, Soumaya, Ben Ayed N, Bouaziz S, Attafik K, Rebai (2013) Investigation of historical seismicity in Tunisia from the records of building damage. In: The large Mediterranean earthquakes from past to present-international symposium, Algiers, 21–23 May 2013, Abstracts vol

Bahrouni N, Bouaziz S, Soumaya A, Ben Ayed N, Attafi K, El Ghali A, Rebai N (2014) Neotectonic and seismotectonics investigation of seismically active regions in Tunisia: a multidisciplinary approach. J Seismolog 18:235–256. https://doi.org/10.1007/s10950-013-9395-y

Ben Ayed N, Zargouni F (1990) Carte sismotectonique de la Tunisie à l'échelle 1/1000 000. Fondation Nationale de la Recherche Scientifique, Tunisia

Benhallou (1985) Les catastrophes séismiques de la région d'Echllef dans le contexte de la séismicité Historique de l'Algérie. Thèse de Doctorat Es-Sciences, IST-USTHB, Alger, Algérie

Benouar D (1994) Materials for investigation of the seismicity of Algeria and adjacent regions during the twentieth century. Ann Geophys 37(4)

Benfedda A, Abbes K, Bouziane D, Bouhadad Y, Slimani A, Larbes S, Haddouche D, Bezzeghoud M (2017) The August 1st, 2014 (Mw 5.3) Moderate earthquake: evidence for an active thrust fault in the bay of Algiers (Algeria), Pure Appl Geophys 174(3):1503–1511. https://doi.org/10.1007/s00024-017-1481-6

Bezzeghoud M, Buforn E (1999) Source parameters of the 1992 Melilla (Spain, Mw = 4.8), 1994 Al Hoceima (Morocco, Mw = 5.8) and Mascara (Algeria, Mw = 5.7) earthquakes and seismotectonic implications. Bull Seismol Soc Am 89(2):359–372

Bezzeghoud M, Dimitrov D, Ruegg JC, Lammali K (1995) Faulting mechanism of the El Asnam (Algeria) 1954 and 1980 earthquake from modeling of vertical movements. Tectonophysics 249:249–266

Bezzeghoud M, Ayadi A, Sébaï A, Aït Messaoud M, Mokrane A, Benhallou H (1996) Seismicity of Algeria between 1365 and 1989: map of maximum observed intensities (MOI). Av Geofís Geod 1:107–114

Bezzeghoud M, Borges JF, Caldeira B, Buforn E, Udias A (2008) Seismic activity in the Azores region in the context of the western part of the Eurasia–Nubia plate boundary. In: International seminar on seismic risk and rehabilitation on the 10th anniversary of the July 9 1998 Azores earthquake, Horta-Azores, 9–13 July, pp 27–31

Bezzeghoud M, Borges JF, Caldeira B (2012) Fontes sísmicas ao longo da fronteira de placas tectónicas entre os Açores e a Argélia: um modelo sismotectónico. In: Dias R, Araújo A, Terrinha P, Kullberg JC (eds) Geologia de Portugal, vol 2. Escolar Editora, pp 747–790. ISBN: 978-972-592-364-1

Bezzeghoud M, Adam C, Buforn E, Borges JF, Caldeira B (2014) Seismicity along the Azores-Gibraltar region and global plate kinematics. J Seismolog 18:205–220. https://doi.org/10.1007/s10950-013-9416-x

Borges JF, Fitas AJS, Bezzeghoud M, Teves-Costa P (2001) Seismotectonics of Portugal and its adjacent Atlantic area. Tectonophysics 337:373–387

Borges JF, Bezzeghoud M, Buforn E, Pro C, Fitas AJS (2007) The 1980, 1997 and 1998 Azores earthquakes and its seismotectonic implications. Tectonophysics 435:37–54

Borges JF, Caldeira B, Bezzeghoud M, Buforn E (2008) Seismicity and seismotectonics of Azores: geodynamic implications. In: Oliveira CS, Costa A, Nunes JC (eds) A book on the 1998 Azores earthquake—10 years after its occurrence, pp 99–110. ISBN 978.989-20-1223-0, 741 p

Bouaziz S, Barrier E, Soussi M, Turki MM, Zouari H (2002) Tectonic evolution of the northern African margin in Tunisia from paleostress data and sedimentary record. Tectonophysics 357:227–253

Bouhadad Y, Nour A, Laouami N, Behai D (2003) The Tachaouaft-Beni-ourtilane fault and seismotectonic aspects of the Babors region (eastern Algeria). J Seismolog 7:79–88. https://doi.org/10.1023/A:1021259931865

Bouhadad Y, Benhamouche A, Bourenane H, Ait ouali A, Chikh M, Guessoum N (2010) The Laalam (Algeria) damaging landslide triggered by a moderate earthquake (Mw = 5.2). J Nat Hazards 54:261–272

Bounif MA, Dorbath C (1998) Three-dimensional velocity structure and relocated aftershocks for the 1985 Constantine, Algeria (Ms = 5.9) earthquake. Ann Geophys 41(1):93–104

Bounif MA, Haessler H, Meghraoui M (1987) The Constantine (Northeast Algeria) earthquake of October 27, 1985; surface ruptures and aftershock study. Earth Planet Sci Lett 85(4):451–460. https://doi.org/10.1016/0012-821X(87)90140-3

Bounif A, Bezzeghoud M, Dorbath L, Legrand D, Deschamps A, Rivera L, Benhallou H (2003) Seismic source study of the 1989, October 29, Chenoua (Algeria) earthquake from aftershocks, broad-band and strong motion records. Ann Geophys 46(4):625–646

Bounif A, Dorbath C, Ayadi A, Meghraoui M, Beldjoudi H, Laouami N, Frogneux M, Slimani A, Alasset PJ, Kharroubi A, Ousadou F, Chikh M, Harbi A, Larbes S, Maouche S (2004) The 21 May Zemmouri (Algeria) earthquake Mw 6.8: relocation and aftershock sequence analysis. Geophys Res Lett 31:L19606. https://doi.org/10.1029/2004gl020586

Braunmiller J, Bernardi F (2005) The 2003 Boumerdes, Algeria earthquake: regional moment tensor analysis. Geophys Res Lett 32(6):L06305. https://doi.org/10.1029/2004GL022038

Buforn E (2008) Seismotectonics of Azores-Tunisia. In: Mendes-Victor L, Sousa Oliveira C, Azevedo J, Ribeiro A (eds) The 1755 Lisbon earthquake revisited, pp 397–410

Buforn E, Udías A, Madariaga R (1991) Intermediate and deep earthquakes in Spain. Pure Appl Geophys 136:375–393

Buforn E, Bezzeghoud M, Udias A, Pro C (2004) Seismic sources on the Iberia-African plate boundary and their tectonic implications. Pure Appl Geophys 161(3):623–646

Buforn E, Udías A, Pro C (2015) Large earthquakes at the Ibero-Maghrebian region. Basis for an EEWS. Pure Appl Geophys 172(2015):2387–2396. https://doi.org/10.1007/s00024-014-0954-0

Buforn E, Pro C, de Galdeano CS, Cantavella JV, Cesca S, Caldeira B, Udías A, Mattesini M (2017) The 2016 south Alboran earthquake (Mw = 6.4): a reactivation of the Ibero-Maghrebian region? Tectonophysics 712:704–715. http://www.sciencedirect.com/science/article/pii/S0040195117302792

De Vicente G, Cloetingh S, Munoz-Martin A, Olaiz A, Stich D, Vegas R, Galindo-Zaldivar J, Fernandez-Lozano J (2008) Inversion of moment tensor focal mechanisms for active stresses around the micro-continent Iberia: tectonic implications. Tectonics 27:TC1009. https://doi.org/10.1029/2006tc002093

Delouis B, Vallée M, Meghraoui M, Calais E, Maouche S, Lammali K, Mahsas A, Briole P, Benhamouda F, Yelles K (2004) Slip distribution of the 2003 Boumerdes-Zemmouri earthquake, Algeria, from teleseismic, GPS, and coastal uplift data. Geophys Res Lett 31

Deschamps A, Gaudemer Y, Cisternas A (1982) The El Asnam, Algeria, earthquake of October 1980: multiple-source mechanism determined from long-period records. Bull Seismol Soc Am 72(4):1111–1128

Deschamps A, Bezzeghoud M, Bounif A (1991) Seismological study of the Constantine (Algeria) earthquake (27 October 1985). In: Mezcua J, Udias A (eds) Seismicity, seismotectonics and seismic risk of the Ibero-Maghrebian region, vol 8. Publ. IGN. Madrid, pp 163–173

Dewey (1990) The 1954 and 1980 Algerian earthquakes: implications for the characteristic-displacement model of fault behavior. Bull Seismol Soc Am 81(2):446–467

El Mrabet T (2005) The great earthquakes in the Maghreb region and their consequences on man and environment, (in Arabic with abstract in English). Edit Centre National de Recherche Scientifique et Technique, Rabat, Morocco

Espinoza AF, Lopez-Arroyo A (1984) Inconsistencies in the modified Mercalli intensities scales, El Asnam, Algeria, earthquake. In: Proc Conf Int Sur la microzonation sismique, Ech Cheliff, Algeria, 10–12 Oct 1984, pp 51–65

Gueddiche M, Harjono H, Ben Ayed N, Hfaiedh M, Diament M, Dubois J (1992) Analyse de la séismicité et mise en évidence d'accidents actifs dans le nord de la Tunisie. Bull Soc Geol Fr 163:415–425

Gutenberg B, Richter CF (1956) Earthquake magnitude, intensity, energy and acceleration. Bull Seismol Soc Am 46:105–145

Gutscher MAS, Dominguez GK, Westbrook P Le, Roy F Ros (2012) The Gibraltar subduction: a decade of new geophysical data. Tectonophysics 574–575(2012):72–91

Harbi A, Maouche S, Ayadi A (1999) Neotectonics and associate seismicity in the eastern Tellian Atlas of Algeria. J Seismolog 3(1):95–104

Harbi A, Peresan A, Panza GF (2010) Seismicity of eastern Algeria: a revised and extended earthquake catalogue. Nat Hazards 54(3):725–747. https://doi.org/10.1007/s11069-009-9497-6

Harbi A, Sebaï A, Benmedjber M, Ousadou F, Rouchiche R, Grigahcene A, Aïni D, Bourouis S, Maouche S, Ayadi A (2015) The Algerian homogenized macroseismic database (267–1989): a deeper insight into the Algerian historical seismicity. Seismol Res Lett 86:6. https://doi.org/10.1785/0220150092

Hatzfeld D (1978) Étude sismotectonique de la zone de collision Ibero-Maghrébine. Ph.D. thesis, Grenoble (France), 281 pp

Heezen BC, Ewing M (1955) Orleansville earthquake and turbidity currents. AAPG Bull 39(12): 2505–2514

Henares J, López Casado C, Sanz de Galdeano C, Delgado J (2003) Stress fields in the Iberian-Maghrebi region. J Seismolog 7(1):65–78. https://doi.org/10.1023/A:1021294015027

Hfaiedh M, Ben Ayed N, Dorel J (1985) Etude néotectonique et sismotectonique de la Tunisie nord-orientale. Note Serv Géol Tunisie n 16, pp 41–56

Houseman G (1996) From mountains to basins. Nature 379:771–772

Kanamori H (1977) The energy release in great earthquakes. J Geophys Res 82(20):2981–2987. https://doi.org/10.1029/JB082i020p02981

King GCP, Vita-Finzi C (1981) Active folding in the Algerian earthquake of 10 October 1980. Nature 292(5818):22–26

Maouche S, Meghraoui M, Morhange C, Belabbes S, Bouhadad Y, Haddoum H (2011) Active coastal thrusting and folding, and uplift rate of the Sahel Anticline and Zemmouri earthquake area (Tell Atlas, Algeria). Tectonophysics 509:69–80. https://doi.org/10.1016/j.tecto.2011.06.003

Martinez Solares JM, Mezcua Rodriguez J (eds) (2002) Catalogo simico de la Peninsula Iberica (880 a.C.-1990). IGN, Madrid, Monografía Núm. 18, 254 pp

McKenzie D (1972) Active tectonics of the Mediterranean region. Geophys J Roy Astron Soc 30 (2):109–185

Meghraoui M (1988) Géologie des zones sismiques du Nord de l'Algérie: Paléosismologie, tectonique active et synthèse sismotectonique. Thèse de doctorat Université Pierre et Marie Curie, France, p 356

Meghraoui M, Maouche S, Chemaa B, Cakir Z, Aoudia A, Harbi A (2004) Coastal uplift and thrust faulting associated with the M (sub w) = 6.8 Zemmouri (Algeria) earthquake of 21 May 2003. Geophys Res Lett 31:L19605. https://doi.org/10.1029/2004GL020466

Mokrane A, Aït Messaoud A, Sebaï A, Menia A, Ayadi A, Bezzeghoud M (1994) Les séismes en Algérie de 1365 à 1992. In: Bezzeghoud M, Benhallou H (eds). Publication du Centre de Recherche en Astronomie Astrophysique et Géophysique, Alger-Bouzaréah, 277 p

Mourabit T, Abou Elenean KM, Ayadi A, Benouar D, Ben Suleman A, Bezzeghoud M, Cheddadi A, Chourak M, ElGabry MN, Harbi A, Hfaiedh M, Hussein HM, Kacem J, Ksentini A, Jabour N, Magrin A, Maouche S, Meghraoui M, Ousadou F, Panza GF, Peresan A, Romdhane N, Vaccari F, Zuccolo E (2014) Neo-deterministic seismic hazard assessment in North-Africa. J Seismolog 18(2):301–318. http://dx.doi.org/10.1007/s10950-013-9375-2

Ousadou F, Dorbath L, Dorbath C, Bounif MA, Benhallou H (2013) The Constantine (Algeria) seismic sequence of 27 October 1985: a new rupture model from aftershock relocation, focal mechanisms, and stress tensors. J Seismolog 17(2):207–222. https://doi.org/10.1007/s10950-012-9320-9

Ousadou F, Dorbath L, Ayadi A, Dorbath C, Gharbi S (2014) Stress field variations along the Maghreb region derived from inversion of major seismic crisis fault plane solutions. Tectonophysics 632(2014):261–280. https://doi.org/10.1016/j.tecto.2014.06.017

Ouyed M, Meghraoui M, Cisternas A, Deschamps A, Dorel J, Frechet J, Gaulon R, Hatzfeld D, Philip H (1981) Seismotectonics of the El Asnam earthquake. Nature 292(5818):26–31

Ouyed M, Yielding G, Hatzfeld D, King GCP (1983) An aftershock study of the El Asnam (Algeria) earthquake of 1980. Geophys J Int 73(3):605–639

Philip H, Meghraoui M (1983) Structural analysis and interpretation of the surface deformations of the El Asnam earthquake of October 10, 1980. Tectonics 2(1):17–49. https://doi.org/10.1029/tc002i001.P00017

Rothé, JP (1950) Les séismes de Kherrata et la sismicité de l'Algérie, Bulletin du Service de la Carte Géologique de l'Algérie, 4e série Géophysique, 3, Mende, France, 40 pp

Rothé JP (1955) Le tremblement de terre d'Orléansville et la sismicité de l'Algérie. Nature (Paris) 3237:1–9

Rothé JP, Lepervier C, Truillet R (1977) Les déformations liées au séisme de 1954 d'El Asnam (ex. Orléansville-Algérie). Bull Soc Geol Fr XIX(3):641–644

Roussel J (1973) Les zones actives et la fréquence des séismes en Algérie (1716–1970). Bull Soc Hist Nat Afr Nord Alger 64(3 and 4):211

Santos R., Caldeira B, Bezzeghoud M, Borges JF (2014) The rupture process and location of the 2003 Zemmouri–Bou- merdes earthquake (Mw 6.8) inferred from seismic and geodetic data. Pure Appl Geophys 172(9):2421–2434

Santos R, Caldeira B, Bezzeghoud M, Borges JF (2015) The rupture process and location of the 2003 Zemmouri–Boumerdes earthquake (Mw 6.8) inferred from seismic and geodetic data. Pure Appl Geophys 172(9):2421–2434. Published online: 23 Nov 2014. http://dx.doi.org/10.1007/s00024-014-0978-5

Shirokova EJ (1967) General features in the orientation of principal stresses in earthquake foci in the Mediterranean Asian seismic belt. Izv Earth Phys 1:12–22

Soumaya A, Ben Ayed N, Delvaux D, Ghanmi M (2015) Spatial variation of present-day stress field and tectonic regime in Tunisia and surroundings from formal inversion of focal mechanisms: geodynamic implications for central Mediterranean. Tectonics 33. https://doi.org/10.1002/2015tc003895

Stich D, Ammon CJ, Morales J (2003) Moment tensor solutions for small and moderate earthquakes in the Ibero-Maghreb region. Geophys Res Let 108(B3):2148, ESE 7-1-7-20

Thio HK, Song X, Saikia CK, Helberger DV, Woods BB (1999) Seismic source and structures estimation in the western Mediterranean using a sparse broadband network. J Geophys Res 104(B1):845–861

Thomas G (1985) Géodynamique d'un bassin intramontagneux : Le bassin du bas Cheliff occidental (Algérie) durant le Mio-Plio-Quaternaire. Thèse ès Sciences, Univ. de Pau, p 594

Van der Woerd J, Dorbath C, Ousadou F, Dorbath L, Delouis B, Jacques E, Tapponnier P, Hahou Y, Menzhi M, Frogneux M, Haessler H (2014) The Al Hoceima Mw 6.4 earthquake of 24 February 2004 and its aftershocks sequence. J Geodyn 77:89–109. https://doi.org/10.1016/j.jog.2013.12.004

Wessel P, Smith WHF (1991) Free software helps map and display data. EOS Trans AGU 72:441

Yelles-Chaouche AK, Djellit H, Beldjoudi H, Bezzeghoud M, Buforn E (2004) The Ain Temouchent (Algeria) earthquake of December 22nd, 1999. Pure Appl Geophys 161(3):607–621

Yielding G, Jackson JA, King GCP, Sinvhal H, Vita-Finzi C, Wood RM (1981) Relations between surface deformation, fault geometry, seismicity, and rupture characteristics during the El Asnam (Algeria) earthquake of 10 October 1980. Earth Planet Sci Lett 56:287–304

Yielding G, Ouyed M, King GCP, Hatzfeld D (1989) Active tectonics of the Algerian Atlas mountains—evidence from aftershocks of the 1980 El Asnam earthquake. Geophys J Int 99(3):761–788

An Overview on 40 Years of Remote Sensing Geology Based on Arab Examples

Jean-Paul Deroin

Abstract Remote sensing techniques are particularly suitable for geological studies in arid regions such as the Arab world. Here, we present an overview of the main applications of satellite sensors to geology for geological mapping, tectonic and structural geology, hydrogeology, mining geology, geoarchaeology, oil and gas prospection, earthquake and seismicity, landslides and coastal erosion. The main characteristics of the sensors are discussed. In addition, three examples of modern remote sensing imagery show geology in Arab countries: The Dead Sea Fault in the Middle East as seen with the large-scale ALOS PALSAR (ScanSAR mode) image, the Bou Azzer Anticline (Morocco) viewed by multispectral Landsat TM and the evolution of the coastal front in Nouakchott (Mauritania), monitored with multi-temporal Landsat data.

Keywords Satellite imagery · Multispectral · Radar · Arab countries
Geosciences · Desert

1 Introduction

Arab countries extend over 8000 km from the Arabian Sea to the East, to the Atlantic Ocean to the West. They include 20 different states: Oman, the United Arab Emirates, Qatar, Bahrain, Yemen, Saudi Arabia, Kuwait, Iraq, Syria, Jordan, Lebanon, Egypt, Sudan, Djibouti, Somalia, Libya, Tunisia, Algeria, Morocco and Mauritania. This large area can be divided into a hot-summer Mediterranean climate (Csa in the Köppen climate classification) to the North, and the largely dominant hot desert climate (Bwh) to the South, corresponding to arid inhospitable deserts (Sahara, Rub Al Khali). This climate setting is particularly suitable for the use of remote sensing techniques in geological mapping, because landforms are remarkably expressed, and often devoid

J.-P. Deroin (✉)
Université de Reims Champagne-Ardenne, Faculty of Science,
GEGENAA EA 3795 and SFR Condorcet FR CNRS 3417, Reims, France
e-mail: jean-paul.deroin@univ-reims.fr

A. Bendaoud et al. (eds.), *The Geology of the Arab World—An Overview*,
Springer Geology, https://doi.org/10.1007/978-3-319-96794-3_12

of vegetation. The area comprises ancient or recent mountain belts (Jabal Akhdar in Oman, Hedjaz in Saudi Arabia, Mount Lebanon, the East African Orogen, Hoggar in southern Algeria, the Atlas Mountains in northern Africa, etc.) and many sedimentary basins often exploited for oil and gas (Central Arabia, Upper Egypt, Murzuq and Ghadames in Libya, Reggane in Algeria, etc.). The present paper aims to document the main trends in remote sensing applied to the geological knowledge of the Arab countries. The main scientific domains considered are: geological mapping, tectonic and structural geology, hydrogeology, mining geology, geoarchaeology, oil and gas prospection, earthquake and seismicity, landslides and coastal erosion.

Bibliometric analysis provides statistical information on scientific activity (e.g. Agapiou and Lysandrou 2015). The Scopus and ScienceDirect databases were used to extract some general information on remote sensing geology in the cited countries (Table 1).

Scopus is one of the largest abstract and citation databases of peer-reviewed literature. All references per country are indicated first (column B), showing that Egypt and Saudi Arabia are the more cited Arab countries. Then, 'country' and 'geology' (column C) were crossed showing that Egypt, Morocco, Saudi Arabia and Algeria provide 46% of all the references (1663/3622). Basically, only 0.6–3.2% (mean of 1.2%) of the total references are related to geology, the maximum rate being observed for Oman, Libya and Mauritania (column D). 'Remote sensing' has been added to the previous query string, showing that 0–19% (with a mean of 7%) of the citations related to geology deal also with remote sensing, particularly in Djibouti, Sudan, Egypt and Mauritania (column F). However, Egypt, Jordan and Sudan provide the greater amount of references in remote sensing geology with 69, 21 and 20, respectively (column E). Note that this raw analysis provides only orders of magnitude.

ScienceDirect is a full-text scientific database provided by Elsevier. The number of references per country is necessarily lower than the Scopus number (column G). We used this database engine to discriminate between structural geology (with the word 'tectonic'), sedimentological and lithological geology (with the word 'sediment') and mapping as a whole (with the word 'mapping'), using only these terms in the title, abstract or keywords of the paper. The results provide the following information. First, tectonics (with a mean of 3.5% of the total references) is more frequently studied in Oman, Djibouti, Morocco and Libya (column H). Sedimentology (with a mean of 5.2% of the total references) is more frequently studied in Mauritania, Oman, Libya and Somalia (column K). Quantitatively speaking, Egypt provides the main set of publications with 322 references in tectonics (second is Morocco with 274) and 527 references in sedimentology (second is also Morocco with 294). The 'remote sensing' term indicates the number of total publications dealing with remote sensing in the country (column N). With only 0.9% in average of the publications dealing with remote sensing, it was hard to decipher between remote sensing applied to geology and to other fields (urban planning, agriculture, etc.). Egypt, Morocco and Saudi Arabia are the three first providers of references in this field with 129, 37 and 35 references, respectively.

This paper proposes an overview of the different methodological aspects leading to the improvement of the geological knowledge from remote sensing approaches,

Table 1 Bibliometric analysis of the Scopus and ScienceDirect databases (March 2017)

A	B	C	D	E	F	G	H	I	J	K	L	M	N	O
		Scopus (number of references)					ScienceDirect (number of references)							
	'Country'	'Country' + 'geology'	%C/B	'Country' + 'geology' + 'remote sensing'	%E/C	'Country'	'Country' + "Tectonic'	%H/G	'Country' + 'sediment'	%J/G	'Country' + 'mapping'	%L/G	'Country' + 'remote sensing'	%N/G
Algeria	15,496	253	1.6	17	6.7	2409	108	4.5	93	3.9	86	3.6	25	1.0
Bahrain	3620	21	0.6	0	0.0	503	1	0.2	17	3.4	4	0.8	1	0.2
Djibouti	838	21	2.5	4	19.0	164	13	7.9	3	1.8	14	8.5	0	0.0
Egypt	50,909	667	1.3	69	10.3	8100	322	4.0	527	6.5	362	4.5	129	1.6
Iraq	25,904	170	0.7	14	8.2	2076	26	1.3	58	2.8	45	2.2	11	0.5
Jordan	28,036	214	0.8	21	9.8	5412	92	1.7	177	3.3	323	6.0	18	0.3
Kuwait	11,871	115	1.0	3	2.6	1844	10	0.5	100	5.4	33	1.8	8	0.4
Lebanon	11,015	92	0.8	9	9.8	1401	37	2.6	33	2.4	51	3.6	10	0.7
Libya	6492	175	2.7	13	7.4	875	55	6.3	74	8.5	40	4.6	15	1.7
Mauritania	2221	60	2.7	6	10.0	397	19	4.8	86	21.7	23	5.8	9	2.3
Morocco	21,312	387	1.8	16	4.1	3779	274	7.3	294	7.8	133	3.5	37	1.0
Oman	10,501	332	3.2	9	2.7	1710	140	8.2	216	12.6	92	5.4	17	1.0
Qatar	6406	68	1.1	3	4.4	789	4	0.5	23	2.9	19	2.4	3	0.4
Saudi Arabia	35,341	356	1.0	16	4.5	5007	84	1.7	133	2.7	105	2.1	35	0.7
Somalia	4945	46	0.9	1	2.2	459	25	5.4	35	7.6	33	7.2	9	2.0
Sudan	19,228	153	0.8	20	13.1	2889	70	2.4	95	3.3	71	2.5	31	1.1
Syria	9305	103	1.1	8	7.8	1078	51	4.7	40	3.7	56	5.2	12	1.1
Tunisia	20,297	222	1.1	11	5.0	3537	174	4.9	200	5.7	113	3.2	17	0.5
U.A.E.	8327	82	1.0	8	9.8	1118	21	1.9	49	4.4	31	2.8	5	0.4
Yemen	5295	85	1.6	7	8.2	662	32	4.8	43	6.5	30	4.5	9	1.4
Total	297,359	3622		255		44209	1558		2296		1664		401	
Mean			1.2		7.0			3.5		5.2		3.8		0.9

including applied and fundamental aspects, from our own experience but also from the literature. Only a few examples among the very large literature are cited. They are chosen for illustrating the main trends. Table 2 is a compilation of the main satellite systems used in geological remote sensing with an indication of the main characteristics of their onboard sensors.

2 Remote Sensing and Geology

2.1 Geological Mapping

Geological mapping chapter comprises remote sensing studies applied to the synthetic geological mapping, to lithological and petrological studies in sedimentary basins, metamorphic belts or in granitic and volcanic areas. It is also concerned by environmental studies particularly in changing environment such as the arid regions. For this purpose, large-scale data (optical or radar) are particularly suitable because each scene covers thousands of square kilometres. If the panchromatic mode (for examples, Landsat ETM+, SPOT-P or IRS) is interesting because of the better ground space resolution, the multispectral range provides specific information related to the spectral signature of minerals, particularly in the mid and shortwave infrared. False colour composites, band rationing, and principal component analysis are the most classical methods used for visualizing geological features. Image processing is frequently associated with laboratory or field spectral reflectance measurements. Radar remote sensing is also interesting, particularly when some configurations are available: low-frequency, high incidence, cross-polarization, etc.

The most popular remote sensing data are provided by Landsat. As early as the 1970s, Landsat MSS allows the first multispectral observations to be done. For example in the Qattara depression, Egypt, El Shazly et al. (1978) have shown that the depression is actually constituted of several small depressions. Jacobberger et al. (1983) investigated the Meatiq Dome, a granitic area in the central Eastern Desert of Egypt, mapping thoroughly contacts between the dome's granite and granite gneiss core. The ophiolitic sequence and the peripheral mafic mylonites located outside the dome were also distinguished. The same area was reinvestigated by the more accurate Landsat TM data (Sultan et al. 1987). These latter demonstrated that increasing amounts of magnetite decrease the ratio of TM band 5 to band 1, increasing amounts of hydroxyl-bearing minerals increase the ratio of TM band 5 to band 7 and increasing amounts of Fe-bearing aluminosilicates increase the product of the ratios: band 5 to band 4 and band 3 to band 4. A synthetic mapping based on Landsat MSS imagery provided the first geologic map of the whole Sinai peninsula and vicinities (Henry and Chorowicz 1986). This study was later on completed by field observation and led to new interpretations (Chorowicz and Lyberis 1989). Qari (1992) used Landsat TM data to map the Jabal Duhayyah area, Saudi Arabia, and particularly the contact between two tectonic zones of the southern Arabian Shield. Alwash and Zilger (1994) used the same kind of dataset to map the Precambrian

Table 2 Main remote sensing satellite systems with their sensors used in geological remote sensing

Vector (satellite)	Sensor	Agency	Swath (km)	Start	End	Alt. (km)	ECT DN	Range 1	Range 2	Range 3	Range 4	Range 5	Range 6	Range 7	Range 8	Range 9	Range 10
								Pan	Coastal	blue	green	red	Near IR	Mid IR	SWIR	Thermal IR	Radar
								0.50-0.90	0.40-0.42	0.40-0.50	0.50-0.60	0.60-0.70	0.90-1.40	1.55-1.75	2.0-2.4	3.0-12.0	
Alos 1		Jaxa		24-01-2006	12-05-2011	695	10:30										
	PRISM		35-70					Pan (2.5m)									
	AVNIR 2		70							B1 (10m)	B2 (10m)	B3 (10m)	B4 (10m)				
	PALSAR 1		30-350														L-band (FP)
Alos 2		Jaxa		24-05-2014	in operation	637	10:30										
	PALSAR 2		25-350														L-band (FP)
CartoSat 1 (IRS P5)		ISRO		05-05-2005	in operation	618	10:30										
	PAN-F		29.4					Pan (2.5m)									
	PAN-A		26.2					Pan (2.2m)									
CartoSat 2	PAN	ISRO	9.6	10-01-2007	in operation	635	09:30	Pan (1m)									
CartoSat 2A	PAN	ISRO	9.6	28-04-2008	in operation	635	09:30	Pan (1m)									
CartoSat 2B	PAN	ISRO	9.6	12-07-2010	in operation	637	09:30	Pan (1m)									
EnviSat		ESA		28-02-2002	09-05-2012	800	10:00										
	MERIS		1.150						B1 (300m)	(300m)	(300m)	(300m)	(300m)				
	ASAR		100-400														C-band AP/CP
EO 1		Nasa		21-11-2000	06-01-2017	705	10:00										
	ALI							Pan (10m)	B1' (30m)	B1 (30m)	B2 (30m)	B3 (30m)	B4,4',5' (30m)	B5 (30m)	B7 (30m)		
	Hyperion										Hyperspectral (220 bands, resolution 30m)						
EOS-TERRA		Nasa		18-12-1999	in operation	705	10:30										
	MODIS		2.330									B1 (250m)	B2 (250m)				
										B3 (500m)	B4 (500m)		B5 (500m)	B6 (500m)	B7 (500m)		
									B8 (1km)	B9-10 (1km)	B11-12 (1km)	B13-14 (1km)	B15-19 et 26 (1km)			B20-25, 27-36 (1km)	
	ASTER		60							B1 (15m)		B2 (15m)	B3 (15m)	B4 (30m)	(30m)	(90m)	
EOS-AQUA	MODIS	Nasa	2.330	04-05-2002	in operation	702	13:30										
ERS		ESA															
ERS 1				17-07-1991	10-03-2000	782											
ERS 2				21-04-1995	05-09-2011	780	10:30										
	AMI		100														C-band VV
FormoSat 2 (ROCSat 2)		NSPO		20-05-2004	22-08-2016	888	10:00										
	RSI		24					Pan (2m)		B1 (8m)	B2 (8m)	B3(8m)	B4 (8m)				
GeoEye 1		DG		06-09-2008	in operation	681	10:30										
	GIS		15.2					Pan (0.41m)		B1 (1.64m)	B2 (1.64m)	B3 (1.64m)	B4 (1.64m)				
GeoEye-2 (see WorldView 4)																	
Ikonos 2		DG		24-09-1999	31-03-2015	681	10:30										
	OSA		11.3					Pan (0.82m)		B1 (3.2m)	B2 (3.2m)	B3 (3.2m)	B4 (3.2m)				
IRS 1A		ISRO		17-03-1988	00/00/1992	904	10:26										
	Liss I		148							B1 (72.5m)	B2 (72.5m)	B3 (72.5m)	B4 (72.5m)				
	Liss II		148							B1 (36m)	B2 (36m)	B3 (36m)	B4 (36m)				
IRS 1B		ISRO		19-08-1991	00/12/1992	904	10:26										
	Liss I		148							B1 (72.5m)	B2 (72.5m)	B3 (72.5m)	B4 (72.5m)				
	Liss II		148							B1 (36m)	B2 (36m)	B3 (36m)	B4 (36m)				
IRS 1C		ISRO		28-12-1995	21-09-2007	817	10:30										
	Liss III		142								B1 (23.5m)	B2 (23.5m)	B3 (23.5m)	B4 (70m)			
	PAN		70					Pan (5.8m)									
	WiFS		804									B1 (188m)	B2 (188m)				
IRS 1D		ISRO		29-09-1997	00/01/2010	974	10:30										
	Liss III		142								B1 (23.5m)	B2 (23.5m)	B3 (23.5m)	B4 (70m)			
	PAN		70					Pan (5.8m)									
	WiFS		804									B1 (188m)	B2 (188m)				
IRS P1 (1E)				20-09-1993	20-09-1993	*mission failed*											
IRS P2				15-10-1994	00/09/1997	817											
	Liss II M		131							B1 (32m)	B2 (32m)	B3 (32m)	B4 (32m)				
IRS P3				21-03-1996	00/01/2006												
	WiFS		804									B1 (188m)	B2 (188m)	B3 (188m)			
	MOS A		195										B1 (1,500m)				
	MOS B		200					Pan (520m)									
	MOS C		192											B1 (550m)			
IRS P4 (OceanSat 1)				26-05-1999	08-08-2010												
	OCM		1.42					Pan (360m)									
IRS P5 (see CartoSat 1)																	
IRS P6 (ResourceSat 1)		ISRO		17-10-2003		817	10:30										
	Liss-IV		70								B1 (5.8m)	B2 (5.8m)	B3 (5.8m)				
	Liss-III		141								B1 (23.5m)	B2 (23.5m)	B3 (23.5m)	B4 (23.5m)			
	AWiFS		740								B1 (56m)	B2 (56m)	B3 (56m)	B4 (56m)			
Jers 1		Jaxa	75	11-02-1992	11-10-1998	568	10:30										
	OPS										B1 (18m)	B2 (18m)	B3,4 (18m)	B5 (18m)	B6,7,8 (18m)		
	SAR																L-band HH
Landsat 1 (ERTS)		Nasa	185	23-07-1972	06-01-1978	910	09:45 a.m										
	RBV				*05-08-1972*						B1 (80m)	B2 (80m)	B3 (80m)				
	MSS										B1 (80m)	B2 (80m)	B3,4 (80m)				
Landsat 2		Nasa	185	22-01-1975	27-07-1983	910	09:45 a.m										
	RBV										B1 (80m)	B2 (80m)	B3 (80m)				
	MSS										B1 (80m)	B2 (80m)	B3,4 (80m)				
Landsat 3		Nasa	185	05-03-1978	07-09-1983	910	09:45 a.m										
	MSS										B1 (80m)	B2 (80m)	B3,4 (80m)				
Landsat 4		Nasa	185	16-07-1982	14-12-1993	705	10:00										
	MSS										B1 (80m)	B2 (80m)	B3,4 (80m)				
	TM									B1 (30m)	B2 (30m)	B3 (30m)	B4 (30m)	B5 (30m)	B7 (30m)	B6 (120m)	
Landsat 5		Nasa	185	01-03-1984	15-01-2013	705	10:00										
	MSS				*00/08/1995*						B1 (80m)	B2 (80m)	B3,4 (80m)				
	TM									B1 (30m)	B2 (30m)	B3 (30m)	B4 (30m)	B5 (30m)	B7 (30m)	B6 (120m)	
Landsat 6		Nasa		05-10-1993	05-10-1993	*mission failed*											
Landsat 7		Nasa	185	15-04-1999	in operation	705	10:00										
	ETM+							B8 (15m)		B1 (30m)	B2 (30m)	B3 (30m)	B4 (30m)	B5 (30m)	B7 (30m)	B6 (60m)	
Landsat 8		Nasa	185	11-02-2013	in operation	705	10:00										
	OLI							B8 (15m)	B1 (30m)	B2 (30m)	B3 (30m)	B4 (30m)	B5 (30m)	B6,9 (30m)	B7 (30m)		
	TIRS															(100m)	
Meteosat Second Generation																	
	SEVIRI	Eumetsat						B12 (1.67km)				B1 (4.8km)	B2 (4.8km)	B3 (4.8km)		B4-11 (4.8km)	
Meteosat 8				28-08-2002	in operation	GEO at 3.5°E											
Meteosat 9				22-12-2005	in operation	GEO at 9.5°E											
Meteosat 10				05-07-2012	in operation	GEO at 0°											
Meteosat 11				15-07-2015	in operation	GEO at 3.4°W											

(continued)

Table 2 (continued)

NOAA		NOAA															
NOAA 15				13-05-1998	in operation	820	07:30										
NOAA 16				21-09-2000	05-06-2014	840	PM										
NOAA 17				24-06-2002	10-04-2013	825	10:00										
NOAA 18				20-05-2005	in operation	869	PM										
NOAA 19				16-02-2009	in operation	869	PM										
	AVHRR 3		2.900								B1 (VR) 1,100m		B2 (1,100m)	B3a (1,100m)		B3b,4,5 (1,100m)	
OrbView 1		OrbImage		03-04-1995	00/04/2000												
OrbView 2 (Seastar)		OrbImage		01-08-1997	00/12/2010	705	12:00										
	SeaWifs		2.8						B1 (1 km)	B2,3 (1 km)	B4,5 (1 km)	B6 (1 km)	B7,8 (1 km)				
OrbView 3		OrbImage		26-06-2003	23-04-2007	452	10:30 a. m.										
	OHRIS		8					Pan (1m)		B1 (4m)	B2 (4m)	B3 (4m)	B4 (4m)				
Pleiades 1A		CNES		17-12-2011	in operation	694	10:30										
	HiRI		20					Pan (0.7m)		B1 (2.8m)	B2 (2.8m)	B3 (2.8m)	B4 (2.8m)				
Pleiades 1B		CNES		02-12-2012	in operation	694	10:30										
	HiRI		20					Pan (0.7m)		B1 (2.8m)	B2 (2.8m)	B3 (2.8m)	B4 (2.8m)				
PROBA-2		ESA		22-10-2001	in operation	670	11:30										
	CHRIS		13						Hyperspectral (19 to 63 bands, resolution 18m to 63m)								
	HRC		5					Pan (8m)									
QuickBird 1		DG		20-11-2000	20-11-2000	*mission failed*											
QuickBird 2		DG		18-10-2001	27-01-2015	450	10:30										
	BGIS 2000		16.5					Pan (0.61m)		B1 (2.4m)	B2 (2.4m)	B3 (2.4m)	B4 (2.4m)				
RADARSAT	SAR	CSA	50 to 500			798	06:00										C-band HH, VV
RADARSAT 1				04-11-1995	09-05-2013												
RADARSAT 2				14-12-2007	in operation												
RapidEye		BlackBridge		29-08-2008		630	11:00	Pan (6.5m)									
	REIS		78							B1 (6.5m)	B2 (6.5m)	B3 (6.5m)	B4-5 (6.5m)				
SEASAT	SAR	Nasa		28-06-1978	10-10-1978	769-799											L-band HH
Sentinel		ESA															
Sentinel 1A				03-04-2014	in operation	693	06:00										
	SAR		250														C-band
Sentinel 1B				22-04-2016	in operation	693	06:00										
	SM		80														C-band DP
	IWS		240														C-band DP
	EWS		400														C-band DP
	WM		20														C-band DP
Sentinel 2A				23-06-2015	in operation	786	10:30										
	MSI		290						B1 (60m)	B2 (10m)	B3 (10m)	B4 (10m)	B8 (10m), 5, 6, 7, 8A (20m), 9 (60m)	B10 (60m), 11 (20m)	B12 (20m)		
SIR A	SAR	Nasa	50	12-11-1981	14-11-1981	222											L-band HH
SIR B	SAR	Nasa		05-10-1984	13-10-1984	232											L-band HH
SIR C - SRL-1		Nasa		09-04-1994	20-04-1994	225											
SIR C - SRL-2		Nasa		30-09-1994	11-10-1994	225											
	SAR		15-90														L-band FP
	SAR		15-90														C-band FP
	X-SAR	DLR	15-60														X-band VV
Spot 1		CNES		22-02-1986	31-12-1990	832	10:30										
	HRV		60					Pan (10m)			B1 (20m)	B2 (20m)	B3 (20m)				
Spot 2		CNES		22-01-1990	30-07-2009	832	10:30										
	HRV		60					Pan (10m)			B1 (20m)	B2 (20m)	B3 (20m)				
Spot 3		CNES		26-09-1993	14-11-1997	832	10:30										
	HRV		60					Pan (10m)			B1 (20m)	B2 (20m)	B3 (20m)				
Spot 4		CNES		23-03-1998	29-06-2013	832	10:30										
	HRVIR		60					Pan (10m)			B1 (20m)	B2 (20m)	B3 (20m)	B4 (20m)			
	VEGETATION-1		2.200							B1 (1,100m)		B2 (1,100m)	B3 (1,100m)	B4 (1,100m)			
Spot 5		CNES		04-05-2002	31-03-2015	832	10:30										
	HRG		60					Pan (5m)			B1 (10m)	B2 (10m)	B3 (10m)	B4 (20m)			
	VEGETATION-2																
Spot 6		CNES		09-09-2012	in operation	694	10:30										
	NAOMI		60					Pan (1.5m)		B1 (6m)	B2 (6m)	B3 (5m)	B4 (6m)				
Spot 7		CNES		30-06-2014	in operation	694	10:30										
	NAOMI		60					Pan (1.5m)		B1 (6m)	B2 (6m)	B3 (5m)	B4 (6m)				
TerraSAR X	SAR	DLR		15-06-2007	in operation	515	06:00										X-band FP
TanDEM-X	SAR	DLR		21-06-2010	in operation	514	06:00										X-band FP
	StaringSpotLight		2.5x6														0.5m
	SpotLight		10x10														1.7-3.5m
	StripMap		50x30														3m
	ScanSar		150x100														18m
	Wide ScanSar		200x270														40m
TRMM		NASA/JAXA															
	VIRS		720	27-11-1997	15-04-2015	403						B1 (2km)		B2 (2km)		B3 to B5 (2km)	
WorldView 1		DG		18-09-2007	in operation	496	10:30										
	WV60		17.6					Pan (0.5m)									
WorldView 2		DG		08-10-2009	in operation	770	10:30										
	WV110																
WorldView 3		DG		13-08-2014	in operation	617	01:30										
	WV 3		13.1					(0.31m)	B1 (1.24m)	B2 (1.24m)	B3 (1.24m)	B4,5 (1.24m	B6,7,8 (1.24m)	B9 to 12 (3.	B13 to 16 (3.7m)		
	CAVIS		13.1						B1 (30m)	B2 (30m)	B3 (30m)	B4,5 (30m)	B6,7,8 (30m)				
WorldView 4 (GeoEye 2)		DG		11-11-2016	in operation	617	10:30										
	SpaceView110		13.1					(0.31m)		B1 (1.24m)	B2 (1.24m)	B3 (1.24m)	B4 (1.24m)				

basement and the volcanic rocks north-west of Al Madinah, also in Saudi Arabia, at a scale of 1:100,000. Al Kwatli et al. (2012) investigated the Quaternary volcanic field of the Al-Lajat plateau, southern Syria, using Landsat ETM+. They proposed colour composites using ratio images (band 5 to band 7, band 5 to band 4, band 3 to band 1 in RGB) for identifying alteration patterns of ferric, ferrous and clay minerals and mapping the main lava flows, and also the classical principal components analysis. This latter revealed high concentrations of ferrous minerals in the third principal component. This study combined remote sensing data and geochronological dating as important tools for mapping the location and timing of past volcanic activity. Gad and Kusky (2006) studied the Barramiya area, central Eastern Desert, Egypt, with two Landsat TM band ratio images: band 5 to band 3, band 5 to band 1, band 7 to band 5 in RGB 1, and band 7 to band 5, band 5 to band 4, band 3 to band 1 in RGB 2. They also developed supervised classification methods and demonstrated that these techniques are effective for lithological mapping of serpentinite rocks, metasediments, metavolcanics and granitoids. From 1986 onward, SPOT provided multispectral data with a narrower spectrum than Landsat TM but with a significant increase in ground space resolution (from 30 to 20 m). This facilitated the development of texture processing methods. Thus, some studies demonstrated the interest of SPOT-XS data for geological mapping, for example Emran et al. (1996) showing the advantage of the variogram analysis for analyzing the famous Bleida mining district in the eastern part of the Bou Azzer anticline, Moroccan Anti-Atlas. They also used SPOT imagery for the mapping of Precambrian ophiolitic occurrences in the Anti-Atlas of Morocco (Emran et al. 1995).

The launch of ASTER in 1999 opened new paths of research in multispectral remote sensing geology. Gad and Kusky (2007) used the ASTER band-ratio image 4/7-4/6-4/10 for mapping rocks in the Wadi Kid metamorphic belt in Egypt, showing improvement over previous techniques in detailing the main rock units. Qiu et al. (2006) investigated the Neoproterozoic Allaqi-Heiani suture, Southern Egypt, also with the help of ASTER data. The spectral analysis demonstrated to be more effective in detecting lithological units than traditional multispectral analysis procedures. Jakobs et al. (2015) improved the existing geological maps in the Eastern Desert, Egypt, by the input of textures in a classification of ASTER and Landsat 8 data. Roy et al. (2009) carried out a HyMap airborne and field hyperspectral survey of the Oman ophiolite. On the basis of spectral signatures, they were able to distinguish three assemblages in the mantle sequence, harzburgites, dunites and harzburgite with interstitial carbonate, respectively. They also recognized four assemblages in the crustal sequence, corresponding to massive gabbros, amphibolized gabbros associated with dykes, wehrlite with high serpentine content, and gabbronorite, respectively.

The interest of synthetic aperture radar data in arid studies has been illustrated for the first time in the Eastern Sahara, Egypt, from the SIR-A and SIR-B missions (McCauley et al. 1982; Schaber et al. 1986). These studies revealed palaeodrainage channels along the Nile valley using their high potential for penetrating sand veneers (McCauley et al. 1986). The potential for geological mapping has been less frequently evaluated, but is known to be better when large wavelengths are used (Schaber et al. 1997). Deroin et al. (1998) and Baghdadi et al. (2005) investigated the potential of radar imagery compared with optical images in geological exploration in Mauritania, in the Atar and Akjoujt region, respectively. The results indicated how radar imagery, particularly when L-band is used, can help not only in detecting surface geological structures such as dykes and veins, but also mapping subsurface structures beneath a shallow layer of sand. Several examples of SAR ERS C-band waves penetrating sands are shown over the Sahara in Dhont et al. (2005a, b). More recently, Deroin et al. (2014) made the detailed geological mapping of the In Ouzzal area, Algeria, emphasizing extensive Archaean dome-and-basin patterns. The main sensor used was the ALOS Palsar compared to Envisat ASAR and to optical imagery (Landsat TM, ETM+ and ASTER).

Figure 1 provides an example of geological remote sensing with optical data. The Landsat TM colour composites cover the Bou Azzer anticline (southern Morocco). The comparison is between the colour composite 7 3 1 (band 7 [SWIR range] in red, band 3 [red range] in green, band 1 [blue range] in blue), and the natural colours (3 2 1). A–F are geological features better perceived with the RGB = 731.

2.2 Tectonics and Structural Geology

Tectonic and structural studies mainly dealt with the use of space-borne techniques to extract geological elements such as active faults, morphotectonic features, large-scale folding, foliation track, etc. Digital elevation models (DEMs) are very useful for morphotectonic approaches. Most of the studies combined remote sensing analysis and detailed field observations.

A Chorowicz team discussed with Butler et al. (1997) the seismic hazards in Beyrouth and the actual location of the Arabian plate boundary in this region on the basis of SAR ERS imagery combined with fieldwork over Lebanon and Syria (Fleury et al. 1999; Chorowicz et al. 2004). Huchon and Khanbari (2003) used SPOT panchromatic data for studying extensional basins along the northern margin of the Gulf of Aden, Yemen, showing that Oligocene–Miocene syn-rift extension trends N.20°E in agreement with the strike of main rift-related normal faults, but oblique to the main trend of the Gulf. Mostafa and Bishta (2005) extracted lineaments from Landsat 7 ETM+ data for the study of Gebel Gharib-Dara, Egypt. They shew that the younger granitic rock units are characterized by a short amplitude of NW trends relative to the NE trends. They also correlated lineament density maps with aeroradiometry and ground radiometric maps and demonstrated that rock units with high radioactivity are characterized by high lineament density, with an implication for new possible uranium targets.

Saadi and Watanabe (2009) studied the Eljufra area in Libya using Landsat ETM +, ERS-2 SAR and SRTM DEM data. They drew the different lithological units and geological lineaments, particularly using the fusion of Landsat ETM+ and ERS-2 SAR data. Nine rock units were plotted in a new geological map. Saadi et al. (2011) integrated SRTM DEM, Landsat ETM+, geologic maps and subsurface data, for the study of the Ghadames Basin, Libya. They indicated that this basin is controlled by four main fault systems and that the basin gradually deepens towards the southwest. Chaabouni et al. (2012) investigated the stratigraphic sequences and structures buried under the Grand Erg Oriental, with the use of SPOT, Landsat and digital

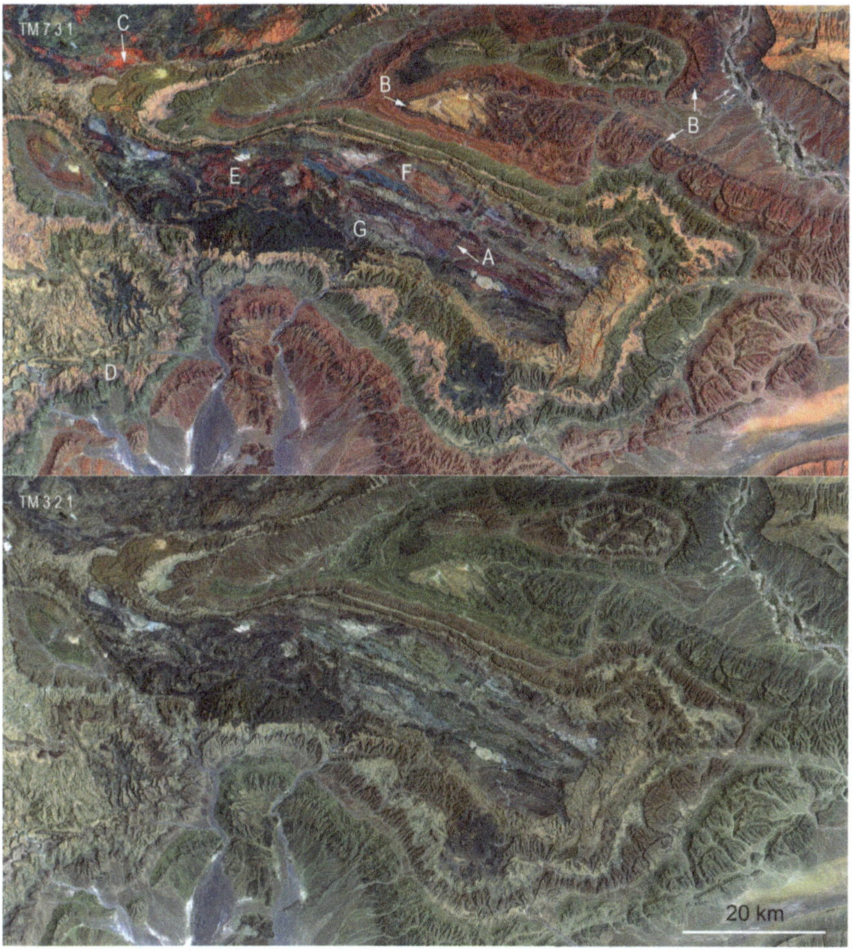

Fig. 1 Example of geological remote sensing with optical data. Landsat TM colour composites on the Bou Azzer anticline (southern Morocco). Comparison of the Band 7 [SWIR range] in red, band 3 [red range] in green, band 1 [blue range] in blue, and the natural colours (3 2 1). Image acquired on 6 June 1987. (A–F) Geological features better perceived with the RGB = 731. (A) Track of the major Jurassic dyke crossing the Bou Azzer anticline. (B) Two layers (in blue in the 731) interbedded within the Upper Cambrian sandstones (reddish) and quartzites (yellow). Note that the ferruginous sandstones appear green. (C) A facies in the Precambrian metamorphic basement (possibly spilites, a kind of altered basalt). (D) Detail of the Cambrian cover of the Bou Azzer anticline (alternating sandstones, schists, carbonates, etc.). (E) Detail of the Lower Cambrian Ouarzazate Fm (ignimbrites, volcanic rocks). (F) Diabases associated with the mineralization (Upper Proterozoic). (G) Detail within the polymetamorphic gneisses (Lower Proterozoic)

elevation models. They obtained a lineament map showing two major directions of faults: the main lineament direction 035–065° (NE–SW) and the secondary lineament direction 110–130° (NW–SE). Tilted blocks delimited by fracture and folding of the Upper Cretaceous series are related to these fractures. Results were compared with surrounding areas such as the Tataouine basin and the Ghadames basin. Chorowicz and Benissa (2016) carried out geological mapping and structural studies based on Landsat ETM+, ERS SAR and SRTM DEMs. They thoroughly investigated the Palaeozoic evolution of the Ghadames and Murzuq Basins to build a new model which is alternative to the classical concept of the Variscan compressional orogen.

Reif et al. (2011) presented a software tool including digital elevation models for interactively mapping and measuring the spatial orientation of the geologic structure in remote and inaccessible areas. This tool was tested in the Zagros belt, Kurdistan, northeast Iraq. The comparison of computed data with the field measurements led to correct results when favourable conditions are present, i.e. moderately dipping planes, strong competence contrast between lithologies, etc. The Alima anticline (Tunisian Atlas) served to test the links between the surface geometry of a fold and its deep architecture (Jabbour et al. 2012). Remote sensing analysis of the SRTM DEM and a set of multispectral images including Landsat ETM+ and ASTER scenes allowed drawing a new geological map. The results were confirmed by a seismic profile. Deffontaines et al. (1992) carried out a neotectonic study in Morocco using morphometric methods. A map of vertical deformation was established and the analysis led to a clear distinction between ancient faults and currently active faults. Chorowicz and Fabre (1997) studied the braided system made of numerous wadis in the Tanezrouft area, southern Algeria. They used Landsat TM, SPOT-P and XS and SAR ERS 1 to emphasize palaeodrainage networks and intracratonic tectonic activity. More recently, Le Béon et al. (2012) investigated the Dead Sea fault using images derived particularly from Landsat ETM+, SPOT-5, very high resolution "Google Earth" imagery, and also DEMs derived from SRTM and ASTER. They emphasized possible maximum offsets of about 5.5 km for the oldest surfaces with a fault slip rate estimated at 5–7 mm year^{-1}. Poujol et al. (2014) carried out a geomorphological and morphotectonic analysis of the northeastern part of the Rif (Morocco) using digital elevation models from SPOT-5 data. The results were consistent with GPS measurements showing a southwestward motion of the Rif belt with respect to stable Africa.

Figure 2 is an example of geological remote sensing with the SAR in the Middle East.

2.3 Hydrogeology

Remote sensing is certainly the backbone of hydrogeological reconnaissance in regions where detailed geological maps are missing. However, even in temperate climate zones where field data are numerous and reliable, remote sensing reveals new features on the structures and lithologies that previously were not sensed. In arid regions, such as the Arab countries, SAR sensors can detect some subsurface

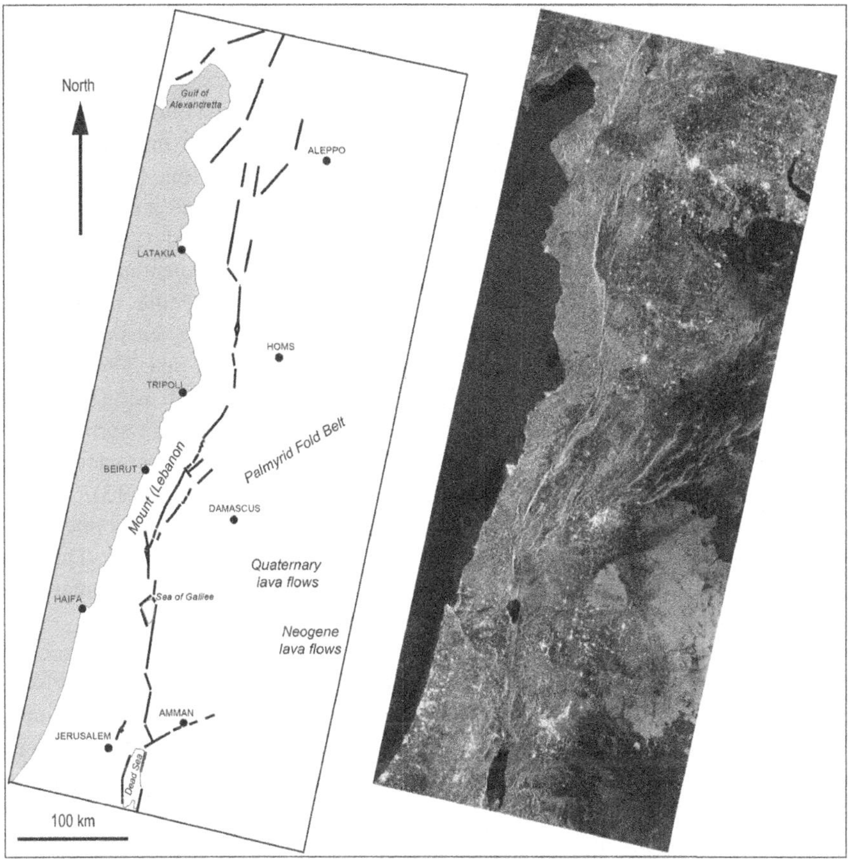

Fig. 2 Example of geological remote sensing with the SAR. Mosaic of two ALOS PALSAR scenes acquired on 20 March 2008 (at about 08:10 UTC). Main lineaments corresponding to the Dead Sea Fault in the Middle East. The space resolution is about 350 m in the ScanSAR mode. The mosaic covers about 600 km × 200 km along the Dead Sea Fault from the East Anatolian Fault to the North to the Dead Sea to the South

features below the surface. The different wavelengths (even if the C-band is the most commonly used) provide detailed information particularly interesting for soil moisture retrieval and for surface roughness retrieval as well. The combination of multi-frequency remote sensing information from the visible to microwave wavelengths should be developed. Some examples revealed the interest of a combined radar and optical/infrared (especially thermal infrared) approach.

As early as 1986, Voûte used Landsat MSS and aerial photographs for optimizing the groundwater exploration in the Kasserine Basin, Tunisia. Mahmood (1996) extracted lineaments from Landsat data in central-north Morocco and shew that they are important guides for groundwater exploration. The lineament has been traced in a granitic pluton and its surrounding metamorphic terrane. It has been observed that the lineament is due to a major fault along the contact between two granitic facies. Quantitative slope analysis has also been carried out. It is concluded that in crystalline basement terranes, deep-seated fractures with the weather rock overburden provide favourable sites for groundwater circulation. New 3D tools where used for volumic imaging of hydrogeologic data on the basis of remote sensing approach (Dhont et al. 2005a, b). Elewa and Qaddad (2011) utilized Landsat ETM+ data in the process to identify the groundwater potential areas in the Sinai Peninsula, Egypt. They indicated that this area is generally of moderate groundwater potentiality. Siebert et al. (2014) carried out a remote sensing approach for studying groundwater fluxes towards the hypersaline Dead Sea. They particularly used the thermal band of Landsat ETM+ and drew a 3D graph using RapidEye and DEM data representing the regional geology and the offshore springs. In a coastal Lebanese watershed, environmental parameters were generated from satellite imageries and digital elevation models by Bou Kheir et al. (2013). The aim of the study was to predict the spatial distribution and concentration of copper using a geographic information system (GIS) and regression-tree analysis. The best regression-tree model combined soil pH and surroundings to waste areas, and explained most of the variability in copper laboratory measurements.

Omar (1990) used SIR-A imagery to study the hydrogeology of an area in southern Iraq. Geological, structural lineament and drainage density maps were constructed and the significant hydrogeological features extracted from them in order to locate areas of high underground water content. Paillou et al. (2010) have shown the ability of the L-band SAR of ALOS PALSAR to detect subsurface features in the Sahara, including palaeo-rivers which are of interest for water resource prospection. Amer et al. (2010, 2013) selected the central Eastern Desert of Egypt for studying groundwater occurrences in shallow fracture zone aquifers using geological, geophysical and remote sensing data (PALSAR, ASTER). A GIS model of groundwater potential was developed indicating that the distribution of the existing water wells is in agreement with the high groundwater potential zones. Milewski et al. (2009) used also the Sinai Peninsula and the Eastern Desert of Egypt as test areas to study the recharge process. They used global and high-resolution remote sensing datasets, particularly the Tropical Rainfall Measuring Mission (TRMM), Landsat TM, NOAA-AVHRR and EO-ASTER. Results demonstrated the enhanced opportunities for groundwater development in the Sinai Peninsula compared to the Eastern Desert. Gaber et al. (2015) mapped and characterized surface sediment in the non-vegetated El-Gallaba Plain of the Western Desert, Egypt. Six optical (ALOS-AVNIR-2, ASTER) or radar (ALOS PALSAR, Radarsat-2) datasets were used for unsupervised and supervised classifications. The best accuracy was obtained with the supervised classification using the full polarimetric Radarsat-2. The low-frequency ALOS PALSAR data revealed a set of linear features that are hidden under windblown sand and are invisible in the field.

In order to map and monitor evaporitic surfaces in the Chott Jerid, Tunisia, optical multisource, multispectral and multidate satellite data have been used, particularly Landsat TM, EO-ASTER and SPOT-6 (Abbas et al. 2018). The medium spectral resolution of the data provided relevant information about the mineralogical nature (mainly halite and gypsum) and distribution of evaporites of the Chott.

2.4 Mining Geology

When metallogeny and mining geology are concerned, spectral aspects play an important role. This is the reason why ASTER is the most frequently used sensor.

As soon as 1984, spectral signatures of ore bodies in Morocco were analyzed in the field and laboratory (Bothorel et al. 1984; Bouchet et al. 1984). The measurements in laboratory correlate with that of outcropping surfaces in the arid environment. Iron caps can be sensed from LANDSAT MSS and SPOT-XS, but spectral signatures of cobalt–nickel in serpentized ophiolite and copper impregnations within red sandstone are not discriminated with these satellites. Massironi et al. (2008) carried out an interesting survey of the Saghro massif, in the Anti-Atlas of Morocco, using all the spectral bands available with ASTER. First, they calculated false colour composites, band ratios, and principal components analyses in the range visible to shortwave infrared. Thus, they recognized major lithological contacts and mineralized faults. Second, they concentrated on the study of plutons using also the thermal infrared range. Gemail et al. (2016) emphasized the combined use of ASTER-SWIR data and airborne geophysical techniques for mineral exploration in the granitic rocks of Gabal Dara area, Egypt. The results provide information on alteration minerals which are valuable for mineral exploration. Othman and Gloaguen (2014) studied the Mawat Ophiolite Complex, Kurdistan, northeast Iraq, also with ASTER data. They detected chromite occurrences within the ophiolite and identified two previously unknown locations having high concentrations of chromite.

Using the more classical Landsat TM data Abdelhamid and Rabba (1994) studied the Khirbet El Nahas area, Jordan, well known for its copper and manganese mineralization. They emphasized the interest of the colour composites ‘RGB’ of bands 7, 4 and 2, and 7, 5 and 2, and also ratios and principal component composites. Ciampalini et al. (2013) described applications of cost-effective remote sensing techniques based on Landsat ETM+ to map areas containing iron deposits in Western Sahara and in Algeria. Lamri et al. (2016) combined Landsat ETM+ and ASTER satellite data with airborne geophysical data for improving the geological map of the Edembo area in the Eastern Hoggar, Algeria. Multispectral band fusion, filtering, band combination, band rationing and principal component analysis were used for better spectral discrimination of the different rocks units.

Airborne magnetic and gamma-ray spectrometry data showed the pervasiveness of large structures beneath the Palaeozoic cover. The aeroradiometric K-range is used for discriminating between high-K granitoids of Djanet from the peraluminous granites of Edembo.

2.5 *Geoarchaeology*

A new path of research is the use of remote sensing for the study of ancient archaeological sites and particularly the ancient mines. Arab countries offer well-adapted settings because remote sensing provides its best results in arid countries where many archaeological artefacts from many old civilizations are generally well preserved. Both large-scale data and very high-resolution data are interesting for a global survey from the one hand, for detailed mapping of the archaeological remains from the other hand. Note that the field method of the ground-penetrating radar (GPR) frequently used in field geoarchaeology does not fit with our objectives which are focused on space-borne remote sensing.

We used a wide range of multispectral sensors (Landsat TM/ETM+, ASTER, ALOS-AVNIR-2) including very high spatial images (QuickBird) to study the ancient silver mines of Jabali, Yemen (Deroin et al. 2011, 2012a). The identification of gossans (iron caps) possibly related to archaeological remains was particularly clear with the red and near-infrared range of Landsat. In Bi'r Samut, Eastern Desert, Egypt, ancient gold mines have been mapped using GeoEye-1 data (Brun et al. 2013). The geological setting, mainly granitic, and the role of N–S to NNE–SSW veins for guiding the gold mineralization have been emphasized. Other archaeological places of interest have been investigated such as alluvial deposits in southern Jordan revealing archaeological remains along wadis with the help of ASTER and SPOT-5 imagery (Löhrer et al. 2013). De Laet et al. (2015) added Corona and QuickBird to the previous dataset to carry out a study of the Dayr al-Barsha area, middle Egypt, emphasizing the routes used at different periods serving quarries, but also cemeteries and settlement logistics. Breeze et al. (2015) used Landsat data and DEM to emphasize a combined method for remotely mapping palaeodrainage and palaeolakes in the whole Arabian Peninsula, particularly in the Nefud desert, Saudi Arabia, in relation to archaeological sites. In the specific case of Byblos, Lebanon, we compared OrbView-3, Ikonos-2 and GeoEye-1 data for mapping the walls and the buildings of the famous archaeological site. Moreover, Corona archive allowed mapping the archaeological site and the coastal environment prior to the beginning of the Lebanese civil war in 1975 (Deroin et al. 2017).

2.6 Oil and Gas Prospection

During an early period, oil companies have shown a considerable scepticism with regard to remote sensing applications to oil exploration. This was easily explained by the great difference between the ground resolution of the first Landsat satellites (80 m) and aerial photographs classically used in field exploration. The new sensors overpass this difficulty and allow thoroughly mapping the geological setting, the industrial facilities or the related pollution. This latter has been recently illustrated by Lacava et al. (2017) in the eastern part of the Mediterranean Sea using MODIS data to quickly detect the presence of oil spills on the sea surface.

Fraser et al. (1997) developed a satellite-based remote sensing approach for generating near surface geological data in central Yemen. This approach allowed mapping faults and structure not apparent in the seismic profiles. The near-surface map highlighted areas of potential closure at reservoir depths. Peña and Abdelsalam (2006) investigated the Dahar Plateau, South Tunisia, which is part of the large Ghadames Basin, using a remote sensing approach to aid exploration for new oil and gas fields. Landsat ETM+, ASTER, Radarsat and SRTM DEMs were used. Both band and band-ratio images have been used for the identification of various lithological units. Radarsat images have been utilized when geological structures are buried under thin sand veneers. The results were successfully compared to in situ investigations including geological mapping and seismic exploration. Cai et al. (2014) integrated remote sensing images, gravity-magnetic data and seismic data to revisit the tectonic evolution of the Cheliff Basin, Algeria. Some areas such as the Tliouanet uplift and the Mostanem uplift were pointed out to be possible targets for hydrocarbon exploration.

2.7 Earthquake and Seismicity

The Arab world is relatively untouched by major earthquakes, which are more frequent in other areas such as Central Asia or the Pacific Ring of Fire. Nevertheless, some destructive events occurred, for example the 551 Beirut earthquake, the 1754 Cairo earthquake or the 1982 North Yemen earthquake, and particularly in northern Africa during the last decades such as the 1960 Agadir earthquake in Morocco or the 2003 Zemmouri (Boumerdès) earthquake in Algeria. Basically, remote sensing imagery applied to seismicity can be used in three ways: First, for estimating the damage after the main shock, second for surveying the deformation along the fault zone also after the main shock, third for 'predicting' the earthquake before the main shock.

Adams et al. (2006) used remote sensing techniques for damage assessment in the aftermath of recent events including the 2003 Boumerdes highly destructive earthquake. Bitelli et al. (2004) presented results on the same earthquake. Pre- and post-event images acquired by different platforms, from medium to very high resolution, panchromatic and multispectral (mainly, IRS and QuickBird) were evaluated and compared. For the same event, Cetin et al. (2012) utilized interferometric synthetic aperture radar (InSAR) technique based on a series of 31 Envisat ASAR images from 2003 to 2010. The analysis revealed relatively strong subsidence phenomena with a maximum rate of 3.5 mm year^{-1} in Boumerdes. The cumulative post-seismic activity during 7 years would be equivalent to an important earthquake. Earlier, Sarti et al. (2003) developed the same methodology using archived ERS data covering the Arava segment of the Dead Sea Rift affected by the 1995 Gulf of Aqaba earthquake. They recognized a mean displacement rate of about 0.4 cm month^{-1} during the 8-month period preceding the earthquake and, in the 3 following years, a measured rate smaller by a factor of ten.

Satellite thermal infrared (TIR) imaging data have recorded short-lived anomalies prior to major earthquakes. Precursor clear local movements were observed, possibly related to groundwater changes before the earthquake event. These anomalies typically appear 4–14 days before the main shock. They affect regions of several to tens of thousands square km and display a positive deviation of 2–4 °C or more, and die out a few days after the event. Some authors have proposed that these signals originate from phenomena associated with pre-seismic processes, causing enhanced TIR emissions. Different causes have been envisaged such as rising fluids that would lead to the emanation of warm gases, CO_2 spreading and causing a local greenhouse effect, radon ionization and heat exchange due to variations of humidity, etc. All imagery data from the geostationary satellites such as MSG-SEVIRI to the polar NOAA-AVHRR, EO-Aqua/Terra-MODIS and Landsat have been used for illustrating thermal anomalies. Whether the anomalies are generally obvious, it is sometimes difficult to separate them from meteorological anomalies. Aliano et al. (2007) presented results in the area affected by the 2003 Boumerdès earthquake. They characterized space–time persistent TIR anomalies about one month before the main shock, overlapping tectonic lineaments and sometimes focusing close to the epicentre.

2.8 Landslides

Landslides are among the major natural hazards related to the geological setting. Landslide survey using remote sensing techniques requires high to very high-resolution data because of the generally limited size of the phenomenon.

In the frame of a GIS, Abdallah et al. (2005) analyzed statistic correlations about terrain parameters and mass movements in Lebanon that were identified from SPOT-4 stereo images at 10 m ground resolution and checked in the field. The main parameters are, in decreasing order: lithology, proximity to fault line, soil type, distance to drainage line, land cover/use, slope aspect, slope gradient. Abdallah et al. (2007) used a multisource approach including Landsat TM, IRS and SPOT-4 data to map mass movement. They emphasized the interest of the anaglyph produced from two panchromatic SPOT-4 stereo-pairs, on the one hand, and the PCA pan-sharpened Landsat TM-IRS image, on the other hand.

Arnous (2011) processed Landsat ETM+ data and combined them with digital elevation models to estimate the landslide hazard zonation in the South Sinai Wadi Watier area, Egypt. Landsat images allowed extracting lineaments, lithology and land-cover types. Youssef et al. (2015) carried out a comprehensive study of landslides in the Jazan region of southwest Saudi Arabia. In this area, landslides are particularly important due to the fast development of uncontrolled urbanization, on the one hand, and the growth of mountain roads to the other hand. An inventory map with 86 landslide occurrences was generated. Landslide susceptibility maps were produced using mainly Landsat ETM+ (15 m) as high-resolution data, and GeoEye (2.5 m) and QuickBird (0.6 m) as very high-resolution data. Othman and Gloaguen (2013) assessed the location and size distribution of landslides in Kurdistan, a highly tectonized region of northeast Iraq. They used an inventory of 3190 landslides mapped from QuickBird images. They proposed an algorithm for automatic landslide extraction. Baer et al. (2002) characterized rates of subsidence along the Dead Sea shores using interferometric synthetic aperture radar technique based on ERS-1 and ERS-2 images. They measured subsidence rates in the range of 0–20 mm year^{-1}, with exceptional rates up to 60 mm year^{-1}. They noticed that the level of the Dead Sea and of the associated groundwater dropped by about 6 m during the same period (1992–1999) and that subsidence is controlled by faults and seaward landslides. In the same region, Dill et al. (2012) utilized Landsat TM, ASTER and SRTM DEM to investigate landslides along the Dead Sea Fault. Li and Zhang (2013) used QuickBird images to apply the change detection method based on the principal component analysis. They demonstrated that landslides and building collapses caused by the Boumerdès earthquake can be recognized. Last but not least, traditional photogrammetry is still used for landslide survey. For example, Laribi et al. (2015) focused on the landslide of El Biar in Algiers, Algeria, using digital photogrammetry based on aerial photographs to detect horizontal displacements between 1995 and 2007. They have shown that the landslide comprises two zones: a peripheral moving at about 5 cm year^{-1} and a central moving at about 10 cm year^{-1}.

2.9 Coastal Erosion

Coastal erosion is an important issue in the Arab countries because each Arab country includes a coastal front. This type of study generally needs a multi-temporal approach and high to very high-resolution data to establish detailed balances of sediment fluxes (loss, i.e. erosion or gain, i.e. sedimentation). One of the major difficulties is due to the fact that whether remote sensing gives a perfect view in the horizontal plane, the estimation of the vertical component (i.e. thickness) needs the use of accurate DEMs.

Hereher (2011) used a set of six Landsat (MSS, TM or ETM+) satellite images to map changes at the famous Rosetta promontory, to the west of the Nile Delta, Egypt, between 1973 and 2008. The results illustrated that this place lost about 12 km^2 of land between 1973 and 2008 and the shoreline withdrew southward about 3.5 km. Coastal protection structures have contributed to reduce coastal erosion but probably promoted downdrift erosion. El Asmar (2002) utilized only three Landsat TM data (1984, 1987, 1991) to study the coastline between Damietta and Port Said, to the east of the Nile Delta. This author recognized erosional and accretionary segments, showing also the strong human influence. In a seminal work, White and El Asmar (1999) studied the Nile Delta as a whole using the same Landsat dataset. They demonstrated that severe erosion affects all the promontories, including the Damietta one, and redistribution of eroded material occurs along the coast. Wu (2007) used SPOT data to study one of the major coastal erosion phenomena along the Atlantic coast, the new harbour of Nouakchott, Mauritania, built in 1987, and its consequences. The coastal dynamics has been deeply altered with a significant littoral geomorphologic change characterized by erosion that caused several suburban and urban floods. This change also concerns the Senegal River estuary further to the south close to the Senegal–Mauritania border, the evolution of which (particularly a human breach made in 2003) has been thoroughly surveyed using Landsat ETM+ data chronicle (2003–2010) and Alos PRISM data (Deroin et al. 2012b). Kermani et al. (2016) analyzed the Bay of Jijel, Algeria. They used multidate aerial photographs (1960, 1973 and 1988) and QuickBird satellite image to reconstruct the coastline variation from the beginning of the 1960s to 2014. After a relatively stable period (to 1973 onward), the coast was clearly eroded with a maximum rate of about 0.9 m $year^{-1}$. The authors indicated this process is probably due to the combined action of stormy climate and human actions along the coast. Louati et al. (2014) explored a series of Landsat data (MSS, TM, ETM+, OLI) to study the shoreline evolution related to the Medjerda delta, northeastern Tunisia, between 1972 and 2013. The main causes of erosion have been linked with the shifting of the river mouth, the damming of the river, the building of a new harbour and the destruction of small bordering dunes. The respective role of longshore currents, coastal subsidence and sand mining has been also emphasized. Deroin et al. (2017) studied the famous archaeological site of Byblos, Lebanon. The comparison of the Corona KH-4B and Ikonos imagery highlighted changes due to the coastal erosion and the cliff degradation from the 1970s onward. The extent of the harbour fittings is associated with erosion to the south and sedimentation to the north due to the interference with the littoral drift. Along this coast, a major landslide occurred in 1922 revealing to the archaeologists the location of the Royal Necropolis.

Figure 3 illustrates the coastal front of Nouakchott, the capital city of Mauritania. This area is characterized by a flat sandy shore exposed to dominant north-west swells. The littoral drift leads the sediments from the north to the south. From 1986 onwards, this drift has been deeply altered when the new harbour, the so-called *Port de l'Amitié*, was built. The comparison of Landsat data acquired in 1984 (before the new harbour), 1995, 2005 and 2016 shows clearly that

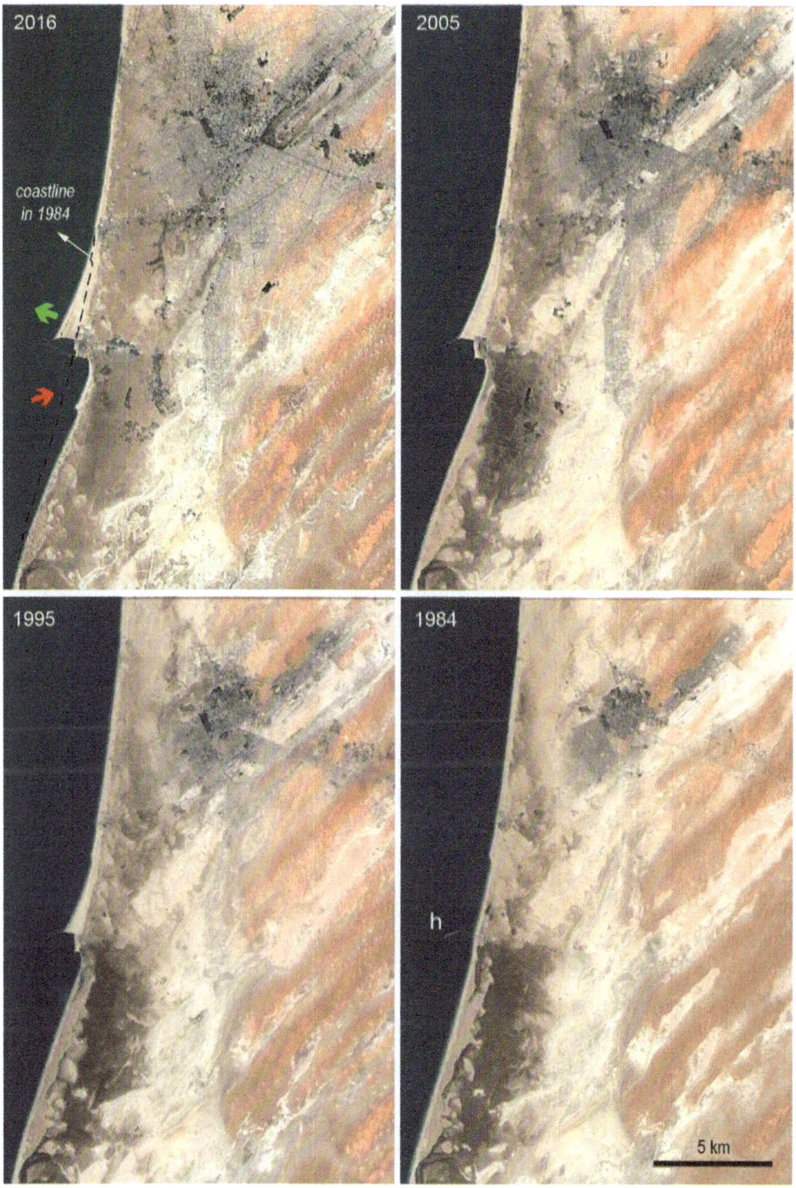

Fig. 3 Example of Landsat data illustrating the coastal front of Nouakchott, the capital city of Mauritania, in 1984, 1995, 2005 and 2016

sedimentation takes place to the north, whereas erosion occurs to the south. It seems that sedimentation to the north decreases from the beginning of the 2000s, whereas erosion remains active at the same rate for 30 years. Note the considerable development of Nouakchott city.

3 Discussion on Remote Sensing Applied to Geology

After a period characterized by large-scale sensors only (1972–end 1990s), the last two decades have seen a lot of very high resolution (VHR) sensors allowing detailed studies. MODIS, MERIS or NOAA-AVHRR allow large-scale mapping (from 1:200,000 to 1:1,000,000) with a ground sample distance of about 300 m. This is also the case with the wide mode of some SAR such as ALOS PALSAR (ScanSAR mode). Here, we illustrate the interest of this kind of data on the Dead Sea fault area covering the Middle East from the Dead Sea to the East Anatolian Fault (Fig. 1).

Landsat ETM+ or TM, and SPOT 1–4 are suitable for intermediate scales (from 1:50,000 to 1:250,000). SPOT 5 allows mapping at 1:20,000–1:50,000, whereas SPOT 6–7 or Formosat allow mapping at 1:5000–1:25,000. Very-high-resolution images (Pléiades, GeoEye, QuickBird, Ikonos, etc.) permit very detailed scale mapping (1:2,000–1:10,000).

3.1 Remote Sensing Geology with Optical Sensors

The most popular remote sensing data are provided by Landsat, the longest running programme for the acquisition of space-borne imagery starting in 1972 with Landsat 1 MSS, improved as early as 1982 by Landsat 4 TM, then by Landsat 7 ETM+ (1999) and Landsat 8 OLI (2013). Multispectral imagery has revealed a great potential in the geological mapping with Landsat but also SPOT XS, IRS, etc. Superspectral imagery (ASTER) and possibly hyperspectral observations, especially those of bare surfaces, are of particular interest. It should be emphasized the specific interest of the shortwave infrared (SWIR between 2.0 and 2.5 μm) for geology. This range is available with Landsat TM/ETM+ or EO-ASTER, but also with modern satellites such as Sentinel 2A or WorldView 3.

Figure 2 provides an example of Landsat TM composites on the area of Bou Azzer in the Anti-Atlas of southern Morocco, a famous Co–Ni–As–Ag–Au district. This place is one of the world most dramatic desert areas for utilizing multispectral sensors, because the rocks are barely weathered. Moreover, structures are particularly well expressed with the large-scale folding of the Bou Azzer Anticline, the magnificent Jebel Sarhro and the Oued Draa river. The contrast is clear between the Proterozoic substrate and the Palaeozoic sedimentary cover (see figure captions for more details).

3.2 Remote Sensing Geology with Hyperspectral Sensors

Hyperion was the first (and currently unique) space-borne hyperspectral sensor covering the range 0.4–2.5 μm. It gathers near-continuous data in 220 narrow bands along the 0.4–2.5 μm spectral range at a 30 m spatial resolution, whereas the CHRIS sensor onboard PROBA covers only the optical range (0.4–1.05 μm). The volume of data collected from Hyperion is about 75 times greater than that for an equivalent area from Landsat ETM+ bands! Basically, the data are relatively difficult to use for scientific purposes (difficult calibration, small coverage, etc.). Unfortunately, no published work has been found regarding Hyperion applications in Arab countries, whereas some Hyperion data cover the Bou Azzer area.

3.3 Remote Sensing Geology with SARs

Since the launch of Seasat in 1978, synthetic aperture radar (SAR) techniques have been widely applied. The last 25 years have shown a considerable increase in the potential for using SAR data with the launches of ERS 1–2, JERS, Radarsat 1–2, Envisat, ALOS PALSAR, etc. New SARs have been launched recently (TerraSAR-X, TanDEM-X, Sentinel 1-1A, etc.). SAR data have been successfully used either for methodological evaluations or for fundamental studies. Radar data help to improve model, particularly for moisture content. Interferometry is probably the most operational property of SAR data, because it allows a detailed survey of soil movement linked with subsidence or other dynamic phenomena, such as the seismicity. In arid regions, SAR sensors can detect some subsurface features up to over one meter below the surface. The different wavelengths (even if the C-band is the most commonly used) provide detailed information particularly interesting for soil moisture retrieval and for surface roughness retrieval as well. The combination of multi-frequency remote sensing information from the visible to microwave wavelengths should be developed.

3.4 Remote Sensing Geology with DEMs

Satellite-derived DEMs have considerably improved the accuracy of topographic information, particularly in poor-known regions. They provide relevant information on the geological and structural setting. DEMs built using C-band SRTM, X-SAR, and ASTER data have been evaluated in addition to the remote sensing dataset, particularly using morphotectonic approaches. DEMs are also useful to evaluate sediment fluxes (erosion or sedimentation).

4 Conclusion

Satellite imagery provides relevant and original information for geological mapping, particularly in arid countries. Basically, remote sensing first advantage is global and synthetic viewing. Remote sensing imageries are sometimes difficult to acquire and process but their value-added is always considerable. Optical data analysis gives multispectral and hyperspectral high-quality results for mapping lithologies and detecting tectonic features. Radar remote sensing is also relevant for revealing surface roughness, subsurface geology, moisture parameters and some structural features. SAR is used for interferometric approaches. Digital elevation models are also interesting to use for geomorphic studies, because Arab countries generally correspond to mountainous areas in a polyphased tectonic context. Methods based on satellite thermal infrared imaging data have recorded short-lived anomalies prior to major earthquakes, possibly related to impending earthquakes such as in the case of the 2003 Boumerdès earthquake. Space-borne remote sensing imagery must be combined with exogenous data (aerial photographs, geophysical survey, field measurements, etc.) for better results. Satellite remote sensing will continue to provide new observation tools in the future. These tools will be applied to geology, hydrogeology, natural hazards, and more generally to the development of the planet Earth. Arab countries located at the edge between the Mediterranean and the arid climate zones are particularly sensitive to evaluate the role of geology in the global impact of the environmental changes.

References

Abbas K, Deroin JP, Bouaziz S (2018) Monitoring of playa evaporites as seen with optical remote sensing sensors: case of Chott El Jerid, Tunisia, from 2003 to present. Arab J Geosci 11(5):1–14

Abdallah C, Chorowicz J, Bou Kheir R, Dhont D (2007) Comparative use of processed satellite images in remote sensing of mass movements: Lebanon as a case study. Int J Remote Sens 28 (19):4409–4427

Abdallah C, Chorowicz J, Bou Kheir R, Khawlie M (2005) Detecting major terrain parameters relating to mass movements' occurrence using GIS, remote sensing and statistical correlations, case study Lebanon. Remote Sens Environ 99(4):448–461

Abdelhamid G, Rabba I (1994) An investigation of mineralized zones revealed during geological mapping, Jabal Hamra Faddan-Wadi Araba, Jordan, using Landsat-TM data. Int J Remote Sens 15(7):1495–1506

Adams BJ, Huyck CK, Gusella L, Wabnitz C, Ghosh S, Eguchi RT (2006) Remote sensing technology for post-earthquake damage assessment—a coming of age. In: 8th US national conference on earthquake engineering, vol 3, pp 1733–1746

Agapiou A, Lysandrou V (2015) Remote sensing archaeology: tracking and mapping evolution in European scientific literature from 1999 to 2015. J Archaeol Sci Rep 4:192–200

Aliano C, Corrado R, Filizzola C, Pergola N, Tramutoli V (2007) Robust satellite techniques (RST) for seismically active areas monitoring: the case of 21st May, 2003 Boumerdes/Thenia (Algeria) earthquake. In: Proceedings of MultiTemp 2007–2007 international workshop on the analysis of multi-temporal remote sensing images, pp 4293042

Al Kwatli MA, Gillot PY, Al Gharib I, Lefèvre JC (2012) Integration of K-Ar geochronology and remote sensing: mapping volcanic rocks and constraining the timing of alteration processes (Al-Lajat Plateau, Syria). Quat Int 251:22–30

Alwash MA, Zilger J (1994) Remote sensing-based geological mapping of the area west of Al Madinah, Saudi Arabia. Int J Remote Sens 15(1):163–172

Amer R, Sultan M, Ripperdan R, Ghulam A, Kusky T (2013) An integrated approach for groundwater potential zoning in shallow fracture zone aquifers. Int J Remote Sens 34 (19):6539–6561

Amer R, Kusky T, Ghulam A (2010) Lithological mapping in the Central Eastern Desert of Egypt using Aster data. J Afr Earth Sci 56:75–82

Arnous MO (2011) Integrated remote sensing and GIS techniques for landslide hazard zonation: a case study Wadi Watier area, South Sinai, Egypt. J Coast Conserv 15(4):477–497

Baer G, Schattner U, Wachs D, Sandwell D, Wdowinski S, Frydman S (2002) The lowest place on earth is subsiding—An InSAR (interferometric synthetic aperture radar) perspective. Bull Geol Soc Am 114(1):12–23

Baghdadi N, Grandjean G, Lahondère D, Paillou P, Lasne Y (2005) The contribution of radar satellite imagery to geological exploration in arid areas. Comptes Rendus Geosci 337(8): 719–728

Bitelli G, Camassi R, Gusella L, Mognol A (2004) Remote sensing imagery for damage assessment of buildings after destructive seismic events. Manag Inf Syst 9:725–734

Bothorel A, Cervelle B, Chorowicz J, Tamain G, Alem EM (1984) Spectral signatures (visible/ near infrared) of rocks and ores; applications to remote sensing of three types of orebodies from S. Morocco. Mod Geol 8:277–294

Bouchet P, Cervelle B, Chorowicz J (1984) Contribution to spectral signature research on ore bodies found in south Morocco at three levels of investigation: satellite, ground and laboratory. Remote Sens Geol Mapp 18:251–266

Bou Kheir R, Greve MH, Deroin J-P, Rebai N (2013) Implementing GIS regression trees for generating the spatial distribution of copper in Mediterranean environments: the case study of Lebanon. Int J Environ Anal Chem 93(1):75–92

Breeze PS, Drake NA, Groucutt HS, Parton A, Jennings RP, White TS, Clark-Balzan L, Shipton C, Scerri EML, Stimpson CM, Crassard R, Hilbert Y, Alsharekh A, Al-Omari A, Petraglia MD (2015) Remote sensing and GIS techniques for reconstructing Arabian palaeohydrology and identifying archaeological sites. Quat Int 382:98–119

Brun J-P, Deroin J-P, Faucher T, Redon B, Téreygeol F (2013) Les mines d'or ptolémaïques. Résultats des prospections dans le district minier de Samut (désert Oriental). Bulletin de l'Institut Français d'Archéologie Orientale (BIFAO) 113:111–141

Butler RWH, Spencer S, Griffiths JM (1997) Transcurrent fault activity on the Dead Sea Transform in Lebanon. J Geol Soc London 154:757–760

Cai S, Xia B, Wan Z-F, Wang L-H (2014) Study on tectonic evolution and target areas for hydrocarbon exploration in Cheliff Basin, Algeria. Nat Gas Geosci 25(4):551–557

Cetin E, Meghraoui M, Cakir Z, Akoglu AM, Mimouni O, Chebbah M (2012) Seven years of postseismic deformation following the 2003 Mw = 6.8 Zemmouri earthquake (Algeria) from InSAR time series. Geophys Res Lett 39(10):L10307

Chaabouni R, Bouaziz S, Peresson H, Wolfgang J (2012) Lineament analysis of South Jenein Area (Southern Tunisia) using remote sensing data and geographic information system. Egypt J Remote Sens Space Sci 15(2):197–206

Chorowicz J, Benissa M (2016) Remote sensing and field analysis of the Palaeozoic structural style in NW Libya: the Qarqaf arch a paleo-transfer fault zone between the Ghadamis and Murzuq basins. J Afr Earth Sci 123:272–293

Chorowicz J, Dhont D, Ammar O, Rukieh M, Bilal A (2004) Tectonics of the Pliocene Homs basalts (Syria) and implications for the Dead Sea fault zone activity. J Geol Soc London 162:259–271

Chorowicz J, Fabre J (1997) Organization of drainage networks from space imagery in the Tanezrouft plateau (Western Sahara): implications for recent intracratonic deformations. Geomorphology 21(2):139–151

Chorowicz J, Lyberis N (1989) Evolution tectonique de la péninsule du Sinai: la formation des plis de l'arc syrien et des golfes de Suez et d'Aqaba. Notes et Mémoires. Compagnie Française des Pétroles 21:199–209

Ciampalini A, Garfagnoli F, Del Ventisette C, Moretti S (2013) Potential use of remote sensing techniques for exploration of iron deposits in Western Sahara and Southwest of Algeria. Nat Resour Res 22(3):179–190

Deffontaines B, Chotin P, Brahim LA, Rozanov M (1992) Investigation of active faults in Morocco using morphometric methods and drainage pattern analysis. Geol Rundsch 81 (1):199–210

De Laet V, van Loon G, Van der Perre A, Deliever I, Willems H (2015) Integrated remote sensing investigations of ancient quarries and road systems in the Greater Dayr al-Barshā Region, Middle Egypt: a study of logistics. J Archaeol Sci 55:286–300

Deroin J-P, Bou Kheir R, Abdallah C (2017) Geoarchaeological remote sensing survey for cultural heritage management. Case study from Byblos (Jbail, Lebanon). J Cult Herit 23:37–43

Deroin J-P, Djemai S, Bendaoud A, Brahmi B, Ouzegane K, Kienast J-R (2014) Integrating geologic and satellite radar data for mapping dome-and-basin patterns in the In Ouzzal Terrane, Western Hoggar, Algeria. J Afr Earth Sci 99:652–665

Deroin J-P, Motti E, Simonin A (1998) Comparison of potential for using optical data and synthetic aperture radar for geological mapping in arid regions (Atar, Western Sahara, Mauritania). Int J Remote Sens 19(6):1115–1132

Deroin J-P, Téreygeol F, Heckes J (2011) Evaluation of very high to medium resolution multispectral satellite imagery for geoarchaeology in arid regions, case study from Jabali, Yemen. J Archaeol Sci 38:101–114

Deroin JP, Téreygeol F, Heckes J (2012) Remote sensing study of the ancient Jabali silver mines (Yemen): from past to present. In: Lasaponara R, Masini N (eds) Satellite remote sensing: a new tool for archaeology. Springer, Berlin, 345 p, pp 231–246

Deroin J-P, Wade S, Dia A (2012b) Impacts of breaching on a bar-built estuary, example of the Senegal River new mouth monitored with satellite imagery from 2003 to 2010. Zeitschrift für Geomorphologie 56:351–370

Dill HG, Hahne K, Shaqour F (2012) Anatomy of landslides along the Dead Sea transform fault system in NW Jordan. Geomorphology 141–142:134–149

Dhont D, Chorowicz J, Collet B, Lichtenegger J, Barbieri M (2005) Spaceborne radar applications in geology. An introduction to imaging radar, and application examples of ERS SAR in Geology and Geomorphology. Edited by European Space Agency, ESA Publications Division, ESTEC, Noordwijk, The Netherland, TM-17, 230 p

Dhont D, Luxey P, Chorowicz J (2005b) 3-D modeling of geologic maps from surface data. AAPG Bull 89(11):1465–1474

El Asmar HM (2002) Short term coastal changes along Damietta-Port Said coast northeast of the Nile Delta, Egypt. J Coast Res 18(3):433–441

Elewa HH, Qaddah AA (2011) Groundwater potentiality mapping in the Sinai Peninsula, Egypt, using remote sensing and GIS-watershed-based modeling. Hydrogeol J 19(3):613–628

El Shazly EM, Abdel Hady MA, El Ghawaby MA, Khawasik SM, El Shazly MM (1978) Landsat satellite imagery of the Qattara Depression area, Egypt. Acta Astronautica 5(10):947–957

Emran A, Chorowicz J, Bassias Y (1995) Analyse tectonique des ophiolites de Tourtit (Anti-Atlas, Maroc) à l'aide de l'imagerie spatiale SPOT, de métamorphisme et des données de terrain: résultats et modèle tectonique. Géo Observateur 6:75–86, Rabat (Maroc)

Emran A, Hakdaoui M, Chorowicz J (1996) Anomalies on geologic maps from multispectral and textural classification: the Bleida mining district (Morocco). Remote Sens Environ 57(1):13–21

Fleury J, Chorowicz J, Somma J (1999) Discussion on transcurrent fault activity on the Dead Sea transform in Lebanon and its implications for plate tectonics and seismic hazards. J Geol Soc London 154:757–760

Fraser A, Huggins P, Rees J, Cleverly P (1997) A satellite remote sensing technique for geological structure horizon mapping. Int J Remote Sens 18(7):1607–1615

Gaber A, Soliman F, Koch M, El-Baz F (2015) Using full-polarimetric SAR data to characterize the surface sediments in desert areas: a case study in El-Gallaba Plain, Egypt. Remote Sens Environ 162:11–28

Gad S, Kusky T (2006) Lithological mapping in the Eastern Desert of Egypt, the Barramiya area, using Landsat Thematic Mapper (TM). J Afr Earth Sci 44(2):196–202

Gad S, Kusky T (2007) Aster spectral ratioing for lithological mapping in the Arabian-Nubian shield, the Neoproterozoic Wadi Kid area, Sinai, Egypt. Gondwana Res 11(3):326–335

Gemail K, Abd-El Rahman NM, Ghiath BM, Aziz RN (2016) Integration of Aster and airborne geophysical data for mineral exploration and environmental mapping: a case study, Gabal Dara, North Eastern Desert, Egypt. Environ Earth Sci 75(592):1–12

Henry C, Chorowicz J (1986) Région du Sinaï. Carte géologique et géomorphologique téléanalytique au 1/500.000. Editions Technip, Paris

Hereher ME (2011) Mapping coastal erosion at the Nile Delta western promontory using Landsat imagery. Environ Earth Sci 64(4):1117–1125

Huchon P, Khanbari K (2003) Rotation of the syn-rift stress field of the northern Gulf of Aden margin, Yemen. Tectonophysics 364(3–4):147–166

Jabbour M, Dhont D, Hervouët Y, Deroin J-P (2012) Geometry and kinematics of fault-propagation folds with variable interlimb angle. J Struct Geol 42:212–226

Jakob S, Bühler B, Gloaguen R, Breitkreuz C, Eliwa HA, El Gameel K (2015) Remote sensing based improvement of the geological map of the Neoproterozoic Ras Gharib segment in the Eastern Desert (NE-Egypt) using texture features. J Afr Earth Sci 111:138–147

Jacobberger PA, Arvidson RE, Rashka DL (1983) Application of Landsat multispectral scanner data and sediment spectral reflectance measurements to mapping of the Meatiq Dome, Egypt. Geology 11(10):587–591

Kermani S, Boutiba M, Guendouz M, Guettouche MS, Khelfani D (2016) Detection and analysis of shoreline changes using geospatial tools and automatic computation: case of Jijelian sandy coast (East Algeria). Ocean Coast Manag 132:46–58

Lacava T, Ciancia E, Coviello I, Di Polito C, Grimaldi CSL, Pergola N, Satriano V, Temimi M, Zhao J, Tramutoli V (2017) A MODIS-based robust satellite technique (RST) for timely detection of oil spilled areas. Remote Sens 9(2):128

Lamri T, Djemai S, Hamoudi M, Zoheirc B, Bendaoudb A, Ouzegane K, Amara M (2016) Satellite imagery and airborne geophysics for geologic mapping of the Edembo area, Eastern Hoggar (Algerian Sahara). J Afr Earth Sci 115:143–158

Laribi A, Walstra J, Ougrine M, Seridi A, Dechemi N (2015) Use of digital photogrammetry for the study of unstable slopes in urban areas: case study of the El Biar landslide, Algiers. Eng Geol 187:73–83

Le Béon M, Klinger Y, Mériaux AS, Al-Qaryouti M, Finkel RC, Mayyas O, Tapponnier P (2012) Quaternary morphotectonic mapping of the Wadi Araba and implications for the tectonic activity of the southern Dead Sea fault. Tectonics 31(5):TC5003

Li C-L, Zhang J-F (2013) Method and application of earthquake damage change detection from remote sensing images based on principal component analysis. Earthquake 33(2):103–108

Löhrer R, Bertrams M, Eckmeier E, Protze J, Lehmkuhl F (2013) Mapping the distribution of weathered Pleistocene wadi deposits in Southern Jordan using ASTER, SPOT-5 data and laboratory spectroscopic analysis. CATENA 107:57–70

Louati M, Saïdi H, Zargouni F (2014) Shoreline change assessment using remote sensing and GIS techniques: a case study of the Medjerda delta coast, Tunisia. Arab J Geosci 8(6):4239–4255

Mahmood A (1996) Lineaments as groundwater exploration guides in hard-rock terranes of arid regions. Can J Remote Sens 22(1):108–116

Massironi M, Bertoldi L, Calafa P, Visoná D, Bistacchi A, Giardino C, Schiavo A (2008) Interpretation and processing of Aster data for geological mapping and granitoids detection in the Saghro massif (eastern Anti-Atlas, Morocco). Geosphere 4(4):736–759

McCauley JF, Breed CS, Schaber GG, McHugh WP, Issawi B, Haynes CV, Grolier MJ, El Kilani A (1986) Paleodrainages of the Eastern Sahara—the radar rivers revisited (SIR-A/B Implications for a Mid-Tertiary Trans-African drainage system). IEEE Trans Geosci Remote Sens 24:624–648

McCauley JF, Schaber GG, Breed CS, Grolier MJ, Haynes CV, Issawi B, Elachi C, Blom R (1982) Subsurface valleys and geoarchaeology of the Eastern Sahara revealed by Shuttle Radar. Science 218:1004–1020

Milewski A, Sultan M, Yan E, Becker R, Abdelhayem A, Soliman F, Gelil KA (2009) A remote sensing solution for estimating runoff and recharge in arid environments. J Hydrol 373(1–2):1–14

Mostafa ME, Bishta AZ (2005) Significance of lineament patterns in rock unit classification and designation: a pilot study on the Gharib-Dara area, northern Eastern desert, Egypt. Int J Remote Sens 26(7):1463–1475

Omar AA (1990) Use of SIR-A interpretation for underground water prospecting in southern Iraq. In: Remote sensing: an operational technology for the mining and petroleum industries. Conference, IMM, London, pp 165–172

Othman AA, Gloaguen R (2014) Improving lithological mapping by SVM classification of spectral and morphological features: The discovery of a new chromite body in the Mawat ophiolite complex (Kurdistan, NE Iraq). Remote Sens 6(8):6867–6896

Othman AA, Gloaguen R (2013) Automatic extraction and size distribution of landslides in Kurdistan region, NE Iraq. Remote Sens 5(5):2389–2410

Paillou P, Lopez S, Farr T, Rosenqvist A (2010) Mapping subsurface geology in Sahara using L-Band SAR: First results from the ALOS/PALSAR imaging radar. IEEE J Sel Top Appl Earth Obs Remote Sens 3(4):632–636

Peña SA, Abdelsalam MG (2006) Orbital remote sensing for geological mapping in southern Tunisia: implication for oil and gas exploration. J Afr Earth Sci 44(2):203–219

Poujol A, Ritz J-F, Tahayt A, Vernant P, Condomines M, Blars P-H, Billant J, Vacher L, Tibari B, Hni L, Idrissi Koulali A (2014) Active tectonics of the Northern Rif (Morocco) from geomorphic and geochronological data. J Geodyn 77:70–88

Qari MYHT (1992) Improving geological mapping in the Arabian Shield using Landsat Thematic Mapper imagery-Jabal Duhayyah area. Int J Remote Sens 13(15):2765–2772

Qiu F, Abdelsalam M, Thakkar P (2006) Spectral analysis of Aster data covering part of the Neoproterozoic Allaqi-Heiani suture, Southern Egypt. J Afr Earth Sci 44(2):169–180

Reif D, Grasemann B, Faber RH (2011) Quantitative structural analysis using remote sensing data: Kurdistan, northeast Iraq. AAPG Bull 95(6):941–956

Roy R, Launeau P, Carrère V, Pinet P, Ceuleneer G, Clénet H, Daydou Y, Girardeau J, Amri I (2009) Geological mapping strategy using visible near-infrared-shortwave infrared hyperspectral remote sensing: application to the Oman ophiolite (Sumail Massif). Geochem Geophys Geosyst 10(2):Q02004

Saadi NM, Abdel Zaher M, El-Baz F, Watanabe K (2011) Integrated remote sensing data utilization for investigating structural and tectonic history of the Ghadames Basin, Libya. Int J Appl Earth Obs Geoinf 13(5):778–791

Saadi NM, Watanabe K (2009) Assessing image processing techniques for geological mapping: a case study in Eljufra, Libya. Geocarto Int 24(3):241–253

Sarti F, Arkin Y, Chorowicz J, Karnieli A, Cunha T (2003) Assessing pre- and post-deformation in the southern Arava Valley segment of the Dead Sea Transform, Israel by differential interferometry. Remote Sens Environ 86(2):141–149

Schaber GG, McCauley JF, Breed CS (1997) The use of multifrequency and polarimetric SIR-C/X-SAR data in geologic studies of Bir Safsat, Egypt. Remote Sens Environ 59:337–363

Schaber GG, McCauley JF, Breed CS, Olhoeft GR (1986) Shuttle imaging radar: physical controls on signal penetration and subsurface scattering in the Eastern Sahara. IEEE Trans Geosci Remote Sens 24:603–623

Siebert C, Rödiger T, Mallast U, Gräbe A, Guttman J, Laronne JB, Storz-Peretz Y, Greenman A, Salameh E, Al-Raggad M, Vachtman D, Ben Zvi A, Ionescu D, Brenner A, Merz R, Geyer S (2014) Challenges to estimate surface- and groundwater flow in arid regions: the Dead Sea catchment. Sci Total Environ 485–486(1):828–841

Sultan M, Arvidson RE, Sturchio NC, Guinness EA (1987) Lithologic mapping in arid regions with Landsat thematic mapper data: Meatiq dome, Egypt. Geol Soc Am Bull 99(6):748–762

Voûte C (1986) Multistage groundwater exploration and satellite remote sensing test area: the Kasserine basin (Tunisia). Photogrammetria 40(4):317–326

White K, El Asmar HM (1999) Monitoring changing position of coastlines using Thematic Mapper imagery, an example from the Nile Delta. Geomorphology 29(1–2):93–105

Wu W (2007) Coastline evolution monitoring and estimation—a case study in the region of Nouakchott, Mauritania. Int J Remote Sens 28(24):5461–5484

Youssef AM, Pradhan B, Jebur MN, El-Harbi HM (2015) Landslide susceptibility mapping using ensemble bivariate and multivariate statistical models in Fayfa area, Saudi Arabia. Environ Earth Sci 73(7):3745–3761

Meteorite Impact Structures in the Arab World: An Overview

M. C. Chabou

Abstract This chapter provides an up-to-date review about confirmed, proposed, and disproven meteorite impact structures in the Arab countries. Among the ~190 confirmed impact structures/sites on Earth, only 13 have been identified in the countries of the Arab world: Agoudal (Morocco), Amguid (Algeria), Aouelloul (Mauritania), BP (Libya), Kamil (Egypt), Oasis (Libya), Ouarkziz (Algeria), Jebel Waqf as Suwwan (Jordan), Saqqar (Saudi Arabia), Talemzane (Algeria), Tenoumer (Mauritania), Tin Bider (Algeria), and Wabar (Saudi Arabia). Thirty-three other structures of alleged impact origin and located in the Arab countries have been proposed in the literature. Based on the examination of satellite images and available field geologic, petrographic, and geochemical documentation, some of these proposed features are not of impact origin. Others are promising impact crater candidates. The record of impact structures in the Arab region is still incomplete considering the great size of the territory and the number of impact craters known in other regions where there have been active programs to study impact structures. Field investigations, including the collection and analyses of rock samples, are required for adding new candidates to the list of confirmed impact craters in the Arab countries. The only criteria that prove evidence for an impact origin of circular geological structures are the occurrence of shock-metamorphic effects in the target rocks (shatter cones and diagnostic shock-metamorphic mineral deformation and transformation phenomena), the discovery of preserved meteorite fragments, or detection of geochemical traces of meteoritic projectiles. Particular attention should be paid during fieldwork and subsequent petrographic studies, as some structures previously identified as impact craters had to be discounted. Nine of these cases of discredited structures are described here.

Keywords Impact structures · Arab world · Shatter cones · Shock metamorphism Simple impact crater · Complex impact crater

M. C. Chabou (✉)
Département Des Sciences de La Terre, Université Ferhat Abbas, Sétif 1, Institut D'Architecture et Des Sciences de La Terre, Campus El-Bez, 19000 Sétif, Algeria
e-mail: charaf.chabou@univ-setif.dz

A. Bendaoud et al. (eds.), *The Geology of the Arab World—An Overview*, Springer Geology, https://doi.org/10.1007/978-3-319-96794-3_13

1 Introduction

Meteorite impacts on Earth are now recognized as a fundamental geological process (French 1998, 2004). There are also several geological structures in the terrestrial impact record that are of particular economic interest (Grieve and Masaitis 1994; Grieve 2005; Reimold et al. 2005). The inventory and description of impact craters on Earth are important primarily for the estimation of the impact cratering rate and its variation with time (Grieve 1984; Grady et al. 1998; Hughes 2000). In the same context, it allows for the estimation of long-term erosion rates on Earth (Hergarten and Kenkmann 2015). On the other hand, observations of the terrestrial impact record are of great relevance in constraining and understanding impact processes (Grieve and Therriault 2004). Numerous papers have been dedicated to the description and scientific investigation of impact sites (including suspected impact structures) in different areas of the world, particularly those where the impact cratering record is low, such as Africa (see, e.g., Koeberl 1994; Reimold and Koeberl 2014) or regions where part or most of the literature is only available in local language (e.g., Russia, Ukraine, and Fennoscandia) (Pesonen 1996; Masaitis 1999; Abels et al. 2002). Currently, there are approximately 190 confirmed impact structures/sites on Earth, and only 13 are currently known in the Arab world, most of them located in the Sahara (Earth Impact Database 2018): Agoudal (Morocco), Amguid (Algeria), Aouelloul (Mauritania), BP (Libya), Kamil (Egypt), Oasis (Libya), Ouarkziz (Algeria), Jebel el Waqf as Suwwan (Jordan), Saqqar (Saudi Arabia), Talemzane (Algeria), Tenoumer (Mauritania), Tin Bider (Algeria), and Wabar (Saudi Arabia). This number is very small considering (i) the large area of this region, with a variety of exposed surfaces suitable for the preservation of impact craters (arid deserts, cratonic areas), (ii) estimates of cratering rates on Earth (Grieve and Dence 1979; Grady et al. 1998), (iii) the large number of impact craters with a diameter between 0.25 and 6 km still waiting to be discovered on the Earth's surface (>340, Hergarten and Kenkmann 2015), (iv) the number of impact craters that remain to be found in Phanerozoic strata on Earth (228 larger than 2.5 km, Stewart 2011), and (v) comparing and contrasting the impact cratering record of the Arab world with that of other regions where active research programs have uncovered many impact structures over the past decades (e.g., in North America, Europe, or Australia) (e.g., Robertson and Grieve 1975; Hodge 1994; Haines 2005). This indicates that a large number of impact structures remain to be discovered in this vast area. In recent years, the status of known and possible impact structures in the Arab countries has been the subject of several conference communications (Chabou 2009, 2011a, 2016; Youbi et al. 2011). The aim of this paper is to review and update our knowledge on the meteorite impact structures in the Arab world based on an up-to-date literature review, and to discuss some problems concerning the discovery of new impact structures in the region.

2 Short Overview of Meteorite Impact Structures

2.1 Different Types of Impact Craters

Impact craters are subdivided into two main morphological groups: simple bowl-shaped and complex impact craters (e.g., Grieve 1987; Grieve and Pilkington 1996; Turtle et al. 2005; Osinski and Pierazzo 2013). Fresh simple impact craters have bowl-shaped depressions with uplifted rims (Fig. 1a). This type of crater shows, if preserved, an overturned flap of target strata on the rim with inverted stratigraphy. This is overlain by ejecta that forms an ejecta blanket around the crater structure (e.g., as seen at Meteor Crater, Arizona, USA; e.g., Kring 2017). The cavity is filled with an allochthonous breccia lens that comprises unshocked and shocked target rock and can be rich in impact melt fragments. These allochthonous materials are underlain by autochthonous fractured and locally brecciated target rocks of the crater floor. Very young and well-preserved simple impact craters may also show an ejecta ray structure as in the case of the Kamil crater in Egypt (Folco et al. 2010) (Fig. 4e).

Complex impact craters exhibit a structurally uplifted central area surrounded by an annular depression partly or largely filled with allochthonous material (breccias and/or impact melt rocks) (Fig. 1b). Inner crater rim sections display downfaulted terraces. The simple-to-complex transition for impact craters on Earth occurs at diameters of ~2 km in sedimentary targets and ~4 km in crystalline rocks (Grieve 1987).

Finally, the largest impact structures are surrounded by multiple concentric uplifted rings and are known as multiring basins (e.g., Spudis 1993; French 1998).

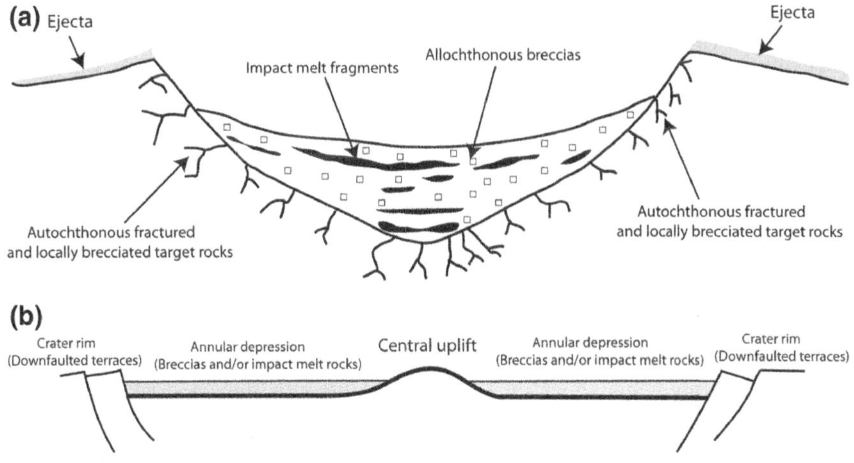

Fig. 1 **a** Schematic cross section of a simple, bowl-shaped impact crater (modified after Grieve 1987). **b** Schematic cross section of a complex impact structure (modified after Grieve and Pilkington 1996)

The three largest known impact structures on Earth (Chicxulub, Sudbury, and Vredefort) have been cited as possible multiring basins (Grieve and Therriault 2000; Grieve et al. 2008).

2.2 How to Recognize an Impact Crater

The accepted criteria for the identification of impact structures on Earth are the occurrence of shock-metamorphic effects in minerals and rocks, the discovery of preserved meteorite fragments, or the detection of traces (chemical or isotopic) of a meteoritic projectile (e.g., French and Koeberl 2010). It is important to emphasize that remote sensing and geophysical studies alone only provide insufficient evidence of the possible existence of impact structures (e.g., Koeberl 2004; Schmieder et al. 2013; Reimold and Koeberl 2014).

Shock-metamorphic effects are subdivided into macroscopic and microscopic features. Shatter cones are the only macroscopic diagnostic shock indicator. They are represented by a range of curved to curvilinear fractures penetrating shocked target rock, decorated with more or less diverging striations that radiate from an apex of a conical feature or from a narrow apical area (e.g., Dietz 1947; Baratoux and Reimold 2016) (Fig. 2). Shatter cones are thought to form in the shock pressure range of 2–30 GPa but the exact mechanisms of their formation are still not fully understood. Among the 13 confirmed impact structures and sites in the Arab world, only 2 have so far yielded shatter cones (Table 1 in Baratoux and Reimold 2016): Jabal Waqf As Suwaan in Jordan and the Agoudal site in Morocco.

Diagnostic microscopic shock features include planar microdeformation features (planar fractures and planar deformation features), isotropization (diaplectic glass), and phase changes in minerals (high-pressure phases; melting) (e.g., French and Koeberl 2010).

Planar microstructures in quartz are the most widely applied criteria for identifying suspected impact structures (Stöffler and Langenhorst 1994).

Planar fractures (PFs) are parallel sets of open subplanar fissures in quartz grains with spacings of about 20–30 μm (Langenhorst and Deutsch 2012). They are characteristic of low shock pressure (<10 GPa). PFs are not considered a unique impact criterion, as they can also occur in quartz grains deformed in nonimpact environments (French 1998; French and Koeberl 2010). However, planar fractures and twin structures in zircon and monazite are of shock-diagnostic value (e.g., Erickson et al. 2016; Timms et al. 2017).

Planar deformation features (PDFs) in quartz (or other rock-forming minerals) are the most convincing shock effect in recognizing new impact structures (French and Koeberl 2010). PDFs are extremely narrow spaced, planar features of amorphous material or fluid inclusion trails, that can occur in single or multiple sets per host grain. Spacings between individual features are in the range of 2–10 μm and the width of individual PDF is typically <2–3 μm (French 1998). PDFs develop in quartz at pressures between ~8 and 35 GPa (Langenhorst and Deutsch 2012).

Fig. 2 Shatter cones in limestone from the Steinheim structure, Germany (Natural History Museum, Vienna)

At low shock degree (between ~8 and 15 GPa), PDFs generally occur as one set per host grain (when oriented perpendicular to (0001), they are mechanical Brazil twins); at higher shock pressures, two or multiple sets become prominent. These sets are oriented parallel to specific crystallographic planes within the quartz lattice. At pressure > 35 GPa, quartz grains are converted to diaplectic glass (shock isotropization). PDFs are considered unique impact indicators (e.g., French 1998).

In porous sedimentary rocks such as sandstone, shock classification based on seven progressive shock stages (SR-S1 to SR-S7) has recently been revised (Stöffler et al. 2018 and references therein). The first three shock stages are defined by the progressive removal of porosity, whereas the division into shock stages SR-SR4 to SR-SR6 is based on the proportions of quartz, coesite, stishovite, and glass. The occurrence of a vesicular (pumiceous) melt rock with vesiculated and flow-textured glass (lechatelierite) is a characteristic of the last shock stage. Based on experimental observation and numerical modeling, Kowitz et al. (2016) showed that the formation of shock deformation features in porous materials strongly depends on the primary porosity and/or water content and proposed new shock classification and calibration scheme for porous, quartz-rich rocks.

Table 1 List of confirmed impact craters/sites in the Arab world

Name	Country	Latitude	Longitude	Diameter (km)	Age (Ma)
Agoudal[a]	Morocco	31°59′ N	05°30′ W	?	<170
Amguid	Algeria	26°05′ N	04°23′ E	0.45	≤0.1
Aouelloul	Mauritania	20°15′ N	12°41′ W	0.36	3.1 ± 0.3
B. P. Structure	Libya	25°19′ N	24°20′ E	2.8	<120
Jebel Waqf as Suwwan	Jordan	31°03′ N	36°48′ E	5.5	≤30
Kamil	Egypt	22°01′ N	26°05′ E	0.045	Quaternary
Oasis	Libya	24°35′ N	24°24′ E	11.5	<120
Ouarkziz	Algeria	29°00′ N	07°33′ W	3.5	<70
Saqqar[b]	Saudi Arabia	29°35′ N	38°42′ E	34	70–410 Ma
Talemzane	Algeria	33°19′ N	04°02′ E	1.75	<3
Tenoumer	Mauritania	22°55′ N	10°24′ W	1.9	2.5 ± 0.5
Tin Bider	Algeria	27°36′ N	05°07′ E	6	<70
Wabar	Saudi Arabia	21°30′ N	50°28′ E	0.12	0.00029

[a]Impact site
[b]Buried structure

High-pressure mineral phases are often observed at meteorite impact craters. The most important are the high-pressure polymorphs of silica (coesite and stishovite), formed and preserved at pressures between 30 and 60 GPa for coesite, and between 12 and 45 GPa for stishovite (Stöffler and Langenhorst 1994). Impact diamonds have also been identified at a number of terrestrial impact structures. Since these polymorphs occur naturally in deep-seated terrestrial rocks, they are excellent impact indicators when found in upper crustal rocks. Preserved meteorite fragments associated with impact craters are very rare and are seemingly restricted to small, simple impact craters (individuals or in crater fields). This association is observed in very young small craters (<0.5 Ma) that formed by an iron meteorite projectile (French and Koeberl 2010). Two impact sites in the Arab world are associated with iron meteorites: Kamil crater in Egypt and the Wabar crater field in Saudi Arabia. It should be noted, however, that there is a possibility of a coincidental association of meteorite fragments with an impact structure as is the case for the Zerga meteorite found immediately outside the rim of the Aouellouel crater (Mauritania) (Fudali and Cressy 1976), or the Agoudal iron meteorite strewn field located near the Agoudal shatter cone site (Chennaoui Aoudjehane et al. 2016).

Finally, the detection of extraterrestrial components in impactites has been used to confirm an impact origin of a crater structure (e.g., Tagle and Hecht 2006; French and Koeberl 2010; Goderis et al. 2013). An excellent example is provided by the Aouelloul crater, where the isotopic analysis of impact glasses showed the admixture of a meteoritic component in the glass (Koeberl et al. 1998). These results constitute the only evidence for the impact origin of this crater, as no unambiguous mineralogical evidence of shock metamorphism has been found in rocks from Aouelloul to date.

2.3 Economic Importance of Meteorite Impact Craters

Several scientific works have been devoted to the economic potential of terrestrial impact structures (Masaitis 1989, 1992; Grieve and Masaitis 1994; Reimold 1995; Grieve 2005, 2013; Reimold et al. 2005). Grieve and Masaitis (1994) estimated that approximately 25% of the then known impact structures in the world were associated with economic resources, and 12% were then actively exploited. Reimold et al. (2005) inventoried 56 impact structures (32% of the then listed terrestrial impact structures) with a particular economic interest, including ore resources, hydrocarbon deposits, building materials and other useful substances, water and hydropower reservoirs, fishing, educational and recreational activities, and tourism values.

Ore and hydrocarbon deposits related to impact cratering are subdivided into three types according to their time of formation (Masaitis 1989; Grieve and Masaitis 1994; Grieve 2005): progenetic, syngenetic, and epigenetic.

Progenetic mineralizations are linked to impact structures that were formed in already ore-bearing target rocks. The impact processes led to the redistribution, preservation, or improved accessibility of ore material. Examples for this type are the gold and uranium deposits of the 2.02 Ga old Vredefort structure (Witwatersrand basin, South Africa), the iron ore deposits and uranium at the Ternovka structure (Ukraine), and the uranium deposits of the Carswell impact structure (Canada) (Reimold et al. 2005 and references therein; Grieve 2013).

Syngenetic ore mineralization is formed during and/or immediately after the impact event—as a direct consequence of impact processes. This includes the postimpact hydrothermal deposits that result from the instantaneous shock-heating, partial melting, and subsequent cooling of the target rock due to the impact process (Grieve 2005, 2013). Examples for this type are the Ni, Cu, PGE of the Sudbury structure (Canada) (Ames et al. 2008) and the impact diamonds of the Popigai impact structure (Russia) (Masaitis 1998).

Epigenetic deposits are formed after the impact event and result from the formation of enclosed basins, with restricted sedimentation, or the flow of fluids into structural traps associated with impact structures. This type is represented by hydrocarbon reservoirs associated with impact structures. Examples of this type are the Viewfield, Red Wing Creek and Ames oil pools (USA) (Donofrio 1981; Johnson and Campbell 1997), and the 66 Ma old Chicxulub impact structure (Mexico) (Grajales-Nishimura et al. 2000; see the review by Reimold et al. 2005).

In some cases, the impact generated lithologies are used as building materials. Examples are the churches and castles in Nördlingen (Germany) and Rochechouart (France) that were constructed using suevite and other impact breccias of the Ries and Rochechouart impact craters. Some depressions that result from impact cratering form permanent lakes or water reservoirs such as the Bosumtwi (Ghana) and Lonar (India) lakes, and Chesapeake Bay (USA) that represents an important hydrological structure regarding the water supply for millions of consumers. The lakes of the Manicouagan (Canada) and Puchezh-Katunki (Russia) impact structures are important reservoirs for hydroelectric power generation. Finally, some impact craters provide sites for educational and recreational activities (Bosumtwi, Rochechouart, Tswaing meteorite crater in South Africa, Vredefort, Chesapeake Bay in USA) and/or serve as tourist attractions (the museums at Meteor Crater in the USA, at the Ries crater and Steinheim basin crater in Germany, at Rochechouart, and at the Tswaing crater) (Reimold et al. 2005).

3 Meteorite Impact Structures/Sites in the Arab World

3.1 Confirmed Impact Structures/Sites (Figs. 3, 4, 5)

As of March 2018, 13 impact structures or sites have been identified in the countries of the Arab world (Earth Impact Database 2018). Eleven of them are exposed impact structures, one (Saqqar, Saudi Arabia) is a buried impact crater, and one (Agoudal, Morocco) is probably an eroded impact site (remnant of an impact crater). An up-to-date summary of these confirmed meteorite impact sites is given below, while further detailed information about the structures located in the Arab countries of Africa can be found in Reimold and Koeberl (2014). The basic data for confirmed impact structures in the Arab world are summarized in Table 1, and their locations are indicated in Fig. 3.

3.1.1 Agoudal, Morocco (31°59′ N–05°30′ W)

On the basis of the discovery of several occurrences of shatter cones in Middle Jurassic marly limestone (Sadilenko et al. 2013), a new impact site was identified near Agoudal in the High Atlas Mountains of Morocco. As no impact structure was found in association with the shatter cone bearing rock (Chennaoui Aoudjehane et al. 2016), the site has been regarded as a remnant of a deeply eroded impact crater. This is the first discovery of an impact site in Morocco, which joins the six Arab countries (Algeria, Egypt, Jordan, Libya, Mauritania and Saudi Arabia) that also have confirmed impact structures. A coeval origin of the strewn field of the Agoudal iron meteorite (Chennaoui-Aoudjehane et al. 2013) and the Agoudal relic impact structure was suggested by Lorenz et al. (2015) following the discovery of fragments of iron which had the same chemical composition as that of the Agoudal meteorite in apparently autochthonous breccia. Schmieder et al. (2015) arrived at the same conclusion following the discovery of primary and oxidized meteoritic matter on a shatter cone surface from the Agoudal site. However, Chennaoui-Aoudjehane et al. (2016) excluded a genetic link between the Agoudal meteorite fall and the Agoudal impact site based on detailed field observations and structural data. The authors concluded that the observed shatter cones belong to an ancient impact structure 1–3 km in diameter that has been almost entirely eroded.

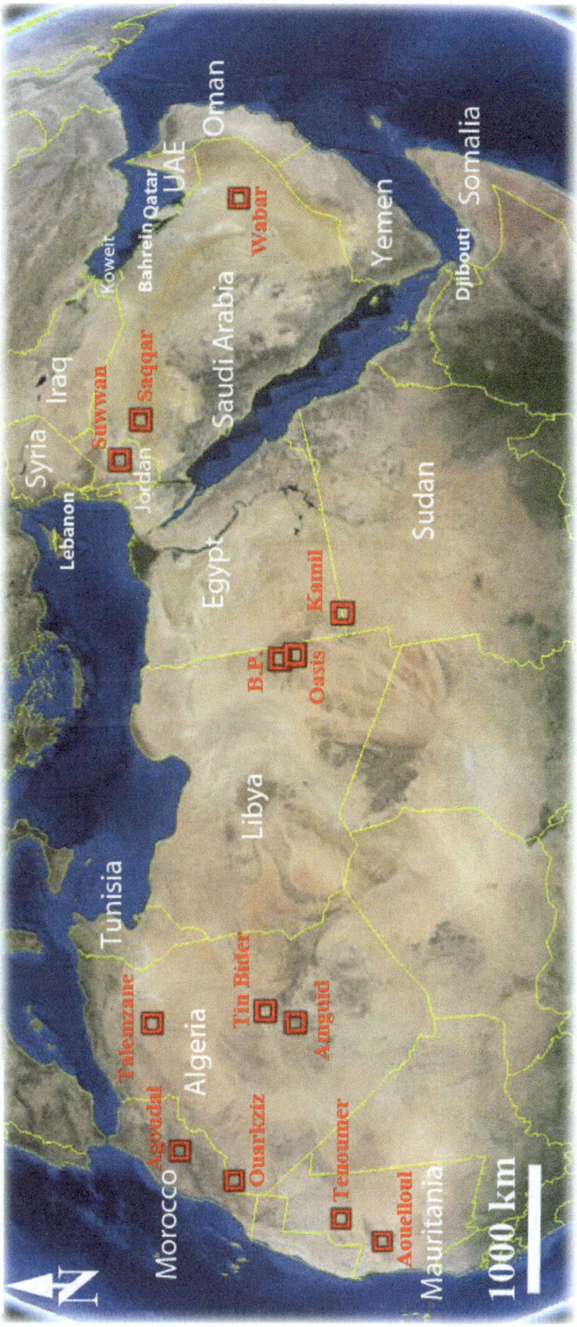

Fig. 3 Location of the 13 currently known meteorite craters and impact structures in the Arab world on a Google Earth Satellite image

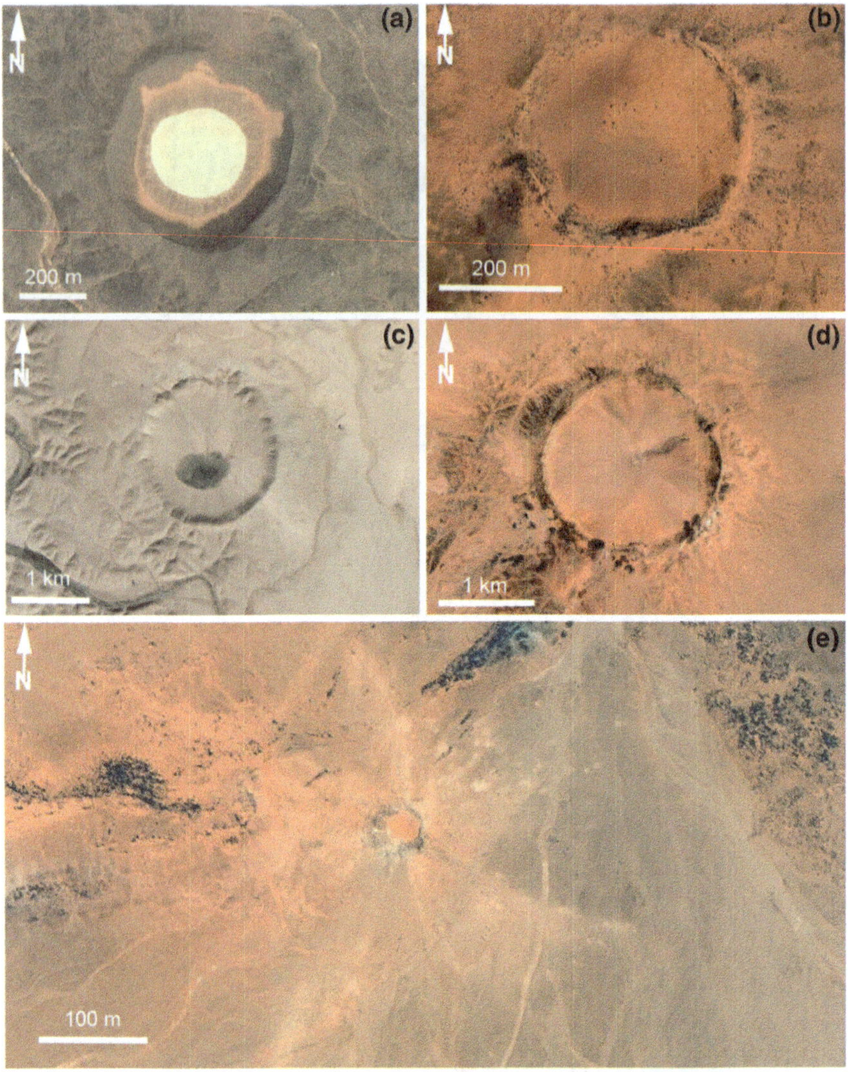

Fig. 4 Satellite views (Google Earth images) of 5 confirmed, simple bowl-shaped impact craters of the Arab world. **a** Amguid (Algeria). **b** Aouelloul (Mauritania). **c** Talemzane (Algeria). **d** Tenoumer (Mauritania). **e** Kamil (Egypt)

They also emphasize that the alleged impact breccias are in fact calcrete formations. Recently, Nachit et al. (2017) suggested the presence of small impact crater of about 500–600 m in diameter that was hidden by a limestone block interpreted as the product of a landslide. This discovery was made using 2D electrical resistivity tomography and must be checked by other investigation methods. The genetic link between the Agoudal meteorite and the Agoudal crater was again suggested by

Nachit et al. (2017) on the basis of the overlap of the meteorite strewn field with the shatter cones. Even though this site is cited in recent publications (e.g., Reimold and Koeberl 2014; Chennaoui-Aoudjehane et al. 2016) and in the Earth Impact Database of the University of New Brunswick, Canada, as a confirmed impact structure, there is no definitive evidence that Agoudal is an in situ impact site. There is a possibility that the site only has a block with shatter cones derived from an impact structure (U.W. Reimold, personal communication).

3.1.2 Amguid, Algeria (26°05′ N–04°23′ E) (Fig. 4a)

The Amguid crater in Algeria was described by Lefranc (1969), who was the first to suggest an impact origin based on the regular circular shape of the feature, its raised rim, and the uniqueness of this structure in the Mouydir area. Amguid is a fresh, circular cavity about 450 m in diameter, 30 m deep, with a raised rim, overturned strata at the rim crest, and an ejecta blanket. The crater lies in Lower Devonian sandstones. The flat center of the crater is filled with fine-grained compacted eolian silts which appear as bright disk on aerial photographs and satellite images (Fig. 4a). In Space Radar Laboratory (SRL) imagery, ejecta deposits appear as a coherent lobe extending outwards from the northern half of the crater (McHone et al. 2002). Evidence for the impact origin of this crater was given by Lambert et al. (1980), who found planar deformation features in quartz grains of samples from the crater wall. The age of Amguid crater was suggested to be between 10,000 and 100,000 years based only on the apparent freshness of the structure (Lambert et al. 1980).

3.1.3 Aouelloul, Mauritania (20°14′ N–12°40′ W) (Fig. 4b)

The Aouelloul crater was first described by Monod and Pourquié (1951), who suggested an impact origin after discovering nickel-rich silica glass in and around the crater. The crater has a diameter of about 390 m, and a rim that rises 53 m above the crater floor. The target rocks consist of Ordovician Zli sandstone and Oujeft quartzite. Glass fragments found around the crater rim have been interpreted as impact glass. Evidence for the impact origin of this crater was provided by Re–Os isotope analyses of impact glasses and sandstones, which showed the presence of a meteoritic component in the glass (Koeberl et al. 1998). Fission track and K–Ar dating of the impact glass suggested an age of 3.1 ± 0.1 Ma for the Aouelloul crater (Fudali and Cressy 1976; Storzer and Wagner 1977).

3.1.4 B. P. Structure, Libya (25°19′ N–24°20′ E) (Fig. 5a)

The B.P. structure was first described by Martin (1969) as three near circular, concentric rock outcrops (*sic*). It consists of three concentric and eroded ridges

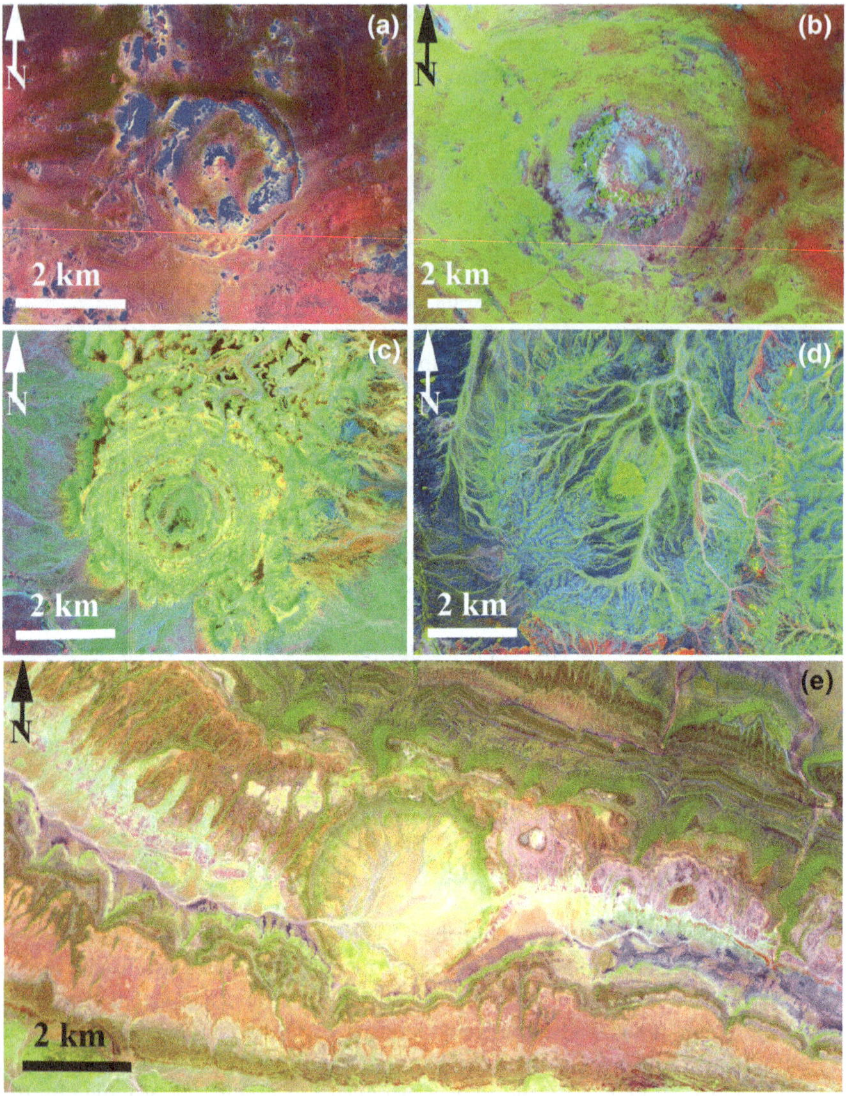

Fig. 5 Sentinel 2A principal component analysis (PCA) images of 5 confirmed complex impact structures of the Arab world. **a** BP (Libya). **b** Oasis (Libya). **c** Tin Bider (Algeria). **d** Jebel Waqf as Suwwan (Jordan). **e** Ouarkziz (Algeria)

(Koeberl et al. 2005a): the "inner ring", which is interpreted as a small central uplift (ca. 425 m in diameter); the "middle ring" with a diameter of about 2 km, and a discontinuous "outer ring", ca. 2.8 km in diameter. The target rocks are quartz sandstone, siltstone and conglomerate of the Jurassic to Lower Cretaceous Nubia Group (Abate et al. 1999). Evidence for the impact origin of this crater was given by French et al. (1974), who found multiple sets of planar deformation features in quartz grains of sandstones from the crater. The age of the structure is less than 120 Ma (Koeberl et al. 2005a).

3.1.5 Jebel El Waqf as Suwwan, Jordan (31°03′ N–36°48′ E) (Fig. 5d)

The Jebel Waqf as Suwwan structure was first described and interpreted as a possible impact structure by Salameh et al. (2006). An earlier radarsat investigation of the structure by Abdelhamid (2001) had described four concentric circles, with two perfect ones of about 6 and 1 km in diameter. The author indicated that this feature greatly resembles an ideal impact structure but concluded that the structure was probably of volcanic origin. Jebel Waqf as Suwwan is an eroded complex impact structure that shows a well-exposed central uplift (innermost "ring") of ca. 900 m diameter and an outer ring structure 6.1 km in diameter (Fig. 5d). The structure is hosted by Cretaceous (Albian) to Paleogene (Eocene) sedimentary strata of various lithology (sandstone, limestone, marl, chalk, and chert) (Kenkmann et al. 2010). Geophysical investigations (Heinrichs et al. 2014) and precise structural studies (Kenkmann et al. 2017) of the structure indicate respectively that the crater is moderately (300–420 m) eroded and was formed by a very oblique impact (~20°). Evidence for the impact origin of this structure was first provided by the occurrence of shatter cones in the outer parts of the central uplifted area, and the detection of multiple sets of planar deformation features and planar fractures in a quartz grain in a sandstone sample from the central uplift (Salameh et al. 2008). Later, PDFs and other microdeformation features were described by Schmieder et al. (2011) from a chert nodule from the center of the Jebel Waqf as Suwwan structure. The age of the Jebel Waqf as Suwwan impact is stratigraphically constrained to post-Eocene (30 Ma or younger) (Salameh et al. 2008).

3.1.6 Kamil, Egypt (22°01′ N–26°05′ E) (Fig. 4e)

In July 2010, Folco et al. (2010) reported the discovery of a new well-preserved impact crater named Kamil in southern Egypt. It is a small bowl-shaped impact crater of 45 m diameter, with a depth of 10 m and a well-preserved ejecta blanket, also displaying some well-developed ejecta rays (Fig. 4e). This highlights the exceptional freshness of the structure, for which an estimated age of <5 ka (Folco et al. 2011; Urbini et al. 2012) was proposed. The target rocks are sandstones of the Gilf Kebir Formation (Lower Cretaceous). The meteorite impact origin for the Kamil crater is based on morphological observations, shock metamorphism (including shatter cones, planar deformation features in quartz, the presence of coesite, stishovite, and diamond), and the preservation of iron meteorite fragments (D'Orazio et al. 2011; Fazio et al. 2016). Thermoluminescence dating of the Kamil crater constrained the impact age between 1600 and 400 BC (Sighinolfi et al. 2015).

3.1.7 Oasis, Libya (24°35′ N–24°24′ E) (Fig. 5b)

The Oasis structure is an eroded, ring-shaped feature with a prominent ring of discontinuous hills (approximately 100 m high) that is 5.1 km in diameter (Abate et al. 1999). The structure lies in Jurassic to Lower Cretaceous sandstones and conglomerates of the Nubia Group and underlying Upper Carboniferous clastic sedimentary rocks (Gibson et al. 2011). A diameter of 11.5 km or even 18 km (from radar images) was suggested for this structure (Koeberl et al. 2005a), while a recent study proposed a diameter of 15.6 ± 0.5 km from high-resolution radar imagery (Van Gasselt et al. 2017). Evidence for the impact origin of this crater was supplied by French et al. (1974) and Gibson et al. (2011), who found multiple sets of planar deformation features in quartz grains of sandstones from the structure. The age of the Oasis structure is not well constrained, with proposed ages ranging from 5.5 to 90–120 Ma (Gibson et al. 2011; Van Gasselt et al. 2017).

3.1.8 Ouarkziz, Algeria (29°00′ N–07°33′ W) (Fig. 5e)

Ouarkziz is a deeply eroded circular ring structure about 3.5 km in diameter. Fabre et al. (1970) described three concentric zones: (i) an outer ring with uplifted and outward dipping formations, (ii) an inner ring with intense folding and inward-dipping strata and (iii) a central zone showing a circular uplift feature with vertically dipping strata. Fabre et al. (1970) reported the occurrences of breccia in the inner ring zone and in the central part of the structure. These authors also reported «planar features» in quartz from the inner part of the rim and in reworked quartz grains. However, no further details were given about these planar features.

3.1.9 Saqqar, Saudi Arabia (29°35′ N–38°42′ E)

In a recent paper, Kenkmann et al. (2015) reported petrographic evidence from drill core for the existence of an impact structure in the Northwest of Saudi Arabia, which they termed the Saqqar structure. With 34 km in diameter, it is the largest confirmed impact structure in the Arab world, and the first buried impact structure recognized among the confirmed impact structures in the Arab countries. Saqqar was first reported by Neville et al. (2014) from a geophysical investigation, who described "a circular structure with an uplifted central high surrounded by a concentric, moat-like syncline, which is, in turn, bounded by a fault system". The structure was formed in Cambrian–Devonian siliciclastics and is unconformably overlain by ~400 m of undeformed Cretaceous and Paleogene sediments. It displays a positive gravity anomaly restricted to the center of the structure (Neville et al. 2014). Thin sections of sandstone samples from cores drilled in the center of the structure show planar deformation features, planar fractures, and feather features (Kenkmann et al. 2015). The age of the Saqqar impact structure is not well constrained but, based on stratigraphic considerations, must lie between the Late Devonian and Late Cretaceous.

3.1.10 Talemzane (Maâdna), Algeria (33°19′ N–04°02′ E) (Fig. 4c)

Talemzane (also called Maâdna structure) was the first impact crater to be discovered and described in Algeria (Karpoff 1953). It is a simple, bowl-shaped depression of about 1750 m diameter, with a raised rim and large ejecta blocks outside the rim. Target rocks are Upper Cretaceous to Eocene limestones. Shuttle imaging radar mission (SIR-A) imagery shows an annular drainage system surrounding the crater, which indicates the probable presence of a weakly developed ring syncline concentric to the crater (McHone and Greeley 1987). Evidence for the impact origin of this crater was alleged by Lambert et al. (1980), who found "planar elements" (PDFs) in quartz grains of a polymict breccia collected at the base of the crater wall. The occurrence of carbonate impact melt at Talemzane crater was recently suggested by Sahoui et al. (2016). Geological field evidence suggested an age between 0.5 and 3 Ma (Lambert et al. 1980), whereas paleomagnetic and rock magnetic investigations indicate that a Late Miocene to Early Pliocene age for the impact event is more probable (Lamali et al. 2016). Finally, it must be noted that a recent geological and geophysical study of the Talemzane crater failed to confirm an impact origin for this structure and suggested a dissolved diapir with inverted relief as an alternative to the impact hypothesis (Lamali et al. 2016). However, the absence of impact-diagnostic evidence following a single field investigation is not sufficient to disprove the impact origin of Talemzane structure. Further detailed field works and collection of more samples for petrographic examination are required.

3.1.11 Tenoumer, Mauritania (22°55′ N–10°24′ E) (Fig. 4d)

Tenoumer crater is a circular depression of about 1.9 km diameter, which has a well-preserved rim rising about 100 m above the crater floor. The target rocks are Precambrian granite, granitic gneiss, and mafic rocks of the Reguibat Shield (Pratesi et al. 2005). Rare occurrences of fused igneous materials outside the crater were interpreted as impact melt rocks (French et al. 1970; Fudali 1974; Pratesi et al. 2005; Schultze et al. 2016). Tenoumer crater was first explored by Richard-Molard (1948b, 1949), who thought it was a crypto-explosion structure. Monod and Pomerol (1966) suggested that the crater is a volcanic feature (a subsidence caldera). Evidence for the impact origin of this crater was given by French et al. (1970), who found up to eight different sets of planar features in quartz grains from granitic basement rock inclusions in melt rock samples. K/Ar dating of impact melt rock indicated the age of the Tenoumer crater as 2.5 ± 0.5 Ma (French et al. 1970), whereas fission track analysis of apatite from melt rocks yielded an age of 21.4 ± 9.7 ka (Storzer et al. 2003). Recently, an $^{40}Ar/^{39}Ar$ age of 1.57 ± 0.14 Ma was reported for the Tenoumer impact event (Schultze et al. 2016), which is now considered the preferred impact age.

3.1.12 Tin Bider, Algeria (27°36′ N–05°07′ E) (Fig. 5c)

Tin Bider is a multiring structure with at least three perfectly concentric annular ridges; an innermost ridge of about 2 km diameter is surrounded by two annular ridges 3.5 and 6 km in diameter. The structure was first investigated by Guillemot (1962), who described a central core surrounded by two annular folds that "curiously recalls the wave train formed when a stone is dropped in water". Detailed field descriptions were provided in Lambert et al. (1981) and Attar (1981), who suggested an impact origin. The structure is located in Upper Cretaceous clays and limestones, and Lower Cretaceous massive continental sandstones are exposed in the central zone (Lambert et al. 1981; Attar 1981). The vertical extent of the central area is at least 500 m. An unusually complex and widespread development of ductile deformation characterizes this feature—reflecting the low-strength nature of the target materials. Evidence for an impact origin of this structure was reported by Lambert et al. (1981), who found up to eight sets of planar deformation features in quartz grains from sandstones in the central zone. The age of the Tin Bider structure is not well constrained; however, an early Tertiary age is likely (Lambert et al. 1981).

3.1.13 Wabar, Saudi Arabia (21°30′ N–50°28′ E)

The Wabar crater field consists of at least three craters, where elevated rims or parts thereof are visible, i.e., that are not covered by encroaching dunes. These craters, first described by Philby (1933), are approximately 11, 64, and 116 m in diameter (Wynn and Shoemaker 1998). An ejecta field is associated with these craters and contains small fragments of white quartz sand, many coated with a black glass, shocked sandstone fragments, and a scattering of fragments of the original meteorite (an iron–nickel meteorite) (Gnos et al. 2013). A meteorite impact origin for these craters was confirmed by the occurrence of impact glasses and fragments of iron–nickel meteorite on the site. Hamann et al. (2013) show that these glasses result from mixing iron meteorite matter with highly siliceous target rocks. Luminescence dating of the Wabar impactite yielded an age of 290 $\pm$ 38 years (Prescott et al. 2004).

3.2 *Possible Impact Structures of the Arab World (Figs. 6, 7, 8, 9, 10, 11, 12)*

During the last few decades, several structural features located in Arab countries have been reported as possible impact structures in the literature or in databases on the internet. A review of these structures is given below, where only possible

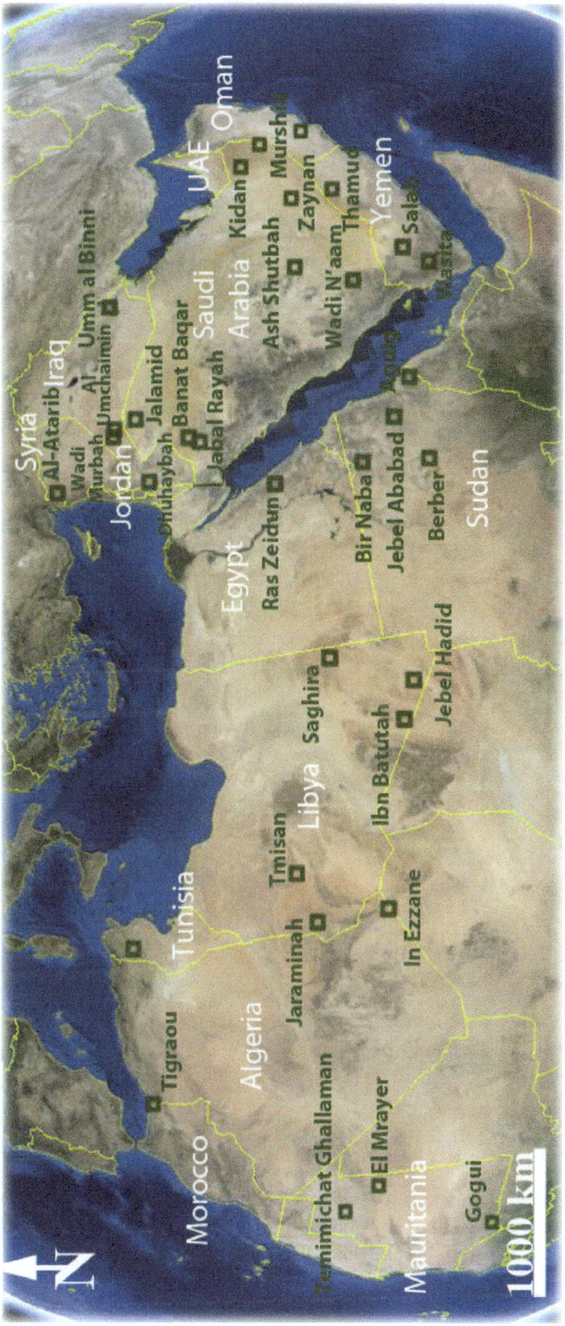

Fig. 6 Location of the possible impact structures in the Arab world on a Google Earth Satellite image

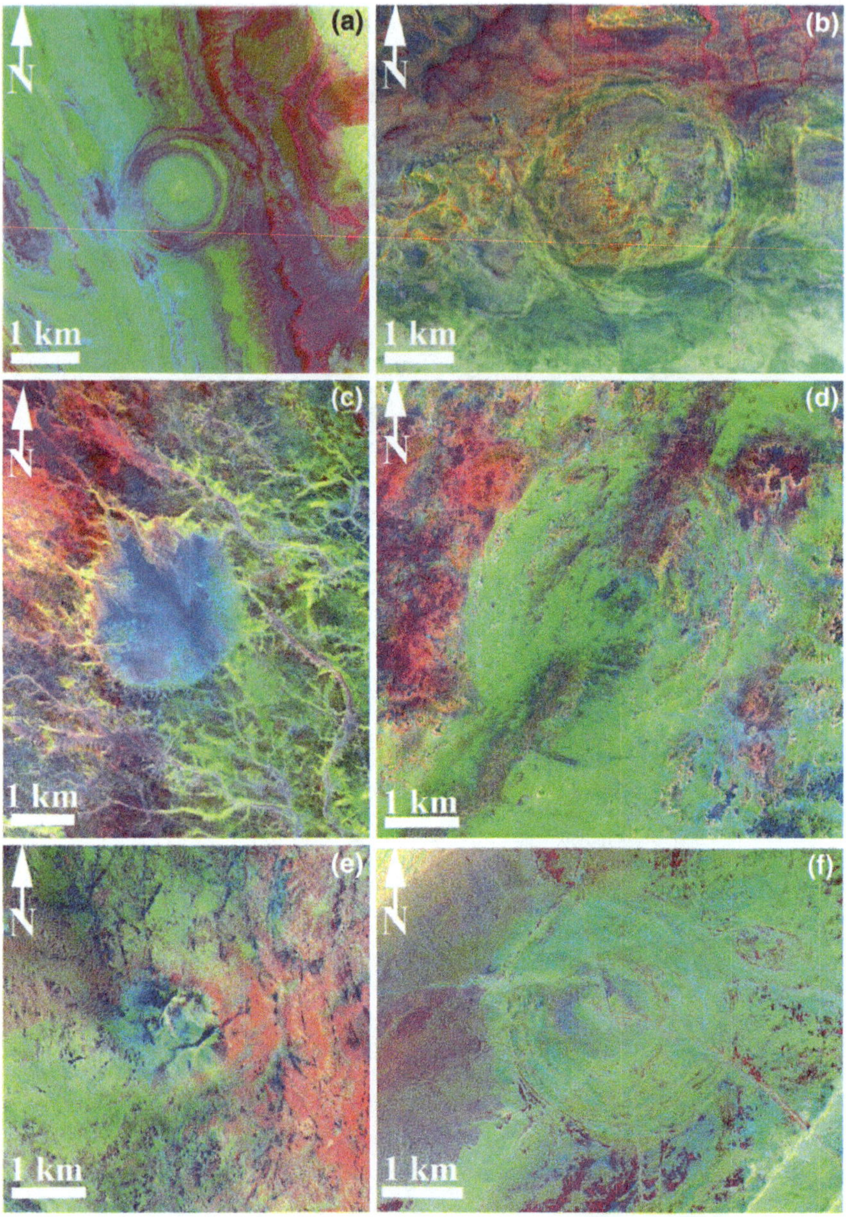

Fig. 7 Sentinel 2A principal component analysis (PCA) images of 6 possible impact structures of the Arab world. **a** Jaraminah (Libya). **b** Tmisan (Libya). **c** Ibn-Batutah (Libya). **d** Jebel Hadid (Libya). **e** Temimichat-Ghallaman (Mauritania). **f** El Mrayer (Mauritania)

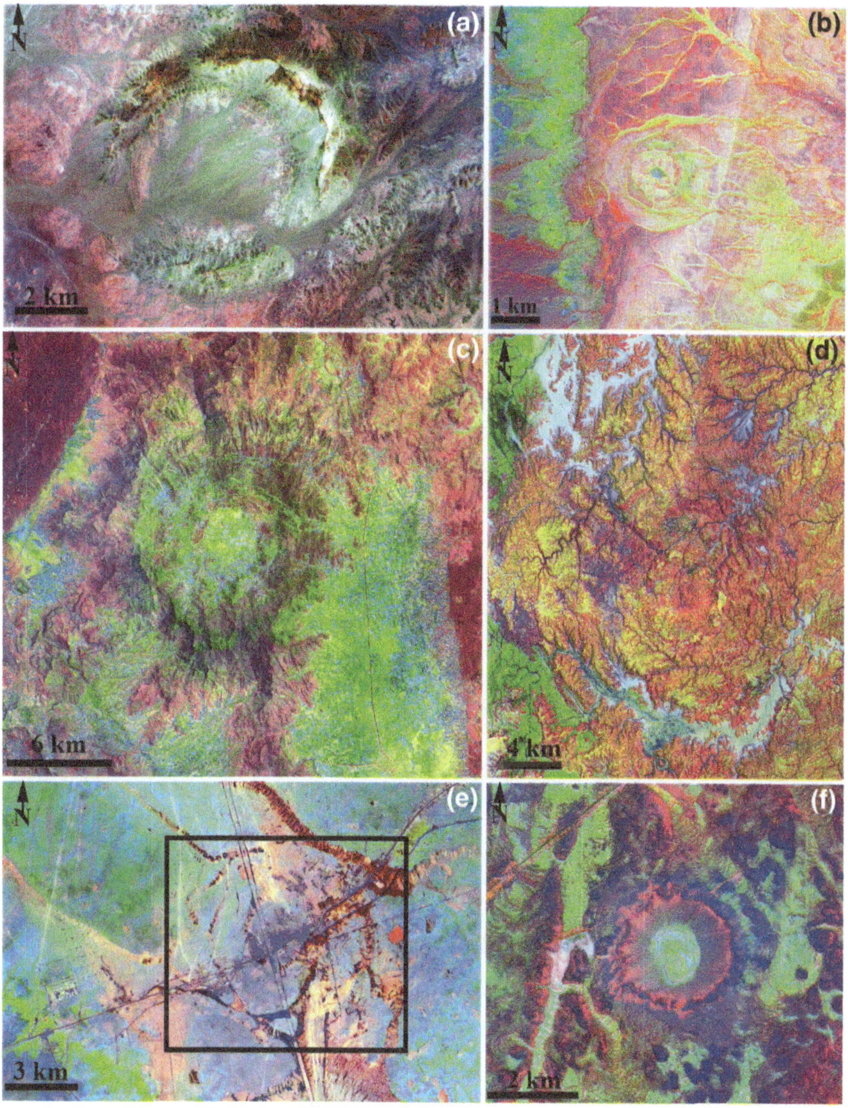

Fig. 8 Sentinel 2A principal component analysis (PCA) images of 6 possible impact structures of the Arab world. **a** Jabal Rayah (Saudi Arabia). **b** Ash Shutbah (Saudi Arabia). **c** Wasita (Yemen). **d** Wadi Na'am (Saudi Arabia). **e** Wadi el Murbah (Jordan/Iraq) (Black rectangular box). **f** Al Umchaimin (Iraq)

Table 2 List of proposed impact structures in the Arab world

Name	Country	Latitude	Longitude	Diameter (km)	Age (Ma)
Adh Dhuhaybah	Jordan	31°30′ N	35°46′ E	0.75	1450–3000 BC
Agig	Sudan	17°55′ N	37°55′ E	6	–
Al-Atarib	Syria	36°08′ N	36°49′ E	0.08	1822?
Al Umchaimin	Iraq	32°36′ N	39°25′ E	2.75	–
Ash Shutbah	Saudi Arabia	21°37′ N	45°39′ E	2.3	<165 Ma
Banat Baqar[a]	Saudi Arabia	29°07′ N	37°36′ E	12	450–470 Ma
Berber	Sudan	18°03′ N	33°30′ E	10	–
Bir Naba	Sudan	21°17′ N	33°59′ E	6–7	–
El Mrayer	Mauritania	22°43′ N	07°18′ W	3.5	–
Gogui	Mauritania	15°33′ N	11°18′ W	0.5	–
Hamidan[a]	Saudi Arabia	20°36′ N	54°44′ E	16	0–50 Ma
Ibn-Batutah	Libya	21°34′ N	20°50′ E	2.5	<144
In Ezzane (5 structures)	Algeria	23°04′ N	11°04′ E	4–9	Quaternary
Jabal Rayah (Al-Madafi)	Saudi Arabia	28°40′ N	37°11′ E	5	<360
Jalamid[a]	Saudi Arabia	31°27′ N	39°35′ E	19	470–480 Ma
Jaraminah	Libya	26°32′ N	10°35′ E	2.2	–
Jebel Ababad	Sudan	19°13′ N	35°59′ E	3	–
Jebel Al Bukrah	Tunisia	35°45′ N	09°08′ E	5	–
Jebel Hadid	Libya	20°52′ N	22°42′ E	4.7	–
Kidan	Saudi Arabia	22°07′ N	53°26′ E	0.1	–
Murshid[a]	Oman	18°10′ N	54°55′ E	2.5	Albian
Ras Zeidun	Egypt	25°37′ N	33°52′ E	7	–
Saghira	Libya	24°40′ N	24°27′ E	2.7	–
Salab	Yemen	15°39′ N	44°41′ E	2.2	–
Temimichat-Ghallaman	Mauritania	24°15′ N	09°39′ W	0.75	<0.006
Thamud	Yemen	18°09′ N	50°04′ E	0.77	~0.002?
Tigraou	Algeria	35°02′ N	01°54′ W	1	–
Tmisan	Libya	27°25′ N	13°24′ E	3.2	–
Umm al Binni	Iraq	31°08′ N	47°04′ E	3.4	–
Wadi el Murbah	Iraq/Jordan Border	32°44′ N	39°00′ E	7.5	–
Wadi Na'am	Saudi Arabia	19°08′ N	44°00′ E	17	Precambrian
Wasita	Yemen	14°54′ N	44°12′ E	14	–
Zaynan[a]	Saudi Arabia	20°23′ N	50°08′ E	5	180–230 Ma

[a]Buried structures

impact structures suggested in scientific literature or conference papers are listed. I shall not include in this review several structures listed in some databases on the internet (e.g., David Rajmon Global impact crater database, http://www.datapages.com/gis-map-publishing-program/gis-open-files/geothematic/david-rajmon-global-impact-crater-gis-project), for which no substantiating information is available. We will propose names to some currently unnamed possible impact structures following the current procedure for nomenclature of impact craters (name of nearby city, mountain, river, etc.). The basic data concerning possible impact structures in the Arab world are summarized in Table 2, and the locations are shown in Fig. 6.

3.2.1 Adh Dhuhaybah, Jordan (31°30′ N–35°46′ E) (Fig. 12a)

Reported by Robert S. Fritzius on 17 October 2010 (Fritzius website 2017, http://www.datasync.com/~rsf1/fun/moab-cr.htm), the Ad Dhuhaybah crater is a small circular feature with a diameter of 750 m. The northeastern portion of the structure is missing (Fig. 12a). The crater lies in the Late Cretaceous Amman Silicified Unit and Al-Hisa Phosphorite Unit (Khirfan et al. 2011). Strata in the crater walls and rims are subhorizontal (Khirfan et al. 2011) suggesting that this crater may not be of meteorite impact origin.

3.2.2 Agig, Sudan (17°55′ N–37°55′ E) (Fig. 11a)

Di Achille (2005) investigated a nearly circular feature in northeastern Sudan using remote sensing images. Agig is proposed as the name of this structure in reference to the nearest town located 40 km to the northeast of the structure. The Agig structure, approximately 6 km in diameter, exhibits a bowl-shaped appearance, a well-preserved rim, and a central uplift that may represent the remnant of a possible central peak. It lies in Neoproterozoic volcano-sedimentary rocks of the Nubian Shield and sharply interrupts the N-S trending regional fabric (Di Achille 2005). The maximum floor-rim height is about 350 m and the central elevated area rises up to ~75 m above the surrounding terrain. The structure is covered by sediments and alluvium of two channels that drain the interior of the structure. Parts of eastern and western rims of the structure appear to have collapsed which, assuming an impact origin, is interpreted as the postimpact crater wall collapse. Based on geomorphological and topographic evidences, Di Achille (2005) interprets this feature as a complex impact structure that has been moderately eroded.

3.2.3 Al-Atarib, Syria (36°08′ N–36°49′ E)

Al-Atarib is a small structure of about 80 m diameter and 50 m depth. Basurah (2009) discussed this structure as a possible excavation from the fall of a meteorite on 22 August 1822, as mentioned in a historical document. A photo of the crater published in Basurah (2009) indicates that it is probably not an impact crater, as the structure strongly resembles a sinkhole.

3.2.4 Al Umchaimin, Iraq (32°36′ N–39°25′ E) (Fig. 8f)

The Al Umchaimin was first recognized as a possible impact structure in 1957 (Merriam and Holwerda 1957). One year later, Mitchell (1958) suggested a volcano-tectonic origin of the feature. It is a nearly circular depression of 2.75 km diameter and 32–42 m depth. The center of the depression is filled with fine-grained silty clay that shows mud cracks. The country rocks are chert-bearing beds of limestone of the Dammam Formation of middle to late Eocene age (Underwood 1994). The radar images (SIR-A and SIR-C) of the structure show an intensely bright ring corresponding to a rim and features that were interpreted to indicate the former existence of possible thin ejecta deposits (Greeley et al. 1987; Berlin 1989; McHone and Greeley 1997). As no impact evidence has yet been reported, the origin of this structure is still unsettled: volcanic collapse, impact crater, or large sinkhole (Underwood 1994). The latter alternative (solution collapse depression) seems more plausible according to geological and geophysical studies of the structure done by Iraqi geologists in the 1970s (works cited in Underwood 1994).

3.2.5 As Shutbah, Saudi Arabia, (21°37′ N, 45°39′ E) (Fig. 8b)

The As Shutbah circular structure was first reported as a possible meteorite impact structure by geologists of the Saudi Aramco company in the 1970 and 1980s (Stewart 2015, and references therein). The structure was then identified by Schmieder et al. (2009b) on satellite images. Gnos et al. (2014) investigated it and concluded that this structure was probably a strongly eroded complex impact crater, although no unequivocal evidence for impact was found by them (with the possible exception of one poorly defined possible shatter cone in sandstone from the central uplifted domain). As Shutbah is a circular feature approximately 2.1 km in diameter, with a central dark area, 400 m wide. The dark color is due to black, desert-varnish-covered debris of a massive orthoquartzite (Gnos et al. 2014). The structure is formed in undeformed limestone beds of the Tuwaiq Mountain Limestone of Jurassic age. The central area consists of strongly folded, calcareous sandstone of the Jurassic Dhruma Formation (Gnos et al. 2014). The age of the structure is constrained stratigraphically as post-Callovian (<165 Ma) (ibid).

3.2.6 Banat Baqar Structure, Saudi Arabia (29°07′ N, 37°36′ E)

Banat Baqar is a subsurface structure of 12 km diameter that is located at 2 km below the surface (Neville et al. 2014). The structure is entirely contained in upper Cambrian to Upper Ordovician strata. A central high is apparent within the structure

Fig. 9 Landsat 8 image of the alleged (Bonin et al. 2011) In Ezzane nested structures

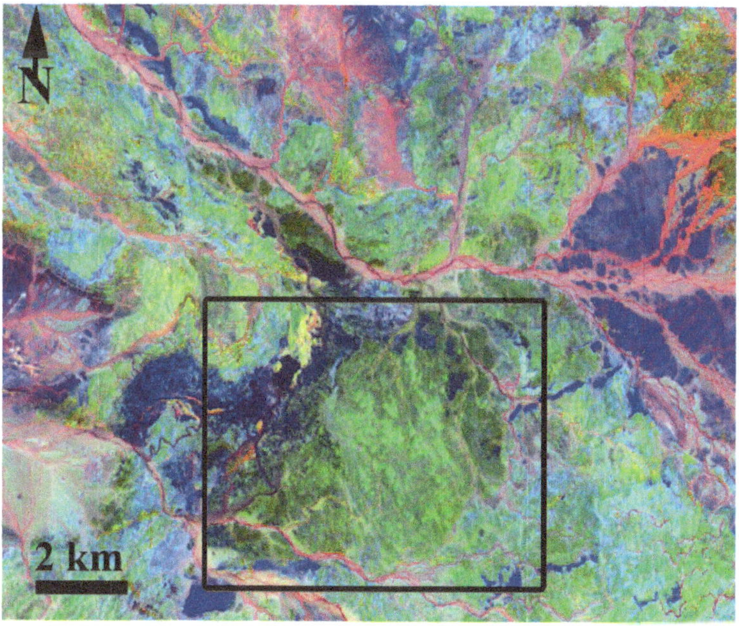

Fig. 10 Sentinel 2A principal component analysis (PCA) image of the alleged Ras Zeidun structure (Black rectangular box)

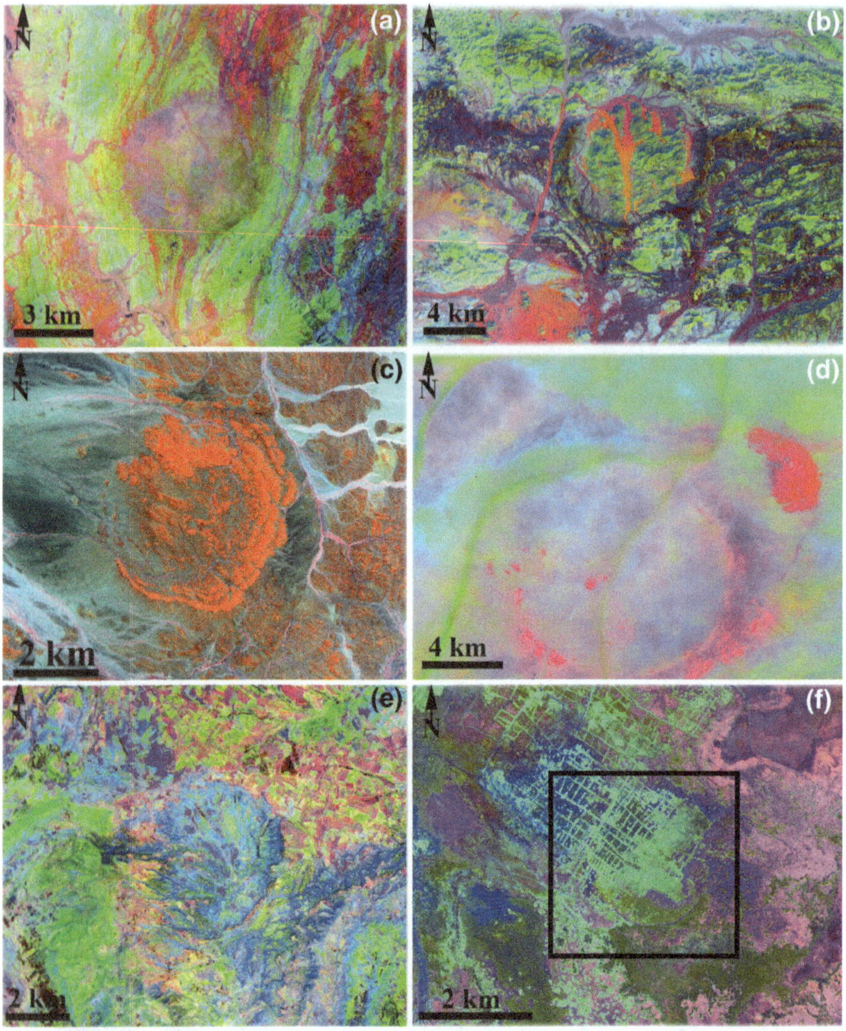

Fig. 11 Sentinel 2A principal component analysis (PCA) images of 6 proposed impact structures of the Arab world. **a** Aguig (Sudan). **b** Bir Naba (Sudan). **c** Jebel Ababad (Sudan). **d** Berber (Sudan). **e** Jebel Al Bukrah (Tunisia). **f** Umm el Binni (Iraq) (Black rectangular box)

on seismic sections, and a dome structure is observed in the strata above the crater. The central high rises about 170 m above the crater floor. The structure displays a positive gravity anomaly of about 2 mGal. The general form and geometric attributes of the structure support a possible impact origin (Neville et al. 2014).

3.2.7 Berber, Sudan (18°03′ N–33°30′ E) (Fig. 11d)

A circular feature located in Sudan was reported by Sparavigna (2010a) as a possible complex impact crater. This structure, 10 km in diameter, is located near the town of Berber in the Bayuda Desert. However, after simple consultation of available scientific papers, there is no doubt that this circular feature corresponds to the so-called Singeir ring complex, which consists of arcuate intrusions of sodic amphibole granite dated at 432 Ma (Vail 1985; Wooley 2001). This structure is one of the 25 alkaline ring complexes identified in the Bayuda Desert and Nile Valley in Sudan (Vail 1985).

3.2.8 Bir Naba, Sudan (21°17′ N–33°59′ E) (Fig. 11b)

A circular structure located near Bir Naba in Northeastern Sudan was also reported by Sparavigna (2010b). This feature, 6 or 7 km in diameter, displays a well-defined, radially oriented drainage pattern and bears a strong morphological resemblance to the Gweni Fada impact structure in northern Chad (Koeberl et al. 2005b). Sentinel 2A principal component analysis of the structure (Fig. 11b) does not indicate the presence of lithologies within the structure that could be different from the surrounding host rock, which may mitigate against a volcanic origin of this structure. Thus, a possible impact origin of the Bir Naba structure cannot be ruled out at this time.

3.2.9 El Mrayer Structure, Mauritania (22°43′ N–7°18′ W) (Fig. 7f)

The El Mrayer structure was described by Rossi (2002) as an about 2-km-wide feature, which consists of a subcircular inner crater bounded by an apparently concentric deformed region. The target rocks are Cambro-Ordovician formations of the Taoudenni basin. A careful look at the satellite images of the structure indicates that this feature appears as a multiringed structure with at least three concentric ridges. The subcircular inner ridge is about 1.5 km wide, while the outermost ridge measures 3.5 km in diameter. Therefore, the total diameter of the structure is at least 3.5 km. The El Mrayer structure is cut by a number of dolerite dykes of the Taoudenni dyke swarm. These dolerites belong to the Triassic-Jurassic (201 Ma) Central Atlantic Magmatic Province (Verati et al. 2005). Therefore, the formation of this structure predates the Triassic-Jurassic period. Rossi (2002) mentioned the possible existence of a small secondary crater 2 km N of the El Mrayer structure. In satellite images, this feature of 150 m diameter seems to be a ring dyke rather than an impact feature. It cannot be excluded that the entire El Mrayer structure could be a volcanic ring complex. Moreover, this structure was studied in the field (Orti et al. 2008), and no evidence of impact-related or -affected rocks was found. According to Orti et al. (2008), the ridges of the structure are composed of sandstones cut by a network of fractures that show Fe–Mn oxide/hydroxide mineralizations. Also, these authors emphasize the geological and morphological similarities between the El Mrayer structure and the crater-like features in the Gilf Kebir area (Egypt).

3.2.10 Gogui Crater, Mauritania (15°33′ N–11°18′ W) (Fig. 12c)

Gogui is a nearly perfectly circular structure of about 450 m diameter. It has a distinct circular rim and a rather flat bottom (Rossi 2002). Around the crater, possible ejecta can be seen on satellite images (Fig. 12c). A radial and locally concentric drainage pattern extends outward from the crater rim. Given its excellent state of conservation, this structure is probably young. The morphology of the crater and its uniqueness in the region make it a very good candidate for a small bowl-shaped impact crater, which obviously must be resolved by ground-based investigations.

3.2.11 Hamidan, Saudi Arabia (20°36′ N–54°44′ E)

The Hamidan structure is a subsurface structure of 16 km diameter that is buried at a depth of 100 m below surface (Neville et al. 2014). Three wells traversing the structure encountered disrupted strata in the Paleogene and upper Cretaceous, while seismic data indicate that this disturbed area is bounded by inward facing extensional faults that penetrate into the Paleozoic strata (Neville et al. 2014). A central uplift is imaged by seismic data. Furthermore, the structure displays positive gravity and magnetic anomalies. Nevertheless, the sources of these anomalies are not well understood. In the Paleogene to uppermost Cretaceous intervals, strata are structurally disrupted—broken, locally steeply inclined, and chaotic. The lithology is described as "mixed" and paleontology as "contaminated", leading to unresolvable stratigraphy within the Paleogene section. According to Neville et al. (2014), observations at Hamidan suggest a possible impact origin. In this interpretation, the disturbed Paleogene section would be explained as crater floor and infill, and the seismically imaged faults that extend into the Paleozoic would represent the faults and terraces along the edge of a complex crater structure (Neville et al. 2014).

3.2.12 Ibn-Batutah, Libya (21°34′ N–20°50′ E) (Fig. 7c)

The Ibn-Batutah structure was discovered by E. M. Ghoneim on satellite imagery. It is a circular structure of 2.5 km diameter and $\sim$25 m depth, with a bowl-like form and an elevated circular rim that rises up to 5 m above the surrounding terrain (Ghoneim 2009). Ibn-Batutah is formed in sandstones of the Cretaceous Nubia Formation and its age is constrained to <144 Ma (Ghoneim 2009). This structure displays some structural features known from simple impact structures and resembles the Talemzane impact crater in Algeria (Ghoneim 2009), but to date, no confirming evidence has been unveiled.

3.2.13 In Ezzane, Algeria (23°01′ N–11°04′ E for the Largest Structure) (Fig. 9)

Bonin et al. (2011) suggested an impact origin for five nested circular features that are visible on satellite images of the region near the Algeria-Niger border. These circular structures, with diameters between 4 and 9 km, are aligned along a north–south direction (Fig. 9). They formed in the volcanic lava flows of the In Ezzane Quaternary volcanic province and the underlying Ordovician Tassili sandstones. A Quaternary age for these structures was suggested by Bonin et al. (2011). However, close examination of these features on satellite images rules out any resemblance with young impact structures of moderate size. The eastern sides of these alleged structures seem to be purely erosional features, and the semicircular (arc)-shaped western sides are dolerite dykes already shown on the Geological Map of the Central Sahara at 1/2,000,000 (1962). These dolerite dykes are related to Paleozoic mafic magmatism that outcrops in Southeastern Algeria and Northeastern Niger along the western border of the Murzuq basin (Denis et al. 2007; Derder et al. 2016). Brecciated sandstones with dark veins of goethite in the western rims of the structures (Bonin et al. 2011) are typical of rocks affected by fluid migration related to the emplacement of subvolcanic intrusives (dykes). The alleged arc-shaped rims of the In Ezzane structures are similar to the arcuate fractures that define the rims of some volcanic, crater-like features of the Gilf Kebir area in Egypt (Orti et al. 2008).

3.2.14 Jabal Rayah (Al Madafi), Saudi Arabia (28°40′ N–37°11′ E) (Fig. 8a)

The Jabal Rayah structure, also known as Al Madafi, is a promising possible impact structure (Janjou et al. 1997). It was first recognized in 1971 by Blodget as a near circular, eroded dome (Garvin and Blodget 1986). Stewart (2015) described the structure in detail and concluded that Jabal Rayah was a probable impact structure. Jabal Rayah represents a 5-km-diameter circular ridge involving hills ranging in height from 150 to 250 m above the surroundings, forming a topographic ring structure enclosing an inner rim terrace about 3 km in diameter. The rocks of the Jabal Rayah structure belong to the siliciclastics Silurian Sharawra Formation and Devonian Tawil Formation. These strata are bound by inward-dipping normal faults and within of the ring fault system, the Sharawra and Tawil Formations are folded (Stewart 2015). The structure is not associated with any obvious gravity or magnetic anomaly. According to Stewart (2015), there is no reasonable alternative to the impact hypothesis, but the degree of exhumation of the structure indicates that any rocks with impact-related features would have probably been removed by erosion.

Stewart (2015) estimates the depth of erosion relative to the time of impact of about 2.5 km based on the strain state observed at the outcrop in the center of Jabal Rayah, i.e., intense folding but with no grain-scale high-pressure deformation effects present. The age of the structure is <360 Ma (Grieve et al. 1988).

3.2.15 Jalamid, Saudi Arabia (31°27′ N–39°35′ E)

The Jalamid structure is a subsurface feature, 19 km in diameter, buried 4–5 km below the surface (Neville et al. 2014). The structure was recognized on seismic data. It is characterized by a synclinal moat bounded by extensional faults, and a central uplifted area, ca. 6 km in diameter and at least 800 m in height measured from its base (based on the interpretation of seismic section). The structure involves Lower Ordovician sandstones (Saq Formation) which are interpreted to form the crater floor, and older strata down to at least the uppermost Neoproterozoic (Neville et al. 2014). Neville et al. (2014) interpreted this structure as a possible impact crater of Early Ordovician age since the structure and central uplifted area are truncated by Middle Ordovician strata.

3.2.16 Jaraminah, Libya (26°32′ N–10°35′ E) (Fig. 7a)

The Jaraminah structure was discovered by A. Dunford on satellite imagery. It is a ring-like feature with a diameter of about 2.2 km. The center of the structure is characterized by an uplifted region of 300–400 m width (Dunford and Koeberl 2009). The strata within the structure are highly deformed. Preliminary petrographic examination of thin sections of rock samples from the Jaraminah structure did not reveal PDFs in quartz grains (Dunford and Koeberl 2009), so that this structure remains in the category of possible impact craters.

3.2.17 Jebel Ababad, Sudan (19°13′ N–35°59′ E) (Fig. 11c)

Using Google Earth images, Sparavigna (2010b) reported a circular feature near Jebel Ababad in Sudan. The Jabal Ababad structure appears as a ca. 3-km-wide ring. The country rocks belong to the Nakasib suture zone of late Proterozoic age. No further information is available about this alleged structure. The Sentinel 2A principal component analysis of the structure (Fig. 11c) shows concentric rings composed of rocks having a lithology different from the surrounding rocks, which is characteristic of igneous ring complexes, as they occur abundantly in that region. Thus, an impact origin of the Jabal Ababad structure is not impossible but considered unlikely.

3.2.18 Jebel Al Bukrah, Tunisia (35°45′ N–09°08′ E) (Fig. 11e)

Tomlinson (1989) identified a possible impact structure in Tunisia on Landsat satellite images. This structure was named Jebel Al Bukrah by Youbi et al. (2011) after the name of the landform that forms the eastern side of the structure. Jebel Al Bukrah is a bowl-shaped depression of about 5 km diameter, with a well-defined rim along its northern and eastern sides. Neither a central structural or stratigraphic uplift, nor outer concentric zones are visible, which is uncommon for an impact structure of this dimension. The country rocks are Middle Cretaceous marine sediments.

3.2.19 Jebel Hadid, Libya (20°52′ N–22°42′ E) (Fig. 7d)

The Jebel Hadid structure was first investigated by Schmieder et al. (2009a) using remote sensing imagery. It is a circular feature about 4.7 km in diameter, with a set of five concentric annular ridges. The country rocks are probably part of the Early Cretaceous Nubian sandstone. This structure bears a strong morphological resemblance to the Tin Bider impact structure in Algeria. Schmieder et al. (2009a) suggested that the Jebel Hadid structure might represent an eroded, complex impact structure. The age of the structure could be post-Early Cretaceous to pre-Pliocene (Schmieder et al. 2009a).

3.2.20 Kidan (Shaybah), Saudi Arabia (22°07′ N–53°26′ E)

The Kidan structure was first described by Stewart (2015), who reported that it had first been proposed as a crater of meteorite impact origin in 1967 by geologists of the Saudi Aramco. This structure was also reported by McHone and Dietz (1988) under the name Shaybah, as a small fresh crater with an uplifted rim and possible ejecta. The Kidan structure is a subcircular feature approximately 120 × 90 m in diameter, with an elevated rim approximately 7 m above the surrounding surface and a 3 m maximum depth relative to the rim. The crater lies in a Quaternary Sabkha. No impactite material has been found in or around the structure, which led Stewart (2015) to favor a fluid escape structure, which is also supported by reflection seismic imaging that shows the existence of a small fault underlying the depression. Nevertheless, the uniqueness of this depression in a large sabkha does not totally rule out a possible impact origin for Kidan crater.

3.2.21 Murshid, Oman (18°10′ N–54°55′ E)

Murshid is a 2.5 km diameter possible crater that was interpreted from seismic data. It is buried to a depth of 680 m and displays a raised rim that is 10–15 m high (Levell et al. 2002). The structure has a maximum depth of about 60 m; it shows a complex morphology with a central peak and terraces on the northern and western sides. The central peak is 500 m wide at its base and about 40 m high. The Murshid structure has been interpreted as a possible impact crater on the basis of its morphology (its circular form, raised rim, and central peak) and the absence of other satisfying explanations (Levell et al. 2002). The structure is contained within Albian sediments, which constraints its age to this period.

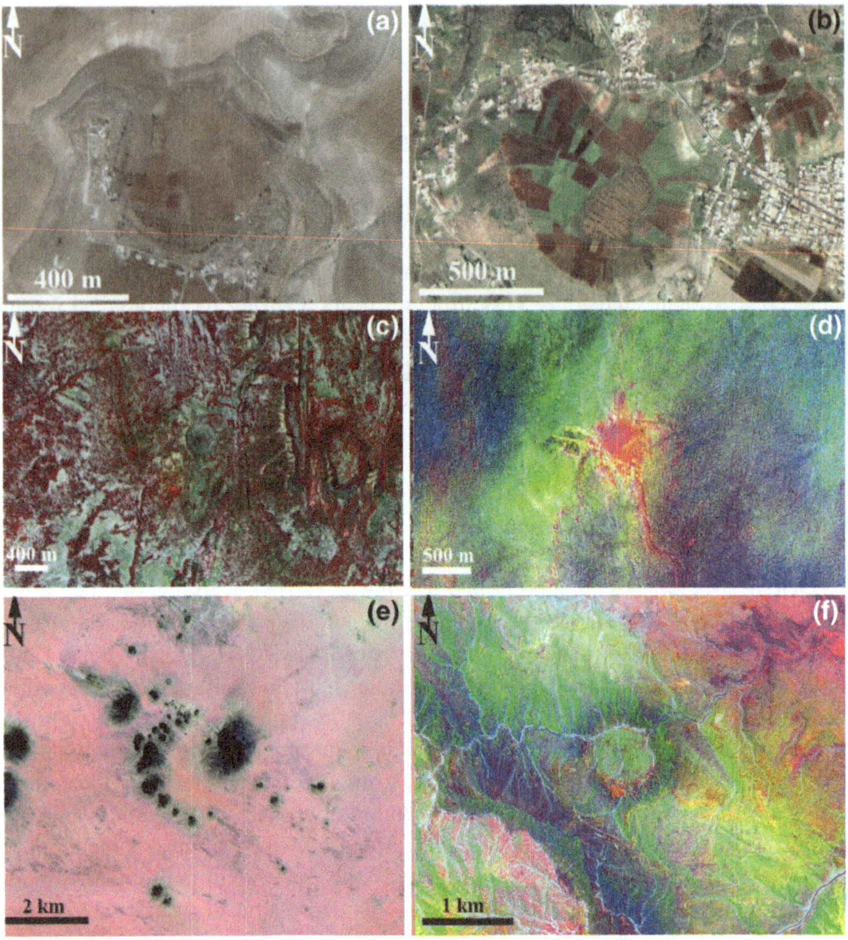

Fig. 12 Google Earth images (**a** and **b**) and Sentinel 2A principal component analysis (PCA) images (**c–f**) of 6 proposed impact structures of the Arab world. **a** Ad Dhuhayba (Jordan). **b** Tigraou (Algeria). **c** Gogui (Mauritania). **d** Thamud (Yemen). **e** Saghira (Libya). **f** Salab (Yemen)

3.2.22 Ras Zeidun, Egypt (25°37′ N–33°52′ E) (Fig. 10)

Barakat (2011) suggested the existence of a possible eroded impact structure of 7 km diameter within the Late Proterozoic basement of the central Eastern Desert of Egypt. This structure is located at Ras Zeidun, where unusual jasper deposits occur as huge concentric plugs several meters in diameter. According to Barakat (2011), petrographic investigations indicate that the quartz grains of the Jasper deposit show features characteristic of meteorite impact and the Ras Zeidun site matched the general character of eroded multiring impact structures. However, no evidence concerning these claims was presented, and no circular feature is visible on satellite images at Ras Zeidun. The geographic coordinates of the site (25°37′15″ N, 33°52′ 35″ E) are mentioned in Soliman and Barakat (2004). Optical and radar remote sensing investigations of the area do not show any unusual structural feature. Sentinel 2A principal component analysis (PCA) shows a change in lithology in the Ras Zeidun area (Fig. 10), but no circular feature is visible. Therefore, the existence of this alleged complex impact structure at Ras Zeidun is viewed with extreme caution.

3.2.23 Saghira, Libya (24°40′ N–24°27′ E) (Fig. 12e)

The Saghira structure was first identified on an Apollo-Soyuz photograph by El-Baz and Mitchell (1976), who named it "Twin Site". It was reported a second time by El-Baz and Issawi (1981), who mentioned it as a probable impact crater. Finally, El-Baz and Ghoneim (2007) named this crater "Saghira", an Arabic word meaning "small", given its small diameter (2.7 km in diameter) in comparison with the so-called "Kebira" (31 km in diameter) structure (Kebira meaning big in Arabic), which had been described in the same area. Saghira is a partial ring feature 2.7 km in diameter, located 10 km north-northeast of the Oasis impact structure. It is very likely that this partial ring structure has nothing to do with an impact crater. It belongs to the field of circular or rounded knobs of sandstone that occur north and northeast of the Oasis structure (Koeberl et al. 2005a). Based on field investigations, Koeberl et al. (2005a) interpreted these plug-like structures as remnants of fold cores.

3.2.24 Salab, Yemen (15°39′ N–44°41′ E) (Fig. 12f)

The Salab structure was first cited by D. Rajmon in his Global Impact crater database (http://www.datapages.com/gis-map-publishing-program/gis-open-files/geothematic/david-rajmon-global-impact-crater-gis-project). It was recently investigated by Daneshvar and Nejadsoleymani (2017) using remote sensing. The structure shows two annular features: an interior "ring" ca. 1.13 km in diameter, which could be interpreted as a central uplift feature, and an exterior annular ridge with a diameter of about 2.2 km. The host rocks are Jurassic limestone of the Amran Group. In the absence of another explanation, Daneshvar and Nejadsoleymani (2017) interpreted the Salab structure as a possible impact crater. Unfortunately, field-based study of this structure is currently impossible due to the civil war in Yemen.

3.2.25 Temimichat-Ghallaman, Mauritania (24°15′ N–09°39′ W) (Fig. 7e)

Temimichat-Ghallaman is a 700 m diameter crater-like structure with a slightly hexagonal shape. The structure is moderately eroded, with a rim height ranging from a few meters to a few tens of meters (Rossi et al. 2003). The country rocks are Precambrian granitic gneiss and gabbros of the Reguibat Shield and the interior of the depression is covered by recent eolian sediments (Rossi et al. 2003). The alleged crater was first noted by Richard-Molard (1948b), and briefly described by Monod and Pourquié (1951). A volcanic origin of the structure was suggested by Pomerol (1967) after the discovery of basaltic fragments inside the crater. However, these volcanic materials were not found in place. More recently, Rossi et al. (2003) found dark to greenish veins of glassy material with fluidal texture inside the granitoid gneisses of the rim. These structures, referred to by the authors as "pseudotachylite veins", do not occur in the surrounding basement suggesting an origin related to the formation of the structure. No evidence for shock metamorphism has yet been found, but this is a good candidate for further exploration work.

3.2.26 Thamud, Yemen (18°09′ N–50°04′ E) (Fig. 12d)

Thamud was discovered during analysis of Landsat satellite images of eastern Yemen for an archaeological project. It is a circular feature about 770 m in diameter, with a shallow rim of only about 1 m height (Blom et al. 1998). The depression is filled with eolian sediment and evaporites, and is bisected by a wadi. The country rocks are early Tertiary carbonates. No overturned rim section, shatter cones, or meteoritic material were observed during field reconnaissance of the structure (Blom et al. 1998). However, the authors felt that an impact origin of this feature was more plausible than other explanations, such as a sinkhole or a volcanic maar crater (Blom et al. 1998). The feature is older than about 6000 years (Blom et al. 1998).

3.2.27 Tigraou, Algeria (35°02′ N–01°54′ W) (Fig. 12b)

The Tigraou structure was first described by Kock (1901), who interpreted this crater feature to be of volcanic origin. Sadran (1958) noted the lack of volcanic products associated with this structure and described it as a "crypto-explosion" structure. Chabou (2011b) reinvestigated this structure and concluded that the Tigraou crater was a good candidate for a simple impact crater based on its morphology (raised rim, overturned strata at the rim crest) and the occurrence of monomict and polymict breccias, and possible melt breccia exposed in one outcrop. The crater of Tigraou has a diameter of about 1 km and is emplaced in Early Jurassic limestones. The southwestern rim of the crater rises 75 m above the crater floor, whereas the northern and eastern rim sectors are eroded. The flat center of the actual crater floor is filled by eolian deposits. At the top of the southwestern rim, the limestone strata become nearly vertical and overturned limestone beds occur at the top in front of a quarry on the southern rim of the crater. Breccia outcrops were observed along the base of the rim near the crater floor. They comprise monomict and polymict breccias mainly composed of angular limestone clasts. Still closer to the crater floor occurs an outcrop of likely melt breccia. This melt breccia consists of limestone fragments, commonly surrounded by brownish flow-banded glass (cored inclusions), in a matrix of fluidal glassy materials. Quartz grains are common in the matrix, with possible occurrence of silica glass as fluidal inclusions. Further work is necessary to confirm the impact origin of this promising structure.

3.2.28 Tmisan, Libya (27°25′ N–13°24′ E) (Fig. 7b)

The Tmisan structure was discovered—together with Jaraminah (see above)—by A. Dunford as features visible on satellite images. Morphologically, the structure consists of an outermost set of undulatory annular ridges with 500–700 m widths, an annular inner plain with an average width of 600–700 m, and an uplifted central region, which forms a plateau with a diameter of 600–800 m (Dunford and Koeberl 2009). The structure has an outermost diameter of about 3.2 km and occurs in siliciclastic rocks of the B'ir al Qasr Formation of Eifelian (387–393 Ma) age (Dunford and Koeberl 2009). Shatter cone-like features were observed in the field. Preliminary petrographic examination of thin sections of 40 samples from Tmisan did not reveal any PDFs in quartz (Dunford and Koeberl 2009). According to these authors, this structure remains in the category of possible impact craters.

3.2.29 Umm Al Binni, Iraq (31°08′ N–47°04′ E) (Fig. 11f)

Master (2001) discovered a circular structure near the Tigris-Euphrates confluence in southern Iraq. This structure was identified on a regional map as the Umm al Binni Lake (Master 2002). It was described as a 3.4 km diameter, near circular structure, showing a roughly polygonal outline and an apparently raised rim surrounded by a ~500 m wide annulus (Master 2001, 2002; Master and Woldai 2007). The Umm al Binni lake belongs to the Al' Amarah marshes of the Mesopotamian Plains composed of recent sediments deposited in the last 5000 years. According to Master and Woldai (2007), an origin of the Umm al Binni structure by salt doming, karst dissolution, interference folding, or igneous intrusion could be ruled out. On the other hand, a possible meteorite impact origin of the structure might account for the simple bowl-shaped geometry with a markedly polygonal outline, and the apparent rim and annulus around the structure (Master and Woldai 2007). A young age of this structure (<5000 years) was proposed by Master (2001, 2002), which led the latter to speculate about the relationship between the alleged Umm al Binni impact event and some catastrophic events related in historical legends of Mesopotamia.

3.2.30 Wadi El Murbah, Iraq/Jordan Border (32°44′ N, 39°01′ E) (Fig. 8e)

The Wadi el Murbah structure was discovered in 1970 by Heimbach on aerial photographs (Master 2009). Theilen-Willige (1987) briefly described the structure as a possible complex crater named Triebel. McHone and Dietz (1988) and McHone and Greeley (1997) referred to the site as a suspected impact feature named El-Mirba. Finally, a detailed remote sensing study of the structure was done by Master (2009), who named the site Wadi el Murbah after the Jordanian border post and highway airstrip of that name, where the structure is located. Wadi el Murbah is a multiring feature highlighted by two concentric arcs identified as white sand-filled drainage streams with an outer diameter of ca. 7.5 km. The inner ring has a ca. 2.5 km diameter. The bedrock comprises Paleocene and middle to late Eocene limestones, with chert layers belonging to the Dammam Formation. The valleys that define the outer and inner rings in the structure are filled with unconsolidated Quaternary sediments of terrestrial fluviatile and lacustrine origin (Master 2009). According to Master (2009), the geometry and topography of the Wadi el Murbah structure are consistent with it being an eroded complex impact structure that is partly buried by younger sediments. The presence of breccias in a borehole drilled into the structure and a positive gravity anomaly associated with the center of the structure seem to support the hypothesis of an impact origin of the structure (Master 2009). A middle Eocene age of the structure was proposed by Master (2009).

3.2.31 Wadi Na'am, Saudi Arabia (19°08′ N–44°00′ E) (Fig. 8d)

Wadi Na'am is a possible complex impact structure with a diameter of 17 km and several ring-like structures (Al-Mishwat 2003). This near circular structure is masked by the overlying regionally horizontal Cambro-Ordovician Wajid sandstone. In the central area are several small outcrops of gabbro which are seemingly detached from the basement below the sandstone. These gabbros are interpreted as part of the top of a possible central uplift. Within the structure, the sandstone displays radial variation in dip, showing progressively shallower dip outward from the center. In contrast, these sandstones are horizontal outside the Wadi Na'am Structure. Arcuate and continuous wadi drainage defines the outer edge of the structure along the entire southern half of its perimeter. Finally, folds and circular fractures are common in the sedimentary cover. Al-Mishwat (2003) interpreted the Wadi Na'am structure as a possibly concealed impact feature of Precambrian age that is now buried under Lower Paleozoic sedimentary cover rocks.

3.2.32 Wasita, Yemen (14°54′ N–44°12′ E) (Fig. 8c)

McHone and Dietz (1988) reported the existence of a circular depression about 4 km wide and surrounded by a 14 km diameter concentric structure near the town of Wasita in Yemen. According to these authors and volcanologists that conducted field investigations in that area, this site is not considered as one of the volcanic features in the region. There is no other information about this medium-sized circular feature.

3.2.33 Zaynan, Saudi Arabia (20°23′ N–50°08′ E)

The Zaynan structure is a subsurface feature, 5 km in diameter and 190 m deep, buried to 3800 m below the surface (Neville et al. 2014). It appears in 2D seismic lines as a structural depression in the upper layers of Triassic strata and shows a broad central uplift. The crater-like structure affects the Triassic Jilh and Sudair formations and is filled with Jurassic carbonates of the Arab to Dhruma formations. According to Neville et al. (2014), the geometry and regional uniqueness of the Zaynan structure are compatible with a meteorite impact origin.

3.3 *Discredited Impact Craters (Figs. 13, 14)*

A number of disproven impact structures are given below; this list is limited to the structures for which an impact origin has been proposed in scientific or conference papers, with an impact in the media for a few of them. The proposed impact origin was later discredited based on more detailed scientific investigations. This listing does not contain any of the circular structures that had previously been investigated for a possible impact origin in Algeria, Morocco and Saudi Arabia, and for which an impact origin was ruled out earlier (see Lambert et al. 1980, 1981; Chaabout et al. 2015; Stewart 2015).

3.3.1 Aflou Structure, Algeria (34°00′ N–02°03′ E) (Fig. 13e)

The Aflou structure is an oval depression, 3 × 5 km in size, which is elongated in SW to NE direction. It was first described by Marks et al. (1972), who assigned it a probable impact origin based on the presence of "vitreous" rocks at the center of the structure, which were interpreted as impact melt lithologies. Lambert et al. (1980) noted a lack of evidence for impact and described it as an erosional and/or dissolution feature. They related the igneous rocks to magmatic activity that affected the area during the late Tertiary or Quaternary. Reimold and Koeberl (2014) reviewed the current knowledge about this structure, listing it as a disproven impact crater in Africa. Chabou and Laghouag (2014) reinvestigated this structure and interpreted it

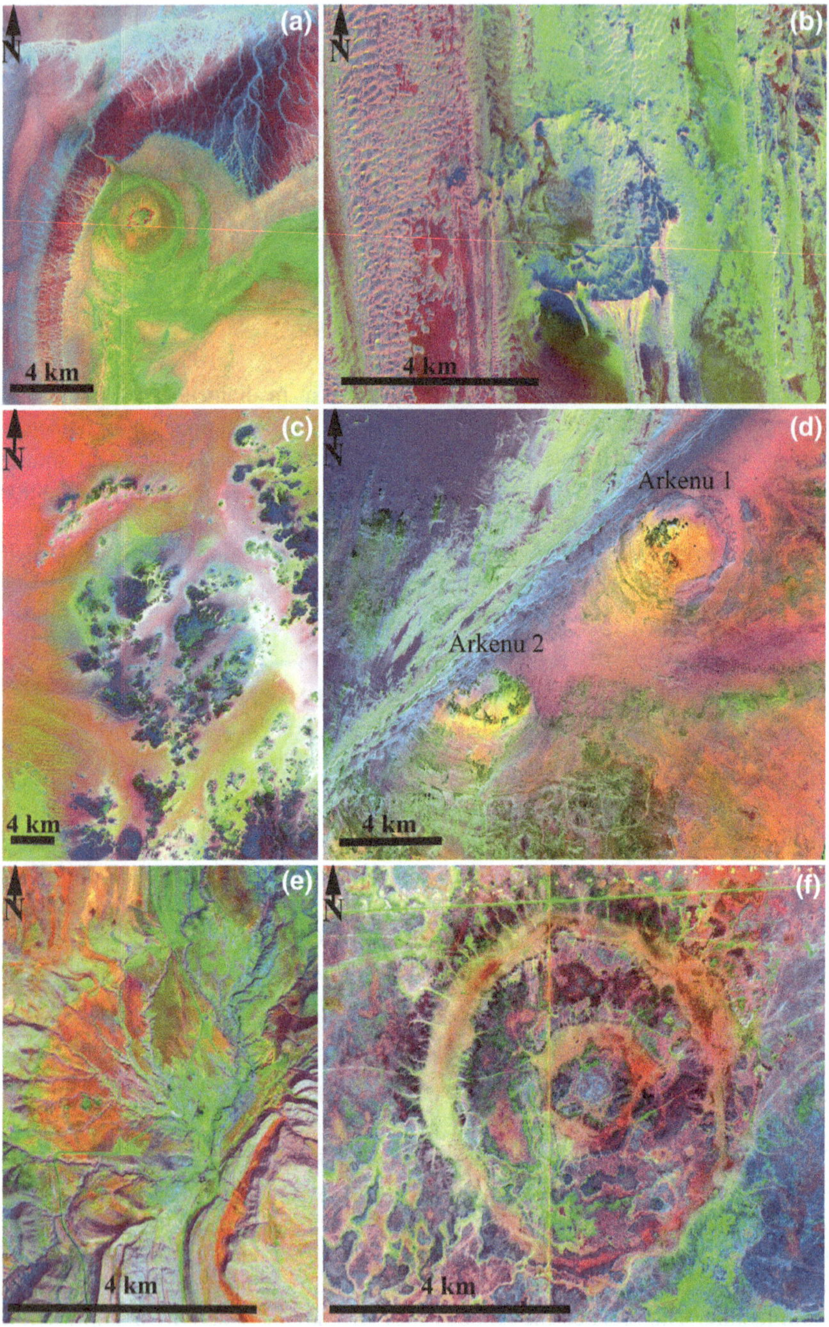

Fig. 13 Sentinel 2A principal component analysis (PCA) images of 6 disproven impact structures of the Arab world. **a** Foum Taguentour (Algeria). **b** El Baz (Egypt). **c** Kebira (Libya/Egypt). **d** Arkenu (Libya). **e** Aflou (Algeria). **f** Habhab (Yemen)

Fig. 14 **a** Google Earth image of the area around lakes Isli and Tislit (High Atlas Mountains, Morocco). **b** Google Earth image of the Richat structure (Mauritania)

as a Triassic diapir. The igneous rocks that outcrop in the area are intraformational Triassic/Jurassic ophites.

3.3.2 Arkenu Double Crater, Libya (22°06′ N–23°45′ E for Arkenu 1 and 22°02′ N–23°43′ E for Arkenu 2) (Fig. 13d)

Paillou et al. (2003) reported the discovery of two impact craters in the southeastern part of the Libyan desert: a northeastern crater named Arkenu 1 of 6.8 km diameter, and a southwestern crater named Arkenu 2, 10.3 km in diameter. The evidence supporting the impact hypothesis proposed by these authors was the observation of shatter cone-like structures in the field and impact breccias with planar fractures in quartz grains. Reimold (2007), Di Martino et al. (2008), and Cigolini et al. (2012) contested these interpretations: the alleged shatter cones are in fact wind ablation features in sandstones, and the reported planar fractures are not truly planar. Detailed investigations performed by Cigolini et al. (2012) failed to detect any evidence of shock metamorphism in the samples outcropping in the areas of the Arkenu structures. However, there is strong evidence supporting that their endogenous origin is related to the emplacement of "paired" subvolcanic stocks of syenitic composition accompanied by hydrothermal degassing (Di Martino et al.

2008; Cigolini et al. 2012). The Arkenu structures have been referred to in the past as carbonatite plugs (Baegi 2008) and probably correspond to the two circular intrusions of carbonatitic rocks described in Woolley (2001) (page 145).

3.3.3 El Baz, Egypt (24°12′ N–26°24′ E) (Fig. 13b)

Discovered on Landsat images, the El Baz crater is a circular feature of 4 km in diameter, which shows a complex structure with a flat floor, a sharp and crenulated rim crest, a terraced wall and a discontinuous inner structure of approximately 1.6 km diameter and that has been interpreted as a central uplift (El-Baz 1981). El Baz (1981) suggested that this structure was of impact origin based on its morphological characteristics. However, a field investigation showed that the El Baz crater is of volcanic origin and was formed as a result of basaltic intrusion into the Nubia sandstone (Barakat 2001).

3.3.4 Foum Taguentour, Algeria (26°14′ N–02°57′ E) (Fig. 13a)

Foum Taguentour is an 8-km-diameter concentric ring feature containing a central depression surrounded by a slightly oval ridge (1 × 1.5 km in diameter). The structure was first noticed by Guillemot (1962) and recognized as of probable impact origin by McHone and Dietz (1978) based on its perfect circularity and its morphology. Lambert et al. (1981) visited the structure and found no evidence of shock metamorphism. They suggested a diapiric origin to explain the characteristics of the structure. Reimold and Koeberl (2014) proposed to give the status of possible impact structure to Foum Taguentour given the limited information available about this perfectly circular structure. However, Haddoum (2009) emphasizes the major role of tectonic activity and argilokinesis in the formation of circular structures in the area, in particular, that of Foum Taguentour. In that respect, and given the relationship between this feature and the surrounding structures (Foum Taguentour structure is linked to a contractional horsetail splay at the end of a major N-S strike-slip fault) (Haddoum 2009), an impact origin for Foum Taguentour is unlikely.

3.3.5 Gilf Kebir Field, Egypt (23°10′–23°40′ N and 26°50′–27°35′ E)

Paillou et al. (2006) reported the occurrence of more than 1300 small crater-like structures distributed over an area of 40,000 km^2 in the Western Egyptian Desert, close to the Gilf Kebir plateau. It was suggested that these structures could be the result of meteorite impacts based on the presence of alleged shatter cones and planar deformation features in quartz at several of the Gilf Kebir structures (Paillou et al. 2004). Orti et al. (2008) conducted detailed studies of the Gilf Kebir crater field and found no evidence supporting an impact origin. They concluded that the circular structures of the Gilf Kebir area are related to endogenic geological processes typical of hydrothermal vent complexes in volcanic areas, which may reflect the emplacement of subvolcanic intrusives that did not reach the surface and crystallized at depth undergoing degassing.

3.3.6 Habhab (Oman Ring), Oman (19°55′ N–57°00′ E) (Fig. 13e)

The Habhab structure was discovered in 1975 by Dietz et al. (1975) on Landsat images. The structure appears as two concentric dark rings, 6 and 2 km in diameter, respectively. The inner ring was interpreted as a central domal uplift (Dietz et al. 1975). Habhab has been mentioned in the literature several times as a possible or suspected impact structure (Grieve et al. 1988; McHone and Dietz 1988; McHone and Greeley 1997). However, Levell et al. (2002) noted, based on seismic surveys, that the structure was formed by a salt dome rising from deep below the surface. A detailed field study carried out by Hofmann et al. (2003) did not yield any evidence for impact, and these authors concluded that the Habhab structure is most likely caused by salt dissolution/salt tectonics related to a near-surface salt plug.

3.3.7 Isli and Tislit Lakes, Morocco (32°13′ N–05°33′ W for Isli and 32°12′ N–05°38′ W for Tislit) (Fig. 14a)

Ibhi et al. (2013) reported the discovery of dual impact structures in the High Atlas Mountains of Morocco (Lakes Isli and Tislit) based on the occurrence of meteorite fragments near the lakes, monomict and polymict breccias, planar structures in quartz grains and radial fractures in rocks. They estimated the age of these twin structures to 40 ka. These allegations received a considerable amount of Moroccan press coverage. However, the results of field investigations summarized in Chaabout et al. (2013) do not favor an impact origin of the twin lakes. Neither evidence of shock metamorphism nor structural features indicating an impact origin were found. Finally, Ibouh et al. (2014) presented strong arguments in favor of a tectonic-karstic origin of Lake Isli. According to these authors, the lake formed during the Lower–Middle Pleistocene in a small Pliocene pull-apart basin through collapse due to karst phenomena in the underlying limestones.

3.3.8 Kebira, Libya (24°39′ N–24°58′ E) (Fig. 13c)

El Baz and Ghoneim (2007) suggested the existence of a $\sim$31 km impact structure in the Cretaceous Nubia sandstones near the border between Libya and Egypt. They described this structure as a complex impact crater composed of an eroded outer rim measuring 31 km in diameter, a well-preserved inner ring of ca. 17 km diameter, and a central peak.

The authors, moreover, speculated about the possible relationships between this alleged structure and the Libyan Desert Glass found in southwestern Egypt. Despite the substantial media coverage about this alleged discovery, evidence for impact at this structure is poor (Reimold 2007; Reimold and Koeberl 2014). The author fully supports the opinion of Reimold and Koeberl (2014) to include this structure in the list of disproven impact craters because there is no remote sensing evidence for the existence of a circular feature associated with the so-called "Kebira crater". The only feature visible on satellite imagery is a semicircular outcrop that emerges from the desert pavement in the northwestern part of the alleged structure, which seems to be purely erosional. Also, Google Earth shows quite clearly that the strata at the center of the structure are in fact horizontal, something that is unexpected in a complex crater with a central uplift.

3.3.9 Richat Structure, Mauritania (21°07′ N–11°24′ W) (Fig. 14b)

The Richat structure, also called the "Eye of Africa", is a large eroded dome at least 40 km in diameter that is located within the Late Proterozoic to Ordovician sequence of the Taoudenni basin (Matton et al. 2005; Matton and Jebrak 2014). The structure was first described by Richard-Molard (1948a), who interpreted it to be of endogenic origin. Cailleux et al. (1964) suggested that this structure was of meteorite impact origin based on the discovery of brecciated cherts and quartzites in the center of the dome, and the presence of coesite in some of the quartzite breccias. The site was later visited by Dietz et al. (1969), who failed to find any evidence of shock-metamorphic effects in the host rocks. The alleged coesite was recognized by Fudali (1969) as a misidentification of barite. Finally, Matton et al. (2005) interpreted the Richat structure and its breccia core as the superficial expression of a Cretaceous alkaline complex with an exceptionally well-preserved hydrothermal karst infilling at its summit.

4 Discussion and Conclusion

Only 13 impact structures or sites are currently known in the Arab world, and numerous craters still await to be discovered according to recent predictions of yet-to-find impact craters on Earth (Stewart 2011; Hergarten and Kenkmann 2015). Several possible impact structures were reported in the literature based on remote sensing results but need detailed field studies to either prove or dismiss an impact origin. Thirty-three possible impact structures were briefly described in this review. Based on the examination of satellite images, the description of the structures given by the authors and the available geological documentation, some of these proposed features are probably not of impact origin (Al-Atarib, Berber, In Ezzane, Jebel Ababad, Ras Zeidun, and Saghira) (Table 3). Others are very promising impact crater candidates. Among these features, the best candidates for meteorite impact

Table 3 List of structures in the Arab world that have been discredited as impact structures

Name	Country	Latitude	Longitude	Explanation
Aflou	Algeria	31°59′ N	05°30′ W	Diapir
Arkenu 1 and 2	Libya	26°05′ N	04°23′ E	Volcanic origin
El Baz	Egypt	20°15′ N	12°41′ W	Volcanic origin
Foum Taguentour	Algeria	25°19′ N	24°20′ E	Tectonic origin
Gilf Kebir structures	Egypt	31°03′ N	36°48′ E	Volcanic origin
Habhab	Oman	22°01′ N	26°05′ E	Salt diapir
Isli and Tisli Lakes	Morocco	24°35′ N	24°24′ E	Karstic origin
Kebira	Libya/Egypt	29°00′ N	07°33′ W	No structure
Richat	Mauritania	21° 07′ N	11°24′ W	Exhumed and eroded magmatic body
From this study				
Al-Atarib	Syria	33°19′ N	04°02′ E	Sinkhole
Berber	Sudan	18°03′ N	33°30′ E	Igneous ring complex
In Ezzane (5 structures)	Algeria	22°55′ N	10°24′ W	Ring dykes
Jebel Ababad	Sudan	19°13′ N	35°59′ E	Igneous ring complex
Ras Zeidun	Egypt	27°36′ N	05°07′ E	No structure
Saghira	Libya	21°30′ N	50°28′ E	Tectonic origin

structures are Temimichat-Ghallaman located in Mauritania and the Jabal Rayah and Ash Shutbah structures located in Saudi Arabia.

It is crucial to constantly reiterate that the only criteria that are generally accepted for assigning an impact origin to a suspected terrestrial structure are the occurrence of shock-metamorphic effects, such as shatter cones and planar deformation features in quartz, the discovery of preserved meteorite fragments, or the detection of traces of a meteoritic projectile (e.g., French and Koeberl 2010). Consequently, field investigations, including the collection and analysis of rock samples, are required. Unfortunately, in many countries of the Arab region, field work is impossible due to political instability, civil war, and/or the presence of landmines. For example, the Ibn-Batutah and Jebel el Hadid structures in Southern Libya were recognized as two suspected impact structures based on remote sensing observations. Fieldwork at both sites is currently impossible in this area due to the security situation at the Libyan–Chad border and the presence of landmines. The same is true for structures located in Iraq, Syria or Yemen. Another aspect of the problems in recognizing impact craters is the misidentification of shock-metamorphic features (e.g., shatter cones can be easily confused with cone-in-cone structures or wind abrasion features, especially in desert regions that characterize much of the Arab territories, and some reports of planar deformation features in quartz were shown to be, in fact, not planar and of tectonic origin) (e.g., Reimold 2007; Hamers and Drury 2011; Reimold et al. 2014). Most of the structures recently proposed as impact structures—based on misidentification of shock-metamorphic features—are located in the Arab world (Schmieder et al. 2014). For example, detailed studies of the Arkenu double "craters" (Libya) and the Gilf Kebir "crater field" (Egypt) have shown no evidence supporting an impact origin of these structures (Orti et al. 2008; Di Martino et al. 2008) despite the early reports by Paillou et al. (2003, 2004) of the presence of shatter cones, planar fractures and planar deformation features (PDFs) in quartz. As noted by Reimold (2007), it is an unfortunate fact that these announcements were often relayed by the consumer press and distributed widely in the Internet. Conversely, revision of these claims is rarely reported in the media. For example, the announcement in 2006 of the discovery of the so-called "Kebira crater" near the border between Libya and Egypt (El Baz and Ghoneim 2007) and its relationship with the Libyan Desert glass was sensationalized in the media despite the fact that this information was not supported by field or analytical evidence. This information is still being broadcast on serious websites such as the NASA earth observatory site (https://earthobservatory.nasa.gov/IOTD/view.php?id=6351). Another example is the announcement of the discovery of dual impact structures in Morocco (Lakes Isli and Tislit) (Ibhi et al. 2013) that received a considerable amount of Moroccan press coverage, and which was widely broadcast on the Internet. This alleged discovery was refuted by independent teams (Chaabout et al. 2013; Ibouh et al. 2014). It should be noted that the article of Ibhi et al. (2013) was published in a journal with no peer review control. With the proliferation of such journals in the Internet, the list of alleged impact structures in the Arab world without convincing evidence will probably grow in the future. It is therefore necessary to sensitize journalists about the reliable sources of scientific information.

On another issue, it should be noted that the geological community in the Arab countries is still, to a large part, unfamiliar with the study of impact craters, with the result that there is a lack of research programs and funding in the fields of impact cratering. Efforts have been made in recent years to promote impact cratering and astrogeology in the countries of the Arab world by the organization of 3 Arab Impact Cratering and Astrogeology Conferences (AICAC) in Jordan, Morocco, and Algeria (Reimold 2010; Baratoux et al. 2012; Belhai et al. 2017). More recently, an African initiative for planetary and space sciences including the Arab countries of Africa was launched (Baratoux et al. 2017). This initiative will certainly help elevate planetary and space science across the entire African continent and may be extended to the rest of the Arab world.

In terms of economic importance, none of the confirmed impact craters in the Arab world seem to provide deposits of particular interest. A few studies addressed the issue of economic significance of impact structures in the Arab countries. Schmieder et al. (2009a) suggested that the possible impact structure of Jebel Hadid in Libya might have provided pathways for hydrocarbon migration, as well as possible local hydrocarbon traps. Belhai et al. (2006) also stressed the possibility that the Algerian complex impact craters (Tin Bider and Ouarkziz) could form possible local hydrocarbon traps. On the other hand, the isolated geographical locations of most confirmed impact craters in the Arab world would make their development into geotouristic areas challenging. Nonetheless, plans are underway by the local authorities to establish the Talemzane crater in Algeria as a tourist attraction, and Jordanian authorities are looking into developing the Jebel Waqf as Suwwan site.

In summary, the record of impact structures in the Arab region is still incomplete considering the great size of the territory and the number of impact craters known in other regions where there have been active programs to study impact structures. It is important to not only promote impact crater studies in the Arab world but also to train the Arab scientific community in recognizing and documenting the macro- and microscopic shock-metamorphic criteria diagnostic for impact structures.

Acknowledgements I thank Martin Schmieder for comments and suggestions on the first version of this manuscript. I appreciate the detailed and helpful comments and reviews by Wolf Uwe Reimold that strongly improved the final version of the manuscript. Abderrahmane Bendaoud is thanked for editorial handling of this chapter. I am grateful to Zakaria Hamimi for providing copies of papers of the Annals of the Geological Survey of Egypt. This paper is a contribution to the Africa Initiative for Planetary and Spaces Sciences (AFIPS).

Appendix

Most of the images of the structures presented in this paper were processed in principal component analysis (PCA) from Sentinel 2A images for better visibility of these structures. Sentinel-2 is ESA's medium spatial resolution super-spectral instrument with a large field of view (290 km), a high revisit capability (5 days with two satellites), a high resolution (10, 20 and 60 m), and a moderately large band set (13 spectral bands) aimed at ensuring data continuity, following the Landsat and SPOT ages, for global land surface monitoring (van der Meer et al. 2014). The Sentinel-2A satellite was successfully launched on 23 June 2015. The Sentinel-2A MSI images were downloaded from the Sentinel's Scientific Data Hub (https://scihub.copernicus.eu/). PCA processing was applied to the four bands at 10 m resolution (B2, B3, B4, and B8, corresponding to the blue, green, red, and near infrared spectral segments). The Landsat 8 OLI satellite image of the In Ezzane structures (Fig. 9) was downloaded from the website http://earthexplorer.usgs.gov/.

References

Abate B, Koeberl C, Kruger FJ, Underwood JR Jr (1999) BP and Oasis impact structures, Libya, and their relation to Libyan Desert Glass. In: Dressler BO, Sharpton VL (eds) Large meteorite impacts and planetary evolution, geological society of America, special paper 339, pp 177–192

Abdelhamid G (2001) Radarsat investigation of a circular feature in east central Jordan. Int J Remote Sens 22:2231–2239

Abels A, Plado J, Pesonen LJ, Lehtinen M (2002) The impact cratering record of fennoscandia—a close look at the database. In: Plado J, Pesonen LJ (eds) Impacts in precambrian shields. Springer, Berlin, pp 1–58

Al-Mishwat AT (2003) Wadi Na'am structure: a possible concealed impact feature from Central Southwest Saudi Arabia. In: 34th lunar and planetary science conference. Abstract #1049

Ames DE, Davidson A, Wodicka N (2008) Geology of the giant Sudbury polymetallic mining camp, Ontario, Canada. Econ Geol 103:1057–1077

Attar A (1981) La structure de Tin Bider (plateau du Tinhert, Algérie). Un impact météoritique. Unpublished report, Sonatrach, p 15

Baegi M (2008) Carbonatites versus meteorite impact origin of two circular structures outcrop at eastern Al-Awaynat area, SE Libya. In: Large meteorite impacts and planetary evolution IV conference, Vredefort, South Africa. Lunar and Planetary Institute, Abstract #3036

Barakat AA (2001) El-Baz crater: basaltic intrusion versus meteorite impact crater. Ann Geol Surv Egypt 24:167–177

Barakat AA (2011) Ras Zeidun, Eastern desert of Egypt: a possible astrobleme. In: Abstracts, 2nd Arab impact cratering and astrogeology conference (AICAC II). Hassan II Casablanca University, Casablanca, Morocco, p 10

Baratoux D, Reimold WU (2016) The current state of knowledge about shatter cones: introduction to the special issue. Meteorit Planet Sci 51:1389–1434

Baratoux D, Reimold WU, Chennaoui-Aoudjehane H (2012) The second Arab impact cratering and astrogeology conference, Casablanca, 14–20 Nov 2011—A bridge between geoscientists and astronomers. Meteorit Planet Sci 47:1098–1103

Baratoux D, Chennaoui-Aoudjehane H, Gibson R, Lamali A, Reimold WU, Sapah MS, Chabou MC, Habarulema JB, Jessell MW, Mogessie A, Benkhaldoun Z, Nkhonjera E, Mukosi NC, Kaire M, Rochette P, Sickafoose A, Martínez-Frías J, Hofmann A, Folco L, Rossi AP, Faye G, Kolenberg K, Tekle K, Belhai D, Elyajouri M, Koeberl C, Abdeen MM (2017) Africa initiative for planetary and space sciences. EOS 98(11):16–23. https://eos.org/opinions/africa-initiative-for-planetary-and-space-sciences

Basurah HM (2009) Al-Atarib crater, Syria. In: Abstracts, the first Arab impact cratering and astrogeology conference, Amman, Jordan, pp 3–4

Belhai D, Merle O, Vincent P, Devouard B, Afalfiz AH (2006) Etat des connaissances et mise au point sur les cratères météoritiques du Sahara algérien, des indicateurs de pièges à hydrocarbures? Bulletin du Service Géologique de l'Algérie 17:95–112

Belhai D, Chennaoui-Aoudjehane H, Baratoux D, Ferrière L, Lamali A, Sahoui R, Lambert P, Ayadi A (2017) The fourth Arab impact cratering and astrogeology conference (AICAC IV), Apr 9–12, 2017, Algiers (Algeria). Meteorit Planet Sci 52:2067–2071

Berlin GL (1989) SIR-A signature of Al Umchaimin crater, western Iraq. Int J Remote Sens 10:451–452

Blom RG, McHone JF, Crippen RE (1998) Possible 770-meter diameter impact crater detected by satellite; Yemen Arab Republic. In: 29th lunar and planetary science conference. Abstract #1559

Bonin B, Ouabadi A, Liégeois JP, Bruguier O, Passchier C, Guillou H, Abdallah N, Fezaa N, Laouar R, Boissonnas J (2011) Possible impact craters within the In Ezzane quaternary volcanic province, Eastern Hoggar, Algeria. In: Abstracts, 2nd Arab impact cratering and astrogeology conference (AICAC II). Hassan II Casablanca University, Casablanca, Morocco, p 15

Cailleux A, Guillemaut A, Pomerol C (1964) Présence de coésite, indice de hautes pressions, dans l'accident circulaire des Richat (Adrar mauritanien). Comptes Rendus de l'Académie des Sciences Paris 258:5488–5490

Chaabout S, Chennaoui Aoudjehane H, Reimold WU, Aboulahris M, Aoudjehane M (2013) Evidence of non-impact cratering origin of Imilchil (Morocco) lakes (Isli and Tislit). In: Large meteorite impacts and planetary evolution V conference, Sudbury, Canada, Lunar and Planetary Institute, Abstract #3074

Chaabout S, Chennaoui-Aoudjehane H, Reimold WU, Baratoux D, Youbi N (2015) Prospecting for possible impact structures in Morocco. J Afr Earth Sci 112:339–352

Chabou MC (2009) Meteorite impact craters in the Arab world: an overview. In: Abstracts, The first Arab impact cratering and astrogeology conference (AICAC), 9–11 Nov 2009, Amman, Jordan, pp 7–8

Chabou MC (2011a) On the problems of the identification of new impact structures in the Arab world. Abstracts, The second Arab impact cratering and astrogeology Conference (AICAC II), 13–20 Nov 2011. Casablanca, Morocco, pp 23–24

Chabou MC (2011b) Tigraou structure: a possible new impact crater in northwestern Algeria. Abstracts, The second Arab impact cratering and astrogeology conference (AICAC II), 13–20 Nov 2011. Casablanca, Morocco, pp 25–26

Chabou MC (2016) An updated inventory of meteorite impact structures in the Arab world. In: Abstract, first Arab geosciences union (ArabGU) international conference (AIC-1), 17–18 Feb 2016, USTHB, Algiers, Algeria, p 174

Chabou MC, Laghouag MY (2014) On the origin of the Aflou structure (Algeria). In: 77th annual meeting of the meteoritical society, 7–12 Sept 2014, Casablanca, Morocco. Abstract #5368

Chennaoui Aoudjehane H, Garvie LAJ, Herd CDK, Chen G, Abouhlaris M (2013) Agoudal: the most recent iron meteorite from Morocco. In: 76th annual meeting of the meteoritical society, July 29–Aug 7 2013, Edmonton, Canada, Abstract #5025

Chennaoui Aoudjehane H, El Kerni H, Reimold WU, Baratoux D, Koeberl C, Bouley S, Aoudjehane M (2016) The Agoudal (High Atlas Mountains, Morocco) shatter cone conundrum: a recent meteorite fall onto the remnant of an impact site. Meteorit Planet Sci 51:1497–1518

Cigolini C, Di Martino M, Laiolo M, Coppola D, Rossetti P, Morelli M (2012) Endogenous and nonimpact origin of the Arkenu circular structures (al-Kufrah basin-SE Libya). Meteorit Planet Sci 47:1772–1788

D'Orazio M, Folco L, Zeoli A, Cordier C (2011) Gebel Kamil: the iron meteorite that formed the Kamil crater (Egypt). Meteorit Planet Sci 46:1179–1196

Daneshvar MRM, Nejadsoleymani H (2017) Remote sensing analysis of unknown origin of a crater in western Yemen. Spat Inf Res 25:575–583

Denis M, Buoncristiani JF, Konaté M, Ghienne JF, Guiraud M (2007) Hirnantian glacial and deglacial record in SW Djado Basin (NE Niger). Geodin Acta 20:177–195

Derder MEM, Maouche S, Liegeois JP, Henry B, Amenna M, Ouabadi A, Bellon H, Bruguier O, Bayou B, Bestandji R, Nouar O, Bouabdallah H, Ayache M, Beddiaf M (2016) Discovery of a Devonian mafic magmatism on the western border of the Murzuq basin (Saharan metacraton): paleomagnetic dating and geodynamical implications. J Afr Earth Sci 115:159–176

Di Achille G (2005) A new candidate impact site in north-eastern Sudan detected from remote sensing. In: 36th lunar and planetary science conference. Mar 14–18 2005, League City, Texas, USA, Abstract #1606

Di Martino M, Cigolini C, Orti L (2008) Non-impact origin of the Arkenu craters (Libya). In: Large meteorite impacts and planetary evolution IV conference, Vredefort, South Africa, Lunar and Planetary Institute, Abstract #3012

Dietz RS (1947) Meteorite impact suggested by the orientation of shattercones at the Kentland, Indiana, disturbance. Science 105:42–43

Dietz RS, Fudali R, Cassidy WA (1969) Richat and Semsityat Domes, Mauritania: not astroblemes. Geol Soc Am Bull 80:1367–1372

Dietz RS, McHone JF, Short NM (1975) Oman ring: suspected astrobleme. Meteoritics 10 (393):509

Donofrio RR (1981) Impact craters: implications for basement hydrocarbon production. J Pet Geol 3:279–302

Dunford A, Koeberl C (2009) Tmisan and Jaraminah, two possible impact craters in western Libya. In: Abstracts, the first Arab impact cratering and astrogeology conference, Amman, Jordan, pp 18–20

Earth Impact Database 2018. http://www.passc.net/EarthImpactDatabase/index.html. Accessed Mar 2018

El Baz F (1981) Circular feature among dunes of the great sand sea, Egypt. Science 213:439–440

El Baz F, Ghoneim EM (2007) Largest crater shape in the Great Sahara revealed by multi- spectral images and radar data. Int J Remote Sens 28:451–458
El-Baz F, Issawi B (1981) Crater forms in the Uweinat region. Ann Geol Surv Egypt 11:79–89
El-Baz F, Mitchell DA (1976) Earth observations and photography experiment MA-136. In: Apollo-Soyuz test project preliminary science report. NASA, Johnson Space Center, Houston, Texas, TMX-58173, pp 10-1–10-64
Erickson TM, Cavosie AJ, Pearce MA, Timms NE, Reddy SM (2016) Empirical constraints on shock features in monazite using shocked zircon inclusions. Geology 44:635–638
Fabre J, Kazi-Tani N, Megartsi M (1970) Le «rond» de l'Ouarkziz (Sahara nord-occidental): un astroblème? Comptes Rendus de l'Académie des Sciences, Paris 270:1212–1215
Fazio A, D'Orazio M, Cordier C, Folco L (2016) Target-projectile interaction during impact melting at Kamil crater, Egypt. Geochim Cosmochim Acta 180:33–50
Folco L, Di Martino M, El Barkooky A, D'Orazio M, Lethy A, Urbini S, Nicolosi I, Hafez M, Cordier C, van Ginneken M, Zeoli A, Radwan AM, El Khrepy S, El Gabry M, Gomaa M, Barakat A, Serra R, El Sharkawi M (2010) The Kamil crater in Egypt. Science 329:804
Folco L, Di Martino M, El Barkooky A, D'Orazio M, Lethy A, Urbini S, Nicolosi I, Hafez M, Cordier C, van Ginneken M, Zeoli A, Radwan AM, El Khrepy S, El Gabry M, Gomaa M, Barakat A, Serra R, El Sharkawi M (2011) Kamil crater (Egypt): ground truth for small-scale meteorite impacts on Earth. Geology 39:179–182
French BM (1998) Traces of catastrophe: a handbook of shock-metamorphic effects in terrestrial meteorite impact craters. Lunar and Planetary Institute, Houston, TX, Contribution CB-954. 120 p. Available online at: https://www.lpi.usra.edu/publications/books/CB-954/CB-954.intro.html
French BM (2004) The importance of being cratered: the new role of meteorite impact as a normal geological process. Meteorit Planet Sci 39:169–197
French BM, Koeberl C (2010) The convincing identification of terrestrial meteorite impact structures: what works, what doesn't, and why. Earth Sci Rev 98:123–170
French BM, Hartung JB, Short NM, Dietz RS (1970) Tenoumer crater, Mauritania: age and petrologic evidence for origin by meteorite impact. J Geophys Res 75:4396–4406
French BM, Underwood JR Jr, Fisk EP (1974) Shock metamorphic features in two meteorite impact structures, Southeastern Libya. Geol Soc Am Bull 85:1425–1428
Fritzius RS website (2017). http://www.datasync.com/~rsf1/fun/moab-cr.htm
Fudali RF (1969) Coesite from the Richat Dome, Mauritania: a misidentification. Science 166:228–230
Fudali RF (1974) Genesis of the melt rocks at Tenoumer Crater, Mauritania. J Geophys Res 79:2115–2121
Fudali RF, Cressy PJ (1976) Investigation of a new stony meteorite from Mauritania with some additional data on its find site: Aouelloul crater. Earth Planet Sci Lett 30:262–268
Garvin JB, Blodget HW (1986) Suspected impact crater near Al Madafi, Saudi Arabia. Meteoritics 21:366–367
Ghoneim EM (2009) Ibn-Batutah: a possible simple impact structure in southeastern Libya, a remote sensing study. Geomorphology 103:341–350
Gibson RL, Reimold WU, Baegi M, Crosta AP, Shbeli E, Eshwedi A (2011) The Oasis structure, southeastern Libya: new constraints on size, age and mechanism of formation. In: 42th lunar and planetary science conference. March 07–11 2011, The Woodlands, Texas, USA, Abstract #1024
Gnos E, Hofmann BA, Halawani A, Tarabulsi Y, Hakeem M, Al Shanti M, Greber ND, Holm S, Alwmark C, Greenwood RC, Ramseyer K (2013) The wabar impact craters, Saudi Arabia, revisited. Meteorit Planet Sci 48:2000–2014
Gnos E, Hofmann BA, Schmieder M, Al-Wagdani K, Mahjoub A, Al-Solami AA, Habibullah SN, Matter A, Alwmark C (2014) Ash Shutbah: a possible impact structure in Saudi Arabia. Meteorit Planet Sci 49:1902–1914

Goderis S, Paquay F, Claeys P (2013) Projectile identification in terrestrial impact structures and ejecta material. In: Osinski GR, Pierazzo E (eds) Impact cratering: processes and products. Wiley-Blackwell, pp 223–239

Grady MM, Hutchison R, McCall JGH, Rothery DA (eds) (1998) Meteorites: flux with time and impact effects. Geological society of America special publications 140, 280 p

Grajales-Nishimura JM, Cadillo-Pardo E, Rosales-Domínguez C, Moráan-Zenteno D, Alvarez W, Claeys P, Ruíz-Morales J, García-Hernández J, Padilla-Avila P, Sánchez-Ríos A (2000) Chicxulub impact: the origin of reservoir and seal facies in the southeastern Mexico oil fields. Geology 28:307–310

Greeley R, Christensen PR, McHone JF (1987) Radar characteristics of small craters: implications for venus. Earth Moon Planet 37:89–111

Grieve RAF (1984) The impact cratering rate in recent time. J Geophys Res 89(Supplement): B403–B408

Grieve RAF (1987) Terrestrial impact structures. Annu Rev Earth Planet Sci 15:245–270

Grieve RAF (2005) Economic natural resource deposits at terrestrial impact structures. In: McDonald I, Boyce AJ, Butler IB, Herrington RJ, Polya DA (eds) Mineral deposits and earth evolution. Geological society of America special publications 248, pp 1–29

Grieve RAF (2013) Economic deposits at terrestrial impact structures. In: Osinski GR, Pierazzo E (eds) Impact cratering: processes and products. Wiley-Blackwell, pp 177–193

Grieve RAF, Dence MR (1979) The terrestrial cratering record. II: the crater production rate. Icarus 38:230–248

Grieve RAF, Masaitis VL (1994) The economic potential of terrestrial impact craters. Int Geol Rev 36:105–151

Grieve RAF, Pilkington M (1996) The signature of terrestrial impacts. ASGO J Aust Geol Geophys 16:399–420

Grieve RAF, Therriault M (2000) Vredefort, Sudbury, Chicxulub: three of a kind? Annu Rev Earth Planet Sci 28:305–338

Grieve RAF, Therriault M (2004) Observations at terrestrial impact structures: their utility in constraining crater formation. Meteorit Planet Sci 39:199–216

Grieve RAF, Wood CA, Garvin JB, McLaughlin G, McHone JF (1988) Possible impact craters. In: Astronaut's guide to terrestrial impact craters. LPI technical report 88-03, pp 75–82

Grieve RAF, Reimold WU, Morgan J, Riller U, Pilkington M (2008) Observations and interpretations at Vredefort, Sudbury, and Chicxulub: towards an empirical model of terrestrial impact basin formation. Meteorit Planet Sci 43:855–882

Guillemot A (1962) Structures anticlinales circulaires d'origine mal connue. Photo interprétation, Technip Editions, fascicule 4

Haddoum H (2009) Les structures hercyniennes dans la couverture sédimentaire paléozoïque de l'Ahnet occidental et de Bled el Mass (N.O. du Hoggar, Algérie): une conséquence du rejeu des failles panafricaines. Bulletin du Service Géologique National 20:221–243

Haines PW (2005) Impact cratering and distal ejecta: the Australian record. Aust J Earth Sci 52:481–507

Hamann C, Hecht L, Ebert M, Wirth R (2013) Chemical projectile–target interaction and liquid immiscibility in impact glass from the Wabar craters, Saudi Arabia. Geochim Cosmochim Acta 121:291–310

Hamers MF, Drury MR (2011) Scanning electron microscope-cathodoluminescence (SEM-CL) imaging of planar deformation features and tectonic deformation lamellae in quartz. Meteorit Planet Sci 46:1814–1831

Heinrichs T, Salameh E, Khoury H (2014) The Waqf as Suwwan crater, Eastern Desert of Jordan: aspects of the deep structure of an oblique impact from reflection seismic and gravity data. Int J Earth Sci 103:233–252

Hergarten S, Kenkmann T (2015) The number of impact craters on Earth: any room for further discoveries? Earth Planet Sci Lett 425:187–192

Hodge PW (1994) Meteorite craters and impact structures of the earth. Cambridge University Press, Cambridge, 124 p

Hofmann BA, Gnos E, Al-Kathiri A, Al-Busaidi SA, Al-Rajhi A, Al-Batashi M, Terken J (2003) The Habhab structure of central Oman: not an impact crater. Meteorit Planet Sci, 38 (Suppl.), Abstract #5096

Hughes DW (2000) A new approach to the calculation of the cratering rate of the Earth over the last 125 ± 20 Myr. Mon Not R Astron Soc 317:429–437

Ibhi A, Nachit H, Abia EH, Touchnt AA, Vaccarto C (2013) Isli and Tislit: The first dual impact crater discovered in Morocco. Int J Astron Astrophys 3:1–4

Ibouh H, Michard A, Charrière A, Benkaddour A, Rhoujjati A (2014) Tectonic–karstic origin of the alleged "impact crater" of Lake Isli (Imilchil district, High Atlas, Morocco). CR Geosci 346:82–89

Janjou D, Halawani MA, Brosse JM, Al-Muallem MS, Becq-Giraudon JF, Dagain J, Genna A, Razin P, Roobol MJ, Shorbaji H, Wyns R (1997) Explanatory notes to the geologic map of the Tabuk Quadrangle, Kingdom of Saudi Arabia. Geoscience Map GM-137, scale 1:250,000, sheet 28B

Johnson KS, Campbell JA (eds) (1997) Ames structure in Northwest Oklahoma and similar features: origin and petroleum production. Oklahoma geological survey circular 100, Oklahoma Geological Survey, Norman, 396 p

Karpoff R (1953) The meteorite crater of Talemzane in southern Algeria. Meteoritics 1:31–38

Kenkmann T, Reimold WU, Khirfan M, Salameh S, Khoury H, Konsul K (2010) The complex impact crater Jebel Waqf as Suwwan in Jordan: effects of target heterogeneity and impact obliquity on central uplift formation. In: Gibson RL, Reimold WU (eds) Large meteorite impacts and planetary evolution, Geological society of America special paper 465, pp 471–487

Kenkmann T, Afifi AM, Stewart SA, Poelchau MH, Cook DJ, Neville AS (2015) Saqqar: a 34 km diameter impact structure in Saudi Arabia. Meteorit Planet Sci 50:1925–1940

Kenkmann T, Sturm S, Krüger T, Salameh S, Al-Raggad M, Konsul K (2017) The structural inventory of a small complex impact crater: Jebel Waqf as Suwwan, Jordan. Meteorit Planet Sci 52:1351–1370

Khirfan M, Fritzius RS, Reimold WU (2011) A small possible impact structure South of Amman, Jordan. In: Abstracts, the second Arab impact cratering and astrogeology conference, Casablanca, Morocco, p 58

Kock AB (1901) Volcan éteint de Tigraou. Bulletin de la Société de Géographie et d'Archéologie d'Oran 21:99–101

Koeberl C (1994) African meteorite impact craters: characteristics and geological importance. J Afr Earth Sci 18:263–295

Koeberl C (2004) Remote sensing studies of impact craters: how to be sure? CR Geosci 336:959–961

Koeberl C, Reimold WU, Shirey SB (1998) The Aouelloul crater, Mauritania: on the problem of confirming the impact origin of a small crater. Meteorit Planet Sci 33:513–517

Koeberl C, Reimold WU, Plescia J (2005a) BP and Oasis impact structures, Libya: remote sensing and field studies. In: Koeberl C, Henkel H (eds) Impact tectonics, impact studies series 8. Springer, Berlin, pp 161–190

Koeberl C, Reimold WU, Copper G, Cowan D, Vincent PM (2005b) Aorounga and Gweni Fada impact structures, Chad: remote sensing, petrography, and geochemistry of target rocks. Meteorit Planet Sci 40:1455–1471

Kowitz A, Güldemeister N, Schmitt RT, Reimold WU, Wünnemann K, Holzwarth A (2016) Revision and recalibration of existing shock classifications for quartzose rocks using low-shock pressure (2.5–20 GPa) recovery experiments and mesoscale numerical modeling. Meteorit Planet Sci 51:1741–1761

Kring DA (2017) Guidebook to the geology of barringer meteorite crater, Arizona. Lunar and Planetary Institute, Houston, TX, Contribution n 2040, 270 pp. Available online at: https://www.lpi.usra.edu/publications/books/barringer_crater_guidebook/

Lamali A, Rochette P, Merabet N, Abtout A, Maouche S, Gattacceca J, Ferrière L, Hamoudi M, Team ASTER, Meziane EH, Ayache M (2016) Geophysical and magneto-structural study of the Maâdna structure (Talemzane, Algeria): Insights on its age and origin. Meteorit Planet Sci 51:2249–2273

Lambert P, McHone JF, Dietz RS, Houfani M (1980) Impact and impact-like structures in Algeria, part 1, four bowl-shaped depressions. Meteoritics 15:158–178

Lambert P, McHone JF, Dietz RS, Briedj M, Djender M (1981) Impact and impact-like structures in Algeria. Part 2, multi-ringed structures. Meteoritics 16:203–227

Langenhorst F, Deutsch A (2012) Shock metamorphism of minerals. Elements 8:31–36

Lefranc JP (1969) Reconnaissance du cratère météoritique d'Amguid (Mouydir, Sahara Central). Comptes Rendus de l'Académie des Sciences, Paris 268:900–902

Levell B, Richard P, Hoogendijk F (2002) A possible Albian impact crater at Murshid, Southern Oman. GeoArabia 7:721–730

Lorenz CA, Ivanova MA, Artemieva NA, Sadilenko DA, Chennaoui Aoudjehane H, Roshina IA, Korochantsev AV, Humayunm M (2015) Formation of a small impact structure discovered within the Agoudal meteorite strewn field, Morocco. Meteorit Planet Sci 50:112–134

Marks P, Van Enk R, Groen C (1972) Sur l'origine météoritique probable de l'accident cratériforme près d'Aflou (Sud algérien). Verhandelingen der Koinklijke Akademie von Wetenschappen te Amsterdam, ser. B 75:348–355

Martin AJ (1969) Possible impact structure in southern Cyrenaica, Libya. Nature 223:940–941

Masaitis VL (1989) The economic geology of impact craters. Int Geol Rev 31:922–933

Masaitis VL (1992) Impact craters: are they useful? Meteoritics 27:21–27

Masaitis VL (1998) Popigai crater: origin and distribution of diamond-bearing impactites. Meteorit Planet Sci 33:349–359

Masaitis VL (1999) Impact structures of northeastern Eurasia: the territories of Russia and adjacent countries. Meteorit Planet Sci 34:691–711

Master S (2001) A possible Holocene impact structure in the Al'Amarah Marshes, near the Tigris-Euphrates confluence, southern Iraq. Meteorit Planet Sci 36(9), Suppl.:A124

Master S (2002) Umm al Binni lake, a possible Holocene impact structure in the marshes of southern Iraq: geological evidence for its age, and implications for Bronze-age Mesopotamia. In: Leroy S, Stewart IS (eds) Environmental catastrophes and recovery in the Holocene, Abstr. Vol., Dept Geogr., Brunel University, Uxbridge, West London, U.K., Aug 29–2 Sept 2002, pp 56–57. http://atlasconferences.com/cgi-bin/abstract/caiq-15

Master S (2009) Wadi El Murbah, a possible 7.5 km diameter buried impact structure on the Jordan-Iraq border: geological setting and remote sensing. In: Abstracts, the first Arab impact cratering and astrogeology conference, Amman, Jordan, pp 58–60

Master S, Woldai T (2007) Umm al Binni Structure, Southern Iraq, as a postulated late holocene meteorite impact crater. In: Rickman H (ed) Bobrowsky PT. Comet/Asteroid Impacts and Human Society (I), Springer, pp 89–103

Matton G, Jébrak M (2014) The "eye of Africa" (Richat dome, Mauritania): an isolated Cretaceous alkaline–hydrothermal complex. J Afr Earth Sc 97:109–124

Matton G, Jébrak M, Lee JKW (2005) Resolving the Richat enigma: doming and hydrothermal karstification above an alkaline complex. Geology 33:665–668

McHone JF, Dietz RS (1978) Foum Taguemtour: probable astrobleme in Algerian Sahara. Meteoritics 13:557–560

McHone JF, Dietz RS (1988) Arabian penninsula: known and suspected impact structures. Meteoritics 23:288–289

McHone JF, Greeley R (1987) Talemzane: Algerian impact crater detected on SIR-A orbital imaging radar. Meteoritics 22:253–264

McHone JF, Greeley R (1997) More impact and impact-like structures on SIR-C radar: Europe, Africa and Arabian Peninsula. In: 28th lunar and planetary science conference. Abstract #1149

McHone JF, Greeley R, Williams KK, Blumberg DG, Kuzmin RO (2002) Space shuttle observations of terrestrial impact structures using SIR-C and X-SAR radars. Meteorit Planet Sci 37:407–420

Merriam R, Holwerda JG (1957) Al Umchaimin, a crater of possible meteoritic origin in western Iraq. Geogr J 123:231–233

Mitchell RC (1958) The Al Umchaimin crater, western Iraq. Geogr J 124:578–580
Monod Th, Pourquié A (1951) Le cratère d'Aouelloul (Adrar, Sahara occidental). Bulletin de l'Institut Français d'Afrique Noir 13:292–302
Monod Th, Pomerol C (1966) Le cratère de Tenoumer et ses laves. Bulletin de la Société Géologique de France S7-VIII:165–172
Nachit H, Abia EH, Bonadiman C, Di Martino M, Vaccaro C (2017) The preservation of the Agoudal impact crater, Morocco, under a landslide: indication of a genetic link between shatter cones and meteorite fragments. Geomorphology 295:76–83
Neville AS, Cook DJ, Afifi AM, Stewart SA (2014) Five buried crater structures imaged on reflection seismic data in Saudi Arabia. GeoArabia 19:17–44
Orti L, Di Martino M, Morelli M, Cigolini C, Pandeli E, Buzzigoli A (2008) Non-impact origin of the crater-like structures in the Gilf Kebir area (Egypt): Implications for the geology of eastern Sahara. Meteorit Planet Sci 43:1629–1639
Osinski GR, Pierazzo E (2013) Impact cratering: processes and products. In: Osinski GR, Pierazzo E (eds) Impact cratering: processes and products. Wiley-Blackwell, pp 1–20
Paillou Ph, Rosenqvist A, Malézieux JM, Reynard B, Farr T, Heggy E (2003) Discovery of a double impact crater in Libya: the astrobleme of Arkenu. CR Geosci 335:1059–1069
Paillou Ph, El Barkooky A, Barakat A, Malézieux JM, Reynard B, Dejax J, Heggy E (2004) Discovery of the largest impact crater field on Earth in the Gilf Kebir region, Egypt. Comptes Rendus Geosci 336:1491–1500
Paillou PH, Reynard B, Malezieux JM, Dejax J, Heggy E, Rochette P, Reimold WU, Michel P, Baratoux D, Razin P, Colin JP (2006) An extended field of crater-shaped structures in the Gilf Kebir region, Egypt: observations and hypotheses about their origin. J Afr Earth Sci 46:281–299
Pesonen LJ (1996) The impact cratering record of Fennoscandia. Earth Moon Planet 72:377–393
Philby HSJ (1933) The empty quarter. Constable and Company Ltd, London, p 433
Pomerol C (1967) Découverte de laves basaltiques dans le cratère du Temimichat (Ghallaman, Mauritanie). Comptes Rendus de l'Académie des Sciences, Paris 265:1164–1166
Pratesi G, Morelli M, Rossi AP, Ori GG (2005) Chemical compositions of impact melt breccias and target rocks from the Tenoumer impact crater, Mauritania. Meteorit Planet Sci 40:1653–1672
Prescott JR, Robertson GB, Shoemaker C, Shoemaker EM, Wynn J (2004) Luminescence dating of the Wabar meteorite craters, Saudi Arabia. J Geophys Res 109:E01008. https://doi.org/10.1029/2003JE002136
Reimold WU (1995) Impact cratering—a review, with special reference to the economic importance of impact structures and the Southern African impact crater record. Earth Moon Planet 70:21–45
Reimold WU (2007) The impact crater bandwagon (some problems with the terrestrial impact cratering record). Meteorit Planet Sci 42:1467–1472
Reimold WU (2010) The first Arab impact cratering and astrogeology conference, Amman, Jordan. 9–11 Nov 2009—An appreciation. Meteorit Planet Sci 45:157–160
Reimold WU, Koeberl C (2014) Impact structures in Africa: a review. J Afr Earth Sci 93:57–157
Reimold WU, Koeberl C, Gibson RL, Dressler BO (2005) Economic mineral deposits in impact structures: a review. In: Koeberl C, Henkel H (eds) Impact tectonics, impact studies series 8. Springer, Berlin, pp 479–552
Reimold WU, Ferrière L, Deutsch A, Koeberl C (2014) Impact controversies: impact recognition criteria and related issues. Meteorit Planet Sci 49:723–731
Richard-Molard J (1948a) La boutonnière du Richat en Adrar mauritanien. Comptes Rendus de l'Académie des Sciences Paris 227:142–143
Richard-Molard J (1948b) Le cratère d'explosion de Tenoumer et l'existence probable d'une grande fracture rectiligne au Sahara occidental. Comptes Rendus de l'Académie des Sciences Paris 227:213–214
Richard-Molard J (1949) Un cratère d'explosion dans le Sahara occidental: le Guelb Tenoumer. Bulletin de la Société de géographie de Lyon et de la région lyonnaise 24:309–318

Robertson PB, Grieve RAF (1975) Impact structures in Canada—their recognition and characteristics. J Roy Astron Soc Can 69:1–21

Rossi AP (2002) Seven possible new impact structures in western Africa detected on Aster imagery. In: 33th lunar and planetary science conference. Abstract #1309

Rossi AP, Baliva A, Piluso E (2003) New evidences of an impact origin for Temimichat crater, Mauritania. In: 34th lunar and planetary science conference. Abstract #1882

Sadilenko DA, Lorenz CA, Ivanova MA, Roshchina IA, Korochantsev AV (2013) A new small impact crater in the High Atlas, in the Agoudal iron strewnfield. In: 76th annual meeting of the meteoritical society, July 29–Aug 7 2013, Edmonton, Canada, Abstract #5215

Sadran G (1958) Les formations volcaniques tertiaires et quaternaires du Tell oranais. Bulletin du Service de la Carte Géologique de l'Algérie 18, 533 p

Sahoui R, Belhai D, Jambon A (2016) Impact-generated carbonate melts in the Tamelzane impact structure (Laghouat, Algeria). Arab J Geosci 9:641

Salameh E, Khoury H, Schneider W (2006) Jebel Waqf as Suwwan, Jordan, a possible impact crater–a first approach. Zeitschrift der deutschen Gesellschaft für Geowissenschaften 157:319–325

Salameh E, Khoury H, Reimold WU, Schneider W (2008) First large meteorite impact structure discovered in the Middle East: Jebel Waqf as Suwwan, Jordan. Meteorit Planet Sci 43:1681–1690

Schmieder M, Buchner E, Le Heron DP (2009a) The Jebel Hadid structure (Al Kufrah Basin, SE Libya)—a possible impact structure and potential hydrocarbon trap? Mar Pet Geol 26:310–318

Schmieder M, Buchner E, Hofmann BA, Gnos E (2009b) The ash Shutbah circular feature—A suspected meteorite impact site in Saudi Arabia. In: Abstracts, the first Arab impact cratering and astrogeology conference, Amman, Jordan, pp 84–85

Schmieder M, Reimold WU, Buchner E, Khirfan M, Salameh E, Khoury H (2011) Shock-metamorphic microfeatures in chert from the Jebel Waqf as Suwwan impact structure, Jordan. Meteorit Planet Sci 46:574–586

Schmieder M, Seyfried H, Gerel O (2013) The circular Uneged Uul structure (East Gobi Basin, Mongolia)—geomorphic and structural evidence for meteorite impact into an unconsolidated coarse-clastic target? J Asian Earth Sci 64:58–76

Schmieder M, Buchner E, Salameh E, Khoury H (2014) Comment on "Some observations facing the interpretation of Waqf as Suwwan structure, SE desert of Jordan, as an impact crater", by A. M. Abed et al. (2014), Arab J Geosci 8:6459–6463

Schmieder M, Chennaoui Aoudjehane H, Buchner E, Tohver E (2015) Meteorite traces on a shatter cone surface from the Agoudal impact site, Morocco. Geol Mag 152:751–757

Schultze DS, Jourdan F, Hecht L, Reimold WU, Schmitt RT (2016) Tenoumer impact crater, Mauritania: impact melt genesis from a lithologically diverse target. Meteorit Planet Sci 51:323–350

Sighinolfi GP, Sibilia E, Contini G, Martini M (2015) Thermoluminescence dating of the Kamil impact crater. Meteorit Planet Sci 50:204–213

Soliman KA, Barakat AA (2004) A note on new jasper deposit in the Eastern desert. Ann Geol Surv Egypt 27:599–604

Sparavigna AC (2010a) Crater-like landform in Bayuda Desert (A processing of satellite images). http://www.arxiv.org/abs/1008.0500

Sparavigna AC (2010b) Craters and ring complexes of the North-East Sudanese Country. http://www.arxiv.org/ftp/arxiv/papers/1008/1008.3976.pdf

Spudis PD (1993) The geology of multi-ring impact basins. Cambridge University Press, Cambridge, 280 p

Stewart SA (2011) Estimates of yet-to-find impact crater population on Earth. J Geol Soc, London 168:1–14

Stewart SA (2015) Circular geological structures outcropping in the sedimentary basins of Saudi Arabia. J Asian Earth Sci 106:95–118

Stöffler D, Langenhorst F (1994) Shock metamorphism of quartz in nature and experiment: I basic observation and theory. Meteoritics 29:155–181

Stöffler D, Hamann C, Metzler K (2018) Shock metamorphism of planetary silicate rocks and sediments: proposal for an updated classification system. Meteorit Planet Sci 53:5–49
Storzer D, Wagner GA (1977) Fission-track dating of meteorite impacts. Meteoritics 12:368–369
Storzer D, Sélo M, Latouche L, Fabre J (2003) The age of Tenoumer crater, Mauritania, revisited. In: 34th lunar and planetary science conference, Abstract #1183
Tagle R, Hecht L (2006) Geochemical identification of projectiles in impact rocks. Meteorit Planet Sci 41:1721–1735
Theilen-Willege B (1987) The use of airborne and spaceborne radar images for the detection and investigation of impact structures. In: Pohl J (ed) Research in terrestrial impact structures. Vieweg & Sohn, Braunschweig, Germany, pp 115–130
Timms NE, Erickson TM, Pearce MA, Cavosie AJ, Schmieder M, Tohver E, Reddy SM, Zanetti MR, Nemchin AA, Wittmann A (2017) A pressure-temperature phase diagram for zircon at extreme conditions. Earth Sci Rev 165:185–202
Tomlinson WD (1989) A possible impact structure in Tunisia. In: 20th lunar and planetary science conference, pp 1122–1123
Turtle EP, Pierazzo E, Collins GS, Osinski GR, Melosh HJ, Morgan JV, Reimold WU (2005) Impact structures: what does crater diameter mean? In: Kenkmann T, Hörz F, Deutsch A (eds), Large meteorite impacts III. Geological society of America special paper 384, pp 1–24
Underwood JR (1994) Al Umchaimin depression, western Iraq: an impact structure? Large meteorite impacts and planetary evolution. In: Dressler BO, Grieve RAF, Sharpton VL (eds) Large meteorite impacts and planetary evolution. Geological society of America special paper 293, pp 259–263
Urbini S, Nicolosi I, Zeoli A, El Khrepy S, Lethy A, Hafez M, El Gabry M, El Barkooky A, Barakat A, Gomaa M, Randwan AM, El Sharkawi M, D'Orazio M, Folco L (2012) Geological and geophysical investigation of Kamil Crater, Egypt. Meteorit Planet Sci 47:1842–1868
Vail JR (1985) Alkaline ring complexes in Sudan. J Afr Earth Sc 3:51–59
van der Meer F, Van der Werff HMA, Van Ruitenbeek FJA (2014) Potential of ESA's Sentinel-2 for geological applications. Remote Sens Environ 148:124–133
Van Gasselt S, Kim JR, Choi YS, Kim J (2017) The Oasis impact structure, Libya: geological characteristics from ALOS PALSAR-2 data interpretation. Earth, Planets Space 69:35
Verati C, Bertrand H, Feraud G (2005) The farthest record of the Central Atlantic Magmatic Province into West Africa craton: precise $^{40}A/^{39}Ar$ dating and geochemistry of Taoudenni basin intrusives (northern Mali). Earth Planet Sci Lett 235:391–407
Woolley AR (2001) Alkaline rocks and carbonatites of the world, Part 3: Africa. Geological Society of London, London, p 384
Wynn JC, Shoemaker EM (1998) The day the sands caught fire. Sci Am 79:36–45
Youbi N, Ziadi A, Doblas M, Madeira J, Martins LT, Mata J, Chennaoui Aoudjehane H, Chaabout S, Bahir M, El Mountassir M, Ghanmi M, Bensalah MKh, Chellai EH, El Hachimi H, R'Kha KC, Nahid A (2011) Toward a more complete database of impact craters of the Arab world. Abstracts, 2nd Arab impact cratering and astrogeology conference (AICAC II). Hassan II Casablanca University, Casablanca, Morocco, pp 90–92
Dietz RS, Fudali R, Cassidy W (1969) Richat and Semsiyat Domes (Mauritania): Not Astroblemes. Geol Soc Am Bull 80:1367-1372

Holocene Climate Development of North Africa and the Arabian Peninsula

S. Lüning and F. Vahrenholt

Abstract Holocene climate in North Africa and on the Arabian Peninsula has undergone major changes. In this contribution, we review hydroclimate and temperature changes in the region over the past 15,000 years by correlating and integrating all available case studies. A pronounced wet period corresponding to the 'Green Sahara' and its equivalent in Arabia commenced between 15,000 and 9000 years BP and ended sometime between 6500 and 3500 years BP, followed by arid conditions throughout the late Holocene. Start and end dates vary between locations, depending on local factors, climate amplifiers and chosen climate proxies, leading to a spatially and temporally complex distribution. Some studies show gradual transitions between the hydroclimatic states while other locations evidenced abrupt changes. The humid phase was triggered by a northward migration of the Intertropical Convergence Zone (ITCZ) due to orbital precession. The northernmost parts of North Africa and the Arabian Peninsula were not reached by the ITCZ. Here, increased early Holocene humidity may have been due to an intensification of southern Mediterranean winter rain and its deeper penetration southwards into the present-day desert areas. The early Holocene 'Green Sahara' forms part of a long series of wet periods that have occurred over the past hundred thousand to million years in North Africa and Arabia. Notably, climate models are still unable to match the observed hydroclimatic changes in a quantitative way. Simulated rainfall during the African Humid Period over the Sahara is not sufficient to sustain vegetation at a level seen in the palaeo record, indicating that processes such as vegetation and dust feedbacks still need to be refined. Sea surface temperatures in North Africa and Arabia during the early Holocene were generally one to several degrees C warmer than during the late Holocene. Warming began around 12,000 years BP and ended around 5000 years BP. The warm period generally coincided with the early Holocene wet phase in the region and is linked to the Holocene Thermal Maximum, an early Holocene period during which temperatures

S. Lüning (✉)
Independent Scientist, Lisbon, Portugal
e-mail: luening@uni-bremen.de

F. Vahrenholt
Department of Chemistry, University of Hamburg, Hamburg, Germany
e-mail: fritz.vahrenholt@chemie.uni-hamburg.de

A. Bendaoud et al. (eds.), *The Geology of the Arab World—An Overview*,
Springer Geology, https://doi.org/10.1007/978-3-319-96794-3_14

were globally elevated. The review suggests that the Holocene climate history of North Africa and Arabia is closely linked to the global development and that significant temperature changes have also occurred in subtropical climate belts.

Keywords Holocene Thermal Maximum · Green Sahara · Hydroclimate Palaeoclimate

1 Introduction

Global climate has seen some major changes over the industrial period comprising the past one and a half centuries. It is generally understood that the observed shifts are driven by a combination of anthropogenic and natural forcings. In order to be better able to distinguish between the respective contributions of these two groups of climate control mechanisms, it is important to have a good understanding of the pre-industrial climate evolution and, therefore, the full past range of natural variability. The field of palaeoclimatology has made major progress over the past 15 years and a large number of case studies have been published, making use of a diverse and sophisticated set of climate proxies. In this contribution, we are attempting to synthesize the current understanding of the climate development in North Africa and on the Arabian Peninsula during the past 15,000 years. The region is located in a climatically sensitive area encompassing some of the driest parts of the world. Rainfall is sourced from both the Mediterranean in the north and the Sahel zone in the south, whereby changes in the position of atmospheric circulation systems have major impacts on precipitation in the region. In this baseline study, we investigate long-term trends of natural pre-industrial change in precipitation and temperature in North Africa and on the Arabian Peninsula and set them in a global context.

2 Materials and Methods

This regional synthesis is based on an intense iterative literature screening process during which a large number of published North African and Arabian palaeoclimate case studies were evaluated towards their temporal coverage, types of climate information and data resolution (Tables 1 and 2). Relevant climate curve data were collected in digital form for flexible plotting and curve correlation. The tabulated data was retrieved (a) from palaeoclimate online data repositories (e.g. Pangaea, NOAA's National Climatic Data Center NCDC and data supplements of papers), (b) from the authors by email request and (c) by digitizing and vectorising using WebPlotDigitizer (https://automeris.io/WebPlotDigitizer/). The climate curve data were loaded into the database and correlation software IC^{TM}, which was originally developed for geological well correlations but can handle climate logs effectively as

Table 1 Key Holocene palaeoclimate studies in northern Africa

Country	Locality	Proxies	Coverage	Studies
Mauritania	ODP 658C	Inorganic and organic geochemistry	Last 25,000 years	Zhao et al. (1995), deMenocal et al. (2000a, b), Adkins et al. (2006)
Morocco/ WS	GC49	Grain size, dust flux	Last 22,000 years	McGee et al. (2013)
Morocco	GeoB 5546	K as fluvial input proxy	9–3 k years BP	Kuhlmann et al. (2004a, b)
Morocco	GeoB 6007	K as fluvial input proxy	Last 10,000 years	Kuhlmann et al. (2004a, b), Kim et al. (2007)
Morocco	Lake Sidi Ali	Diatoms, elemental geochemistry, TOC, carbonate, ostracod stable isotopes, pollen	Last 12,000 years	Lamb et al. (1999), Zielhofer et al. (2017)
Morocco	Lake Hachlaf	Pollen, C/N, $\delta^{13}C$	Last 6000 years	Nourelbait et al. (2016)
Morocco	Lakes Ifrah	Grain size, petrography, magnetic susceptibility	Last 29,000 years	Damnati et al. (2016), Rhoujjati et al. (2010)
Morocco	Lake Tigalmamine	Pollen	Last 10,000 years	Cheddadi et al. (1998)
Morocco	GC27	Leaf wax isotopes	Last 25,000 years	Tierney et al. (2017)
Morocco	Ait Ichou Swamp	Pollen, charcoal, grain size, geochemistry	Last 25,000 years	Tabel et al. (2016)
Algeria	Gueldaman Cave	$\delta^{13}C$, $\delta^{18}O$	6–3 k years BP	Ruan et al. (2016)
Algeria	I-n-Atei palaeolake	$\delta^{13}C$, diatoms	11–7 k years BP	Lécuyer et al. (2016)
Tunisia	Sebkha Boujmel	Pollen, sediment petrography	Last 8000 years	Jaouadi et al. (2016)
Tunisia	Medjerda floodplain	Fluvial C14 age dating	Last 14,000 years	Zielhofer et al. (2008)
Tunisia	MD04-2797CQ	Pollen, mineralogy, grain size, geochemistry, planktic foraminifera	Last 30,000 years	Desprat et al. (2013), Bout-Roumazeilles et al. (2013), Essallami et al. (2007)
Chad	Lake Yoa	Pollen, diatoms, lithology, Chironomidae	Last 6000 years	Kröpelin et al. (2008), Eggermont et al. (2008)
Chad	Lake Megachad	Age dating of shoreline	Last 7500 years	Drake and Bristow (2006)

(continued)

Table 1 (continued)

Country	Locality	Proxies	Coverage	Studies
Egypt	MC78, Red Sea	$\delta^{18}O$ in planktic foraminifera	Last 6000 years	Badawi (2016)
Egypt	Burullus Lagoon	Pollen	Last 8000 years	Bernhardt et al. (2012)
Egypt	Lake Qarun	Lithology, diatoms, $\delta^{18}O$, $\delta^{13}C$ in mollusc shells	Last 10,000 years	Baioumy et al. (2010), Hassan et al. (2012)
Egypt	MD04-2726	Geochemistry	Last 10,000 years	Revel et al. (2015)
Egypt	MS27PT	Geochemistry	Last 21,000 years	Revel et al. (2014, 2015)
Egypt	MS21PCC, Nile Delta	$\delta^{18}O$, $\delta^{13}C$	Last 13,000 years	Hennekam et al. (2015)
Egypt	P362/2-33	Inorganic and organic geochemistry, grain size	Last 10,000 years	Blanchet et al. (2013, 2014)
Egypt	Nile catchment	Fluvial OSL and C14 age dating	Last 12,000 years	Macklin et al. (2015)
Egypt	KL17VL, N Red Sea	Tex_{86} SST	Last 15,000 years	Trommer et al. (2010)

Table 2 Key Holocene palaeoclimate studies on the Arabian Peninsula and nearby locations

Country	Locality	Proxies	Coverage	Studies
Israel, Palestine	GA-110 and 112	$\delta^{18}O$, $\delta^{13}C$	Last 4000 years	Schilman et al. (2001), Martín-Puertas et al. (2010)
Israel, Palestine	Tel Akko	pollen grains, fern spores, micro-charcoal fragments, and dinoflagellate cysts	Last 6000 years	Kaniewski et al. (2013)
Israel, Palestine	Mishqafaim Cave	Wood, cave geometry	Last 7000 years	Frumkin et al. (1991)
Israel, Palestine	Dead Sea	Shoreline age dating, pollen	Last 10,000 years	Enzel et al. (2003), Litt et al. (2012)
Israel, Palestine	PS009PC	Foraminifera, $\delta^{18}O$	Last 13,000 years	Mojtahid et al. (2015), Hennekam et al. (2014, 2015)
Israel, Palestine	GeoB 7702-3	Uk'37, TEX_{86}	Last 27,000 years	Castañeda et al. (2010, 2016)
Israel, Palestine	GeoTü SL112	Grain size, sediment petrography, $\delta^{18}O$, $\delta^{13}C$	Last 27,000 years	Hamann et al. (2008), Kuhnt et al. (2008)
Israel, Palestine	Soreq Cave	C and O isotopes	Last 60,000 years	Bar-Matthews et al. (1999), Schilman et al. (2002), Orland et al. (2009), Bar-Matthews and Ayalon (2011)

(continued)

Table 2 (continued)

Country	Locality	Proxies	Coverage	Studies
Jordan	GeoB 5804-4	$\delta^{18}O$, sediment petrography	Last 8000 years	Arz et al. (2003a), Lamy et al. (2006)
Lebanon	Jeita Cave	$\delta^{18}O$, $\delta^{13}C$	Last 20,000 years	Verheyden et al. (2008), Cheng et al. (2015)
Syria	Jableh	Pollen	Last 1100 years	Kaniewski et al. (2008, 2011)
Oman	Hoti Cave	$\delta^{18}O$	Last 10,000 years	Fleitmann et al. (2007)
Oman	Cores RC2730, RC2735 and Site 723A	Planktonic foraminifera *Globigerina bulloides*	Last 12,000 years	Gupta et al. (2003), Anderson et al. (2010)
Oman	SL163	Foraminifera and geochemistry	6–8.5 k years BP	Munz et al. (2016)
Oman	Qunf Cave	$\delta^{18}O$	Last 10,000 years	Fleitmann et al. (2003)
Oman	RC-27-42	Alkenone temperatures	28–5 k years BP	Pourmand et al. (2007)
Saudi Arabia	GeoB5844-2	Foraminifera	Last 10,000 years	Arz et al. (2003a, b)
Yemen	Hoq Cave	$\delta^{13}C$, $\delta^{18}O$	Last 9500 years	Van Rampelbergh et al. (2013)
Sudan	MC93	Planktic foraminifera	Last 6000 years	Edelman-Furstenberg et al. (2009)

well. Climate correlation panels were compiled and the early Holocene climate anomaly visualized by peak colouring.

3 Modern Climate Systems of North Africa and Arabia

Present-day climate in the study area consists of three main climate zones. Warm Mediterranean climate occurs along the coasts of northern Morocco, Algeria, Tunisia, parts of Libya and the Levant. The greatest part of North Africa and the Arabian Peninsula lies in warm desert climate, including the Sahara and Rub' al Khali (Empty Quarter). The warm semi-arid climate of the Sahel zone occurs to the south of the Sahara. Rain is supplied by two systems: (1) Mediterranean winter rain brings precipitation to the northern part of the region. (2) The seasonal northward shift of the Intertropical Convergence Zone (ITCZ) brings moisture-laden air to the southern parts of North Africa and Arabia in the northern hemispheric summer (Fig. 1). Case studies suggest that climatic temperature changes influence both the location and width of the ITCZ (Byrne and Schneider 2016; Sachs et al. 2009; Schneider et al. 2014).

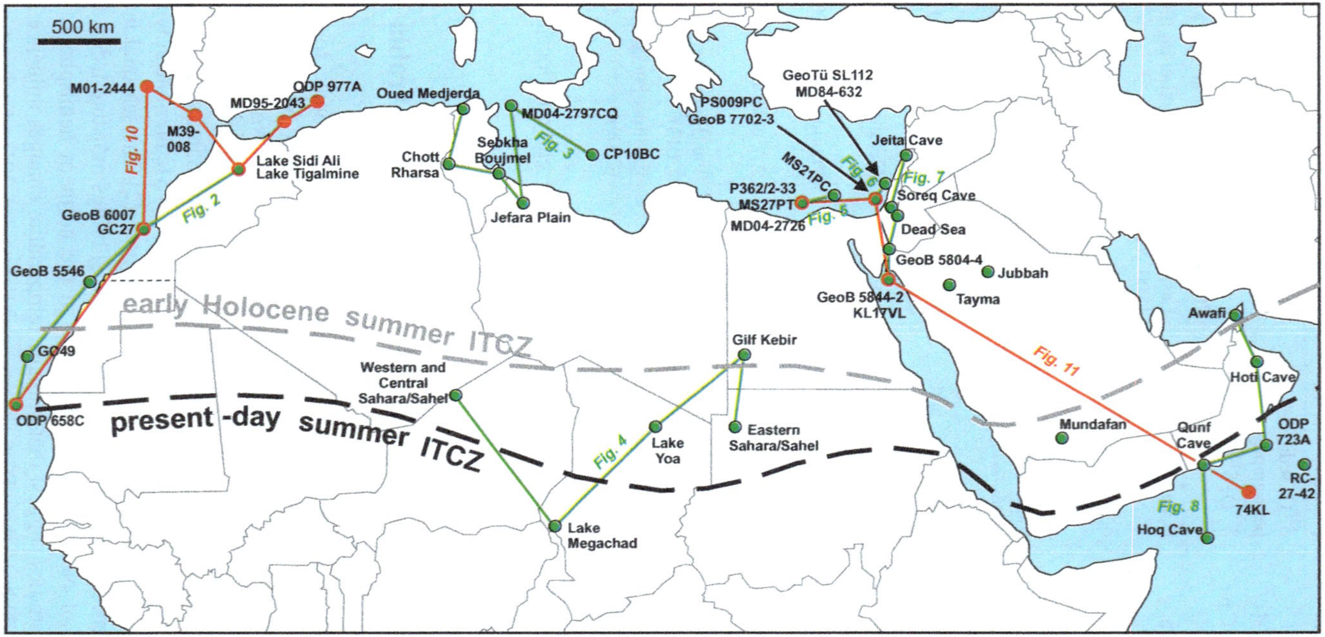

Fig. 1 Location Map. Green transects mark hydroclimate correlation panels, red transects represent temperature correlation panels; respective figure numbers are indicated. Early Holocene ITCZ after Revel et al. (2015) and Fleitmann et al. (2007)

The Northwest African Atlantic coast is dominated by the southward-flowing Canary Current, which forms part of the clockwise-rotating North Atlantic Gyre. Most of the Moroccan Atlantic margin lies in a weak permanent upwelling zone (26–35°N) (Cropper et al. 2014). To the south lies a sector with strong permanent annual upwelling which extends into northern Mauritania (21–26°N). A strong upwelling regime also exists along the western coast of the Arabian Sea off Oman and Yemen.

4 Holocene Hydroclimate of North Africa and the Arabian Peninsula

During the first half of the Holocene, the Sahara and Arabian Peninsula have received significantly more rainfall, a phase that has been termed 'The Green Sahara', 'African Humid Period' (AHP) or Early Holocene Humid Period (EHHP) (e.g. Hoelzmann et al. 2004; Lézine et al. 2011; Robinson et al. 2006; Tierney et al. 2017). The main driver behind this is the northward shift of the ITCZ which brought rain into core parts of the modern deserts. The humid phase extended over several thousand years and coincides roughly with the Holocene Thermal Maximum (HTM), a major global warm phase (Ljungqvist 2011; Marcott et al. 2013). In the following, we will review key palaeohydroclimatic evidence for this phase on a region-by-region basis, starting in Mauritania and then continuing eastwards.

4.1 *Mauritania to Morocco*

ODP658C Ocean Drilling Program (ODP) Hole 658C was cored during Leg 108 in 2263 m water depth off Cap Blanc, Mauritania (Fig. 1). The eolian terrigenous fraction was studied by deMenocal et al. (2000a, b) and shows significantly reduced values between 11,500 and 5500 years BP (Fig. 2). The reduction in eolian supplies during this time is interpreted by the authors as reflecting a period of greatly increased rainfall and vegetation cover in the now hyperarid Saharan desert. Aridification around 5500 years BP subsequently led to shrinking vegetation and greater amounts of dust.

GC49 This gravity core has been taken from 2303 m water depth off El Argoub during the 2007 CHEETA cruise. McGee et al. (2013) carried out a grain size analysis of the core and found a significantly reduced eolian dust flux during the interval 10,000–5000 years BP (Fig. 2). According to the authors, the decline in dust flux likely reflects a combination of decreased wind gustiness and increased precipitation associated with a northward shift of the ITCZ at this time. McGee et al. (2013) studied the dust flux in several other cores along the NW African margin which yielded generally similar results.

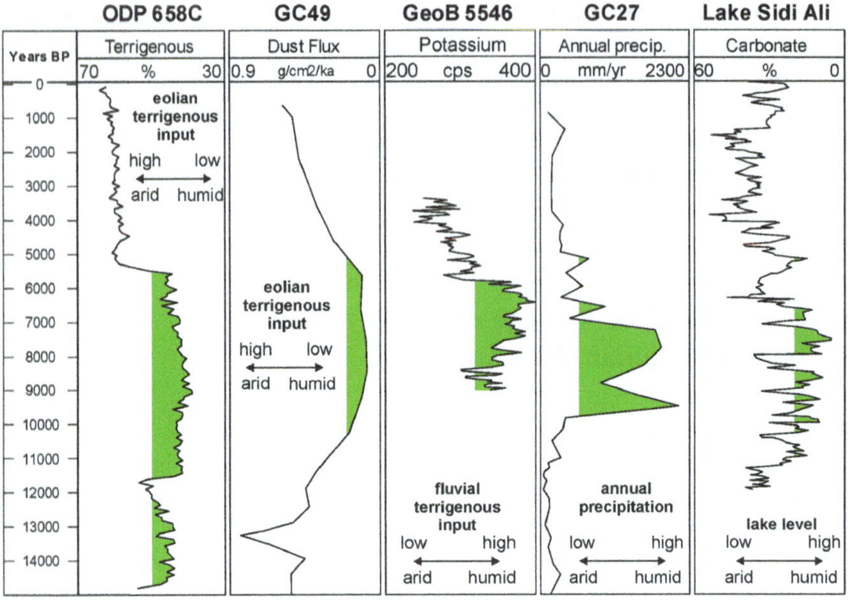

Fig. 2 Development of hydroclimate in NW Africa during the past 15,000 years based on climate proxies. Curves: ODP 658C (deMenocal 2000a, b), GC49 (McGee et al. 2013), GeoB 5546 (Kuhlmann et al. 2004b), GC27 (Tierney et al. 2017), Lake Sidi Ali (Zielhofer et al. 2017). Green peaks mark phases with more humid conditions. Location map in Fig. 1

GeoB 5546 This core has been taken off Cape Juby in southern Morocco, east of the Canary Islands in 1071 m water depth (Fig. 1). Kuhlmann et al. (2004a, b) measured the potassium (K) content as an indicator for fluvial terrigenous input by means of X-ray fluorescence (XRF) core scanner analysis for the time period 9000–3000 years BP. High K values from the base of the dataset up to 5800 years BP reflect intensified fluvial activity associated with the African Humid Period (AHP). The subsequent interval 5800 to nearly 3000 years BP appears to have been significantly drier.

GC27 Core GC27 has been retrieved off Cape Ghir, north of Agadir from 1258 m water depth (Fig. 1). The location forms part of a series of gravity cores along the NW African Atlantic coast studied by Tierney et al. (2017) for which they carried out a quantitative reconstruction of western Saharan annual precipitation derived from leaf wax isotopes. For core GC27, rainfall was found to be significantly elevated during 10,000–5000 years BP, with annual precipitation 10 times higher during 10,000–7000 years BP compared to the drier phases preceding and following the AHP (Fig. 2). The other cores GC37, GC49, GC68 to the south yielded a similar hydroclimatic history, although with slight differences in individual timing.

Lake Sidi Ali The lake lies in the Middle Atlas in Morocco at an altitude of 2080 m a.s.l. (Fig. 1). Zielhofer et al. (2017) carried out a multi-proxy study which found a significant anti-correlation between total organic carbon (TOC) and carbonate contents. The authors interpret low carbonate content as reflecting a high lake level associated with humid and warm conditions. During such times, high productivity conditions are thought to have prevailed, resulting in high TOCs and an anaerobic setting, related to reducing processes driven by organic matter decomposition. Carbonate content was generally low between 11,000 and 4000 years BP, suggesting a wet (and warm) climate for this period (Fig. 2). The humid phase was interrupted by a drier millennium during 6500–5500 years BP when the carbonate content was higher and TOC lower.

Other Locations in Morocco The general decrease in annual precipitation over the past 6000 years in the Middle Atlas has also been documented by Nourelbait et al. (2016) from **Lake Hachlaf** (1700 m a.s.l.) based on pollen analysis. The authors managed to differentiate seasonal trends in precipitation and showed that the aridification was mostly driven by a reduction in the summer rain, while the amount of winter precipitation slightly increased.

Cheddadi et al. (2016) presented a pollen-based Holocene climate reconstruction for **Bab El Karn**, a mire outlet located in the western Rif Mountains. *Cedrus atlantica* shows a pronounced abundance peak 7500–5500 years BP interpreted as peak humidity. Interestingly, the distribution of *Quercus canariensis* shows an inverse relation compared to *C. atlantica*.

Notably, Tabel et al. (2016) describe a different hydroclimatological development from **Ait Ichou Swamp** in the southern Middle Atlas (1560 m a.s.l.), based on pollen, micro-charcoals, grain size and geochemical analysis. In their paper, the authors suggest a rather dry early Holocene and an increase in rainfall after the mid-Holocene, based on the disappearance of the drought-tolerant species *Artemisia* and the aquatics development. However, there may be alternative interpretations of the dataset. In particular noteworthy is a pronounced abundance peak of the arboreal oak pollen *Quercus canariensis* 11,500–4000 years BP which coincides surprisingly well with the African Humid Period: increased humidity during this period may have boosted *Q. canariensis* distribution because rainfall usually has a positive impact on deciduous tree development. Generally, elevated temperatures during the early Holocene may have further contributed to the abundance peak.

Likewise, palaeohydroclimatic interpretations published by Cheddadi et al. (1998) from **Lake Tigalmine** in the Middle Atlas (1626 m a.s.l.) suggested a dry early Holocene and a wetter middle Holocene. The rainfall reconstruction may be worth being re-evaluated, especially because a previous pollen analysis by Lamb et al. (1989) from the same lake identified an early Holocene *Q. canariensis* abundance peak similar to the one in Ait Ichou Swamp (Tabel et al. 2016).

4.2 Tunisia and Northern Libya

Oued Medjerda is the main perennial watercourse in northern Tunisia and adjacent northeastern Algeria (Fig. 1). Its catchment covers approx. 23,500 km^2.

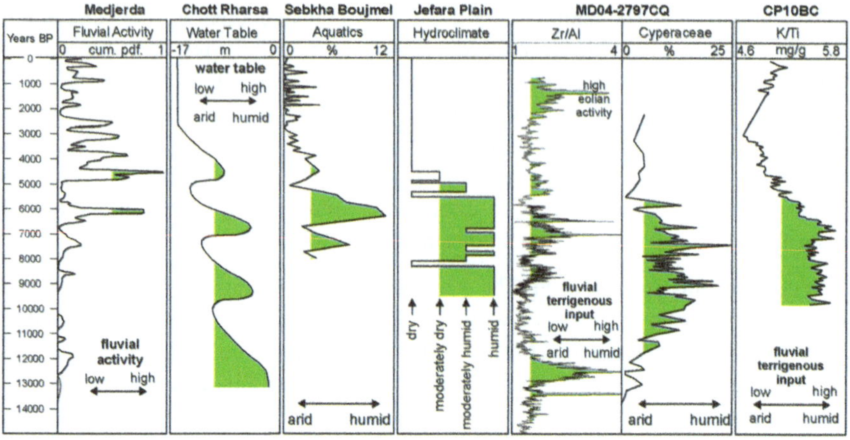

Fig. 3 Development of hydroclimate in Tunisia and northern Libya during the past 15,000 years based on climate proxies. Curves: Oued Medjerda (Zielhofer et al. 2008), Chott Rharsa (Swezey et al. 1999), Sebkha Boujmel (Jaouadi et al. 2016), Jefara Plain (Giraudi et al. 2013), MD04-2797CQ (Zr/Al: Bout-Roumazeilles et al. 2013; Cyperaceae: Desprat et al.2013), CP10BC (Wu et al. 2016). Location map in Fig. 1

The northern catchment includes the Mediterranean sub-humid south-facing slopes of the Tunisian Tell region. Zielhofer et al. (2008) reconstructed the Holocene flood history based on ^{14}C age dating of coarse beds in 12 Medjerda key profiles of the well-stratified cohesive overbank floodplain. The plots of the cumulative probability density functions (pdf) document a phase of generally increased fluvial activity during the period 6500–2500 years BP, composed of a series of millennial-scale activity cycles (Fig. 3). Peak fluvial activity in Oued Medjerda occurred around 4500 years BP. Subsequently, Medjerda River activity declined but kept flowing at rather high levels until 2500 years BP. Abundant vegetation during the early Holocene may have prevented the development of major floods. Reduced vegetational cover during the middle Holocene at the end of the African Humid Period might have resulted in lowered soil protection, which facilitated erosion and the formation of flood deposits. Subsequent aridification during the late Holocene would then have led to declining fluvial activity. Notably, the area did not entirely dry up at this time, probably reflecting the Mediterranean rather than Saharan–Sahelian influence in the Medjerda catchment area.

Chott Rharsa is located in western Central Tunisia and is a large playa lakes along the northern margin of the Sahara. Swezey et al. (1999) carried out field mapping of sedimentary facies, in particular, measuring elevation ranges of the eolian units, age-dated by thermoluminescence (TL) methods. The authors reconstructed three humid phases of high water table during the early and middle Holocene, with an overall falling trend. Water table stayed low after 4000 years BP, marking longer term aridification.

Sebkha Boujmel is situated in East Central Tunisia on the coast of the Jefara Plain about 40 km south of the island of Djerba (Fig. 1). The supratidal sebkhas form part of the paralic complex of Bahiret el Bibane. Jaouadi et al. (2016) undertook pollen and clay mineralogical analyses covering the period of the past 8000 years and found a succession of five wet–dry oscillations. Based on the aquatics distribution, climate has been predominantly wet during the period 8000–5000 years BP, with signs of a strong aridification after 4000 years BP (Fig. 3).

Jefara Plain, NW Libya The Jefara Plain extends from Tunisia eastwards into northwest Libya (Fig. 1), where Giraudi et al. (2013) and Mercuri et al. (2015) studied pollen and fossil molluscs in palustrine and eolian sediments for the past 10,000 years. A period of higher humidity began around 9400 years BP and ended around 5000 years BP (Fig. 3). Climate gradually transitioned into arid conditions during over 1000 years between 5500 and 4500 years BP.

MD04-2797CQ The core MD04-2797CQ was taken during the PRIVILEGE-PRIMAROSA cruise in 2004 at 771 m water depth in the Sicilian–Tunisian Strait. Bout-Roumazeilles et al. (2013) studied elemental and mineralogical composition as well as the grain size distribution of the core over the past 20,000 years. For the early and middle Holocene, the authors found two major phases of high Zr/Al ratios (13,000–12,000 and 9000–5000 years BP) which are interpreted as periods of enhanced river supply. In contrast, a third peak of high Zr/Al occurring during 2000–1000 years BP is thought to represent a phase of enhanced eolian activity because it coincides with an increase of the Ti/Al and high supply in quartz. The pollen flora of the core was analysed by Desprat et al. (2013). A major increase in the abundance of Cyperaceae occurs 11,500–5500 years BP, marking higher moisture availability in a more humid climate during this period.

CP10BC This core has been collected from the margin of the Libyan shelf in the southwestern Ionian Sea in 1501 m water depth during the RV Pelagia cruise CORTADO in 2011 (Fig. 1). Wu et al. (2016) studied the elemental geochemical composition of the core for the past 10,000 years and identified a pronounced peak in the fluvial indicator K/Ti ratio for the period 10,000–5000 years BP (Fig. 3). Based on this proxy, riverine input appears to have gradually decreased over several thousand years spanning the interval 6300–3000 years BP (Wu et al. 2017).

4.3 Sahara and Sahel

Western and Central Sahara/Sahel The appearance and eventual desiccation of lakes in the western and Central Sahara/Sahel have been summarized in a series of regional synopses (Gasse 2002; Hoelzmann et al. 2004; Schütt and Krause 2009). The lakes in this region generally formed between 12,000 and 9500 years BP and dried out between 6000 and 3500 years (Fig. 4) (e.g. Callot and Fontugne 2008). The early Holocene wet period has also been described from southern Algeria from

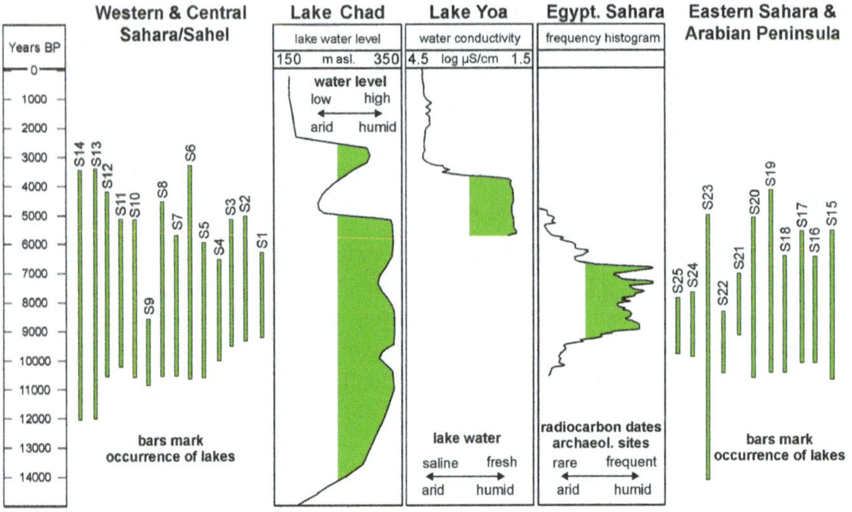

Fig. 4 Development of hydroclimate in the Sahara and Sahel during the past 15,000 years based on climate proxies. Green bars on left and right side of figure mark occurrence of lakes in the Western & Central Sahara/Sahel and Eastern Sahara & Arabian Peninsula, respectively. Location identifiers and charts from Hoelzmann et al. (2004). Lake Chad (Armitage et al. 2015), Lake Yoa (Eggermont et al. 2008), Egyptian Sahara (Riemer et al.2017). Location map in Fig. 1

an area south of the Hoggar where Lécuyer et al. (2016) studied the **I-n-Atei palaeolake**. The site was a lake from 11,000 to 7400 years BP after which it dried out and left place to swamps. Another prominent study location is **Ine Sakane** sand sea in northeastern Mali that was studied by Hillaire-Marcel et al. (1983). The authors reconstructed a lacustrine period during the time 9500–6500 years BP during which numerous lakes occupied depressions formed in the Cretaceous limestones and between the sand ridges. Between ca. 6500 and 5500 years BP the lakes dried up. A second lacustrine episode is dated between ca. 5500 and 4500 years BP and corresponds to the establishment of numerous Neolithic settlements in the area. Arid conditions have developed since about 4500 years BP to attain the conditions of the present day. From the **Acacus Mountains** in Southwest Libya, Cremaschi et al. (2010) reported calcareous tufa deposits that formed during a wet phase 9600–8100 years BP. At that time, specialized hunter-gatherers colonized the area (Cancellieri and di Lernia 2014).

Lake Megachad During the early to middle Holocene Lake Chad (Fig. 1) was much larger than today with a lake level that was more than 150 m higher. Armitage et al. (2015) reconstructed the lake level history based on optically stimulated luminescence (OSL) dating of sedimentary deposits. Lake level began to rapidly rise 15,000 years ago and stayed high until 5000 years BP (Fig. 4) after which the lake shrank significantly. At its peak around 6000 years ago, Palaeolake Mega-Chad was the largest freshwater lake on Earth, with an area of 360,000 km^2 (Amaral et al. 2013; Armitage et al. 2015; Drake and Bristow 2006). The general drying trend was interrupted by a wet phase that lasted 3500–2500 years BP during which lake level rose markedly.

Lake Yoa This lake lies 380 m above sea level and occupies a Pleistocene deflation basin in the desert oasis of Ounianga Kebir, situated halfway between the Tibesti and Ennedi mountains in northern Chad (Fig. 1). Eggermont et al. (2008) studied changes in fossil chironomid assemblages from the lake for the past 6000 years. They identified a fairly abrupt change at 4000 years BP from fresh to saline conditions, reflecting a strong reduction in precipitation in the region (Fig. 4). The aridification is also reflected in the palynology as documented by Kröpelin et al. (2008), who reported strong reductions in tropical trees after 4800 years BP and a sharp increase in Saharan plant taxa at 2500 years BP.

Eastern Sahara Hoelzmann et al. (2004) compiled the ages of lakes in the Eastern Sahara and on the Arabian Peninsula. The majority of lakes formed around 10,000 years BP and dried out sometime between 7500 and 4000 years BP (Fig. 4). This general timing is confirmed by radiocarbon data from archaeological excavations in the Eastern Sahara (Bubenzer and Riemer 2007; Kuper and Kröpelin 2006; Linstädter and Kröpelin 2004; Manning and Timpson 2014). Riemer et al. (2017) studied rock art in the **Gilf Kebir-Jebel Uweinat** region in SE Libya, NW Sudan and SW Egypt and suggest that the vast majority of the region's rock art was produced after 8500 years BP during the early Holocene humid phase. Arid conditions returned to the region around 5500 years BP. From Jebel Uweinat, Marinova et al. (2014) reported carbonate structures dated 8100 and 9400 years BP that line the walls of two valleys in Gebel Uweinat and which were likely formed in standing water. A palaeogeographic map showing the vegetational belts and distribution of main rivers and lakes in North Africa during the peak phase of the African Humid Period (AHP) was compiled by Larrasoaña et al. (2013). From NW Sudan, Hoelzmann et al. (2010) described palaeolake Gureinat which existed from 10,900 years BP until at least 7100 years BP.

4.4 Nile Delta and Levant Offshore

MS27PT is a core that has been taken in the offshore Nile Delta in 1389 m water depth, 90 km outward of the Rosetta mouth of the Nile (Fig. 1). The elemental geochemical distribution of the core was studied by Revel et al. (2015), who identified the ratio between titanium and calcium as a proxy for the terrigenous fluvial output of the Nile. The Log Ti/Ca shows a pronounced positive peak during the period 14,000–8000 years BP after which the ratio steadily decreases until 3000 years BP, marking a long gradual transition over several thousand years from the African Humid Period to the subsequent dry phase. The humid climate of the early Holocene is generally confirmed by the $\delta^{18}O$ values of planktonic foraminifera *Globigerinoides ruber* var. *alba* which show a negative peak 12,000–6000 years BP (Revel et al. 2014) (Fig. 5). The $\delta^{18}O_{ruber}$ traces freshwater

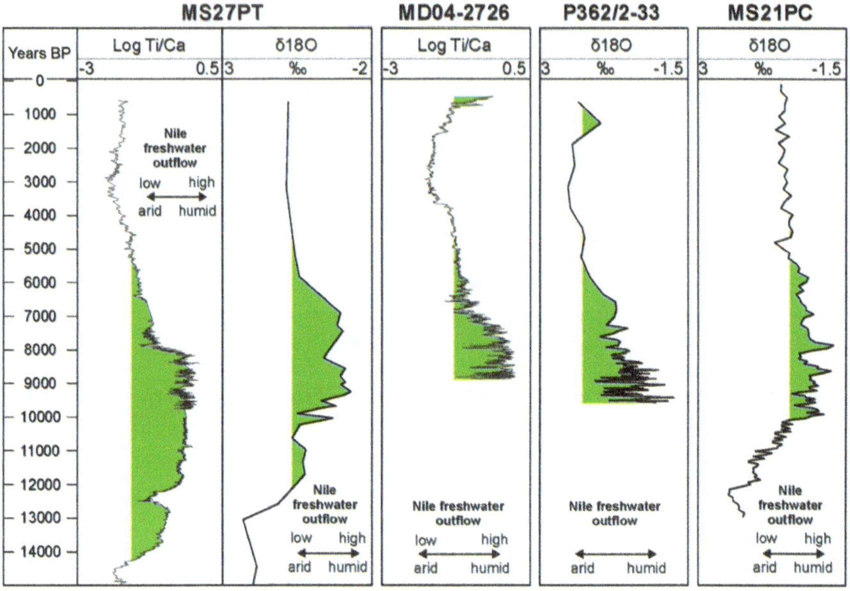

Fig. 5 Development of hydroclimate in the Nile Delta, offshore Egypt, during the past 15,000 years based on climate proxies. Curves: MS27PT (Log Ti/Ca: Revel et al. 2015; d18O: Revel et al. 2014), MD04-2726 (Revel et al. 2015), P362/2-33 (Blanchet et al. 2014), MS21PC (Hennekam et al. 2015). Location map in Fig. 1

variability, because Nile water has a more depleted $\delta^{18}O$ than seawater. The foraminifer *G. ruber* generally lives in the top ~50–100 m of the water column and is able to thrive in freshwater lenses (Hennekam et al. 2015). A recent study by Jalali et al. (2017) generally confirms the earlier hydroclimatic findings.

MD04-2726 This core has been retrieved from the offshore Nile Delta in 1058 m water depth, about 40 km east of core MS27PT on the Rosetta upper slope (Fig. 1). The Nile freshwater output proxy Ti/Ca was analysed by Revel et al. (2015) and shows a positive peak from the start of the dataset around 9000 years BP until 7000 years BP, followed by a gradual decline until 3000 years BP (Fig. 5). See also similar findings of Jalali et al. (2017) for this core.

P362/2-33 The core has been taken in the offshore Nile Delta in 700 m water depth, in the vicinity of the most active Rosetta Canyon, upslope of MD04-2726 (Fig. 1). Stable oxygen isotopes ($\delta_{18}O$) were measured on *G. ruber* by Blanchet et al. (2014) (Fig. 5). The trends match those from the MS27PT core described above. Ti/Ca and grain size data presented in Blanchet et al. (2013) from the same core support the hydroclimatic timing of the wettest phase ending around 7000 years BP and a gradual switch from wet to arid at around 5000 years BP.

MS21PC This core originates from the Central Province of the Nile delta and was recovered from 1022 m water depth (Fig. 1). Hennekam et al. (2015) published the

$\delta^{18}O$ values of *G. ruber* which show a negative peak during the period 10,000–5000 years BP, reflecting the African Humid Period (AHP) (Fig. 5).

PS009PC The core has been taken in the southeast Levantine Basin, approximately 100 km southwest of Haifa, Israel, from 552 m water depth (Fig. 1). Stable oxygen isotope data ($\delta^{18}O_{ruber}$) measured by Hennekam et al. (2014) show a marked switch towards more negative values around 12,000 years BP, marking the beginning of the AHP. Peak humidity is reached 10,000–5000 years BP, represented by a broad negative trough in the oxygen isotopes, followed by a slow and gradual increase of values until modern times (Fig. 6). The AHP is also reflected by an interval of light $\delta^{13}C_{ruber}$ and $\delta^{13}C_{org}$, and an increase in *Quercus* (oak), corresponding to the inflow of depleted terrestrial $\delta^{13}C_{DIC}$, indicating a vegetated 'Green' Nile catchment with more abundant tree vegetation in the river plain (Hennekam et al. 2015). An enhanced size of *Globigerinoides ruber* during the AHP is interpreted as a response to environmental stress caused by low-saline waters leading to delayed reproduction and larger growth due to prolonged maturity (Mojtahid et al. 2015) (Fig. 6).

GeoB 7702-3 This core has been collected on the continental slope off Israel in 562 m water depth (Fig. 1). Castañeda et al. (2016) reconstructed the variability in the amount of precipitation in the Nile catchment area in tropical Africa based on

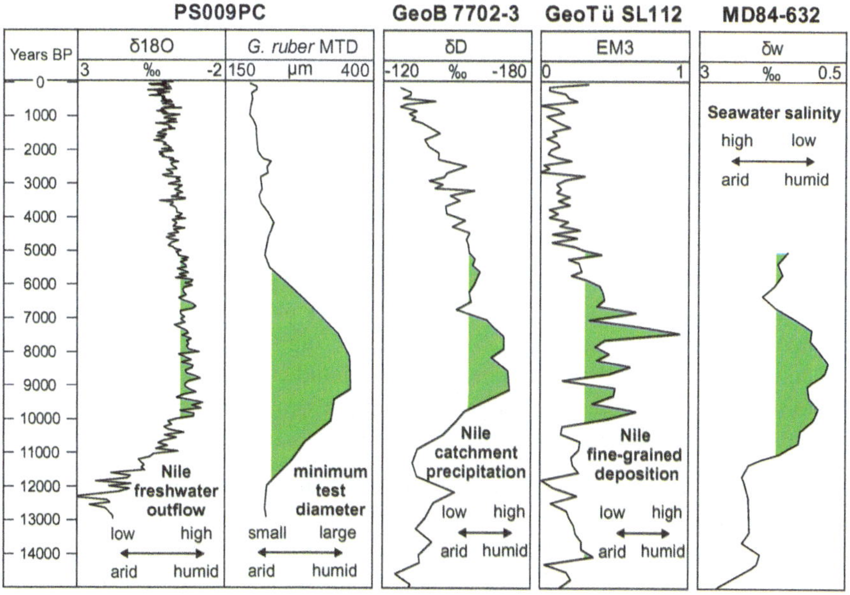

Fig. 6 Development of hydroclimate along offshore Levant during the past 15,000 years based on climate proxies. Curves: PS009PC (δ18O: Hennekam et al. 2014; minimum test diameter G. ruber white s.s.: Mojtahid et al. 2015), GeoB 7702-3 (Castañeda et al. 2016), GeoTü SL112 (Hamann et al. 2008), MD-84-632 (Essallami et al. 2007). Location map in Fig. 1

the deuterium isotopic composition of plant leaf waxes in the core. The authors found a pronounced wet phase 10,000–5000 years BP as indicated by more negative deuterium isotopes (Fig. 6).

GeoTü SL112 The gravity core was recovered from the southeastern Levantine Sea off Israel from 892 m water depth (Fig. 1). Hamann et al. (2008) analysed grain size distributions of the terrigenous fraction of the core and found a marked increase of the fine-grained EM3 fraction between 10,000 and 5000 years BP (Fig. 6) which is interpreted as the enhanced influx of Nile river water.

MD84-632 This core has been retrieved from the Levantine basin 50 km west of Haifa from 1425 m water depth (Fig. 1). Essallami et al. (2007) determined oxygen isotopes in planktonic foraminifera to derive changes in the oxygen isotope composition of seawater (δ_w), a proxy for salinity, to examine the alteration of freshwater input/evaporation budget. The study identified a period of less saline waters 11,000–7000 years, marking increased humidity during the AHP (Fig. 6). *G. ruber* oxygen isotopes presented by Essallami et al. (2007) show a negative anomaly from 11,000 to 5000 years BP (where the studied core ends), suggesting a continuation of the wet phase up to this point.

Qarun Lake The Holocene development of Qarun Lake in the Faiyum Oasis (Egypt) was studied by Marks et al. (2017) and Zhao et al. (2017) in core FA-1. The freshwater lake formed around 10,000 years BP and turned brackish 6200 years BP. Rapid onset of hyper-aridification occurred at 4200 years BP. The evolution is in general agreement with the hydroclimatic evolution as reconstructed in the various Nile Delta offshore sediment cores (see above).

4.5 Onshore Levant to Northern Red Sea

Jeita Cave This cave is located 15 km northeast of Beirut in central Lebanon, about 100 m above sea level (Fig. 1). Stable oxygen isotopes of stalagmites were studied by Verheyden et al. (2008) and Cheng et al. (2015). A negative $\delta^{18}O$ anomaly during the phase 11,000–5000 years BP suggests an increase in the precipitation–evaporation (P-E) balance, i.e. more humid conditions (Fig. 7). An early Holocene wet phase with higher rainfall than at present has also been reported by Willcox et al. (2009) from northern Syria.

Soreq Cave The cave is located in the Judean Mountains in central Israel, 40 km east of the Mediterranean Sea, 400 m above sea level and 10–50 m below the surface (Fig. 1). Oxygen isotopes have been measured by Schilman et al. (2002) and Bar-Matthews et al. (2003) and document a phase of more negative $\delta^{18}O$ values 11,000–5000 years BP which is interpreted as indicating high rainfall in the Eastern Mediterranean region at these times (Fig. 7).

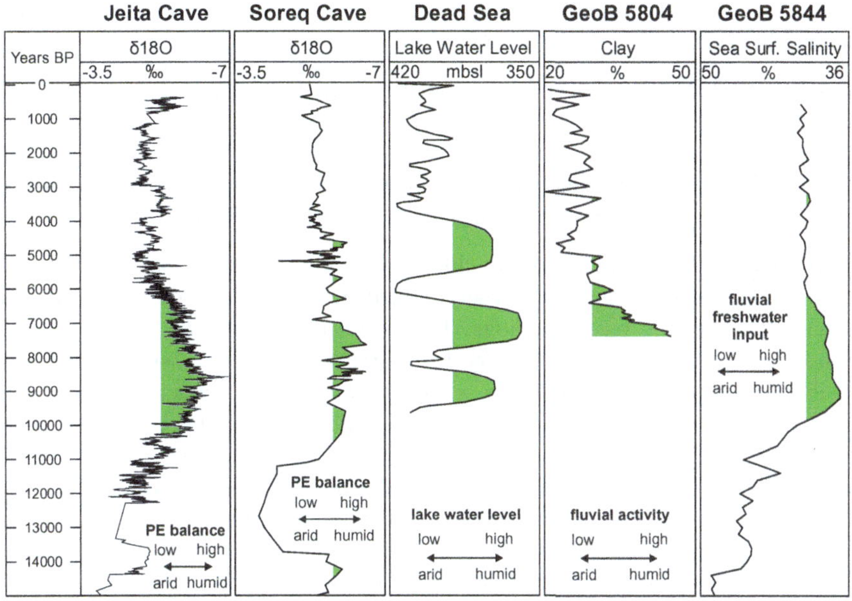

Fig. 7 Development of hydroclimate onshore Levant to northern Red Sea during the past 15,000 years based on climate proxies. Curves: Jeita Cave (Cheng et al. 2015; PE=precipitation-evaporation), Soreq Cave (Bar-Matthews et al. 2003), Dead Sea (Enzel et al. 2003; lake water level below sea level, mbsl), GeoB 5804 and GeoB 5844 (Arz et al.2003a). Location map in Fig. 1

Dead Sea The Dead Sea (Fig. 1) is a terminal lake of one of the largest hydrological systems in the Levant and its lake level may, thus, be viewed as a large rain gauge for the region (Enzel et al. 2003). Over the years, the lake level has been reconstructed by various teams and methods (e.g. Bartov et al. 2007; Enzel et al. 2003; Litt et al. 2012; Liu et al. 2013; Migowski et al. 2006; Robinson et al. 2006). Resulting lake level curves differ from each other when compared in detail. However, all authors commonly agree that there have been phases of high lake level during the early to middle Holocene (Fig. 7), matching well with the concept of a more humid climate at this time. Parts of the Pleistocene seem to have been even more humid, as evidenced by even higher lake levels (Liu et al. 2013; Robinson et al. 2006). This may explain why caves of the northern Negev Desert in southern Israel show good speleothem growth during repeated long wet phases of the Pleistocene but lack any speleothem deposition during the Holocene when the required wetness threshold was apparently not reached (Vaks et al. 2006, 2007).

GeoB 5804-4 This sediment core has been taken in the northernmost part of the Gulf of Aqaba from 464 m water depth (Fig. 1). Arz et al. (2003a) reconstructed palaeosalinities based on $\delta^{18}O$ values of *G. ruber* and clay content as fluvial input proxy. High clay content and more negative $\delta^{18}O_{ruber}$ occur from the base of the studied core at 8000 years BP until 5000 years BP, marking a more humid climate during this period (Fig. 7).

GeoB 5844-2 This core has been collected from the northern Red Sea off the southern tip of the Sinai Peninsula in 963 m water depth (Fig. 1). Sea surface salinities were reconstructed by Arz et al. (2003a) based on foraminiferal oxygen isotopes. Salinities start rising sharply at 11,000 years BP and reach peak values at 9000 years BP. The high salinity phase ends around 6000 years BP (Fig. 7).

Northern Saudi Arabia Holocene climate research in northern Saudi Arabia has been intensified over the past decade, nevertheless, there is still a lack of fully quantitative climatic reconstructions. The continental sabkha of **Tayma** in NW Saudi Arabia (Fig. 1) is one of very few North Arabian sites, where substantial landscape changes are recorded in a laminated lacustrine sequence. During the period 9200–8000 years BP, a perennial lake with a surface area of up to 22 km^2 formed in Tayma, marking early Holocene peak humidity. Newly formed grasslands were surrounding the lake at the time. The development of this lake has been reconstructed by means of shoreline age dating, palynology and ostracods (Dinies et al. 2015; Engel et al. 2012; Ginau et al. 2012). The lake contracted abruptly after 8000 years BP, and the grasslands were replaced by more drought-resistant dwarf-shrublands (Dinies et al. 2015).
The **Jubbah** palaeolake is located about 300 km east of Tayma and may have a similar age as the lake in Tayma (Fig. 1). Crassard et al. (2013b) dated the lake deposits as 9000–8000 years BP based on calibrated radiocarbon ages. Prehistoric rock art near the lake documents human occupation during wet periods (Jennings et al. 2013). Interestingly, Hilbert et al. (2014) described two different lake phases from the Jubbah Basin based on optically stimulated luminescence dating, the first occurring during the last part of the Terminal Pleistocene prior to c. 12,200 years BP, and the second after 6600 years BP. It is unclear why these dates are incongruous with records from elsewhere in Arabia.

The early Holocene humid phase does not appear to have been wet enough to allow speleothem growth in Saudi Arabian caves. Fleitmann et al. (2004) studied stalagmites from five caves in the northeast and central east Saudi Arabia. Hardly any of them were younger than 400,000 years, and none of them were of Holocene age.

4.6 SE Arabian Peninsula

UAE Rub' al Khali Lakes The Awafi palaeolake lies in the Ra's al-Khaimah Emirate of the United Arab Emirates (UAE) at the northeastern edge of the Rub' al Khali desert (Fig. 1). The lake existed during the period 8500–4100 years BP (Parker et al. 2004, 2006). From 5900 years BP, the water level started declining and vegetation became reduced, probably with an intermittent seasonal water body (Preston et al. 2012). From this time, the region's semi-nomadic, herder-gatherer

populations abandoned much of the landscape and concentrated in selected environmental refugia, such as along the northern Omani coast (Parker and Goudie 2008; Preston et al. 2012).
Atkinson et al. (2013) studied two additional UAE Rub' al Khali sites, **Al Ain and Tawi Asmar**, in which they confirmed the humid early Holocene episode and the onset of dry conditions at around 6000 years BP. Gebel et al. (1989) documented a lake phase at At Ain Faida/Al Ain during the period 11,000–5800 years BP. The timing of this wet phase corresponds well with that from other lakes in the Rub' al Khali desert. Parker et al. (2006) presented a probability density function (pdf) of radiocarbon ages of various Rub' al Khali lakes that show a maximum lake occurrence during 10,000–6000 years BP, coinciding with the early Holocene wet phase (Fig. 8).

Hoti Cave This cave is located in northern Oman about 800 m above sea level (Fig. 1). Fleitmann et al. (2007) studied oxygen isotopes in a stalagmite from this cave and found an interval of more negative values for the period 10,000–6500 years BP, reflecting increased precipitation during this time (Fig. 8).

The early Holocene wet period has been also documented 200 km east of Hoti Cave, in playa-like sediments from the **Hajar Mountain range** in northern Oman. Based on optical stimulated luminescence (OSL) sediment dating, Fuchs and

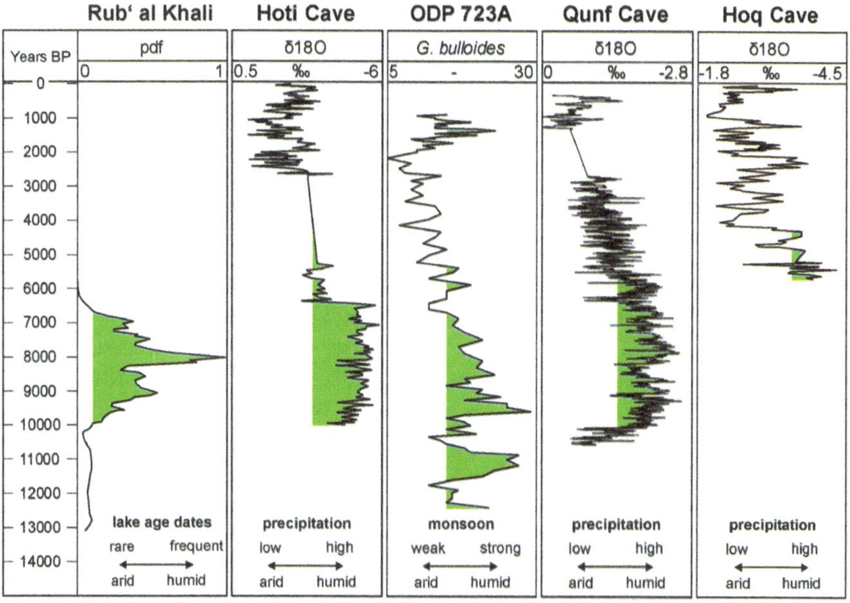

Fig. 8 Development of hydroclimate of the southeastern Arabian Peninsula during the past 15,000 years based on climate proxies. Curves: Rub Al-Khali lakes (Parker et al. 2006), Hoti Cave, N-Oman (Fleitmann et al. 2007), ODP 723A (Gupta et al. 2005), Qunf Cave, S-Oman (Fleitmann et al. 2003), Hoq Cave, Socotra Island, Yemen (Van Rampelbergh et al. 2013). Location map in Fig. 1

Buerkert (2008) found an abrupt increase in sedimentation rates in the playa at 10,500 years BP indicating the onset of the early Holocene humid period. Rainfall in the study location reached its maximum at 9000–8000 years BP. A decreasing sedimentation rate after 8000 years BP characterizes the beginning of arid conditions of the middle to late Holocene.

ODP 723A Ocean Drilling Program (ODP) Hole 723A was cored off the southern coast of Oman in the northwestern Arabian Sea in 808 m water depth in the centre of an Oxygen Minimum Zone (OMZ), providing a high-resolution sedimentary record of variations in the summer monsoon-induced upwelling (Fig. 1). Gupta et al. (2005) studied abundance variations of the planktic foraminifer *Globigerina bulloides* which are an upwelling-loving species and serves as a proxy of the summer monsoon intensity and associated rainfall. *G. bulloides* shows a characteristic abundance peak during the phase 12,000–5500 years BP, evidencing strong monsoon activity resulting in humid conditions (Fig. 8).

Qunf Cave This cave is located in southern Oman and lies 650 m above sea level (Fig. 1). The stalagmites oxygen isotopes were measured by Fleitmann et al. (2003) and show a significant negative anomaly during 10,000–5500 years BP, marking a phase with increased precipitation (Fig. 8).

The general Holocene hydroclimatic timing in southern Oman was confirmed by Cremaschi et al. (2015), who studied cavities of **Jebel Qara** in the Dhofar region. The authors found increasing humidity since 12,000 years BP which eventually allowed a large community of land snails to form that was exploited by Early Holocene hunter-gatherers who lived in the rock shelters 10,500–9500 years BP and left consistent accumulations of land shells (*escargotières*). The maximum of Holocene humidity was reached between 9000 and 8000 years BP when the regional aquifer was recharged. Calcareous tufa was deposited at the entrance of caves up to 4500 years BP, followed by a transition towards arid conditions (Cremaschi et al. 2015).

Hoq Cave The cave is located 335 m above sea level in the eastern part of Socotra Island. The island belongs to Yemen and is situated between the Horn of Africa and the Arabian Peninsula (Fig. 1). Van Rampelbergh et al. (2013) measured oxygen isotopes in various speleothems of the cave. Stalagmite STM1 comprises the last 5500 years and captures the very end of the early Holocene wet period as reflected by a negative $\delta^{18}O$ anomaly that abruptly ends 4000 years BP when isotope values increased sharply (Fig. 8). Stalagmite Hq1 covers an even longer time period of more than 9000 years. However, this speleothem is less suitable for hydroclimatic reconstructions because of the stalagmite's location near the cave entrance. Here, stronger air circulation has brought significant evaporation effects on the stalagmite that led to an isotopic disequilibrium of the calcite precipitation. In contrast, stalagmite STM1 was sampled deeper in the cave where variations in oxygen isotopes are dominated by changes in precipitation and effective moisture (Van Rampelbergh et al. 2013).

Other Study Locations in Yemen and Somalia The early to middle Holocene wet phase has been also documented in **Mukalla Cave** in mainland southern Yemen where Fleitmann et al. (2011) carried out a lower resolution study covering the past 350,000 years. Likewise, similarly aged humid conditions were also reported from al-Hawa, a large flat basin at 710 m above sea level located in the inland desert of Yemen, the Ramlat as-Sabatayn, which is the southernmost extension of the Rub' al Khali hyperarid sand sea. Lézine et al. (2007) identified a freshwater lake phase that occurred here between 12,000 and 7500 years BP. The Holocene aridification history was also reconstructed for a core off Somalia. Jung et al. (2004) studied Sr–Nd isotope ratios of the dust fraction from **Core 905**. The highly variable Sr isotope record is interpreted to reflect mainly changes in the evaporation/precipitation balance over NE Africa. It shows that the first aridification step occurred at 8500 years BP followed by an unstable transitional period up to 6000 years BP. A second aridification step began at 6000 years BP and ceased at 3800 years BP when modern-day dry climate was established (Jung et al. 2004).

Southern and Central Saudi Arabia The **Mundafan palaeolake** situated in the Rub' al-Khali Desert in SW Saudi Arabia (Fig. 1) represents an important study location for Holocene hydroclimate in the region. The lake existed during the pluvial period, 10,500–6000 years BP and hosted a benthic foraminifera fauna that was probably imported from the sea by birds (Gennari et al. 2011). Neolithic Occupations around the Mundafan lake testify of suitable living conditions at the time (Crassard et al. 2013a). The early Holocene wet phase at Mundafan forms part of a series of four discrete humid periods over the past 150,000 years (Rosenberg et al. 2011).

The early Holocene humid phase also led to a reactivation of the trans-Arabian **Wadi ad Dawasir** fluvial system in central Saudi Arabia, which was last active during certain times of the Pleistocene (Matter et al. 2016). The absence of Holocene lakes in most of the Rub' al-Khali implies that trans-Arabian rivers were mainly fed by precipitation in the Asir Mountains. Matter et al. (2016) carried out radiocarbon and luminescence age dating and noted a systematic discrepancy between the two data types. All radiocarbon ages were between 7000 and 9000 years BP, but the luminescence ages of the same strata were mainly several thousand years younger (up to 4000 years BP), indicating the additional need for research to achieve clarification.

4.7 Discussion Hydroclimate

Onset and Termination of Humid Phase All case studies available from North Africa and the Arabian Peninsula document a pronounced wet period that commenced between 15,000 and 9000 years BP and ended sometime between 6500 and 3500 years BP. Start and end dates vary between locations, depending on local factors, climate amplifiers and chosen climate proxies leading to a spatially and temporally complex distribution (Lézine et al. 2014). Some studies show gradual transitions between the hydroclimatic states, while other locations evidenced abrupt changes. It has to be noted that many case studies reflect the hydroclimate within a distance of a few 100 s kilometres to the study location, whereas other hydroclimate reconstructions represent long-distance climate signals. For example, variations in the Nile freshwater influx react to changes in rainfall in the Nile catchment in areas as far as East Africa, more than 6000 km away from the Nile Delta (Revel et al. 2015). Lézine et al. (2011) suggested that widespread AHP humidity in the Sahara and Sahel started several millennia earlier to the east than to the west. Considering the totality of all data covered in this review, it is unclear if this east–west trend is robust. A more important trend for the arid–wet transition appears to be of latitudinal nature. Reconstructions by Tierney et al. (2017) along the NW African margin indicate that the wet period started earlier and lasted longer in lower latitudes while it was more restricted in extent at higher latitudes. Likewise, the onset of aridity started in the north, with the monsoon rains weakening progressively later at lower latitudes (deMenocal 2015; Fleitmann et al. 2003; Shanahan et al. 2015). A similar situation is found on the Arabian Peninsula, where southern Arabia was already under monsoonal influence during the earliest Holocene while the rains migrated into the Gulf region only a few thousand years later (Parker and Goudie 2008). In the Sahara, the wet–arid transition appears to have occurred in the east earlier than in the west (Lézine et al. 2011).

Triggers of the Humid Phase The early Holocene wet phase of North Africa and Arabia was triggered by a northward migration of the ITCZ due to orbital precession that gradually aligned boreal summer solstice with the perihelion (deMenocal et al. 2000b). Summer insolation in the Northern Hemisphere at that time had risen to peak levels about 8% greater than today (Fig. 11). The ITCZ migration resulted in a northward shift and expansion of the monsoonal summer rains which in a time transgressive manner for a few thousand years reached the core areas of the present-day Sahara (Drake and Bristow 2006; Shanahan et al. 2015) as well as the Arabian Peninsula (Cremaschi et al. 2015; Fuchs and Buerkert 2008; Groucutt and Petraglia 2012) (Fig. 1). Aridification of North Africa and the Arabian Peninsula commenced when the ITCZ slowly moved back southwards and monsoonal rains could no longer reach the area (deMenocal 2015; Lézine et al. 2017). The ITCZ migration model provides a good explanation for the observed spatio-temporal patterns associated with the early Holocene humid period in the region. The exact position of the ITCZ on the Arabian Peninsula is still a matter of debate. The most widely accepted model is probably from Fleitmann et al. (2007), who assume the early Holocene ITCZ running in a line linking Jeddah in southern Saudi Arabia to Dubai (UAE) (Fig. 1) (see also Fleitmann and Matter 2009). Other authors proposed a slightly more southward location (Fuchs and Buerkert 2008). Recently, Enzel et al. (2015) challenged that the ITCZ may not have moved at all in Arabia which resulted in a controversial debate (Engel et al. 2017; Enzel et al. 2017).

The northernmost parts of North Africa and the Arabian Peninsula were not reached by the ITCZ (Arz et al. 2003a; Guagnin et al. 2016), therefore, the early

Holocene wet conditions documented from here require a different explanation. The early Holocene appears to have been the wettest phase of the last 25,000 years across much of the Levant and Eastern Mediterranean (Robinson et al. 2006). The most cited possibility is an intensification of southern Mediterranean winter rain and its deeper penetration southwards into North Africa and the Arabian Peninsula (Arz et al. 2003a; Desprat et al. 2013; Engel et al. 2012). Higher sea surface temperatures (SST) may have contributed by intensifying evaporation in the Mediterranean, which, in turn, could have brought more rainfall (Giraudi et al. 2013). Increased fluvial input off Libya and Tunisia could also be related to long-distance rivers transporting water from the interior of the Green Sahara under ITCZ rain influence (Wu et al. 2016), similar as in the case for the Nile in Egypt.

Pleistocene-Tertiary Context The early Holocene humid Phase forms part of a long series of wet periods that have occurred over the past hundred thousands to million years in North Africa and Arabia (Castañeda et al. 2009; Ehrmann et al. 2017; Fleitmann et al. 2011; Fleitmann and Matter 2009). Larrasoaña et al. (2013) have identified 230 of these recurrent wet phases associated with astronomically forced insolation changes that have driven monsoon dynamics. The authors suggest that these humid events have played an important role in migration and evolution of hominins.

Climate Models Various groups have attempted to simulate the early Holocene humid phase of North Africa and Arabia by means of climate models. While the general increase in humidity as a consequence of precessional ITCZ migration was successfully replicated, the models are still unable to match the observed hydro-climatic changes in a quantitative way. Namely, simulated AHP rainfall over the Sahara is not sufficient to sustain vegetation at a level seen in the palaeo record (Boos and Korty 2016; Hopcroft et al. 2017). According to Tierney et al. (2017), the climate models require improvements with regard to vegetation and dust feedbacks without which climate models systematically fail to reproduce the Green Sahara. Skinner and Poulsen (2016) have suggested that part of the missing rainfall could be related to a trans-Saharan moisture-sucking mechanism. A low pressure zone that forms during fall over NW Africa may have drawn tropical moisture in a wide corridor northwards across the western part of the Sahara. Another issue that climate models can still not fully replicate is the nature of the wet-to-dry transition at the end of the AHP (Renssen et al. 2006).

5 Holocene Temperature Development of North Africa and the Arabian Peninsula

5.1 Northwest Africa

ODP 658C This core from the Ocean Drilling Program has been taken off Cap Blanc in northern Mauritania from 2263 m water depth (Fig. 1). The well has been widely studied by various groups using a large number of techniques.

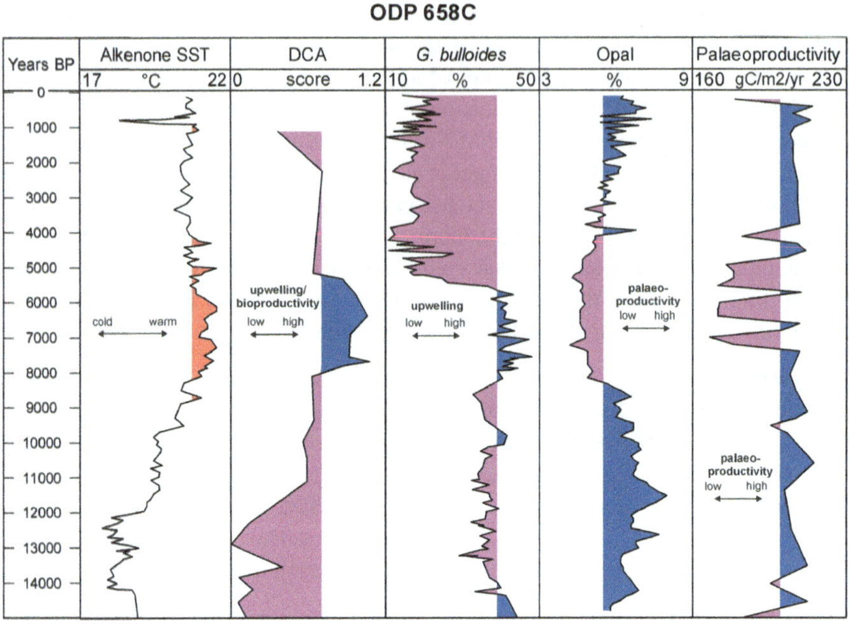

Fig. 9 Development of temperature and productivity markers in ODP core 658C off Cape Blanc during the past 15,000 years. Curves: Alkenone sea surface temperature, SST (Adkins et al. 2006), radiolarian detrended correspondence analysis, DCA (Haslett and Smart 2006), Globigerina bulloides (Adkins et al. 2006), biogenic opal (deMenocal et al. 2000a), diatom palaeoproductivity (Knaack 1997). Red peaks= warm, blue peaks=high productivity, purple peaks=low productivity. Location map in Fig. 1

Alkenone-derived temperatures begin to rise steadily around 12,000 years BP with peak warmth 9000–4000 years BP, followed by a gentle cooling (Adkins et al. 2006; Zhao et al. 1995, 2000) (Fig. 9). Originally, deMenocal et al. (2000b) had suggested an inverse temperature development for this core, with a cold early Holocene and a warm late Holocene. It appears, however, that the foraminifera transfer function has essentially failed in this case, disturbed by an increase in abundance and size of *Globigerina bulloides* during the early Holocene that was independent of sea surface temperature, but probably rather related to enhanced river nutrient supply during the African humid period (Adkins et al. 2006) (Fig. 9). In addition, dissolution effects may have reduced the preserved abundance of *G. bulloides* in the late Holocene part of the section (Haslett and Smart 2006).

Initially, it was assumed that Holocene temperature changes documented in this core could be associated with variations in the upwelling intensity in the ODP 658C region, whereby cold upwelling water would have cooled the sea surface (deMenocal et al. 2000b) and lack of upwelling would have led to warming (Adkins et al. 2006). Comparison with other offshore temperature reconstructions, however, suggests that the long-term temperature trends observed in the ODP 658C core are largely in line with neighbouring Atlantic studies (see discussion below, Fig. 10). Changes in palaeoproductivity may, therefore, not have played a dominant role for the temperature development here, possibly just modifying it. Furthermore, the upwelling and bioproductivity history off Cap Blanc is far from fully understood. Unfortunately, different palaeoproductivity and fertility proxies yield contradictory trends. Radiolarians (Haslett and Smart 2006) and planktic foraminifera (Adkins et al. 2006; deMenocal et al. 2000b) indicate peak upwelling conditions for the early Holocene, coinciding with peak sea surface warming (Fig. 9). On the other hand, diatoms (Knaack 1997), associated biogenic opal (Adkins et al. 2006; Bradtmiller et al. 2016) and molecular biomarkers (Zhao et al. 2006) suggest low productivity conditions for the same period.

GeoB 6007-2 This Moroccan core was cut off Cape Ghir in 582 m water depth (Fig. 1). Alkenone sea surface temperatures were measured by Kim et al. (2007) and show a pronounced warm phase from the start of the dataset at 10,000 years BP until 6000 years BP, followed by a general cooling until 2000 years BP (Fig. 10). The authors also studied oxygen isotopes of *G. bulloides,* which they interpret as indicator for changes in upwelling intensity. Significant $\delta^{18}O$ fluctuations occur, however, the early Holocene warm phase does not coincide with generally weaker upwelling, therefore, lack of upwelling cannot be the main driver of the observed warming here.

MD01-2444 This core has been retrieved offshore southern Portugal, about 150 km SW of Lisbon from 2637 m water depth (Fig. 1). Alkenone-derived sea surface temperatures were published for this core by Martrat et al. (2007). Abrupt warming occurred around 12,000 years BP, with peak temperatures at 8000 years BP (Fig. 10). The early Holocene warm phase ended around 5000 years BP, followed by a gentle long-term cooling until present. A nearby core MD01-2443 described by the same authors shows a similar development.

M39-008 This core originates from the northeastern Gulf of Cadiz in Spain from 576 m water depth (Fig. 1). Similar as in the Portuguese core MD01-2444, sea surface temperatures rise abruptly at around 12,000 years BP with the warm phase lasting until 5000 years BP, as evidenced by alkenone data measured by Cacho et al. (2001) (Fig. 10).

Lake Tigalmamine The lake is located in the Middle Atlas in Morocco at an elevation of 1626 m (Fig. 1). Cheddadi et al. (1998) reconstructed January temperatures based on pollen and documented a warm phase that extended from the base of the sediment core at 10,000 years BP until 5500 years BP (Fig. 10). Notably, this dataset represents the only available land temperature reconstruction for the Holocene in onshore North Africa and Arabia. Targeted research efforts are needed to add further land temperature records for the past ten millennia in the region.

MD95-2043 This core has been retrieved in 1841 m water depth from the Alboran Sea, about 90 km south of Almeria (Fig. 1). Cacho et al. (1999) published alkenone

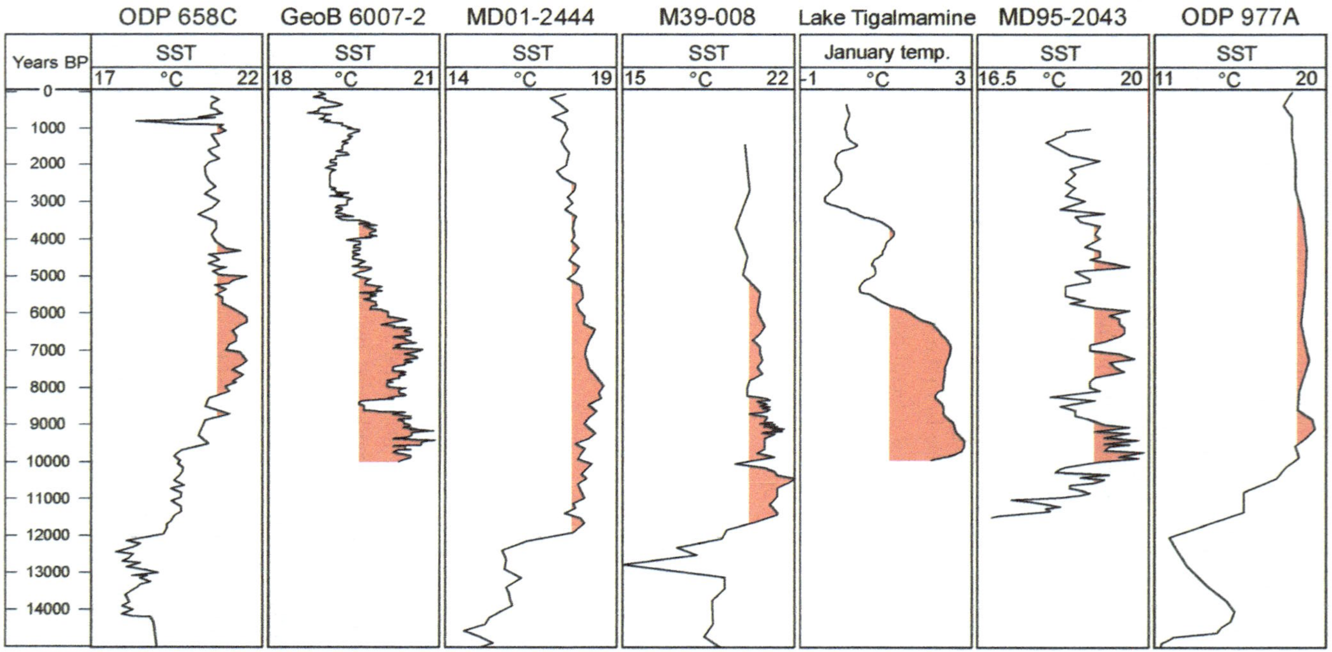

Fig. 10 Temperature development in NW Africa and SW Europe during the past 15,000 years. Curves: ODP 658C (Adkins et al. 2006), GeoB 6007-2 (Kim et al. 2007), MD01-2444 (Martrat et al. 2007), M39-008 (Cacho et al. 2001), Lake Tigalmine (Cheddadi et al. 1998), MD95-2043 (Cacho et al. 1999), ODP 977A (Martrat et al. 2004). Red peaks mark warming. Location map in Fig. 1

sea surface temperature data that document a warm phase comprising 10,000–6000 years BP that was briefly interrupted by a colder episode 8500–7500 years BP (Fig. 10).

OPD 977A Also, this core was taken from the Alboran Sea, from a location east of MD-95-2043 in 1984 m water depth (Fig. 1). Alkenone sea surface temperatures were measured by Martrat et al. (2004) and show an abrupt warming around 12,000 years, like in the other neighbouring Portuguese and Spanish cores described above (Fig. 10). Peak temperatures were reached 9000 years BP. The warm phase lasted until 3000 years BP, followed by a long-term gentle cooling until the Little Ice Age 300 years BP.

MD04-2797 This core was retrieved from the Siculo–Tunisian Strait in 771 m water depth, about 50 km east off Cape Bon in NE Tunisia and 110 km off Sicily (Fig. 1). Essallami et al. (2007) reconstructed sea surface temperatures based on alkenones and planktonic foraminiferal Modern Analog Technique (MAT). Unfortunately, data for the last 2000 years is missing. Alkenones show high temperatures 11,000–8500 years BP and 5500–2500 years BP, separated by a colder phase. The MAT technique did not yield any clear Holocene long-term trend in this core.

Other Parts of the Mediterranean Warm early Holocene climate has also been documented in lakes in the Tuscan–Emilian Apennines in Italy. Samartin et al. (2017) reconstructed July temperatures based on fossil midge assemblages and identified a period of 9000–5000 years BP during which they estimate temperatures that have been 1–2 °C warmer than the recent pre-industrial period.

5.2 NE Africa and Arabian Peninsula

P362/2-33 This core was taken from the western Nile Delta from 700 m water depth, about 50 km north of Alexandria (Fig. 1). Alkenone sea surface temperatures were measured by Blanchet et al. (2014) which indicate generally warm conditions from the base of the core at 9500 years BP until about 5000 years BP (Fig. 11). The late Holocene is colder, except during renewed warming of the last 1000 years.

GeoB 7702-3 The core has been cut from the continental slope offshore Gaza in 562 m water depth (Fig. 1). Castañeda et al. (2016) examined branched Glycerol dialkyl glycerol tetraethers (GDGTs) using BAYSPAR calibration to reconstruct mean annual air temperature. The authors found a pronounced warm phase 10,500–4500 years BP, followed by sharp cooling (Fig. 11). A series of shorter warm phases occurred from 2000 years BP onwards. The results supersede previously published Tex_{86} temperatures published by Castañeda et al. (2010) that originally suggested overall similar trends but 4 °C lower temperatures throughout the Holocene than the new GDGT curve. Alkenone-derived temperatures in Castañeda

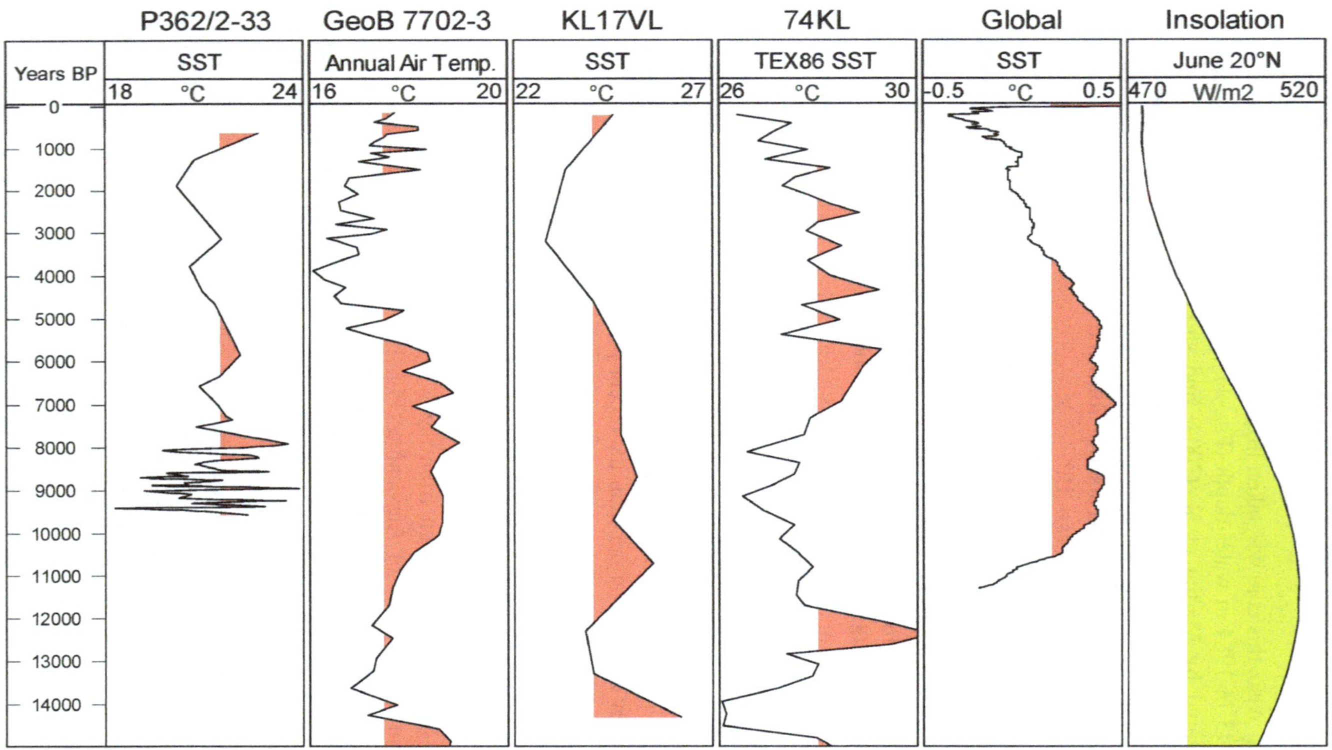

Fig. 11 Temperature development offshore Nile Delta (P362/2-33), Levant (GeoB 7702-3), northern Red Sea (KL17VL) and Oman (74KL) during the past 15,000 years. Curves: P362/2-33 (Blanchet et al. 2014), GeoB 7702-3 (Castañeda et al. 2016), KL17VL (Trommer et al. 2010), 74KL (Huguet et al. 2006). Location map in Fig. 1

et al. (2010) showed a rather stable temperature throughout the Holocene and do not really match the GDGT data. Temperature reconstructions based on alkenones and planktonic foraminifera assemblages from core MD84-632 offshore northern Israel did not yield any clear Holocene temperature trends (Essallami et al. 2007).

KL17VL The core was taken from the northern Red Sea in 1014 m water depth (Fig. 1). Trommer et al. (2010) measured sea surface temperatures based on TEX_{86} and documented a warm phase that lasted from 11,000 to 5000 years BP, followed by a cooler period (Fig. 11). Temperatures increased again in the last millennium BP, similar as seen in cores P362/2-33 and GeoB 7702-3.

RC-27-42 The core was retrieved from the western Arabian Sea in 2040 m water depth, about 350 km from the Oman continental margin (Fig. 1). Pourmand et al. (2007) measured alkenone-derived sea surface temperatures and documented a cold phase from 28,000 to 14,000 years BP, followed by a warm period until 5000 years BP at which point the core study ended, unfortunately.

74KL This core was obtained 400 km off the southern Omani coast from the East Sheba Ridge in 3212 m water depth (Fig. 1). Sea surface temperatures based on TEX_{86} and Uk'37 were studied by Huguet et al. (2006). The curves yield very different trends and also differ markedly from the pattern observed in other parts of North Africa and Arabia. TEX_{86} data suggests that the early Holocene was cold with low temperatures 12,000–7000 years BP (Fig. 11). A subsequent warm phase occurred 7000–2000 years BP, followed again by lower temperatures until present. In contrast, Uk'37-derived temperatures show a continuous warming from 16,000 years BP until present. Due to the discrepancy, the temperature results from this core have to be used with caution.

5.3 Discussion Temperature Development

The early to mid-Holocene Thermal Maximum (HTM) is a well-recognized feature in the global postglacial climate development. Ljungqvist (2011) analysed 60 different reconstructions of annual mean temperature from locations around the globe and found a marked statistical warming 8000–4000 years BP. A similar study by Marcott et al. (2013) based on 73 sea surface temperature reconstructions dated the peak HTM warming at 10,000–5000 years BP (Fig. 11). The HTM end dates suggested by the global syntheses of Ljungqvist (2011) and Marcott et al. (2013) generally match those from North Africa and Arabia, while warming appears to have commenced a few thousand years earlier than on global average. HTM warming is thought to be driven by northern hemisphere high summertime insolation associated with orbital precession.

A global temperature synthesis including onshore case studies in a systematic way does, unfortunately, not exist yet. Changes in onshore temperatures are likely to have significantly exceeded those in the marine realm. For example, the HTM appears to have been only mildly warmer than the late Holocene in some offshore parts of NW Africa (Fig. 10), matching the concept of Marcott et al. (2013) who assume a temperature amplitude of less than 1 °C over the past 11,000 years. However, half of the offshore temperature reconstructions show HTM warming of several degrees centigrade, indicating that the global HTM amplitude may actually be larger than thought.

The global dataset of Marcott et al. (2013) includes eight temperature records from North Africa and Arabia. For the northern Mauritanian ODP core 658C, the authors have unfortunately chosen the meanwhile outdated foraminifera-based curve of deMenocal et al. (2000b) which erroneously suggested early Holocene cooling, while the now accepted alkenone-derived temperatures document early Holocene warming (Fig. 10). Another region of inconsistency is the northern Red Sea. Here, Marcott et al. (2013) chose the Uk'37 temperature reconstruction of Arz et al. (2003b) that suggested a cool early Holocene compared to a warmer late Holocene. These data, however, have meanwhile been superseded by more recent TEX_{86} temperatures published by Trommer et al. (2010) from the nearby core KL17VL. This newer data show pronounced HTM warming, in line with other data from North Africa and Arabia (Fig. 11). Finally, Marcott et al. (2013) included the Uk'37-derived temperature data from core 74KL off Oman (Huguet et al. 2006) in their global temperature average, for which two important issues have to be considered. First, the Uk'37 and Tex86 temperature reconstructions for the same core differ greatly from each other (see above), whereby the chosen Uk'37 data show a warm late Holocene, contrasting with a cold late Holocene in the Tex86 curve. Second, core 74KL may be considered unsuitable altogether for global temperature reconstructions because its temperature signal is controlled by local upwelling dynamics rather than regional trends.

The review has shown that large discrepancies may exist in some cores where the temperature history has been analysed based on several independent techniques. Results may, therefore, not in all cases be fully reliable. A regional correlation approach such as in this contribution can help to better interpret regional trends and identify outliers which warrant further discussion. Such a systematic regional approach also allows distinguishing between the dominant drivers of the observed warming: Of particular importance is the identification of local (e.g. upwelling) versus and regional/global effects. In this contribution, we focused on long-term Holocene climate change. Notably, characteristic climate fluctuations have also been identified on shorter, i.e. millennial- and centennial-scale (e.g. deMenocal et al. 2000b; Gupta et al. 2005; Hennekam et al. 2014; Neff et al. 2001). This will be the subject of a separate paper.

6 Conclusions

Holocene climate in North Africa and on the Arabian Peninsula has undergone major changes. In this contribution, we review hydroclimate and temperature changes in the region over the past 15,000 years by correlating and integrating all available case studies. A pronounced wet period corresponding to the 'Green Sahara' and its equivalent in Arabia commenced between 15,000 and 9000 years BP and ended sometime between 6500 and 3500 years BP, followed by arid conditions throughout the late Holocene. Start and end dates vary between locations, depending on local factors, climate amplifiers and chosen climate proxies, leading to a spatially and temporally complex distribution. Some studies show gradual transitions between the hydroclimatic states while other locations evidenced abrupt changes. The humid phase was triggered by a northward migration of the ITCZ due to orbital precession.

The northernmost parts of North Africa and the Arabian Peninsula were not reached by the ITCZ. Here, increased early Holocene humidity may have been due to an intensification of southern Mediterranean winter rain and its deeper penetration southwards into the present-day desert areas. The early Holocene 'Green Sahara' forms part of a long series of wet periods that have occurred over the past hundred thousand to million years in North Africa and Arabia. Notably, climate models are still unable to match the observed hydroclimatic changes in a quantitative way. Simulated AHP rainfall over the Sahara is not sufficient to sustain vegetation at a level seen in the palaeo record.

Sea surface temperatures in North Africa and Arabia during the early Holocene were generally one to several degrees warmer than during the late Holocene. Warming began around 12,000 years BP and ended around 5000 years BP. The warm period generally coincided with the early Holocene wet phase in the region and is linked to the Holocene Thermal Maximum, an early Holocene period during which temperatures were globally elevated. The review suggests that the Holocene climate history of North Africa and Arabia is closely linked to the global development and that significant temperature changes have also occurred in subtropical climate belts.

Acknowledgements The authors would like to thank Nathalie Combourieu-Nebout, Viviane Bout-Roumazeilles, Rick Hennekam, Miryam Bar-Matthews, Marie-Alexandrine Sicre, Sophie Verheyden, Hai Cheng, Jessica Tierney, Jiao-yang Ruan, Stéphanie Desprat, Cecile Blanchet, Jalal Tabel and Meryem Mojtahid for the kind provision of tabulated data. Lydia Gerullis (MARUM, Univ. Bremen) is thanked for her valuable support with regards to the PANGAEA database. We are grateful to Jurgis Klaudius and Lloyd's Register for providing the database and correlation software IC^{TM} for this project. Special thanks go to the technical IC^{TM} team who kindly helped initiating the database and charts. Project support by Jens Kröger is greatly acknowledged. We thank Mariusz Gałka (Adam Mickiewicz University, Poznan) and Christoph Zielhofer (Leipzig University) for valuable discussions. We extend our thanks to the editor of this book, Abderrahmane Bendaoud (University of Science and Technology Houari Boumedienne, Algiers), for the invitation to write this chapter. We are grateful to the two reviewers, whose valuable suggestions helped to improve the manuscript.

References

Adkins J, deMenocal P, Eshel G (2006) The "African humid period" and the record of marine upwelling from excess 230Th in Ocean Drilling Program Hole 658C. Paleoceanography 21(4)

Amaral PGC, Vincens A, Guiot J, Buchet G, Deschamps P, Doumnang JC, Sylvestre F (2013) Palynological evidence for gradual vegetation and climate changes during the African humid period termination at 13°N from a Mega-Lake Chad sedimentary sequence. Clim Past 9(1): 223–241

Anderson DM, Baulcomb CK, Duvivier AK, Gupta AK (2010) Indian summer monsoon during the last two millennia. J Quat Sci 25(6):911–917

Armitage SJ, Bristow CS, Drake NA (2015) West African monsoon dynamics inferred from abrupt fluctuations of Lake Mega-Chad. Proc Natl Acad Sci 112(28):8543–8548

Arz HW, Lamy F, Pätzold J, Müller PJ, Prins M (2003a) Mediterranean moisture source for an early-Holocene humid period in the northern Red Sea. Science 300(5616):118–121

Arz HW, Pätzold J, Müller PJ, Moammar MO (2003b) Influence of Northern Hemisphere climate and global sea level rise on the restricted Red Sea marine environment during termination I. Paleoceanography 18(2)

Atkinson OAC, Thomas DSG, Parker AG, Goudie AS (2013) Late Quaternary humidity and aridity dynamics in the northeast Rub' al-Khali, United Arab Emirates: implications for early human dispersal and occupation of eastern Arabia. Quatern Int 300:292–301

Badawi A (2016) Late Holocene hydrographic settings of the northern Red Sea. Egypt J Aquat Res 42(1):41–48

Baioumy HM, Kayanne H, Tada R (2010) Reconstruction of lake-level and climate changes in Lake Qarun, Egypt, during the last 7000 years. J Great Lakes Res 36(2):318–327

Bar-Matthews M, Ayalon A (2011) Mid-Holocene climate variations revealed by high-resolution speleothem records from Soreq Cave, Israel and their correlation with cultural changes. Holocene 21(1):163–171

Bar-Matthews M, Ayalon A, Kaufman A, Wasserburg GJ (1999) The Eastern Mediterranean paleoclimate as a reflection of regional events: Soreq cave, Israel. Earth Planet Sci Lett 166(1–2): 85–95

Bar-Matthews M, Ayalon A, Gilmour M, Matthews A, Hawkesworth CJ (2003) Sea–land oxygen isotopic relationships from planktonic foraminifera and speleothems in the Eastern Mediterranean region and their implication for paleorainfall during interglacial intervals. Geochim Cosmochim Acta 67(17):3181–3199

Bartov Y, Enzel Y, Porat N, Stein M (2007) Evolution of the Late Pleistocene–Holocene Dead Sea basin from sequence statigraphy of fan deltas and lake-level reconstruction. J Sediment Res 77(9):680–692

Bernhardt CE, Horton BP, Stanley J-D (2012) Nile Delta vegetation response to Holocene climate variability. Geology 40(7):615–618

Blanchet CL, Tjallingii R, Frank M, Lorenzen J, Reitz A, Brown K, Feseker T, Brückmann W (2013) High- and low-latitude forcing of the Nile River regime during the Holocene inferred from laminated sediments of the Nile deep-sea fan. Earth Planet Sci Lett 364:98–110

Blanchet CL, Frank M, Schouten S (2014) Asynchronous changes in vegetation, runoff and erosion in the Nile River watershed during the Holocene. PLoS ONE 9(12):e115958

Boos WR, Korty RL (2016) Regional energy budget control of the intertropical convergence zone and application to mid-Holocene rainfall. Nat Geosci 9(12):892–897

Bout-Roumazeilles V, Combourieu-Nebout N, Desprat S, Siani G, Turon JL, Essallami L (2013) Tracking atmospheric and riverine terrigenous supplies variability during the last glacial and the Holocene in central Mediterranean. Clim Past 9(3):1065–1087

Bradtmiller LI, McGee D, Awalt M, Evers J, Yerxa H, Kinsley CW, deMenocal PB (2016) Changes in biological productivity along the northwest African margin over the past 20,000 years. Paleoceanography 31(1):185–202

Bubenzer O, Riemer H (2007) Holocene climatic change and human settlement between the central Sahara and the Nile Valley: archaeological and geomorphological results. Geoarchaeology 22(6):607–620

Byrne MP, Schneider T (2016) Energetic constraints on the width of the intertropical convergence zone. J Clim 29(13):4709–4721

Cacho I, Grimalt JO, Pelejero C, Canals M, Sierro FJ, Flores JA, Shackleton N (1999) Dansgaard-Oeschger and Heinrich event imprints in Alboran Sea paleotemperatures. Paleoceanography 14(6):698–705

Cacho I, Grimalt JO, Canals M, Sbaffi L, Shackleton NJ, Schönfeld J, Zahn R (2001) Variability of the western Mediterranean Sea surface temperature during the last 25,000 years and its connection with the Northern Hemisphere climatic changes. Paleoceanography 16(1):40–52

Callot Y, Fontugne M (2008) Les sites lacustres d'âge holocène dans l'est du Grand Erg occidental (nord-ouest du Sahara algérien): interprétation géomorphologique et paléoclimatique. Géomorphol Relief Process Environ 3:187–200

Cancellieri E, di Lernia S (2014) Re-entering the central Sahara at the onset of the Holocene: a territorial approach to Early Acacus hunter–gatherers (SW Libya). Quatern Int 320:43–62

Castañeda IS, Mulitza S, Schefuß E, Lopes dos Santos RA, Sinninghe Damsté JS, Schouten S (2009) Wet phases in the Sahara/Sahel region and human migration patterns in North Africa. Proc Natl Acad Sci 106(48):20159–20163

Castañeda IS, Schefuß E, Pätzold J, Sinninghe Damsté JS, Weldeab S, Schouten S (2010) Millennial-scale sea surface temperature changes in the eastern Mediterranean (Nile River Delta region) over the last 27,000 years. Paleoceanography 25(1)

Castañeda IS, Schouten S, Pätzold J, Lucassen F, Kasemann S, Kuhlmann H, Schefuß E (2016) Hydroclimate variability in the Nile River Basin during the past 28,000 years. Earth Planet Sci Lett 438:47–56

Cheddadi R, Lamb HF, Guiot J, van der Kaars S (1998) Holocene climatic change in Morocco: a quantitative reconstruction from pollen data. Clim Dyn 14(12):883–890

Cheddadi R, Bouaissa O, Rhoujjati A, Dezileau L (2016) Environmental changes in the Moroccan western Rif Mountains over the last 9,000 years. Quaternaire 27(1):15–25

Cheng H, Sinha A, Verheyden S, Nader FH, Li XL, Zhang PZ, Yin JJ, Yi L, Peng YB, Rao ZG, Ning YF, Edwards RL (2015) The climate variability in northern Levant over the past 20,000 years. Geophys Res Lett 42(20):8641–8650

Crassard R, Petraglia MD, Drake NA, Breeze P, Gratuze B, Alsharekh A, Arbach M, Groucutt HS, Khalidi L, Michelsen N, Robin CJ, Schiettecatte J (2013a) Middle Palaeolithic and Neolithic occupations around Mundafan palaeolake, Saudi Arabia: implications for climate change and human dispersals. PLoS ONE 8(7):e69665

Crassard R, Petraglia MD, Parker AG, Parton A, Roberts RG, Jacobs Z, Alsharekh A, Al-Omari A, Breeze P, Drake NA, Groucutt HS, Jennings R, Régagnon E, Shipton C (2013b) Beyond the Levant: first evidence of a pre-pottery Neolithic incursion into the Nefud Desert, Saudi Arabia. PLoS ONE 8(7):e68061

Cremaschi M, Zerboni A, Spötl C, Felletti F (2010) The calcareous tufa in the Tadrart Acacus Mt. (SW Fezzan, Libya): an early Holocene palaeoclimate archive in the central Sahara. Palaeogeogr Palaeoclimatol Palaeoecol 287(1–4):81–94

Cremaschi M, Zerboni A, Charpentier V, Crassard R, Isola I, Regattieri E, Zanchetta G (2015) Early–Middle Holocene environmental changes and pre-Neolithic human occupations as recorded in the cavities of Jebel Qara (Dhofar, southern Sultanate of Oman). Quatern Int 382:264–276

Cropper TE, Hanna E, Bigg GR (2014) Spatial and temporal seasonal trends in coastal upwelling off Northwest Africa, 1981–2012. Deep Sea Res Part I 86:94–111

Damnati B, Etebaai I, Benjilani H, El Khoudri K, Reddad H, Taieb M (2016) Sedimentology and geochemistry of lacustrine terraces of three Middle Atlas lakes: paleohydrological changes for the last 2300 cal BP in Morocco (western Mediterranean region). Quatern Int 404(Part B):163–173

deMenocal PB (2015) Palaeoclimate: end of the African humid period. Nat Geosci 8(2):86–87 deMenocal P, Ortiz J, Guilderson T, Adkins J, Sarnthein M, Baker L, Yarusinsky M (2000a) Abrupt onset and termination of the African humid period: rapid climate responses to gradual insolation forcing. Quatern Sci Rev 19(1–5):347–361

deMenocal P, Ortiz J, Guilderson T, Sarnthein M (2000b) Coherent high- and low-latitude climate variability during the Holocene warm period. Science 288(5474):2198–2202

Desprat S, Combourieu-Nebout N, Essallami L, Sicre MA, Dormoy I, Peyron O, Siani G, Bout Roumazeilles V, Turon JL (2013) Deglacial and Holocene vegetation and climatic changes in the southern Central Mediterranean from a direct land–sea correlation. Clim Past 9(2):767–787

Dinies M, Plessen B, Neef R, Kürschner H (2015) When the desert was green: grassland expansion during the early Holocene in northwestern Arabia. Quatern Int 382:293–302

Drake N, Bristow C (2006) Shorelines in the Sahara: geomorphological evidence for an enhanced monsoon from palaeolake Megachad. Holocene 16(6):901–911

Edelman-Furstenberg Y, Almogi-Labin A, Hemleben C (2009) Palaeoceanographic evolution of the central Red Sea during the late Holocene. Holocene 19(1):117–127

Eggermont H, Verschuren D, Fagot M, Rumes B, Van Bocxlaer B, Kröpelin S (2008) Aquatic community response in a groundwater-fed desert lake to Holocene desiccation of the Sahara. Quatern Sci Rev 27(25–26):2411–2425

Ehrmann W, Schmiedl G, Beuscher S, Krüger S (2017) Intensity of African humid periods estimated from Saharan dust fluxes. PLoS ONE 12(1):e0170989

Engel M, Brückner H, Pint A, Wellbrock K, Ginau A, Voss P, Grottker M, Klasen N, Frenzel P (2012) The early Holocene humid period in NW Saudi Arabia—sediments, microfossils and palaeo-hydrological modelling. Quatern Int 266:131–141

Engel M, Matter A, Parker AG, Parton A, Petraglia MD, Preston GW, Preusser F (2017) Lakes or wetlands? A comment on 'The middle Holocene climatic records from Arabia: reassessing lacustrine environments, shift of ITCZ in Arabian Sea, and impacts of the southwest Indian and African monsoons' by Enzel et al. Glob Planet Change 148:258–267

Enzel Y, Bookman R, Sharon D, Gvirtzman H, Dayan U, Ziv B, Stein M (2003) Late Holocene climates of the Near East deduced from Dead Sea level variations and modern regional winter rainfall. Quatern Res 60(3):263–273

Enzel Y, Kushnir Y, Quade J (2015) The middle Holocene climatic records from Arabia: reassessing lacustrine environments, shift of ITCZ in Arabian Sea, and impacts of the southwest Indian and African monsoons. Glob Planet Change 129:69–91

Enzel Y, Quade J, Kushnir Y (2017) Response to Engel et al. (in press): Lakes or wetlands? A comment on "The middle Holocene climatic records from Arabia: reassessing lacustrine environments, shift of ITCZ in Arabian Sea, and impacts of the southwest Indian and African monsoons" by Enzel et al. (2015). Glob Planet Change 148:268–271

Essallami L, Sicre MA, Kallel N, Labeyrie L, Siani G (2007) Hydrological changes in the Mediterranean Sea over the last 30,000 years. Geochem Geophys Geosyst 8(7)

Fleitmann D, Matter A (2009) The speleothem record of climate variability in Southern Arabia. C R Geosci 341(8–9):633–642

Fleitmann D, Burns SJ, Mudelsee M, Neff U, Kramers J, Mangini A, Matter A (2003) Holocene forcing of the Indian monsoon recorded in a stalagmite from southern Oman. Science 300 (5626):1737–1739

Fleitmann D, Matter A, Pint JJ, Al-Shanti MA (2004) The speleothem record of climate change in Saudi Arabia. Saudi Geological Survey, Open File Report SGS-OF-2004-8, pp 1–40

Fleitmann D, Burns SJ, Mangini A, Mudelsee M, Kramers J, Villa I, Neff U, Al-Subbary AA, Buettner A, Hippler D, Matter A (2007) Holocene ITCZ and Indian monsoon dynamics recorded in stalagmites from Oman and Yemen (Socotra). Quatern Sci Rev 26(1–2):170–188

Fleitmann D, Burns SJ, Pekala M, Mangini A, Al-Subbary A, Al-Aowah M, Kramers J, Matter A (2011) Holocene and Pleistocene pluvial periods in Yemen, southern Arabia. Quatern Sci Rev 30(7–8):783–787

Frumkin A, Magaritz M, Carmi I, Zak I (1991) The Holocene climatic record of the salt caves of Mount Sedom Israel. Holocene 1(3):191–200

Fuchs M, Buerkert A (2008) A 20 ka sediment record from the Hajar Mountain range in N-Oman, and its implication for detecting arid–humid periods on the southeastern Arabian Peninsula. Earth Planet Sci Lett 265(3–4):546–558

Gasse F (2002) Diatom-inferred salinity and carbonate oxygen isotopes in Holocene waterbodies of the western Sahara and Sahel (Africa). Quatern Sci Rev 21(7):737–767

Gebel HG, Hanss C, Liebau A, Raehle W (1989) The Late Quaternary environments of ‘Ain al-Faidha/Al-‘Ain, Abu Dhabi Emirate. Archaeol U A E 5:9–48

Gennari G, Rosenberg T, Spezzaferri S, Berger J-P, Fleitmann D, Preusser F, Al-Shanti M, Matter A (2011) Faunal evidence of a Holocene pluvial phase in southern Arabia with remarks on the morphological variability of Helenina anderseni. J Foramin Res 41(3):248–259

Ginau A, Engel M, Brückner H (2012) Holocene chemical precipitates in the continental sabkha of Tayma (NW Saudi Arabia). J Arid Environ 84:26–37

Giraudi C, Mercuri AM, Esu D (2013) Holocene palaeoclimate in the northern Sahara margin (Jefara Plain, northwestern Libya). Holocene 23(3):339–352

Groucutt HS, Petraglia MD (2012) The prehistory of the Arabian Peninsula: deserts, dispersals, and demography. Evol Anthropol Issues News Rev 21(3):113–125

Guagnin M, Jennings R, Eager H, Parton A, Stimpson C, Stepanek C, Pfeiffer M, Groucutt HS, Drake NA, Alsharekh A, Petraglia MD (2016) Rock art imagery as a proxy for Holocene environmental change: a view from Shuwaymis, NW Saudi Arabia. Holocene 26(11):1822–1834

Gupta AK, Anderson DM, Overpeck J (2003) Abrupt changes in the Asian southwest monsoon during the Holocene and their links to the North Atlantic Ocean. Nature 421:354–357

Gupta AK, Das M, Anderson DM (2005) Solar influence on the Indian summer monsoon during the Holocene. Geophys Res Lett 32(L17703)

Hamann Y, Ehrmann W, Schmiedl G, Krüger S, Stuut JB, Kuhnt T (2008) Sedimentation processes in the Eastern Mediterranean Sea during the Late Glacial and Holocene revealed by end-member modelling of the terrigenous fraction in marine sediments. Mar Geol 248(1–2):97–114

Haslett SK, Smart CW (2006) Late Quaternary upwelling off tropical NW Africa: new micropalaeontological evidence from ODP Hole 658C. J Quat Sci 21(3):259–269

Hassan FA, Hamdan MA, Flower RJ, Keatings K (2012) The oxygen and carbon isotopic records in Holocene freshwater mollusc shells from the faiyum paleolakes, Egypt: their paloenvironmental and paleoclimatic implications. Quatern Int 266:175–187

Hennekam R, Jilbert T, Schnetger B, de Lange GJ (2014) Solar forcing of Nile discharge and sapropel S1 formation in the early to middle Holocene eastern Mediterranean. Paleoceanography 29(5):343–356

Hennekam R, Donders TH, Zwiep K, de Lange GJ (2015) Integral view of Holocene precipitation and vegetation changes in the Nile catchment area as inferred from its delta sediments. Quatern Sci Rev 130:189–199

Hilbert YH, White TS, Parton A, Clark-Balzan L, Crassard R, Groucutt HS, Jennings RP, Breeze P, Parker A, Shipton C, Al-Omari A, Alsharekh AM, Petraglia MD (2014) Epipalaeolithic occupation and palaeoenvironments of the southern Nefud desert, Saudi Arabia, during the Terminal Pleistocene and Early Holocene. J Archaeol Sci 50:460–474

Hillaire-Marcel C, Riser J, Rognon P, Petit-Maire N, Rosso JC, Soulie-Marche I (1983) Radiocarbon chronology of Holocene hydrologic changes in Northeastern Mali. Quatern Res 20(2):145–164

Hoelzmann P, Gasse F, Dupont LM, Salzmann U, Staubwasser M, Leuschner DC, Sirocko F (2004) Palaeoenvironmental changes in the arid and sub arid belt (Sahara-Sahel-Arabian Peninsula) from 150 kyr to present. In: Battarbee RW, Gasse F, Stickley CE (eds) Past climate variability through Europe and Africa. Springer Netherlands, Dordrecht, pp 219–256

Hoelzmann P, Schwalb A, Roberts N, Cooper P, Burgess A (2010) Hydrological response of an east-Saharan palaeolake (NW Sudan) to early-Holocene climate. Holocene 20(4):537–549

Hopcroft PO, Valdes PJ, Harper AB, Beerling DJ (2017) Multi vegetation model evaluation of the Green Sahara climate regime. Geophys Res Lett 44(13):6804–6813

Huguet C, Kim J-H, Sinninghe Damsté JS, Schouten S (2006) Reconstruction of sea surface temperature variations in the Arabian Sea over the last 23 kyr using organic proxies (TEX86 and U37K′). Paleoceanography 21(3)

Jalali B, Sicre M-A, Kallel N, Azuara J, Combourieu-Nebout N, Bassetti M-A, Klein V (2017) High-resolution Holocene climate and hydrological variability from two major Mediterranean deltas (Nile and Rhone). Holocene 27(8):1158–1168

Jaouadi S, Lebreton V, Bout-Roumazeilles V, Siani G, Lakhdar R, Boussoffara R, Dezileau L, Kallel N, Mannai-Tayech B, Combourieu-Nebout N (2016) Environmental changes, climate and anthropogenic impact in south-east Tunisia during the last 8 kyr. Clim Past 12(6):1339–1359

Jennings RP, Shipton C, Al-Omari A, Alsharekh AM, Crassard R, Groucutt H, Petraglia MD (2013) Rock art landscapes beside the Jubbah palaeolake, Saudi Arabia. Antiquity 87(337): 666–683

Jung SJA, Davies GR, Ganssen GM, Kroon D (2004) Stepwise Holocene aridification in NE Africa deduced from dust-borne radiogenic isotope records. Earth Planet Sci Lett 221(1–4): 27–37

Kaniewski D, Paulissen E, Van Campo E, Al-Maqdissi M, Bretschneider J, Van Lerberghe K (2008) Middle East coastal ecosystem response to middle-to-late Holocene abrupt climate changes. Proc Natl Acad Sci 105(37):13941–13946

Kaniewski D, Van Campo E, Paulissen E, Weiss H, Bakker J, Rossignol I, Van Lerberghe K (2011) The medieval climate anomaly and the little Ice Age in coastal Syria inferred from pollen-derived palaeoclimatic patterns. Glob Planet Change 78:178–187

Kaniewski D, Van Campo E, Morhange C, Guiot J, Zviely D, Shaked I, Otto T, Artzy M (2013) Early urban impact on Mediterranean coastal environments. Sci Rep 3:3540

Kim J-H, Meggers H, Rimbu N, Lohmann G, Freudenthal T, Müller PJ, Schneider RR (2007) Impacts of the North Atlantic gyre circulation on Holocene climate off northwest Africa. Geology 35(5):387–390

Knaack J (1997) Eine neue Transferfunktion zur Rekonstruktion der Paläoproduktivität aus Gemeinschaften mariner Diatomeen [A new transfer function to reconstruct paleoproductivity from marine diatom assemblages]. PhD thesis, Berichte Geologisch-Paläontologisches Institut und Museum, Kiel, vol 83, pp 1–119

Kröpelin S, Verschuren D, Lézine A-M, Eggermont H, Cocquyt C, Francus P, Cazet J-P, Fagot M, Rumes B, Russell JM, Darius F, Conley DJ, Schuster M, von Suchodoletz H, Engstrom DR (2008) Climate-driven ecosystem succession in the Sahara: the past 6000 years. Science 320 (5877):765–768

Kuhlmann H, Freudenthal T, Helmke P, Meggers H (2004a) Reconstruction of paleoceanography off NW Africa during the last 40,000 years: influence of local and regional factors on sediment accumulation. Mar Geol 207(1–4):209–224

Kuhlmann H, Meggers H, Freudenthal T, Wefer G (2004b) The transition of the monsoonal and the N Atlantic climate system off NW Africa during the Holocene. Geophys Res Lett 31(22)

Kuhnt T, Schmiedl G, Ehrmann W, Hamann Y, Andersen N (2008) Stable isotopic composition of Holocene benthic foraminifers from the Eastern Mediterranean Sea: past changes in productivity and deep water oxygenation. Palaeogeogr Palaeoclimatol Palaeoecol 268(1–2): 106–115

Kuper R, Kröpelin S (2006) Climate-controlled Holocene occupation in the Sahara: motor of Africa's evolution. Science 313(5788):803–807

Lamb HF, Eicher U, Switsur VR (1989) An 18,000-year record of vegetation, lake-level and climatic change from Tigalmamine, Middle Atlas, Morocco. J Biogeogr 16(1):65–74

Lamb H, Roberts N, Leng M, Barker P, Benkaddour A, van der Kaars S (1999) Lake evolution in a semi-arid montane environment: response to catchment change and hydroclimatic variation. J Paleolimnol 21(3):325–343

Lamy F, Arz HW, Bond GC, Bahr A, Pätzold J (2006) Multicentennial-scale hydrological changes in the Black Sea and northern Red Sea during the Holocene and the Arctic/North Atlantic Oscillation. Paleoceanography 21(1)

Larrasoaña JC, Roberts AP, Rohling EJ (2013) Dynamics of green Sahara periods and their role in hominin evolution. PLoS ONE 8(10):e76514

Lécuyer C, Lézine A-M, Fourel F, Gasse F, Sylvestre F, Pailles C, Grenier C, Travi Y, Barral A (2016) I-n-Atei palaeolake documents past environmental changes in central Sahara at the time of the "Green Sahara": charcoal, carbon isotope and diatom records. Palaeogeogr Palaeoclimatol Palaeoecol 441(Part 4):834–844

Lézine A-M, Tiercelin JJ, Robert C, Saliège J-F, Cleuziou S, Inizan M-L, Braemer F (2007) Centennial to millennial-scale variability of the Indian monsoon during the early Holocene from a sediment, pollen and isotope record from the desert of Yemen. Palaeogeogr Palaeoclimatol Palaeoecol 243(3–4):235–249

Lézine A-M, Hély C, Grenier C, Braconnot P, Krinner G (2011) Sahara and Sahel vulnerability to climate changes, lessons from Holocene hydrological data. Quatern Sci Rev 30:3001–3012

Lézine A-M, Bassinot F, Peterschmitt J-Y (2014) Orbitally-induced changes of the Atlantic and Indian monsoons over the past 20,000 years: new insights based on the comparison of continental and marine records. Bull Soc Geol Fr 185(1):3–12

Lézine A-M, Ivory SJ, Braconnot P, Marti O (2017) Timing of the southward retreat of the ITCZ at the end of the Holocene humid period in Southern Arabia: data-model comparison. Quatern Sci Rev 164:68–76

Linstädter J, Kröpelin S (2004) Wadi Bakht revisited: Holocene climate change and prehistoric occupation in the Gilf Kebir region of the Eastern Sahara, SW Egypt. Geoarchaeology 19(8): 753–778

Litt T, Ohlwein C, Neumann FH, Hense A, Stein M (2012) Holocene climate variability in the Levant from the Dead Sea pollen record. Quatern Sci Rev 49:95–105

Liu T, Broecker WS, Stein M (2013) Rock varnish evidence for a Younger Dryas wet period in the Dead Sea basin. Geophys Res Lett 40(10):2229–2235

Ljungqvist FC (2011) The spatio-temporal pattern of the mid-Holocene thermal maximum. Geogr–Sb ČGS 116(2):91–110

Macklin MG, Toonen WHJ, Woodward JC, Williams MAJ, Flaux C, Marriner N, Nicoll K, Verstraeten G, Spencer N, Welsby D (2015) A new model of river dynamics, hydroclimatic change and human settlement in the Nile Valley derived from meta-analysis of the Holocene fluvial archive. Quatern Sci Rev 130:109–123

Manning K, Timpson A (2014) The demographic response to Holocene climate change in the Sahara. Quatern Sci Rev 101:28–35

Marcott SA, Shakun JD, Clark PU, Mix AC (2013) A reconstruction of regional and global temperature for the past 11,300 years. Science 339(6124):1198–1201

Marinova MM, Meckler AN, McKay CP (2014) Holocene freshwater carbonate structures in the hyper-arid Gebel Uweinat region of the Sahara Desert (Southwestern Egypt). J Afr Earth Sci 89:50–55

Marks L, Salem A, Welc F, Nitychoruk J, Chen Z, Blaauw M, Zalat A, Majecka A, Szymanek M, Chodyka M, Tołoczko-Pasek A, Sun Q, Zhao X, Jiang J (2017) Holocene lake sediments from the Faiyum Oasis in Egypt: a record of environmental and climate change. Boreas

Martín-Puertas C, Jiménez-Espejo F, Martínez-Ruiz F, Nieto-Moreno V, Rodrigo M, Mata MP, Valero-Garcés BL (2010) Late Holocene climate variability in the southwestern Mediterranean region: an integrated marine and terrestrial geochemical approach. Clim Past 6(6):807–816

Martrat B, Grimalt JO, Lopez-Martinez C, Cacho I, Sierro FJ, Flores JA, Zahn R, Canals M, Curtis JH, Hodell DA (2004) Abrupt temperature changes in the Western Mediterranean over the past 250,000 years. Science 306(5702):1762–1765

Martrat B, Grimalt JO, Shackleton NJ, de Abreu L, Hutterli MA, Stocker TF (2007) Four climate cycles of recurring deep and surface water destabilizations on the Iberian margin. Science 317 (5837):502–507

Matter A, Mahjoub A, Neubert E, Preusser F, Schwalb A, Szidat S, Wulf G (2016) Reactivation of the Pleistocene trans-Arabian Wadi ad Dawasir fluvial system (Saudi Arabia) during the Holocene humid phase. Geomorphology 270:88–101

McGee D, deMenocal PB, Winckler G, Stuut JBW, Bradtmiller LI (2013) The magnitude, timing and abruptness of changes in North African dust deposition over the last 20,000 yr. Earth Planet Sci Lett 371–372:163–176

Mercuri AM, Florenzano A, Giraudi C, Garcea EAA (2015) Humans and water in desert "Refugium" areas: palynological evidence of climate oscillations and cultural developments in early and mid-Holocene Saharan edges. Interdiscip Archaeol Nat Sci Archaeol 6(2):151–160

Migowski C, Stein M, Prasad S, Negendank JFW, Agnon A (2006) Holocene climate variability and cultural evolution in the Near East from the Dead Sea sedimentary record. Quatern Res 66(3):421–431

Mojtahid M, Manceau R, Schiebel R, Hennekam R, de Lange GJ (2015) Thirteen thousand years of southeastern Mediterranean climate variability inferred from an integrative planktic foraminiferal-based approach. Paleoceanography 30(4):402–422

Munz PM, Steinke S, Böll A, Lückge A, Groeneveld J, Kucera M, Schulz H (2016) Decadal resolution record of Oman margin upwelling indicates persistent solar forcing of the Indian summer monsoon after the early Holocene summer insolation maximum. Clim Past Discuss 2016:1–31

Neff U, Burns SJ, Mangini A, Mudelsee M, Fleitmann D, Matter A (2001) Strong coherence between solar variability and the monsoon in Oman between 9 and 6 kyr ago. Nature 411:290–293

Nourelbait M, Rhoujjati A, Benkaddour A, Carré M, Eynaud F, Martinez P, Cheddadi R (2016) Climate change and ecosystems dynamics over the last 6000 years in the Middle Atlas, Morocco. Clim Past 12(4):1029–1042

Orland IJ, Bar-Matthews M, Kita NT, Ayalon A, Matthews A, Valley JW (2009) Climate deterioration in the Eastern Mediterranean as revealed by ion microprobe analysis of a speleothem that grew from 2.2 to 0.9 ka in Soreq Cave, Israel. Quatern Res 71(1):27–35

Parker AG, Goudie AS (2008) Geomorphological and palaeoenvironmental investigations in the southeastern Arabian Gulf region and the implication for the archaeology of the region. Geomorphology 101(3):458–470

Parker AG, Eckersley L, Smith MM, Goudie AS, Stokes S, Ward S, White K, Hodson MJ (2004) Holocene vegetation dynamics in the northeastern Rub' al-Khali desert, Arabian Peninsula: a phytolith, pollen and carbon isotope study. J Quat Sci 19(7):665–676

Parker A, Davies C, Wilkinson T (2006) The early to mid-Holocene moist period in Arabia: some recent evidence from lacustrine sequences in eastern and south-western Arabia. Proc Semin Arab Stud 36:243–255

Pourmand A, Marcantonio F, Bianchi TS, Canuel EA, Waterson EJ (2007) A 28-ka history of sea surface temperature, primary productivity and planktonic community variability in the western Arabian Sea. Paleoceanography 22(4)

Preston GW, Parker AG, Walkington H, Leng MJ, Hodson MJ (2012) From nomadic herder-hunters to sedentary farmers: the relationship between climate change and ancient subsistence strategies in south-eastern Arabia. J Arid Environ 86:122–130

Renssen H, Brovkin V, Fichefet T, Goosse H (2006) Simulation of the Holocene climate evolution in Northern Africa: the termination of the African humid period. Quatern Int 150(1):95–102

Revel M, Colin C, Bernasconi S, Combourieu-Nebout N, Ducassou E, Grousset FE, Rolland Y, Migeon S, Bosch D, Brunet P, Zhao Y, Mascle J (2014) 21,000 years of Ethiopian African monsoon variability recorded in sediments of the western Nile deep-sea fan. Reg Environ Change 14(5):1685–1696

Revel M, Ducassou E, Skonieczny C, Colin C, Bastian L, Bosch D, Migeon S, Mascle J (2015) 20,000 years of Nile River dynamics and environmental changes in the Nile catchment area as inferred from Nile upper continental slope sediments. Quatern Sci Rev 130:200–221

Rhoujjati A, Cheddadi R, Taïeb M, Baali A, Ortu E (2010) Environmental changes over the past c. 29,000 years in the Middle Atlas (Morocco): a record from Lake Ifrah. J Arid Environ 74(7): 737–745

Riemer H, Kröpelin S, Zboray A (2017) Climate, styles and archaeology: an integral approach towards an absolute chronology of the rock art in the Libyan Desert (Eastern Sahara). Antiquity 91(355):7–23

Robinson SA, Black S, Sellwood BW, Valdes PJ (2006) A review of palaeoclimates and palaeoenvironments in the Levant and Eastern Mediterranean from 25,000 to 5000 years BP: setting the environmental background for the evolution of human civilisation. Quatern Sci Rev 25(13–14):1517–1541

Rosenberg TM, Preusser F, Fleitmann D, Schwalb A, Penkman K, Schmid TW, Al-Shanti MA, Kadi K, Matter A (2011) Humid periods in southern Arabia: windows of opportunity for modern human dispersal. Geology 39(12):1115–1118

Ruan J, Kherbouche F, Genty D, Blamart D, Cheng H, Dewilde F, Hachi S, Edwards RL, Régnier E, Michelot JL (2016) Evidence of a prolonged drought ca. 4200 yr BP correlated with prehistoric settlement abandonment from the Gueldaman GLD1 Cave, Northern Algeria. Clim Past 12(1):1–14

Sachs JP, Sachse D, Smittenberg RH, Zhang Z, Battisti DS, Golubic S (2009) Southward movement of the Pacific intertropical convergence zone AD 1400–1850. Nat Geosci 2(7):519–525

Samartin S, Heiri O, Joos F, Renssen H, Franke J, Bronnimann S, Tinner W (2017) Warm Mediterranean mid-Holocene summers inferred from fossil midge assemblages. Nat Geosci 10(3):207–212

Schilman B, Bar-Matthews M, Almogi-Labin A, Luz B (2001) Global climate instability reflected by Eastern Mediterranean marine records during the late Holocene. Palaeogeogr Palaeoclimatol Palaeoecol 176(1–4):157–176

Schilman B, Ayalon A, Bar-Matthews M (2002) Sea-land paleoclimate correlation in the Eastern Mediterranean region during the late Holocene. Israel J Earth Sci 51(3–4):181–190

Schneider T, Bischoff T, Haug GH (2014) Migrations and dynamics of the intertropical convergence zone. Nature 513(7516):45–53

Schütt B, Krause J (2009) Comparison of proxy-based palaeoenvironmental reconstructions and hindcast modelled annual precipitation—a review of Holocene palaeoenvironmental research in the Central Sahara. In: Baumhauer R, Runge J (eds) Holocene palaeoenvironmental history of the Central Sahara. CRC Press, Bota Raton, pp 23–37

Shanahan TM, McKay NP, Hughen KA, Overpeck JT, Otto-Bliesner B, Heil CW, King J, Scholz CA, Peck J (2015) The time-transgressive termination of the African humid period. Nat Geosci 8(2):140–144

Skinner CB, Poulsen CJ (2016) The role of fall season tropical plumes in enhancing Saharan rainfall during the African humid period. Geophys Res Lett 43(1):349–358

Swezey C, Lancaster N, Kocurek G, Deynoux M, Blum M, Price D, Pion J-C (1999) Response of aeolian systems to Holocene climatic and hydrologic changes on the northern margin of the Sahara: a high-resolution record from the Chott Rharsa basin, Tunisia. Holocene 9(2):141–147

Tabel J, Khater C, Rhoujjati A, Dezileau L, Bouimetarhan I, Carre M, Vidal L, Benkaddour A, Nourelbait M, Cheddadi R (2016) Environmental changes over the past 25 000 years in the southern Middle Atlas, Morocco. J Quat Sci

Tierney JE, Pausata FSR, deMenocal PB (2017) Rainfall regimes of the Green Sahara. Sci Adv 3(1)

Trommer G, Siccha M, Rohling EJ, Grant K, van der Meer MTJ, Schouten S, Hemleben C, Kucera M (2010) Millennial-scale variability in Red Sea circulation in response to Holocene insolation forcing. Paleoceanography 25(3)

Vaks A, Bar-Matthews M, Ayalon A, Matthews A, Frumkin A, Dayan U, Halicz L, Almogi-Labin A, Schilman B (2006) Paleoclimate and location of the border between Mediterranean climate region and the Saharo–Arabian Desert as revealed by speleothems from the northern Negev Desert, Israel. Earth Planet Sci Lett 249(3–4):384–399

Vaks A, Bar-Matthews M, Ayalon A, Matthews A, Halicz L, Frumkin A (2007) Desert speleothems reveal climatic window for African exodus of early modern humans. Geology 35(9):831–834

Van Rampelbergh M, Fleitmann D, Verheyden S, Cheng H, Edwards L, De Geest P, De Vleeschouwer D, Burns SJ, Matter A, Claeys P, Keppens E (2013) Mid- to late Holocene Indian Ocean Monsoon variability recorded in four speleothems from Socotra Island, Yemen. Quatern Sci Rev 65:129–142

Verheyden S, Nader F, Cheng H, Edwards L, Swennen R (2008) Paleoclimate reconstruction in the Levant region from the geochemistry of a Holocene stalagmite from the Jeita cave, Lebanon. Quatern Res 70:368–381

Willcox G, Buxo R, Herveux L (2009) Late Pleistocene and early Holocene climate and the beginnings of cultivation in northern Syria. Holocene 19(1):151–158

Wu J, Böning P, Pahnke K, Tachikawa K, de Lange GJ (2016) Unraveling North-African riverine and eolian contributions to central Mediterranean sediments during Holocene sapropel S1 formation. Quatern Sci Rev 152:31–48

Wu J, Liu Z, Stuut J-BW, Zhao Y, Schirone A, de Lange GJ (2017) North-African paleodrainage discharges to the central Mediterranean during the last 18,000 years: a multiproxy characterization. Quatern Sci Rev 163:95–113

Zhao M, Beveridge NAS, Shackleton NJ, Sarnthein M, Eglinton G (1995) Molecular stratigraphy of cores off northwest Africa: sea surface temperature history over the last 80 ka. Paleoceanography 10(3):661–675

Zhao M, Eglinton G, Haslett SK, Jordan RW, Sarnthein M, Zhang Z (2000) Marine and terrestrial biomarker records for the last 35,000 years at ODP site 658C off NW Africa. Org Geochem 31(9):919–930

Zhao M, Mercer JL, Eglinton G, Higginson MJ, Huang C-Y (2006) Comparative molecular biomarker assessment of phytoplankton paleoproductivity for the last 160 kyr off Cap Blanc, NW Africa. Org Geochem 37(1):72–97

Zhao X, Liu Y, Salem A, Marks L, Welc F, Sun Q, Jiang J, Chen J, Chen Z (2017) Migration of the Intertropical Convergence Zone in North Africa during the Holocene: evidence from variations in quartz grain roundness in the lower Nile valley, Egypt. Quatern Int 449 (Supplement C):22–28

Zielhofer C, Faust D, Linstädter J (2008) Late Pleistocene and Holocene alluvial archives in the Southwestern Mediterranean: changes in fluvial dynamics and past human response. Quatern Int 181(1):39–54

Zielhofer C, Fletcher WJ, Mischke S, De Batist M, Campbell JFE, Joannin S, Tjallingii R, El Hamouti N, Junginger A, Stele A, Bussmann J, Schneider B, Lauer T, Spitzer K, Strupler M, Brachert T, Mikdad A (2017) Atlantic forcing of Western Mediterranean winter rain minima during the last 12,000 years. Quatern Sci Rev 157:29–51

The manufacturer's authorised representative in the EU is Springer
Nature Customer Service Centre GmbH, Europaplatz 3, 69115 Heidelberg,
Germany. If you have any concerns regarding our products, please
contact ProductSafety@springernature.com

Printed and bound by CPI Group (UK) Ltd, Croydon, CR0 4YY
30/09/2026
02225658-0015